Methods in Enzymology

Volume 101

RECOMBINANT DNA

Part C

METHODS IN ENZYMOLOGY

EDITORS-IN-CHIEF

Sidney P. Colowick   Nathan O. Kaplan

Methods in Enzymology

Volume 101

Recombinant DNA

Part C

EDITED BY

Ray Wu
SECTION OF BIOCHEMISTRY
MOLECULAR AND CELL BIOLOGY
CORNELL UNIVERSITY
ITHACA, NEW YORK

Lawrence Grossman
DEPARTMENT OF BIOCHEMISTRY
THE JOHNS HOPKINS UNIVERSITY
SCHOOL OF HYGIENE AND PUBLIC HEALTH
BALTIMORE, MARYLAND

Kivie Moldave
DEPARTMENT OF BIOLOGICAL CHEMISTRY
COLLEGE OF MEDICINE
UNIVERSITY OF CALIFORNIA
IRVINE, CALIFORNIA

1983

ACADEMIC PRESS

A Subsidiary of Harcourt Brace Jovanovich, Publishers

New York London
Paris San Diego San Francisco São Paulo Sydney Tokyo Toronto

ACADEMIC PRESS, INC.
111 Fifth Avenue, New York, New York 10003

United Kingdom Edition published by
ACADEMIC PRESS, INC. (LONDON) LTD.
24/28 Oval Road, London NW1 7DX

Library of Congress Cataloging in Publication Data

Main entry under title:

Recombinant DNA.

Pt. edited by Ray Wu, Lawrence Grossman, Kivie Moldave.
Includes bibliographical references and indexes.
1. Recombinant DNA. I. Wu, Ray. II. Grossman, Lawrence, Date . III. Moldave, Kivie, Date .
IV. Series: Methods in enzymology ; v. 68, etc.
[DNLM: 1. DNA, Recombinant. W1 ME9615K v. 68, etc. /
QU 58 R312 1979]
QP601.M49 vol. 68 574.1'925'08s 79-26584
[QH442] [574.8'732]
ISBN 0-12-182001-7 (v. 101)

PRINTED IN THE UNITED STATES OF AMERICA

83 84 85 86 9 8 7 6 5 4 3 2 1

Table of Contents

Section I. New Vectors for Cloning Genes

Section II. Cloning of Genes into Yeast Cells

Section III. Systems for Monitoring Cloned Gene Expression

A. Intact Cell Systems

B. Introduction of Genes into Mammalian Cells

C. Cell-Free Systems; Transcription

D. Cell-Free Systems; Translation

Contributors to Volume 101

Article numbers are in parentheses following the names of contributors.
Affiliations listed are current.

GUSTAV AMMERER (11), *Zymos Corporation, Seattle, Washington 98103*

CARL W. ANDERSON (41), *Biology Department, Brookhaven National Laboratory, Upton, New York 11973*

WAYNE M. BARNES (5), *Department of Biological Chemistry, Washington University School of Medicine, St. Louis, Missouri 63110*

LESLIE BARNETT (1), *MRC Laboratory of Molecular Biology, Cambridge CB2 2QH, England*

KENNETH A. BARTON (33), *Cetus Madison Corporation, Middleton, Wisconsin 53562*

MICHAEL D. BEEN (4), *Department of Microbiology and Immunology, School of Medicine, University of Washington, Seattle, Washington 98195*

MICHAEL BEVAN (5), *Plant Breeding Institute, Trumpington, Cambridge CB2 2LQ, England*

DAVID BOTSTEIN (9), *Department of Biology, Massachusetts Institute of Technology, Cambridge, Massachusetts 02139*

SYDNEY BRENNER (1), *MRC Laboratory of Molecular Biology, Cambridge CB2 2QH, England*

JAMES R. BROACH (21), *Department of Microbiology, State University of New York, Stony Brook, New York 11794*

NATHAN BROT (45), *Department of Biochemistry, Roche Institute of Molecular Biology, Nutley, New Jersey 07110*

PATRICIA A. BROWN (18), *Rosenstiel Basic Science Research Center, Brandeis University, Waltham, Massachusetts 02154*

JOHN CARBON (20), *Department of Biological Sciences, University of California, Santa Barbara, California 93106*

YVES CENATIEMPO (45), *Laboratoire de Biologie Moleculaire, University Lyon, 69622 Villeurbanne, France*

M. CHAMBERLIN (34), *Department of Biochemistry, University of California, Berkeley, California 94720*

GLENN H. CHAMBLISS (37), *Department of Bacteriology, University of Wisconsin, Madison, Wisconsin 53706*

JAMES J. CHAMPOUX (4), *Department of Microbiology and Immunology, School of Medicine, University of Washington, Seattle, Washington 98195*

HUI-ZHU CHEN (44), *Fairchild Center for Biological Sciences, Columbia University, New York, New York 10027*

MARY-DELL CHILTON (33), *Department of Biology, Washington University, St. Louis, Missouri 63130*

FORREST CHUMLEY (13), *Department of Biology, Massachusetts Institute of Technology and the Whitehead Institute for Biomedical Research, Cambridge, Massachusetts 02139*

JOSEPHINE E. CLARK-CURTISS (23), *Department of Microbiology, University of Alabama in Birmingham, Birmingham, Alabama 35294*

LOUISE CLARKE (20), *Department of Biological Sciences, University of California, Santa Barbara, California, 93106*

LAWRENCE COHEN (43), *Dana Farber Cancer Institute, Harvard Medical School, Boston, Massachusetts 02115*

GRAY F. CROUSE (3), *Basic Research Program—LBI, Frederick Cancer Research Facility, Frederick, Maryland 21701*

ROY CURTISS III (23), *Department of Microbiology, University of Alabama in Birmingham, Birmingham, Alabama 35294*

A. DEVERA (34), *Department of Biochemistry, University of California, Berkeley, California 94720*

JOHN D. DIGNAM (36), *Department of Biochemistry, University of Mississippi Medical Center, Jackson, Mississippi 39216*

BERNARD S. DUDOCK (41), *Department of Biochemistry, State University of New York, Stony Brook, New York 11794*

GERALD R. FINK (13), *Department of Biology, Massachusetts Institute of Tech-*

nology and the Whitehead Institute for Biomedical Research, Cambridge, Massachusetts 02139

ANDREW FIRE (35), *Center for Cancer Research, Massachusetts Institute of Technology, Cambridge, Massachusetts 02139*

ITZHAK FISCHER (40), *Department of Biological Chemistry, College of Medicine, University of California, Irvine, California 92717*

ANNEMARIE FRISCHAUF (3), *European Molecular Biology Laboratory, Postfach 102209, 6900 Heidelberg, Federal Republic of Germany*

EUGENIUSZ GASIOR (42), *Department of Molecular Biology, Institute of Microbiology and Biochemistry, University of Marie Curie-Sklodowska, Lublin, Poland*

M. GILMAN (34), *Department of Biochemistry, University of California, Berkeley, California 94720*

JOSEPH GLORIOSO (27), *Unit for Laboratory Animal Medicine, University of Michigan, Ann Arbor, Michigan 48109*

ALAN L. GOLDIN (27), *Department of Human Genetics, University of Michigan Medical School, Ann Arbor, Michigan 48109*

JON W. GORDON (28), *Department of Obstetrics and Gynecology, Mount Sinai School of Medicine, New York, New York 10029*

A. GRAESSMANN (30, 31), *Institut für Molekularbiologie und Biochemie der Freien Universität Berlin, D-1000 Berlin 33, Federal Republic of Germany*

M. GRAESSMANN (30), *Institut für Molekularbiologie und Biochemie der Freien Universität Berlin, D-100 Berlin 33, Federal Republic of Germany*

LEONARD GUARENTE (10), *Department of Biology, Massachusetts Institute of Technology, Cambridge, Massachusetts 02139*

J. B. GURDON (25), *MRC Laboratory of Molecular Biology, Cambridge CB2 2QH, England*

MARK S. GUYER (24), *Department of Molecular Genetics, GENEX Corporation, Gaithersburg, Maryland 20877*

TINA M. HENKIN (37), *Department of Bacteriology, University of Wisconsin, Madison, Wisconsin 53706*

EDGAR C. HENSHAW (39), *University of Rochester Cancer Center, Rochester, New York 14642*

YEN-SEN HO (6), *Department of Molecular Genetics, Smith Kline and French Laboratories, Philadelphia, Pennsylvania 19101*

PETER M. HOWLEY (26), *Laboratory of Pathology, National Cancer Institute, National Institutes of Health, Bethesda, Maryland 20205*

CHU-LAI HSIAO (20), *Central Research and Development Department, E. I. DuPont de Nemours and Company, Experimental Station, Wilmington, Delaware 19898*

JUNGI HUANG (29), *Institute of Economic Crops, Jiangsu Academy of Agricultural Sciences, Nanjing, Peoples Republic of China*

JONATHAN KARN (1), *MRC Laboratory of Molecular Biology, Cambridge CB2 2QH, England*

R. KINGSTON (34), *Center for Cancer Research, Massachusetts Institute of Technology, Cambridge, Massachusetts 02139*

MING-FAN LAW (26), *Laboratory of Pathology, National Cancer Institute, National Institutes of Health, Bethesda, Maryland 20205*

HANS LEHRACH (3), *European Molecular Biology Laboratory, Postfach 102209, 6900 Heidelberg, Federal Republic of Germany*

JUDITH M. LEVENTHAL (37), *Department of Bacteriology, University of Wisconsin, Madison, Wisconsin 53706*

MYRON LEVINE (27), *Department of Human Genetics, University of Michigan Medical School, Ann Arbor, Michigan 48109*

GUILING LIU (29), *Institute of Economic Crops, Jiangsu Academy of Agricultural Sciences, Nanjing, Peoples Republic of China*

A. LOYTER (31), *Department of Biological Chemistry, Institute of Life Sciences, The Hebrew University of Jerusalem, 91904 Jerusalem, Israel*

VIVIAN L. MACKAY (22), *Waksman Institute of Microbiology, Rutgers University, The State University of New Jersey, New Brunswick, New Jersey 08903, and Zymos Corporation, Seattle, Washington 98103*

JAMES L. MANLEY (35), *Department of Biol-*

ogy, Columbia University, New York, New York 10027

PAUL L. MARTIN (36), *Laboratory of Biochemistry and Molecular Genetics, The Rockefeller University, New York, New York 10021*

WILLIAM C. MERRICK (38), *Department of Biochemistry, Case Western Reserve University, Cleveland, Ohio 44106*

JOACHIM MESSING (2), *Department of Biochemistry, University of Minnesota, St. Paul, Minnesota 55108*

LUIS MEZA-BASSO (45), *Universidad Austral de Chile, Casilla 567, Valdivia, Chile*

JACQUELINE S. MILLER (43), *Department of Biological Chemistry, Harvard Medical School, Boston, Massachusetts 02115*

KIVIE MOLDAVE (40, 42), *Department of Biological Chemistry, College of Medicine, University of California, Irvine, California 92717*

ANDREW W. MURRAY (16), *Dana Farber Cancer Institute and The Committee on Cell and Developmental Biology, Harvard Medical School, Boston, Massachusetts 02115*

BRIAN P. NICHOLS (8), *Department of Biological Sciences, University of Illinois at Chicago, Chicago, Illinois 60607*

TERRY L. ORR-WEAVER (14), *Dana Farber Cancer Institute and Department of Biological Chemistry, Harvard Medical School, Boston, Massachusetts 02115*

RICHARD PANNIERS (39), *University of Rochester Cancer Center, Rochester, New York 14642*

DEMETRIOS PAPAHADJOPOULOS (32), *Cancer Research Institute and Department of Pharmacology, University of California, San Francisco, California 94143*

BRUCE M. PATERSON (43), *Laboratory of Biochemistry, National Cancer Institute, National Institutes of Health, Bethesda, Maryland 20205*

SIYING QIAN (29), *Institute of Economic Crops, Jiangsu Academy of Agricultural Sciences, Nanjing, Peoples Republic of China*

A. RAZIN (31), *Department of Cellular Biochemistry, Hebrew University–Hadassah Medical School, 91000 Jerusalem, Israel*

ROBERT P. RICCIARDI (43), *The Wistar Institute of Anatomy and Biology, Philadelphia, Pennsylvania 19104*

NIKOLAOS ROBAKIS (45), *Department of Microbiology, Hoffmann-La Roche Inc., Nutley, New Jersey 07110*

BRYAN E. ROBERTS (43), *Department of Biological Chemistry, Harvard Medical School, Boston, Massachusetts 02115*

ROBERT G. ROEDER (36), *Laboratory of Biochemistry and Molecular Genetics, The Rockefeller University, New York, New York 10021*

MARK ROSE (9), *Department of Biology, Massachusetts Institute of Technology and Whitehead Institute for Biomedical Research, Cambridge, Massachusetts 02139*

MARTIN ROSENBERG (6), *Department of Molecular Genetics, Smith Kline and French Laboratories, Philadelphia, Pennsylvania 19101*

RODNEY J. ROTHSTEIN (12, 14), *Department of Microbiology, UMDNJ–New Jersey Medical School, Newark, New Jersey 07103*

STEPHANIE W. RUBY (16), *Dana Farber Cancer Institute and Department of Biological Chemistry, Harvard Medical School, Boston, Massachusetts 02115*

FRANK H. RUDDLE (28), *Department of Biology and Human Genetics, Yale University, New Haven, Connecticut 06511*

MARK SAMUELS (35), *Center for Cancer Research, Massachusetts Institute of Technology, Cambridge, Massachusetts 02139*

ROZANNE M. SANDRI-GOLDIN (27), *Department of Human Genetics, University of Michigan Medical School, Ann Arbor, Michigan 48109*

NAVA SARVER (26), *Laboratory of Pathology, National Cancer Institute, National Institutes of Health, Bethesda, Maryland 20205*

PHILLIP A. SHARP (35), *Center for Cancer Research, Massachusetts Institute of Technology, Cambridge, Massachusetts 02139*

BARKUR S. SHASTRY (36), *Laboratory of Biochemistry and Molecular Genetics, The Rockefeller University, New York, New York 10021*

ALLAN SHATZMAN (6), *Department of Mo-*

lecular Genetics, Smith Kline and French Laboratories, Philadelphia, Pennsylvania 19101

PAULA H. SON (5), *Department of Biological Chemistry, Washington University School of Medicine, St. Louis, Missouri 63110*

JOHN I. STILES (19), *Department of Botany, Hawaii Institute of Tropical Agriculture and Human Resources, University of Hawaii, Honolulu, Hawaii 96822*

ROBERT M. STRAUBINGER (32), *Cancer Research Institute and Department of Pharmacology, University of California, San Francisco, California 94143*

J. WILLIAM STRAUS (41), *Department of Biochemistry, State University of New York, Stony Brook, New York 11794*

NEAL SUGAWARA (17), *Dana Farber Cancer Institute and Department of Biological Chemistry, Harvard Medical School, Boston, Massachusetts 02115*

JACK W. SZOSTAK (14, 15, 16, 17, 18), *Dana Farber Cancer Institute and Department of Biological Chemistry, Harvard Medical School, Boston, Massachusetts 02115*

A. VAINSTEIN (31), *Department of Biological Chemistry, The Hebrew University of Jerusalem, 91904 Jerusalem, Israel*

HERBERT WEISSBACH (45), *Department of Biochemistry, Roche Institute of Molecular Biology, Nutley, New Jersey 07110*

JIAN WENG (29), *Shanghai Institute of Biochemistry, Academia Sinica, Shanghai 200031, Peoples Republic of China*

M. P. WICKENS (25), *Department of Biochemistry, University of Wisconsin, Madison, Wisconsin 53706*

J. WIGGS (34), *Department of Biochemistry, University of California, Berkeley, California 94720*

FRED WINSTON (13), *Department of Biology, Massachusetts Institute of Technology and the Whitehead Institute for Biomedical Research, Cambridge, Massachusetts 02139*

CHARLES YANOFSKY (8), *Department of Biological Sciences, Stanford University, Stanford, California 94305*

GEORGE H. YOAKUM (7), *Laboratory of Human Carcinogenesis, National Cancer Institute, National Institutes of Health, Bethesda, Maryland 20205*

YISHEN ZENG (29), *Shanghai Institute of Biochemistry, Academia Sinica, Shanghai 200031, Peoples Republic of China*

GUANG-YU ZHOU (29), *Shanghai Institute of Biochemistry, Academia Sinica, Shanghai 200031, Peoples Republic of China*

GEOFFREY ZUBAY (44), *Fairchild Center for Biological Sciences, Columbia University, New York, New York 10027*

Preface

Exciting new developments in recombinant DNA research allow the isolation and amplification of specific genes or DNA segments from almost any living organism. These new developments have revolutionized our approaches to solving complex biological problems and have opened up new possibilities for producing new and better products in the areas of health, agriculture, and industry.

Volumes 100 and 101 supplement Volumes 65 and 68 of *Methods in Enzymology*. During the last three years, many new or improved methods on recombinant DNA or nucleic acids have appeared, and they are included in these two volumes. Volume 100 covers the use of enzymes in recombinant DNA research, enzymes affecting the gross morphology of DNA, proteins with specialized functions acting at specific loci, new methods for DNA isolation, hybridization, and cloning, analytical methods for gene products, and mutagenesis: *in vitro* and *in vivo*. Volume 101 includes sections on new vectors for cloning genes, cloning of genes into yeast cells, and systems for monitoring cloned gene expression.

RAY WU
LAWRENCE GROSSMAN
KIVIE MOLDAVE

METHODS IN ENZYMOLOGY

EDITED BY

Sidney P. Colowick and Nathan O. Kaplan

VANDERBILT UNIVERSITY
SCHOOL OF MEDICINE
NASHVILLE, TENNESSEE

DEPARTMENT OF CHEMISTRY
UNIVERSITY OF CALIFORNIA
AT SAN DIEGO
LA JOLLA, CALIFORNIA

METHODS IN ENZYMOLOGY

EDITORS-IN-CHIEF

Sidney P. Colowick   Nathan O. Kaplan

VOLUME VIII. Complex Carbohydrates
Edited by ELIZABETH F. NEUFELD AND VICTOR GINSBURG

VOLUME IX. Carbohydrate Metabolism
Edited by WILLIS A. WOOD

VOLUME X. Oxidation and Phosphorylation
Edited by RONALD W. ESTABROOK AND MAYNARD E. PULLMAN

VOLUME XI. Enzyme Structure
Edited by C. H. W. HIRS

VOLUME XII. Nucleic Acids (Parts A and B)
Edited by LAWRENCE GROSSMAN AND KIVIE MOLDAVE

VOLUME XIII. Citric Acid Cycle
Edited by J. M. LOWENSTEIN

VOLUME XIV. Lipids
Edited by J. M. LOWENSTEIN

VOLUME XV. Steroids and Terpenoids
Edited by RAYMOND B. CLAYTON

VOLUME XVI. Fast Reactions
Edited by KENNETH KUSTIN

VOLUME XVII. Metabolism of Amino Acids and Amines (Parts A and B)
Edited by HERBERT TABOR AND CELIA WHITE TABOR

VOLUME XVIII. Vitamins and Coenzymes (Parts A, B, and C)
Edited by DONALD B. MCCORMICK AND LEMUEL D. WRIGHT

VOLUME XIX. Proteolytic Enzymes
Edited by GERTRUDE E. PERLMANN AND LASZLO LORAND

VOLUME XX. Nucleic Acids and Protein Synthesis (Part C)
Edited by KIVIE MOLDAVE AND LAWRENCE GROSSMAN

VOLUME XXI. Nucleic Acids (Part D)
Edited by LAWRENCE GROSSMAN AND KIVIE MOLDAVE

VOLUME XXII. Enzyme Purification and Related Techniques
Edited by WILLIAM B. JAKOBY

VOLUME XXIII. Photosynthesis (Part A)
Edited by ANTHONY SAN PIETRO

VOLUME XXIV. Photosynthesis and Nitrogen Fixation (Part B)
Edited by ANTHONY SAN PIETRO

VOLUME XXV. Enzyme Structure (Part B)
Edited by C. H. W. HIRS AND SERGE N. TIMASHEFF

VOLUME XXVI. Enzyme Structure (Part C)
Edited by C. H. W. HIRS AND SERGE N. TIMASHEFF

VOLUME XXVII. Enzyme Structure (Part D)
Edited by C. H. W. HIRS AND SERGE N. TIMASHEFF

VOLUME XXVIII. Complex Carbohydrates (Part B)
Edited by VICTOR GINSBURG

VOLUME XXIX. Nucleic Acids and Protein Synthesis (Part E)
Edited by LAWRENCE GROSSMAN AND KIVIE MOLDAVE

VOLUME XXX. Nucleic Acids and Protein Synthesis (Part F)
Edited by KIVIE MOLDAVE AND LAWRENCE GROSSMAN

VOLUME XXXI. Biomembranes (Part A)
Edited by SIDNEY FLEISCHER AND LESTER PACKER

VOLUME XXXII. Biomembranes (Part B)
Edited by SIDNEY FLEISCHER AND LESTER PACKER

VOLUME XXXIII. Cumulative Subject Index Volumes I-XXX
Edited by MARTHA G. DENNIS AND EDWARD A. DENNIS

VOLUME XXXIV. Affinity Techniques (Enzyme Purification: Part B)
Edited by WILLIAM B. JAKOBY AND MEIR WILCHEK

VOLUME XXXV. Lipids (Part B)
Edited by JOHN M. LOWENSTEIN

VOLUME XXXVI. Hormone Action (Part A: Steroid Hormones)
Edited by BERT W. O'MALLEY AND JOEL G. HARDMAN

VOLUME XXXVII. Hormone Action (Part B: Peptide Hormones)
Edited by BERT W. O'MALLEY AND JOEL G. HARDMAN

VOLUME XXXVIII. Hormone Action (Part C: Cyclic Nucleotides)
Edited by JOEL G. HARDMAN AND BERT W. O'MALLEY

VOLUME XXXIX. Hormone Action (Part D: Isolated Cells, Tissues, and Organ Systems)
Edited by JOEL G. HARDMAN AND BERT W. O'MALLEY

VOLUME XL. Hormone Action (Part E: Nuclear Structure and Function)
Edited by BERT W. O'MALLEY AND JOEL G. HARDMAN

VOLUME XLI. Carbohydrate Metabolism (Part B)
Edited by W. A. WOOD

VOLUME XLII. Carbohydrate Metabolism (Part C)
Edited by W. A. WOOD

VOLUME XLIII. Antibiotics
Edited by JOHN H. HASH

VOLUME XLIV. Immobilized Enzymes
Edited by KLAUS MOSBACH

VOLUME XLV. Proteolytic Enzymes (Part B)
Edited by LASZLO LORAND

VOLUME XLVI. Affinity Labeling
Edited by WILLIAM B. JAKOBY AND MEIR WILCHEK

VOLUME XLVII. Enzyme Structure (Part E)
Edited by C. H. W. HIRS AND SERGE N. TIMASHEFF

VOLUME 62. Vitamins and Coenzymes (Part D)
Edited by DONALD B. MCCORMICK AND LEMUEL D. WRIGHT

VOLUME 63. Enzyme Kinetics and Mechanism (Part A: Initial Rate and Inhibitor Methods)
Edited by DANIEL L. PURICH

VOLUME 64. Enzyme Kinetics and Mechanism (Part B: Isotopic Probes and Complex Enzyme Systems)
Edited by DANIEL L. PURICH

VOLUME 65. Nucleic Acids (Part I)
Edited by LAWRENCE GROSSMAN AND KIVIE MOLDAVE

VOLUME 66. Vitamins and Coenzymes (Part E)
Edited by DONALD B. MCCORMICK AND LEMUEL D. WRIGHT

VOLUME 67. Vitamins and Coenzymes (Part F)
Edited by DONALD B. MCCORMICK AND LEMUEL D. WRIGHT

VOLUME 68. Recombinant DNA
Edited by RAY WU

VOLUME 69. Photosynthesis and Nitrogen Fixation (Part C)
Edited by ANTHONY SAN PIETRO

VOLUME 70. Immunochemical Techniques (Part A)
Edited by HELEN VAN VUNAKIS AND JOHN J. LANGONE

VOLUME 71. Lipids (Part C)
Edited by JOHN M. LOWENSTEIN

VOLUME 72. Lipids (Part D)
Edited by JOHN M. LOWENSTEIN

VOLUME 73. Immunochemical Techniques (Part B)
Edited by JOHN J. LANGONE AND HELEN VAN VUNAKIS

VOLUME 74. Immunochemical Techniques (Part C)
Edited by JOHN J. LANGONE AND HELEN VAN VUNAKIS

VOLUME 75. Cumulative Subject Index Volumes XXXI, XXXII, and XXXIV-LV
Edited by EDWARD A. DENNIS AND MARTHA G. DENNIS

VOLUME 76. Hemoglobins
Edited by ERALDO ANTONINI, LUIGI ROSSI-BERNARDI, AND EMILIA CHIANCONE

VOLUME 77. Detoxication and Drug Metabolism
Edited by WILLIAM B. JAKOBY

VOLUME 78. Interferons (Part A)
Edited by SIDNEY PESTKA

VOLUME 79. Interferons (Part B)
Edited by SIDNEY PESTKA

VOLUME 80. Proteolytic Enzymes (Part C)
Edited by LASZLO LORAND

VOLUME 81. Biomembranes (Part H: Visual Pigments and Purple Membranes, I)
Edited by LESTER PACKER

VOLUME 82. Structural and Contractile Proteins (Part A: Extracellular Matrix)
Edited by LEON W. CUNNINGHAM AND DIXIE W. FREDERIKSEN

VOLUME 83. Complex Carbohydrates (Part D)
Edited by VICTOR GINSBURG

VOLUME 84. Immunochemical Techniques (Part D: Selected Immunoassays)
Edited by JOHN J. LANGONE AND HELEN VAN VUNAKIS

VOLUME 85. Structural and Contractile Proteins (Part B: The Contractile Apparatus and the Cytoskeleton)
Edited by DIXIE W. FREDERIKSEN AND LEON W. CUNNINGHAM

VOLUME 86. Prostaglandins and Arachidonate Metabolites
Edited by WILLIAM E. M. LANDS AND WILLIAM L. SMITH

VOLUME 87. Enzyme Kinetics and Mechanism (Part C: Intermediates, Stereochemistry, and Rate Studies)
Edited by DANIEL L. PURICH

VOLUME 88. Biomembranes (Part I: Visual Pigments and Purple Membranes, II)
Edited by LESTER PACKER

VOLUME 89. Carbohydrate Metabolism (Part D)
Edited by WILLIS A. WOOD

VOLUME 90. Carbohydrate Metabolism (Part E)
Edited by WILLIS A. WOOD

VOLUME 91. Enzyme Structure (Part I)
Edited by C. H. W. HIRS AND SERGE N. TIMASHEFF

VOLUME 92. Immunochemical Techniques (Part E: Monoclonal Antibodies and General Immunoassay Methods)
Edited by JOHN J. LANGONE AND HELEN VAN VUNAKIS

VOLUME 93. Immunochemical Techniques (Part F: Conventional Antibodies, Fc Receptors, and Cytotoxicity)
Edited by JOHN J. LANGONE AND HELEN VAN VUNAKIS

VOLUME 94. Polyamines (in preparation)
Edited by HERBERT TABOR AND CELIA WHITE TABOR

VOLUME 95. Cumulative Subject Index Volumes 61–74 and 76–80 (in preparation)
Edited by EDWARD A. DENNIS AND MARTHA G. DENNIS

VOLUME 96. Biomembranes (Part J: Membrane Biogenesis: Assembly and Targeting (General Methods; Eukaryotes)) (in preparation)
Edited by SIDNEY FLEISCHER AND BECCA FLEISCHER

VOLUME 97. Biomembranes (Part K: Membrane Biogenesis: Assembly and Targeting (Prokaryotes, Mitochondria, and Chloroplasts)) (in preparation)
Edited by SIDNEY FLEISCHER AND BECCA FLEISCHER

VOLUME 98. Biomembranes (Part L: Membrane Biogenesis (Processing and Recycling)) (in preparation)
Edited by SIDNEY FLEISCHER AND BECCA FLEISCHER

VOLUME 99. Hormone Action (Part F: Protein Kinases) (in preparation)
Edited by JACKIE D. CORBIN AND JOEL G. HARDMAN

VOLUME 100. Recombinant DNA (Part B)
Edited by RAY WU, LAWRENCE GROSSMAN, AND KIVIE MOLDAVE

VOLUME 101. Recombinant DNA (Part C)
Edited by RAY WU, LAWRENCE GROSSMAN, AND KIVIE MOLDAVE

VOLUME 102. Hormone Action (Part G: Calmodulin and Calcium-Binding Proteins) (in preparation)
Edited by ANTHONY R. MEANS AND BERT W. O'MALLEY

VOLUME 103. Hormone Action (Part H: Neuroendocrine Peptides) (in preparation)
Edited by P. MICHAEL CONN

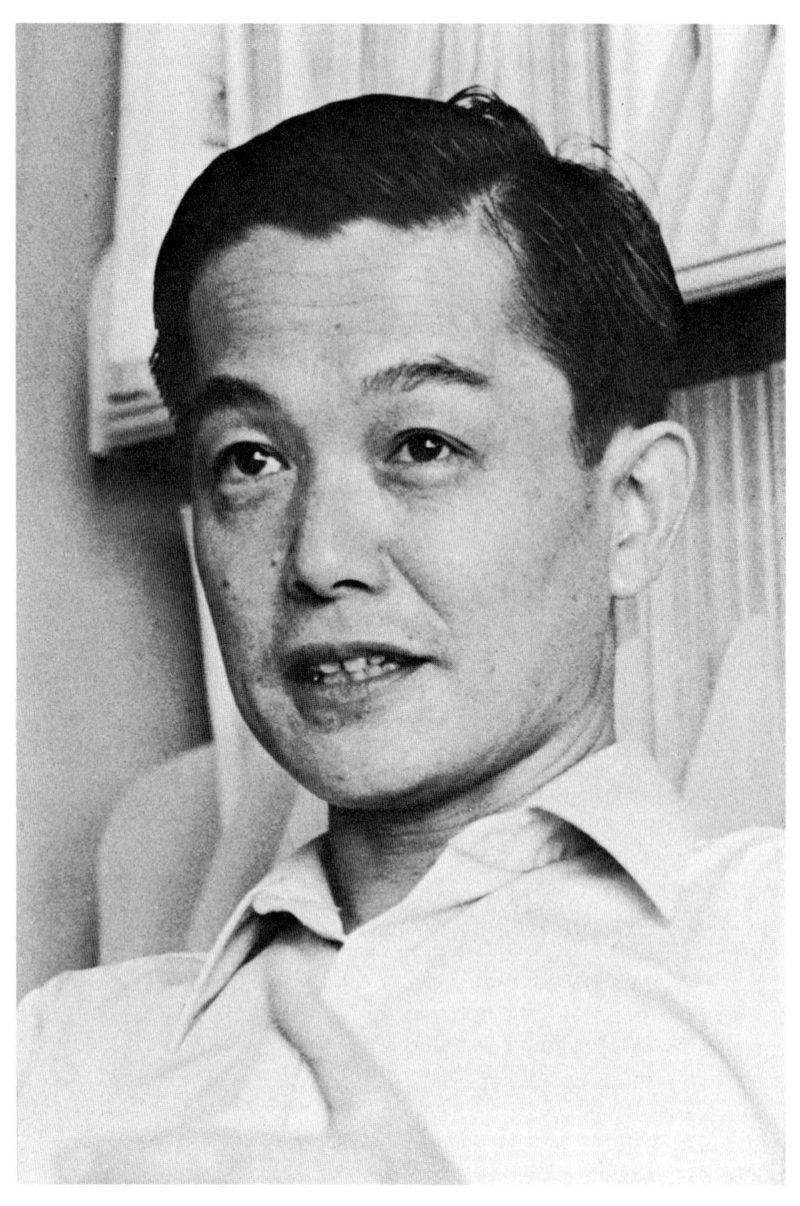

Reiji Okazaki
1930–1975

Reiji Okazaki (1930–1975)

Reiji Okazaki has been memorialized by the nascent DNA replication fragments that bear his name. His discovery of the Okazaki fragments in the discontinuous synthesis of DNA at the replication fork helped solve a perplexing problem: how DNA polymerases with an invariant unidirectional mode of synthesis can copy the oppositely oriented strands of the duplex chromosome. Those of us who knew him do not require the adjectival use of his name to keep his memory alive. We retain the image of a scientist utterly dedicated to understanding the molecular basis of biology.

Reiji Okazaki was born in Hiroshima in 1930 and received his Ph.D. training in developmental biology under Tsuneo Yamada at Nagoya University.[1] In seeking systems simpler than sea urchins to study cell proliferation, he chose *Lactobacillus* and *Escherichia coli* in which he discovered thymidine diphosphate rhamnose, the coenzyme of lipopolysaccharide synthesis. With J. L. Strominger in St. Louis in 1960–1961 he worked out the enzymatic synthesis of this coenzyme. In my laboratory, the following year, he purified thymidine kinase of *E. coli* and demonstrated the allosteric regulation of this key salvage enzyme. On returning to Nagoya, as Professor of Molecular Biology, he initiated the series of elegant studies of phage T4 DNA replication that led to his key discovery of discontinuous replication.

His bibliography of some thirty papers (1966–1977) can be consulted for the innovative approaches he introduced to solve fundamental questions of DNA replication. His research style, less readily gleaned from the literature, is illustrated by two incidents which are vivid in my memory.

One I call an Okazaki maneuver. In purifying thymidine kinase, he used a heating step: the enzyme was held in a test tube at 70° for 5 minutes. When he decided to prepare a large amount of enzyme, going from a scale of 10 milliliters to several liters, he simply repeated the same heating procedure, this time using *236* test tubes. I was embarrassed to report such an unsophisticated procedure. But then I realized that he was able to complete this step in a few hours, and saw no point in wasting precious days and material learning how to do the heating in a big beaker or flask. Recently, one of my colleagues purified the single-strand DNA binding protein with a heating step. When he came to scale-up the procedure from

[1] An appreciative obituary by Sakaru Suzuki appeared in *Trends in Biochemical Sciences* **1,** N39 (Feb. 1976).

3 milliliters to 6 liters he was guided by the Okazaki maneuver; he heated 2000 test tubes each containing 3 milliliters. Others who tried heating larger volumes of enzyme lost the preparation in a thick coagulum.

A second incident I call Okazaki courage. It had been customary in my laboratory when characterizing an enzyme to set up protocols containing 10 to 20 assay tubes. Rarely, some ambitious person might do a 24-tube assay. Reiji set a record that may never be broken. He performed a 128-tube assay of thymidine kinase, even though each assay included a laborious electrophoretic separation of the product from the substrate. Because the pure enzyme was rather labile he felt it essential to measure at once all the substrate, effector, inhibitor, and other parameters. The successful completion of this experiment was a feat of courage, concentration, skill, and enterprise unique in my experience.

Okazaki died of leukemia in 1975, a sudden and cruel loss to his wife and co-worker, Tuneko, to his devoted students, and to the worldwide scientific community. The continued productivity of his laboratory by his students under Tuneko Okazaki's direction is a tribute to its scientific prowess and to Reiji Okazaki's inspirational legacy.

ARTHUR KORNBERG
Department of Biochemistry
Stanford University School of Medicine
Stanford, California

Section I

New Vectors for Cloning Genes

[1] New Bacteriophage Lambda Vectors with Positive Selection for Cloned Inserts

By JONATHAN KARN, SYDNEY BRENNER, and LESLIE BARNETT

Molecular cloning methods eliminated the necessity for physical fractionation of DNA and permitted, for the first time, the isolation of eukaryotic structural genes.[1-7] In principle, any eukaryotic gene may be isolated from a pool of cloned fragments large enough to give sequence representation of an entire genome. A simple multicellular eukaryote such as *Caenorhabditis elegans* has a haploid DNA content of approximately 8×10^7 bp.[8] Assuming random DNA cleavage and uniform cloning efficiency, a collection of 8×10^4 clones with an average length of 10^4 bp will include any genomic sequence with greater than 99% probability. Similarly, the human genome with 2×10^9 bp will be represented by 10^6 clones of 10^4 bp length.[6] Clones of interest are then identified in these genome "libraries" by hybridization and other assays, and flanking sequences can be obtained in subsequent "walking" steps.

Bacteriophage lambda cloning vectors offer a number of technical advantages that make them attractive vehicles for the construction of genome libraries.[9] DNA fragments of up to 22 kb may be stably maintained, and recombinants in bacteriophage lambda may be efficiently recovered by *in vitro* packaging. The primary pool of clones may be amplified without significant loss of sequences from the population by limited growth of the phage. Subsequently the entire collection may then be stored as bacteriophage lysates for long periods. Finally, bacteriophage plaques from the amplified pools may readily be screened by the rapid and sensitive

[1] P. C. Wensink, D. J. Finnegan, J. E. Donelson, and D. Hogness, *Cell* **3,** 315 (1974).
[2] M. Thomas, J. R. Cameron, and R. W. Davis, *Proc. Natl. Acad. Sci. U.S.A.* **71,** 4579 (1974).
[3] L. Clarke, and J. Carbon, *Cell* **9,** 91 (1976).
[4] S. M. Tilghman, D. C. Tiemeier, F. Polsky, M. H. Edgell, J. G. Seidman, A. Leder, L. W. Enquist, B. Norman, and P. Leder, *Proc. Natl. Acad. Sci. U.S.A.* **75,** 725 (1978).
[5] S. Tonegawa, C. Brach, N. Hozumi, and R. Scholler, *Proc. Natl. Acad. Sci. U.S.A.* **74,** 3518 (1977).
[6] T. Maniatis, R. C. Hardison, E. Lacy, J. Lauer, C. O'Connell, and D. Quon, *Cell* **15,** 687 (1978).
[7] F. R. Blattner, A. E. Blechl, K. Denniston-Thompson, M. E. Faber, J. E. Richards, J. L. Slighton, P. W. Tucker, and O. Smithies, *Science* **202,** 1279 (1978).
[8] J. E. Sulston, and S. Brenner, *Genetics* **77,** 95 (1974).
[9] N. E. Murray, *in* "The Bacteriophage Lambda II," Cold Spring Harbor Laboratory, Cold Spring Harbor, New York.

ISBN 0-12-182001-7

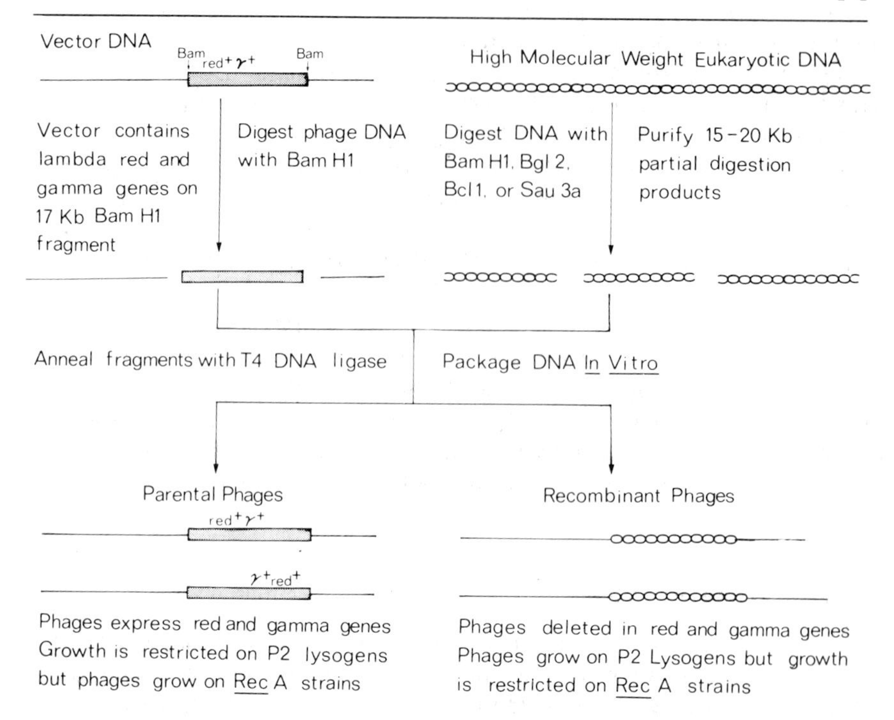

FIG. 1. Schematic diagram outlining the construction of recombinants using the λ1059 vector.

plaque hybridization method of Benton and Davis,[10] genetic selections,[11,12] or immunological assays[13-16] that take advantage of the high levels of transcription that may be achieved with clones in bacteriophage.

Most bacteriophage vectors are substitution vectors that require internal filler fragments to be physically separated from the vector arms before insertions of foreign DNA.[2,6,7,9,17] This step is inefficient and leads to the contamination of the recombinant phage pools with phages harboring one

[10] W. D. Benton, and R. W. Davis, *Science* **196,** 180 (1977).

[11] B. Seed, unpublished results.

[12] M. Goldfarb, K. Shimizu, M. Pervcho, and M. Wigler, *Nature* (*London*) **296,** 404 (1982).

[13] B. Sanzey, T. Mercereau, T. Ternynck, and P. Kourilsky, *Proc. Natl. Acad. Sci. U.S.A.* **73,** 3394 (1976).

[14] A. Skalka, and L. Shapiro, *in* "Eucaryotic Genetics Systems" (*ICN-UCLA Symp. Mol. Cell. Biol.* **8**), p. 123. Academic Press, New York, 1977.

[15] S. Broome, and W. Gilbert, *Proc. Natl. Acad. Sci. U.S.A.* **75,** 2746 (1978).

[16] D. J. Kemp, and A. F. Cowman, *Proc. Natl. Acad. Sci. U.S.A.* **78,** 4520 (1981).

[17] N. E. Murray, and K. Murray, *Nature* (*London*) **251,** 476 (1974).

or more of these fragments.[6,7] Some years ago we developed a bacteriophage lambda *Bam*HI cloning vector, lambda 1059, with a positive selection for cloned inserts.[18] This feature allows construction of recombinants in lambda without separation of the phage arms. A schematic diagram outlining the strategy we have adopted for cloning in bacteriophage 1059 (and derivative strains) is shown in Fig. 1. Genomic DNA is partially digested with restriction endonucleases to produce a population of DNA fragments from which molecules 15–20 kb long are purified by agarose gel electrophoresis. The size-selected fragments are ligated with T4 DNA ligase to the arms of the phage vector cleaved with an appropriate enzyme. Viable phage particles are recovered by *in vitro* packaging of the ligated DNAs, and a permanent collection of recombinant phages is then established by allowing the phages harboring inserts to amplify through several generations of growth on a strain that restricts the growth of the original vector. Clones of interest are then identified by hybridization with specific probes.

Principle of the Method

Our selection scheme for inserts is based on the spi phenotype of lambda. Spi^- derivatives of phage lambda are phages that can form plaques on *Escherichia coli* strains lysogenic for the temperate phage P2. This phenomenon was first described by Zissler *et al.*,[19] who demonstrated that concomitant loss of several lambda early functions at the *red* and *gamma* loci was required for full expression of the phenotype. We reasoned that if the *red* and *gamma* genes were placed on a central fragment in a bacteriophage lambda vector, then recombinants that substituted foreign DNA for this fragment should be spi^- and distinguished from the parent vector by plating on P2 containing strains. In order to ensure that the *red* and *gamma* genes were expressed in either orientation of the central fragment, we placed these genes under pL control, and specific *chi* mutations[20,21] were introduced into the vector arms in order to assure good growth of the recombinant phages. Selection for the spi phenotype alone does not distinguish between phages that harbor foreign DNA fragments and phages that have simply deleted the central fragment. We took advantage of lambda's packaging requirements to complete the selection

[18] J. Karn, S. Brenner, L. Barnett, and G. Cesareni, *Proc. Natl. Acad. Sci. U.S.A.* **77,** 5172 (1980).

[19] J. Zissler, E. R. Signer, and F. Schaefer, *in* "The Bacteriophage Lambda" (A. D. Hershey, ed.), p. 455. Cold Spring Harbor Laboratory, Cold Spring Harbor, New York, 1971.

[20] F. W. Stahl, J. M. Craseman, and M. M. Stahl, *J. Mol. Biol.* **94,** 203 (1975).

[21] D. Henderson, and J. Weil, *Genetics* **79,** 143 (1975).

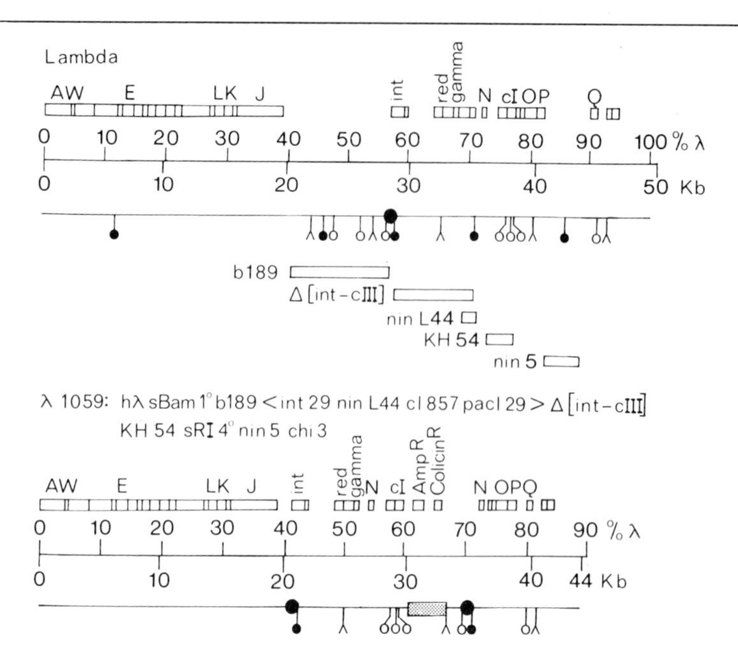

FIG. 2. Structure of λ1059. The top panel shows the *Bam*HI (●), *Eco*RI (/\), and *Hin*dIII (○) restriction maps of lambda, and the position of many of the known lambda genes. The bars underneath the lambda map indicate the map positions of the deletions used in the construction of λ1059. A restriction map of λ1059 is shown here. The left arm of the phage carries the λ structural genes *A*–*J*. The *s*baml° mutation and the *b*189 deletion remove the *Bam*HI sites from this arm. The central fragment carries the sequence from the first *att* site (△●P′, shown on the map as a large filled circle) to the *Bgl* site at coordinate 745 in the *cro* gene. At this juncture sequences from the mini ColE1 plasmid pACL29 (stippled region) are introduced. This plasmid introduces the β-lactamase gene (*Amp*^{R}) and colicin immunity gene (*Colicin*^{R}). The central fragment terminates in a duplicated λ*att* site (P●P′). This sequence is present in wild-type lambda from the *Eco*RI site at coordinate 543 to the *Bam*HI site at 578. The *Bam*HI site at 714 has been removed from the central fragment by the *nin*L44 deletion. The short right arm carries a deletion Δ[*int-c*III] originally made *in vitro* by removing DNA from between the two *Bam*HI sites at 580 and 714, the KH54 deletion, which removes the *rex* and *c*I genes, and the *nin*5 deletion. Substitution of the central fragment produces a spi^{-} phage with a *b*189 arm, a single λ*att* site, a 9–22 kb insert cloned between the *Bam*HI sites at 580 and 714, and an immunity arm with the KH54 and *nin*5 deletions. The growth of these phages is enhanced by the *chi* D mutation present on the right arm of the vector.

scheme. Lambdoid phages require genome sizes of between 0.7 and 1.08 of the wild-type DNA properly to fill the phage heads,[22,23] yet all the essential functions required for lambda growth and maturation can be obtained on DNA fragments of approximately 0.6 the genome size. By using

[22] J. Weil, R. Cunningham, R. Martin III, E. Mitchell, and B. Bolling, *Virology* **50,** 373 (1972).

[23] N. Sternberg, and R. Weisberg, *Nature* (*London*) **256,** 97 (1975).

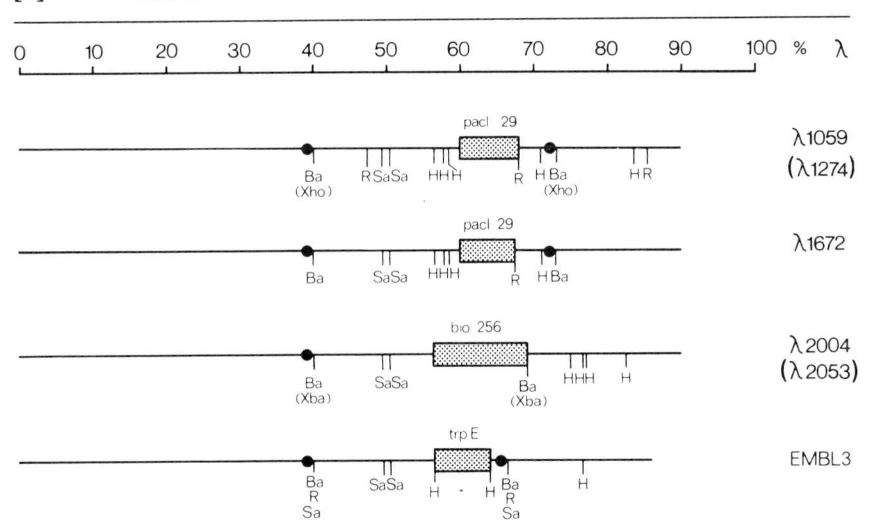

FIG. 3. Restriction endonuclease cleavage maps of λ1059, λ1274, λ1672, λ2004, λ2053, EMBL3. Sites of cleavage for *Bam*HI (Ba), *Hin*dIII (H), *Eco*RI (R), *Sal*I (Sa), *Xba*I (Xba), and *Xho*I (Xho) are indicated. Genotypes are given in Table II. In 2004 and 2053 the pACL29 plasmid has been replaced by an *Eco*RI-*Bam*HI fragment from the *bio*256 substitution in Charon 4a. This removed the *c*I857 gene and the *Hin*dIII sites from the central fragment of the phage. In EMBL3 a *Hin*dIII fragment carrying the *trpE* gene replaces pACL29.

a set of naturally occurring[24–27] and enzymically generated[28] deletions, we were able to construct vectors with appropriately short arms.[29] It should be noted that the packaging requirement places both an upper and a lower limit on the size of DNA fragments to be cloned in the bacteriophage and that this must be taken into account when designing cloning experiments.

Structure of the Bacteriophage Vectors

The restriction endonuclease cleavage maps of our original vector, lambda 1059, is given in Fig. 2, and a number of derivative strains are shown in Fig. 3 and in Table I. The phages are each composed of three fragments, separable by cleavage with an appropriate restriction enzyme: a 19.6 kb left arm carrying the genes for the lambda head and tail proteins, a 12–14 kb central fragment carrying the *red* and *gamma* genes under pL

[24] R. W. Davis, and J. S. Parkinson, *J. Mol. Biol.* **56,** 403 (1971).

[25] J. S. Salstrom, M. Fiandt, and W. Szybalski, *Mol. Gen. Genet.* **168,** 211 (1979).

[26] F. R. Blattner, M. Fiandt, K. K. Hass, P. A. Twose, and W. Szybalski, *Virology* **62,** 458 (1974).

[27] D. Court, and K. Sato, *Virology* **39,** 348 (1969).

[28] L. Enquist, and R. A. Weisberg, *J. Mol. Biol.* **111,** 97 (1979).

[29] S. Brenner, G. Cesareni, and J. Karn, *Gene* **17,** 27 (1982).

TABLE I
LAMBDA CLONING VECTORS WITH POSITIVE SELECTION FOR INSERTS

Strain	Genotype	*Chi*	Cloning sites	Capacity (kb)
1059	*h*λ*s*bam1°*b*189⟨*int*29*nin*L44*c*I857pACL29⟩ Δ[*int*-*c*III]KH54*s*RI4°*nin*5	D	*Bam*HI	9–22
1672	*h*λ*s*bam1°*b*189⟨*int*29*s*RI3°*nin*L44*c*I857 pACL29⟩Δ[*int*-*c*III]KH54*s*RI4°*nin*5 *s*RI5°*sHin*dIII6°	C	*Bam*HI	9–22
2004	*h*λ*s*bam1°*b*189*att int29s*RI3°ninL44 Δ[*sHin*dIII3-*sHin*dIII5]Σ*bio*256Δ[*int*-*c*III]*c*I857*s*RI4°*nin*5*s*RI5°	C	*Bam*HI	7–20
1259	*h*λ*s*bam1°E*am*2001K*am*424*b*189 ⟨*int*29*nin*L44pACL29⟩Δ[*int*-*c*III] KH54*s*RI4°*nin*5	D	*Bam*HI	9–22
1274	*Xho*I linker in 1059 *Bam*HI sites	D	*Bam*HI, *Xho*I	9–22
2053	*Xba*I linker in 2004 *Bam*HI sites	C	*Xba*I	7–20
2149	*h*λ*s*bam1°*b*189*att int* (*Xba*I)*nin*L44 Δ[*sHin*dIII3-*sHin*dIII5]Σ*bio*256*int* (*Xba*I) [*int*-*c*III]*c*I857	C	*Xba*I	5–18
EMBL3,4	*h*λ*s*bam1°*b*189⟨*int*(linker)*nin*L44 Δ[*sHin*dIII3-*sHin*dIII5]Σ*trp*E⟩*int*(linker) Δ[*int*-*c*III]KH54*s*RI4°*nin*5*s*RI5°	D	*Eco*RI, *Bam*HI, *Sal*I	9–22

control, and a 9–11 kb right arm carrying the lambda replication and lysis genes from which the *red* and *gamma* genes have been deleted. The two arms of the vector contain all the essential functions required for lambda replication and maturation in a DNA sequence less than 65% of the wild-type length. Viable phages are produced when these arms are annealed with internal DNA fragments between 5 and 22 kb; however, the two arms together do not produce viable phages. The left arms of all our phages carry the b189 deletion (17.5%)[24] and the *s*baml° mutation[30] removing the *Bam*HI site in the *D* gene. The right arms are all deleted between the *Bam*HI sites in the lambda *int* gene and the *c*III gene (13.1%)[28] and have defined *chi* sites (either *chi* C or *chi* D) that have been introduced to ensure efficient growth of the recombinant spi phages.[20,21]

Most of the vectors we have constructed are "phasmid" vectors and carry a ColE1 type plasmid (pACL29) on the central fragment.[29] This proved to be a disadvantage in some experiments since commonly used ColE1 plasmid probes such as pBR322[31] will cross-hybridize with these

[30] B. Klein, and K. Murray, *J. Mol. Biol.* **133,** 289 (1979).

[31] F. Bolivar, R. L. Rodriguez, P. J. Greene, M. C. Betlach, H. L. Heyneker, H. W. Boyer, J. H. Crosa, and S. Falkow, *Gene* **2,** 95 (1977).

sequences and detect those parental phages that survive the spi selection procedure. Some derivatives of 1059 have therefore been constructed that substitute other DNA fragments for the plasmid component. We cloned a fragment of biotin operon from a *bio*256[32,33] phage between the first *Hin*dIII site in 1672 (in the *c*I gene) and the *Bam*HI site on the right arm of lambda 1129.[29] (Note that the fragment is inverted compared with normal *bio*256 transducing phages and that one lambda *att* site is deleted.) Lehrach *et al.*[34] have prepared similar derivatives that substitute a *Hin*dIII fragment carrying the *E. coli trpE* gene for the plasmid component (EMBL 3,4).

Other derivatives of 1059, which introduce defined amber mutations or alter the restriction enzyme sites on the vector, have also been constructed.[35] The *Xho*I and *Xba*I vectors were prepared by cloning synthetic oligonucleotide linkers into the *Bam*HI sites of parental phages. These linkers, were decamers composed of G-A-T-C followed by the relevant restriction site (i.e., G-A-T-C-C-T-C-G-A-G and G-A-T-C-T-C-T-A-G-A). These self-anneal to yield double-stranded hexanucleotide sequences with G-A-T-C sticky ends, which may be cloned directly into the *Bam*HI sites. The *Xho*I linker maintains the *Bam*HI site, whereas the *Xba*I linker destroys it. The derivatives with amber mutations are of use in genetic selection experiments (see below) as well as providing biological containment.

We now describe the use of these vectors in detail.

Growth of Bacteriophage

Media

- CY broth: 10 g of Difco casamino acids, 5 g of Difco Bacto yeast extract, 3 g of NaCl, 2 g of KCl adjusted to pH 7.0. For most experiments this is supplemented with 10 m*M* Tris-HCl, pH 7.4, and 10 m*M* $MgCl_2$
- Lambda dil: 10 m*M* Tris-HCl, pH 7.4, 5 m*M* $MgSO_4$, 0.2 *M* NaCl, 0.1% gelatin
- Lambda agar: 10 g of Difco Bacto-tryptone, 2.5 g of NaCl, 12 g of agar (bottom) or 6 g of agar (top) per plate

[32] E. R. Signer, K. F. Manly, and M. Brunstetter, *Virology* **39,** 137 (1969).

[33] F. R. Blattner, B. G. Williams, A. E. Blechl, K. Denniston-Thompson, H. E. Faber, L. A. Furlong, D. J. Grunwald, D. O. Kiefer, O. D. Moore, J. W. Schumm, E. L. Sheldon, and O. Smithies, *Science* **196,** 161 (1977).

[34] H. Lehrach, and N. Murray, in preparation.

[35] J. Karn, H. Mattes, M. Gait, L. Barnett, and S. Brenner, *Gene,* in press (1983).

Bacterial Strains

Lambda 1059 and its derivative strains will grow on any lambda-sensitive host. The stringency of the spi selection scheme varies markedly with different strains. In general, *E. coli* C strains harboring P2 are more stringent than the corresponding K strains; however, we routinely work with K strains that are derivatives of C600 (the Q series strains, Table II), since we have found that recombinants grow considerably better on these strains. It is important to use strains that are restriction-deficient in the initial plating of bacteriophage clone collections to prevent loss of recombinants that introduce unmodified restriction sites. Accordingly, we have introduced the hsr^{-}_{K} hsm^{+}_{K} alleles into our set of isogenic plating strains. Derivatives of 1059 harboring amber mutations must be plated on hosts carrying the appropriate supressor mutations. Table II lists the genotypes and origins of the bacterial strains. The P2 lysogens will segregate on long-term storage in stabs, and it is advisable to keep master stocks as glycerinated cultures at $-70°$.

Phage DNA Preparation

Recombinants in lambda 1059 and related strains grow well, and titers of 10^9 to 10^{10} PFU per milliliter of lysate may be expected. Bacteriophage were grown as liquid lysates on Q358 bacteria using CY medium supplemented with 25 m*M* Tris-HCl, pH 7.4, and 10 m*M* $MgCl_2$. Early log-phase cultures were inoculated with the phage from a single purified plaque. Occasionally these starter cultures fail to lyse after 5–7 hr of growth and the bacteria approach saturation. Tenfold dilution of the cul-

TABLE II
BACTERIAL STRAINS

Strain	Relevant features	Source
EQ82	su_{II}^{+} su_{III}^{+} hsr_{K}^{-} hsm_{K}^{+}	N. Murray
Q276	*rec*A1 su_{II}^{+}	Cambridge
Q342	*rec*A1 su_{II}^{+} su_{III}^{+}	Cambridge
Q358	su_{II}^{+} hsr_{K}^{-} hsm_{K}^{+}	Cambridge
Q359	su_{II}^{+} hsr_{K}^{-} hsm_{K}^{+} P2	Cambridge
Q360	su_{II}^{+} P2	Cambridge
Q364	su_{II}^{+} hsr_{K}^{-} hsm_{K}^{+} P2 Δ[*lac-pro*]	Cambridge
CQ6	*E. coli* C, P2	G. Bertani
WR3	*rec*A1 $su°$	M. Gottesman
D91	Δ[*lac-pro*]	Cambridge
WX71	su_{III}^{+} P2	I. Herskowitz

tures with fresh media allows renewed growth of the bacteria, and lysis usually ensues after 3–4 hr. DNA was prepared from 1-liter cultures inoculated with 2–5 ml of the primary lysate. The phages were recovered from lysates by precipitation with 70 g of polyethylene glycol (PEG-6000) per liter and purified by two cycles of the CsCl density gradient centrifugation.[36] DNA was extracted from concentrated, dialyzed, phage suspensions by phenol extraction and stored at a concentration of 0.5–2.5 mg/ml in 10 m*M* Tris-HCl, 10 m*M* NaCl, 0.1 m*M* EDTA.

Amplifying the Clone Collection

Recombinant phage were plated at a density of approximately 2000 plaques per 10-cm dish of Q359 bacteria. Plate stocks were prepared as follows: 5 ml of lambda dil were added to each dish, and the top agar was scraped off. The agar suspension was vortexed, and bacteria, agar, and debris were removed by centrifugation at 5000 rpm for 10 min in a Sorvall GLC centrifuge. The extracted phage, which typically had titers of 10^9 per milliliter, were stored over chloroform at 4°.

Preparation of DNA Fragments

Random Fragments

Genomic DNA suitable for insertion into the spi vectors (Table I) may be prepared with a variety of enzymes. Vectors with *Bam*HI sites can accommodate fragments prepared with *Bam*HI, *Bgl*II, *Bcl*I, *Sau*3a, or *Mbo*I. Vectors with *Xho*I sites can accommodate fragments prepared with either *Sal*I or *Xho*I. Cleavage of the DNA with a restriction enzyme with a four base-pair recognition sequence, such as *Sau*3a, produces a nearly random population of fragments, whereas cleavage to completion with restriction enzymes with larger recognition sequences allows purification of particular sequences. *Sau*3a cleaves at the sequence G-A-T-C and leaves a tetranucleotide extension.[37,38] These fragments may therefore be cloned directly into *Bam*HI sites (G-G-A-T-C-C-) without linker addition.[18,38,39] The *Sau*3a sites should occur once every 256 bp in DNA with 50% G+C, and only $\frac{1}{80}$th of these sites need to be cleaved to produce

[36] K. R. Yamamoto, B. M. Alberts, R. Benzinger, L. Hawthorne, and C. Treiber, *Virology* **46,** 734 (1970).

[37] J. S. Sussenbach, C. H. Monfoort, R. Schipof, and E. C. Stobberingh, *Nucleic Acids Res.* **3,** 3193 (1976).

[38] R. J. Roberts, *CRC Crit. Rev. Biochem.* **4,** 123 (1976).

[39] G. A. Wilson, and F. E. Young, *J. Mol. Biol.* **97,** 123 (1975).

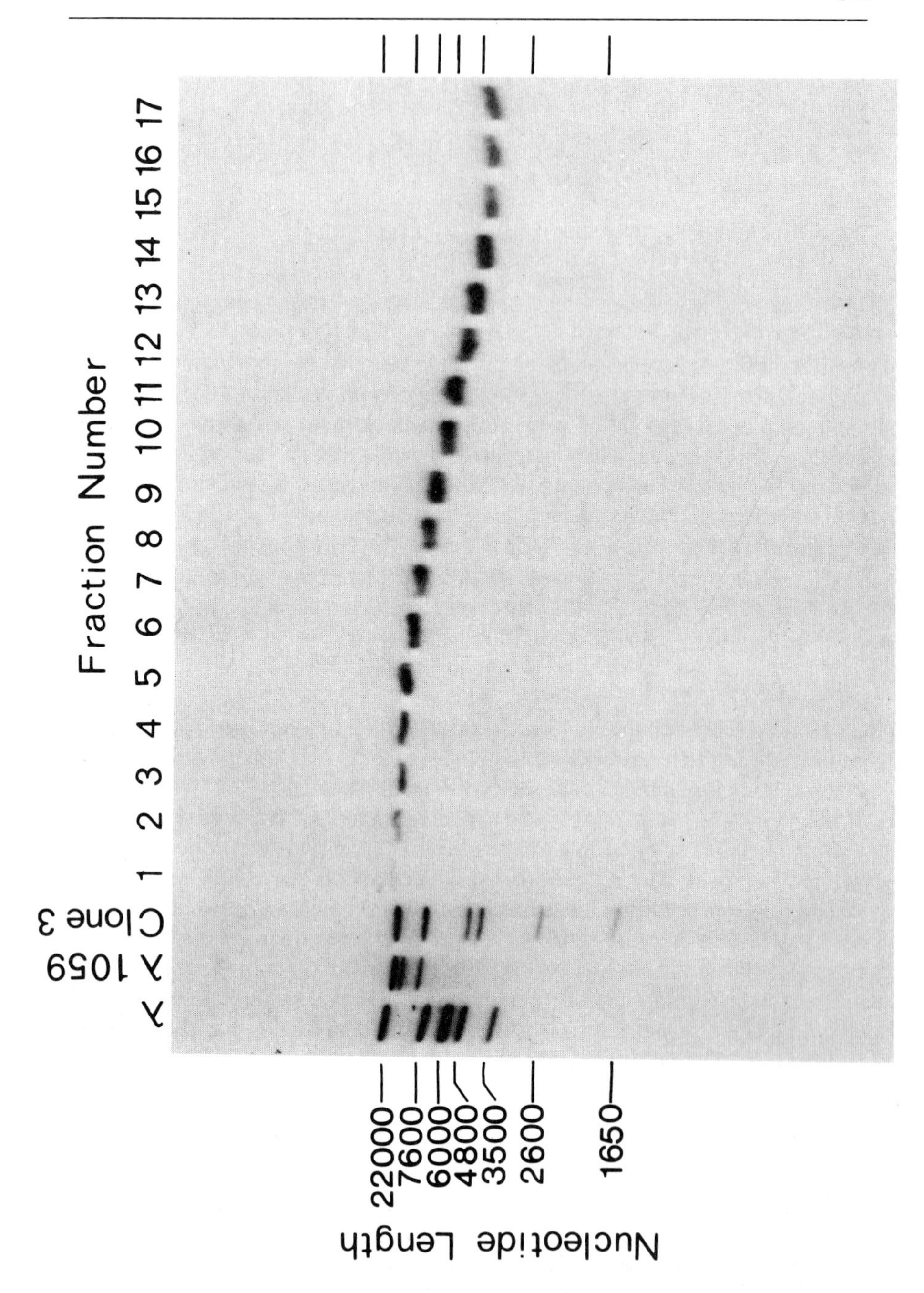
Fraction Number
λ
λ 1059
Clone 3
1
2
3
4
5
6
7
8
9
10
11
12
13
14
15
16
17
22000
7600
6000
4800
3500
2600
1650
Nucleotide Length

FIG. 4. Fractionation of partially digested nematode DNA. Nematode DNA (N2 DNA) was prepared from frozen animals purified by flotation on sucrose.[8] The worms were pulverized by grinding in a mortar chilled with liquid nitrogen. DNA was released from the disrupted worms by suspending the animals in 1% SDS, 100 m*M* Tris-HCl, pH 7.4, 1 m*M* EDTA using 100 ml of buffer per 5 g wet weight of worms. The viscous suspension was extracted with phenol and then phenol–chloroform–isoamyl alcohol (25:24:1), and crude high molecular weight DNA was precipitated by addition of 2 volumes of ethanol. This preparation was further purified by CsCl density gradient centrifugation. Purified DNA was stored at 500 μg/ml in 10 m*M* Tris-HCl, pH 7.4, 10 m*M* NaCl, 0.1 m*M* EDTA at 4°.

Analysis of this material on neutral agarose gels showed the DNA to be greater than 100 kb. N2 DNA was digested with *Bam*HI or *Sau*3a for 1 hr at 37° in a buffer containing 10 m*M* Tris-HCl, pH 7.4, 10 m*M* $MgCl_2$, 10 m*M* 2-mercaptoethanol, 50 m*M* NaCl. Aliquots of 20 μg of DNA were digested in 100-μl reactions containing 0.1, 0.2, 0.5, 1.0, and 2.0 units of *Sau*3a or 1, 2, 5, 10, and 20 units of *Bam*HI. The reaction mixes prepared with each enzyme were pooled, and an aliquot containing 1 μg of DNA was end-labeled by incubation with 0.1 unit of *E. coli* DNA polymerase I large subunit (Boehringer) in a 10-μl reaction mix containing 10 μCi of [α-^{32}P]dATP (350 mCi/mmol) 500 μ*M* dCTP, 500 μ*M* dGTP, 500 μ*M* dTTP, 10 m*M* Tris-HCl, pH 7.4, 10 m*M* $MgCl_2$, 0.1 m*M* DTT, 50 m*M* NaCl. After incubation for 20 min at 25°, the reaction was terminated by heat inactivation of the polymerase at 70° for 5 min. The labeled DNA was mixed with the remaining DNA, and the sample was extracted with phenol and then ether. Residual phenol and unincorporated triphosphates were removed by chromatography of the sample on small columns of Sepharose 4B equilibrated with 10 m*M* Tris-HCl, pH 7.4, 10 m*M* NaCl, 0.1 m*M* EDTA.

The excluded peak was concentrated by ethanol precipitation and redissolved at a final DNA concentration of 500 μg/ml. Aliquots containing 50 μg of labeled, digested DNA were fractionated by electrophoresis on columns of 0.5% low melting temperature agarose (BRL). Gels were cast in 1.5 × 20 cm tubes sealed at one end by a piece of dialysis tubing fixed with an elastic band. A flat upper surface was obtained by overlayering the melted agarose with a small layer of butan-2-ol. Samples were applied in 0.3% agarose containing 0.01% bromophenol blue and 0.01% xylene cyanole fast tracking dyes. Electrophoresis was for approximately 18 hr at 150 V, after which time the xylene cyanole dye had moved approximately 15 cm. Both the gel and the electrophoresis buffer contained 40 m*M* Tris-acetate, pH 8.3, 20 m*M* sodium acetate, 2 m*M* EDTA (TAE buffer), and 2 μg of ethidium bromide per milliliter.

After electrophoresis, fractions were cut from the gel with a sterile razor blade. DNA was recovered from the agarose gel slices by melting the agarose at 70° for 5 min. The melted agarose slice was diluted with 10 volumes of H_2O and transferred to a 37° water bath. This was loaded on 300-μl columns of phenyl neutral red polyacrylamide affinity absorbent (Boehringer product No. 275, 387) equilibrated with 0.1 × TAE buffer. The columns were washed with 10 ml of 0.1 × TAE, and the DNA was eluted with 2 *M* $NaClO_4$ in 1.0 × TAE. One-drop fractions were collected, and fractions containing radioactive DNA were pooled. The eluted DNA was concentrated by ethanol precipitation and redissolved at 10 m*M* Tris-HCl, pH 7.4, 10 m*M* NaCl, 0.1 m*M* EDTA. After phenol extraction and subsequent ethanol precipitation, the DNA was redissolved in 10 m*M* Tris-HCl, pH 7.4, 10 m*M* NaCl, 0.1 m*M* EDTA at a final concentration of 500 μg/ml and stored at −20°. Recovery of DNA from agarose gel varied from 50 to 70%.

The autoradiograph depicted in the figure shows fractions of *Sau*3a-digested nematode DNA prepared as described, analyzed by electrophoresis on a 1% agarose gel. The gel was cast in 0.1 × 1.8 × 20 cm slabs. Electrophoresis was at 100 mA for 4 hr using TAE buffer containing 2.0 μg of ethidium bromide per milliliter. Nick-translated *Eco*RI-cutλDNA, *Bam*HI-cut 1059 DNA, and a clone of *unc*54 DNA cut with *Bam*HI were included as size markers. Fractions 3, 4, and 5 contain 15–20 kb *Sau*3a fragments suitable for insertion into λ1059.

DNA fragments 20 kb long. The frequency of *Sau*3a sites does not vary appreciably with changes in base composition. In DNA with 67% G+C, the sites should occur once every 324 bp. In practice, however, we minimize the possibility of obtaining abnormal distributions of fragments by a single digestion condition and routinely digest genomic DNA in several reactions in which the concentration of enzyme is varied over a 20-fold range (Fig. 4).

Specific Fragments

In some experiments it may be desirable to clone specific restriction fragments produced by limit digestion with a six-base pair enzyme. For example, most of the *unc*54 myosin heavy-chain gene coding sequence is present on a 8.3 kb *Xba* fragment.[40,41] In order to isolate this fragment from strains carrying *unc*54 mutations quickly, we have constructed an *Xba* vector (2149) with slightly extended arms to accommodate this fragment. Since the distribution of *Xba* sites in the nematode genome is nonrandom, a considerable sequence enrichment is obtained simply by purifying a single size fraction from a limit enzyme digest. It should be noted that the use of more than one restriction enzyme in succession would provide additional sequence purification.

Size Fractionation

Rigorous size fractionation of the DNA to be cloned is essential to avoid spurious linkage produced by multiple ligation events. If fragments greater than 14 kb are ligated to the vector arms, any dimers or multimers formed during the ligation reactions will exceed the 22 kb cloning capacity of the phage and will not appear in the recombinant phage population. Fragments less than 12 kb are frequently cloned as multiples. We have found that preparation of DNA fragments by agarose gel electrophoresis is more satisfactory than purification of fragments by velocity sedimentation on sucrose or NaCl density gradients. Any method of recovery of DNA from gels that yields ligatable DNA is satisfactory. In most of our recent experiments we have recovered DNA from low melting temperature agarose gels by phenol extraction. Usually the DNA is sufficiently pure after ethanol precipitation, without additional purification. Figure 4 shows nematode DNA fragments prepared by *Sau*3a partial digestion and size-fractionated by agarose gel electrophoresis. Fractions 2, 3, and 4 contain 15–20 kb DNA fragments suitable for cloning.

[40] A. R. MacLeod, J. Karn, and S. Brenner, *Nature (London)* **291,** 386 (1981).
[41] J. Karn, and L. Barnett, *Proc. Natl. Acad. Sci. U.S.A.*, in press (1983).

Preparation of Recombinants

Enzymes and In Vitro Packaging

Successful and efficient cloning requires highly purified restriction enzymes and active DNA ligase. Commercial preparations have improved markedly in recent years and most are satisfactory; however, we have found it convenient to prepare our own enzymes in order to have large quantities of calibrated materials. T4 DNA ligase was prepared from a lysogen of a lambda-T4 gene 30 recombinant originally prepared by Murray *et al.*[42] Restriction enzymes were prepared by standard methods. A number of *in vitro* packaging systems have been developed, and each gives much the same packaging efficiencies. In our experiments we have used extracts prepared from NS 428 supplemented with partially purified protein A, following the method of Sternberg[43] and Becker and Gold[44] as modified by Blattner[7] and ourselves.[18] There should be no incompatibility between our vectors and other packaging extracts.

Yield of Recombinants

We routinely monitor the yield of religated vector molecules and recombinants by plating on Q358 and Q359. Figure 5 shows a cloning experiment in which lambda 1059 DNA was cleaved with *Bam*HI and 2-μg aliquots were religated with T4 DNA ligase in the presence of 0–0.6 μg of 18 kb fragments produced by *Bam*HI or *Sau*3a cleavage of nematode DNA. Cleavage and religation of the vector DNA in the absence of nematode DNA produces more than 1×10^6 phage particles per microgram of phage DNA. These phages grow on Q358, but fewer than 2×10^3 PFU are detectable on Q359. This background is reduced to less than 2×10^2 PFU per microgram on the more stringent selective strain, CQ6. Cleavage and relegation of 1059 in the presence of nematode DNA fragments produce recombinant phages that are detected by plating on Q359. The yield of recombinants is linear with the amount of nematode DNA added as long as the DNA concentration is low. The ligation reaction is saturated with a greater than 2.0-fold molar excess of insert DNA to vector DNA (0.5 μg insert DNA per 1.0 μg of vector). The yield of recombinants in this experiment ranged from 2.4×10^5 to 5.4×10^5 per microgram of 15–20 kb nematode DNA. This yield is approximately 10-fold higher than the yield reported by Maniatis *et al.*[6] using Charon 4a vectors. At saturation of the ligation reaction with nematode DNA, approximately

[42] N. E. Murray, S. A. Bruce, and K. Murray, *J. Mol. Biol.* **132,** 493 (1979).

[43] N. Sternberg, D. Tiemeier, and L. Enquist, *Gene* **1,** 255 (1977).

[44] A. Becker, and M. Gold, *Proc. Natl. Acad. Sci. U.S.A.* **72,** 581 (1975).

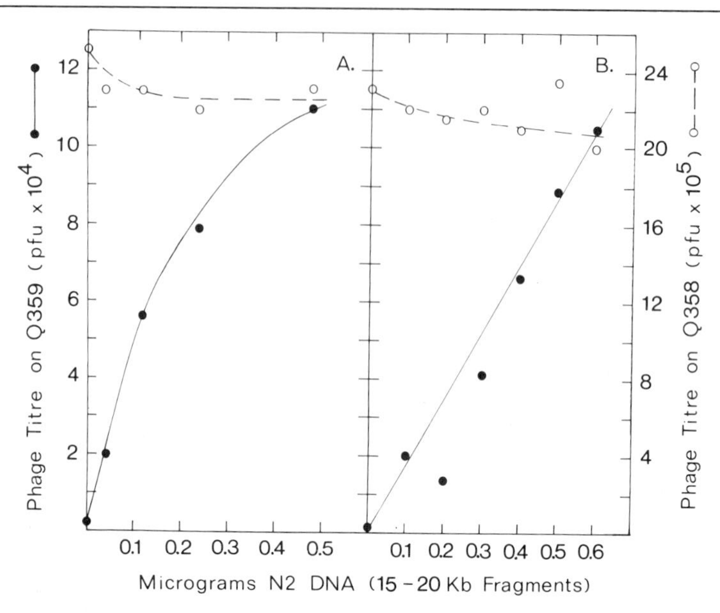

FIG. 5. Insertion of nematode DNA into 1059 arms. 1059 DNA was digested with a threefold excess of *Bam*HI as described in the legend to Fig. 4. The reaction was terminated by incubation at 70° for 5 min. Aliquots of 2.0 μg of cleaved 1059 DNA were ligated in the presence of 0–0.6 μg of 15–20 kb nematode DNA prepared as described in the legend to Fig. 4. In 20-μl reactions containing 0.1 Weiss unit of T4 DNA ligase, 10 m*M* Tris-HCl, pH 7.4, 10 m*M* $MgCl_2$, 50 m*M* NaCl, 0.1 m*M* ATP. Incubation was at 4° for 18 hr. The ligated DNAs were packaged *in vitro* as follows, using extracts of heat-induced NS 428. Extracts were prepared by lysing 10 g of induced cells in 50 ml of 50 m*M* Tris-HCl, pH 8.0, 3 m*M* $MgCl_2$, 10 m*M* 2-mercaptoethanol, 1 m*M* EDTA in a French pressure cell operated at 1000 psi. Cellular debris was removed by centrifugation of the extract for 30 min, 35,000 rpm in a T60 rotor, and aliquots of the supernatant were stored at −70°. Extracts prepared in this manner are active in *in vitro* packaging when supplemented with partially purified protein A prepared as described by Blattner *et al.*[7] Packaging was performed in 150-μl reactions containing 50 μl of extract, 10 μl of protein A, 20 m*M* Tris-HCl, pH 8.0, 3 m*M* $MgCl_2$, 0.05% 2-mercaptoethanol, 1 m*M* EDTA, 6 m*M* spermidine, 6 m*M* putrescene, 1.5 m*M* ATP, and 2.0 μg of cleaved and religated 1059 DNA. After incubation for 60 min at 20°, the extracts were diluted to 1 ml with λdil (10 m*M* Tris-HCl, pH 7.4, 5 m*M* $MgSO_4$, 0.2 *M* NaCl, 0.1% w/v gelatin), and titered on Q358 and Q359 bacteria. Panel A: Yield of total phage (PFU on Q358, ○---○) and recombinant phage (PFU on Q359, ●——●) genomes obtained by religation of *Bam*HI-cleaved 1059 in the presence of 0–0.5 μg of 15–20 kb fragments of *Bam*HI-cleaved N2 DNA. Panel B: Yield of total phage (PFU on Q358, ○---○) and recombinant phage (PFU on Q359, ●——●) genomes obtained by ligation of *Bam*HI-cloned 1059 in the presence of 0–0.6 μg of 15–20 kb fragments of *Sau*3a-cleaved N2 DNA.

10% of the total phages produced harbor inserts. The total yield of phages decreases somewhat upon addition of nematode DNA to the ligation reaction. This may be due to the addition of trace quantities of inhibitors of the T4 ligase or the result of sequestering of vector arms by broken nematode fragments.

Identification of Specific Clones

Mean Length of DNA Inserts

The distribution of DNA sizes in a lambda phage population can be determined by measuring the density of phages on CsCl density gradients.[45] Since the amount of protein in the phage particles is constant, the buoyant density of a phage is a function of the DNA-to-protein ratio. Changes in the length of lambda DNA of as little as 500 bp may be detected by this method. Figure 6 shows the results of a density gradient analysis of the clone collections prepared in the experiment shown in Fig. 4. An *h*434*c*I857*nin*5 phage (46.1 kb) as well as 1059 were used as size markers. The recombinant phages varied in size from 46 to 44 kb with an average of 45 kb. This corresponds to an average insert size of 15 kb. The half-maximal bandwidth of the density distribution of the recombinant phage population was approximately twice that of the marker phage, demonstrating that the recombinants contain DNA inserts with limited heterogeneity.

Plaque Hybridization

In most of our work we have used probes made from the mp series of M13 vectors.[46,47] Originally we used nick-translated RF DNA, but more recently we have been using probes made by priming on M13 single-stranded DNA to the 5′ sides of the clone insert and hybridizing with the partially double-stranded material.[48,49] A slight background of hybridization of M13 DNA to *lac* DNA sequences from the host strains was encountered in our early experiments. This can be eliminated by the addition of 20 μg of M13 vector DNA per milliliter as a competitor when

[45] N. Davidson, and W. Szybalski, *in* "The Bacteriophage Lambda" (A. D. Hershey, ed.), pp. 45–82. Cold Spring Harbor Laboratory, Cold Spring Harbor, New York, 1971.

[46] J. Messing, B. Gronenborn, B. Müller-Hill, and P. H. Hofschneider, *Proc. Natl. Acad. Sci. U.S.A.* **74,** 3642 (1977).

[47] J. Messing, R. Crea, and H. Seeburg, *Nucleic Acids Res.* **9,** 309 (1981).

[48] N-t. Hu, and J. Messing, *Gene* **17,** 271 (1982).

[49] D. Brown, J. Frampton, P. Goelet, and J. Karn, *Gene* **20,** 139 (1982).

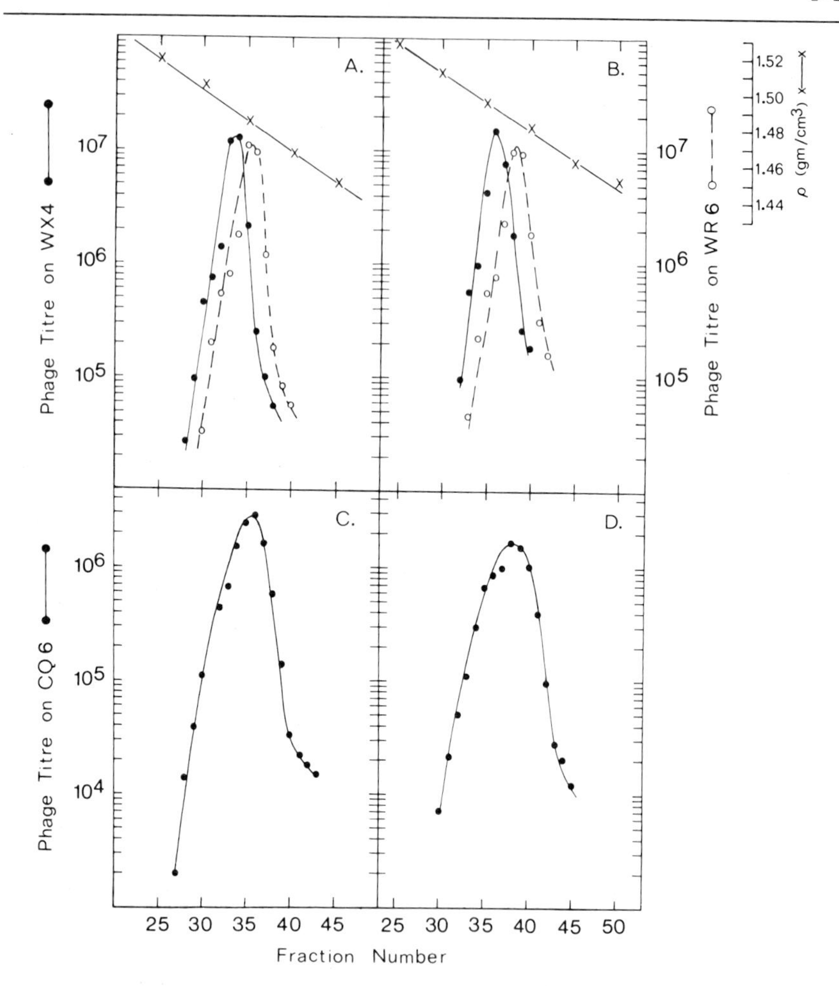

FIG. 6. Analysis of recombinant phage collections by CsCl density gradient centrifugation. 1059 (44 kb) and 308: *h*434*c*I857*s*RI 4° *nin*5 *s*RI 5° *Sam*7 (46.1 kb) were included as density markers. Approximately 10^5 of each marker phage and 10^8 phages from recombinant phage pools were added to 5 ml of 100 m*M* Tris-HCl, pH 7.4, 10 m*M* $MgCl_2$. Solid CsCl was added to a final refractive index of 1.3810, and the phage were banded by centrifugation in an SW60 rotor at 40,000 rpm for 24 hr. After centrifugation, one drop fractions were collected by puncturing the bottom of the centrifuge tubes with a needle. The refractive index of every fifth fraction was measured, and 20 μl aliquots of each fraction were added to 1 ml of dil. Each fraction was titered on WX4 (λ^R), ●——●; WR6, (*rec*A, $h434^R$), ○---○; and CQ6 ($h434^R$ P2), ●——●, to determine the position of the *h*434, 1059, and recombinant phages, respectively. Panels (A) and (C) plot the distribution of *Bam*HI-generated recombinant phage (panel C) and marker phages (panel A) included in the same gradient. Panels B and D plot the results of a similar analysis of the recombinants generated with *Sau*3a fragments.

working with the single-stranded DNA probes or by using strains with deletions of the *lac* region for plating (D91 or Q364).

Plasmid probes may also be used to screen libraries, but it is preferable to construct the library in a vector that lacks the plasmid insert (2004, 2139, EMBL4) in order to avoid false positives arising from hybridization to parental phages that escape the spi selection.

Immunological Assay

Recombinants of the spi vectors have a number of features that are of use when immunological detection of cloned sequences is planned.[13–16] Each of the phages is constructed so that DNA is inserted in the *Bam*HI site at 714 in the leftward promoter. This segment is efficiently transcribed from pL and any inserted fragment that contains an intact gene and ribosome binding site will be expressed at a high level when cloned in this site. We cloned the T4 DNA ligase gene 30 into 1059 and found that during lytic infection the protein was produced at the same levels as in the original phages constructed by Murray,[42] which placed this gene under the control of the rightward promoter. Additionally, clones in 1059 and its derivatives retain a single lambda attachment site. All the recombinants may therefore be inserted efficiently into the *E. coli* chromosome with helper phage supplying integrase and repressor.

Genetic Selections

Most genetic selection schemes involve suppression of amber mutations in the vector arms by cloned suppressor tRNA genes. Seed *et al.*[11] have found that plasmids carrying both the tRNA su^{+}_{III} gene and a cloned insert may be inserted in specific lambda clones through *rec*-mediated "lifting" events. If the vector has amber mutations in essential functions, then only "phasmids"[29] carrying the plasmid and the suppressor gene will grow on su^{-} strains. A second selection scheme was developed by Goldfarb *et al.*[12] following a suggestion by one of us. DNA is cotransformed into mammalian cells together with su^{+}_{III} DNA to provide a selective marker. Larger DNA fragments carrying transforming DNA and the su^{+}_{III} DNA are then selected in a lambda 1059 derivative carrying the *Sam7* mutation. The vector 1259 would also be suitable for this experiment.[50]

[50] We have recently constructed another vector, λ2001, which is a derivative of 2053 and carries a Δ[*int-c* III] KH54 *s* RI 4° *nin5* *s* RI 5° *sHin* dIII 6° *chi* C right arm and has the polylinker sequence TCTAGAATTCAAGCTTGGATCCTCGAGCTCTAGA cloned into the *Xba* sites. This phage is a vector for *Eco* RI, *Hin* dIII, *Bam* HI, *Xho* I, *Sac* I, and *Xba* I.

[2] New M13 Vectors for Cloning

By JOACHIM MESSING

In the summer of 1978 I began to circulate a manual for the use of M13mp2 and its derivatives that has been used in the Minnesota M13 laboratory courses and is the basis for this article. The revisions for this chapter include an updated list of single-stranded DNA phage (M13, fl, fd) vectors, their relevant data, and new methods of hybridization and mutagenesis based on these vectors. The strategies of M13 cloning have been extended. A discussion of computer software to aid in shotgun DNA sequencing and designing primers for *in vitro* mutagenesis has also been added.

Principle of the Method

Life Cycle of M13 and Use in Cloning

An impressive groundwork of today's technology in molecular biology is based on studies using the bacteriophage of *Escherichia coli;* this also furnishes the basic tools for genetic engineering. The transducing properties of the bacteriophage presented a model for the development of a scheme to dissect and combine the segments of DNA separated by evolutionary barriers.[1] Bacteriophage lambda was the first phage converted into a general transducing system.[2,2a,3] Besides the large-genome phage like the T or lambda phage, a group of small phage has also played an important role in the understanding of gene organization, e.g., overlapping genes[4] or DNA synthesis.[5] In contrast to the larger phage, the phage of this group usually contain circular DNA molecules.[6] The study of the DNA of these phage particles and the subsequent physical studies of circular molecules in general[7,8] furnish the understanding needed to use the

[1] D. A. Jackson, R. H. Symons, and P. Berg, *Proc. Natl. Acad. Sci. U.S.A.* **69,** 2904 (1972).
[2] N. E. Murray and K. Murray, *Nature* (*London*) **251,** 476 (1974).
[2a] A. Rambach and P. Tiollais, *Proc. Natl. Acad. Sci. U.S.A.* **71,** 3927 (1974).
[3] M. Thomas, J. R. Cameron, and R. W. Davis, *Proc. Natl. Acad. Sci. U.S.A.* **71,** 4579 (1974).
[4] B. G. Barrell, G. M. Air, and C. A. Hutchison III, *Nature* (*London*) **264,** 34 (1976).
[5] A. Kornberg, W. H. Freeman, San Francisco, 1980.
[6] W. Fiers and R. L. Sinsheimer, *J. Mol. Biol.* **5,** 408 (1962).
[7] A. Burton and R. L. Sinsheimer, *J. Mol. Biol.* **14,** 327 (1965).
[8] W. Bauer and J. Vinograd, *J. Mol. Biol.* **33,** 141 (1968).

METHODS IN ENZYMOLOGY, VOL. 101

ISBN 0-12-182001-7

plasmid vectors as a second vehicle for recombinant DNA techniques.[9] A particularly important feature of this group of phage is that their circular DNA molecule is single-stranded.[10]

Two families of phage containing single-stranded DNA have been studied extensively, the isometric phage ϕX174 or G4[11] and the F-specific rod-shaped filamentous phage (Ff) fl, fd, and M13.[12,13] The latter group has two important properties that aid their use as recombinant DNA vehicles: they do not lyse their host cells and they allow the packaging of larger than unit-length viral DNA.[14]

The two major components of the viral coat, the protein products of gene *III* and gene *VIII*, are encoded by the phage genome.[15] About 2700 subunits of the major coat protein (gene *VIII*), in combination with the viral DNA, form the rod-shaped viral particle.[16] If larger recombinant molecules are incorporated, the length of the protein coat is extended proportionally. The amount of gene *VIII* protein in infected cells seems to be in excess; up to seven times the unit length of the viral genome can be packaged.[17] Four copies of the other coat protein (gene *III*) are located at one end of the rod[18] and involved in the adsorption of the phage to the F pilus of the host cell.[19] Other viral genes, such as *I, IV, VI, VII,* and *IX*, furnish minor protein components for the assembly of the viral particles.

The rod-shaped virus penetrates the F pilus and is stripped of its major coat protein in the cell membrane. The other coat protein (gene *III*) seems to guide the virus in the infection process[20] and is considered to be a "pilot" protein. The viral single-stranded DNA (ss-DNA) is converted into a double-stranded circular form (RF DNA) without the synthesis of any viral product. The synthesis of the first RF DNA (parental RF) is an interesting step, because by this means the virus passes from a passive into an active life form. Part of this transition results from the fact that the

[9] S. N. Cohen, A. C. Y. Chang, H. W. Boyer, and R. B. Helling, *Proc. Natl. Acad. Sci. U.S.A.* **71,** 1743 (1973).

[10] R. L. Sinsheimer, *in* "Phage and the Origins of Molecular Biology" (J. Cairns *et al.*, eds.), p. 258. Cold Spring Harbor Laboratory, Cold Spring Harbor, New York, 1966.

[11] R. L. Sinsheimer, *Prog. Nucleic. Acid Res. Mol. Biol.* **8,** 115 (1968).

[12] D. Pratt, *Annu. Rev. Genet.* **3,** 343 (1969).

[13] D. A. Marvin and B. Hohn, *Bacteriol. Rev.* **33,** 172 (1969).

[14] W. O. Salivar, T. J. Henry, and D. Pratt, *Virology* **32,** 41 (1967).

[15] D. Pratt, H. Tzagoloff, and W. S. Erdahl, *Virology* **39,** 42 (1969).

[16] D. A. Marvin and E. J. Wachtel, *Philos. Trans. R. Soc. London Ser. B* **276,** 81 (1976).

[17] J. Messing, *in* "Recombinant DNA. Proceedings of the Third Cleveland Symposium on Macromolecules" (A. G. Walton, ed.), p. 143. Elsevier, Amsterdam, 1981.

[18] R. Marco, *Virology* **68,** 280 (1975).

[19] T.-C. Lin and I. J. Bendet, *Biochem. Biophys. Res. Commun.* **72,** 369 (1976).

[20] S. M. Jaswinski, A. A. Lindberg, and A. Kornberg, *Virology* **66,** 283 (1975).

viral DNA is always the same strand, indicated as the (+) strand. The complementary strand, the (−) strand, serves two important functions. It represents the sense strand and contains all the coding information; all transcription occurs in the same direction. It also provides the template for the synthesis of progeny phage DNA. Therefore, both viral gene expression and viral DNA replication require the synthesis of the complementary strand.

Although transcriptional starts precede most viral genes, there are not an equal number of transcriptional stops. Transcription is terminated only in the two intergenic regions. This leads to a cascade form of transcription. Viral genes like gene *VIII* are transcribed very frequently, whereas gene *III* is transcribed in small amounts.[21]

The viral protein products of gene *II*[22] and gene *V*[23] contribute to the asymmetric fashion in which the progeny viral DNA is synthesized. Only after the completion of the synthesis of the complementary strand and the formation of the parental double-stranded replicative form is the gene *II* product synthesized. This protein initiates (+) strand synthesis by specifically introducing a nick into the (+) strand of the RF molecule.[24] The free 3′-OH end of the (+) strand is extended by DNA polymerase while the 5′ end is displaced. Thus, a new copy of the (+) strand is synthesized using the (−) strand as the template. After the (+) strand synthesis has migrated once around the intact (−) strand circle, the gene *II* product cleaves the (+) strand again to separate the parental (+) strand from the newly synthesized (+) strand. The (−) strand continues to serve as a template for new progeny (+) strands while the parental (+) strand is circularized.[25] The displaced parental (+) strand can then be converted into an RF form. Therefore the RF replication can be divided into stages, RF to ss and ss to RF. About 100–200 copies of the RF molecules accumulate in the cell before a second viral gene product (gene *V*) involved in viral replication reaches high levels. The gene *V* product interferes with the second stage of RF replication by binding to the newly synthesized (+) strands so that they cannot be converted into the double-stranded form.

[21] H. Schaller, E. Beck, and M. Takanami, *in* "The Single-Stranded DNA Phages" (D. T. Denhardt, D. Dressler, and D. S. Ray, eds.), p. 139. Cold Spring Harbor Laboratory, Cold Spring Harbor, New York, 1978.

[22] N. S.-C. Lin and D. Pratt, *J. Mol. Biol.* **72,** 37 (1972).

[23] J. S. Salstrom and D. Pratt, *J. Mol. Biol.* **61,** 489 (1971).

[24] T. F. Meyer, K. Geider, C. Kurz, and H. Schaller, *Nature* (*London*) **278,** 365 (1979).

[25] W. L. Staudenbauer, B. E. Kessler-Liebscher, P. K. Schneck, B. van Dorp, and P. H. Hofschneider, *in* "The Single-Stranded DNA Phages" (D. T. Denhardt, D. Dressler, and D. S. Ray, eds.), p. 369. Cold Spring Harbor Laboratory, Cold Spring Harbor, New York, 1978.

This protein–ss-DNA complex then moves to the periplasmic space, where the DNA binding protein is replaced and the ss-DNA associates with the coat proteins. After phage assembly, the mature particles are released without dissolution of the bacterial cell wall. This release is different from the entry mechanism and does not require the presence of F-specific pili. Therefore phages can be produced from infected colonies in which phage is contained like a plasmid.

A second route can be taken to introduce the phage DNA into the cell. Double-stranded or single-stranded M13 phage DNAs are not infectious and do not penetrate the F pili. If cells are rendered competent for the uptake of DNA by $CaCl_2$,[26] phage DNA can be introduced in the same way as plasmids[27]; the ss-DNA with somewhat lower efficiency than the RF form. Because this transfection does not require the F pili, the choice of host cells would appear to be broad. However, the phage released are barred from entry into other cells, as these cells have no pili, so that cells not transfected are free of phage DNA, rapidly overgrowing those that are infected. This problem can be overcome by counterselection using lactose[27] or a drug resistance.[28]

The scheme of entry and exit of the various forms of M13 phage into and from the bacterial cell is summarized in Fig. 1. From this scheme three important facts of technological importance are seen.

1. The single-stranded form of the phage DNA, as well as the double-stranded form, can be introduced into the cell.
2. The double-stranded form is converted into the single-stranded form.
3. The single-stranded form is released from the intact bacterial cell in a filamentous phage particle that can easily be separated from all bacterial and intracellular viral components.

All viral genes in M13 are essential, however, and therefore cannot be replaced by cloned DNA as was done with the phage lambda cloning vehicles. Although the packaging mode that results in the mature M13 phage particle allows for the extension of the length of viral DNA packaged, this does not clarify how foreign DNA should be added to the viral DNA. Two intergenic regions, a small one of less than 100 bp between genes *VIII* and *III*, and a larger one of about 500 bp between genes *IV* and *II*, have been

[26] S. N. Cohen, A. C. Y. Chang, and L. Hsu, *Proc. Natl. Acad. Sci. U.S.A.* **69,** 2110 (1972).

[27] J. Messing, B. Gronenborn, B. Müller-Hill, and P. H. Hofschneider, *Proc. Natl. Acad. Sci. U.S.A.* **74,** 3642 (1977).

[28] R. Herrmann, K. Neugebauer, H. Zentgraf, and H. Schaller, *Mol. Gen. Genet.* **159,** 171 (1978).

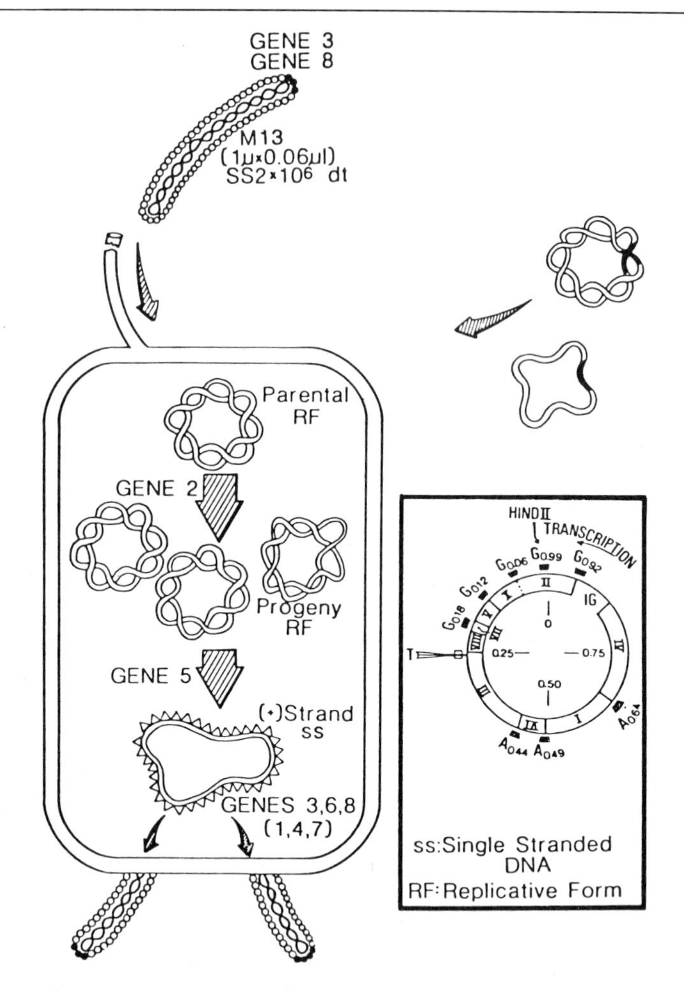

FIG. 1. Life cycle of filamentous single-stranded bacteriophage. The flow of steps is given in detail in the text. Viral genes are designated with roman numerals, and their map position is indicated in the physical map with the original *Hin*cII site as a reference point. The viral gene products are designated with arabic numerals and their role in the various steps during the life cycle is indicated by associating them with the particular intermediate viral form.

described.[21,29] They do not code for proteins, but have important functions; the starts of (−) and (+) strand synthesis are in a region of about

[29] K. Horuishi, G. F. Vovis, and P. Model, *in* "The Single-Stranded DNA Phages" (D. T. Denhardt, D. Dressler, and D. S. Ray, eds.), p. 113. Cold Spring Harbor Laboratory, Cold Spring Harbor, New York, 1978.

100 nucleotides within the larger intergenic region. The start of transcription of gene *II* and the termination of the transcription of gene *IV* both reside within this region. The second intergenic region contains the central rho-independent transcription termination signal as well as the start for the transcription of gene *III*.

Therefore, viral sequences that can be interrupted by insertion have been selected by insertion mutagenesis. Although insertion mutants derived by the *in vivo* transposition of transposons have been isolated,[28] most Ff vectors are based on a different insertion mutagenesis procedure. If the replicative form is linearized by partial digestion with a restriction endonuclease that cleaves at many different positions, and if it is fused to a DNA fragment containing a marker like the *E. coli lac in vitro,* the number and location of the insertions can be controlled. The insertion mutant that has been recovered from transformed cells has been shown to contain the *lac* DNA in the larger intergenic space.[27]

Two Constant and One Variable Primer

Recombinant DNA in single-stranded form is useful for exploring gene structure and function and constructing synthetic genes. This was shown by a number of techniques that have been developed with the ss-DNA phage ϕX174. The DNA sequencing technique based on a primer extension reaction[30,31] was used to determine the complete nucleotide sequence of the ϕX174 genome.[32] The DNA–DNA hybridization technique, which is the basis of Southern blot hybridization,[33] has been developed with ϕX174 DNA.[34] The primer extension reaction has been used both for labeling DNA and for synthesis of the first DNA with biological activity.[35] The marker rescue scheme developed for the construction of a genetic map of ϕX174[36] is the basic strategy for site specific mutagenesis.[37] Because it has been demonstrated that the replicative form, RF, of the ss-DNA phage M13 can be recombined *in vitro* with another restriction fragment to yield recombinant ss-DNA phages,[27] all the techniques applicable to ϕX174 are also applicable to any DNA cloned into M13.

[30] F. Sanger and A. R. Coulson, *J. Mol. Biol.* **94,** 441 (1975).

[31] F. Sanger, S. Nicklen, and A. R. Coulson, *Proc. Natl. Acad. Sci. U.S.A.* **74,** 5463 (1977).

[32] F. Sanger, G. M. Air, B. G. Barrell, N. L. Brown, A. R. Coulson, J. C. Fiddes, C. A. Hutchison III, P. M. Slocombe, and M. Smith, *Nature (London)* **265,** 687 (1977).

[33] E. M. Southern, *J. Mol. Biol.* **98,** 503 (1975).

[34] D. T. Denhardt, *Biochem. Biophys. Res. Commun.* **23,** 641 (1966).

[35] M. Goulian, A. Kornberg, and R. L. Sinsheimer, *Proc. Natl. Acad. Sci. U.S.A.* **58,** 2321 (1967).

[36] C. A. Hutchison III and M. H. Edgell, *J. Virol.* **8,** 181 (1971).

[37] C. A. Hutchison III, S. Phillips, M. H. Edgell, S. Gillam, P. Jahnke, and M. Smith, *J. Biol. Chem.* **253,** 6551 (1978).

The application of these techniques to a double-stranded DNA (ds-DNA) requires the separation of the two strand of the DNA. The single-stranded bacteriophage offers a biological scheme for strand separation. The preparation of pure ϕX174 phage particle yields pure ss-DNA in large amounts. This avoids the technical difficulties of DNA strand separation by alkaline CsCl gradients,[38] poly(UG) CsCl gradients,[39] polyacrylamide gels,[40,41] or exonuclease treatment.[42] Although strand separation techniques as well as the purification of DNA from other cellular DNA after cell lysis can be improved, it is difficult to match the purity the biological scheme provides.

If the recombinant DNA in the single-stranded form is to be used as a template for the *in vitro* synthesis of DNA, the start of DNA synthesis has to be defined by a primer.[5] In most cases synthesis of the complementary strand has to occur in a nonrandom way at a specific site. This can be achieved by use of either a restriction endonuclease fragment[36] or a synthetic primer.[43] In order to hybridize the primer to the template strand, the primer must be single-stranded. The complementary strand from the restriction fragment is obtained by denaturation of the ds-DNA. The separation of the strands of the restriction fragment can also be obtained by treating the ds-DNA with exonuclease III.[42] The synthetic primer is prepared as a short piece of ss-DNA that is complementary to the site of initiation of DNA synthesis.

The primers can be used for three purposes: (*a*) to direct the synthesis of the cloned DNA[44]; (*b*) to direct the synthesis of the vector DNA[17,45]; and (*c*) to introduce site-specific changes into the cloned DNA.[46] The first two primers can be of an universal character if they represent sequences that flank a unique cloning site of the single-stranded phage vector. Thus, any improvements made with this cloning site broadens the application of the universal primers.[17,44]

Site-specific mutagenesis may be improved by the combined use of a

[38] J. Vinograd, J. Morris, N. Davidson, and W. F. Dove, *Proc. Natl. Acad. Sci. U.S.A.* **49,** 12 (1963).

[39] W. Szybalski, H. Kubinski, Z. Hradecna, and W. C. Summers, this series, Vol. 21, p. 383.

[40] A. M. Maxam and W. Gilbert, *Proc. Natl. Acad. Sci. U.S.A.* **74,** 560 (1977).

[41] A. A. Szalay, K. Grohmann, and R. L. Sinsheimer, *Nucleic Acids Res.* **4,** 1569 (1977).

[42] R. Wu, C. D. Tu, and Padmanabhan, *Biochem. Biophys. Res. Commun.* **55,** 1092 (1973).

[43] F. Sanger, J. E. Donelson, A. R. Coulson, H. Kossel, and H. Fischer, *Proc. Natl. Acad. Sci. U.S.A.* **70,** 1209 (1973).

[44] G. Heidecker, J. Messing, and B. Gronenborn, *Gene* **10,** 69 (1980).

[45] N.-T. Hu, and J. Messing, *Gene* **17,** 271 (1982).

[46] M. Smith and S. Gillam, *in* "Genetic Engineering," Vol. 3 (J. Setlow and A. Hollaender, eds.), Vol. 3, p. 1. Plenum, New York, 1981.

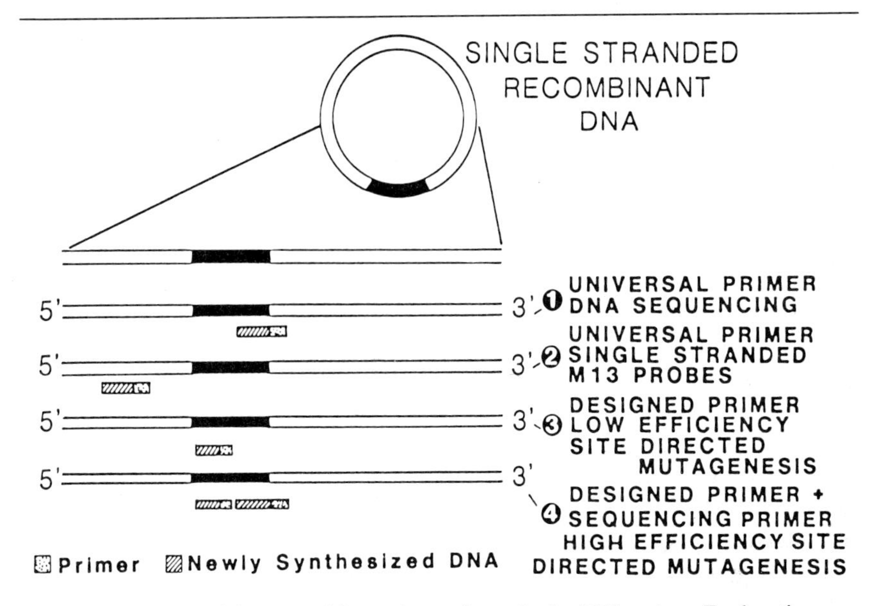

FIG. 2. Scheme of the uses of the various primers in the M13 system. Explanations are given in the text.

custom-designed primer with the universal sequencing primer.[47] Because gaps in the complementary strand are properly filled in *in vivo,* the site-specific mutagenesis does not depend on the completion of the complementary strand *in vitro.* Repair by nick translation of the two mismatching strands will not occur *in vivo* unless the 5′ end is too close to the mismatch.[47] The easiest way to extend the 5′ end of an oligonucleotide serving as a "mutagen" is the use of the sequencing primer at the same time in the presence of DNA polymerase and DNA ligase. If the large fragment of DNA polymerase (Klenow fragment) is used *in vitro,* then synthesis of the complementary strand extended from the sequencing primer will stop at the 5′ end of the mutagen primer, and the nick can be sealed by the DNA ligase. At the same time the 3′ end of the mutant primer is extended so that the site of mismatch is located far from either end of the complementary strand. A summary outline of the various primer extension experiments is given in Fig. 2.

Primer extension is not the only aspect of recombinant DNA technology that can be studied with the M13 cloning system. Other aspects, such as the overproduction of gene products, have recently been explored.

[47] J. Messing, R. Crea, and P. H. Seeburg, *Nucleic Acids Res.* **9,** 309 (1981).

[47a] P. Slocombe, A. Easton, Boseley, and D. C. Burke, *Proc. Natl. Acad. Sci. U.S.A.* **79,** 5455 (1982).

Purification of nucleic acids by single-stranded M13 DNA immobilized on a solid phase may prove to be useful. The circular single-stranded form of M13mp7 can be linearized at a specific site because of the inverted repeat of the multiple cloning site[47] (Heidecker and Messing, unpublished). Thus, one could use the ss-DNA instead of the RF for cloning full-length RNA molecules, if ligation techniques can be improved.

Materials and Reagents

Use distilled water for media (dH_2O), and double-distilled water (ddH_2O) for enzyme reactions.

Growth Media

M9 salts (10×): Na_2HPO_4, 60 g; KH_2PO_4, 30 g; NaCl, 5 g; NH_4Cl, 10 g. These quantities, for 1 liter of 10× solution, are dissolved in distilled water and autoclaved. To make 1 liter of medium, 100 ml of 10× M9 salts, 1 ml of a 1 *M* $MgSO_4 \cdot 7\ H_2O$, 10 ml of 20% glucose, 1 ml of 1% vitamin B_1, and 10 ml of 0.01 *M* $CaCl_2$ are added under sterile conditions to distilled water to make up a solution of 1 liter. If plates are to be prepared, autoclave 20 g of Difco agar in 900 ml dH_2O and all the other solutions separately, and combine them afterward. Use a 2-liter Erlenmeyer flask, let the solution cool in a 55–65° water bath, and pour into standard petri dishes. If the solution is hot when poured the plates will be too wet.

2× YT medium: Bacto tryptone, 16 g; Bacto yeast extract, 10 g; NaCl, 5 g. Add to 1 liter of dH_2O and autoclave. For plates add 20 g of Difco agar per liter; for soft agar, 6 g of Difco agar per liter. The soft agar after autoclaving is kept in a 55° water bath.

B broth: Bacto tryptone, 10 g; NaCl, 8 g. Add 1 liter of dH_2O and 1 ml of 1% vitamin B_1 solution and autoclave. For plates add 20 g of Difco agar per liter, and for soft agar add 6 g of Difco agar per liter. The soft agar can be kept in a 55° water bath once melted.

The B broth medium is better for achieving a deeper blue color for the M13 plaques, possibly because of the lack of catabolite repression. In this respect, it may be worth noting again that the *lac* regulatory region in M13 does not contain the *uv5* mutation.[27]

All the media may be obtained from Difco Laboratories, the chemicals from Mallinckrodt Inc. or Sigma Chemicals Co.

Other Materials

IPTG, isopropylthiogalactoside (Sigma): Dissolve in dH_2O to prepare 100 m*M* solution and sterilize by filtration through 0.22 μm Millipore filters.

Xgal, 5-dibromo4-chloro3-indolylgalactoside (Sigma or Bachem): Dissolve in dimethylformamide (Eastman Chemicals) to make a 2% solution. Wrap aluminum foil around the tube to avoid damage by light, and keep refrigerated.

The M13mp vectors, the plasmids derived from M13 (the pUC plasmids) or the plasmids containing the biological primers and their hosts are made available by Bethesda Research Laboratories as a courtesy to the scientific community. Other firms supplying these strains have agreed to observe the same obligation, if strains are requested by interested researchers.

Replicative Form (*RF*) *Preparation*

Growth media and chemicals are as described above.

Sucrose, 25% in 0.05 *M* Tris-HCl, pH 8.0, 0.01 *M* EDTA

Lysozyme (Sigma), 5 mg/ml in 0.05 *M* Tris-HCl, pH 8.0, 0.01 *M* EDTA. Store at 4°. Do not use a lysozyme solution older than 2 weeks.

Tris-HCl, 0.25 *M*, pH 8.0, 0.25 *M* EDTA

Pancreatic RNase (Sigma): Dissolve in 0.01 *M* sodium acetate buffer (pH 5.0) to give a final concentration of 10 mg/ml, boil for 2 min, and store at −20° in small aliquots.

Tris-HCl, 0.05 *M*, pH 8.0; 0.01 *M* EDTA, 2% Triton X-100 (Sigma)

Ethidium bromide (Sigma): Dissolve in distilled H_2O to give a concentration of 10 mg/ml. Handle only with gloves. Avoid any skin contact or inhalation of powder; treat as a mutagen.

CsCl (Kawecki Berylco Industries, Inc.), technical grade.

Light mineral oil (Mallinckrodt)

Supercoiled DNA: In the presence of ethidium bromide, the buoyant density is 1.54 g/cm^3 in CsCl solution at 20°; this corresponds to a refractive index of 1.385.

n-butanol (Mallinckrodt): Extraction depends on the saturation of the alcohol with salt, therefore do not dialyze before extraction of the dye. CAUTION: In the presence of light, ethidium bromide introduces nicks into DNA.

Dialysis tubing, obtained from Spectrum Medical Industries and boiled for 10 min in 10 m*M* EDTA, pH 8.3. The EDTA solution is replaced with fresh solution and the tubing is stored wet at 4° until use.

Sodium acetate, 3 *M*: adjust the pH with glacial acetic acid to 4.5–5.0. It takes a lot of acid; final adjustment to the correct molarity can be made with double-distilled (dd) H_2O.

Sucrose, 5%: Dissolve 5 g of sucrose in 100 ml 0.05 *M* Tris-HCl, pH 8.0, 1 *M* NaCl, 0.005 *M* EDTA.

Sucrose, 20%: dissolve 20 g of sucrose in 100 ml of 0.05 *M* Tris-HCl, pH 8.0, 1 *M* NaCl, 0.005 *M* EDTA.

Phenol (bought from Mallinckrodt): Redistill, collect under dH_2O in smaller bottles covered with argon gas, and store in the dark at −20°. Just before use it is thawed, it is saturated with buffer (10 m*M* Tris-HCl, pH 7.5, 1 m*M* EDTA) and adjusted to a pH of 7.5–8 with NaOH. To phenol-extract, it is mixed with an amount equal to that of the aqueous phase and vortexed for 10 sec. The phases are separated by centrifugation. A second extraction is done with 1:1 mixture of phenol–chloroform (Mallinkrodt) as the organic phase.

Low Tris buffer: 10 m*M* Tris-HCl, pH 7.5, 0.1 m*M* EDTA

Cloning Strategies

Growth media and chemicals are as described above.

The enzymes used are restriction endonucleases, ligase (only T4 DNA ligase is used), exonuclease III, DNA polymerase (only the large fragment or Klenow fragment[5] is used), DNA polynucleotide kinase, *Bal*31, DNase I (electrophoretic grade), and bacterial alkaline phosphatase. They are bought from one of the following companies: Bethesda Research Laboratories, Biotech, Boehringer-Mannheim, New England BioLabs, New England Nuclear, P-L Biochemicals, Sigma, Worthington Biochemicals. Dilution buffers, incubation, and storage conditions are as recommended by the manufacturers. Modifcations of the manufacturers recommendations are made from time to time as pointed out in the text.

Polypropylene tubes (1.5 ml, 0.5 ml, and 0.25 ml) and yellow tips for pipetmen are obtained from Evergreen Scientific.

Agarose, acrylamide, bisacrylamide, TEMED, ammonium persulfate, BSA, DTT, nucleotide triphosphates, yeast tRNA and other chemicals are obtained from Bethesda Research Laboratories, Eastman Chemicals, Mallinckrodt, P-L Biochemicals, or Sigma.

Acrylamide (Eastman) was recrystallized from chloroform. The purified acrylamide was used to prepare a 40% stock of acrylamide and bisacrylamide (38:2). Acrylamide was dissolved in double-distilled water overnight and filtered through Whatman 3 M paper. The filtered solution was stored in the dark at 4°.

A 6% polyacrylamide gel is prepared as follows: Glass plates are washed with detergent, rinsed with water, then ethanol, and dried. They are assembled with the appropriate spacers, clamped with book binders, and sealed with 1% melted agarose. Acrylamide, 4.5 ml of the 40% stock solution, is mixed with 3 ml of borate buffer, 0.3 ml of fresh 10% ammonium persulfate solution, and 22.2 ml of ddH_2O. Before pouring, 30 μl of TEMED are added and mixed with the acrylamide mixture.

Other Materials

Borate buffer (10×): Dissolve in 3 liters of ddH_2O, 363.3 g of Tris-base, 185.5 g of boric acid, and 17.5 g of EDTA.

Loading buffer: 0.05% bromophenol blue, 0.2 *M* EDTA, pH 8.3, 50% glycerol

Ethidium bromide staining bath: 0.5 μg of ethidium bromide per milliliter in 1 m*M* EDTA pH 8.3, containing 1 μg of RNase per milliliter (diluted from the stock solution as described above)

SSC, (1×): 0.15 *M* NaCl, 0.015 *M* sodium citrate

Transformation

Growth media and chemicals are as described above.

Identification of Recombinants

Growth media, chemicals, and enzymes are as described above.

SDS, 2%: 2 g of SDS (Bio-rad) in 100 ml of ddH_2O

Submerged agarose gels: electrophorese horizontally using borate buffer as the running buffer. For a 1% agarose gel, 1 g of agarose is placed in 100 ml of borate buffer in a round-bottom flask. The flask is placed in a heating mantle. A condenser is attached to the flask. The melted agarose solution is allowed to cool to about 50° and swirled from time to time before it is poured to cast the gel.

Nitrocellulose filter paper (S&S BA 85) for hybridization experiments (Schleicher & Schuell). Before use it is floated on both sides on ddH_2O.

Radiochemicals from Amersham, New England Nuclear, and ICN

H buffer: 100 m*M* Tris-HCl, pH 7.9, 600 m*M* NaCl, 66 m*M* $MgCl_2$.

Denhardt solution: 0.02% Ficoll (Pharmacia), 0.02% poly(vinylpyrrolidone) (Sigma), 10 mg of BSA per milliliter

Prehybridization solution: 5× Denhardt, 5× SSC, 50% formamide (MCB), 50 m*M* phosphate buffer, pH 6.8; 1% glycine (Sigma), 250 μg of salmon sperm DNA per milliliter (P-L Biochemicals, sonicated a few times in 10 m*M* Tris-HCl, pH 7.5, 1 m*M* EDTA, boiled for 15 min, and chilled on ice), 100 μg of poly(A) (Sigma) per milliliter, and 0.1% SDS (Bio-Rad). If M13 probes are used, add cold M13ss.

Hybridization solution: 1× Denhardt, 5× SSC, 50% formamide, 20 m*M* phosphate buffer, pH 6.8, 100 μg of salmon sperm DNA, 100 μg of poly(A) per milliliter, and 0.1% SDS

SSCP: 120 m*M* NaCl, 15 m*M* sodium citrate, 13 m*M* KH_2PO_4, 1 m*M* EDTA; adjusted to pH 7.2 with 1 *M* NaOH

Washing solution (room temperature): 2× SSCP, 0.1% Sarkosyl (ICN)

Washing solution (50°): 0.2× SSCP, 0.1% Sarkosyl

Exposure: Kodak XR2 film at room temperature or at −70° with a Cronex Lighting Plus intensifying screen (DuPont).

STET buffer: 50 m*M* Tris-HCl, pH 8.0, 8% sucrose, 5% Triton X-100, 50 m*M* EDTA.

In Vitro DNA Synthesis

Growth media, chemicals, and enzymes are as described above.

PEG solution: 3.3 *M* NaCl, 27% PEG-6000 (Union Carbide); or in two separate solutions at 5 *M* NaCl and 40% PEG-6000.

Ampicillin (5 mg/ml in water) and tetracycline-HCl (5 mg/ml in water) are filter-sterilized. Chloramphenicol (20 mg/ml) is dissolved in 70% ethanol.

Custom-made oligonucleotides may be obtained from BioLogicals, Collaborative Research, New England BioLabs, or P-L Biochemicals.

TCA solutions: 1 mg of salmon sperm DNA per milliliter, 5% sodium pyrophosphate, 50 m*M* EDTA
 10% TCA
 5% sodium pyrophosphate, 50 m*M* EDTA
 5% TCA

Sequencing gel: 19 g of urea (BRL) are dissolved in a solution containing 8 ml of 40% acrylamide (38:2), 4 ml of 10× borate buffer, 0.4 ml of fresh 10% ammonium persulfate, and ddH_2O to make up a volume of 40 ml. The solution is filtered through Whatman 3 M paper, and 20 μl of TEMED are added before pouring the solution into the gel cast.
 Synthetic primers are available from Bethesda Research Laboratory, BioLogicals, Collaborative Research, New England BioLabs, New England Nuclear, and P-L Biochemicals.

Deoxy Mixes

Mix[a]	G′	A′	T′	C′
0.5 m*M* dGTP	1	7.5	10	10
0.5 m*M* dTTP	10	7.5	1	10
0.5 m*M* dCTP	10	7.5	10	1
H buffer	7.5	7.5	7.5	7.5

[a] 0.5 m*M* dNTPs are made up from 10 m*M* stock solutions, frozen in 12-μl aliquots, and kept only for 2 weeks.

Dideoxy solutions: 1 m*M* ddGTP, 0.25 m*M* ddATP, 2 m*M* ddTTP, 1 m*M* ddCTP. Stock solutions are 10 or 20 m*M* and neutralized with 10 m*M* Tris-HCl, pH 7–8. Solutions are made in 25-μl aliquots, frozen, and thawed only once for immediate use.
Chase solution: 40 μl each of 10 m*M* dGTP, 10 m*M* dATP, 10 m*M* dTTP, 10 m*M* dCTP, ddH_2O
Stop solution: 0.05% each of xylene cyanole FF and bromophenol blue, 99% deionized formamide, 10 m*M* EDTA, 10 m*M* NaOH

Computer Requirements

Hardware
- Apple II plus 48K memory
- Apple language system (PASCAL)
- 2 Apple II disk drives
- Sanyo monitor, 12 inch
- M & R Super Term. 80 column board
- Printer (Silentype or Epson MX-80F/T with Grappler Interface)
- D.C. Hayes Micromodem for Apple II

Software
- Communications system for Micromodem
- Apple II software for DNA sequencing
- The ready-to-use-code for both the sequencing programs and communication program are available through the Department of Biochemistry, University of Minnesota, St. Paul, Minnesota 55108

Methods

1. Maintenance and Growth of Phage

The infective phage particles are stable for many years, if frozen at −20°. A culture of infected cells is cleared of the cells by a low speed centrifugation (6000–8000 *g*, 10 min). The supernatant containing the free phage can be frozen without the addition of glycerol. Infected cells have a tendency to lose the F episome that is required for the infection of the phage. Therefore, noninfected cells are needed for plaque formation. Usually the term plaque is used to indicate the lysis of host cells by the bacteriophage. However, a plaque formed by M13-infected cells, represents a zone of infected cells within a lawn of noninfected cells. Infected cells are distinguished from noninfected cells by their slower rate of growth, which is up to twice as long as for noninfected cells. This is in contrast to cells harboring a plasmid which does not influence growth rate significantly. The differential growth rate also causes host cells to cure Ff phage rapidly if selection as described below is withdrawn or the infection mechanism is blocked.

The infection process and the extrusion mechanism are independent. Phage can be formed without infection by virions. A host system resistant to natural infection can be transformed with phage DNA, but once transformed these cells need to be kept under permanent selection to maintain the phage. If phage are propagated as plasmids by colony formers, derivatives with a selectable marker, such as a drug resistance, or an auxotrophic marker should be used (Table I).

The packaging properties of the Ff phage creates a second type of selection. The growth rate of cells is decreased not only because of phage infection, but also because the size of the phage genome is increased. This can be screened for by plaque morphology, since phage genomes a third larger than unit length form smaller phaques. Nevertheless, an insert of about 40 kb has been cloned into M13mp2.[17] Over a broad range of phage DNA length (from about seven times larger than unit length to one-third of the unit length), DNA is not barred from packaging.[17,48,49] Defective or mini phage, however, grow only in the presence of a helper phage. Thus, there is no selection for phage smaller than unit length. As any cloned DNA is not essential for phage propagation, large recombinant DNA-containing phage are selected against because of their slower growth rate. Therefore, the cloning of large DNA fragments into Ff vectors is not a suitable way to generate a clone library, as has been done with the lambda packaging system.[50] Cloning into Ff phages require selection and maintenance of single transformants.

A third precaution is related to the phage host system. The NIH guidelines place Ff vectors in the K12/P1 category. *Escherichia coli* K12 strains deficient in conjugation must be used as a host. Since Ff phages need conjugal F pili for infection, a host carrying an F factor mutation such as *traD* or *traI*[51] must be used. These mutations reduce the promotion of conjugation, but allow Ff phage infection. Because conjugation cannot be used as a means to maintain the F episome in the host, a selective nutrient marker has to be used to maintain the mutant F episome.

A special F episome is required for *E. coli* infection by the M13mp

[48] A. S. Grandis and R. E. Webster, *Virology* **55,** 14 (1973).
[49] J. Griffith and A. Kornberg, *Virology* **59,** 139 (1974).
[50] N. Sternberg, D. Tiemeier, and L. Enquist, *Gene* **1,** 255 (1977).
[51] M. Achtman, N. Willets, and A. J. Clark, *J. Bacteriol.* **106,** 529 (1971).

phages. The *tra* mutations were originally introduced into a F′lac.[51] This episome complements lactose prototrophy in a host with a deletion in the lactose operon on the chromosome. The M13mp phages make use of a *lac* phenotype as a direct screening assay for cloning. Therefore, the *traD36* mutation had to be crossed to a F episome carrying *proAB* so that its maintenance can be based on proline prototrophy.[52] The hosts (JM101: *Δlacpro, supE, thi,* F′ *traD36, proAB, lac*I^q *ZΔM15;* or JM103: *Δlacpro, supE, thi, strA, sbcB15, endA, hspR4,* F′ *traD36, proAB, lac*I^q *ZΔM15*) should be streaked out on glucose, B_1, minimal salt medium (Materials and Reagents) without the addition of amino acids.

The assay for the expression of the *lac* DNA of the M13mp phages in JM101 or JM103 is based on an *in situ* color reaction of Xgal. The compound itself, does not induce the *E. coli lac* operon, but is cleaved by active β-galactosidase to produce deep blue dibromodichloroindigo. IPTG is added as an inducer and is not cleaved by active β-galactosidase.[53] Thus the expression of β-galactosidase in infected cells can be screened for in a lawn on agar plates in the presence of Xgal and IPTG. Upon induction by IPTG, a defective β-galactosidase is synthesized under the control of the F′ episome, which is complemented by the β-galactosidase fragment produced by the phage. This results in a blue plaque if the Xgal assay is used. If the phage *lac* DNA is interrupted by a cloned DNA fragment and the β-galactosidase complementing fragment is not produced, plaques stay colorless.[27]

For the propogation of any Ff phage, it is important to determine the titer or the phage concentration. To do this, streak host cells (JM101 or JM103) out on an agar plate containing M9 medium with glucose as a carbon source (Materials and Reagents) and grow at 37° overnight. Take a single colony and inoculate into 2 ml of 2× YT (Materials and Reagents) and incubate for 7 hr at 37°. Make a number of 10-fold serial dilutions (up to 10^{10}) of the phage suspension. Take 0.1 ml from each dilution and mix with 0.01 ml of 100 m*M* IPTG, 0.05 ml of 2% Xgal in DMF, 0.2 ml of the freshly grown host cells, and 3 ml of soft agar; plate the mixture on B broth agar plates (Materials and Reagents). Let the soft agar harden and incubate the plates at 37°. At temperatures below 34° F pili are not properly formed and infection is blocked. Therefore, it is critical that the incubation is carried out at 37°, not at lower temperatures. Plaques can be seen after about 4 hrs, the plaque color can be determined after a few more hours of incubation. It should be stressed again that what appear as turbid plaques are in fact zones of retarded growth of infected cells.

[52] J. Messing, *Recomb. DNA Tech. Bull.* (NIH Publ. No. 79-99, 2) **2,** 43 (1979).

[53] J. Miller, Cold Spring Harbor Laboratory, Cold Spring Harbor, New York, 1972.

TABLE I
LIST OF SINGLE-STRANDED DNA PHAGE VECTORS[a]

Phage	Parents	Insertion	Cloning sites	Phenotype	P mode[a]	Reference[b]
M13mp1	M13, pMG1106	5868	*Pvu*I, *Bgl*I	lac	P/C	27
M13mp2	M13mp1	5868	mp1 + *Eco*RI	lac	P/C	56
M13mp5	M13mp2	5868	mp2 + *Hin*dIII	lac	P/C	52
M13mp6	M13mp2	5868	mp2	lac, −*Bam*HI	P/C	52
M13mp61	M13mp6, M13amI,II	5868	mp2	lac, amI,II	P/C	47
M13mp62	M13mp61	5868	mp2	lac, −*Hin*cII	P/C	47
M13mp63	M13mp62	5868	mp2	lac, −*Acc*I	P/C	47
mWU43	M13mp2	5868	mp2 + *Bam*HI	lac	P/C	59
mWJ22	M13mp2	5868	mp2 + *Hin*dIII	lac	P/C	59
M13mp71	MWJ43	5868	mWJ43 + *Sal*I, *Pst*I	lac	P/C	47
M13mp7	M13mp71, M13mp63	5868	mp71 + *Hin*cII, *Acc*I	lac	P/C	47
M13mp8	M13mp7, pUC8	5868	mp7 + *Hin*dIII, *Sma*I, *Xma*I,	lac	P/C	76
M13mp9	M13mp7, pUC9	5868	mp8, but opposite polarity	lac	P/C	76
M13mp10	M13mp8c, pUC12	5868	mp8 + *Xba*I, *Sst*I	lac	P/C	101[c]
M13mp11	M13mp8c, pUC13	5868	mp9 + *Xba*I, *Sst*I	lac	P/C	101[c]
M13Ho176	M13, λh80dhis	5727	*Eco*RI, *Sal*I, *Xho*I, *Kpn*I	his	C	59

M13Goril	M13, G4	5565	*Eco*RI, *Sst*I, *Xho*I, *Kpn*I	—	P	62
M13bla	M13, Tn3	5565	*Pst*I	ap	C	62
M13blacat	M13, Tn3, pACYC184	5565	*Eco*RI, *Pst*I	ap, cm	C	62
R199	f1	5725	*Eco*RI	—	P	60
R208	R199	5725	*Hin*dIII, *Pst*I, *Sal*I	ap, tc	C	60
R209	f1	5868	*Eco*RI	—	P	60
R229	f1	5614	*Eco*RI	—	P	66
fd11	fd, pKB252	5830	*Eco*RI	—	P	61
fd101	fd, pACYC177	5565	*Pst*I, *Hin*dIII, *Sma*I	ap, km	C	61
fd103	fd, pACYC184/177	5565	*Pst*I, *Eco*RI	ap, cm	C	61
fd104	fd, pACYC177	5565	*Xho*I, *Hin*dIII, *Sma*I	km	C	61
fd106	fd, pACYC184, 177	5565	fd104 + *Eco*RI	cm, km	C	61
fd109	fd, pBR322	5565	*Pst*I, *Hin*dIII, *Eco*RI	ap	C	61
fd107	fd, pBR322	5644	fd109 + *Sal*I	ap	C	61
fdtet	fd, Tn10	5644	*Eco*RI, *Hin*dIII	tc	C	63
fKN16	fdtet	5644	fdtet	tc, Δ gene*III*	C	64

[a] The name of phage vectors, their ancestors with the parental phage first, the map position of the insertion to yield the vector phage, the cloning sites of the vector phage, useful phenotypes of the vector phage, the propagation mode (P mode) of the vector phage as plaque (P) or colony (C) formers, and their sources are presented.

[b] Numbers refer to text footnotes.

[c] J. Messing, J. Vieira, J. Norrander, T. Kemps, and G. Heidecker, manuscript in preparation.

Pick some infected cells from a single blue plaque by piercing the plaque with a toothpick, put them in 2 ml of 2 × YT medium (titer will be around 10^7 to 10^8 PFU/ml), and let them grow for 3 hr. During the 3-hr growth of infected cells, also grow a culture of noninfected cells from a colony transferred from a M9 glucose plate to 10 ml of 2 × YT liquid medium. These two cultures should provide phage from the infected cells with a titer of 10^{10} to 10^{11} PFU/ml and noninfected cells with an OD_{600} = 0.3. The noninfected cells should then be infected with a multiplicity of infection of 1 to 10. Incubation is continued at 37° for another 4–5 hr. The titer should reach 10^{11} to 5×10^{12}/ml.

Phage particles numbering 3×10^{11} are equivalent to 1 μg of single-stranded M13 DNA. Since there are about 100 copies of RF and about 20 times as many phage per cell, the theoretical yield of a 10-ml culture grown to about 2×10^9/ml is 12 μg of RF and 120 μg of ss-DNA. Depending on the experiment, the volume of the culture may be scaled down (one dideoxy sequencing reaction or a primer–mutagenesis reaction requires about 0.6 μg of ss-DNA, but the preparation of hybridization probes requires only 0.05 μg of ss-DNA), or scaled up for larger amounts of RF DNA for cloning (1 μg of RF gives rise to about 10^6 transformants) or ss-DNA for preparative work (the purification of large amounts of a specific mRNA).

2. *Isolation of the Replicative Form* (*RF*)

The amount of RF DNA required for cloning depends upon how many different cloning sites of the same RF DNA are to be used (see also Cloning Strategies). A 40-ml culture of infected cells should yield about 20–40 μg of RF DNA; this is sufficient for most projects. Since phage vectors allowing double-digest cloning are prepared as pairs (see Forced Cloning): two 40-ml cultures, one for each of the vectors, are conveniently handled with standard laboratory equipment.

Growth and infection are described above, and a culture of infected cells is grown to saturation at 37°. The following procedure, originally developed for plasmid preparation by Radloff *et al.*[54] and modified as described below, is used to purify the RF DNA. Infected cells are collected in polypropylene tubes in the Sorvall SS34 rotor, or any equivalent rotor, by centrifugation at 6000 rpm for 5 min at 4°. The supernatant fraction can be saved to prepare ss-DNA (see Preparation of Template, ss-DNA). The pellet is resuspended in sucrose buffer (Materials and Reagents) to a volume of 1.0 ml. Keep the suspension in the plastic Sorvall tube on ice at all times. Add 0.3 ml of lysozyme solution, gently mix, and wait for 5 min.

[54] R. Radloff, W. Bauer, and J. Vinograd, *Proc. Natl. Acad. Sci. U.S.A.* **57,** 1514 (1967).

Then add 0.6 ml of Tris–EDTA and subsequently 0.025 ml of 10 mg/ml RNase solution. After a 5-min incubation the lysis is completed by incubation of the mixture with 2.5 ml of Tris–EDTA–Triton for a further 10 min. All solutions are mixed very gently by slowly revolving and tipping the tube. The lysis mixture is cleared by centrifugation in the Sorvall SS34 rotor at 12,000 rpm for 20 min at 4°. After the centrifugation, turn the brake off at 1000 rpm to avoid loosening of the pellet.

The supernatant is diluted to a volume of 8.5 ml with a solution containing 0.16 ml of 10 mg/ml ethidium bromide and 8 g of CsCl (Materials and Reagents). The two lysates are enough for two nitrocellulose or polyallomer tubes. The tubes are balanced with light mineral oil and centrifuged opposite each other in the Ti 50 or type 65 Beckman rotor at 40,000 rpm for 40 hr at 15°.

The RF isolation can be speeded up by using a vertical tube centrifuge rotor. The cleared lysate is centrifuged as before, and the supernatant fraction is diluted with distilled H_2O to a volume of 5 ml containing 0.1 ml of ethidium bromide and used to dissolve 4.8 g of CsCl (Materials and Reagents). The mixture is left in the dark for about an hour to allow the protein to float to the top, or it is centrifuged at 6000 rpm for 5 min. The cleared solution is removed from underneath the protein film and transferred to a Beckman VTi65 ultracentrifuge tube and centrifuged at 60,000 rpm for 16 hr at 20°. Because the gradient needs to reorient after the run, deceleration is reduced by turning the brake off at 1000 rpm. Critical to this procedure are the RNase step and the removal of the bulk protein.

After the run the DNA can be visualized by illumination with UV, 300 nm (usually the yield will be sufficient so that a band can be seen even without UV). The denser or the lower band is removed by puncturing the tube from the side with a syringe and a needle and withdrawing fluid. Avoid removing any of the upper, less dense band. Occasionally a third band can be seen with an intermediate density. This represents ss-DNA and should be left behind. The recovered fraction is mixed with equal amounts of *n*-butanol to extract the dye (Materials and Reagents). Repeat the extraction two or three times, until the aqueous phase is clear. Any interphase is left behind. Dialyze for 6 hr against 2 liters of low-Tris buffer (Materials and Reagents), with several changes. The DNA is precipitated with 0.1 volume of 3 *M* NaAc and 2.5 volumes of ethanol (Materials and Reagents), washed once with 70% ethanol, dried under vacuum, and redissolved in 200 μl of low-Tris buffer (Materials and Reagents). This stock solution should have a concentration of about 100 ng of DNA per microliter.

The purity of the RF solution at this point may be satisfactory for cloning: it can be tested as described below (Cloning Strategies). The RF preparation obtained using the vertical rotor centrifugation procedure usually still contains protein that is removed by a phenol extraction. The stock is mixed with an equal volume of phenol–chloroform (1:1) mixture, vortexed for 10 sec, and cleared by centrifugation in a microfuge; the aqueous phase is removed with a Pipetman, and the DNA is concentrated as described above.

If the RF preparation is contaminated by *E. coli* DNA, or RNA, or mini phage RF,[49] another purification step using a sucrose gradient centrifugation can be carried out. In high salt concentrations the RF has a high *s* value[7] and can be easily separated from the other nucleic acids. Prepare a 5 to 20% sucrose gradient with 18.5 ml of 5% and 18.5 ml of 20% sucrose in high salt (Materials and Reagents) in Beckman SW 27 nitrocellulose tubes, and load 0.2 ml of the RF solution. Tubes are balanced with light mineral oil and spun at 23,000 rpm for 15 hr at 4°. Fractions of 1.5 ml are collected from the bottom of the tube after puncture with a needle, and their optical density is determined. The peak with the RF will be the fastest sedimenting material, and the DNA is concentrated as described above but taken up in half the volume of low-Tris buffer to adjust for the losses that occur during this purification step.

3. *Cloning Strategies*

a. Single-Stranded DNA Phage Cloning Vectors

The first Ff cloning experiment described[27] was very inefficient because RF could not be cleaved by a restriction endonuclease in a single site. However, a number of more efficient Ff cloning vectors have been developed that are based on a marker inactivation system, such as the use of fragments of the *lac* operon,[17,47,55–59] the histidine operon,[60] fragments

[55] B. Gronenborn and J. Messing, *Hoppe-Seyler's Z. Physiol. Chem.* **358,** 1208 (1977).
[56] B. Gronenborn and J. Messing, *Nature* (*London*) **272,** 375 (1978).
[57] J. Messing and B. Gronenborn, *in* "The Single-Stranded DNA Phages" (D. T. Denhardt, D. Dressler, and D. S. Ray, eds.), p. 449. Cold Spring Harbor Laboratory, Cold Spring Harbor, New York, 1978.
[58] R. J. Rothstein, L. F. Lau, C. P. Bahl, S. A. Narang, and R. Wu, this series, Vol. 68, p. 98.
[59] R. J. Rothstein and R. Wu, *Gene* **15,** 167 (1981).
[60] W. Barnes, *Gene* **5,** 127 (1979).

conferring drug resistances,[61–65] or synthetic sites without a marker system.[65,66] A partial list of the different Ff cloning system is given in Table I.

b. Selection Schemes of Recombinant Phage

A marker inactivation system, however, often relies on using the Ff cloning system as a colony-forming system rather than a plaque-forming one. The former concept can be very time consuming, as replica plating of transformants to screen them introduces an additional step. This is still necessary even if the cloning efficiency is enhanced by treating the cut RF of the vector with bacterial alkaline phosphatase.[67] Furthermore, recombinants must be maintained as infected cultures and therefore undergo continuous growth. Any change in growth rate would select against bacteria harboring recombinant phage DNA. Thus if a deletion occurs within the recombinant DNA even if at a low frequency, it leads to an increase in the growth rate of those host cells and results in a dilution of the original recombinant DNA with every cell division. Recombinants propagated as plaques and stored as phage can be kept as purified clones.

The *lac* operon marker inactivation system allows for the use of a plaque assay system without the need for replica plating. By using as the cloning site for M13 a piece of *lac* DNA, a blue plaque can be produced under appropriate plating conditions.[27] M13*lac* phage therefore, not only can be selected for on lactose minimal plates, but also screened for by a color reaction *in situ* in the absence of any selection, making it a system that can be used directly to identify recombinants. Insertion of DNA into the *lac* region of the phage vector leads to the reduction or inactivation of α-complementation (see also Maintenance and Growth of Phage, above), and either light blue or colorless plaques are produced.[56] If recircularization of the cut vector is prevented by treating the 5′ ends with alkaline phosphatase (see below) or using a double-digest cloning scheme (see below), recombinants can be produced with high efficiency and the resid-

[61] R. Herrmann, K. Neugebauer, E. Pirkl, H. Zentgraf, and H. Schaller, *Mol. Gen. Gent.* **177,** 231 (1980).

[62] J. C. Hines and D. S. Ray, *Gene* **11,** 207 (1980).

[63] A. N. Zacher, C. A. Stock, J. W. Golden, and G. P. Smith, *Gene* **9,** 127 (1980).

[64] F. K. Nelson, S. M. Friedman, and G. P. Smith, *Virology* **108,** 338 (1981).

[65] J. D. Boeke, G. F. Vovis, and N. D. Zinder, *Proc. Natl. Acad. Sci. U.S.A.* **76,** 2699 (1979).

[66] J. D. Boeke, *Mol. Gen. Genet.* **181,** 288 (1981).

[67] A. Ulrich, J. Shine, J. Chirgwin, R. Pictet, E. Tischer, W. Rutter, and H. Goodman, *Science* **196,** 1313 (1977).

ual empty vectors sorted out directly in the initial transformation experiment.

c. Controls in Ff Cloning

Despite the use of a marker inactivation scheme for cloning, problems arise in the detection of recombinant molecules. Contamination of the enzymes involved in the cloning experiment can cause deletions and thus produce marker inactivation without insertions. Also, DNA fragments can be cloned without disruption of the structural gene, by an in-frame insertion, for instance. Marker activity can be sustained even at a reduced level allowing recombinants to not be identified. Contamination of the RF with *E. coli* DNA or the restriction endonuclease with DNA will result in nonspecific insertions and subsequent inactivation of the marker. Some of these problems are alleviated by treating the cut vector with alkaline phosphatase. Insertions in the *lac* DNA can give rise to blue plaques, an observation that eventually led to the construction of an in-frame multiple cloning site without loss of the *lac* phenotype.[17,47,52] The solutions to these problems depend on the experimental goal and are discussed below.

In general terms, any cloning experiment should be accompanied by the following controls:

1. The efficiency of opening a cloning site in the vector by the restriction enzyme is tested by transforming host cells with equal amounts of cleaved and uncleaved vector DNA. The ratio should be at least 1:100.
2. Transformants from uncleaved vector should be scored for their marker activity. There should be less than 1 colorless plaque per 100 blue ones, and they should normally not occur at all, if the RF preparation has been started with a single blue plaque.
3. Cleaved RF should be treated with alkaline phosphatase as described below. The efficiency of the treatment should be checked by ligating the RF to itself before and after alkaline phosphatase treatment. Ligation products are tested by transforming competent host cells. The treatment of RF with alkaline phosphatase should give a differential recircularization efficiency as tested by transformation. Compared to circular RF, the number of transformants of religated RF should not be higher than 1% after alkaline phosphatase treatment and between 20 and 50% before phosphatase treatment.
4. Every transformation experiment should include a control where competent host cells are plated with no DNA added and another control where RF is ligated in the absence of donor DNA.

d. Shotgun Cloning

Lambda packaging allows propagation of a mixed population of recombinant and vector phage. This plays an important role in constructing large eukaryotic genomic libraries, when phage is propagated batchwise. Because of the differential growth rate of recombinant and Ff vector phage, they have to be propagated by plaque or colony formation. Less complex libraries of the order of 10,000 different recombinants (e.g., single eukaryotic chromosomes), however, may be constructed with Ff vectors if transformants are plated on agar plates in a nonconfluent manner.

Most current molecular genetic work is dedicated to small sections of eukaryotic or prokaryotic DNA. Large viruses, mitochondrial or chloroplast DNA, multigene families, or functionally related genes have complexities of only 10–500 kb. They have been purified from genomic libraries by using their gene products as molecular probes or by virtue of their compartmentization, as is the case with virus particles or mitochondria.

Detailed studies of these genetic entities must be done on even smaller segments of the DNA. Structural and functional analyses require the dissection of the DNA into smaller fragments. DNA sequencing is done discontinuously on fragments of 200–400 nucleotides. Analysis of the relationships of DNA sequences using hybridization depends on the specificity of the probe, and therefore on using short restriction fragments as probes. The fastest way to dissect a DNA into small fragments and to separate them in pure form is shotgun cloning. The order of these small fragments can be determined by sequencing. To do this, the DNA has to be dissected in overlapping fragments, and the sequence information is used to join two fragments of DNA to a larger fragment and eventually into the total DNA molecule. This approach is used not only to generate information, but also to define material for the experiments that can be done following the determination of the primary structure.[68]

There are now four ways in which a DNA molecule can be dissected into small overlapping fragments and cloned in Ff vectors: (*a*) the combination of various restriction endonucleases with addition of a linker technique[44,69]; (*b*) the direct cloning using various restriction endonucleases without the aid of linkers[47]; (*c*) breakage by shear, and (*d*) DNase I treatment. Procedures (c) and (d) result in DNA fragments with protruding ends that have to be converted into blunt ends for cloning by either

[68] J. Messing, *in* "Genetic Engineering" (J. Setlow and A. Hollaender, eds.), Vol. 4, p. 19. Academic Press, New York, 1982.

[69] F. Sanger, A. R. Coulson, B. G. Barrell, A. J. H. Smith, and B. A. Roe, *J. Mol. Biol.* **143**, 161 (1980).

*Bal*31[47,70] or DNA polymerase.[71] Despite the specificity of the cleavage site of a restriction endonuclease used, in combination they are sufficient to determine the complete sequence of cauliflower mosaic virus DNA.[72] Since this allows the ligation of DNA fragments directly, it should be preferred over any other method and used if possible. The DNA to be sequenced should be checked with various enzymes and agarose gel electrophoresis of the fragments to determine whether sufficient sites are available.

Whichever cleavage procedure is used, a size selection of the DNA fragments is necessary to ensure that every sequencing step results in maximal sequence information. In addition, the shotgun library may be generated in all cases from a recombinant plasmid containing the DNA of interest. Although the cauliflower mosaic virus sequences were separated from pBR322 before subcloning into M13,[72] this step can be omitted.[70] A procedure to identify rapidly M13 subclones containing only pBR322 sequences before the sequencing step is described below. The plasmid containing the sequence of interest is prepared as described in Section 6b.

Restriction Endonuclease Treatment. After using various restriction endonucleases to cleave the DNA, the resulting fragments can be ligated directly in to the M13mp7, M13mp8, and M13mp9 cloning system (see Tables I and II) without additional DNA modifying steps. The enzymes that are the easiest to use for shotgun sequencing are those that recognize sites containing four nucleotides and produce the appropriate ends for cloning into the Ff vector.

About 3 μg of the purified plasmid DNA to be cloned are digested with one of the following restriction enzymes: *Sau*3A, *Msp*I, *Taq*I, *Sci*NI (this has not been tested yet), *Alu*I, *Rsa*I, *Hae*III, or *Fnu*DII, according to procedures suggested by their supplier. An aliquot (1 μg of total DNA) of the reaction is analyzed by polyacrylamide gel electrophoresis. A 6% polyacrylamide gel in Tris-borate buffer (Materials and Reagents) is prepared, and samples are run at 150 V until the bromophenol blue dye front runs off. M13mp2 RF cleaved with *Hpa*II is included as size markers [1596, 829, 818, 652, 545, 543, 472, 454, 357, 250, 176, 156, 129, 123, 60, 18, 18 (if M13mp7, M13mp8, or M13mp9 RF is used, the 250 bp fragment will be cleaved because of the *Hpa*II site between *Eco*RI and *Bam*HI]. After the run, the gel is stained with ethidium bromide and a picture is taken for documentation of the cleavage patterns. The remainder of each reaction is

[70] J. Messing and P. H. Seeburg, *in* "Developmental Biology Using Purified Genes" (*ICN-UCLA Symp. Mol. Cell. Biol.*), Vol. 23, p. 659. Academic Press, New York, 1981.

[71] S. Anderson, *Nucleic Acids Res.* **9,** 3015 (1981).

[72] R. C. Gardner, A. J. Howarth, P. Hahn, M. Brown-Luedi, R. J. Shepherd, and J. Messing, *Nucleic Acids Res.* **9,** 2871 (1981).

pooled into one of the three mixtures. The *Sau*3A fragments are kept separately; the *Msp*I, *Taq*I, and *Sci*NI fragments are combined for the second mixture, and the *Alu*I, *Rsa*I, *Hae*III, and *Fnu*DII fragments are combined for the third mixture. These three mixtures of restriction fragments are concentrated by ethanol precipitation as described above (RF Isolation). They are then taken up in 20 μl of low Tris buffer and 3 μl of loading buffer (Materials and Reagents) and are ready for preparative gel electrophoresis. In addition to these enzymes, *Eco*RI may be very useful as recently described.[72,72a]

Shearing. The plasmid DNA to be cloned is precipitated with ethanol and taken up in 6× SSC to give a final concentration of 12 μg/ml. The DNA is converted into the relaxed form by holding the solution for 10 min in boiling water[34] and then slowly cooling it for 10 min at 65° followed by 10 min at room temperature to allow renaturation. The solution is then placed on ice and degassed with nitrogen or argon; the microtip of the sonicator is inserted into the solution as deep as possible. The amplitude of sonication has to be determined empirically with every experiment. For example, the treatment could consist of pulses of 5 sec with 30-sec intervals of cooling, using a quarter of the maximal output. Aliquots of the time course are analyzed by polycrylamide gel electrophoresis as described above. *Hpa*II-cleaved M13mp2 RF DNA can be included as size markers. The residual volume of the solution with fragments that are shown to be in the size range of 500 to 1200 nucleotides (5 μg of total DNA) are kept, concentrated with ethanol as described (RF isolation) and taken up in 20 μl of low-Tris buffer. The sample is then ready for preparative gel electrophoresis.

DNase I Treatment. The plasmid containing the DNA that is to be sequenced is prepared as described in the preceding paragraph. After molecules are relaxed and concentrated by ethanol, the DNA is taken up in 50 m*M* Tris-HCl, pH 7.5 to give a concentration of 1 mg/ml. Electrophoretically pure DNase I is dissolved in 0.01 *N* HCl to give a concentration of 1 mg/ml and stored in 0.05-ml aliquots at −20°.[73] About 50 μl of DNA, 10 μl of DNase I (diluted 1:1000 in incubation buffer just before use), and 10 μl of DNase 10× buffer (500 m*M* Tris-HCl, pH 7.5, 10 m*M* $MnCl_2$, 1 mg of BSA per milliliter) are combined with double-distilled H_2O in a final volume of 100 μl.[70,71,74] Incubation is carried out at room temperature. Aliquots of 25 μl are removed every 5 min and inactivated by the addition of 1 μl of 0.5 *M* EDTA, pH 8.3. An aliquot of 10 μl of every sample is diluted with 10 μl of low-Tris buffer and 3 μl of loading buffer. The

[72a] R. C. Gardner, A. J. Howarth, J. Messing, and R. J. Shepherd, *DNA* **1,** 109 (1982).
[73] P. W. J. Rigby, M. Dieckmann, C. Rhodes, and P. Berg, *J. Mol. Biol.* **113,** 237 (1977).
[74] T. E. Shenk, J. Carbon, and P. Berg, *J. Virol.* **18,** 664 (1976).

digestion is checked by electrophoresis through a 6% polyacrylamide gel. M13mp2 RF DNA cleaved with *Hpa*II is used as size markers. The sample showing a size distribution of 500–1200 bp is saved, the protein is removed by the standard phenol extraction, and the DNA is concentrated as described (RF isolation). The DNA is taken up in 50 μl of low-Tris buffer and is ready for preparative gel electrophoresis.

Preparative Gel Electrophoresis. A 7% polyacrylamide gel is prepared (Materials and Reagents). Samples from the restriction cleavage, shearing, or DNase I treatment are diluted in 20 μl of low-Tris buffer to give a final concentration of 50 μg/ml. Loading buffer (10 μl) is added, and samples are run at 100 V until the dye marker (bromophenol blue) runs off the gel. M13 size markers should be run alongside as described above. The diluted samples will show only a weak staining. On the other hand, smaller molecules are better separated and not trapped under the larger-size molecules. The region of the gel that corresponds to the 800 bp M13 *Hpa*II fragments is cut out with razor blade. The piece of polyacrylamide is placed into dialysis tubing that has been filled with a 1:10 dilution of the running buffer.[44] The tubing (0.3–0.4 ml of buffer volume) is closed with closure clips and place into an elution box containing 1:10 borate buffer between two electrodes. The DNA is electrophoresed for 1 hr at 100 V and for a further 5 min with the polarity reversed. The aqueous solution is then transferred to a 1.5-ml polypropylene tube; 3 *M* sodium acetate (40 μl) and 2.5 volumes of ethanol are added to fill the tube for the concentration of the DNA fragments as described for RF isolation. Small amounts of polyacrylamide that go into solution act as a carrier. The samples are combined and taken up in 20 μl of low-Tris buffer. The restriction fragments are ready for cloning; the sheared and the DNase I fragments, however, have to be treated with *Bal*31[75] before use.

Bal*31 Treatment.* About 300–500 ng of the DNA fragments should be recovered from the preparative gel electrophoresis. The DNA is treated with 0.1 unit of *Bal*31 in a volume of 30 μl at 30° for 1 min as recommended by the supplier. The reaction is stopped after the dilution of the sample in 0.3 ml of low-Tris buffer by the addition of phenol. A standard phenol extraction is carried out, 5 μg of tRNA are added, and the DNA is concentrated by ethanol precipitation. The DNA is taken up in 20 μl of low-Tris buffer and is ready for the ligation.

Alkaline Phosphatase Treatment. Three samples of M13mp8 RF DNA are prepared. The M13mp8 RF is cleaved with *Bam*HI for the *Sau*3A library, with *Acc*I for the *Taq*I, *Msp*I, *Sci*NI mixture, and with *Sma*I or *Hinc*II for blunt-ended fragment mixtures. About 5 μg of RF should be

[75] R. H. Legerski, J. L. Hodnett, and H. B. Gray, *Nucleic Acids Res.* **5,** 1445 (1978).

cleaved with the appropriate amount of the particular enzyme as recommended by the manufacturer. The reaction is terminated by adding twice the amount of EDTA needed to titrate the $MgCl_2$, the enzyme is removed with phenol, and the DNA is concentrated as described above (RF isolation). The cleaved RF is taken up in 100 μl of 100 m*M* Tris-HCl, pH 7.5, and incubated with 75 units of bacterial alkaline phosphatase at 65° for 60 min. The reaction is terminated by a phenol extraction; the DNA is concentrated by alcohol precipitation and taken up in 25 μl of low-Tris buffer for the ligation reaction.

Ligation. The ligation for the *Sau*3A library is carried out at 15° overnight and terminated with EDTA. About 200 ng of RF are ligated to about 10 ng of *Sau*3A produced fragments in a volume of 100 μl. An aliquot of 10 μl should be used to transform 0.3 ml of competent JM103 cells. If too many plaques are formed as a consequence of the transformation, the transformation should be repeated. The reaction mixture is diluted so that not more than 100–300 plaques are formed per agar plate.

About 10 ng of the *Taq*I, *Msp*I, *Sci*NI mixture is ligated to 200 ng of RF cleaved with *Acc*I at 8° overnight in a volume of 100 μl with the required amount of ligase as reommended by the supplier and terminated with EDTA. The transformation is handled as for the *Sau*3A library.

About 20 ng of blunt-ended fragments and 200 ng of cleaved RF are ligated in 100 μl at room temperature overnight. Most suppliers recommend about a 20 to 100-fold higher amount of ligase for blunt ligation as compared to sticky-end ligation. The reaction is terminated with a phenol extraction. The DNA is concentrated and taken up in 20 μl of low-Tris buffer for the transformation. The efficiency with which the ligation products will transform competent cells is about a factor of 5 lower than with the sticky-end ligation products.

All ligation experiments should include one control reaction using the cleaved RF and another with phosphatase-treated RF DNA by itself. Further details are included in the transformation procedure, as well as in Section 3,c, controls in Ff cloning.

e. Forced Cloning

Shotgun cloning leads to the rapid accumulation of a large number of different recombinant phage. In many cases, however, a different approach is necessary. If only a small number of clones are needed or a small segment has to be analyzed, it is desirable to circumvent the analysis of a large collection of shotgun clones. One alternative to shotgun cloning, therefore, is the forced cloning scheme. If a vector is cleaved with two different restriction endonucleases to produce two different cohesive ends or one cohesive and one blunt end, the vector cannot be recircularized again unless a second DNA with compatible ends is added. Because this determines the orientation in which a DNA fragment is inserted into the RF, only one of the two strands of the cloned DNA can be produced in single-stranded form. Therefore, a pair of vectors with the cloning sites in the two possible orders in respect to the *lac* promoter has to be used so that double-digest forced cloning in both orientations can be conducted. M13mp8 and M13mp9 are such a pair of vectors. They contain a set of unique restriction sites with opposite polarities in the *lac* DNA.[76]

The appropriate M13 RF is cleaved with two restriction endonucleases that provide the appropriate cloning sites (see Section 3,f, Cloning Guide, for planning), extracted with phenol, and concentrated by ethanol precipitation. The small synthetic insert that is cleaved out of the RF does not go back easily under ligation conditions; background of self-ligated vector should be low.

Transformation efficiency of host cells with double-cut RF should drop at least to 1% compared to uncleaved RF. Self-ligation should result in only a slight increase in the number of transformants. Include these two controls when testing the insertion of the fragments of interest. Insertion of fragments should lead to a large increase of colorless plaques.

The strategy of forced cloning can be improved if a particular double-digest fragment can be selected out of a number of fragments cleaved with the same two enzymes. For instance, if one wishes to clone a DNA segment that can be excised with *Sau*3A and *Msp*I and does not contain any *Hae*III site, the insertion of other *Sau*3A/*Msp*I fragments can be reduced by cleaving the mixture with the third nuclease producing a blunt end. For a detailed study of the different combination of restriction site see the next section.

A ratio of vector to insert of 3:1 in ligation reactions has worked well and avoids obtaining a number of clones with multiple inserts. A final concentrationof 1–2 μg/ml of vector ina 20–30 μl reactionvolume is recommended.

f. Cloning Guide

The basic principle that has guided the construction of the multiple cloning site in the M13mp phage is the provision of a number of sticky ends and a blunt end for cloning that are contained within unique recognition sequences on the phage genome. Because the recognition sequences are longer than the sticky ends (6 nucleotides versus 4 nucleotides), en-

[76] J. Messing and J. Vieira, *Gene* **19**, 263 (1982).

TABLE II
RESTRICTION FRAGMENT FAMILIES OF THE M13mp7 CLONING SYSTEM[a]

M13mp7 cloning site	Shotgun fragment family
*Hin*dII	*Alu*I
GTC ↓ GAC	AG ↓ CT
	*Rsa*I
	GA ↓ TC
	*Fnu*DII
	CG ↓ CG
	*Hae*III
	GG ↓ CC
*Acc*I	*Taq*I
GT ↓ CGAC	T ↓ CGA
	*Msp*I
	C ↓ CGG
	*Sci*NI
	G ↓ CGC
*Bam*HI	*Sau*3A
G ↓ GATCC	↓ GATC

[a] Explanations are given in the text.

zymes with different specificity can be used to cleave the donor DNA at different sites, but still produce the same sticky ends. This combination of different sites allows us to use a small number of cloning sites for a large variety of differently cleaved DNAs.[77] Because sticky-end cloning is not only more efficient but also very selective, it is a preferable cloning method. A list of sites and corresponding enzymes has been provided in Table II; on the left is a list of the few cloning sites in the vector M13mp7,[47] and on the right are the various enzymes with their matching sites. In addition, the sequences of the various multiple cloning sites in the M13mp phage vectors and in their pUC plasmid derivatives are presented in Fig. 3.

The formation of recombinant molecules in the ligation reaction are also dependent on the concentration of the DNA fragments added. The concentrations that have been given above should work well in most cases, but the yield of recombinants may be improved by calculating the amounts of the DNA fragments as described.[77a]

[77] R. J. Roberts, *Nucleic Acids Res*. **9,** r75 (1981).

[77a] A. Dugaiczyk, H. W. Boyer, and H. M. Goodman, *J. Mol. Biol.* **96,** 171 (1975).

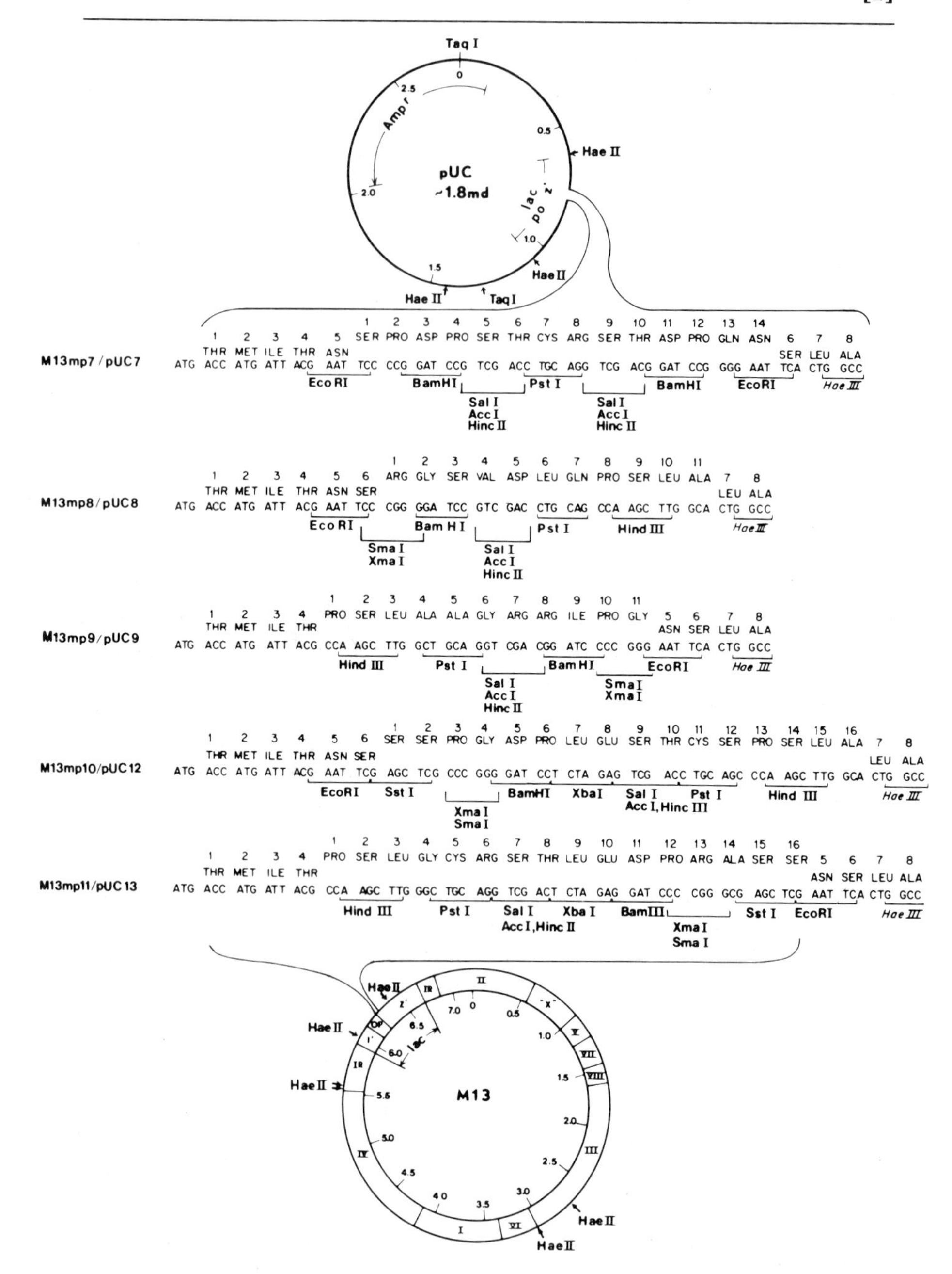

FIG. 3. The multiple cloning sites of M13mp7, 8, 9, 10, 11 and pUC7, 8, 9, 12, 13. The *lac* regions containing the multiple cloning sites, which are the basis for the cloning strategies as

4. Transformation

Besides infection, the most important route for the introduction of Ff DNA into host cells is transformation. Both RF and deproteinized single-stranded viral DNA can be used to transform host cells. From 1 ng of circular RF molecules, about 1000 transformants can be obtained. The same amount of ss-DNA gives about a 10-fold lower efficiency. On the average, one of every 50,000 molecules successfully transforms a target cell. The transformation efficiency with the same amount of *Eco*RI-cleaved RF drops to about zero. Ligated *Eco*RI-cleaved RF gives a transformation efficiency of up to 50% that of uncleaved molecules. A plate of bacteria transfected with circular RF should show only blue plaques. If not, the starting culture was contaminated and needs to be plaque-purified. Thus the only important step to control is the purity of the single plaque that is used to generate the stock; this is also true for any recombinant phage. The standard controls that should accompany any cloning experiment have already been discussed in the cloning section.

The volume of competent cells needed depends on the number of transformation mixtures. Normally, for each 2 ng of vehicle DNA, 0.3 ml of competent cells (10^9) and one agar plate are used. Larger quantities of DNA will produce a lawn in which the plaques are too densely packed; cross-contamination can occur. Even with the quantity recommended above, single plaques may have to be restreaked on a freshly poured plate with uninfected host cells for further purification. Because competent cells are concentrated 10-fold out of the growth volume, 3 ml of culture medium are needed per transformation experiment.

A loopful of cells of JM101 or JM103 are transferred from an M9 glucose minimal plate to a culture flask containing the appropriate volume of 2× YT (Materials and Reagents). Cells are grown to a density of 0.6–0.7 OD (660 nm). Cells are collected in a Sorvall SS34 at 6000 rpm for 5 min at 4°. Cells are resuspended immediately into ice-cold 0.05 *M* $CaCl_2$ (half of the growth volume) and kept in suspension for 20 min on ice. Cells

described in the text, are presented in their primary structure. The sequences are presented with the recognition sites for the various restriction endonucleases as described in the first column of Table II in the reading frame of the amino terminus of the β-galactosidase, starting with the ATG codon at the left and ending with the *Hae*III site and codon 8, alanine, to the right. The additional codons of the modified *lac* sequence in the multiple cloning sites are set up and numbered. The map position of the cloning sites with respect to their replicon, the M13mp phage[47,76] (see also J. Messing, J. Vieira, J. Norrander, T. Kempe, and G. Heidecker, manuscript in preparation) and the pUC plasmids [J. Vieira and J. Messing, *Gene* **19,** 259 (1982); and J. Messing, J. Vieira, J. Norrander, T. Kempe, and G. Heidecker, manuscript in preparation] is given by a circular map, respectively.

are centrifuged again and resuspended in a volume of iced 0.05 *M* $CaCl_2$ equal to one-tenth of growth volume. About 2 ng of RF are added to 0.3 ml of competent cells in a 13 × 100 mm glass test tube, and the mixture is kept on ice for 40 min more. Uptake of DNA is induced by a heat pulse at 42° for 2 min or 37° for 15 min. It seems to be that JM103 is more sensitive to heat than JM101. This may not be critical as long as 42° is not exceeded. After the heat pulse the mixture is transferred to room temperature; 0.01 ml of IPTG (100 m*M*), 0.05 ml of Xgal (2% in DMF) (Materials and Reagents), 0.2 ml of fresh exponentially growing JM101 or JM103 cells, and 3 ml of soft agar are added. The solution is vortexed for 10 sec and plated on B broth plates (Materials and Reagents). After the agar solidifies, the plates are incubated at 37°.

If larger DNA fragments (>2500 bp) are cloned, the diffusion of phages is somewhat lower, and the size of recombinants can be screened for by plaque morphology (smaller and clearer plaques). Sometimes plaques are simply smaller because of late growth. Plaque morphology can be used as a screen only by repeated plating. The JM101 or JM103 culture for plating may be obtained by adding an appropriate amount of fresh 2× YT to the culture flask used to obtain the cells for the $CaCl_2$ treatment, because it still contains some growing bacteria. During the time the cells are being made competent, this new culture will grow to the right density (late logarithmic phase) to plate the transformation mix.

The preparation of competent cells would be more convenient, if they could be stored. Although cells can be aliquoted and frozen at −80° for future transformation experiments,[78] the transformation efficiency is usually not as high as with cells made competent without storage. This may be still satisfactory if the amount of DNA to be cloned is not limiting. For the storage of competent cells a few additional steps are necessary. One wash with 0.1 *M* $MgCl_2$ (half the growth volume) precedes the first $CaCl_2$ wash for 5–10 min. The second wash is also done with 0.1 *M* $CaCl_2$ instead of 0.05 *M*. The cells are finally resuspended in one-tenth of the growth volume of 0.1 *M* $CaCl_2$ containing 14% glycerol. Frozen cells are allowed to thaw in an ice-water bath for 10 min, RF is added, and incubation is continued for 30 min. A heat pulse for 2 min at 42° follows, and the plating procedure including fresh exponentially growing cells is as described above.

If cells are made competent from a strain that is used for transformation experiments using plasmids like pBR322, they can also be used for the transformation experiment involving M13 phage. This is particularly useful if the competent strain is lacking an F factor and does not allow

[78] D. A. Morrison, this series, Vol. 68, p. 326.

infection or plaque formation of M13. As long as the standard host is added after the heat pulse under the standard plating conditions as described above, plaque and color formation can be observed. The $CaCl_2$-treated cells grow very poorly, if plated directly after the heat pulse. The transfected phage DNA matures and is secreted out of the host without being able to infect other cells. If, however, fresh exponentially growing standard cells (JM101 or JM103) are added, the lawn is formed by the latter strain, which can be infected and induced to produce the color reaction. This scheme may also be useful for special types of cloning experiments.

5. *Identification of Recombinants*

In any cloning system the inactivation of a marker by insertion of a cloned fragment is a helpful step in the identification of recombinants. The direct test for β-galactosidase activity (blue colonies or plaques) is a fast screen for recombinants. However, there are other biochemical methods available to characterize the nature of the recombinant molecules. They can be carried out either with the intracellular replicative form or with the extracellular single-stranded form of M13 DNA. Because in a shotgun DNA sequencing approach you will have already prepared the phage, the screening procedure listed below for single-stranded DNA may be employed without much effort. In addition, the *in situ* hybridization technique helps to prevent the loss of recombinants where insertions do not lead to the formation of colorless plaques (see discussion in Section 3,c, Controls in Ff cloning). If the background of empty vectors has been suppressed by the alkaline phosphatase treatment (see also Section 3,d, Shotgun Cloning), the numbers of additional transformants to test is relatively small.

a. Direct Gel Electrophoresis (DIGE)

If the DNA fragments to be cloned are larger than 200 bp, the insertions are large enough to be detected by direct agarose gel electrophoresis (DIGE) of DNA from an SDS-disrupted culture. A JM103 culture is prepared as described for the transformation procedure. At an $OD_{600} = 0.1$ for every clone, 2-ml aliquots are transferred to long test tubes (30 ml). Use 8 cm-long sterile sticks and transfer individual plaques to the 2-ml culture. Tubes are vigorously rotated on a roller drum in a 37° chamber for 6–7 hr. The infected cells are transfered to 1.5-ml Eppendorf tubes and centrifuged, 20 μl of supernatant are withdrawn with a sterile Pipetman tip, mixed with 1 μl of 2% SDS and 3 μl of loading buffer (Materials and Reagents). Samples are electrophoresed through a 0.7% agarose gel in

Tris–borate buffer (Materials and Reagents) at 100 V for 5 hr or at 70 V (0.7–0.8 V/cm) overnight. Vector phage DNA should be included as a standard. The gel is stained with ethidium bromide (Materials and Reagents), and a picture is taken in the presence of UV illumination. The mobility of the ss-DNA of recombinant phages decreases proportionally to the length of inserted DNA. Although a difference of 200 bp in 7000 bp hardly changes the mobility, a difference of 400–500 as advisable for shotgun DNA sequencing will cause a clearly detectable change in the mobility.

This sizing experiment is important for the characterization of recombinants for use in the rapid sequencing procedure; larger inserts should be selected in order that each primer reaction gives the maximum amount of sequence information (see discussion under shotgun cloning). Another important piece of information that can be obtained by the DIGE test is a rough estimate of the quantity of single-stranded DNA produced. This should be known before the phage are concentrated and their viral coat is removed by the phenol extraction as described in Section 6,a, (Preparation of Template, ss-DNA). Therefore, the relative intensity of the fluorescent ethidium bromide of the size standard included in the DIGE can serve also as a quantitating standard.

b. C-Test—DNA Cloned in Both Orientations

Because the (+) strand is always contained in the viral particle, two recombinant phage with a DNA fragment cloned in opposite orientations into the RF will hybridize only via the insert region. To test whether two recombinant phages have DNA complementary to each other, 20 μl of two supernatants are mixed with SDS and loading buffer as described above. After overlaying the mixture with a small amount of light mineral oil, the sample is incubated in a 65° water bath for at least 1 hr. The sample is electrophoresed through an agarose gel as described above. If the two viral DNAs hybridize via their insert region, they form a figure eight-like structure and migrate slower in the gel than the single viral circles. A sample of the mixture before the incubation at 65° is included as a control.

This procedure has been useful not only to determine the polarity of inserts in Ff vectors, but also to screen smaller insects. If hybridization is performed with a recombinant phage containing a large insert (2000–3000 bp), other recombinant phage containing smaller parts of this 2000–3000 bp region from a shotgun cloning experiment, for instance, can be identified if these inserts have the opposite polarity.[72,79]

[79] A. J. Howarth, R. C. Gardner, J. Messing, and R. J. Shepherd, *Virology* **112,** 678 (1981).

c. B-Test—Hybridization to Viral DNA by Blotting

Transfer from Agarose Gels. After visualization of the viral DNA by ethidium bromide staining, the gel is soaked in 1 *M* Tris-HCl, pH 7.5, 1.5 *M* NaCl for 30 min, and the DNA is transferred to a nitrocellulose filter overnight using the Southern procedure.[33] The procedure uses a tray filled with 10× SSC, a glass plate across the tray, Whatman 3 M paper lying across the glass plate with both ends floating in the 10× SSC, the gel, the nitrocellulose filter, Whatman No. 1 filter, stack of paper towels, and a rigid plate (the last three layers fit the size of the gel). A 2-liter Erlenmeyer flask containing 1 liter of water is placed on top. The nitrocellulose filter is first wetted by floating on water, soaked in 10× SSC, and then placed on the gel. The next morning the nitrocellulose filter is washed briefly in 2× SSC (1 min), placed on dry Whatman 3 M paper, and air dried. The filter is then baked for at least 2 hr in a vacuum oven at 80°.

Phage Dots. To screen a larger number of recombinants in a single hybridization experiment, 10 μl of the phage supernatant to be tested are spotted directly on a sheet of nitrocellulose filter.[45] Up to 120 clones can be spotted on a 15 × 18 cm sheet of nitrocellulose filter. The paper is wetted with a solution of 0.1 *N* NaOH, 1.5 *M* NaCl for 5 min, air dried, and then floated on a solution of 0.5 *M* Tris-HCl, pH 7, 3 *M* NaCl for 5 min. The paper is then baked in a vacuum oven for 2 hr or more as in the procedure described above. The buffer treatment can also be omitted and the filter baked directly after applying the phage solution on the filter.

Transfer from Agar Plates. The phage DNA screen described above is very effective for a shotgun library where recombinants represent more than two-thirds of the total transformants. It can be tedious, however, if a particular clone is represented only up to a few percent. For this, the Benton and Davis[80] procedure can be applied. The soft agar should be hardened by placing the agar plate in the refrigerator at 4° for 60 min. A nitrocellulose filter is placed on top of the soft agar with Ff plaques. The contact should be tight, leaving no air bubbles between the filter and the soft agar layer. The filter should darken evenly and be stained by the blue color of the plaques as it adheres to the wet surface. The nitrocellulose filter is allowed to remain for 5–10 min and then carefully lifted from the agar surface. The filter is wetted with the alkaline solution as described in the preceding paragraph. After baking, the filter is ready as in the other two preceding procedures.

Preparation of Probes. A new procedure for preparing M13 single-

[80] W. D. Benton and R. W. Davis, *Science* **196,** 180 (1977).

strand specific probes is described in Section 6,c, Preparation of Single-Strand Specific Probes. Alternatively, the plasmid DNA that has been subcloned into M13 can be labeled. The labeled probes should be made from both parts of the plasmid, the vector and its cloned DNA as previously described.[70]

This type of hybridization does not require a very sensitive hybridization probe. A typical reaction is as follows[44]: 1 μl of the restriction fragment (100 ng) is mixed with 1.5 μl of H buffer (Materials and Reagents), 7 μl of double distilled H_2O, and 5 units of exonuclease III. The reaction is carried out at room temperature for 30 min. After the inactivation of exonuclease III by boiling for 3 min, the DNA is renatured by incubation at 65° for 30 min. The solution is chilled on ice and transferred to an Eppendorf tube containing 5 μCi of [α-^{32}P]dATP (400 Ci/mmol) that has been dried down; 1 μl of 0.1 *M* DTT, 0.5 unit of DNA polymerase I (Klenow fragment), and 10 μl of a mixture of dGTP, dCTP, and dTTP (0.1 m*M*) in 10 m*M* Tris-HCl (pH 7.5), 8 m*M* $MgCl_2$, 2 m*M* DTT are added. The final volume is adjusted to 25 μl, and the reaction is carried out at room temperature for 30 min. The reaction is terminated by boiling for 3 min followed by quenching on ice. The DNA has a specific label of 10^7 to 10^8 cpm/μg. The probe should not be renatured before the hybridization step.

Hybridization and Filter Processing. After preparing the filter and the probe, the filter is prehybridized at 43° for 2 hr in a solution containing 5× Denhardt solution, 5× SSC, 50% formamide, 50 m*M* phosphate buffer (pH 6.8), 250 μg of sonicated salmon sperm DNA and 100 μg of poly(A) per milliliter, and 0.1% SDS. This solution is then replaced by the hybridization solution containing 1× Denhardt solution, 5× SSC, 50% formamide, 20 m*M* phosphate buffer (pH 6.8), 100 μg of sonicated salmon sperm DNA and 100 μg of poly(A) per milliliter, 0.1% SDS, and the radioactive probe. The nitrocellulose filter is sealed by heat into plastic freezer bags containing 10 ml of the described solutions. Hybridization is allowed to proceed in the presence of the probe at 43° overnight on a rocking table. The nitrocellulose filter is washed afterward three times for 15 min with 2× SSCP, 0.1% Sarkosyl (Materials and Reagents) at room temperature and then four times for 15 min at 50° in 0.2× SSCP, 0.1% Sarkosyl. The filter is then air-dried and exposed to X-ray film. The film is usually exposed overnight at room temperature without an intensifying screen and developed the next morning.

d. Mini Screen—Quick Isolation of Small Amounts of RF

In a few cases it may be desirable to analyze the recombinant phage for certain restriction cleavage sites or to cut out a particular fragment and subclone it by forced cloning using M13mp8 and M13mp9. Such an experiment may not require going through the RF isolation procedure described above. This mini screen also makes use of the infected cells collected at the same time the phage are prepared for the DIGE test described above. The following procedure is a modification of one of the many plasmid purification methods described.[81] The bacterial pellet is taken up in 80 μl of STET buffer and 5 μl of the lysozyme solution are added (Materials and Reagents). The tube is vortexed for 10 sec and placed in boiling water for 50–60 sec. The sample is then centrifuged in a microfuge B (Beckman) to remove cell debris. The supernatant, which is slightly yellow, is withdrawn with a Pipetman, and an equal volume of iced isopropanol is added. The solution is mixed on a vortex mixer and kept at $-20°$ for 10 min. The DNA is precipitated by centrifuging the solution in the microfuge in a cold room for 10 min. The pellet is washed once with 70% ethanol, dried under low pressure and taken up in 350 μl of low-Tris buffer. Any insoluble material is removed by a short spin in the microfuge, and the clear supernatant is saved. For cloning experiments at this point a standard phenol extraction is advised, and the DNA should be concentrated as described under RF isolation. The DNA is taken up in 30 μl of low-Tris buffer and is ready for further use of the RF DNA. The yield should be 0.3–1 μg of RF.

e. Marker Rescue

Homologous recombination in *E. coli* between M13 recombinants and the *E. coli* chromosome, the F factor, or plasmids can be used to rescue mutant sequences. The rescue of mutant sequences may be useful either analytically or preparatively. For instance, homology exists between the F factor carrying the *lac* operon with the U118 mutation (e.g., Cold Spring Harbor strain CSH 34[53] and the M13 *lac* phage. The U118 mutation leads to an early termination of the synthesis of β-galactosidase unless it is rescued by the M13 phage. If the standard host (JM101) is infected with the mutant phage, a colorless plaque will be formed. Although the rescue may occur at a relatively low frequency (0.1%; Heidecker and Messing, unpublished), a single colorless plaque is seen easily against a background of blue plaques under the appropriate plating conditions.[27] The same rescue occurs probably between the F factor carrying the M15 deletion (JM101) and the M13mp phage and gives rise to colorless plaques (see Section 1, Maintenance and Growth of Phage).

[81] D. S. Holmes and M. Quigley, *Anal. Biochem.* **114,** 193 (1981).

In Vitro DNA Synthesis

A number of ways have been given for how double-stranded DNA can be cloned into a Ff vector system and the recombinant phage characterized. One of the major applications of the recombinant DNA in single-stranded form is to serve as a template in an *in vitro* DNA synthesis reaction. A scheme of the various aspects of these applications is represented in Fig. 2 and is discussed above. Some of these methods are described in a general way elsewhere without the Ff cloning system, like the DNA sequencing procedure[31] or the site-specific mutagenesis.[46] In the following, however, all methods are described as an integrative part of the Ff cloning system.

a. PREPARATION OF TEMPLATE, SS-DNA

Using the Ff vector system, the preparation of the recombinant DNA in single-stranded form is reduced to two steps: (*a*) the concentration of phage by PEG precipitation; and (*b*) the removal of the viral coat protein by phenol extraction.

The following procedure, which can be scaled up, is usually sufficient for preparation of enough template for all three of the applications: DNA sequencing, single-strand specific probes, and site-directed mutagenesis. The supernatant that has been saved from the 2-ml cultures for the characterization of recombinants (DIGE, Section 5,a) can be processed for template preparation. To reduce to a minimum the number of templates that have to be prepared, three analytical criteria are used: the sample should contain enough DNA as judged by the ethidium bromide staining, the recombinant DNA should have the appropriate size increase, and the insert should be related to a known sequence by hybridization. The latter step may be omitted, if the number of possible insertions is relatively small: the sequencing procedure will be faster than carrying out the hybridization, and more accurate.

If the supernatant that has been kept in 1.5-ml Eppendorf tubes (see DIGE, Section 5,a) has become turbid during storage in the cold, it should be cleared again by centrifugation. A new 1.5-ml tube is filled with 200 μl of 27% PEG-6000 in 3.3 *M* NaCl (Materials and Reagents). The tube is then filled with the clear supernatant containing the phage. This procedure is based on the PEG procedure described by Yamamoto *et al.*[82] The tube is shaken and left in the cold room for 60 min. The solution becomes turbid and is then cleared by centrifugation. The supernatant is removed

[82] K. R. Yamamoto, B. M. Alberts, R. Benzinger, L. Lonhorne, and G. Treiber, *Virology* **40,** 734 (1970).

with a Pasteur pipette to avoid any disruption of the soft pellet. Any residual fluid is removed with a Kimwipe. The pellet is then resuspended in 0.65 ml of low-Tris buffer. The solution should then be clear. If not, insoluble material should be removed by centrifugation. To the clear solution 40 μl of 40% PEG-6000 and 80 μl of 5 *M* NaCl are added, mixed, and again left at room temperature for further 30 min. The solution once more becomes turbid and is again cleared by centrifugation. The soft pellet is saved as described above and resuspended in 300 μl of low-Tris buffer. The viral coat protein is removed by two organic solvent extractions, and the viral DNA is concentrated with ethanol as described for the RF isolation. The first extraction should be done with buffered phenol only, the second with a 1:1 mixture of buffered phenol–chloroform. Any interphase should be left behind. The DNA is taken up in 10 μl; the yield should be 5–10 μg.

b. PREPARATION OF MASTER PRIMERS

Obtaining the Two Universal Primers from Amplifiable Plasmids. The two universal primers, the first for sequencing and the second for making single-strand-specific probes, can be prepared from two plasmids that are made available through Bethesda Research Laboratories. They are purified as *Eco*RI fragments and serve as an alternative to commercially available synthetic primers. Because the primer for site-directed mutagenesis has to be individually designed and synthesized, oligonucleotide synthesis, however, will become a more important technique for a modern biochemical laboratory.

In the meantime it is useful to have a biological source for the universal primers. They have been constructed from a 109 bp-long *Alu*I restriction fragment that contains the *Eco*RI cloning site of M13mp2 vector. If this fragment is cleaved by *Eco*RI, two fragments of 17 bp and 92 bp are produced. The (−) strand of the 92 bp fragment serves as a universal primer for the synthesis of the complementary strand of the cloned DNA (Fig. 2). This is a prerequisite for the shotgun DNA sequencing approach.[44]

The second universal primer or probe primer is the (−) strand of the 17 bp fragment, which serves as the primer for the synthesis of the complementary strand of the vector DNA. If this synthesis is controlled so that the 3′ end of the newly synthesized DNA does not extend into the region of the cloned DNA, then the cloned DNA remains single-stranded and is now attached to a highly labeled molecule.[17,45] Because all the M13mp phage vectors have their cloning sites at the same map position, only one set of master primers is required for all of them.

Using a fill in reaction and a blunt-end ligation, both primer fragments have been converted into *Eco*RI fragments of 96 bp and 21 bp.[44] For amplification they have been introduced into pBR325 resulting in the new plasmids pHM232[44] and pHM235.[17] Both fragments have been inserted into the *Eco*RI site of the region conferring chloramphenicol resistance. Two copies of the 21 bp fragment have been inserted into the plasmid vector, but only one copy of the 96 bp fragment. Both plasmids have been introduced into the *E. coli* K12 strain CSH26,[53] which lacks the *lac* operon to minimize plasmid instability.

Preparation of the Biological Primers. The plasmids containing the primers can be selected by resistance to ampicillin and tetracycline and can be amplified with chloramphenicol. Cells are grown in 500 ml of 2 × YT in the presence of 20 μg of ampicillin per milliliter. When the culture reaches an OD (600 nm) of 1, 200 μg of chloramphenicol per milliliter are added for amplification, and incubation is continued overnight at 37°.[83] The cells are collected by low speed centrifugation at 4°, and the pellets are resuspended in 5 ml of ice cold 25% sucrose (Materials and Reagents). The cell suspension is then distributed into four 40-ml Sorvall centrifuge tubes and kept on ice at all times. The remainder of the purification scheme is identical to the RF isolation described above. After concentration of the plasmid DNA by ethanol precipitation, the dried pellet is taken up in 100 μl of *Eco*RI buffer and cleaved with *Eco*RI. The reaction is terminated by heating the DNA solution for 10 min at 65°. In the case of the probe primer, the DNA is now ready for use. The sequencing primer should be further purified by a sucrose gradient centrifugation as described for the RF isolation. The gradient fractions containing the small *Eco*RI fragment are easily separated from the remaining plasmid DNA, pooled, dialyzed against water, and concentrated by lyophilization. The DNA is taken up in 300 μl of water, dialyzed against low-Tris buffer, and concentrated by ethanol precipitation. The dried pellet is finally taken up in low-Tris buffer (Materials and Reagents) and adjusted to a concentration of 20 μg of purified primer per milliliter. The probe primer yields 1–2 μg of DNA, sufficient for 500–1000 labeling experiments. The sequencing primer yields 5–10 μg, sufficient for 250–500 sequencing reactions.

Preparation of Single-Strand Specific Probes

Single-strand specific hybridization probes can be used to determine the polarity of RNA transcripts and of M13 recombinant viral DNA. If they can be highly labeled, they may be used in general as probes for hy-

[83] V. Hershfield, H. W. Boyer, C. Yanofsky, M. A. Lovett, and D. R. Helinski, *Proc. Natl. Acad. Sci. U.S.A.* **71,** 3455 (1974).

bridization experiments. If the recombinant single-stranded DNA is immobilized to a solid phase, hybridization to complementary RNA or DNA can be used to purify related DNA sequences.

Hybridization Probes for "Walking." In respect to the DNA sequencing strategy, the Ff shotgun library can be rapidly mapped using the phage-dot hybridization procedure as described in (see Section 5,c, B-Test). To "walk" along a sequence in larger steps than just from one subclone overlapping with the next subclone, a larger probe can be prepared that hybridizes to a number of subclones. Using the forced cloning procedure with M13mp8 and M13mp9, double-digested DNA fragments of 2000–3000 nucleotides are readily cloned in both orientations (see Section 3, Cloning Strategies) and can be used as a source for preparing probes for "walking."

Subclones Prepared for DNA Sequencing: a Source for Hybridization Probes. Once the primary structure of each of the DNA subclones used in the shotgun DNA sequencing procedure has been determined, a number of questions remain that require the use of hybridization experiments. Which part of the DNA sequenced is transcribed, and in which direction? How is the transcript processed? How well are certain sequences conserved when compared to other related sequences? Whatever subclone is needed to answer one of these questions, it is now available in template form from the shotgun DNA sequencing project. These types of hybridization experiments as well as the comparison of the primary structure of DNA having related sequences play an important role in the understanding of gene structure and function.

Labeling Procedure for Single-Strand-Specific Probes. For the preparation of the probe, the 21 bp primer DNA before use is boiled for 2 min and then chilled on ice; 1 μl of primer (about 2 ng), 1 μl of template (50 ng), 1.5 μl of H buffer, 1 μl of 0.1 M DDT, and 4.5 μl of double-distilled H_2O are mixed and incubated for 15 min at 65°. After cooling to room temperature, 10 μCi of [α-^{32}P]dATP (2000–3000 Ci/mmol) that has been dried down is taken up in the template–primer mixture. One microliter of a mixture containing 500 μM each of dGTP, dCTP, dTTP, and 1 μl (0.5 unit) of DNA polymerase (Klenow) are combined. Incubation is continued for 60 min at room temperature. The reaction is terminated by adding 1 μl of 0.25 M EDTA, pH 8.3. It is important to realize that, after labeling, the probe is not to be heated to denature the complementary DNA strands. The labeled probe is ready for immediate use; keep it on ice and use it as described above under B-test. Although the primer DNA has not been separated from small amounts of *E. coli* DNA or pBR325, all contaminating DNA at this stage is double-stranded and does not hybridize to the single-stranded DNA immobilized to filter paper. Interference with this hybridization, however, occurs, if the (−) strand of M13 (from RF, for instance) or the pertinent *E. coli lac* DNA has been transferred to the filter and immobilized. Little, if any newly synthesized labeled M13 (−) strand is released from the probe under the described hybridization conditions.[45]

If a synthetic primer is to be used, 1 μl (10-fold molar excess for an oligonucleotide of 13 bases, 2-fold for a 15-mer) is mixed directly with the template and H buffer; the mixture is not boiled. The annealing mixture is heated, however, to 55° for 5 min and then transferred to room temperature. DNA polymerase and the dXTPs are added after the mixture has cooled for 20 min, and the reaction is continued as described for the biological primer. Under these conditions probes with a specific activity of 1.5 to 3 × 10^8 cpm/μg have been obtained.

Highly Labeled Probes with a Set of Probe Primers. A limiting parameter in obtaining probes of high specific activity is the quality of the elongation reaction. A short elongation reaction, however, can be overcome by a higher frequency of initiation of DNA synthesis. Therefore, nick translation has been a very useful labeling technique[73] because the number of nicks (starts) can be controlled by treating the DNA with DNase I. In the procedure described above, only one start for DNA synthesis is provided. On the other hand, no strand displacement or exonucleoytic removal of the 5′ end of a DNA has to occur during the elongation reaction. To obtain probes of higher specific labeling, it may be useful to synthesize a set of probe primers allowing multiple starts on the viral strand of the vector. These primers would have to be designed so that a maximal elongation with the radioactive labeled precursor occurs. The elongation should be limited as to produce as much of the (−) strand of the vector as possible without copying the insert. Probes with a specific activity of 10^9 cpm/μg would be obtained. This would allow the detection of a single gene copy in a genomic blot or very small amounts of transcripts.

d. Site-Specific Mutagenesis

This procedure will be described only as it is integrated into the M13 cloning system. For a detailed protocol of annealing reactions of synthetic primers or primer elongation reactions see this series, Vol [32].

Restriction Cleavage in Site-Specific Mutagenesis. Approaches to site-specific mutagenesis, which have been developed along with the construction of M13 vectors, may be of general use. One approach uses mutagenesis by random insertion, which is illustrated by the original M13 cloning experiment. The *lac* fragment was inserted into restriction sites. Whether or not this led to an interruption of an important stretch of sequence was tested by determining the viability of these insertion mutants.[27] Instead of a marker like the *lac* fragment, a fragment containing synthetic restriction cleavage sites can also be used if their introduction creates a unique site in the receptor chromosome.[84]

Another way restriction endonuclease cleavages can be used in mutagenesis is as selective agents. Chemicals used in random mutagenesis modify bases and cause mispairing. To select for site-specific mutations with a restriction endonuclease, the particular change has either to create or to abolish a unique restriction cleavage site.[47,56] Random chemical mutagenesis with M13 cloned DNA will accumulate the mutations in the cloned portion of the DNA; mutations within the M13 vector DNA are often nonviable.[56] If bisulfite is used as a single-strand-specific mutagen,[85] mutagenesis may also be localized to the single-stranded region containing the insert by hybridizing the complementary strand from the RF of the vector to the (+) strand of the recombinant phage. This would leave the cloned DNA single-stranded; the vector part has been converted preferentially into a double-stranded form. The phage treated with a chemical mutagen can then be used to transform competent host cells, and individual mutants can be recovered. This procedure can actually be conducted in a way that mutants can be selected for as described below in the Ff life cycle subsection.

The Variable Primer as a Mutagen. A new procedure of designing genes with defined insertions, deletions, or point mutations has been developed.[37] The advantage of the new technique is that an exact change can be designed. Only small polynucleotide segments containing the change have to be chemically synthesized.[86] Owing to improvements of solid phase-supported oligonucleotide synthesis,[87] any short polynucleotide segment can be produced rapidly. To test these small chemically synthesized oligonucleotides functionally, the appropriate flanking sequences have to be added.

Elongation of the Variable Primer Based on the Ff Cloning System. The single-stranded DNA phage cloning system can be used to reconstitute a DNA sequence embedding the chemically synthesized oligonucleotide. In such a reconstitution the synthetic oligonucleotide acts as a primer and the recombinant single-stranded DNA of the phage acts as a

[84] F. Heffron, M. So, and. J. McCarthy, *Proc. Natl. Acad. Sci. U.S.A.* **75,** 6012 (1978).

[85] D. Shortle and D. Nathans, *Proc. Natl. Acad. Sci. U.S.A.* **75,** 2170 (1978).

[86] H. G. Khorana, K. L. Agarwal, H. Buchi, M. H. Caruthers, N. K. Gupta, K. Kleppe, A. Kumar, E. Ohtsuka, U. L. RajBhandary, J. H. van de Sande, V. Sgaramella, T. Terao, H. Weber, and T. Yamada, *J. Mol. Biol.* **72,** 209 (1972).

[87] M. H. Caruthers, S. L. Beaucage, J. W. Efcavitch, E. F. Fisher, M. D. Matteucci, and Y. Stabinsky, *Nucleic Acids Res. Symp. Ser.* **7,** 215 (1980).

template to allow the addition of the flanking sequences to the primer by an enzymic reaction. Although a primer derived from a restriction fragment and hybridized to the M13 template is also extended *in vivo* after the transfection of competent *E. coli* cells,[47] a synthetic primer is too short for such an experiment. The enzymic addition of flanking sequences to the mutagen primer, however, raises the stability of the primer–template heteroduplex formation. This requires the extension of both the 3′ and the 5′ ends of the synthetic oligonucleotide; 3′ extension can be easily achieved, 5′ extension is more difficult. This problem has been solved in the ϕX174 system by carrying out a complete round of synthesis and closing the nick in the presence of DNA ligase.[37,88]

A simpler method of extending the 5′ end is the use of a second primer close to the 5′ end of the primer used as a mutagen. As outlined in Fig. 2, the master primer used for DNA sequencing is the closest primer available for those mutagenesis experiments. If both the sequencing primer and the mutagen primer are used in the same *in vitro* DNA synthesis reaction in the presence of DNA ligase, the flanking sequences can be added at the same time to the 3′ and 5′ end of the polynucleotide segment used to carry out the mutagenesis. The product is then used to transform competent host cells, and the remaining gap is properly filled in *in vivo*.[89,90]

Use of the Ff Life Cycle for the Production of Mutant Sequences. The gap filling *in vivo* does not affect the stability of the M13 heteroduplex significantly, unless the noncomplementary region destabilizes the hybridization of the 3′ or 5′ end of the annealed (−) strand.[47] A flow of the life cycle of the M13 heteroduplex is shown in Fig. 4. The parental RF that is formed now contains a heteroduplex region. As a next step, the gene *II* product is formed and produces a specific nick in the (+) strand. This directs the (−) strand, which contains the synthetic oligonucleotide sequence, to be the template for DNA synthesis while the old (+) strand is displaced.[25] Therefore, the population of progeny phage will be a mixture of wild-type and mutant phages.

Alternatively, the sequencing primer can also be replaced by the complete (−) strand of the vector as described above for the bisulfite method. Because the (−) strand is available only by the isolation of RF, its hybridization to the template used in the mutagenesis is always less efficient than the hybridization of the single-stranded synthetic sequencing primer. This handicap, however, may be compensated for somewhat by the following trick. The currently used M13mp vectors carry two *amber* mutations so that it can be propagated only in host cells with the *Su2* suppres-

[88] S. Gillam and M. Smith, *Gene* **8**, 81 (1979).
[89] Peter Seeburg, personal communication.
[90] J. Norrander and J. Messing, manuscript in preparation.

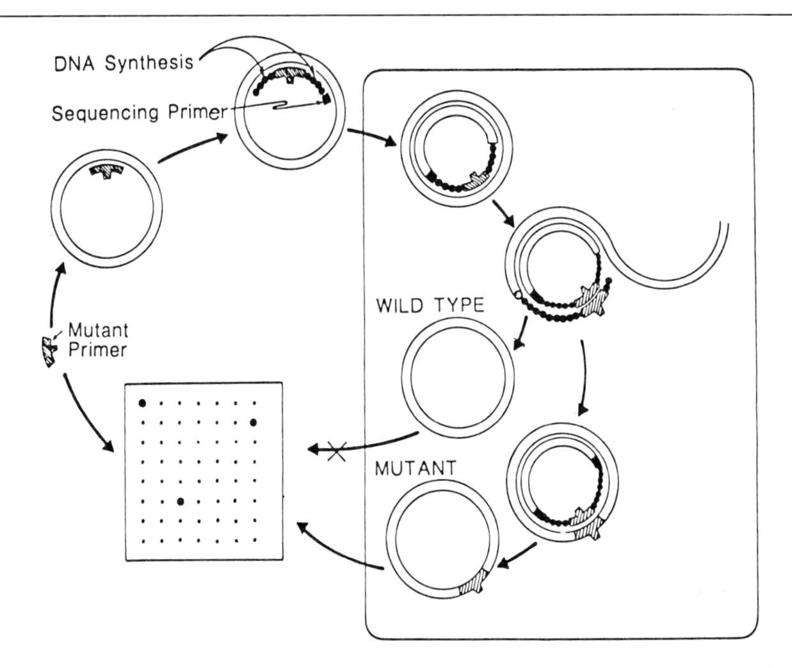

FIG. 4. Cloning scheme of a synthetic mutant sequences. Explanations are given in the text.

sor.[47] If wild-type M13 RF without any *amber* mutations is cleaved with *Hae*III (the *lac* DNA in the M13mp phage has been cloned into a *Hae*III site[27]) and the restriction fragments are heat denatured, the (−) strand of these fragments can be hybridized to the template used in the mutagenesis experiment. When the primer elongation is completed as described above, the (−) strand not only carries the desired mutation, but is also phenotypically distinguishable from the (+) strand because of the two *amber* mutations. This has been useful for transferring certain mutations from one strand to the other strand in the construction of the M13mp vectors.[47] If such a heteroduplex is used to transform a strain without the *Su2* suppressor like JM83,[53] the mutant phage can be enriched.

Finally, the mutagen primer can be used as a hybridization probe to discriminate mutant progeny phage DNA from any background. The clone of the desired mutant phage is then subjected to a rapid DNA sequence analysis; this is the ultimate confirmation of the structure change.

e. DNA SEQUENCING BY CHAIN TERMINATION

DNA Sequencing Techniques. The DNA sequencing method by chemical cleavage of DNA[40] has been widely used. Because the DNA sequencing method by primer extension[31] required the purification of a primer and a template, it has been used less frequently. Both methods require the fragmentation of larger DNA (>300 bp) into smaller fragments because of the limitation in the sequencing reaction and the resolution of the sequencing gels. Although improvements in the separation of DNA chains by polyacrylamide gel electrophoresis have been made,[91] the determination of sequences of over 400 nucleotides is difficult. This problem is met by providing the DNA in overlapping pieces.

End Labeling versus Continuous Labeling of DNA. On the other hand, the primer extension method has some advantages. Primer extension results in the continuous labeling of the newly synthesized DNA chain; the chemical method requires that the DNA be end labeled before the chemical cleavage. It is important, however, that both labeling procedures give equally strong signals for every chain length in the autoradiogram of the ladder gel. This requires that the molarity of the DNA molecules with different chain lengths has to be about the same for the end-labeling procedure, but should decrease with the chain length for the continuous labeling procedure. In the presence of the chain-inhibiting dideoxynucleotidetriphosphates, this requirement for the continuous labeling procedure is met, an excellent exploitation of the properties of the DNA polymerase by Sanger and his co-workers.[31]

Therefore, both steps, DNA labeling and the generation of DNA molecules with varying chain lengths, are combined into one experimental step and allow the use of smaller amounts of radioactivity.

DNA Strand Separation and Cloning as a One-Step Procedure. If a continuous labeling procedure is used for DNA sequencing, however, the complementary strands of the DNA have to be separated. The life cycle of Ff phages provides a biological means for the separation of the DNA strands and the purification of one strand.[27] Purification of DNA by cloning is superior to most physical purification techniques in purity and yield. Combining the separation of DNA strands and the purification of the ss-DNA by cloning in one step is an important aid in DNA sequencing. Consequently, Ff cloning and primer extension by DNA polymerase can be viewed as an integrative strategy for the structural analysis of the genetic information.

The Integration of Ff Cloning into DNA Sequencing. The integration of the primer extension reaction with chain terminators into Ff cloning may be best handled by keeping the isolation of the DNA template as the only variable component. (See under Comments in the section Cloning Strategies, how some situations may have to be handled differently.) This is accomplished by setting the number of primer extension reactions equal

[91] F. Sanger and A. R. Coulson, *FEBS Lett.* **87,** 107 (1978).

to the number of recombinant phage. In this case, for every chain-termination reaction a new template has to be prepared. However, the design of the M13mp cloning system allows for the use of the same primer for every extension reaction. Because the primer is short, it can be chemically synthesized.[47,92] In summary, the flow of steps for the use of the M13 vectors in the chain termination sequencing reaction are outlined in Fig. 5.

Experimental Time Frame. The cloning strategies, the characterization of recombinant phages, the preparation of template and primer have been described above (Sections 3, 4, 5 6,a and 6,b). This section focuses on the primer extension reaction and the subsequent analysis of the extension products.

The sequencing experiments including gel electrophoresis can be conducted during the day, and the radioactive polyacrylamide gel can be exposed to film overnight. Thus, the data can be recorded the following day. This protocol is designed to allow one person to carry out the reactions for 10 different templates at the same time. If desired, the running and drying time of the gels for the first set of reactions can be used for the preparation of another set of 10 templates. The time-limiting aspects of the procedure involve the cloning, the template preparation, the data recording, and the evaluation before and after the biochemistry of the sequencing reaction.

Preparation of the Gel. The experimental day should be started with the preparation of the gels. Glass plates (8 × 14.75 and 8 × 15.5 inches) are washed with detergent, rinsed with water, then with ethanol and air dried. Use two 0.4 mm spacers for the long sides and one for the bottom of the gel. The plates are clamped tightly together with clip binders, and the spacers are sealed with melted agarose. The plates are left flat on the bench. A solution of 8 *M* urea, 8% acrylamide solution, Tris-borate buffer, 1% of a fresh 10% ammonium persulfate solution, and 0.05% TEMED (Materials and Reagents) is used to make the gel. While keeping the glass plate assembly slightly tilted, pour the acrylamide solution slowly into the open slot of the top. After insertion of the comb (20 slots, each 5 mm wide), leave the gel flat on the bench during polymerization.

After polymerization the clamps are removed, and the gel is attached to the gel apparatus. We use an apparatus that utilizes two sets of gel plates to form a tank that holds the upper buffer (Shadel, Inc., San Francisco, California). The "ears" of the smaller glass plate, which contain

[92] S. A. Narang, R. Brousseau, H. M. Hsiung, W. Sung, R. Scarpulla, G. Ghangas, L. Lau, B. Hess, and R. Wu, *in* "Proceedings of the International Symposium on Chemical Synthesis of Nucleic Acids" Nucleic Acids Symp. No. 7 (H. Koster, ed.), p. 377. IRL Press, Oxford and Washington, D.C., 1980.

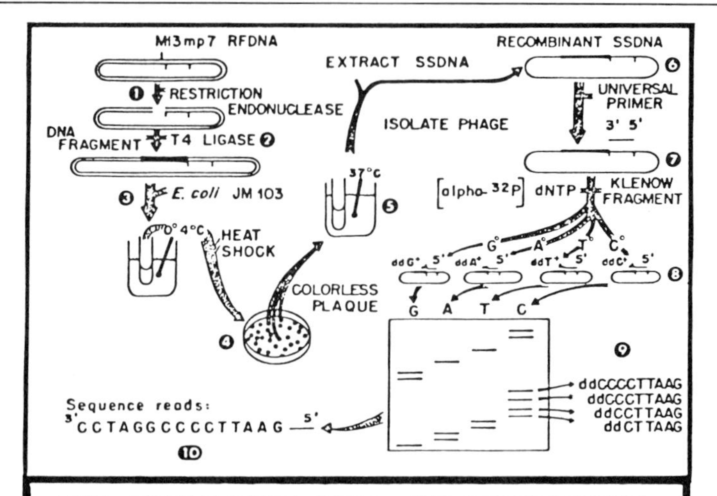

M13 SHOTGUN DNA SEQUENCING

1. CLEAVE M13mp7 RF WITH A RESTRICTION ENDONUCLEASE TO PRODUCE STICKY OR BLUNT ENDS AS NEEDED TO CLONE THE PARTICULAR FRAGMENT FAMILY.
2. ADD FRAGMENT FAMILY AND START LIGATION REACTION.
3. PREPARE COMPETENT HOST CELLS (JM103), ADD LIGATION MIX, AND HEAT SHOCK CELLS.
4. PLATE TRANSFORMED CELLS IN THE PRESENCE OF IPTG, Xgal, AND EXPONENTIALLY GROWING CELLS. CULTURE CELLS OVERNIGHT AT 37°C.
5. PICK COLORLESS PLAQUES AND CULTURE THEM INDIVIDUALLY IN 1.5 ml VOLUMES.
6. ISOLATE PHAGE AND EXTRACT SINGLE STRAND DNA(ssDNA)
7. ANNEAL EACH TEMPLATE WITH UNIVERSAL PRIMER.
8. FOR EACH TEMPLATE/PRIMER MIX, RUN 4 REACTIONS INCLUDING ONE OF THE 4 DIFFERENT CHAIN TERMINATORS, ddG, ddA, ddT, ddC.
9. ELECTROPHORESE REACTIONS THROUGH A DENATURING POLYACRYLAMIDE GEL LOADING IN THE ORDER GATC. AFTER RUN EXPOSE TO X-RAY FILM.
10. READ SEQUENCE FROM AUTORADIOGRAM. PROCESS DATA USING A COMPUTER ASSISTED PROGRAM.

FIG. 5. Scheme showing flow of steps in shotgun DNA sequencing.

the buffer to the gel, tend to be broken easily. We find that small pieces of rubber can substitute for these "ears," which are sealed with agarose. The gel plates have to be heated during the run to avoid compression of bands arising from undenatured DNA chains. Usually the temperature rises during the run owing to the high voltage used. In the apparatus described, the upper buffer conducts the heat evenly over the glass plates so that compression and also "smiling" is avoided. Because the initial temperature of the buffer is not very high, even if the gel is prerun, the desirable temperature of the upper buffer of 60–70° can be controlled by a heating system during the entire electrophoresis.

Reaction Tubes. To handle a large number of templates at the same time, the reactions are carried out in standard 250-μl plastic test tubes that fit the appropriate adapters of the Beckman microfuge B. Eight of the 250-μl plastic test tubes are held by one of these adapters, for a total of 48 per rotor head. Volumes are measured with a calibrated 20-μl Pipetman (it is used to dispense the small quantities needed for the chain-termination reaction). It should be noted that not all yellow tips available for the Pipetman are suited for this purpose. They must be slim enough to reach to the bottom of the tube. Although these tips can be autoclaved, the 250-μl plastic test tubes are heat sensitive. If the tubes are taken from a sealed bag, they are usually clean enough to carry out the reaction; no problems have been observed. The DNA polymerase and the thawed nucleotide solutions (Materials and Reagents) are placed in an ice bucket.

Use of the Biological Primer in the Sequencing Reaction. One of the major differences between the chemically synthesized primer and the biological primer is that the latter is provided in the double-stranded form. Although the double-stranded primer can be denatured into its single strands, the renaturation of these strands interferes with the annealing of the primer to the template. This is avoided by converting the double-stranded *Eco*RI fragment from pHM232 into a single-stranded form by exonuclease III treatment.[42] The M13 template is then the only complementary sequence to the primer. When the primer is elongated during the sequencing reaction, the cloning site region of the vector becomes double-stranded so that the primer can be cleaved off by the corresponding restriction endonuclease.[44] This additional step resets the 5′ end of the newly synthesized chains and reduces the length of the primer to the shortest possible length.

Annealing of Template and Biological Primer. The 0.5-ml polypropylene tubes that are heat resistant are used for the denaturation of the biological primer. Combine 1 μl of primer (20 μg/ml) purified as described in Section 6,b, 1.5 μl H-buffer, 7 μl of double-distilled H_2O, and 3 units of exonuclease III (0.5 μl) and incubate at room temperature for 15 min. The reaction is stopped by placing the tube in boiling water for 3 min. The tube is then placed in ice, 1 μl of template (500 μg/ml) is added, and the volume is adjusted to 13 μl with double distilled H_2O. The primer is hybridized to the template by incubating the mixture at 65° for 30 min. The tube is subsequently removed from the heat block and is ready for the primer extension reaction.

Annealing of Template and Synthetic Primer. Because there is a sequence in gene II similar to the *lac* sequence (see also Section 7, Computer Software), most of the synthetic sequencing primers commercially available should not be used in high molar excess. Combine 1 μl of primer (0.25 pmol of a 15-mer) with 1 μl of template (500 ng/ml), 1.5 μl of H buffer, and 9.5 μl of ddH_2O, and heat to 55° for 5 min. The tube is removed from the heat block and is ready for the primer extension reaction.

Synchronizing Pulse, Chase and Stop. The tubes for the primer extension reaction made for 10 templates at a time are best arranged in a tray. The plastic test tubes (250 μl) are marked G1, A1, T1, C1 for the first template and G2 and so on for the next template. Groups of four tubes for the sequencing reaction of each template are placed into a centrifuge adapter for the Beckman microfuge B. The centrifuge adapters can each hold one or two sets of templae reactions. The loaded adapters are placed in the tray so that the tubes are supported at a low angle and can be easily entered with a Pipetman tip. The tubes with the template primer mixes can be placed in the wells of Gilson fraction collector plates at room temperature.

Each template primer complex is mixed with 15 μCi of [α-^{32}P]dATP (600 Ci/mmol) that has been dried down, 1 μl of 0.2 *M* DTT, and 1 μl of DNA polymerase (Klenow, 500 units/ml) and then put on ice. From each template mixture, 3 μl are pipetted into the set of four tubes lying out in the tray as described in the preceding paragraph.

The corresponding deoxynucleotide and dideoxynucleotide triphosphate solutions (Materials and Reagents) are mixed 1:1 (e.g., G′ and ddGTP), and 2 μl of the G-specific mixture are added to the G tubes of each template set in the tray. The A, T, and C specific mixtures are added to the other tubes accordingly. Care is taken that the two solutions in each tube are not mingled together during the pipetting.

After all reactions are pipetted, the centrifuge holders are placed into the microfuge. All 40 reactions can now be started at the same time by combining the solution of the template–primer mix with the precursor mix by means of a short centrifugation. The tube holders containing the reaction tubes are placed in the tray while the reactions are allowed to proceed for 15 min at room temperature. During this incubation time, 1 μl of chase solution is placed near the upper tube lip. When the 15 min incubation time has elapsed, the tube holders are placed in the microfuge and the chase reaction is started by a short centrifugation. Afterward the tube holders are placed in the tray for 15 min at room temperature. During this incubation time 14 μl of stop solution (Materials and Reagents) are added. The stop solution is then added to the reactions synchronously by centrifugation.

If the biological 96 bp primer has been used, the reaction has to be split in half after the pulse. To one the chase is added as described; to the other, the chase is added containing the appropriate amount of *Eco*RI enzyme. Reactions are terminated as described for the synthetic primer. In this case, however, it may be difficult to do the pipetting for 10 templates at the same time without getting confused.

Gel Electrophoresis and Multiple Loadings. The reactions to be loaded are first placed in a heat block and incubated for 3 min at 95°. The comb is carefully removed from the gel (see the section on preparation of the gel, above) after the buffer reservoir has been filled with the Tris-borate buffer (Materials and Reagents). The slots are thoroughly washed by swirling the buffer in each slot with a syringe and a needle held between the glass plates. This has to be done before loading the samples to remove the urea. Then 3 μl of each of four reactions for a given template are loaded in the order GATC. Because each gel has 20 slots, 5 templates are loaded per gel. A Hamilton syringe is used, which is washed in the buffer between each loading. The samples are run into the gel for 5 min by electrophoresis. Then the urea is washed out of the slots of the second gel, and the reaction mixtures for the other 5 templates are loaded. The gel is run at 1600 V, and electrophoresis is terminated after 2–2.5 hr or when the bromophenol blue dye reaches the bottom of the 8% polyacrylamide gel. A second run using the same samples can be done on a 6% polyacrylamide gel. This run is also done at 1600 V and proceeds until the xylene cyanole dye front reaches the bottom of the gel. These two runs should allow about 250–300 nucleotides to be read. The sequences should contain an overlap and the sequence of the cloning site. If the biological primer was used and cleaved off with *Eco*RI, then the cleaved samples would be run on a 6% gel. The uncleaved samples should be run through an 8% gel until the xylene cyanole has migrated about two-thirds of the way down the gel.

Gel Drying, Exposure to X-Ray Film. After the run, the buffer is removed and the gel apparatus is disassembled. The smaller of the two glass plates is lifted by placing a spatula in the middle between the "ears." Although we do not siliconize the plates, the gel will stay attached usually to the lower glass plate, which is resting on the bench. Next, Whatman 3 M paper cut to the size of the gel is placed on top of the gel. The gel will wet and adhere to the paper. The paper with the gel and the glass plate is turned over and placed on top of another clean glass plate of the same size. The very bottom of the gel is wetted with water where it is adhering to the glass plate. The gel can then be peeled off the plate adhering to the filter paper. The same procedure is done with the second gel. Both gels can be placed on a Hoefer slab gel dryer SE 1140 after cutting off the first 5 cm of each gel (nearest to the slots), which are not important. The filter paper is now wetted thoroughly with water, and the paper-gel sheet is placed gel side up on a stack of paper towels on the dryer. The gels are covered with Saran wrap and dried in about 30 min. After drying, the standard XR-2 14 x 17 inches X-ray film can be placed on both gels in an appropriate film holder. The film and the gels should be held close to one another to ensure good contact. This avoids fuzzy bands. Exposure is at room temperature, usually overnight.

Gel Reading. The X-ray films are developed using the solutions recommended by the manufacturer. They are read on a milky light box. It is best if two persons process the data. The first person reads, and the second enters the data directly from a keyboard into the computer. Then they should switch, and the process should be repeated. The two sets of data can then be compared. If discrepancies emerge, the pertinent section of the autoradiogram can be inspected again. It should be determined whether a mistake has occurred due to a false reading or to a mistake in recording the data, or if the interpretation of the gel at that particular position is ambiguous. The data can then be edited, or both sets of data are kept, each with its ambiguities. Clarification of data has then to rely on additional data; for example, data involving complementary or overlapping sequences. From time to time in every sequencing project regions of the gel patterns occur that are difficult to interpret. Gingeras and Roberts[93] have developed a semiautomatic gel reading and recording device integrated into the computer system. The investigator may actually supervise the computer while it reads the gel. The instrumentation is meant to ease and speed up the procedure. However, the final decision will still have to come from the investigator.

7. *Computer Software*

The storage, evaluation, and design of the long strings of the nucleotide bases in DNA or RNA sequences have necessitated the use of a computer. In the last 5 years a number of programs have been developed[93-95]

[93] T. R. Gingeras and R. J. Roberts, *in* "Genetic Engineering" (J. Setlow and A. Hollaender, eds.), Vol. 3, p. 319. Plenum, New York, 1981.

[94] L. J. Korn, C. L. Queen, and M. N. Wegman, *Proc. Natl. Acad. Sci. U.S.A.* **74,** 4401 (1977).

[95] R. Staden, *Nucleic Acids Res.* **8,** 3673 (1980).

to provide the software for handling nucleic acid sequence data. Any software is limited, because it can be used only with the compatible computer hardware. The adaptation of software to other computers is costly and slows down the exchange of improved versions among different researchers. Software written for the larger computers may also generate too large a financial burden for the average molecular biology laboratory. Expensive hardware may have to be purchased or leased, or may be used as a time-sharing system. The running costs for such a system may be high, and user time is limited. All these expenses may be lessened by the purchase of a microcomputer that can also serve as a terminal for any larger computer. A complete package, including even a modest printer with graphics capabilities, may not exceed the cost of an ultracentrifuge rotor, a typical accessory instrument in a molecular biology laboratory. With this type of equipment many investigators can use the same software and are able to exchange improved versions.

To compensate for the rather small memory space of a microcomputer, we have selected among the different languages one that seems to be well suited for handling large strings of characters in the smallest possible memory space. The UCSD PASCAL programs are preferred over other standard languages like BASIC or FORTRAN. The Apple II computer can be equipped with a PASCAL language card (Materials and Reagents). Therefore, we have written a number of programs for handling sequence data for the Apple II in PASCAL language. To speed up the running time of some of the programs, Assembly language has been used in addition.[96,96a]

We have designed programs to handle two components of the Ff cloning system, the DNA sequencing and the design of primers. First, with respect to DNA sequencing the program can be used to reconstruct a large sequence out of small overlapping sequences and to check artifacts of gel runs and find mistakes in entering data. This program depends on matching overlapping and complementary sequence data. The computer and investigator have to interact to decide whether two pieces of sequence information should be joined to a larger block of sequence. Therefore, the program must allow the investigator to open files while carrying out the comparisons, reviewing mismatches, editing mistakes, storing ambiguous data. To open a file while the program is running, however, cannot be carried out with many large computers like the Burroughs, which we have used with the Pl1 programs written by Korn *et al.*[94] An example of the output generated by this program is shown in Fig. 6, in which two overlapping sequences from a sequence project are compared.

Second, the design of oligonucleotides to serve as mutagens is also

[96] R. Larson and J. Messing, *Nucleic Acids Res.* **10,** 39 (1982).

[96a] R. Larson and J. Messing, *DNA* **2,** 31 (1983).

```
A COMPARISON OF Z4ALU3A AND Z4RSA78
 SEARCHING FOR COMPLEMENTARY HOMOLOGIES

THE UPPER SEQUENCE IS  Z4RSA78
THE LOWER SEQUENCE IS Z4ALU3A


        10        20        30        40        50        60        70
CATAAGCATGAGAGTTCAATGCTGCCAGTTGGTTGAATGGAAGAAATTGCCGGGGGTAGGCAGCAG
 *  *          ***************************************************
AACTAAGCATGAGAGTCAATGCTGCCAGTTGGTTGAATGGAAGAAATTGCCGGGGGTAGGCAGCAG
        35        45        55        65        75        85        95

        10        20        30        40        50        60        70
CATAAG
**  **
CAGCAG
        95
```

Fig. 6. A computer search to map two overlapping shotgun clones. The option 4 of the program described[96] has been executed to compare a new gel reading to an archive of sequences. The lower sequence is matched with one of the archived sequence. Despite the long stretch of overlap, ambiguities are indicated that have to be checked as explained in the text.

handled by a comparison program. Because the synthetic oligonucleotide is used as a primer in DNA synthesis as well as a hybridization probe to discriminate between mutant and wild-type sequences, it must specifically hybridize to the templae at a unique location. This approach can be used to design a better synthetic sequencing primer. As an example, by comparing the sequence of the primer we have synthesized and used so far[47] to the (+) strand of M13mp vectors, we find that a second site the primer is nearly complementary (Fig. 7). The second site differs from the primary target by two nucleotides. This lowers the affinity of the primer to this site, but not enough so that if the primer is in molar excess compared to the template it will hybridize to both sites.

The sequence of M13mp8 has been recorded with the Apple II program and is available with these programs as described in Materials and Reagents upon request. The option 3, suboption M[96] can be used to print out the complete sequence with the name of the restriction enzymes written above the cleavage sites. The option 3, suboption A can be used to list the map position of all restriction cleavage sites as well as their map distances for the analysis of restriction fragments pattern in gel electrophoresis. A simple map with the detailed structure of the multiple cloning sites is given in Fig. 3.

The M13 wild-type sequence has been determined by Wezenbeek *et al.*[97] As a reference point the original *Hin*cII site, which has been altered in M13mp8, is used. The C, position 3, has been converted into a T, which

[97] P. Wezenbeek, T. J. M. Hulsebos, and J. G. G. Schoenmakers, *Gene* **11,** 129 (1980).

```
 A COMPARISON OF SP1 AND M13MP7
 SEARCHING FOR DIRECT HOMOLOGIES

THE UPPER SEQUENCE IS  M13MP7
THE LOWER SEQUENCE IS SP1


      797       807       817       827       837       847       857
AACGTCCTGACTGGTATAATGAGCCAGTTCTTAAAATCGCATAAGGTAATTCACAATGATTAAAGTTGAAATTAAA
****** *******
AACGTCGTGACTGGG
       10        20


 THERE WERE NO MATCHES IN M13MP7

 A COMPARISON OF SP1 AND M13MP7
 SEARCHING FOR DIRECT HOMOLOGIES

THE UPPER SEQUENCE IS  M13MP7
THE LOWER SEQUENCE IS SP1


     6305      6315      6325      6335      6345      6355      6365
AACGTCGTGACTGGGAAAACCCTGGCGTTACCCAACTTAATCGCCTTGCAGCACATCCCCCTTCGCCAGCTGGCG
***************
AACGTCGTGACTGGG
       10        20
```

FIG. 7. A homology search between a synthetic primer and the template. The sequencing primer SP1 is compared with M13mp7. The position at 6300 is the site close to the cloning site, the other position at 800 is the region of gene *II*. Although there are two mismatching base pairs, the 3′ end lines up perfectly and can contribute to some background priming.

eliminates the *Hin*cII site, the G, 2220, into an A eliminating the *Bam*HI site, and the C, 6917, into a T eliminating the *Acc*I and introducing the *Bgl*II site. The *lac* DNA that has been inserted in the position 5868 (Table I) is 789 nucleotides long.[98] It contains the 36 bp-long multiple cloning site (Table I) at map position 6231, increasing the length of the *lac* DNA by 33 nucleotides. The modified *lac* DNA (5869–6690) gives M13mp8 a total length of 7229 nucleotides. Two *amber* codons, one in gene *I* and one in gene *II*, have not been mapped[99] and not considered in the nucleotide sequence.

The catalog of restriction endonucleases in the Apple II program has been taken from a published list.[77]

8. Comments

Maintenance and Growth of Phage. A low yield of phage is probably due to one of two reasons: the F factor of the host has been lost or the cells have received poor aeration before and after infection. The maintenance of the F factor is assured by requiring proline prototrophy, and the

[98] A. M. Maxam, W. Gilbert, N. Chapman, G. Copenhaver, H. Donis-Keller, N. Rosenthal, D. States, and W. Herr, personal communication.

[99] D. Pratt, personal communication.

aeration is improved by vigorous shaking of a small volume in a relatively large culture flask (e.g., 200 ml of medium in a 2-liter Erlenmeyer).

Poor plaque formation occurs if the bacterial lawn is thin; to correct this, prepare and use a fresh plate. It helps to infect a culture with phage taken from a growing plaque from a fresh plate.

Isolation of the Replicative Form (RF). Low yield is due to the same reasons as described for the phage. If the growth volume is scaled up (1 liter), aeration of the host cells and the selection of the F factor are even more important.

If you do not get a clear lysate after centrifugation, use a swing-out rotor such as HB4. The volume of the lysate should be only one-tenth of the volume of the centrifuge tube to avoid high shear forces during centrifugation. The brake should be turned off at about 500 rpm.

DNA solutions that contain ethidium bromide should not be exposed to light because of the danger of introducing nicks into the supercoiled form of the RF or plasmid.

Cloning Strategies. If subclones are prepared for DNA sequencing, it is important that these clones be readily obtained. If certain sequences are difficult to get as subclones for DNA sequencing, much effort may be spent to generate particular M13 recombinants. This may arise especially in situations where convenient restriction sites are missing. In this case, I would recommend not to pursue a subcloning experiment, but rather to prepare another primer. This, in the future, may be a very attractive solution when restriction fragments as primers are replaced by synthetic primers. Oligonucleotide synthesis may help not only in site-directed mutagenesis, but also in filling gaps of sequence information in a shotgun DNA sequencing project.

If large fragments are to be cloned, they should be separated from competing smaller fragments by preparative polyacrylamide or agarose (low melting point, BRL) gel electrophoresis.

The concentration of the DNAs used in the ligation reaction should be carefully checked.[77a] Concatemers do not transform cells, and multiple inserts are not desirable.

Transformation. A poor yield of transformants can result from the presence of too much ligase in the DNA solution or too high concentrations of DTT. No more than 10 μl of a ligation reaction should be used to transform about 0.3 ml of competent cells. If necessary, DNA should be concentrated by ethanol precipitation and taken up in a small volume of low-Tris buffer.

Too high a heat shock temperature (above 42°) as well as aged cells reduce the transformation efficiency. Cells should be grown for at least 3 hr, but harvested in the early logarithmic growth phase.

Identification of Recombinants. Instead of growing cultures in test tubes, they can be prepared in Microtiter wells that hold 0.1–0.2 ml. There are adapters for table-top centrifuges in which one can spin these Microtiter plates. The culture volume is big enough for first screens, both DIGE and dot blot. The latter test is simplified by having the cultures arranged with the Microtiter plate in a similar way as they appear on the dot blot. There are also Pipetmen, which pipette eight samples, i.e., one row of the Microtiter plate at a time. These plates can also be used directly to keep phage recombinants frozen as a master storage system without taken up too much space. For template preparation, larger volumes have to be grown up as described.

If M13 single-stranded DNA is electrophoresed through an agarose gel, sometimes two bands show up instead of one. The second band can contain either random linearized circles or mini phage DNA.

If a number of discrete bands show up, the plaque that has been picked contains a heterologous population of phage. Diffusion of plaques through soft agar can cause cross contamination of plaques. In this case one must repeat plaque purification.

Another reason for the appearance of several discrete bands can be the occurrence of spontaneous deletions. The same care is taken as described above.

In Vitro DNA Synthesis. Template preparation or a bad batch of radioactively labeled precursor are usually the main source of problems for the primer extension reaction. Prepare template only from an infected bacterial culture that produces a normal yield of phage. This should be apparent from the DIGE-test. We do not observe any difference with various culture medias. Small amounts of PEG do not inhibit a primer extension experiment. If a problem occurs, it seems not to be solved by additional purification steps. It is easier to prepare a new phage batch.

A weak preparation of polymerase enzyme causes nonspecific stops in all tracks. The amount of polymerase that is included in the reaction should be empirically determined. Although higher amounts would normally abolish nonspecific stops in all four tracks, there is the danger of increasing the amounts of unspecific nucleases; this can cause the same problem. If a synthetic primer is annealed to the template, the polymerase should be added to the template–primer solution only shortly before the reaction. The exonuclease in the Klenow fragment may attack the 3′ end of the primer. The shorter primer does not adhere stably to the template. As a result, the radioactive signals are equally reduced in the latter gel.

If deoxynucleotide solutions are a problem, incorporation first becomes poor, then nonspecific stops occur in all four tracks.

If dideoxynucleotide solutions are a problem, false stops will occur in individual tracks. The ratio of dideoxynucleotides to deoxynucleotides has to be adjusted in a few initial experiments to allow chain determination to occur over the range of chain lengths that are separated by the sequencing gel. The concentrations of dideoxynucleotides given in Materials and Reagents serve as a guideline.

Weak radioactive signals in general in all tracks may occur, and strong signals may remain at the top of the gel if the DNA is not properly denatured after the elongation reaction. The pH of the stop buffer needs to be checked. You may need to make it more alkaline.

Acknowledgments

I wish to thank colleagues who have helped me in discussions and suggestions to work out such a detailed manual, especially Steve Anderson, Jurgen Brosius, John Carlson, Richard Gardner, Rich Gelinas, Bruno Gronenborn, Shirley Halling, Gisela Heidecker, Nien-Tai Hu, Jan Norrander, Dave Pratt, Rich Roberts, Irwin Rubenstein, Peter Seeburg, Mike Smith, Jeff Vieira, Huber Warner, and Mark Zoller. I thank Kris Kohn for her aid in preparing the manuscript.

I was supported during this work by the Deutsche Forschungsgemeinschaft, the College of Agriculture, University of California at Davis, the U.S. Department of Agriculture, SEA 5901-9-0386, the Experiment Station of the University of Minnesota, MN 15030, and the U.S. Department of Energy, DOE-AC02-81ER10901.

[3] An Integrated and Simplified Approach to Cloning into Plasmids and Single-Stranded Phages

By GRAY F. CROUSE, ANNEMARIE FRISCHAUF, and HANS LEHRACH

A particular DNA fragment that is to be cloned is often a part of a mixture of DNA fragments. To obtain the desired fragment, either one must first isolate the fragment from the mixture before cloning it into the vector, or one must clone the entire mixture and then screen for the desired fragment. Both procedures are time consuming and can require a significant amount of DNA.

In this chapter, we describe several methods that, taken together, greatly facilitate the cloning and selection of DNA fragments. The most important of these methods is the ligation and transformation of fragments taken directly from gels made with low-melting-point agarose. This method has been described before[1]; however, in this chapter we describe the protocol in more detail and show its usefulness for single-stranded

[1] A. M. Frischauf, H. Garoff, and H. Lehrach, *Nucleic Acids Res.* **8,** 5541 (1980).

ISBN 0-12-182001-7

phage vectors such as M13. The convenient and direct cloning of DNA fragments from gels allows several other simple techniques to be used. For example, because a vector DNA band can be taken from a gel, contamination of the vector with small amounts of chromosomal DNA or single-stranded phage DNA is unimportant and banding of the vector in CsCl is unnecessary. The combination of these methods requires very little DNA and is rapid and suitable for either plasmid or single-stranded phage vectors.

Materials

Reagents

TE: 10 m*M* Tris-HCl, pH 8.0, and 1 m*M* EDTA

Tris-acetate buffer for gel electrophoresis: 40 m*M* Tris, 1 m*M* EDTA, adjusted to pH 8.2 with glacial acetic acid. 25 × Tris acetate buffer is made up, per liter, of 121 g of Tris base, 11.7 g of EDTA (acid form), and ~26.5 ml of glacial acetic acid.

ATP, 0.1 *M*. Adjust to pH 7 with Tris base and store at −20° in aliquots.

Spermidine, 0.1 *M*. Adjust to pH 7 with Tris base and store in aliquots at −20°. A stock with a concentration of 0.01 *M* is also useful.

L broth: 10 g of tryptone, 5 g of yeast extract, and 5 g of NaCl per liter

2× TY: 16 g of tryptone, 10 g of yeast extract, and 5 g of NaCl per liter

Buffer-saturated phenol[2]: To 2.27 kg of phenol (5 lb) (Mallinckrodt), add 1 liter of 1 *M* Tris, pH 7.8, 125 ml of H_2O, 125 ml of *m*-cresol, 5 ml of 2-mercaptoethanol, and 2.5 g of 8-hydroxyquinoline. The phenol does not need to be distilled, and the solution is stable indefinitely at room temperature when stored in a brown bottle.

Ligase buffer: 30 m*M* Tris, pH 7.6, 10 m*M* $MgCl_2$, and 5 m*M* dithiothreitol (DTT). A 10× stock of ligase buffer can be stored in aliquots at −20° indefinitely.

T4 DNA polymerase buffer: 30 m*M* Tris, pH 7.6, 10 m*M* $MgCl_2$, 5 m*M* DTT, and 0.05 m*M* each dCTP, dGTP, dTTP, and dATP. A 10× stock of this buffer can be stored in aliquots at −20°.

Kinase buffer: 30 m*M* Tris, pH 8.0, 10 m*M* $MgCl_2$, 5 m*M* DTT. A 10× stock can be stored at −20° in aliquots.

Isopropyl-β-D-thiogalactopyranoside (IPTG). Dissolve in H_2O (20 mg/ml) and store at 4° or −20°.

5-Bromo-4-chloro-3-indolyl-β-D-galactoside (Xgal). Dissolve in dimethylformamide (20 mg/ml) and store at 4°.

[2] N. Blin and D. W. Stafford, *Nucleic Acids Res.* **3,** 2303 (1976).

Lysozyme solution: 50 mM glucose, 10 mM EDTA, and 25 mM Tris-HCl, pH 8.0. Store at 4°.

NaOH–SDS: 0.2 N NaOH and 1% sodium dodecyl sulfate (SDS). This preparation should be made weekly from stocks of 1 N NaOH and 10% SDS, and can be stored at room temperature.

NaAc, 3 M, pH 4.8: Dissolve 3 mol of NaAc in a small amount of H_2O. Add a large amount of glacial acetic acid to adjust the pH to 4.8, and then add H_2O to make 1 liter of solution. This solution can be stored at room temperature.

STET: 8% sucrose, 5% Titron X-100, 50 mM EDTA, and 50 mM Tris-HCl, pH 8.0

Enzymes

Most of the enzymes we have used were purchased from either Bethesda Research Laboratories or New England BioLabs. The T4 polynucleotide kinase was obtained from Biogenics Research Corporation. The purity of the T4 polynucleotide kinase is critical, because traces of 3′ exonuclease could easily prevent the ligation of an oligonucleotide linker.

Calf alkaline phosphatase (Boehringer) comes as a suspension in $(NH_4)_2SO_4$. In order to remove the $(NH_4)_2SO_4$, which can inhibit some enzymes, and to provide more convenient storage for the enzyme, the suspension is centrifuged for 1 min at 12,000 g. The pellet is resuspended in 0.1 M glycine, pH 9.5, 1 mM $MgCl_2$, 0.1 mM $ZnCl_2$, and 50% glycerol at a concentration of 200 units/ml and stored at −20°.

Lysozyme (Sigma), is dissolved at either 10 mg/ml or 50 mg/ml in 10 mM Tris-HCl, pH 7.5, and aliquots are immediately quick-frozen in liquid N_2 and stored at −20°. Aliquots should be thawed just before being used.

RNase A at 20 mg/ml in 0.15 M NaCl is boiled for 15 min and stored at −20°. A working stock at 100 μg/ml is also stored at −20°. The RNase at 1 μg/ml should be tested to ensure that it has no DNA-nicking activity when incubated with plasmid supercoils at 37°.

Methods

Alkaline DNA Extraction

The procedure we use for extraction of either plasmid or M13 replicative form DNA is that of Birnboim and Doly[3] with some modifications. Cells containing plasmids are grown overnight in L broth plus antibiotic.

[3] H. C. Birnboim and J. Doly, *Nucleic Acids Res.* **7**, 1513 (1979).

To obtain the replicative form of M13 and other single-stranded phages, 5–10 μl of high-titer phage stock are added to 5 ml of cells at $A_{550} = 0.5$ in 2× TY, and the cells are grown for 3–4 hr. The extraction procedure for both types of culture is as follows: Spin down 5 ml of culture. Resuspend the M13 cells in 5 ml of medium, and respin to wash out phages. Resuspend the pellet of cells in 180 μl of lysozyme solution. Transfer 192 μl to a microfuge tube. Add 8 μl of lysozyme (50 mg/ml) freshly thawed from a frozen stock. Let sit for 5 min at room temperature, and then chill the suspension on ice. Add 400 μl of NaOH–SDS, and gently vortex. The solution should clear. Let the solution sit for 5 min on ice. Add 300 μl of 3 *M* NaAc, pH 4.8, and gently vortex. A clot of DNA should form. Let this preparation sit for at least 10 min on ice and then spin it for 15 min in a microfuge. Place 750 μl of supernatant into a microfuge tube. Add 450 μl of isopropanol, mix well, and spin for 5 min. (It is not necessary to chill the mixture before spinning.) Wash the pellet with cold 70% ethanol and then remove the ethanol with a finely drawn pipette. Add 200 μl of TE + 0.15 *M* NaCl, and vortex to dissolve. Add 200 μl of buffer-saturated phenol, and vortex. Add 200 μl of chloroform, and vortex. Spin for 2 min in a microfuge. Carefully remove the aqueous layer and to it add 500 μl of ethanol. Chill for 5 min in a dry ice–ethanol bath. Spin for 5 min and then wash the pellet with 70% ethanol. Dissolve the DNA pellet in 100 μl of TE + 1 μg of RNase per milliliter.

Five microliters of the DNA should suffice for one restriction digest. This method can be adjusted for larger volumes: 1 liter of plasmid, amplified with chloramphenicol, can be prepared using 20 ml of the lysozyme solution and proportionate amounts of the other reagents. Fifty milliliters of M13 can be prepared with 1 ml of lysozyme solution. The DNA may not react efficiently with alkaline phosphatase or polynucleotide kinase because of competition from the ribooligonucleotides remaining from the RNase digestion. These can be removed by chromotography of the DNA on a BioGel A-5m (Bio-Rad) column.

Transformation Directly from Agarose Gels

Step 1. Restrict the DNA according to the enzyme supplier's specifications. If a vector fragment is to be treated with phosphatase, add the phosphatase after restriction and incubate for the desired time [e.g., 150 units of bacterial alkaline phosphatase (Bethesda Research Laboratories) incubated for 60 min at 65° or 0.2–0.4 unit of calf alkaline phosphatase (Boehringer) incubated for 60 min at 37°]. Prepare the samples for electrophoresis by precipitation with ethanol. Removal or inactivation of the phosphatase is unnecessary because it separates from the DNA on the gel.

The method is sensitive and generally, if one can clearly see DNA by ethidium bromide (EtBr) staining, there is enough to clone. The vector concentration should usually be in the range of 25–100 ng in a 50-μl gel slice, and the fragment to be cloned can be at this concentration or at a greater concentration. If the vector is cleaved with only a single enzyme, the amount of vector should be relatively low and a greater amount of insert fragment should be added in order to minimize the background of reclosed plasmids.

Step 2. Run the samples in Seaplaque agarose (Marine Colloids) in Tris-acetate buffer. (The low-melting-point agarose from Miles Laboratories or Bethesda Research Laboratories will also work.)

Because of the necessary subsequent dilution, it is helpful to have very narrow gel slots and to use a low-concentration gel. We generally run 0.6% gels, but higher concentrations can be used. The 0.6% gel is fragile and will be difficult to handle if it is not supported. The agarose should be cooled to 37° before pouring. It still will require more time for hardening than conventional agarose; the time can be shortened by pouring the gel in a cold room (4°).

Step 3. Stain the gel with 1 μg of EtBr per milliliter for 15–30 min. Destain in H_2O for an equal time.

Step 4. Illuminate the gel with long-wave ultraviolet radiation (365 nm; do not use short wave), and cut out the desired bands. Try to cut out as small a piece of gel as possible. Also, the amounts of DNA can be adjusted by adding only a fraction of a gel slice. Mix together gel slices of fragments to be ligated together. Alternatively, the bands may be removed with glass capillary micropipettes, which can also be used to measure the volume of the removed bands.

Step 5. Determine the approximate gel volume by weighing the slices.

Step 6. Determine the final volume necessary for each sample so that the agarose concentration will be 0.1%. Then calculate the amount of 10× ligase buffer, 0.1 *M* ATP, and ligase necessary to make the sample 1 m*M* ATP and 1× ligase buffer.

Step 7. Make up the remainder of the volume with 10 m*M* Tris, pH 7.6. Add the 10 m*M* Tris to the gel slices, and melt the slices by incubating them at 68° for 10 min.

Step 8. Cool the agarose to 37°, and then add the appropriate amounts of 10× ligase buffer, ATP, and T4 DNA ligase. [We generally use 300 units (New England BioLabs) in a volume of 300–500 μl.] Incubate the tubes for several hours to overnight at 12°.

Step 9. For transformation, vortex the ligated mixture, which should have formed a soft gel, to break up the gel, and mix it with at least an equal volume of competent cells. Transformation should be performed as described in the next section. Transformation of all the ligated mixture is usually not necessary to get the desired clones; 100 μl of the gel mixture usually suffice. The remainder can be stored at 4° for later transformation, if necessary.

Although the ligation is generally left overnight at 12°, it is possible to do a shorter ligation at 12° or 20°. The DNA can be restricted either before or after ligation by adding 10× restriction buffer and enzyme to the melted and diluted gel and incubating it at 37°. Removal of the restriction buffer is not necessary before transformation, although transformation efficiency may be reduced. To remove the restriction buffer before ligation, the DNA (and agarose) is precipitated with ethanol by the addition of NaCl to a final concentration of 0.2 *M* and 2 volumes of ethanol. The pellet is washed with 70% ethanol and then redissolved in 10 m*M* Tris, pH 7.6, at 68° to yield a final agarose concentration of 0.1%.

Preparation of Competent Cells

We prepare competent cells by the technique of Dagert and Ehrlich[4] with an additional step in which the competent cells are frozen for later use. Cells are grown in 500 ml of L broth to A_{600} = 0.2. (We generally use HB101[5]; 4 ml of an overnight culture of HB101 added to 500 ml of L broth require 2–4 hr of growth.) The cells are chilled thoroughly (at least 15–20 min on ice). The cells should be kept at 0–4° throughout the remainder of the procedure. The cells are spun at 3000 rpm for 15 min in a GS3 rotor, and the pellet is resuspended in 200 ml of cold 100 m*M* $CaCl_2$. The cells are left for 20 min on ice and then are spun again for 10 min. The pellet is resuspended in 5 ml of cold 100 m*M* $CaCl_2$ and chilled on ice for 20–24 hr. Cold glycerol is added to make a 10% solution. Aliquots (100–200 μl) are frozen in liquid N_2 and stored at −80°.

For M13 competent cells, a colony of JM103[6] is grown in 2× TY to A_{550} = 0.3. The procedure is identical to the above with the exception that cells are resuspended in 0.5 volume of cold 50 m*M* $CaCl_2$ and then in 0.1 volume of cold 50 m*M* $CaCl_2$ and frozen in 200-μl aliquots. We do not know whether the differences in the two procedures are important.

Plasmid Transformation

The cells are thawed on ice. The cells and the DNA (use an equal volume, or less, of ligated DNA in agarose) are mixed together and are first incubated for 30 min on ice and then incubated for 2 min at 37°. Ten vol-

[4] M. Dagert and S. D. Ehrlich, *Gene* **6,** 23 (1979).

[5] H. W. Boyer and D. Roulland-Dussoix, *J. Mol. Biol.* **41,** 459 (1969).

[6] J. Messing, R. Crea, and P. H. Seeburg, *Nucleic Acids Res.* **9,** 309 (1981).

umes of L broth are added, and the mixture is incubated for 30 min at 37° to allow preexpression of antibiotic resistance. Cells are pelleted (e.g., for 5 min in a clinical centrifuge), resuspended in a small volume of L broth, and plated on a selection plate.

M13 Transformation

The cells are thawed on ice. Less than 0.5 volume of ligated DNA in agarose is added, and the mixture is first incubated for 30 min on ice and then incubated for 2 min at 42°. Three milliliters of top agar + 0.2 ml of an exponential culture of JM103 cells are added, and the culture is plated on glucose minimal plates. If one of the M13 vectors developed by Messing and his collaborators[6] is being used, 30 μl of IPTG and Xgal are added in order to score for *lac* complementation.

Plasmid Boiling Preparation

The following procedure is that of Holmes and Quigley[7] with slight modifications.

Method 1. Streak colonies on an antibiotic plate (up to 12 can be streaked on 1 plate). Pick up a broad swipe of cells with a toothpick and suspend them in 100 μl of cold STET buffer in a microfuge tube by twirling the toothpick and then vortexing the suspension. Keep on ice.

Method 2. Grow the cells overnight in L broth and antibiotic. Spin down 0.5 ml of culture in a microfuge tube (a few seconds is enough). Suspend the cells in 100 μl cold STET buffer (a toothpick can be used for resuspending the cells).

Method 3 (for M13). Stab the plaques with a toothpick into 1 ml of 2× TY; grow for 5–6 hr. Spin down 1 ml of cells for 2 min in a microfuge; save the supernatant. Resuspend the cells in 100 μl of cold STET buffer.

For any of the above methods, continue as follows: Add to the cells 8 μl of 10 mg/ml lysozyme, freshly thawed from a frozen stock. Suspend the cells in a boiling water bath for 90 sec, and then spin in a microfuge for 10 min. Remove the gelatinous pellet with a toothpick and discard it. Add 80 μl of isopropanol and chill for 5 min at −20°. Spin for 5 min and then remove the supernatant, wash with 70% ethanol, and dry. Resuspend in 15 μl of 1 μg/ml RNase in TE. Incubate for 15 min at 68° (to inactivate any remaining nucleases).

The DNA restricts well and there is usually enough DNA for three restrictions. If the samples are not heated at 68° as suggested above, incubations should be kept short (15 min), because a nuclease that is particularly active in high salt is sometimes found in the preparation.

[7] D. S. Holmes and M. Quigley, *Anal. Biochem.* **114,** 193 (1981).

Addition of Synthetic Oligonucleotides (Linkers)

If the ends of the DNA to which a linker is to be added have not been produced by a restriction enzyme that produces blunt ends, or by an enzyme such as BAL-31 that is known to produce blunt ends, the ends should first be made blunt with T4 DNA polymerase.

The DNA is precipitated with ethanol (to remove the previous buffer and to concentrate the DNA) and then is dissolved in 8 μl of 10 m*M* Tris. pH 7.6. Then 1 μl of 10× T4 DNA polymerase buffer and 1 μl of T4 DNA polymerase (1–2 units, Bethesda Research Laboratories) are added. The preparation is incubated for 30 min at 30°.

The linkers are treated with kinase as follows: Mix together 1 μl of linker (0.1–2 μg of linker), 1 μl of 10× kinase buffer, 1 μl of 10 m*M* spermidine, 1 μl of 5 m*M* ATP, 5 μl of H_2O, and 1 μl of T4 polynucleotide kinase (approximately 1 unit, Biogenics). Incubate for 30–60 min at 37°.

The linkers are ligated to the blunt-end DNA as follows: Add the treated linkers to the DNA mix. Add 2 μl of 5 m*M* ATP and 1 μl of T4 DNA ligase (300 units, New England BioLabs). Incubate for up to 24 hr at 15–20°.

Results and Discussion

Transformation from Seaplaque Agarose Gels

Seaplaque agarose has very little effect on transformation and permits efficient ligation of DNA, as can be seen in Table I. The ability to transform cells with DNA taken directly from an agarose gel greatly simplifies the associated cloning procedures. For example, because the vector, as well as the fragment to be cloned, can be taken from the gel, the vector does not need to be pure. We have found that DNA prepared by the alkaline DNA extraction method has little chromosomal DNA contamination and is adequately pure to be used for cloning. The agarose gel can be stained with EtBr, making it possible to adjust the amount of vector and the amount of the fragment to be cloned so that higher transformation efficiencies can be obtained. The addition of synthetic linkers is also facilitated. We usually use a large excess of linker (up to 1 μg per DNA sample). This amount of a 6-bp recognition sequence represents the equivalent of hundreds of micrograms of natural DNA. Therefore, to cleave the ligated linker requires either that a large amount of restriction enzyme be used or that the unattached linkers be removed. The unattached linkers are easily separated from a fragment on a gel, and the linkers ligated to the fragment can be cleaved with restriction enzyme be-

TABLE I
LIGATION AND TRANSFORMATION OF pBR322 IN SEAPLAQUE AGAROSE[a]

DNA	Colonies/ng
pBR322	420
pBR322 in 0.1% agarose	300
pBR322 *Eco*RI cut	0.4
pBR322/*Eco*RI cut and ligated in 270 μl	>400
pBR322/*Eco*RI cut and ligated in 270 μl of 0.1% agarose	>400

[a] Supercoiled pBR322 was used to transform competent HB101 directly (10 ng for 200 μl of frozen competent cells) or it was cleaved with *Eco*RI, the *Eco*RI was inactivated at 68° for 10 min, and then the pBR322 precipitated with ethanol. The *Eco*RI-cleaved pBR322 was then used to transform HB101 without ligation (15 ng for 200 μl of cells), or 80 ng was ligated in a volume of 270 μl, with or without agarose, and half of this ligated DNA was transformed with 200 μl of competent cells.

fore transformation, without requiring the removal of the agarose. Cloning of small fragments is somewhat of a problem because of the low resolution of agarose at lower molecular weights. However, we have successfully cloned a 340-bp fragment from a 0.6% gel (results not shown).

Cloning into M13

These procedures are equally applicable to plasmids or single-stranded phages such as M13. To illustrate the method, we will describe the cloning of two fragments from a plasmid into M13mp8, a cloning vector developed by J. Messing and obtained from Bethesda Research Laboratories. The replicative form of the vector was made by the alkaline extraction procedure, as already described in Methods. In addition to the replicative form of DNA obtained with this procedure, there is another band seen on gels, caused by the single-stranded phage DNA. This band can be reduced, but not eliminated, by washing the cells before lysis. Although this DNA would be infectious, it is easily separated from the cleaved replicative form of DNA on the gel and therefore does not interfere with subsequent steps. Table II shows the result of ligation of M13mp8 with digests of p*dhfr*3.2, a plasmid containing 3.2 kb of a *dhfr*

TABLE II
CLONING OF FRAGMENTS OF pdhfr3.2 INTO M13mp8[a]

DNA	Volume of ligated DNA (μl)	No. of plaques	No. of blue plaques	Clear plaques per ng of vector
*Hin*dIII/*Bam* HI-cleaved M13mp8	10	580	3	130
+ 2.5-kb *Hin*dIII/*Bgl*II p*dhfr*3.2 fragment	100	5300	16	120
*Eco*RI/*Bam* HI-cleaved M13mp8	10	>1000	0	>300
+ 2.0kb-*Eco*RI/*Bgl*II p*dhfr*3.2 fragment	100	8100	1	250

[a] Slices containing the fragments indicated above were ligated in a volume of 300μl and transformed as detailed in Methods. Either 10 μl or 100 μl of the ligated DNA were used to transform 200 μl of frozen competent JM103 cells.

gene cloned into pBR327.[8] mp8 is designed so that vectors without an insert show up blue on Xgal indicator plates, whereas vectors with an insert are clear. The few blue plaques that were noted in Table II were presumably caused by vector molecules that had been cut with one restriction enzyme, but not both, and had subsequently reclosed. The production of clear plaques was very efficient ($> 1 \times 10^5/\mu g$ of vector). In addition, at least 100 μl of the ligation reaction could be used to transform 200 μl of competent cells.

In order to determine whether the clear plaques had inserts, plasmid boiling preparations were made from randomly picked clear plaques. Because the presence of an insert should noticeably change the migration of the supercoiled form of the M13 vector, aliquots of the preparations were electrophoresed without restrictions. Figure 1 shows the result from 12 randomly picked plaques of the 2.5-kb insert. As can be seen, all 12 randomly picked clear plaques have an insert, apparently of identical size. Some of these clones were then restricted and shown to have the expected insert (results not shown).

A General Cloning Strategy

Our routine procedure for cloning is as follows: The vector, either plasmid or M13, is prepared by the alkaline procedure. A 5-ml culture usually yields enough vector for at least 20 experiments. The vector DNA and the DNA to be inserted are restricted with the appropriate enzymes.

[8] G. F. Crouse, R. McEwan, and M. Pearson, *Mol. Cell. Biol.* **3,** 257 (1983).

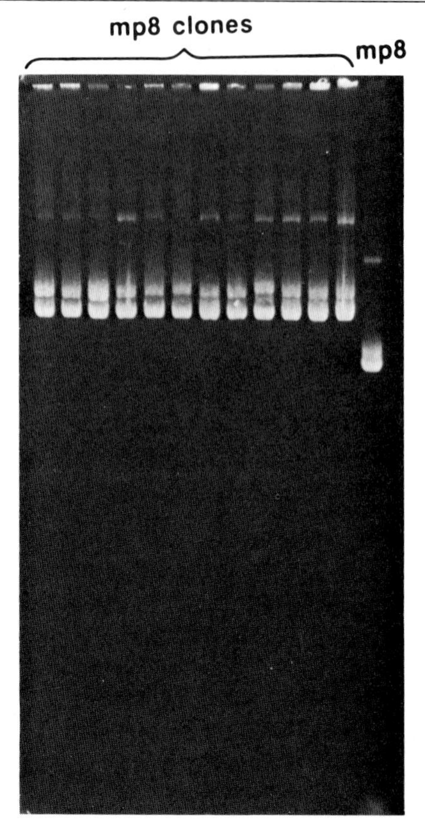

FIG. 1. Screening M13 plaques for inserts. Twelve clear plaques of the 2.5-kb *Hin*dIII/*Bgl*II p*dhrf*3.2 fragment cloned into M13mp8 (see Table II) were picked, and boiling preparations were made as described in Methods. Five microliters of each preparation were electrophoresed on a 0.8% agarose gel in Tris-acetate buffer. M13mp8, prepared by the alkaline procedure, was also run on the same gel as a marker.

Ends are filled in with T4 DNA polymerase, and linkers are added if needed. The samples are electrophoresed through Seaplaque agarose, and the desired fragments are mixed, restricted if necessary (for linkers), ligated, and transformed. Colonies or plaques are initially screened using the boiling preparation; initial screening for larger supercoils is done if the insert is large enough to cause a noticeable shift in supercoil size, and then the candidate clones are tested with one or two restriction digests. Likely candidates are then grown in 5-ml cultures, and an alkaline preparation is made for further characterization of the clones.

Whenever possible, we find it preferable to clone fragments bounded by *different* restriction sites. When such fragments are cloned into appropriately cleaved vectors, the orientation is known and essentially all clones produced will be the desired product, eliminating the need for any preliminary screening. The difficulty in cloning restriction fragments generated by single enzymes is that reclosure of the plasmid competes with the insertion of the fragment. Also the cloned fragment can be in either of two orientations. The reclosure of the plasmid can be greatly reduced (but, in our hands, not eliminated) by pretreating the vector with phosphatase. The ratio of clones with inserts can be increased by increasing the concentration of insert fragment. (This can be accomplished by varying the amounts of the gel slices added). The use of a vector that will give a color difference on an indicator plate or change in antibiotic resistance greatly facilitates the selection of clones with inserts, although this is not always possible.

The procedures described here are quick and simple. They eliminate the need for the banding of vectors with CsCl, and the use of frozen competent cells for transformation saves time and increases reproducibility; transformation from a gel decreases the amount of screening for a desired clone and facilitates the treatment of vectors with phosphatase and the use of linkers in cloning. Vector fragments taken from gels are free of any unrestricted vector that could create serious background problems. The combination of these procedures makes it possible to prepare a vector, to clone fragments into the vector, and to characterize the resulting clones, all within a week.

Acknowledgments

Part of this research was sponsored by the National Cancer Institute, DHHS, under Contract No. NO1-CO1-75380 with Litton Bionetics, Inc. The contents of this publication do not necessarily reflect the views or policies of the Department of Health and Human Services, nor does mention of trade names, commercial products, or organizations imply endorsement by the United States Government.

[4] Cutting of M13mp7 Phage DNA and Excision of Cloned Single-Stranded Sequences by Restriction Endonucleases

By MICHAEL D. BEEN and JAMES J. CHAMPOUX

Type II restriction endonucleases recognize specific sequence in duplex DNA and break both strands of the DNA at a specific position in or near that sequence.[1] Athough a few restriction enzymes can cut sites in single-stranded DNAs,[2-4] apparently by recognizing duplex regions,[4] this is not a property common to most restriction enzymes.

In the course of experiments using single-stranded DNA as substrate for the eukaryotic type 1 topoisomerase, it became desirable to have a method to conveniently excise, label, and purify cloned fragments of single-stranded DNA. The sequence of the M13 derivative, M13mp7,[5] suggested how the above objectives might be met. Messing and co-workers have developed a number of useful single-stranded cloning vehicles primarily for use in DNA sequencing.[5,6] Of those that are in general use, M13mp7 is unique in that the DNA of the multipurpose cloning site contains a sequence in which 21 nucleotides to one side of the unique *Pst*I recognition sequence are complementary to the 21 nucleotides to the other side (Fig. 1). This self-complementarity allows for the formation of a hairpin containing in duplex form the following restriction sites: *Eco*RI, *Bam*HI, *Sal*I, *Hin*cII and *Acc*I. Here we show that *Eco*RI and *Bam*HI will cut the DNA within this hairpin structure and, in addition, that inserts cloned into the *Pst*I, *Sal*I, and *Hin*cII sites can be cut out from the recombinant phage DNA using restriction enzymes. Ricca *et al.*[7] have also demonstrated that the M13mp7 recombinant phage DNA can be cut by restriction enzymes.

[1] D. Nathans and H. O. Smith, *Annu. Rev. Biochem.* **44,** 273 (1975).
[2] K. Horiuchi and N. D. Zinder, *Proc. Natl. Acad. Sci. U.S.A.* **72,** 2555 (1975).
[3] R. W. Blakesley and R. D. Wells, *Nature (London)* **257,** 421 (1975).
[4] R. W. Blakesley, J. B. Dodgson, I. F. Nes, and R. D. Wells, *J. Biol. Chem.* **252,** 7300 (1977).
[5] J. Messing, R. Crea, and P. H. Seeburg, *Nucleic Acids Res.* **9,** 309 (1981).
[6] B. Gronenborn and J. Messing, *Nature (London)* **272,** 375 (1978).
[7] G. A. Ricca, J. M. Taylor, and J. E. Kalinyak, *Proc. Natl. Acad. Sci. U.S.A.* **79,** 724 (1982).

METHODS IN ENZYMOLOGY, VOL. 101

ISBN 0-12-182001-7

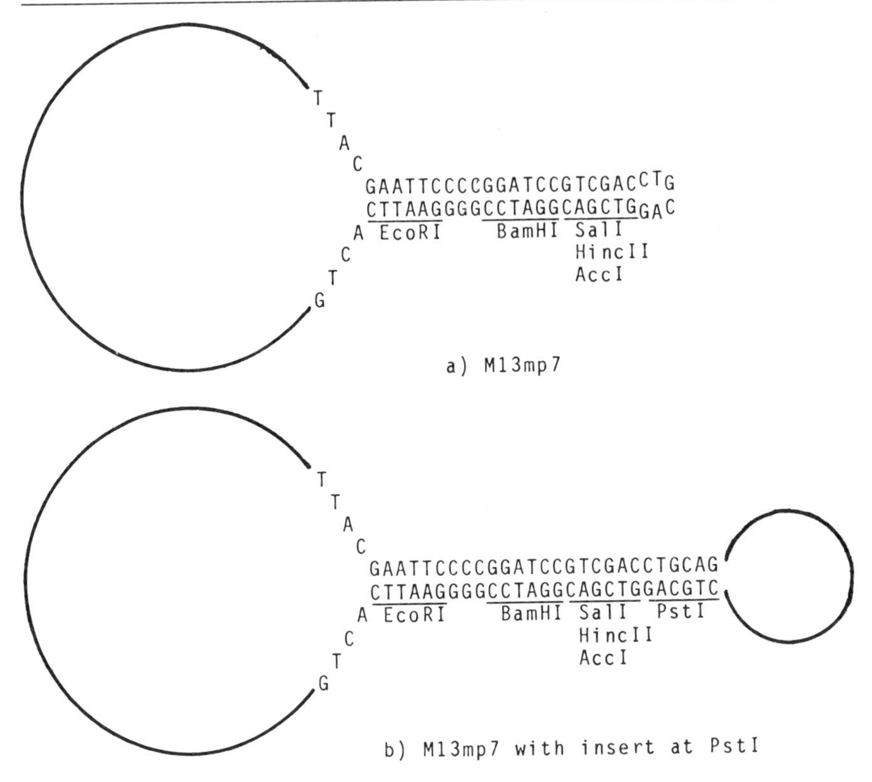

FIG. 1. Sequence of double-stranded regions in M13mp7 (a) and derivatives of M13mp7 containing an insert at the *Pst*I site (b) drawn as the base-paired structures described in the text.

Materials and Methods

Agarose, T4 ligase and all restriction enzymes used were purchased from Bethesda Research Laboratories (BRL). *Escherichia coli* DNA polymerase I (Klenow fragment), additional *Eco*RI and *Bam*HI, and [α-^{32}P]dATP were purchased from New England Nuclear. Phages M13mp2 and M13mp7 and *E. coli* JM103 are those described by Messing *et al.*[5,6] and were obtained from BRL.

Three recombinant DNA phage clones were prepared with DNA fragments of known sequence using methods described.[5] The DNA fragments to be inserted were first purified by acrylamide gel electrophoresis and isolated by electroelution. M13mp7/P contains the 0.31-kb *Hae*III fragment of ϕX174 DNA cloned into the *Hin*cII site. M13mp7/S contains the

1.22-kb *Pst*I fragment of SV40 DNA cloned into the *Pst*I site. M13mp7/M contains the 2.14-kb *Xho*I-*Sal*I fragment from the Moloney MuLV proviral DNA subcloned from pBR322[8] into the *Sal*I site.

M13 phage particles were isolated from the supernatants of overnight cultures of infected cells by the addition of NaCl to 0.5 *M* and PEG to 3% and allowing the phage to precipitate overnight at 4°. The phage were pelleted (8000 rpm for 30 min in a Beckman JA14) and resuspended in 10 m*M* Tris-HCl (pH 7.5), 1 m*M* EDTA. DNA was purified by phenol and chloroform extraction followed by sedimentation in an alkaline 5 to 20% sucrose gradient for 18 hr at 20,000 rpm in a Beckman SW 27 rotor. Twenty-five 1.5-ml fractions were collected, and the DNA was located by agarose gel electrophoresis of 5-μl aliquots. Pooled fractions were adjusted to neutral pH with 1 *M* acetic acid, 2 volumes of ethanol were added, and the DNA precipitated overnight at $-20°$. The DNA was pelleted, resuspended in 0.3 *M* sodium acetate and reprecipitated with ethanol. The DNA pellet was washed with cold 95% ethanol, dried, and suspended in 10 m*M* Tris-HCl (pH 7.5), 1 m*M* EDTA. DNA concentrations were estimated from the optical density taking an A_{260} unit as 40 μg/ml. The replicative form of M13mp7 was prepared from infected cells by CsCl–ethidium bromide banding of a cleared lysate prepared by a lysozyme and Triton X-100 method.[9] The preparation of SV40 DNA and ϕX174 replicative form DNA has been previously described.[10,11]

Restriction enzyme digests were carried out using buffer conditions recommended by the suppliers. The recombinant phage DNA was subjected to a brief denaturation and renaturation step prior to cutting. We routinely heated the DNA in 0.2 *M* NaCl, 10 m*M* Tris-HCl (pH 7.5), 1 m*M* EDTA for 2 min at 100° followed by 5–10 min at 65° but found that the salt and temperature conditions for renaturation were not critical. For 3′-end labeling, each reaction (final volume 10 μl) contained 0.25μg of *Eco*RI cut DNA in 20 m*M* Tris-HCl (pH 7.4), 7 m*M* $MgCl_2$, 50 m*M* NaCl, 10 m*M* 2-mercaptoethanol, 0.1 m*M* dGTP, 1 μ*M* [α-^{32}P]dATP (600 Ci/mmol), and 4 units of Klenow fragment of DNA polymerase I. The reaction was incubated for 30 min at 15°, and the DNA was ethanol precipitated twice from 2 *M* ammonium acetate to remove unincorporated label.[12] The pellets were washed with 95% ethanol and dried.

Agarose gel electrophoresis in Tris-borate EDTA buffer was as pre-

[8] T. M. Shinnick, R. A. Lerner, and J. G. Sutcliffe, *Nature* (*London*) **293,** 543 (1981).

[9] J. H. Crosa, and S. Falkow, *in* "Manual of Methods for General Bacteriology" (P. Gerhardt, ed.), pp. 266–282. American Society for Microbiology, Washington, D.C., 1981.

[10] J. J. Champoux and B. L. McConaughy, *Biochemistry* **14,** 307 (1975).

[11] M. D. Been and J. J. Champoux, *Proc. Natl. Acad. Sci. U.S.A.* **78,** 2883 (1981).

[12] A. M. Maxam, and W. Gilbert, this series, Vol. 65, p. 499.

viously described.[13] Gels were stained in 0.2 μg/ml ethidium bromide and photographed with UV illumination. Agarose gel electrophoresis under alkaline conditions was as described by McDonell *et al.*[14] A vertical gel apparatus was used (gel dimensions: 20 × 15 × 0.22 cm); both gel and reservoirs contained 30 m*M* NaOH and 2 m*M* EDTA. Prior to electrophoresis, all samples were ethanol-precipitated and resuspended in alkaline sample mix containing 30 m*M* NaOH, 2 m*M* EDTA, 7% Ficoll, 0.1% SDS, and 0.04% bromocresol green. Gels were either neutralized in 20 m*M* Tris-HCl (pH 7.5) and stained as above for photographing or dried on a slab gel drier for autoradiography.

Results

Cutting of M13mp7 Phage DNA by EcoRI and BamHI

The conversion of single-stranded circular DNA to a linear species can be monitored using the differential mobility of the two forms during agarose gel electrophoresis. M13mp2 DNA contains one single-stranded *Eco*RI recognition sequence and is not cut by *Eco*RI (Fig. 2, lanes 1 and 2). However, M13mp7 DNA, which contains the self-complementary sequence with an *Eco*RI site[5] (Fig. 1) is cut by *Eco*RI (Fig. 2, lanes 3 and 4). When the *Eco*RI-cut M13mp7 DNA was phosphorylated using [γ-^{32}P]ATP and polynucleotide kinase and the products were analyzed by electrophoresis in an acrylamide gel, a labeled fragment the size of the hairpin was observed (data not shown). These results indicated that the duplex recognition sequence at the base of the hairpin is a substrate for the restriction enzyme. The data in Fig. 2 also show that the restriction enzyme *Bam*HI cuts M13mp7 at the *Bam*HI site contained within the same duplex hairpin. There was little or no cutting of M13mp7 DNA by either *Sal*I or *Hin*cII (data not shown).

Excision of Single-Stranded Cloned Inserts from M13mp7 by EcoRI and BamHI

Insertion of a fragment into M13mp7 RF DNA at the *Pst*I site generates a single-stranded phage DNA with the potential to form the 27-bp duplex structure shown in Fig. 1b. Similarly, cloning into the *Sal*I (also *Hin*cII or *Acc*I) site generates a phage DNA containing a potential intramolecular duplex region 21 bp in length.

As a probe for the presence of such duplex regions in various recom-

[13] M. D. Been and J. J. Champoux, *Nucleic Acids Res.* **8,** 6129 (1980).

[14] M. W. McDonell, M. N. Simon, and F. W. Studier, *J. Mol. Biol.* **110,** 119 (1977).

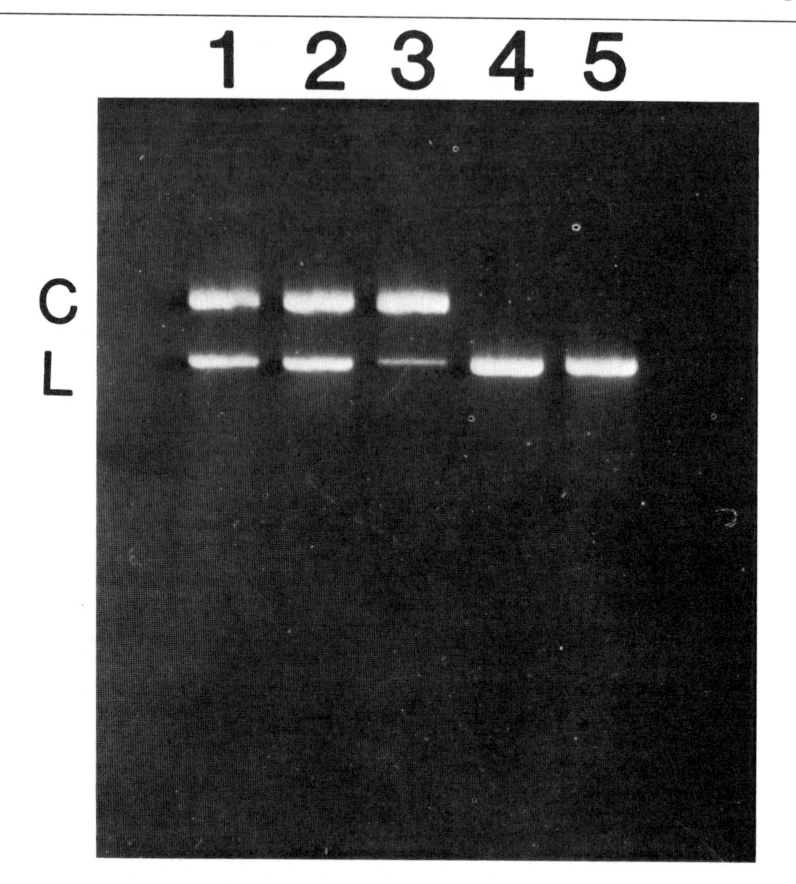

FIG. 2. Cutting M13 derivatives with *Eco*RI or *Bam*HI. Each reaction contained 0.5 μg of DNA (lanes 1 and 2, M13mp2; lanes 3, 4 and 5, M13mp7) in a final volume of 20 μl. Reactions were incubated for 30 min at 37°. After ethanol precipitation, each sample was resuspended in 20 μl of alkaline sample mix, 10 μl of which was electrophoresed in a 1.4% agarose gel in Tris-borate buffer for 2 hr at 100 V. Lanes 1 and 3, no enzyme; lanes 2 and 4, 2.5 units of *Eco*RI; lane 5, 2.5 units of *Bam*HI. Migration is from top to bottom. C indicates position of circles, and L indicates the linears.

binant single-stranded phage DNAs, we tested the DNAs for their sensitivities to the restriction enzymes *Eco*RI and *Bam*HI. Figure 3 shows that inserts of lengths 0.31, 1.22, and 2.14 kb were all efficiently excised from the M13mp7 vector by both *Eco*RI and *Bam*HI. The inserts cut out with *Bam*HI migrated slightly faster than the corresponding inserts cut out with *Eco*RI. The 1.22 kb fragment was cloned into the *Pst*I site whereas the other two were cloned into the position of the *Sal*I site. Thus, fragments cloned at either of these two positions can be excised using *Eco*RI

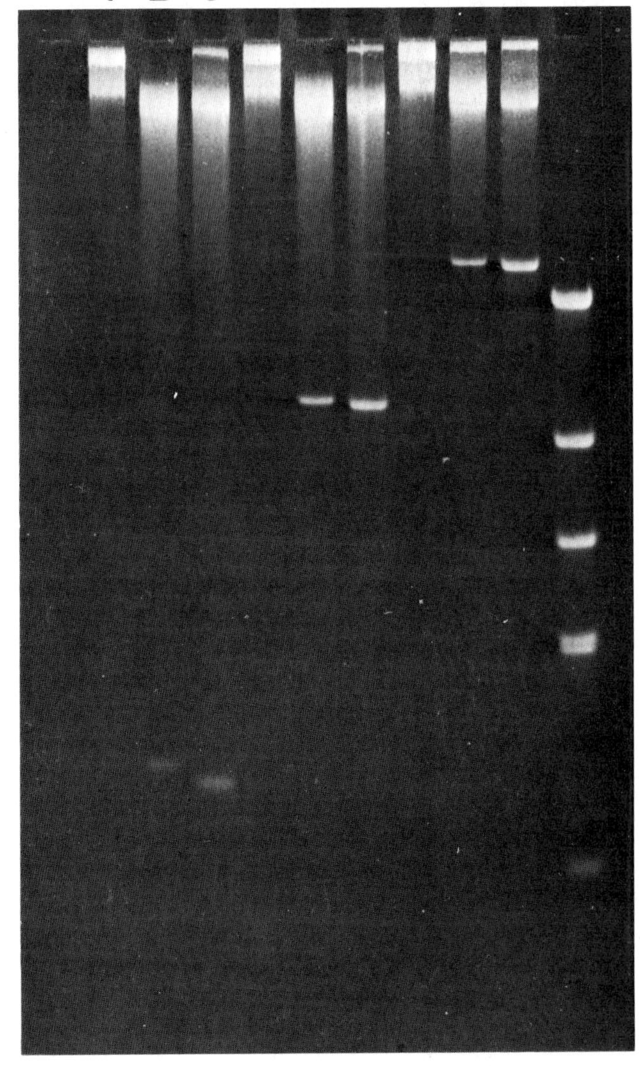

FIG. 3. Excision of inserts from single-stranded DNA. Reactions contained 0.5 μg of M13mp7/P DNA containing the 0.31 kb ϕX fragment (lanes 1-3), M13mp7/S containing the 1.22 kb SV40 fragment (lanes 4–6), or M13mp7/M containing the 2.14 kb Moloney fragment (lanes 7–9). Lanes marked 1, 4, and 7 were incubated under *Eco*RI reaction conditions without enzyme, lanes marked e were cut with *Eco*RI (5 units) and lanes marked b were cut with *Bam*HI (5 units) for 30 min at 37°. Lane 10 contained *Hin*fI digested SV40 DNA as molecular weight markers. All samples were ethanol precipitated, resuspended in 10 μl of alkaline sample mix, and electrophoresed in a 1.6% agarose gel under alkaline conditions for 4 hr at 100 V.

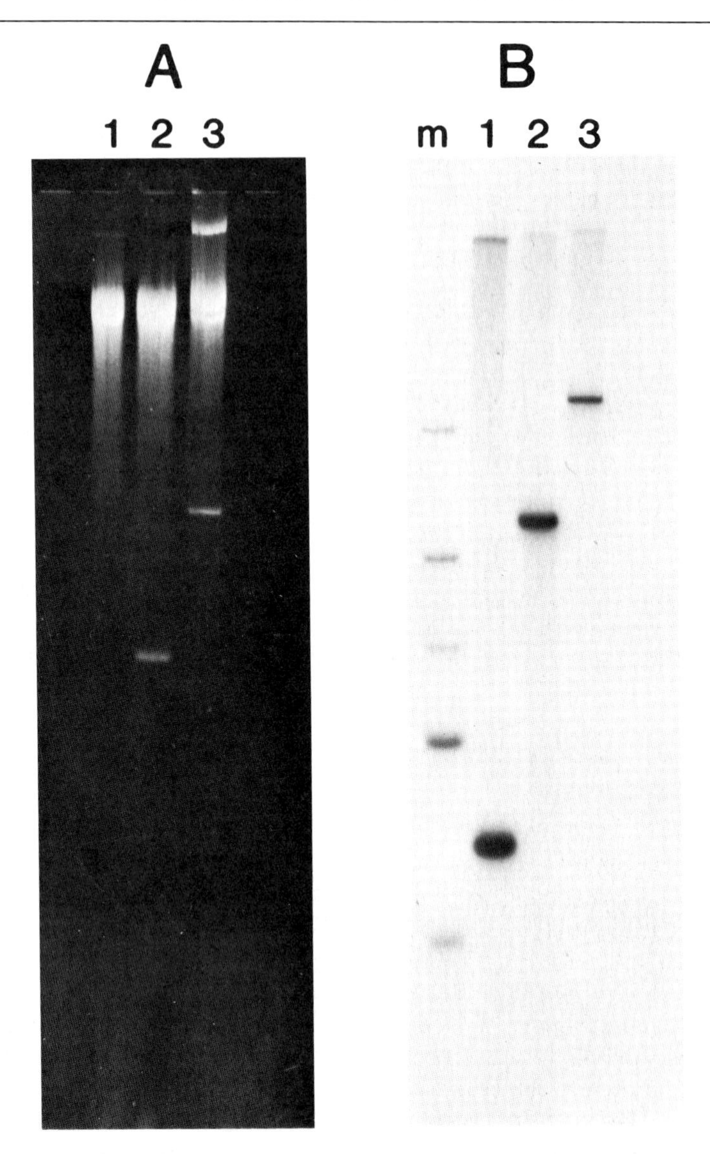

FIG. 4. Labeling excised inserts with ^{32}P at the 3′ terminus. (A) One microgram of DNA from each of the M13mp7 clones was cut with *Eco*RI, the DNA was ethanol precipitated, and half was run in a 1.4% agarose gel under alkaline conditions for 4 hr at 100 V. The gel was stained with ethidium bromide and photographed with UV illumination. (B) The remainder of the DNA was labeled with [^{32}P]dATP using Klenow fragment of DNA polymerase I. Aliquots of labeled DNA were run in a 1.6% alkaline agarose gel. The gel was dried and exposed to X-ray film. Lane 1 contained M13mp7/P (0.31 kb insert); lane 2, M13mp7/S (1.22 kb insert); and lane 3, M13mp7/M (2.14 kb insert). Lane m contained ^{32}P-labeled SV40 *Hin*fI markers.

or *Bam*HI. The recombinant DNA containing an insert at the *Pst*I site could also be cut by *Sal*I, *Hinc*II, and *Pst*I (data not shown). However, excision of the fragment with *Pst*I was less efficient than with the other enzymes. We were not able to investigate the ability of *Hinc*II or *Sal*I to excise fragments cloned into those sites, since the fragments inserted into those positions altered the recognition sequences.

3′-End Labeling of Inserts after Excision by EcoRI

In many instances it would be desirable to label one of the termini of the excised single-stranded fragments for subsequent analyses. An examination of the structure shown in Fig. 1b reveals that after removal of the insert using *Eco*RI, the vector DNA will be a single-stranded linear, but the two ends of the excised fragment can remain base paired. Although polynucleotide kinase can be used to label both structures with high efficiency after dephosphorylation, the Klenow fragment of DNA polymerase I should preferentially label the insert by filling in the *Eco*RI site.

The three single-stranded recombinant phage derivatives described above were cut with *Eco*RI and treated with Klenow fragment of DNA polymerase I in the presence of labeled [α-^{32}P]dATP. The products were separated by alkaline agarose gel electrophoresis and visualized by autoradiography (Fig. 4). The results show that all three of the excised fragments can be labeled by this procedure. Moreover, the ^{32}P label appears to be preferentially incorporated into the excised pieces.

Discussion

We show here that it is possible to cleave single-stranded phage M13mp7 DNA specifically using restriction endonucleases *Eco*RI and *Bam*HI. The ability to cleave this DNA is due to a feature in the sequence of the derivative which generates a duplex hairpin containing these restriction enzyme recognition sequences. One might expect that this DNA could also be cleaved by *Sal*I and *Hinc*II, since the hairpin also contains the recognition sequence for these enzymes. However, we found that there was little or no cutting of M13mp7 by either *Sal*I or *Hinc*II, whereas, recombinant phage DNA with an insert at the *Pst*I site could be cut by both these enzymes. These results suggest that the loop at the end of the hairpin in M13mp7 affects the structure being recognized by *Sal*I and *Hinc*II. Because the DNA sequence to one side of the *Eco*RI site does not allow for conventional base pairing (see Fig. 1), it is noteworthy that *Eco*RI can cleave within the hairpin structure.[15,16]

[15] P. J. Green, M. S. Ponian, A. L. Nussbaum, L. Tobias, D. E. Garfin, H. W. Boyer, and H. M. Goodman, *J. Mol. Biol.* **99,** 237 (1975).

[16] A.-L. Lu, W. E. Jack, and P. Modrich, *J. Biol. Chem.* **256,** 13200 (1981).

Fragments excised by *Eco*RI can be labeled at their 3′ ends using a labeled deoxynucleoside triphosphate and the Klenow fragment of DNA polymerase I. We also note that fragments cleaved out with *Bam*HI can be labeled similarly; T4 DNA polymerase can be used as well as the Klenow fragment of DNA polymerase I, or the 5′ ends can be labeled using T4 polynucleotide kinase.

In combination with preparative alkaline sucrose gradients we have used the methods described here to excise, label, and purify relatively large amounts of single-stranded fragments directly from M13mp7 recombinant phage derivatives without using either agarose or acrylamide gels. Although we show here that *Eco*RI and *Bam*HI suffice for excising the inserted DNA, with the proper recombinant, *Sal*I, *Hin*cII, and possibly PstI could be used.

Acknowledgments

We would like to thank B. Brewer for reviewing the manuscript and K. Spangler for its preparation. This work was supported by National Institute of Health Research Grant GM23224.

[5] Kilo-Sequencing: Creation of an Ordered Nest of Asymmetric Deletions across a Large Target Sequence Carried on Phage M13

By WAYNE M. BARNES, MICHAEL BEVAN, and PAULA H. SON

A heretofore most convenient method for acquisition of large amounts of DNA sequence data involves the random molecular cloning of small [1 kilobase (kb)] pieces of target DNA sequence into M13 phage, adjacent to a primer homology that can be repeatedly used to derive 300–500 nucleotides of sequence data per sequencing experiment[1] (see Messing, this volume [2]. A computer is then required to aid in the identification and ordering of each stretch of data.[2,3] We here describe an organized, nonrandom contrast to this strategy. Our strategy retains the conveniences of the dideoxy method with M13 transducing phage DNA template, with the same

[1] F. Sanger, A. R. Coulson, B. G. Barrell, A. J. H. Smith, and B. A. Roe, *J. Mol. Biol.* **143,** 161 (1980).

[2] R. Staden, *Nucleic Acids Res.* **6,** 2601 (1979).

[3] R. Staden, *Nucleic Acids Res.* **8,** 3673 (1980).

ISBN 0-12-182001-7

peak rate of data acquisition (say 1000 nucleotides per man day), but each experiment spans a predetermined region of the target sequence. To start the procedure, target DNA (3–14 kb) is recloned into our vector so that a unique restriction site is present at one edge, adjacent to the commercially available sequencing primer. The M13 vectors that we have constructed for use in this method are able to carry stably at least 14 kb with little or no tendency to form spontaneous deletions.

By an *in vitro* procedure described in detail below, we create thousands of deletions that start at the unique restriction site adjacent to the sequencing primer and extend various distances across the target DNA. Thus each sequencing experiment will provide sequencing data from a point well within the target sequence, at the boundary of the deletion. Phage carrying the desired size of deletions, that is, phage whose DNA as template will give rise to DNA sequence data in a desired location along the target DNA, are purified by electrophoresis *alive* on an agarose gel. Phage running in the same location on the agarose gel give rise to nucleotide sequence data from the same kilobase of target DNA.

The usefulness of this strategy is not restricted to large sequencing projects: Small regions within large recloned restriction fragments can be accessed with maximum convenience for dideoxy sequencing without the need for more than minimal advance knowledge of the restriction site content of the DNA region. Although we describe the use of the resulting nested set of deletions to sequence DNA, the deletions, which carry a linker restriction site at their boundary, are available to serve purposes besides sequencing, such as the mapping of transcripts, genes, DNA binding sites, promoters. Derivation of this strategy, and an example of its application, have been described.[4,4a]

Summary of the Deletion Procedure (Fig. 1)

Recloning the Target DNA So That a Unique Restriction Site Is on One Side, Adjacent to the Sequencing Primer

This step has basically two approaches. In the first approach one attempts to identify a restriction enzyme that does *not* cut anywhere in the target DNA. Then one clones the target DNA fragment into a site or combination of sites on the other side (*vis-à-vis* the sequencing primer) from that unique restriction site in the linker region of the vector that is lacking in the target DNA. For instance, if the target DNA lacks a *Hin*dIII site, clone the target DNA into a site on the other side of the *Hin*dIII site, such

[4] W. M. Barnes and M. Bevan, *Nucleic Acids Res.* **11,** 349 (1983).
[4a] M. Bevan, W. M. Barnes, and M. D. Chilton, *Nucleic Acids Res.* **11,** 369 (1983).

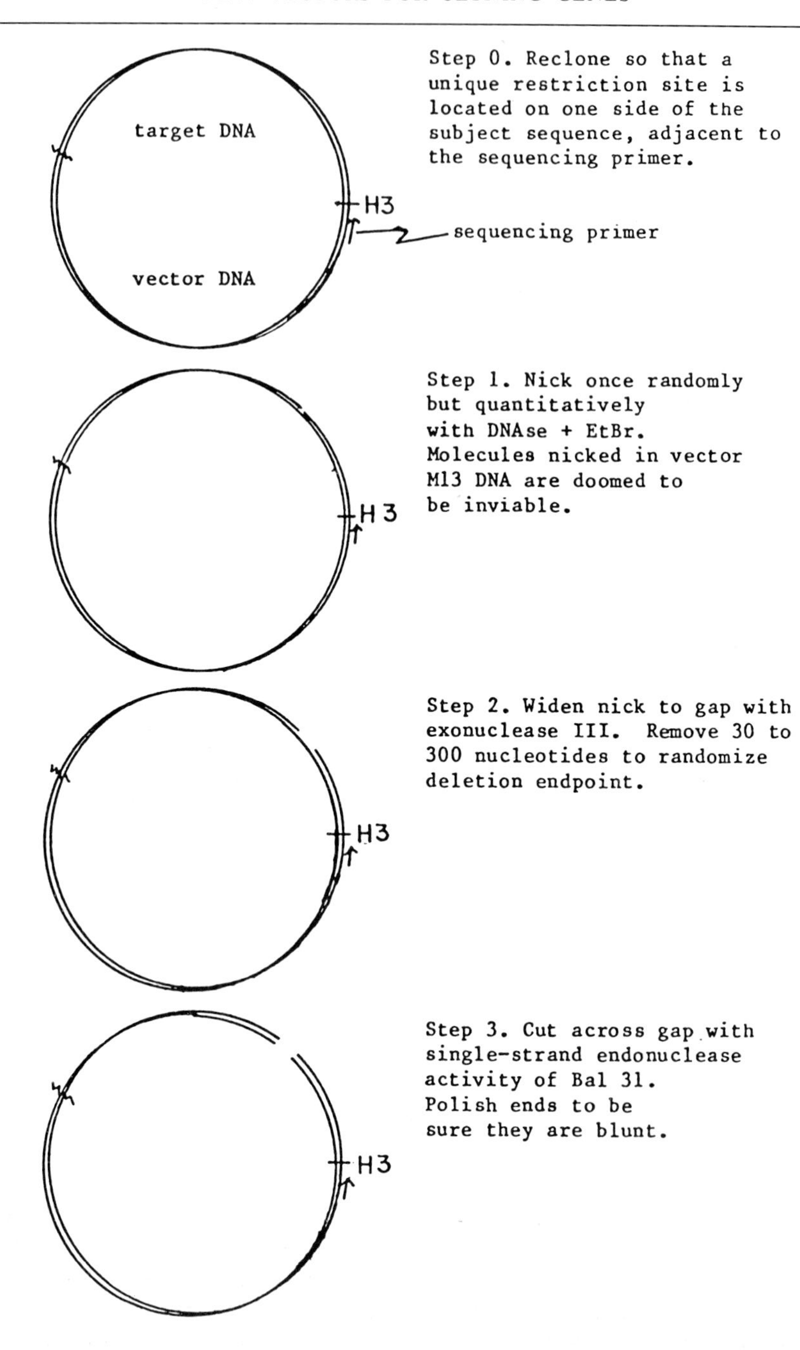
target DNA
vector DNA
H3
sequencing primer
Step 0. Reclone so that a unique restriction site is located on one side of the subject sequence, adjacent to the sequencing primer.
H3
Step 1. Nick once randomly but quantitatively with DNAse + EtBr. Molecules nicked in vector M13 DNA are doomed to be inviable.
H3
Step 2. Widen nick to gap with exonuclease III. Remove 30 to 300 nucleotides to randomize deletion endpoint.
H3
Step 3. Cut across gap with single-strand endonuclease activity of Bal 31. Polish ends to be sure they are blunt.

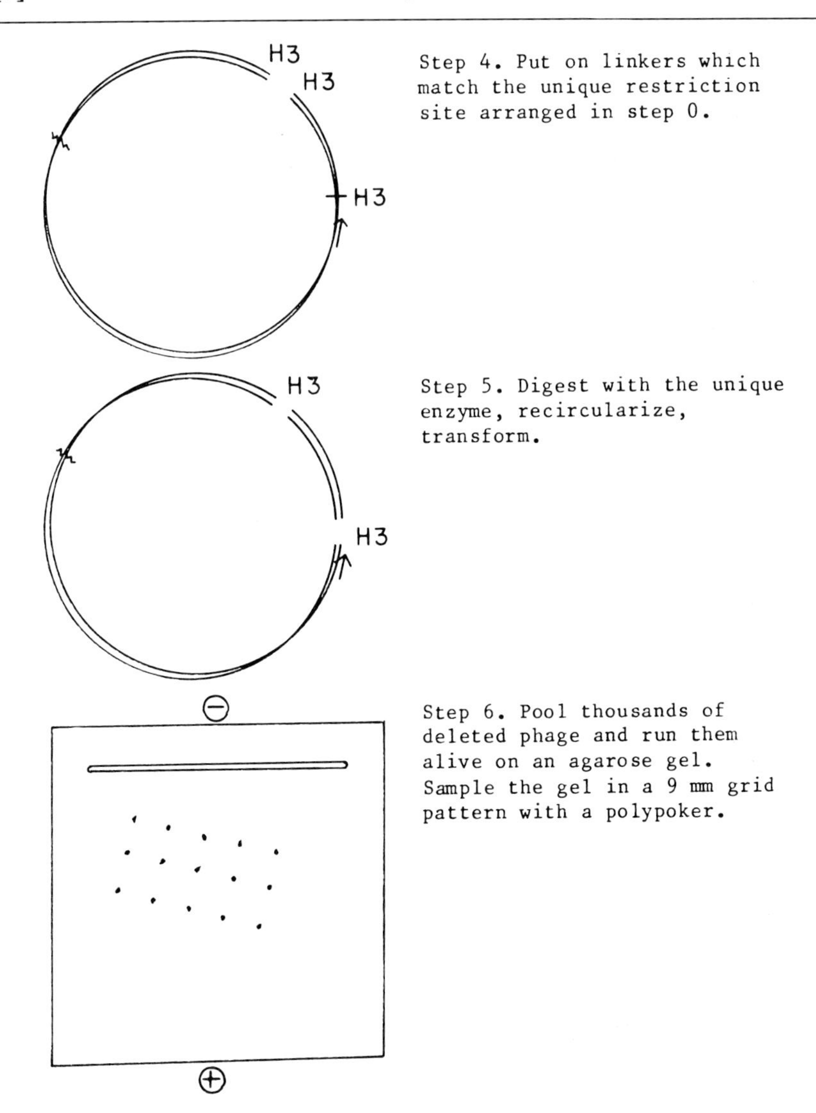

FIG. 1. The steps in the overall plan.

as into the *Pst*I, *Bg*II, *Xba*I, or *Eco*RI site of mWB2344. The ends of the target DNA in this approach can be irrelevant; they can be a partial *Sau*3A digest for insertion into the *Bg*II site of mWB2344, or they can even be a random DNase cleavage product or shear product.

Both orientations will usually be obtainable and desirable, so that the entire deletion and sequencing strategy can derive sequence data on both strands, if necessary. The two possible orientations of the target DNA can be conveniently distinguished by annealling together phage suspensions, followed by agarose gel analysis, as described under Orientation Trick in the Deletion Method section below.

The second approach for recloning is to take advantage of restriction sites in or near the target DNA sequence that have sticky ends matching those in the mWB2341 or mWB2344 polylinker region. In this case the orientation of the target DNA may be forced. Only one sticky end need be used to force the orientation, if one end of both the passenger DNA and target DNA is first polished to a blunt end by treatment with the large fragment of DNA polymerase. If a sticky end used for this cloning approach is on the primer side of the passenger, this restriction site can function as the near-side deletion boundary in the kilo-sequencing strategy.

Vectors and Selection of Recombinants

Figure 2 diagrams the structure, and Fig. 3 shows the cloning region DNA sequence, of our vectors mWB2341, mWB2342, and mWB2344. mWB2341 and mWB2342 are the simpler, as they have only a *Hin*dIII site and an *Eco*RI site. The M13 vectors we have constructed for use with this strategy are able to carry at least 14 kb of passenger DNA with none, or less, of the propensity to form deletions spontaneously that is exhibited by other M13 cloning vectors, such as, the mp vector series of Messing *et al.*[5,6] We do not know why foreign DNA is more stable in our vectors, but it is worth noting that our inserts are in another part of the M13 genome, on the other side of the origins of replication from the insert location for the mp series of vectors. Our constructs always form blue plaques in our host–vector system, even when carrying target DNA, a situation that we have arranged in order to make the plaques easier to see.[7]

We do not use any color change to identify recombinant plaques, since it is more efficient to use phosphatase to select recombinants.[8] Using the phosphatase selection during recloning, 75–95% of the plaques are re-

[5] B. Gronenborn and J. Messing, *Nature* (*London*) **272,** 375 (1978).

[6] J. Messing, R. Crea, and P. H. Seeburg, *Nucleic Acids Res.* **9,** 309 (1981).

[7] The mechanism of induction of the wild-type *lac* operon in the infected host cells is that of out-titration of the *lac* repressor by the many copies of *lac* operator on the replicating RF DNA molecules. This effect has been observed before for plasmids.[26] For the mWB2344 vector series, white plaques are boring deletions of the *lac* operator that result from inefficient ligation or from purposely created deletions that go too far.

[8] A. Ullrich, J. Shine, J. Chirgwin, R. Pictet, E. Tischer, W. J. Rutter, and H. M. Goodman, *Science* **196** 1313 (1977).

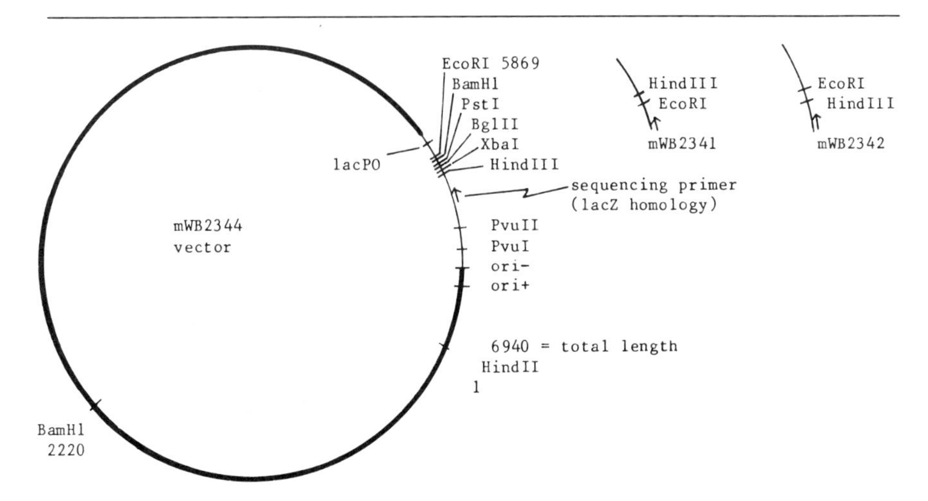

FIG. 2. M13 vectors (W. M. Barnes, in preparation) developed for the kilo-sequencing method. The heavy line of the circular map is M13 DNA, opened at the unique *Sau*96 site. The thin line of the circle is DNA from the first part of the *lac* operon, transferred here from mWJ22 (R. J. Rothstein, L. F. Lau, C. P. Bahl, S. A. Narang, and R. Wu, this series, Vol. 68, p. 98) and later modified to contain the restriction sites shown. For more detail see the sequence printout in Fig. 3. Sites shown are unique, except for the *Bam*HI site, which appears twice. The numbers near the restriction enzyme names are nucleotide position numbers for the total genomes, starting with the unique *Hin*dII site of M13, according to the numbering used for the wild-type sequence.[23a]

combinant. The desired constructs are usually identifiable by the size of their DNA.[9] This identification can and should be confirmed by any of several methods, including hybridization, restriction pattern, phenotype, genotype, and DNA sequence of the first 300 base pairs adjacent to the primer.

Replicative form (RF) DNA is then prepared, and 10 μg are used for the succeeding *in vitro* deletion formation.

Deletion Strategy

The deletions are formed by cleavage first at the far (*vis-à-vis* the sequencing primer) and random side of the deletion, then later at the near side. Random cleavage across both strands of the RF DNA is achieved with the aid of three enzymic treatments. As shown in Fig. 1, step 1, digestion by pancreatic DNase I in the presence of a saturating amount of ethidium bromide is used to create a *quantitative* conversion of RF super-

[9] W. M. Barnes, *Science* **195,** 393 (1977).

```
 M13 vector mWB2344 Cloning site region.  Numbering before lac DNA is from M13 wild-type sequence.

      5610      5620      5630      5640      5650      5660      5670      5680      5690      5700
CTCGCCACGTTCGCCGGCTTTCCCCGTCAAGCTCTAAATCGGGGGCTCCCTTTAGGGTTCCGATTTAGTGCTTTACGGCACCTCGACCCCAAAAAACTTG
           ------          ----                                                    ----

                              .start lac DNA
      5710      5720      5730      5740      5750      5760      5770      5780      5790      5800
ATTTGGGTGATGGTTCACGTAGTGGGCAGTGAGCGCAACGCAATTAATGTGAGTTAGCTCACTCATTAGGCACCCCAGGCTTTACACTTTATGCTTCCGG
    -----          end i gene.----                      ----              -----                 ----

      5810      5820      5830      5840      5850      5860      5870      5880      5890      5900
CTCGTATGTTGTGTGGAATTGTGAGCGGATAACAATTTCACACAGGAAACAGCTATGACCATGATTACGAATTCGCGCGACCGGATCCGGGCAACGTTGT
     <---------------lac operator                    fMet Z gene... ----------- -----------
       probe primer                                                 EcoRI       BamH1**

This is the linker region for mWB2344; see below for 2341 and 2342.
      5910      5920      5930      5940      5950      5960      5970      5980      5990      6000
TGCCATTGCTGCAGGCGCAGAACTGGTAGGTATGGAAGATCTCTAGAAGCTTGTGGAATTAATTCACTGGCCGTCGTTTTACAACGTCGTGACTGGGAAA
     -------                           -------XbaI -----  ----------  ----            <---primer-----
      PstI                              BglII      HindIII                            (N.E.Biolabs)

      6010      6020      6030      6040      6050      6060      6070      6080      6090      6100
ACCCTGGCGTTACCCAACTTAATCGCCTTGCAGCACATCCCCCTTCGCCAGCTGGCGTAATAGCGAAGAGGCCCGCACCGATCGCCCTTCCCAACAGTT
  -----                     -----                  ------          ----------     ------

      6110      6120      6130      6140      6150      6160      6170      6180      6190      6200
GCGTAGCCTGAATGGCGAATGGCGCTTTGCCTGGTTTCCGGCACCAGAAGCGGTGCCGGAAAGCTGGCTGGAGTGCGATCTTCCTGAGGCCGATACTGTC
    -----------    ------   -----   ----                  ----  ----              ---------------

                                                          lac DNA ends.*
      6210      6220      6230      6240      6250      6260      6270      6280      6290      6300
GTCGTCCCCTCAAACTGGCAGATGCACGGTTACGATGCGCCCATCTACACCAACGTGACCTATCCCCCTGATAGACGGTTTTTCGCCCTTTGACGTTGGA
                     -----        -------                                                          --

      6310      6320      6330      6340      6350      6360      6370      6380      6390      6400
GTCCACGTTCTTTAATAGTGGACTCTTGTTCCAAACTTGAACAACACTCAACCCTATCTCGGGCTATTCTTTTGATTTATAAGGGATTTTGCCGATTTCG
F <------------------ F                                     ------                                  -
    ori+ primer
      6410      6420      6430      6440      6450      6460      6470      6480...  total = ...6940
GCCTATTGGTTAAAAAATGAGCTGATTTAACAAAAATTTAACGCGAATTTTAACAAAATATTAACGTTTACAATTTAAAT...     ...TTGGATGTT
---                ----                    ----

.........................................................................................................

mWB2341             5850      5860      5870      5880      5890      5900      5910      5920      5930
cloning region: AAACAGCTATGACCATGATTACGAATTAATTCCACAAGCTTGTGGAATTCACTGGCCGTCGTTTTACAACGTCGTGACTGGGAAA
                       fMet Z gene...               HindIII  EcoRI                   <---primer-----
                                                                                     (N.E.Biolabs)
mWB2342             5850      5860      5870      5880      5890      5900      5910      5920      5930
cloning region: AAACAGCTATGACCATGATTACGAATTCCACAAGCTTGTGGAATTAATTCACTGGCCGTCGTTTTACAACGTCGTGACTGGGAAA
                       fMet Z gene... EcoRI    HindIII                                   <---primer-----
                                                                                     (N.E.Biolabs)
      Total length of mWB2341 and mWB2342 is 6867.
.........................................................................................................

   ** BamH1 site is not unique (yet).
    * note accidental deletion of 8 from M13 ori sequence.
```

FIG. 3. DNA sequence of the cloning site region of vectors mWB2344, mWB2341, and mWB2342. The host bacterial strain that we use is WB373, which is MM294 carrying the tra^- short F plasmid pWB373. MM294 (obtained from M. Innis) is *endA1*, Sm^s, $B1^-$, *hsdR17*, *supE44* [M. Meselson and R. Yuan, *Nature* (*London*) **217,** 1110 (1968)]. This strain exhibits high transformation efficiency. pWB373 is pOX38 [M. Guyer *et al.*, *Cold Spring Harbor Symp. Quant. Biol.* **45,** 135 (1981)], into a *tra* gene of which we have transposed Tn5-DR2 [C. Egner, and D. Berg, *Proc. Natl. Acad. Sci. U.S.A.* **78,** 459 (1981)]. Thus strain WB373 is kan^r, amp^r, tra^-, $M13^s$, in addition to the markers for MM294 listed above.[4] M13 wild-type numbering is from Ref. 23a.

coils to the singly-nicked form II.[10] Since the resulting nick is not significantly sensitive to nuclease S1 or *Bal*31 nuclease (results not shown), the nick is widened to a small gap by a short treatment with exonuclease III. Finally, we utilize the single-strand nuclease to cleave the single strand bridging the gap. We do not use nuclease S1 at this point because the zinc in the S1 reaction buffer interferes with subsequent enzymic steps, and it is not convenient to remove the zinc. S1 may be used if the zinc is removed by dialysis.

Should the location of the resulting double-stranded cleavage not be sufficiently random (and we have obtained the same deletion more than once), the last two enzyme treatments can be adjusted (increased, with various time points), to blur any specificity exhibited by the initial DNase nick. Attempting to digest more than 300 nucleotides with *Bal*31 is not recommended, owing to a low yield, in our experience, of the more digested products (data not shown).

Note that cleavages in the M13 portion of the RF DNA will result in nonviable molecules. Thus, to a first approximation, we have a *selection* for the desired molecules, those cleaved in the desired region, the target passenger DNA to the 3′ side of the sequencing primer.

After addition of linkers that match the unique restriction site we have arranged near the sequencing primer, the DNA is cleaved with that unique enzyme. This creates the desired deletions, which start near the sequencing primer and extend various distances across the target DNA. Standard methodology is used to recircularize the DNA with ligase and transform *E. coli*.

The use of linkers at this point has several advantages over an alternative procedure of blunt-end ligation at the recircularization step. First, the recircularization ligation is easier to carry off efficiently, so that more deleted phages result per microgram of manipulated DNA. Second, the ligation can be performed under more dilute conditions than are required for blunt-end ligation,[11] so that the danger of undesired intermolecular ligation events is reduced. In corollary, any undesired intermolecular ligation events that do result can be recognized, if necessary, since they will create more than one restriction site for the unique enzyme being used. Finally, the restriction site created with the addition of linkers at the boundary of the deletion may be useful in other biological experiments.

Live Phage Gel. One could at this point (and should, to monitor the general success of the procedure) pick plaques at random and determine

[10] L. Greenfield, L. Simpson, and D. Kaplan, *Biochim. Biophys. Acta* **407,** 365 (1975).

[11] A. Sugino, H. M. Goodman, H. L. Heyneker, J. Shine, H. W. Boyer, and N. R. Cozzarelli, *J. Biol. Chem.* **252,** 3987 (1977).

the size by the rapid toothpick assay.[9] However, one more manipulation can allow one to select the sizes of deletions one is interested in: Schaller and co-workers[12] have shown that filamentous phage can be electrophoresed on agarose gels and recovered in their infectious form. We have confirmed this observation and use it to fractionate our deleted phage according to size (Fig. 4).

Approximately 3000 independently deleted phage are obtained per microgram of DNA surviving the above procedure. All the phage plaques are suspended together and electrophoresed on an agarose gel to separate them on the basis of the size of their genome, which corresponds to the size of the deletion, which determines where along the target DNA a dideoxy sequencing experiment will provide data. The gel is sampled with a convenient "polypicker," and gel fractions are stored as phage suspensions in the wells of a Microtiter plate. Each well (gel fraction) may be sampled to obtain sequencing data corresponding to a certain region of the target DNA.

Deletion Method–Detailed Recipe

Stock Solutions and Buffers

Stock Reaction Buffers. Reaction buffers are described as their final concentrations, but they are made up as superconcentrated stocks. Make up 1 or 2 ml at the indicated super concentration (10× or 2×), and store for 1–2 months at −20°. Useful shelf stock solutions, from which all these buffers can be made, are the autoclavable and the not autoclavable stock solutions.

Autoclavable stock solutions: 2 *M* Tris-HCl, pH 7.9, 0.5 *M* $Na_{3\times4}$EDTA, pH 7.9, 1 *M* $MgCl_2$, 5 *M* NaCl.

[12] R. Herrmann, K. Neugebauer, H. Zentgraf, and H. Schaller, *Mol. Gen. Genet.* **159,** 171 (1978).

FIG. 4. Live phage gel for deletions created on mTi23.11, which is mWB2341 carrying the 3300 base-pair *Hin*dIII fragment 23 of the T-DNA of the Ti plasmid pTi T37 *Agrobacterium tumefaciens*. Asymmetric deletions were anchored by the *Eco*RI site, and *Eco*RI linkers were used. The origin is near the top, and the scale may be judged by the 9-mm spacing of the polypoker grid pattern. The heavier band at the top is undeleted phage (due, possibly, to incomplete enzymic reaction at any of the steps of the deletion procedure). The heavy band at the bottom is fully deleted phage (which can be recognized and avoided because the phage form white plaques). Intermediate deletions, phage adhering to pins 6 through 10 of the polypoker, were analyzed further (see Fig. 5) and gave rise to 3000 nucleotides of sequence data (not shown).

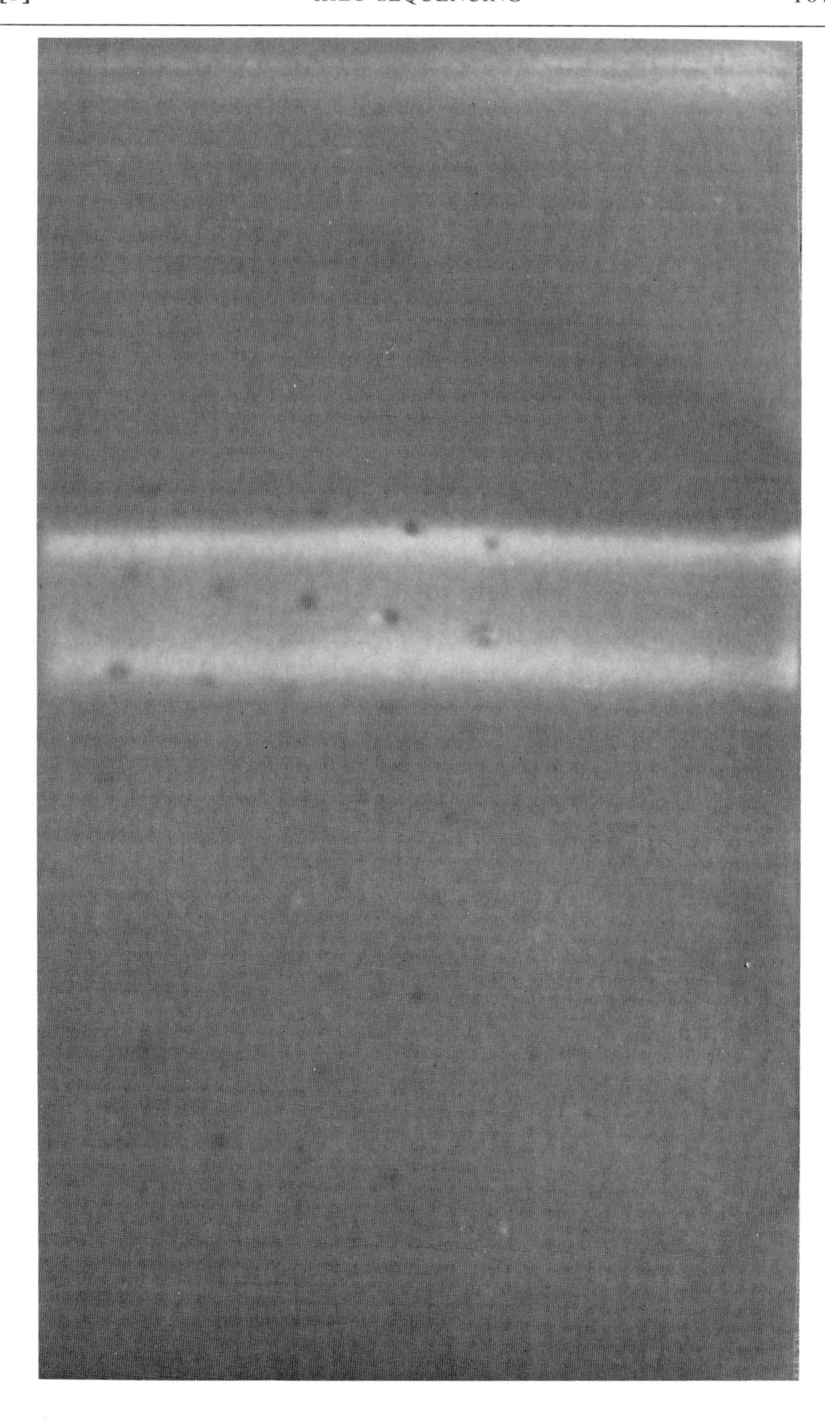

Not autoclavable stock solutions: 1 *M* DTT, 10 m*M* nucleoside triphosphates, 10 m*M* hexamminecobaltic chloride in 10 m*M* HCl. Store at −20°.

NaTMS: 10 m*M* Tris-HCl, pH 7.9, 10 m*M* $MgCl_2$, 50 m*M* NaCl, 10 m*M* 2-mercaptoethanol

Nicking buffer (NB): 0.125 *M* NaCl, 20 m*M* $MgCl_2$, 4 m*M* Tris-HCl, pH 7.9, 60 μg/ml bovine serum albumin (BSA, Miles Pentex). Store as 10×.

Sticky-end T4 ligase buffer: 30 m*M* Tris-HCl, pH 7.9, 4 m*M* $MgCl_2$, 40 μ*M* rATP, 1 m*M* EDTA, and 10 m*M* DTT. Store at 10×. For blunt ends, supplement reaction to 1 m*M* rATP, 1 m*M* hexamminecobaltic chloride, and use 50–100 times more enzyme than is necessary for sticky ends.

*Bal*31 buffer[16]: 12 m*M* $MgCl_2$, 12 m*M* $CaCl_2$, 0.6 *M* NaCl, 25 m*M* Tris-HCl, pH 7.9, 1 m*M* EDTA. Store as 2×.

ExoIII buffer: 66 m*M* Tris-HCl, pH 7.9, 0.66 m*M* $MgCl_2$, 10 m*M* 2-mercaptoethanol. Store as 10×.

Kinase buffer: 5 m*M* Tris-HCl, pH 7.6, 1 m*M* $MgCl_2$, 0.5 m*M* DTT. Store as 10×.

Other Solutions

DNA buffer is 10 m*M* Tris-HCl, pH 7.9, 10 m*M* NaCl, 0.1 m*M* EDTA.

Phenol extraction buffer: 0.3 *M* NaCl, 0.1 *M* Tris-HCl, pH 7.9, 1 m*M* EDTA.

4× GGB (agarose gel electrophoresis buffer): 160 m*M* Tris-acetate, pH 8.3, 80 m*M* sodium acetate, 0.08 m*M* EDTA. Store at 5×, which is labeled 20× GGB.

Formamide-dye mix: 99% formamide (perhaps recrystallized at 0°), 0.1% xylene cyanol (the slow blue), 0.1% bromophenol blue (the fast blue), 5 m*M* EDTA. Store in the dark.

Enzymes

Klenow enzyme (large fragment of DNA polymerase I of *E. coli*,[13,14] was obtained from New England BioLabs. Bad batches of this enzyme exist in the marketplace. These batches are not suitable for dideoxy sequencing. They are most noticeably recognized by too many C bands on the autoradiograph, especially preceding a run of more than one C. The only known cure is to find or isolate a good batch.

[13] H. Klenow, K. Overgaard-Hansen, and S. A. Patkar, *Eur. J. Biochem.* **22**, 371 (1971).

[14] P. Setlow, D. Brutlag, and A. Kornberg, *J. Biol. Chem.* **247**, 224 (1972).

Restriction enzymes, exonuclease III (ExoIII),[15] and *Bal*31 nuclease,[16] were purchased from New England BioLabs. Pancreatic DNase I was from Worthington. T4 DNA ligase was from New England BioLabs or New England Nuclear, or was a gift from Mike Bittner.

Equipment

Electrophoresis equipment, polypoker, and semiautomatic sequencing gel reader are available from Swell Gels [15 Carter Ct., St. Louis, Missouri 62132; (314)994-1587].

Recloning

Calf intestine alkaline phosphatase is the salt-free, lyophilized preparation from Boehringer-Mannheim. The contents of one vial, 1000 units, are suspended in 1 ml of 50% glycerol, 10 m*M* Tris-HCl, pH 7.9, 10 m*M* 2-mercaptoethanol and stored at −20°. Vector RF DNA (10 μg) is cleaved with 10 units of restriction enzyme in NaTMS buffer for 30 mins at 37°. Then 1 unit of phosphatase is added per microgram DNA and incubation is continued for a further 30 min. Extract with phenol,[17] precipitate with ethanol,[18] and resuspend at 0.2 μg/μl in DNA buffer. Control ligations and agarose gel analysis (in parallel with analysis of unphosphatased linear vector DNA), should be carried out to verify the resistance of the phosphatased DNA to ligation, yet its ligatability in the presence of excess target DNA.

Vector DNA (0.2 μg) and target DNA (0.5 μg) may be ligated together with T4 ligase from New England BioLabs or New England Nuclear in a volume of 20 μl, according to the supplier. WB373 may be rendered competent by treatment with 0.1 *M* $CaCl_2$.[19] Commit one-third of the ligase reaction to agarose gel analysis, and transform 0.2 ml of competent

[15] R. R. Richardson, I. R. Lehman, and A. Kornberg, *J. Biol. Chem.* **239,** 251 (1964).

[16] R. J. Legerski, J. L. Hodnett, and H. B. Gray, *Nucleic Acids Res.* **5,** 1445 (1978).

[17] Throughout these recipes, "extract with phenol" means to render the DNA suspension 0.1 *M* in Tris-HCl, pH 7.9, and extract as described for step 9 of the phage DNA isolation procedure.

[18] Throughout these recipes "precipitate with ethanol" means the following: Render the DNA solution about 0.4 *M* in ammonium acetate by adding $\frac{1}{10}$ volume of 4.4 *M* ammonium acetate, pH 5.2. Mix and add 3 volumes of 95% ethanol. Mix and freeze solid in liquid N_2, or store for at least 2 hr at −20°. Thaw if frozen, and spin for 5 min in an Eppendorf microfuge if a small volume, or at 10,000 rpm in a siliconized Corex glass tube if a larger volume. Mark the outside of the tube so you know where to expect the pellet to be. Decant the ethanol or draw it off gently with a pulled and cut Pasteur pipette. Rinse gently, without losing sight of the pellet, with 75% salt-free ethanol. Dry under vacuum.

[19] S. N. Cohen, A. C. Y. Chang, and C. L. Hsu, *Proc. Natl. Acad. Sci. U.S.A.* **69,** 2110 (1972).

WB373 bacteria with 3 μl of the ligase reaction. Be sure to include lawn cells from a freshly saturated culture of WB373 when plating various dilutions to get infected centers (plaques). Each liter of agar plates contains 12 g of agar, 10 g of NZ-amine A or tryptone, 2.5 g of NaCl, 40 mg of XG (no yeast extract, no IPTG).

Toothpick Assay

Recombinant phage plaques can usually be identified by the size of their DNA with the tootpick assay[9] (Fig. 5) as follows:

1. Stick the flat end of a sterile toothpick into each subject plaque, and smear the phage and bacteria onto the surface of a rich agar plate. Make smears about 7 mm in diameter in a replica grid pattern. For edification, pick an "unplaque" too. The more observant will note that recombinant plaques have a subtly different phenotype than do "empty" vector plaques.

2. Incubate overnight at 37°. Note that smears containing infected cells are thinner than uninfected ones, with a few faster-growing papillae in the thin part. The toothpick assay is best performed with bacteria from a fresh such plate, but the plate may be kept at room temperature for a week or so while interesting clones are identified.

3. Using a sterile toothpick, pick up 1–2 mg (1 mm^3) of infected bacteria and put into a 13 × 54 mm glass tube containing 0.2 ml of 4× GGB electrophoresis buffer. Resuspend the bacteria in the buffer by vortexing or by shaking the entire rack of tubes horizontally. Remove toothpicks and reautoclave for next use.

Variation: 0.2 ml of a saturated culture of M13-infected bacteria can be used directly at this point. So can the supernatant of such a culture, if you only want to see the single-stranded virion DNA on the gel. These variations are recommended to monitor all phage infections and preparative isolations of RF or virion DNA.

4. Add two drops of toothpick SDS mix (5% SDS, 50% glycerol, 0.1 *M* EDTA, 0.1% xylene cyanole or bromophenol blue) Do not expect this dye to be visible during the run. Shake to mix. Heat for 10 min at 70°.

5. Cool to room temperature, and load 20–40 μl onto a horizontal 1% agarose, 4× GGB gel. We use the gel conformation described by Shinnick *et al.*[20] except that the gel is covered with Saran wrap instead of Vaseline petroleum jelly, except at the sample wells, which are covered with melted Vaseline petroleum jelly. It is critical to overlay the sample in the

[20] T. M. Shinnick, E. Lund, O. Smithies, and F. R. Blattner, *Nucleic Acids Res.* **2,** 1911 (1975).

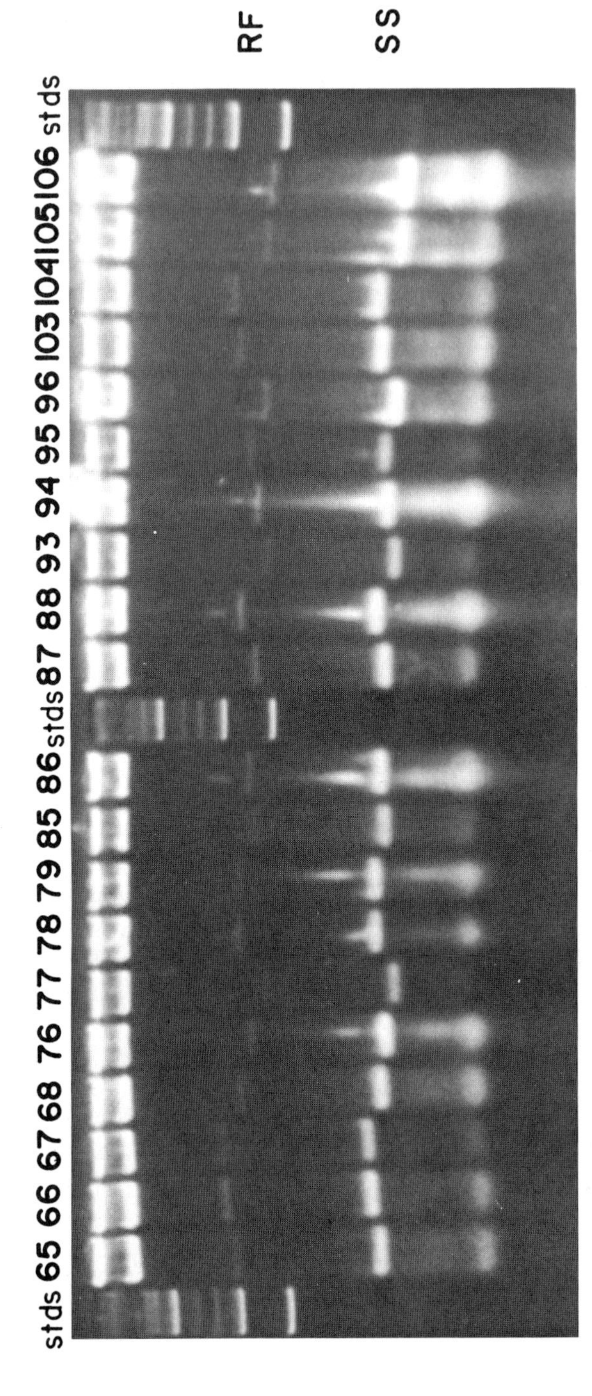

FIG. 5. Toothpick assay to determine genome size for some of the phage from the live phage gel shown in Fig. 4. Numbers in the sixties are randomly chosen phage from pin 6, numbers in the seventies are phage from pin 7, etc. The two smallest RF size standards are mWB2341 (6867 bp) and ml3Hol76 [W. M. Barnes, *Gene* **5**, 127 (1979)] (9738 bp), which thus span an insert size range of 2871 bp.

wells with buffer so that the wells are completely filled before sealing the wells with the melted Vaseline petroleum jelly. For a 2.5 mm thick, 20 cm × 20 cm gel, run for 5 hr at 100 mA, until the bromophenol blue dye in a mix of DNA size standards in their own channel has run 5–7 cm.

Size standards in their own channel will not serve accurately. To avoid a systematic error of 200–300 bp, include RF or infected bacteria size standards within each sample, in amounts that provide a faint grid across the entire gel.

6. Stain the gel for 30–60 min in $\frac{1}{2}$ GGB, 0.5 μg/ml ethidium bromide.

7. *Photography*. Slide the gel directly onto a piece of black (overexposed and developed) x-ray film, a good photographic background. Illuminate from above and to the side with shortwave UV light. Photograph with a 35 mm camera, an orange filter, and tri-X film. Take a few exposures, such as 20 sec, 1 min, 2 min. Do not leave the UV light on or it will ruin the filters in the lamp. Develop film for 10–12 min (ASA 1600).

This method is not guaranteed to be successful (the opposite, even) if the following variations are tried: different electrophoresis buffer, different gel geometry, ethidium in the gel during the run, different photographic geometry, different camera/film system.

Orientation Trick

Size is only one criterion of the correct constructs. In general, each orientation of the inserted DNA will be desired. M13 constructs of the same size and carrying the same DNA fragment may be put into two orientation classes with the orientation trick.[21]

For this trick you may use phage DNA in any of the following forms:[22] (*a*) some of the smear picked up from the replica plate used for the toothpick assay vortexed in 0.2 ml of 4 × GGB in a Eppendorf tube [Pick up a little more than for the toothpick assay, and, since it is probably a day later, vortex extra hard to resuspend the bacteria. Then spin out the bacteria and use the supernatant.]; (*b*) genetic stock (culture supernatant); (*c*) pure, extracted DNA at 1–5 μg/ml.

Render each phage suspension 4× GGB if it is not already, and aliquot 50 μl per Eppendorf tube. Mix together pairs of phage (DNA) that you want to test for carrying the opposite orientation of the same DNA. Be sure, however, to anneal and run each phage DNA by itself, too. Add 1 drop of the toothpick SDS solution, mix, seal, and incubate at 60–70°

[21] W. M. Barnes, *In* "Genetic Engineering (J. K. Setlow and A. Hollaender, eds.), Vol. 2, p. 685. Plenum, New York, 1980.

[22] I do not recommend trying this with total infected bacterial culture.

for 0.5–16 hrs. Load onto agarose gel, run, and photograph exactly as for the toothpick assay above.

Opposite orientation will be signaled by the appearance of two slower-moving bands not present in each DNA by itself.

Preparation of RF DNA

The purified, desired phage construct identified by the toothpick assay and orientation test, and any other tests appropriate for a particular target DNA, may be grown up in 1 liter in quantity as follows. First, grow a 30-ml infected culture as described below for preparation of phage DNA for sequencing, and remove the bacteria by centrifugation. Then infect 1 liter of rich broth with 5 ml of this phage suspension and 5 ml of freshly saturated host bacteria and shake well overnight (16 hr) in a 2-liter flask at 37°.[23] RF DNA may be prepared from the infected cells using almost any procedure that works for plasmid DNA.[24,25]

Creation of Deletions

Do not go through this procedure without checking 0.1–0.2 μg of the DNA on agarose gels after every step. Suggested points for checking these aliquots are marked with asterisks*. In general, each enzyme is used in amounts some threefold above the minimum found necessary to do the job.

Step 1. Nick once with DNase.[10] Mix 5 μg RF DNA, 10 μl of 10× nicking buffer, 10 μl of 5 mg/ml Ethidium bromide (0.5 mg/ml final), HOH to 100 μl; 50 ng of DNase (stored at 4° centigrade at 1 mg/ml in 1 m*M* HCl and diluted appropriately in nicking buffer immediately before use.)

*,*Incubate at 25–30° for 1 hr; render 0.1 *M* Tris-HCl, pH 7.9. Extract with phenol twice, with chloroform twice, and precipitate with ethanol.

Step 2. Widen the nick to a small gap with exonuclease III. Resuspend DNA in 45 μl of DNA buffer; 5 μl of 10× *exo*III*buffer*. And 10–30 units of New England BioLabs exoIII.

Incubate at 25–30° for 10 min. Inactive enzyme by heating at 70° for 10 min.*

Step 3. Cut across gap with the single-strand nuclease activity of

[23] The mWB234 series and constructs grow more slowly than wild-type M13 (or the mp series, which is less prone to deletion if it is not grown this long).

[23a] P. K. G. F. Van Wezenbeck, T. J. M. Hulsoebos, and J. G. G. Schoenmakers, *Gene* **11**, 129 (1980).

[24] L. Katz, D. T. Kingsbury, and D. R. Helinski, *J. Bacteriol.* **114**, 577 (1973).

[25] D. Clewell, and D. R. Helinski, *Biochemistry* **9**, 4428 (1970).

*Bal*31. Use 50 μl of exoIII reaction, 100 μl of 2× bal buffer, 50 μl of HOH, and 1 μl of Bal/10 ($\frac{1}{16}$ unit; diluted from New England BioLabs stock into 0.2 *M* Tris-HCl, pH 7.9, 0.01 *M* BME, 50% glycerol).

Incubate at 25–30° for 5 min, then stop $\frac{1}{2}$ aliquot by adding to excess EDTA.

*,*Stop the rest after a total of 9 min. (Save 0.2 μg of the least and most *Bal*31 digested DNA separate for agarose gel analysis.) Extract with phenol once, with chloroform twice, and precipitate with ethanol.

Step 4. Treat ends[26] with Klenow enzyme fragment to be sure they are blunt. Resuspend DNA in 45 μl of DNA buffer, 6 μl of 10× NaTMS, 6μl of 1 m*M* all 4 dNTPs (final 100 μ*M*), 3 μl (1.5 units) of Klenow (DNA polymerase large fragment).

Incubate at 25–37° for 15–30 min. Extract with phenol once, with chloroform twice, and precipitate with ethanol.

Step 5. Put linkers (which match the unique site near the primer) onto the clean, random cut just made.[27] Use 5 μg of randomly linearized RF, 2 μl of 10× ligase buffer, 2 μl of 10 m*M* rATP, 2 μl of 10 m*M* spermidine or hexamminecobaltic chloride, 2 μl of kinased linkers (0.01 OD, see below) HOH to 19 μl, as needed, 1 μl of T4 ligase, in amount sufficient to ligate blunt ends.

Incubate overnight at 4°. (In a parallel control reaction, ligate 1 μg of DNA *Bsp*I or other blunt-end restriction digest.) Extract with phenol once, with chloroform twice, and precipitate with ethanol.

Step 5b. Separate the genome-length DNA from ATP and the bulk of the polylinkers by any of several methods, such as spermine precipitation,[28] PEG precipitation,[29] or, as described below, gel filtration.

Pack a 1-ml column of BioGel A-15m in a disposable 1-ml pipette, using as column buffer 2.5 m*M* Tris-HCl, 1 m*M* NaCl. After loading the column with a volume of 30–100 μl, monitor the column output with a rate meter, and collect 10–15 two-drop fractions. The fractions containing the desired DNA are just ahead of the radioactivity detectable with the hand monitor. Confirm the identity of the DNA-containing fractions by spotting (along with concentration standards containing amounts in the range 0–50 ng) 3–4-μl aliquots onto the surface of a 1% agarose, 1× GGB, agarose gel containing 0.5 μg of ethidium bromide per milliter. Do not electrophorese, but examine immediately under UV illumination, by photography as described above for the toothpick assay.

[26] K. Backman, M. Ptashne, and W. Gilbert, *Proc. Natl. Acad. Sci. U.S.A.* **73,** 4174 (1976).

[27] If desired, this step can be eliminated, step 4 (Klenow polishing) can be moved to follow step 6, and the recircularization can be carried out under blunt-end ligation conditions.[4]

[28] B. C. Hoopes and W. R. McClure, *Nucleic Acids Res.* **9,** 5493 (1981).

[29] J. T. Lis and R. Schlief, *Nucleic Acids Res.* **2,** 383 (1975).

Step 6. Digest with the unique enzyme. This creates all the desired deletions. It should be necessary to add more than one unit of enzyme per microgram, since much polylinker may be present.*

Extract with phenol once, with chloroform twice, and precipitate with ethanol.

Step 7. Recircularization. Ligate 1 μg in 100 μl overnight at 4°. Ethanol precipitate and resuspend in 15 μl of DNA buffer.*

Step 8. Transform WB373 bacteria; plate 0.9, 0.1, and 0.01 (fractions) of total on XG plates.

Anticipate about 3000 plaques total. Wash them all off the plate in 5 ml of M13 buffer, spin at 10,000 krpm for 15 min to remove bacteria, and heat for 10 min at 70°. Store this phage deletion stock at 4°.

Step 8. Live Phage Gel. Pour a 0.7% high-melting-temperature (type VI) agarose gel in 3 × GGB running buffer. (20 × GGB = 0.8 *M* Tris-acetate, pH 8.3, 0.4 *M* sodium acetate, 4 m*M* EDTA.) Make up gel sample at 60° using 0.3 ml of phage stock, 0.2 ml of blue 2, 0.3 ml of 0.7% agarose. Load wide well of horizontal gel two-thirds full. Load vector and undeleted M13 phage in some side slots, for size standards. Top off with 0.7% agarose, cover with Saran wrap, and run gel at 100 mA (3 V/cm) for 16 hr. Sample gel at 2-mm intervals from 2 to 13 cm migration, using a polypoker, an array of blunt syringe needles (Swell Gels). Rinse gel samples into Microtiter wells with 2 × 0.1 ml of M13 buffer.

Staining gel: Soak $\frac{1}{2}$ hr in 1 liter of 0.1 *N* NaOH to strip phage coats from the DNA. Neutralize by soaking for $\frac{1}{2}$ hr in 3 × GGB. Then soak for 1–16 hr in ethidium stain (0.5 μg/ml in 0.5 × GGB). Photograph as described above for the toothpick assay.[9]

*1% Agarose gel checks[4]:

After

Step 1: Mostly nicked circles; a small amount of linears, no supercoils. (A few time points might be illuminating.)

Step 2: Same as above

Step 3: 100% full-length linears.

Step 4: Same

Step 5: Smooth smear, range of sizes from full length linears to very small. Radioactive.

Step 6: Conversion to much larger sizes than above.

Step 7: Range of sizes of single-stranded DNA from full-sized down to vector-sized.

Kinasing linkers (20 μl):

0.2 OD_{260} of linkers from New England BioLabs

50 μCi gamma rATP
2 μl 10× kinase buffer (10× = 50 mM Tris-HCl, pH 7.6, 10 mM MgC12, 5 mM DTT)
HOH to 18 μl
Heat at 70° for 2 min; cool quickly.
Add 1 μl of T4 polynucleotide kinase (P-L Biochemicals)
Leave for 15 min at 37°
Add cold rATP to 1 mM
Add 1 μl more kinase
Leave for 30 min at 37°
Inactivate enzyme with 70°, 10 min.
Store frozen.

To monitor the reaction, spot 0.5 μl on PEI plate. Develop with 1.5 M NaH_2PO_4.

Preparation of Phage DNA from mWB23 Series Phage Constructs

Since the ready, reliable isolation of phage DNA is critical to the success and convenience of the overall strategy, we present below a modification of the procedure of Sanger *et al.*[1] that we find to be reliable for mWB23 series phage constructs.

1. Inoculate 30 ml of 2× NY with bacteria WB373 (0.2 ml of fresh 7–15 hr culture) and with phage. This phage can be (*a*) 0.2 ml of genetic stock (i.e., infected-culture supernatant); (*b*) a fresh plaque picked up with a Pasteur pipette; or, (*c*) 0.5 ml of phage suspension from an infected-bacteria smear that has just given a pleasing toothpick assay result; pick up a generous amount (3–5 mg) of infected bacteria with a sterile stick, vortex in 1 ml of broth or phage buffer in a sterile Eppendorf tube, spin out bacteria, and (optionally) heat for 10 min at 70°. (This phage suspension can also serve as a genetic stock and as a source of phage for the orientation trick.)

Incubate for 12–16 hr at 37°. Be sure to shake well enough to aerate well. Note that this incubation time is much longer than can (even must) be used for other, faster-growing M13 vectors.[23]

2. Transfer to sterile 40-ml centrifuge tubes and centrifuge for 15 min at 14,000 rpm.

3. Decant carefully into another clear polycarbonate centrifuge tube. If the supernatant is not perfectly clear, centrifuge again. *Save a genetic stock sample:* Remove about 3 ml to a 2-dram vial, heat for 10 min at 70°, and store at 4°.

4. To 30 ml of clear culture supernatant, add 7.5 ml of 15% PEG, 2.5 M NaCl. Shake, and put on ice for 30 min.

5. Noticeable turbidity should develop; if none does, anticipate no yield.

6. Centrifuge for 20 min at 10,000 rpm to pellet PEG-precipitated M13.

7. Drain pellets *well;* even rinse them very gently with 2–3 ml of DNA buffer to remove *all* PEG-containing supernatant.

8. Resuspend pellets in 2 ml of phenol extraction buffer (0.3 *M* NaCl, 0.1 *M* Tris-HCl, 1 m*M* EDTA). (It is convenient to do this by rotary shaking at 37°.) Usually we extract only one-eighth of the phage (the amount from 3–4 ml of infected culture); so, after removing 0.25 ml to a 1.5-ml Eppendorf tube for phenol extraction, store the rest frozen. The viability of phage stored in this manner is sufficient to serve as an emergency genetic stock sample.

9. Phenol extract twice. (Add 0.8 ml of distilled, water-saturated, then frozen, phenol, vortex and shake for 2 min, spin for 1 min in a microfuge, remove phenol layer *without removing any interface material, if present,* since it contains DNA.)

10. Extract twice with $CHCl_3$–isoamyl alcohol, 24:1. Remove any interface at the last extraction.

11. Remove 1% (some 2.5 μl) of the DNA phase to 0.2 ml of diluted blue dye mix, and load 20 μl onto a 1% agarose gel to assay concentration and size vs standards of 20 ng and 40 ng per band. DNA sequencing will not work if there is no template DNA there.

12. Precipitate with ethanol and resuspend at 1 μg/μl, according to the results of the gel concentration assay. Generally, this will be about 19 μl. Store frozen.

DNA Sequencing[30,31]

All mixing is done in siliconized, 13 × 54 mm glass tubes. All reactions are incubated in pulled, cut melting-point capillaries. Template is M13 transducing phage virion DNA, isolated as described above. Primer is single-stranded synthetic DNA from New England BioLabs.

1. Label a siliconized tube with name of the primer–template.
2. Record in book as follows:

Date	*Primer/Template*	*Klenow*	*Chase*	*Stop*

3. Start heating a pan of water.
4. Thaw frozen solutions and vortex.
5. Add to labeled tube and then mix up and down (5×) in capillary: 6 μl of 10× NaTMS, 3 μl of template (1 mg/ml), 15 μl of primer (1 μl New England BioLabs *lac* 15-mer + 14 μl H_2O) (about $\frac{2}{3}$ pmol each).

[30] F. Sanger, S. Nicklen, and A. R. Coulson, *Proc. Natl. Acad. Sci. U.S.A.* **74,** 5463 (1977).
[31] A. J. H. Smith, this series, Vol. 65, p. 560.

TABLE I
ANALOG MIXES FOR THE DIDEOXY SEQUENCING METHOD[a]

		Final 10× (m*M*)	Final 1× (μ*M*)	μl 10 m*M* stock (μl)
ddT (30:1)	dA	0		0
	dG	0.4	40	40
	dC	0.4	40	40
	dT	0.034	3.4	3.4
	ddT	1.0	100	100
	Water	55 *M*		0.82 ml
ddA (17:1)	dA	0[b]	1.8	0
	dG	0.4	40	40
	dC	0.4	40	40
	dT	0.4	40	40
	ddA	0.35	35	35
	Water			0.845 ml
ddG (17:1)	dA	0		0
	dC	0.4	40	40
	dT	0.4	40	40
	dG	0.056	5.6	5.6
	ddG	1.0	100	100
	Water			0.817 ml
ddC (8:1)	dA	0		0
	dG	0.4	40	40
	dT	0.4	40	40
	dC	0.150	15	12.5
	ddC	1.0	100	100
	Water			0.807 ml
ddG (1.7:1) vs dI[c]	dA	0		0
	dC	0.4	40	40
	dT	0.4	40	40
	dI	0.2	20	20
	ddG	0.35	35	35
	Water			0.865 ml

[a] One-milliliter batches of 10× Analog Mixes. Store in water at −20°; it may be better to make up these solutions in 10 m*M* tris-HCl, pH 7.9.

[b] dA is supplied to six 10-μl reactions as 12 μl of [α-^{32}P]dATP (0.5 μ*M* final) and 4 μl of 20 μ*M* cold dATP (1.33 μ*M* final).

[c] The dITP channel is a valuable aid in resolving compressions due to secondary structure in the analyzed DNA (W. M. Barnes, in preparation). All triphosphates are supplied by P-L Biochemicals.

6. Seal capillary, and put into tube full of water.

7. *Anneal:* Place into 70° water bath (about 500 ml, covered), and allow to cool for 30 min (to 45°).

8. Meanwhile, dry down 15 μl (15–30 μCi) [α-^{32} P]dATP (400 mCi/μmol) under vacuum in a siliconized tube.

9. Resuspend A* in 26 μl of H_2O + 4 μl of 0.02 m*M* cold dATP. Mix with annealed DNA.

10. Divide annealed DNA + A + A* into 6 siliconized tubes (~8.5 μl each). ICATGC = tube and gel channel order

11. Add 1 μl of a 10× analog mix (see Table I) to a new, dry capillary, and pop into tube containing the 8.5-μl drop. Do touch the drop with the tip.

12. Fetch 6 aliquots of Klenow enzyme from freezer in a 5-μl capillary. Aliquot 0.2 to 1 μl (0.25 unit) of Klenow enzyme into each of six graduated 5 μl capillaries, and rack them onto a stick of plasticene clay.

13. Add Klenow enzyme aliquot to each reaction capillary. Then mix each reaction up and down 4 or 5 times allowing no bubbles to form. Tilt rack of tubes to assist capillary to attract the entire reaction volume up into the capillary away from air. That is, leave no drop outside the capillary. Incubate (at a tilt) for 15 min in a 37° air incubator.

14. *Chase*. Add 1 μl of chase mix (1 m*M* in each of the four dNTPs) and also another aliquot of Klenow enzyme. (Measure both of these into six more 5-μl capillaries racked onto plasticene clay as before.) Incubate at 37° for 15 min.

15. Add 10 μl of 15 m*M* EDTA to each. (Roll drop down side of tube, then blow out each capillary.)

16. To store overnight, add 2 drops ethanol, cover rack with Saran wrap, and put at −20°.

17. Dry down complete sequencing reactions under vacuum (20–30 min with a good pump). Glycerol will not evaporate.

18. Resuspend each in 11 μl of 99% formamide, 5 m*M* EDTA, 0.1% xylene cyanol, 0.1% bromophenol blue. Vortex well.

19. Heat for 30 sec at 90° and cool to room temperature immediately before loading *or reloading*. Store in dark at room temperature (up to 3 days? sealed with Parafilm) between loadings.

20. Load ~3 μl per gel channel (1–1.5 mm depth of sample).

Sequencing Gels[32,33]

Gel Plates

We recommend gel plates made of Pyrex rolled-surface plate, ⅛ inch thick, 20 cm × 40 cm. Between uses, wash with hot tap water, rinse with distilled water, rinse with 95% ethanol, dry with large Kimwipes. This

[32] F. Sanger and A. R. Coulson, *FEBS Lett*. **87,** 107 (1978).

[33] Equipment source: Swell Gels, 15 Carter Ct., St. Louis, Missouri, 63132; (314)994-1587.

cleaning is most easily and effectively done immediately after use (while the gel is fixing).

Note that Sanger[32] recommends wells no more than 5 mm wide, separated by no more than 1.5 mm.

Assembly. Use no grease or tape.

1. Place the notched plate on a clean bench top or ice bucket.
2. Place long spacers (clear rigid vinyl, 0.015 inch thick) along sides, leaving room at the bottom for the short spacer.
3. Place short spacer strip along bottom, making sure it abuts the bottom ends of the long side spacers. Do not overlap.
4. Carefully emplace the unnotched plate.
5. Put one clamp on bottom, and four on each side. Be sure they are clamping only about $\frac{1}{4}$-inch from the edges of the plates.

Between uses, it is best to cover the notch with Saran wrap and store the gel plates clean and ready to use.

GEL SOLUTION (Table II)

Weigh out ultrapure urea in a clean beaker or suction flask + stir bar. Add deionized 40% acrylamide, deionized 2% bis, water, 10× TBE, and fresh 10% ammonium persulfate while stirring in a warm (tap water) bath. Remove from water bath when urea is dissolved; do not warm past room temperature. Stir until complete dissolution of the urea. Filter if solution is not completely clear. Degas briefly.

Immediately before pouring, add the neat TEMED with rapid stirring. You have 10 min to get the gel(s) poured. Allow to polymerize for 2 hr.

Pouring. Gels are polymerized while flat horizontal and poured at only a slight angle. A few milliliters may leak out of the bottom while you are pouring, but capillary attraction will prevent any leakage once the gel is laid flat. Holding a flask or overfull 25-ml pipette in your dominant hand, pour the gel solution into the far side of the notch. Allow the solution to go down the far side of the gel, across the bottom, and up the near side. Do not allow the water column on the far side to break. Do not allow bubbles to form at any time. If a bubble starts to form, tap it with a pipette. If too many bubbles try to form (or succeed in forming), you did not clean the plates properly. Bubbles in the top fourth of the gel can be swept out with a bubble sweeper (0.01-inch vinyl strip). Bubbles near the edge just leave. If there are bubbles anywhere else, start over. An expert can pour 3, maybe 4 gels at a time with no bubbles.

After pouring the gel, insert well-former only 5 mm, with no bubbles under or between any of the teeth. (Teeth are 5 mm wide, so push them in until a square is below the notch of the gel plate.) Put three clamps across top of gel, clamping directly at the wells.

TABLE II
SEQUENCING GEL SOLUTIONS

10× 3:1 TBE[a]	For 3 liters	For 1 liter
Tris base (g)	486	162
Boric acid (g)	82.5	27.5
Na_2EDTA (g)	27.9	9.3
8% gel[b]	**3–4 gels**	**1–2 gels**
Urea, ultra pure	66 g	33
40% acrylamide	30 ml	15
2% Bis	30 ml	15
H_2O	32.5 ml	16
10× 3:1 TBE	15.3 ml	7.7
10% Ammonium persulfate	0.75 ml	0.37
Degas		
TEMED (neat)	0.075 ml	37 ul
(total)	(150 ml)	(75 ml)

[a] 3:1 TBE is an improvement over TBE [see S. Anderson, *Nucleic Acids Res.* **9,** 3015 (1981).

[b] The 40% acrylamide and 2% Bis solutions below are deionized by stirring for 30 min with a few tablespoons full of Amberlite MB-1 (Mallinckrodt), followed by filtering through fluted filter paper. Store in a dark refrigerator for months. Urea is ultra pure from Schwarz-Mann/Becton-Dickinson, acrylamide is from BDH (not the "specially pure"), and bisacrylamide is from BDH or Sigma.

After some 2 hr, rinse the gel plate assembly under tap water to remove crusted urea and acrylamide solution. Remove all clamps. *Slowly* pull out slot former, and then *immediately* rinse wells with 1× TBE from a squirt bottle to remove any unpolymerized acrylamide solution. Do not forget to remove the bottom spacer. Leave the side spacers in. Dry the outside of the plates.

Fill the lower buffer of the gel apparatus, and slide the gel plate assembly into buffer so that no bubbles are caught under the gel. Clamp at sides of notch, and fill the upper chamber. Use no grease.

Prerun the gel for 15–30 min at full power (34 W). Immediately before loading each set of samples, rinse the urea out of the wells by forcing buffer into them with the aid of a Pasteur pipette with a rubber bulb.

Using a capillary pulled thin enough, load 3–4 μl of just-heated sample in formamide–dye mix. Do not load the sample by allowing it to fall more than 1 mm through the buffer. Place the tip of the capillary at the bottom of the well and force the sample out until the column of sample is

1–1.5 mm in depth. Waste some sample by leaving it in the loading capillary rather than risk blowing a bubble in the well while the sample is in it. Rinse the capillary once with lower reservoir buffer between samples. Do not skip any wells. Run each experiment (set of six channels) into the gel for 5 mins at full power before loading the next set.

Run the gel for a multiple of 2.5 hr, usually. Use 30–34 W constant power.

Autoradiography

Separate gel plates carefully by inserting a spatula near a bottom corner and twisting slowly. It is best if the gel sticks to the unnotched plate. Using a pizza cutter, remove 0.5 inch along the sides (and, if necessary, 2.5 inches along the top) of the gel. This is so that two gels will fit on one film.

Listen to the slow blue dye with a Geiger counter; 10–15 cps may mean a good experiment with 1 or 2 days' exposure time.

You have previously prepared a 1-inch-deep bath of 10% acetic acid made up with tap water. You have also previously placed in this fixing bath a piece of "junk" glass 7 × 16 inches. Now flip the gel plate, with the gel clinging to it, upside down, and lower it into the acid bath. Slosh gently until the gel falls off the gel plate onto the junk glass. After 3 min, or after you have cleaned up and stored your gel plates in a ready-to-pour condition, remove the gel, supported by the junk glass plate, drain a little, and blot excess liquid with many Kimwipes or paper towels.

Cover the wet, fixed gel with Saran wrap. Place sticky labels onto the gel (not off the gel) where they will not cover up any interesting bands. Label with a radioactive pen (^{35}S ink), at least noting the date.

Autoradiograph with single-sided film (Kodak SB5) at room temperature for 2 days, probably.

Acknowledgments

This work was supported by Grant GM24956 from the NIH. W. B. also acknowledges a Faculty Research Award from the American Cancer Society. The *Agrobacterium tumefaciens* DNA work is part of a collaboration with Mary Dell Chilton.

[6] The Use of pKC30 and Its Derivatives for Controlled Expression of Genes

By MARTIN ROSENBERG, YEN-SEN HO, and ALLAN SHATZMAN

There are numerous gene products of biological interest that cannot be obtained in quantities sufficient for detailed study of their structure and function. Over the past few years recombinant technology has offered new approaches to this problem. One such approach has been the development of vector systems designed to achieve efficient expression of cloned genes in bacteria.

In general, the rationale used in the design of these systems involves insertion of the gene of interest into a multicopy vector system (e.g., a plasmid) such that the gene is transcribed from a "strong" bacterial promoter. This usually ensures efficient transcription of the gene, but does not necessarily guarantee its expression. In particular, for those genes that do not naturally contain the proper signals for ribosome recognition and translation initiation in *Escherichia coli,* special procedures must be devised to supply this information. This is done either by fusing the gene to a bacterial ribosome binding site or to the N-terminal coding region of a bacterial gene. In the first case, some difficulties in obtaining efficient translation have been encountered owing to sequence alterations made in the ribosome recognition region prior to or as a result of gene insertion. In the latter case, the gene product is a fusion protein carrying additional peptide information at its N terminus. The fusion products may have physical and functional properties that differ from the normal protein, thereby limiting their value for biological study. In addition to factors such as promoter strength, gene copy number, and translational efficiency, a number of other factors may influence the expression of a cloned gene in bacteria. These include (*a*) reduction of transcription resulting from polarity effects; (*b*) the stability of the mRNA; (*c*) the stability of the gene product; and (*d*) the potential lethality of the product to the growth of the host.

This chapter describes a set of plasmid cloning vehicles that were constructed to achieve efficient expression of cloned genes in *E. coli.* These vectors contain transcriptional and translational signals that derive from the bacteriophage lambda genome. The design of the vectors and the rationale for using these particular regulatory sequences are discussed in relation to each of the above considerations. Procedures for using these systems are described in detail.

ISBN 0-12-182001-7

Rationale of the Method

The system to be described utilizes a plasmid vehicle (a pBR322 derivative) carrying regulatory signals derived from the bacteriophage λ genome. Phage regulatory information was chosen because, in general, these signals tend to be more efficient than their host-derived counterparts. For example, with a vector system designed specifically for studying transcriptional regulatory signals,[1] the phage λ promoter P_L was shown to be 8–10 times more efficient than the bacterial promoter of the lactose operon (P_{lac}). In fact, P_L was as efficient, or more so, as all other bacterial promoters tested.[2,3]

Plasmids carrying P_L are often unstable, presumably owing to the high level of P_L-directed transcription.[4] This problem of instability was overcome by repressing P_L transcription, using bacterial hosts that contain an integrated copy of the λ genome (i.e., bacterial lysogens). In these cells, P_L transcription is controlled by the phage λ repressor protein (*cI*), a product that is synthesized continuously and regulated autogenously in the lysogen.[5] It was demonstrated that certain lysogens synthesize sufficient repressor to inhibit P_L expression completely on the multicopy vector.[6] Thus, the cells can be stably transformed and the vector maintained in these lysogenic hosts. Moreover, by using a lysogen carrying a temperature-sensitive mutation in the λ *cI* gene (cI857),[7] P_L-directed transcription can be activated at any time. Induction is accomplished by simply raising the temperature of the cell culture from 32° to 42°. Thus, cells carrying the vector can be grown initially to high density without expression of the cloned gene (at 32°), and subsequently induced to synthesize the product (at 42°). The ability to control gene expression on the vector, coupled with the rapidity of the induction procedure and the efficiency of P_L, ensures high-level expression of the product in a relatively short time period. These features are particularly useful for the expression of gene products that may be lethal and/or rapidly turned-over in bacteria.

In addition to providing a strong, regulatable promoter, the system

[1] K. McKenney, H. Shimatake, D. Court, U. Schmeissner, C. Brady, and M. Rosenberg, *in* "Gene Amplification and Analysis," Vol. 2: Analysis of Nucleic Acids by Enzymatic Methods" (J. G. Chirikjian and T. S. Papas, eds.), p. 383. Elsevier/North-Holland, Amsterdam, 1981.

[2] M. Rosenberg, K. McKenney, and D. Schümperli, *in* "Promoters: Structure and Function" (M. Chamberlin and R. L. Rodriguez, eds.) p. 387. Praeger, New York, 1982.

[3] A. Shatzman and M. Rosenberg, unpublished data.

[4] R. N. Rao, unpublished data.

[5] M. Ptashne, K. Backman, M. Z. Humayun, A. Jeffrey, R. Maurer, B. Meyer, and R. T. Sauer, *Science* **194,** 156 (1976).

[6] H. Shimatake and M. Rosenberg, *Nature* (*London*) **292,** 128 (1981).

[7] R. Sussman and F. Jacob, *C. R. Hebd. Seances Acad. Sci.* **254,** 1517 (1962).

also ensures that P_L-directed transcription efficiently traverses any gene insert. This is accomplished by providing both the phage λ anti-termination function, *N*, and a site on the P_L transcription unit necessary for *N* utilization (*Nut* site). *N* expression from the host lysogen removes transcriptional polarity, thereby inhibiting termination within the P_L transcription unit.[8,9] Hence, any transcriptional polarity caused by sequences that occur before or within the coding sequence is eliminated by the *N* plus *Nut* system. As demonstrated below, this system leads to a dramatic increase in product yield and also allows much greater flexibility in inserting genes into the vector.

In order to extend this system to the expression of genes lacking *E. coli* translational regulatory information, efficient ribosome recognition and translation initiation sites were engineered into the P_L transcription unit. The site chosen was that of an efficiently translated λ phage gene, *cII*. The entire coding region of this gene was removed, leaving only its initiator fMet codon and regulatory sequences upstream. Neither the sequence nor the position of any nucleotides in the ribosome binding region was altered. Instead, a restriction site for insertion of the desired gene was introduced immediately downstream from the ATG initiation codon. As described below, this system allows direct fusion of any coding sequence to this translational regulatory signal and, furthermore, allows any gene to be adapted for insertion into the vector.

Cloning Prokaryotic Genes into pKC30

The plasmid pKC30 (Fig. 1A) is used to overexpress bacterial genes that contain their own translational regulatory information. This vector contains a unique *Hpa*I restriction site positioned 321 bp downstream from the P_L promoter. Blunt-ended DNA fragments are inserted into this blunt-ended restriction site. The appropriate fragments can be generated: (*a*) directly by restriction; (*b*) subsequent to the removal of single-strand overhanging ends by the action of S_1 or mung bean nuclease; or (*c*) subsequent to the "fill-in" of single-strand overhanging ends by the use of DNA polymerase (Klenow fragment). The *Hpa*I site of pKC30 also can be adapted for the insertion of other DNA fragments by first introducing various synthetic linkers into the site. In addition, the site can be used in combination with other unique restriction sites positioned downstream

[8] S. Heinemann and W. Spiegelman, *Cold Spring Harbor Symp. Quant. Biol.* **35,** 315 (1971).

[9] M. Rosenberg, D. Court, D. L. Wulff, H. Shimatake, and C. Brady, *in* "The Operon" (J. Miller, ed.), p. 345. Cold Spring Harbor Laboratory, Cold Spring Harbor, New York, 1978.

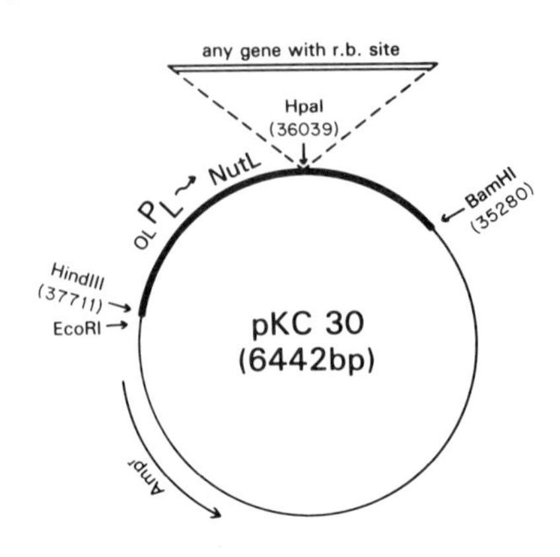

B

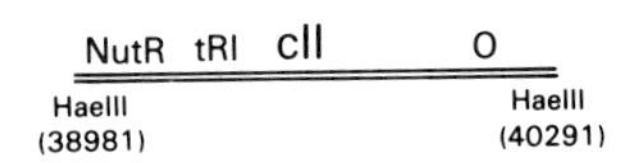

FIG. 1. (A) Schematic diagram of plasmid pKC30. This plasmid is a derivative of pBR322 and contains a 2.4-kb *Hin*dIII-*Bam*HI restriction fragment derived from phage λ inserted between the *Hin*dIII and *Bam*HI restriction sites within the tetracycline gene of pBR322.[4] The λ insert contains the operator (O_L), the promotor signal (P_L), and a site for *N* recognition (*Nut L*). [N. C. Franklin and G. N. Bennett, *Gene* **8,** 197, (1979)]. Fragments are cloned into the *Hpa*I restriction site that occurs within the *N* gene coding region. The numbering system of the λ DNA segment is that of F. Blattner (personal communication). (B) The 1.3 kb *Hae*III restriction fragment of phage λ, which was inserted into the *Hpa*I site of pKC30. This fragment contains the entire *cII* coding region, a site for *N* recognition (*Nut R*), the rho-dependent transcription termination site (*tR1*), and most of the *O* gene region.

from the *Hpa*I site on pKC30 (Fig. 1A). Note that the *Hpa*I cloning site interrupts the coding region of the λ *N* gene, which occurs on pKC30. Hence, insertion into this site also gives rise to *N* protein fusion products that result from *N* gene translation entering the promoter proximal part of the DNA insert. The effect of this translation on the expression of genes cloned into pKC30 has not been examined.

The pKC30 vector has been used to express efficiently several bacterial gene products. The application of the system is probably best exemplified by its use in production of the phage λ transcriptional activator protein *cII*,[6,10] and eight *cII* protein variants that differ by only single amino

[10] Y. S. Ho, M. Lewis, and M. Rosenberg, *J. Biol. Chem.* **257,** 9128 (1982).

acid substitutions.[11,12] Although *cII* is rapidly turned over in *E. coli* and is lethal to cell growth, its insertion and expression in pKC30 allowed levels of synthesis approaching 3–5% of total cellular protein.[6] The following describes in detail the cloning of such a gene fragment into pKC30.

Preparation of HpaI-Restricted pKC30. pKC30 DNA was transformed[13] into an *E. coli* N99λ*cI*$^+$ lysogen, and plasmid DNA was prepared. The yield was approximately 1 μg of DNA per milliliter of cell culture. Plasmid DNA (5–20 μg) was restricted with *Hpa*I (all restriction conditions and enzymes are those of New England BioLabs). The reaction was stopped, and the DNA was recovered by phenol extraction and ethanol precipitation. The DNA was resuspended in 10 m*M* Tris, pH 8.0 (10–50 μl).

Preparation of the Gene Insert. A 1.3-kb *Hae*III fragment from phage λ DNA that carries the *cII* gene (Fig. 1B) was inserted into the *Hpa*I site. Phage DNA (250 μg) was restricted with *Hae*III (200 units) and, after phenol extraction and ethanol precipitation, was dissolved in 600 μl of sample buffer (40 m*M* Tris-HAc, pH 7.9, 2.0 m*M* EDTA, 2.5% glycerol, 0.012% bromophenol blue and xylene cyanol). The sample was loaded into a 5.2 cm wide slot of a preparative 5% polyacrylamide gel (0.15 × 14 × 40 cm^3), and the products were resolved by electrophoresis at 200 V for 18 hr. After electrophoresis, the gel was stained with methylene blue (0.02% for 20 min). The 1.3 kb DNA fragment was cut out of the gel and eluted electrophoretically into 1 ml of 20 m*M* Tris-HAc, pH 8.0. The DNA sample was extracted with phenol and precipitated twice with ethanol. Fragment recoveries vary between 50 and 75%.

Ligation. Fragments are inserted into the *Hpa*I site by blunt-end ligation[14] using T4 DNA ligase (P-L Biochemicals). It is important that reaction volumes be kept small (~25 μl) in order to increase the efficiency of fragment insertion. The ratio of fragment insert to pKC30 vector used in the ligation reaction is ~2:1 (e.g., 0.4 pmol fragment: 0.2 pmol pKC30). The ligation reaction is carried out for 14 hr at 12–15°, a somewhat lower temperature than is usually recommended. This is done to increase the efficiency of blunt-end ligation at A·T rich restriction sites (e.g., *Hpa*I). The ligation reaction is stopped by heating at 70° for 5 min. If the inserted fragment neither recreates nor contains an *Hpa*I restriction site, *Hpa*I (5 units in 50 μl of *Hpa*I restriction buffer) is added to the ligation mixture to

[11] Y. S. Ho, M. Mahony, and M. Rosenberg, in preparation.

[12] D. Wulff, M. Mahony, A Shatzman, and M. Rosenberg, in preparation.

[13] R. W. Davis, D. Botstein, and J. R. Roth, *in* "Advanced Bacterial Genetics." Cold Spring Harbor Laboratory, Cold Spring Harbor, New York, 1981.

[14] A. Ullrich, J. Shine, J. Chirguin, R. Pictet, E. Tischer, W. Rutter, and H. Goodman, *Science* **196,** 1313 (1977).

recut those pKC30 molecules that rejoined without an insert. The reaction mixture is then phenol extracted, ethanol precipitated, dried, and redissolved in 50 μl of H_2O.

Transformation and Clone Analysis. The ligated DNA (25 μl) is then used to transform a λ lysogen carrying a wild-type repressor gene (e.g., *E. coli* N100λ*cI*$^+$). Ampicillin-resistant recombinants are obtained and screened by size and restriction analysis for the presence and orientation of the insert. This is done by preparing rapid-plasmid DNA isolates[13] of individual clones and examining this DNA on 1% agarose gels using appropriate size markers. Since blunt-end ligation can result in the fragment being inserted in either orientation, it is important to select an appropriate restriction analysis to distinguish the different orientations.

Expression of the Gene Product. In order to express the cloned gene, the pKC30 derivative (e.g., pKC30cII) is first transformed into a λ lysogen carrying a temperature-sensitive mutation in its repressor gene. The transformed cells are grown overnight in LB broth containing ampicillin (50 μg/ml) at 32°; 2 ml of this cell culture are then inoculated into 200 ml of LB containing ampicillin (50 μg/ml). The culture is incubated at 32° until A_{650} = 1. At this time, an equal volume of LB, prewarmed to 65° is added with swirling to elevate the temperature rapidly to 42°, and the culture is incubated at 42° for another 60–90 min.[15,16] The cells are harvested

[15] The temperature induction also has been carried out in 50–100-liter volumes using a standard fermentor. Raising the temperature of the culture from 32° to 42° requires approximately 3 min.

[16] The time period of induction resulting in the best yield varies (45–120 min) and appears to depend upon the nature of the gene product, the host cell, and type of phage lysogen. For each cloned gene, expression should be monitored and compared in a number of lysogenic hosts.

FIG. 2. An SDS–polyacrylamide gel analysis of proteins made in a λ*c1857* lysogen carrying pKC30cII before and after heat induction. (A) Autoradiogram of pulse-labeled proteins. Cells were grown at 32° in modified minimal media M56 [M. E. Gottesman and M. B. Yarmolinsky, *J. Mol. Biol.* **31,** 487 (1968)] with ampicillin (50 μg/ml) until A_{650} = 0.6. The cells were harvested by centrifugation and resuspended in minimal media M56, supplemented with 0.2% glucose, 0.2 m*M* amino acid mixture lacking methionine, 1 μg of thiamine and 50 μg of ampicillin per milliliter. Aliquots (200 μl) of these cells were pulse-labeled for 1 min with 20 μCi of [^{35}S]methionine (Amersham, >600 Ci/mmol) at 32° and at 42°, 20 min after incubation at these temperatures. After pulse labeling, 30 μl of 10% SDS and 35% 2-mercaptoethanol were added, and the samples were heated at 95° for 1 min, followed by quick freezing in Dry Ice–ethanol bath, and precipitated with trichloroacetic acid (final concentration 10%). The precipitates were washed twice with cold ethanol, resuspended in sample buffer, heated at 95° for 5 min, and subjected to electrophoresis on a 15% polyacrylamide gel. The gels were finally dried and fluorographed [W. M. Bonner and R. A. Laskey, *Eur. J. Biochem.* **46,** 83 (1974)]. (B) Coomassie brilliant blue-stained SDS–polyacrylamide gel analysis of total cellular protein prepared before and after a 40-min heat induction as described previously.[6]

A

B

32°C 42°C

32°C 42°C

O

cII

by centrifugation, washed once with TMG buffer (Tris–$MgSO_4$–gelatin buffer, pH 7.5) to remove the ampicillin, and then immediately frozen. Using this procedure, cII protein (subunit molecular weight 10,500) was produced as 3–5% of total cellular protein and purified to a final yield of 1 mg of homogeneous protein per gram wet weight of cell culture (Fig. 2).[6,10]

The Use of Antipolarity: *N* + *Nut*

When a gene is cloned into pKC30, sequences upstream of the coding region, which are untranslated or carry transcription termination signals, may be inserted as well. The polar effects on transcription caused by these sequences will reduce gene expression. For example, in the construction of pKC30cII a termination site, *tR1*, positioned immediately upstream of *cII* was inserted along with the gene. The presence of *tR1* could have drastically reduced *cII* expression. To circumvent these problems, we use the phage λ antitermination factor, *N*, which funtions on P_L-directed transcription. *N* is provided in single copy from the host lysogen and its expression is induced by temperature concomitant with P_L transcription on the vector. The regulatory site required for *N* utilization (*Nut* site) is provided on all pKC30 derivatives.

The importance of *N* was first demonstrated with the vector pKC30cII. *cII* production was found to be ~8 times higher in lysogens that provided *N* (N^+) as opposed to those that did not (N^-).[17] In more recent experiments,[3] it was shown that the single-copy lysogen provides sufficient *N* to antiterminate completely all transcription through a terminator on a multicopy plasmid. Moreover, because *N* generally relieves all transcriptional polarity, it can even increase expression of genes cloned into pKC30 that do not have specific terminators preceding them. For example, the *E. coli* galactokinase gene (*galK*) was cloned into the *Hpa*I site of pKC30. Expression was monitored before and after induction of both an N^+ and N^- lysogen. The results (Fig. 3) indicate that, after induction, 3–4 times more *galK* is produced in the N^+ cells than in the N^- cells. No *galK* is detected in either background before induction.

[17] H. Shimatake and M. Rosenberg, unpublished data.

FIG. 3. A Coomassie Brilliant Blue-stained SDS–polyacrylamide gel analysis of total cellular protein obtained from an N^+ and an N^- defective lysogen carrying the plasmid pKC30galK. Protein samples were prepared from cells before (at 32°) and 60 min after thermal induction (at 42°) and analyzed as described in Fig. 2. The position of the galactokinase gene product is indicated by an arrow.

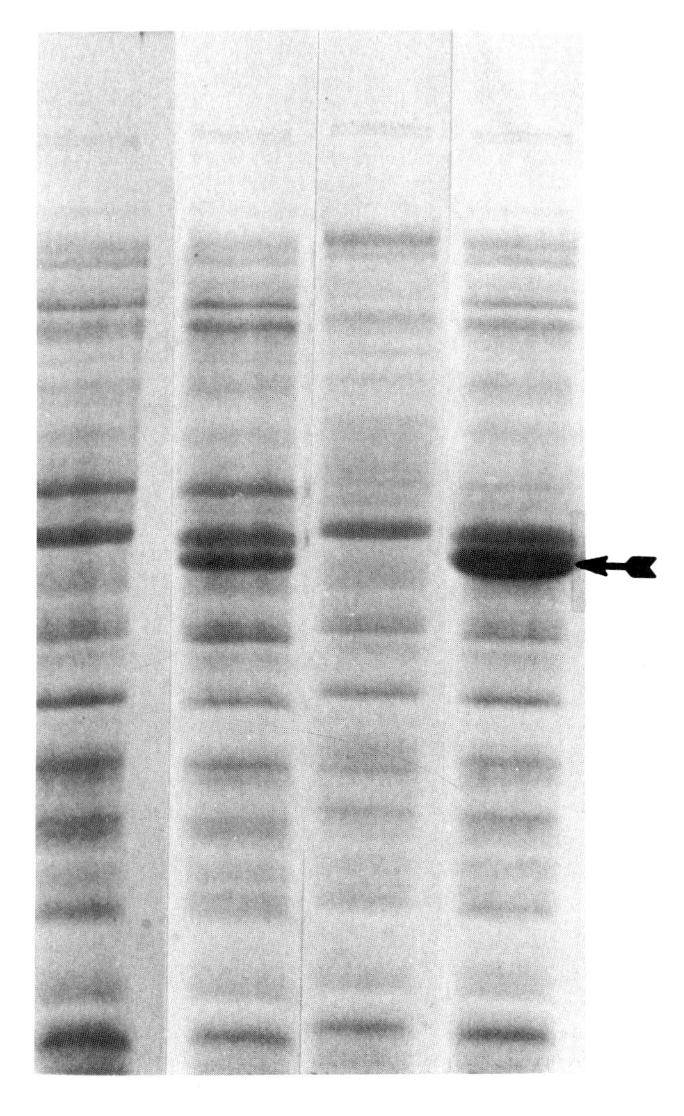
N⁻
N⁺
32° 42° 32° 42°

The effect of *N* on *galK* expression can be attributed primarily to its ability to relieve polarity occurring before or within the *galK* gene. However, it is also possible that *N* may be directly affecting the translation of genes that come under its transcriptional control. Whatever the case, the effect is significant (*galK* is produced as >20% of total cellular protein in the N^+ condition) and eliminates the necessity for manipulating sequences upstream of the gene that might otherwise interfere with its expression.

Adapting pKC30 for the Expression of Eukaryotic Genes

The plasmid constructed for the expression of genes that do not normally carry regulatory signals for their translation in bacteria is shown in

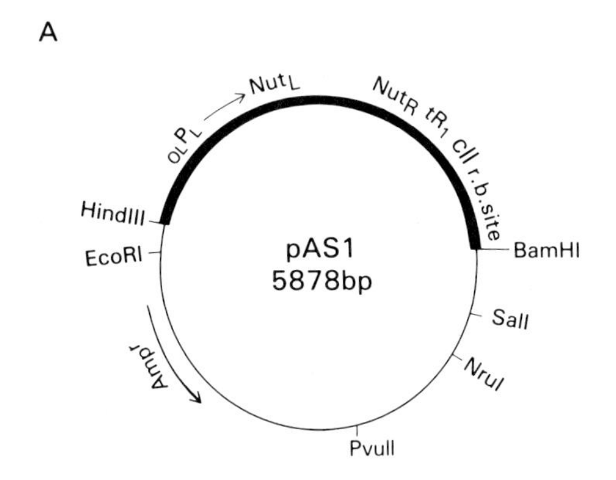

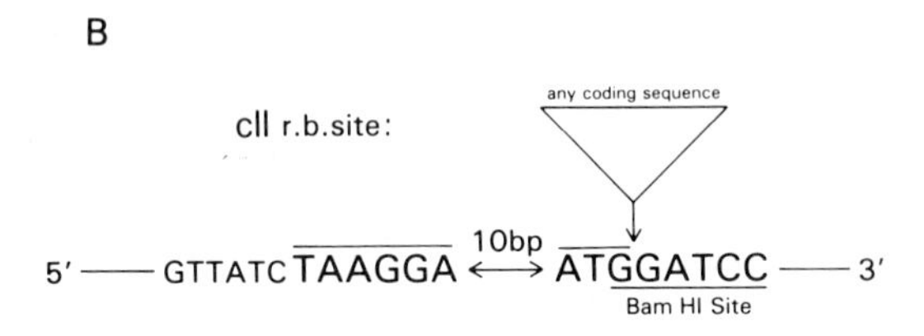

FIG. 4. (A) Schematic diagram of plasmid pAS_1. This vector is a derivative of pKC30 and is made of phage λ sequences (thickened line) inserted between the *Hin*dIII and *Bam*HI restriction sites in pBR322. The region of pKC30 between the *Hpa*I (36060) and *Bam*HI (35301) site has been deleted, and a portion of the fragment shown in Fig. 1B, extending from the *Hae*III site (38981) to and including the ribosome binding site (r.b. site) and ATG initiation codon of *c*II, has been inserted. See text for other details. (B) The region of pAS_1 that surrounds the *cII* translational regulatory information. The ribosome binding site and ATG initiation codon (overscored) of *cII* are indicated, as is the unique *Bam*HI site (underscored), which provides access to this regulatory information. DNA fragments are inserted into this site as detailed in the text.

Fig. 4A. This vector, pAS_1, is essentially identical to pKC30cII except that all λ sequences downstream of the *cII* initiation codon have been deleted. The *Bam*HI site of pBR322 is now fused directly to the *cII* ATG (Fig. 4B). This fusion retains the *Bam*HI site and positions one side of the staggered cut immediately adjacent to the ATG codon permitting ready access to the *cII* translational regulatory information. Eukaryotic and/or synthetic genes, can be adapted and fused to this translation initiation signal. It is most important that all fusions between the gene coding sequence and the *cII* initiation codon maintain the correct translation reading frame. Below, various procedures are described for inserting genes into the pAS_1 vector. Note that all cloning experiments with pAS_1, like those for pKC30, are carried out in a cI^+ lysogen in order to maximize stability of the vector. Expression of the cloned gene takes place in the cI^{ts} lysogen using procedures identical to those described above for pKC30.

Cloning and Expression of Genes in pAS_1

Direct Insertion at the BamHI Site. The only genes that can be fused directly to the *cII* initiation codon are those that contain a *Bam*HI, *Bgl*II, *Sau*3A, or *Bcl*I restriction site[18] at or near their own initiation codon. The necessary restriction site may occur naturally within the gene or be engineered into the gene by recombinant or synthetic techniques. Standard procedures are used for the cloning and for clone analysis and expression.

Two genes have been cloned and expressed in pAS_1 using this technique, the β-galactosidase gene (*lacZ*) of *E. coli* and the metallothionein II gene from monkey. The *lacZ* gene was engineered to contain a unique *Bam*HI site near its 5′ end,[19] whereas the metallothionein gene naturally contained a *Bam*HI site at its second amino acid codon.[20] In both cases direct *Bam*HI ligation of the gene into pAS_1 appropriately positioned the coding sequence in frame with the *cII* ATG codon. Expression of both genes was controlled entirely by transcriptional and translational signals provided by pAS_1. As shown in Fig. 5A, the pAS_1*lacZ* construction results in high level expression of β-galactosidase. Similar results were obtained with the monkey metallothionein gene (not shown).

Adapting pAS_1 for Other Genes. Most genes do not contain the restriction information necessary for their direct insertion into the *Bam*HI site of pAS_1. Thus, it was necessary to provide greater versatility for inserting

[18] These sites all share the common four-base 5′-overhanging end pGATC...

[19] This construction was made and kindly provided by M. Casadaban, M. Casadaban, J. Chou, and S. Cohen, *J. Bacteriol.* **143,** 971 (1980).

[20] Kindly provided by D. Hamer.

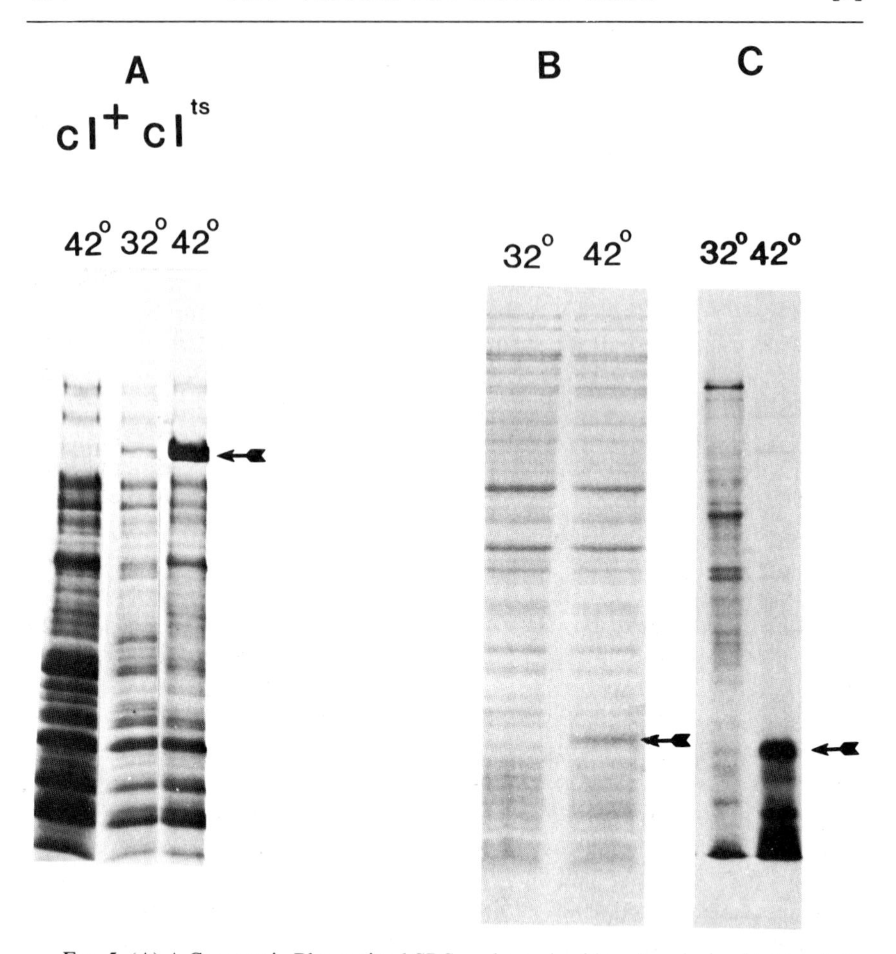

FIG. 5. (A) A Coomassie Blue stained SDS–polyacrylamide gel analysis of total cellular protein obtained from a cI^+ and a cI^{ts} host lysogen carrying the plasmid $pAS_1\beta gal$. Temperature inductions, sample preparations, and gel electrophoresis are the same as described in Figs. 2 and 3. The position of the β-galactosidase gene product is indicated (arrow). (B) Analysis (as in A) of total cellular protein obtained from a cI^{ts} lysogen carrying the plasmid pAS_1t. The position of the SV40 small t antigen is indicated (arrow). (C) An autoradiogram of an SDS–polyacrylamide gel analysis of ^{35}S pulse-labeled proteins synthesized in a cI^{ts} lysogen carrying pAS_1t before and 60 min after heat induction. All procedures are similar to those described in Fig. 2A.

DNA fragments into the vector. This was accomplished by converting the *Bam*HI site of pAS_1 into a blunt-ended cloning site. The four-base 5′-overhanging end of the *Bam*HI cleavage site can be removed using mung bean nuclease, thereby creating a blunt-end cloning site immediately adjacent to the *cII* initiation codon. Mung bean nuclease is used because this enzyme has a high exonucleolytic specificity for single-stranded DNA. We have found this enzyme to be reproducibly better than S1 nuclease for selectively removing the four-base overhang sequence. The following conditions are used for mung bean nuclease removal of the four base overhang generated by *Bam*HI cleavage of pAS_1.

pAS_1 DNA (1.5 pmol; 5 μg) is linearized by *Bam*HI restriction and then treated with mung bean nuclease (25–30 units, P-L Biochemicals) in 50 μl of 30 m*M* NaOAc (pH 4.6), 250 m*M* NaCl, 1 m*M* $ZnCl_2$, 5% glycerol for 30 min at 30°. The reaction is stopped by the addition of SDS (to 0.2%), extracted with phenol, and precipitated twice with ethanol. Approximately 50% of those molecules that have lost their *Bam*HI site are blunt-ended precisely to the correct site.

Any gene containing any restriction site properly positioned at or near its 5′ end can now be inserted into this vehicle. Blunt-ended fragments can be inserted directly, whereas other restriction fragments must first be made blunt-ended. This is accomplished by either removing the 5′- and 3′-overhanging ends with mung bean nuclease (as above) or "filling-in" the 5′-overhanging ends with DNA polymerase.[21] The fill-in reaction is carried out as follows. The DNA (0.3 pmol) is dissolved in 100 μl of 6 m*M* Tris-HCl (pH 7.4), 6 m*M* $MgCl_2$, 60 m*M* NaCl containing 20 μ*M* of each of the four deoxynucleotide triphosphates. DNA polymerase (1 unit; large fragment after Klenow, Boehringer-Mannheim) is added and the reaction incubated at 15° for 3 hr. The reaction is stopped by heat inactivating the polymerase at 65° for 5 min. The reaction mix is extracted with phenol, and the DNA is recovered by ethanol precipitation.

This procedure was used to adapt, insert, and express the metallothionein I gene of mouse in the pAS_1 system. The mouse gene contains a unique *Ava*II restriction site at its second codon (5′ ATG GAC CCC 3′). Cleavage with *Ava*II, followed by fill-in of the three base 5′-overhanging end with DNA polymerase creates a blunt-end before the first base pair of the second codon. This blunt-end fragment was inserted into the mung bean nuclease-treated pAS_1 vector. The resulting fusion places the second codon of metallothionein in-frame with the initiation codon of *cII*. This construction results in high levels of synthesis of the authentic metallothionein gene product in *E. coli*.

Adapting Any Gene for Expression in pAS_1. The procedures described above still limit the use of pAS_1 to those genes that contain appropriate restriction information near their 5′ termini. In order to make the pAS_1 system generally applicable to the expression of any gene, a procedure was developed that allows precise placement of a new restriction site at the second codon (or any other codon) of any gene. Creation of this site

[21] F. Rougeon, *Nucleic Acids Res.* **2**, 2365 (1975).

permits fusion of the gene in-frame to the *cII* initiation codon as pAS_1. The general procedure is outlined below.[22]

1. The gene of interest is cloned into a plasmid vector such that the vector can be opened uniquely at a restriction site just upstream of the gene (e.g., RS1, Fig. 6A).

2. The plasmid, after cleavage at RS1 is digested with the double-stranded exonuclease *Bal*31.[23] Conditions are selected which ensure digestion into the region surrounding the second codon of the gene. For example, a typical reaction will contain 100 μg of plasmid DNA dissolved in 500 μl of 20 m*M* Tris-HCl (pH 8.0), 12 m*M* $CaCl_2$, 12 m*M* $MgCl_2$, 200 m*M* NaCl, 1 m*M* EDTA, and *Bal*31 exonuclease (2–5 units, Bethesda Research Laboratories). The reaction is carried out at 29° and the time of incubation varies depending upon the number of base pairs to be removed. Although the rate of degradation depends upon the particular

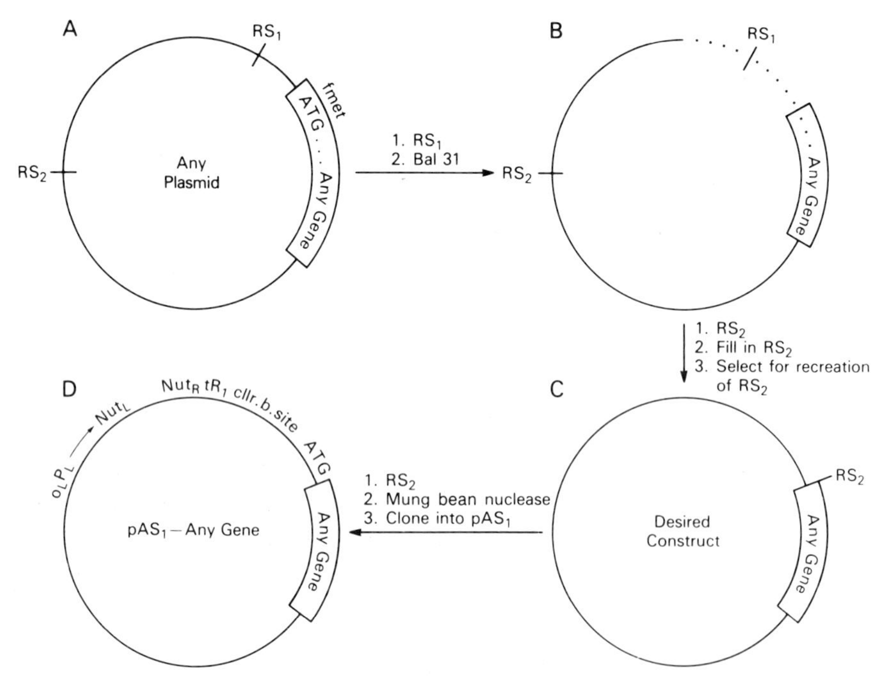

FIG. 6. Schematic diagram of the procedure used to adapt any gene for precise insertion into the pAS_1 expression system. See text for details.

[22] A similar procedure has been used to introduce a *Sal*I site at the initiation codon of the bacteriophage T7 gene *1.1;* N. Panayotatos and K. Truong, *Nucleic Acids Res.* **9,** 5679 (1981).

[23] H. Gray, D. Ostrander, J. Hodnett, R. Legerski, and D. Robberson, *Nucleic Acids Res.* **2,** 1459 (1975).

DNA sequence being digested, a time estimate can be calculated according to Gray *et al.*[23] In order to monitor more precisely the extent of digestion, aliquots (~20 μg) of the reaction are removed at various times and subjected to detailed restriction analysis. In each case the reaction is stopped by addition of EDTA (to 20 m*M*) and phenol. Two successive phenol extractions are carried out, and the DNA is recovered by ethanol precipitation.

3. After digestion by *Bal*31 to the proper extent, the DNA sample is then restricted at a second unique site (RS2, Fig. 6B), positioned well upstream of RS1. This second site should be far enough upstream from RS1 so as not to have been affected by the *Bal*31 digestion. Most important, RS2 should have a six-base pair recognition sequence that gives rise to a four-base 5′-overhanging end and in addition, have as its sixth base pair the same base as the first bp of the second codon of the gene to be expressed (e.g., *Eco*RI, GAATT$\underline{\text{C}}$ or *Bam*HI, GGATC$\underline{\text{C}}$ for codons starting with $\underline{\text{C}}$XX; *Bgl*II, AGATC$\underline{\text{T}}$ or *Hind*III, AAGCT$\underline{\text{T}}$ for codons starting with $\underline{\text{T}}$XX; *Xho*I, CTCGA$\underline{\text{G}}$ or *Xma*I, CCCGG$\underline{\text{G}}$ for codons starting with $\underline{\text{G}}$XX; and *Xba*I, TCTAG$\underline{\text{A}}$ or *Bcl*I, TGATC$\underline{\text{A}}$ for codons starting with $\underline{\text{A}}$XX).

4. The four-base 5′-overhanging end of RS2 is filled-in with DNA polymerae (to recreate five-sixths of the RS2 site) and then blunt-end ligated to the heterogeneous ends created by *Bal*31 digestion (using procedures detailed above).

5. Recombinant plasmids are selected and screened by restriction for those that have recreated RS2 (Fig. 6C). Only those molecules that end in the appropriate base pair will re-form the RS2 site.

6. Those clones containing RS2 are examined by restriction analysis and/or direct sequence analysis. The correct construction has recreated the RS2 site by fusing it to the first base of the second codon of the gene.

7. Restriction of this new construction at RS2, followed by treatment with mung bean nuclease (as above), provides a blunt-ended cloning site adjacent to the first base of the second codon. This blunt end can be fused directly in-frame to the *cII* initiation codon, similarly made blunt-ended by mung bean treatment of the *Bam*HI restricted pAS_1 vector (Fig. 6D).

The above procedure has been used to adapt several genes for insertion and expression into pAS_1. For example, the small t antigen gene of SV40 does not contain an appropriate restriction site at its 5′ end. Using *Bal*31 digestion from an upstream site, the first base pair of the second codon of the small *t* gene ($\overline{\text{ATG}}$ $\overline{\text{GAT}}$···) was fused to an upstream, filled-in *Ava*I restriction site (···$\underline{\text{CCCGA}}$).[24] The fusion, (···$\underline{\text{CCCGAGAT}}$···) recreates the *Ava*I site precisely at the second codon of the small *t* gene.

[24] This construction was made and kindly provided by C. Queen.

Restriction of this vector with *Ava*I followed by mung bean nuclease digestion produces a blunt end that was fused in-frame to the blunt-ended *Bam*HI site of pAS_1. The resulting vector, pAS_1t, expresses authentic SV40 small t antigen from phage regulatory signals. After only a 60-min induction period, small t antigen represents some 10% of the total cellular protein (Fig. 5B). ^{35}S-pulse labeling experiments indicate that small t is the major product synthesized in these bacteria (Fig. 5C). Apparently, the pKC30-pAS_1 vector system offers the potential for efficiently expressing essentially any gene in *E. coli*.

Acknowledgments

We thank Gail Taff for typing and editing the manuscript.

[7] Amplification of DNA Repair Genes Using Plasmid pKC30

By GEORGE H. YOAKUM

A primary goal of many recombinant DNA research projects includes the construction of plasmids that increase the level of gene product for specific genes or pathways. This is usually motivated by the preponderance of biologically important genes expressed at levels that are low enough to make it difficult or impossible to isolate and characterize the gene product. Since product amplification is a generalized goal with special problems for each gene and pathway, a number of specialized vectors have been developed for application to this problem.[1-14] The features of

[1] K. Backman, M. Ptashne, and W. Gilbert, *Proc. Natl. Acad. Sci. U. S. A.* **73,** 4174 (1976).
[2] K. Backman and M. Ptashne, *Cell* **13,** 65 (1978).
[3] J. Hedgpeth, M. Ballivet, and H. Eisen, *Mol. Gen. Genet.* **163,** 197 (1978).
[4] T. M. Roberts, R. Kachich, and M. Ptashne, *Proc. Natl. Acad. Sci. U. S. A.* **76,** 760 (1979).
[5] L. Guarente, G. Lauer, T. M. Roberts, and M. Ptashne, *Cell* **20,** 543 (1980).
[6] T. Taniguchi, L. Guarente, T. M. Roberts, D. Kimelan, J. Donham, and M. Ptashne, *Proc. Natl. Acad. Sci. U. S. A.* **77,** 5230 (1980).

ISBN 0-12-182001-7

gene amplification vehicles include the following: (*a*) a highly efficient promoter from which transcription can be regulated by chemical or thermal sigals; (*b*) a downstream restriction site(s) for insertion of gene segments to be amplified; and (*c*) an antibiotic resistance marker and appropriate host strain(s) for selection of recombinant plasmids and eventual amplification of successful constructs. In addition to this type of amplification vector, two plasmid systems[12,13] carrying a regulable promoter for high-level transcription have been combined with the ability to deregulate plasmid replication and increase the number of plasmid copies during the amplification process.

Simply constructing a plasmid that successfully fuses the gene to be amplified to a plasmid carrying a strong promoter does not ensure high yields of gene product.[15] When choosing a gene amplification system for a project, some consideration should be given to the known genetic and physiological properties that regulate expression of the native gene and describe the biological response expected when the gene is expressed at high levels. A system that utilizes a regulable promoter offers the possibility to amplify genes that may be lethal or damaging to the cell during the amplification procedure. Although most currently available gene amplification systems offer this feature, other factors that must be considered when selecting a vector include the construction of appropriate recipient strains to permit rapid isolation of recombinant DNA plasmids potentially capable of amplifying the gene insert. Consideration should also be given to the method used to identify recombinant plasmid isolates capable of optimal amplification for each particular gene to be amplified.

In this chapter a method is presented that was devised to permit construction of *uvr*/pKC30 hybrid plasmids[16] for optimal amplification of *Escherichia coli* DNA repair genes on this vector. This vector was selected

[7] H. U. Bernard, E. Remant, V. M. Hershfield, H. K. Das, D. R. Helinski, C. Yanofsky, and N. Franklin, *Gene* **5,** 59 (1979).

[8] D. M. Williams, R. G. Schoner, E. J. Duvall, L. H. Preis, and P. S. Lovett, *Gene* **16,** 199 (1980).

[9] R. A. Hallewell and S. Emtage, *Gene* **9,** 27 (1980).

[10] J. Gilbert, G. Khoury, A. K. Seth, and E. Jay, *Proc. Natl. Acad. Sci. U. S. A.* **78,** 5543 (1981).

[11] R. M. Rao and S. G. Rogers, *Gene* **3,** 247 (1978).

[12] M. Bittner and D. Vapnek, *Gene* **15,** 319 (1981).

[13] J. J. Shinsky, B. E. Uhlin, P. Gustafsson, and S. N. Cohen, *Gene* **16,** 275 (1981).

[14] H. Shimatake and M. Rosenberg, *Nature* (*London*) **292,** 128 (1981).

[15] T. M. Roberts and G. D. Lauer, this series, Vol. 68, p. 473.

[16] G. H. Yoakum, A. T. Yeung, W. B. Mattes, and L. Grossman, *Proc. Natl. Acad. Sci. U. S. A.* **79,** 1766 (1982).

because (*a*) it was readily available for use; (*b*) it was relatively easy to construct recipient strains carrying a lysogenized lambda genome capable of regulating the P_L promoter of the vector (pKC30) thereby permitting genetic selection of potentially interesting *uvr*/pKC30 hybrid plasmids; and (*c*) the system was easily adaptable to use in a modified "maxicell"[17] labeling system that permits rapid screening of hybrid plasmids for regulable production of amplified gene product.[16] The general features of this method may be useful for construction and screening pKC30 hybrid plasmids that amplify other genes.

Methods

Construction of Lambda Lysogens

A series of *recA*, *uvr* strains were selected for construction of lambda lysogens with a mutant thermal inducible *c*I repressor (λ*c*I857*sam*7) or wild-type *c*I repressor (λKan) to permit *uvr*/pKC30 hybrid plasmid transformation into strains that could be readily screened for the presence of pKC30 plasmids carrying functional *uvr*-gene segments. Lysogens were prepared as follows: Overnight cultures of *recA uvr* strains grown in tryptone broth with 0.2% maltose[18] were diluted 1:20 into fresh medium. Cultures were incubated at 37° until they reached an approximate density of 2×10^8 cells/ml. Approximately 0.2 ml of each cell culture was placed in sterile tubes, and cultures were incubated at the appropriate temperature for obtaining *c*I-repressed lysogens (32° for λ*c*I857*sam*7, and 37° for λKan). After adding approximately 10^5 phage to each culture, the infected cells were allowed to continue incubation overnight without shaking. The following day, cultures infected with cI857sam7 were diluted to approximately 10^7 cells/ml and infected with 10^8 PFU λcI90. Mutation of the *c*I gene of λcI90 assures that the only potential outcome of λcI90 infection is lysis, since this mutant is incapable of lysogeny. The culture was incubated for 2–4 hr before dilution and plating at 32°. Lysogens carrying λ*c*I857*sam*7 were identified by their inability to grow when plated at 42° and resistance to λcI90 in streak tests on plates. Lysogens carrying λKan were identified by plating cells on Luria agar plates (L plates) containing 20 μg of kanamycin per milliliter and an immunity to λ *c*I90. (Phage stocks were kind gifts from Dr. R. McMacken, JHU.)

[17] A. Sancar, A. Hack, and W. D. Rupp, *J. Bacteriol.* **137,** 692 (1979).
[18] N. Sternberg and R. Weisberg, *J. Mol. Biol.* **117,** 733 (1977).

Construction of uvr/pKC30 Hybrid Plasmids

DNA fragments carrying the *uvr* genes were prepared for fusion with the P_L promoter of pKC30 by endonuclease restriction, digestion with mixed exonuclease *Bal*31, and purification before being ligated at high DNA concentration with *Hpa*I-digested pKC30 DNA (Fig. 1). The intended objectives during hybrid plasmid construction include (*a*) "tailoring" of DNA fragments to remove elements that might interfere with ex-

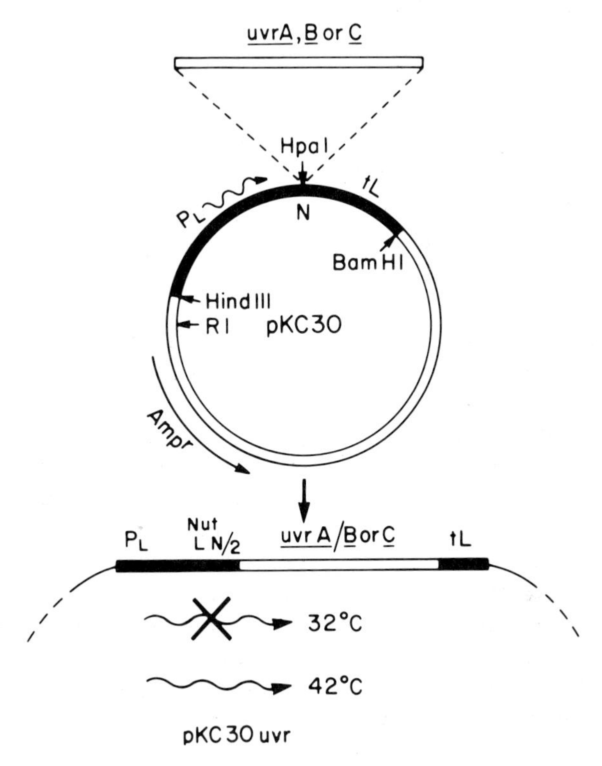

FIG. 1. The construction rationale for amplification of *uvr* proteins is very similar to the rationale applied to the utilization of pKC30,[14] and other P_L-promoter vectors,[7] for amplified expression of inserted genes. The exact hybrid plasmid structure that is most appropriate for amplification of the inserted gene cannot be physically predicted. To express optimally a gene from a novel promoter it is most practical to prepare mixtures of hybrid plasmids with random-size gene fragments after *Bal*31 nuclease digestion.[15] By inserting random-size fragments carrying the structural gene (*uvr* fragments) at the *Hpa*I site of pKC30, the following criteria are met: (*a*) The gene fragment is placed downstream from the P_L-promoter; (*b*) the *Bal*31 nuclease digestion of *uvr* fragments will result in a population of molecules with various distances between the start site for P_L transcription and *uvr*-gene product translation.

pression of *uvr* genes from the P_L promoter of pKC30/*uvr*-gene hybrid plasmids by producing random-length fragments during *Bal*31 nuclease digestion; (*b*) the production by mixed nuclease digestion also of blunt ends capable of ligation to *Hpa*I blunt ends in the downstream site selected for fusion of genes to be amplified on pKC30.[14] Therefore, it is useful to know the location of the structural gene to be amplified on the restriction fragment being fused to pKC30 to provide estimates of the range of *Bal*31 nuclease[19,20] digestion products most likely to result in gene-size fragments for ligation to the vector (Fig. 1).

Plasmid DNA carrying *uvr* genes was usually digested with two restriction endonucleases: (*a*) one enzyme selected to inactivate the replicative capacity of the plasmid providing the *uvr*-gene fragment; and (*b*) a second enzyme(s) selected to release the *uvr* gene on a fragment that is nearest "gene size." Plasmid DNA was prepared from overnight L-broth cultures amplified by chloramphenicol at 2×10^8 cells/ml. The cultures were lysed, and plasmid DNA was banded on CsCl gradients as described.[16] Restriction digests of 100–200 μg of *uvr*-plasmid DNA were incubated at 37° for 2 hr in reaction mixtures with 1.5-fold excess endonuclease activity based on 1-hr units. Reactions were terminated by addition of SDS (sodium lauryl sulfate) to 0.2% final concentration, and 15 min of incubation at 70°. Restriction digests were then placed at 45°, and Pronase CB (Sigma) was added to a final concentration of 50 μg/ml. Pronase reactions were incubated for 45 min, and samples were placed on ice.

Restriction endonuclease-digested *uvr*-plasmid DNA fragments were prepared for *Bal*31 reactions by phenol extraction and ethanol precipitation. Phenol used for the preparation of DNA was redistilled in the laboratory, neutralized by addition of 2 *M* Tris base until the aqueous phase was pH 6.5–7.0, bubbled with argon, and stored at −20°. Restriction digests were diluted by the addition of an equal volume of TE buffer (10 m*M* Tris, pH 8.0–1 m*M* EDTA, ethylenediaminetetraacetic acid) before extraction with two volumes of phenol. Phases were separated by centrifugation at 5000 rpm in a Sorvall SS34 rotor at 4°, and the extractions were carried out twice. The phenol was back-extracted once with 0.25 volume of TE buffer to reduce loss of DNA during the extraction procedure.

The pooled aqueous phase was brought to a final concentration of 0.3 *M* LiCl by addition of a 10 *M* filtered (0.22 μm) and autoclaved stock solution. Purified DNA was precipitated by addition of 2 volumes of −20° 95% ethanol as follows: (*a*) incubation of DNA–ethanol solutions for

[19] P. P. Lau and H. B. Gray, *Nucleic Acids Res.* **6,** 331 (1979).

[20] H. B. Gray, D. A. Ostrander, J. L. Hodwett, R. J. Legerski, and D. L. Robberson, *Nucleic Acids Res.* **2,** (1975).

45 min in a Dry Ice ethanol bath; or (*b*) by placing DNA–ethanol solutions at −20° overnight. Precipitated DNA was then collected by centrifugation at 10,000 rpm for 30 min in a Sorvall SS34 rotor at 4°. The DNA pellets were washed by addition of an equal volume of 70% ethanol to remove residual phenol and salt that remain in the ethanol clinging to the centrifuge tube. The DNA pellet that remains after 5 min centrifugation as indicated above, is covered with a Kimwipe and vacuum dried to remove residual ethanol.

Purified, restriction product is now ready for hydrolysis by reaction with mixed nuclease *Bal*31.[19,20] The DNA pellet is dissolved in a minimal amount (100–300 μl) of *Bal*31 nuclease reaction buffer (BN buffer: 600 m*M* NaCl, 20 m*M* Tris, pH 8.1, 12 m*M* $CaCl_2$, 12 m*M* $MgCl_2$, 1 m*M* EDTA). An aliquot of the solution is used to estimate the DNA concentration by optical density at 260 nm, and BN buffer is added to yield no less than 250 μg of DNA per milliliter in *Bal*31 nuclease reaction buffer. Nuclease reactions were initiated by the addition of 5-μl aliquots of diluted *Bal*31 nuclease (dilution buffer: 50% glycerol, 100 m*M* NaCl, 200 m*M* Tris, pH 8.1, vacuum degassed, and stored at −20°) to reaction mixtures that were prewarmed to 30°. The specific dilution of *Bal*31 nuclease added to each reaction was determined as follows: In the DNA concentration ranges used for reaction (250–650 μg/ml) the enzyme preparation was capable of removing approximately 1 μg of DNA per minute per unit of activity. Therefore, addition of 2.5 units of *Bal*31 nuclease per milliliter to reaction mixes containing 250 μg of DNA per milliliter results in a hydrolysis rate that removes approximately 1% of the fragment length per minute of incubation. If it is necessary to trim the fragment length by more than 30%, additional aliquots of enzyme must be added, since the reaction remains linear for only 30 min. Nuclease reactions were terminated by placing aliquots in 2 volumes of phenol on ice after appropriate incubation times. After phenol extraction and ethanol precipitation of nuclease reaction products (as described above), the DNA samples would be ready for blunt-end ligation to *Hpa*I-digested pKC30. Two sources of *Bal*31 nuclease were used for the experiments described here with no notable difference in experimental performance: (*a*) New England BioLabs (32 Tozer Rd. Beverly, Massachusetts 01915); and (*b*) Bethesda Research Laboratories (BRL; P.O. Box 577 Gaithersburg, Maryland 20760).

In some cases it was desirable to mix the various nuclease reaction products and prepare one ligation mixture with potential inserts representing the entire range of *Bal*31 hydrolysis used in the experiment. If the position of the gene is well known relative to the starting point for *Bal*31 hydrolysis, ligation of each product separately may permit early detection of any problem elements that influence the final performance of hybrid plasmids. The *Bal*31 hydrolysis of DNA used to amplify the *uvrA* gene on pKC30 was adjusted to remove approximately 10–800 nucleotides from the *uvrA* fragment.[16] The reaction was stopped after (*a*) 30 sec (less than 1% hydrolysis); (*b*) 5 min (ca 5% hydrolysis; (*c*) 15 min (ca 15% hydrolysis); and (*d*) 25 min (ca 25% hydrolysis). It is best to use a set of *Bal*31 reaction conditions representing a relatively wide range of products, since variations in DNA concentration and differences in dilution of the enzyme preparation can vary by approximately a factor of 2 in the efficiency of hydrolysis between samples run on different days. Since the *uvrA* gene is regulated by *lexA* repressor,[21,22] we had to consider the possibility that the *lexA* binding site might interfere with P_L-directed transcription of the *uvrA* gene on constructs carrying the *uvrA* gene appropriately positioned for expression from P_L. In addition, the fragment carrying the *uvrA* gene probably included the native promoters for the *uvrA* and *ssb* genes,[16,22] which might also interfere with expression from the P_L promoter of pKC30. Therefore, it was essential to run separate ligations for each extent of *Bal*31 hydrolysis and have a genetic selection to prescreen potentially useful constructs from those that were unable to complement the *uvrA* mutation.

Blunt-end ligation of DNA fragments into the *Hpa*I site of pKC30 requires reaction conditions that include high DNA concentrations due to a very high apparent K_m for blunt-end ligation (ca 80 mg/ml) and of nonlinear reaction kinetics.[23-26] The reactions used to construct the *uvrA*/pKC30 hybrid plasmid were run at DNA concentrations ranging from 300 to 500 μg of *uvr*-DNA per milliliter. After modification by various extents of hydrolysis with *Bal*31 nuclease, the DNA fragments were mixed with approximately 5 ng of *Hpa*I-digested pKC30 DNA. The reaction volume used was 5 μl for each blunt-end ligation. The *Bal*31 nuclease-digested *uvr*-DNA fragments were estimated to be 3000–4000 base pairs in length. Therefore, *Hpa*I-digested pKC30 DNA was present at approximately 100-fold molar ratio less than the *uvr*-DNA fragments. This assured that many of the blunt-end ligation products would include *uvr* fragment inserts, and that few of the recombinant plasmids produced

[21] C. J. Kenyon and G. C. Walker, *Proc. Natl. Acad. Sci. U. S. A.* **77,** 2819 (1980).

[22] A. Sancar, R. B. Wharton, S. Seltzer, B. M. Kacinski, N. D. Clarke, and W. D. Rupp, *J. Mol. Biol.* **148,** 45 (1981).

[23] P. Modrich and I. R. Lehman, *J. Biol. Chem.* **245,** 3626 (1970).

[24] K. V. Dengan and J. H. van de Sande, *Biochemistry* **17,** 723 (1978).

[25] M. Mottes, C. Morandi, S. Cremaschi, and V. Sgaramella, *Nucleic Acids Res.* **4,** 2467 (1977).

[26] A. Sugino, H. M. Goodman, H. L. Heyneker, J. Shine, H. W. Boyer, and N. R. Cozzarelli, *J. Biol. Chem.* **252,** 3987 (1977).

would include multimolecular products. If it is important to convert a high ratio of blunt-end products to recombinant plasmids, the vector must be treated with bacterial alkaline phosphatase (BAP) to remove the 5′-phosphate and prevent recircularization[27] during the ligase step.

The ethanol precipitated DNA pellets from *Bal*31 reactions were dried in small Eppendorf tubes (40 μl), and the reaction products were dissolved in 3 μl of sterile deionized water (DiH_2O). The reaction mixture was completed by adding 1 μl of a 5× stock of ligase reaction buffer (5× L buffer: 100 m*M* Tris, pH 7.5, 50 m*M* MgCl, 50 m*M* dithiothreitol, 3.0 m*M* ATP, stored in one-use aliquots at −70°). The reaction was initiated by adding 1 unit[27] (in 1 μl) of T4 DNA ligase. The exact number of blunt-end ligation activity units was not known for this preparation; however, this is frequently 10- to 50-fold less than the number of units based on ligation of overlapping ends.[24] The ligation reactions were incubated at 22° for approximately 16 hr. The source of ligase for these reactions was Boehringer Mannheim Biochemicals (7941 Castelway Dr., P.O. Box 50816, Indianapolis, Indiana 46250).

Genetic Selection of uvrA/pKC30 Hybrid Plasmids

The benefits provided by designing a genetic selection system for screening transformation products carrying recombinant plasmids constructed *in vitro* include the ability to quickly narrow the search for useful new plasmids carrying the intact *uvr* gene among many plasmids that are capable of replication in the host, and resistant to the appropriate antibiotic. This eliminates the need to physically purify gene fragments that are potentially useful for amplification before ligation to the appropriate vector. By allowing the selection of colonies from the population that carry plasmids with complete gene inserts from those that do not, the percentage of potentially useful plasmids and the probability of rapidly finding the optimal construct for any gene on an amplification vector is increased.

The series of strains used to construct the lambda lysogens for isolation of *uvr*/pKC30 hybrid plasmids is available from the *E. coli* Genetic Stock Center (Department of Human Genetics, Yale University School of Medicine, 333 Ceder St. New Haven, Connecticut 06510). This service provides free strains carrying a wide variety of potentially useful mutations in many economically and academically valuable genes, some of which can be used to construct the necessary lysogens for use in genetic selection of hybrid pKC30 plasmids.

To select the *uvrA*/pKC30 hybrid plasmids a *recA*1, *uvrA*6, λ*c*I857*sam*7

[27] B. Weiss, A. Jacquemin-Sablon, T. R. Live, G. C. Fareed, and C. C. Richardson, *J. Biol. Chem.* **243**, 4543 (1968).

strain[16] (GHY3600) was transformed with recombinant DNA mixtures by a variation of the $CaCl_2$ method, described in detail by S. R. Kushner[28] allowed 2 hr of incubation at 30° to express plasmid genes, and treated with a dose of ultraviolet light (UV) that is approximately 90% lethal for *recA*1, *uvrA*6 double mutants. The UV-irradiated cells were plated on L plates containing 50 μg of ampicillin (Amp) per milliliter and incubated for 48 hr at 30°. Colonies were transfered by toothpick to Microtiter wells containing 0.1 ml of L broth with 50 μg of Amp per milliliter, incubated for 24 hr at 30°, and tested for Amp^R (ampicillin resistance), and UV^R (UV resistance). Those colonies which were both Amp^R and UV^R were considered to be potential sources of *uvr*A/pKC30 hybrid plasmids.

Therefore, AMp^R/UV^R colonies were selected and used to innoculate 2-ml cultures in L broth containing 100 μg of Amp per milliliter. The cultures were grown overnight at 30°, and small amounts of plasmid DNA were isolated by the alkaline lysis method of Birnboim and Doly[29] for comparison with pKC30 on 0.8% agarose gels and subsequent transformation into GHY3600. The reason for secondary transformation is to confirm genetically the Amp^R/UV^R phenotype of the plasmid before more extensive analysis of the physical and gene-amplification properties of hybrid plasmids. During the construction of pGHY5003 *uvrA*/pKC30 (Fig. 2), approximately 400 colonies were initially screened for ampicillin and UV resistance, and of these, 24 or approximately 6% were of the appropriate phenotype for additional characterization. After confirming that these plasmids were *uvrA*/pKC30 hybrid plasmids by genetic test and by determining their physical size on screening gels, several were selected for additional characterization by isolating approximately 100–200 μg of plasmid DNA by the method of Vapnek *et al.*[30] This DNA was used to test for the proper orientation of the *uvrA* gene insert to the P_L promoter of pKC30 by determining the product size after digestion with restriction endonucleases: *Sal*I, and *Bgl*II/*Kpn*I (Fig. 2).[16]

A group of plasmids were identified that contained the *uvr*A gene downstream from the P_L-promoter of pKC30 with the start signal for the *uvr*A-gene in proximal position relative to the P_L promoter. This group of plasmids was considered appropriate for additional testing to determine whether any of the plasmid products isolated by this method were capable of expressing the *uvrA* gene from the P_L promoter.

[28] S. R. Kushner, *in* "Genetic Engineering" (H. W. Boyer and S. Nicosia, eds.), p. 17. Elsevier, Amsterdam, 1978.

[29] H. C. Birnboim and J. Doly, *Nucleic Acids Res.* **7,** 1513 (1979).

[30] D. Vapnek, N. K. Alton, C. L. Basset, and S. R. Kushner, *Proc. Natl. Acad. Sci. U. S. A.* **74,** 3508 (1977).

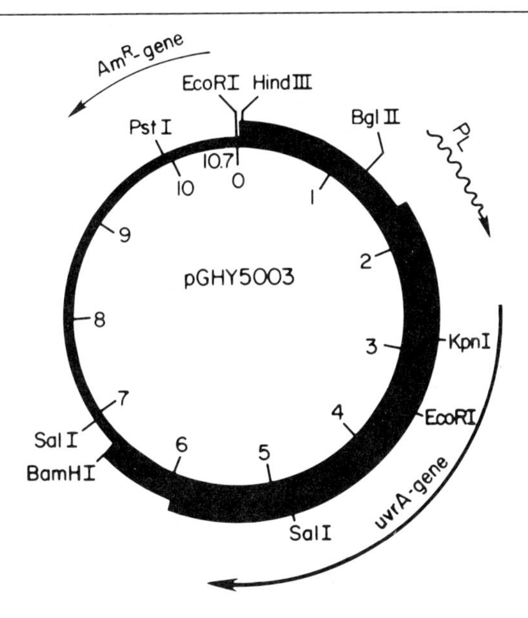

FIG. 2. The structure of hybrid plasmid pGHY5003 consists of pKC30 containing a 4.2 kbp insert with the *uvr*A gene of *Escherichia coli* in transcriptive phase with the P_L promoter.[16] The location and orientation of the *uvr*A gene was deduced from the size of fragments produced by *Bgl*II/*Kpn*I digests, and *Sal*I digests. The segment of the map from 1.7 to 5.9 kbp shows the fragment of *E. coli* DNA blunt-end ligated at the *Hpa*I site of pKC30. The segments from the site of ligation to *Hin*dIII and *Bam*HI are segments of lambda DNA, and the region from 6.7 to 10.7 kbp is the pBR322 portion of pKC30. Mapping pGHY5003 indicates that approximately 80–150 bp have been removed from the P_L-proximal end of the *uvr*A-gene insert, and approximately 250–350 bp were deleted from the P_L-distal end of the *uvrA*-gene insert.[16]

Testing for Inducible Expression of uvrA-Polypeptide Synthesis

The gene product to be amplified from a regulable promotor is often difficult to assay in cell extracts. The degree of this problem varies for each gene and pathway; however, the lack of a reliable quantitative assay for *uvr*-gene products presented an obstacle to rapid screening of *uvrA*/pKC30 hybrid plasmids for P_L-controlled expression of *uvrA* gene-product. The need to screen a number of hybrid plasmids rapidly for gene-product amplification may be partially resolved by comparing cell extracts from induced cultures on SDS–polyacrylamide gels to determine whether the polypeptide band identified as the gene product responds to the induction signal. The interpretation of SDS–polyacrylamide gels after induction of cultures is always difficult, since each particular hybrid plasmid and host strain produces a range of new polypeptide products that may obscure identification. To circumvent these problems, a system for specific labeling of plasmid-encoded gene products[17] was adapted for use in testing hybrid plasmids for inducible expression of genes under P_L control.[16]

The "maxicell" system for specifically labeling plasmid-encoded gene products utilizes the "reckless" DNA-degradation phenotype of *recA* mutants to eliminate the expression of host genes.[17] The *recA* host will convert approximately 85% of chromosomal DNA to acid-soluble hydrolysis products after UV-light treatments that introduce as few as 20–50 pyrimidine dimers per chromosome. The small size relative to the host chromosome, and multiple copies of plasmid molecules per host cell, make plasmid genes much more resistant than host genes to UV treatment in *recA* hosts for post-UV polypeptide radiolabeling.[17] The maxicell procedure includes: (*a*) UV treatment of *recA* host strains carrying the plasmid; (*b*) overnight incubation (12–18 hr) of UV-treated cultures in L broth[17] with D-cycloserine (200 μg/ml) and Geopen (100 μg/ml) to amplify plasmids and degrade the host chromosome; and (*c*) subsequent radiolabeling of plasmid-encoded gene products in media supplemented with [^{35}S]methionine (50 μCi/ml; specific activity 1500 Ci/mmol.)[11] Initial attempts to apply this procedure to compare induction of P_L expression of *uvrA* on hybrid *uvrA*/pKC30 plasmid failed to yield any net incorporation of radiolabel into polypeptide, even though we had successfully used this method to label *uvr* genes on pBR322 vectors in other experiments.[31,32] The possibility that overnight incubation had caused the loss of *c*I repressor (since the chromosomal *c*I gene was hydrolyzed during the post-UV incubation) suggested the testing for inducible expression after shorter time periods of post-UV incubation. Experiments revealed that a 2-hr post-UV incubation of GHY3633 (*recA*1, *uvrA*6, *c*I857*sam*7) carrying hybrid *uvrA*/pKC30 plasmids was optimal for P_L-induced expression of *uvrA* polypeptide for radiolabeling without host gene expression interference.[16]

Plasmid-specific *uvrA* polypeptide synthesis, after induction of GHY3633 strains carrying *uvrA*/pKC30 hybrid plasmids under P_L control, was determined by the following procedure. Overnight cultures of GHY3633 (*recA*1, *uvrA*6, *c*I857*sam*7/pGHY5003) were grown in 25 ml of K medium[18,30] supplemented with 0.1% casamino acids at 30°, and cell pellets were washed once with 10 ml of chilled (4°) M9 buffer.[30] Washed cells were treated with 50 J/m^2 UV light from a 15-W germicidal lamp, resuspended in K medium with 0.1% casamino acids, and incubated at 30° for 90 min. The post-UV culture was chilled for 5 min with shaking to

[31] G. H. Yoakum, S. R. Kushner, and L. Grossman, *Gene* **12**, 243 (1980).

[32] G. H. Yoakum and L. Grossman, *Nature* (*London*) **229**, 171 (1981).

maintain aerobic conditions. The cells were then pelleted by centrifugation at 5000 rpm in a Sorvall SS34 rotor at 4° for 5 min. The cell pellet was resuspended in 10 ml of chilled (4°) K-labeling medium, washed, and pelleted by centrifugation (as above). The washed cells were resuspended in 10.0 ml of K-labeling medium that contains, per milliliter, 2 μg of each of the 20 L-amino acids with the exception of L-methionine (absent), and L-cysteine (0.001 μg/ml). The culture was then incubated at 30° for 30 min in a 125-ml Erlenmeyer flask to complete the post-UV incubation. Induction of expression from the P_L promoter of pKC30 was initiated by moving the flask to a 42° water bath for continued incubation. The rate of *uvrA* polypeptide synthesis was estimated at various times after shifting the culture.[16] To "pulse-label" the induced cultures, a 1.0-ml sample was removed from the flask culture and placed in tubes containing 0.1 ml of K-labeling medium containing 50 μCi of [^{35}S]methionine (Amersham; 1500 μCi/m*M*). The radiolabeling "pulse-incubation" was continued for 45 sec at 42° and terminated by the addition of 0.1 ml of 1.0% L-methionine (unlabeled). Radiolabeled cultures were transferred to ice after the addition of unlabeled L-methionine and stored until the end of the period tested for P_L induction of *uvr*A gene product. The radiolabeled cells taken from cultures incubated at 42° for 0, 15, 30, 45, and 75 min were pelleted by centrifugation at 5000 rpm in a Sorvall SS34 rotor for 15 min at 4°. The K-labeling medium was removed for disposal as radioactive waste, and the cells were washed by resuspension in 2.0 ml of chilled (4°) K medium containing 0.1% casamino acids and 0.1% L-methionine, followed by centrifugation.

Cell lysates were prepared for analysis by SDS–polyacrylamide gel electrophoresis by resuspending radiolabeled cell pellets in 0.2 ml of SDS–gel lysis buffer: 75 m*M* Tris, pH 6.8, 2.0% SDS, 5% 2-mercaptoethanol. The SDS–gel lysis buffer was prepared by adding fresh 2-mercaptoethanol from 14 *M* stock solution to Tris–SDS buffer just before use. The 2-mercaptoethanol stock was bubbled with Argon (or N_2) and stored in a bottle sealed with Parafilm at 4°. Frequently used 2-mercaptoethanol stock solutions were not stored for periods in excess of 2–3 months. Radiolabeled cell pellets were resuspended in lysis buffer by pipeting, and cell lysates were prepared by placing the samples in a boiling water bath for 3 min. An aliquot (50 μl) of each lysate was loaded on 10–20% polyacrylamide gradient–SDS gels.[16,33] The radiolabeled polypeptide products were visualized after gel electrophoresis and fluorography[16,32] (Fig. 3). The radiolabeled samples from peak-level induction points can be used to confirm the identity of the amplified gene product by running a separate

[33] U. Laemmli, *Nature (London)* **227,** 680 (1970).

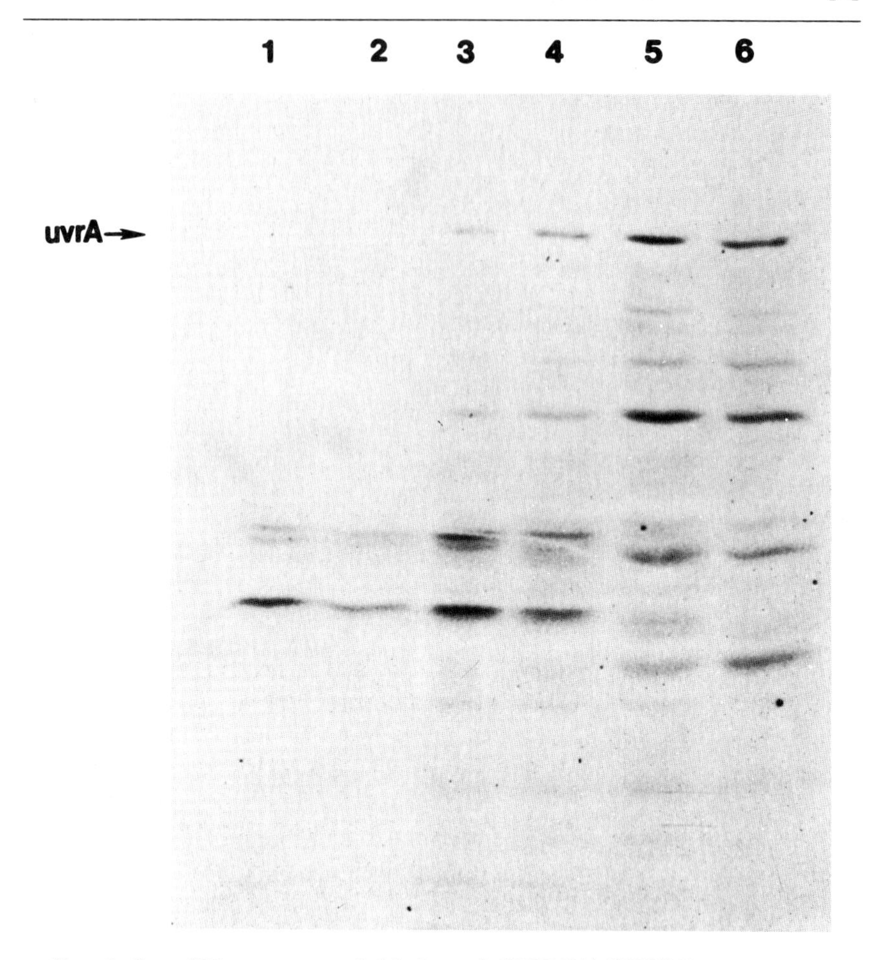

FIG. 3. Post UV-treatment radiolabeling of GHY3633/pGHY5003 (*recA*1, *uvrA*6, *c*I857*sam*7). Induction of *uvrA*-polypeptide synthesis at 42° assayed by pulse-labeling and fluorography of SDS–polyacrylamide gel electrophoresis products from cell lysates.[16] Identification of the *uvrA* band was confirmed by comigration with [^{35}S]methionine-labeled *uvrA*-polypeptide from maxicell labeling[17] experiments with *uvrA*$^{+}$ pGHY3243.[31] Time of "induction-incubation" at 42° was: lane 1, 0 min; lane 2, 15 min; lane 3, 30 min; lane 4, 45 min; lane 5, 60 min; lane 6, 75 min.

gel for comparison to the migration properties of radiolabeled plasmid gene products identified by the maxicell method on pBR322 plasmids.[16,32]

To estimate the gross yield of *uvrA* polypeptide in terms of total protein after thermal induction of hybrid *uvrA*/pKC30 plasmids, pGHY5003 was transformed into a lysogen designed for amplification of large-scale cultures.[14] This strain, which is wild type with respect to *recA*, carries a lysogen with inducible *c*I repressor and mutations affecting the replication, excision, and lethal functions of lambda. This strain (N99) was provided for use in these experiments by M. Rosenberg.[14] The strain produced by transformation with pGHY5003 was grown at 32° to approximately 5×10^8 cells/ml in L broth; cultures were then shifted to 42° for P_L amplification. After 42° incubation periods between 0 and 6 hr, samples were removed for lysis and analysis by SDS–polyacrylamide gel electrophoresis. These gels were stained with Coomassie Blue to obtain estimates of the total yields of *uvrA* polypeptide after various periods of thermal induction (Fig. 4).[16] After selecting several samples from the peak time period for induction of *uvrA*-polypeptide (Fig. 4), the distribution of polypeptide products on SDS–polyacrylamide gels was scanned for densitometric analysis of the relative quantity of the major protein bands (Fig. 5). The area under the peak at the migration position of *uvrA*-polypeptide was compared to the area under the remaining spectrum of Coomassie Blue-stained bands resolved on SDS–polyacrylamide gels.

Results

One of the primary advantages offered by the pKC30 system for amplification of genes is the application of easily designed genetic selection to the problem of hybrid plasmid construction and of identification methods for rapid screening of appropriately constructed plasmids for amplified gene expression even when the gene product may be difficult to assay by direct means. The minimal requirements that must be met in order to utilize pKC30 in amplification projects similar to those described here are that (*a*) the gene to be amplified must have been isolated and mapped on a plasmid that provides at least 100–200 μg of purified DNA (i.e., pBR322) to use for hybrid plasmid construction; (*b*) a series of mutant strains that can be complemented by the gene must be used to construct lambda lysogens capable of regulating P_L expression of the gene for selection of successful hybrid plasmid constructs; (*c*) a *recA* lambda lysogen should be used to provide an appropriate storage strain for hybrid pKC30 plasmids; (*d*) a strain carrying a lambda lysogen capable of thermal regulation of P_L expression but defective in many of the deleterious functions of lambda (N99, from M. Rosenberg) should also be used. Since these conditions can be met for many genes, the pKC30 vector may prove to be useful for application to many gene amplification projects.

Testing the thermal inducibility of hybrid *uvrA*/pKC30 plasmids by screening several potentially useful constructs is best demonstrated by the results in Fig. 4. Adapting the maxicell radiolabeling procedure to screen potentially useful hybrid *uvrA*/pKC30 plasmids for expression of

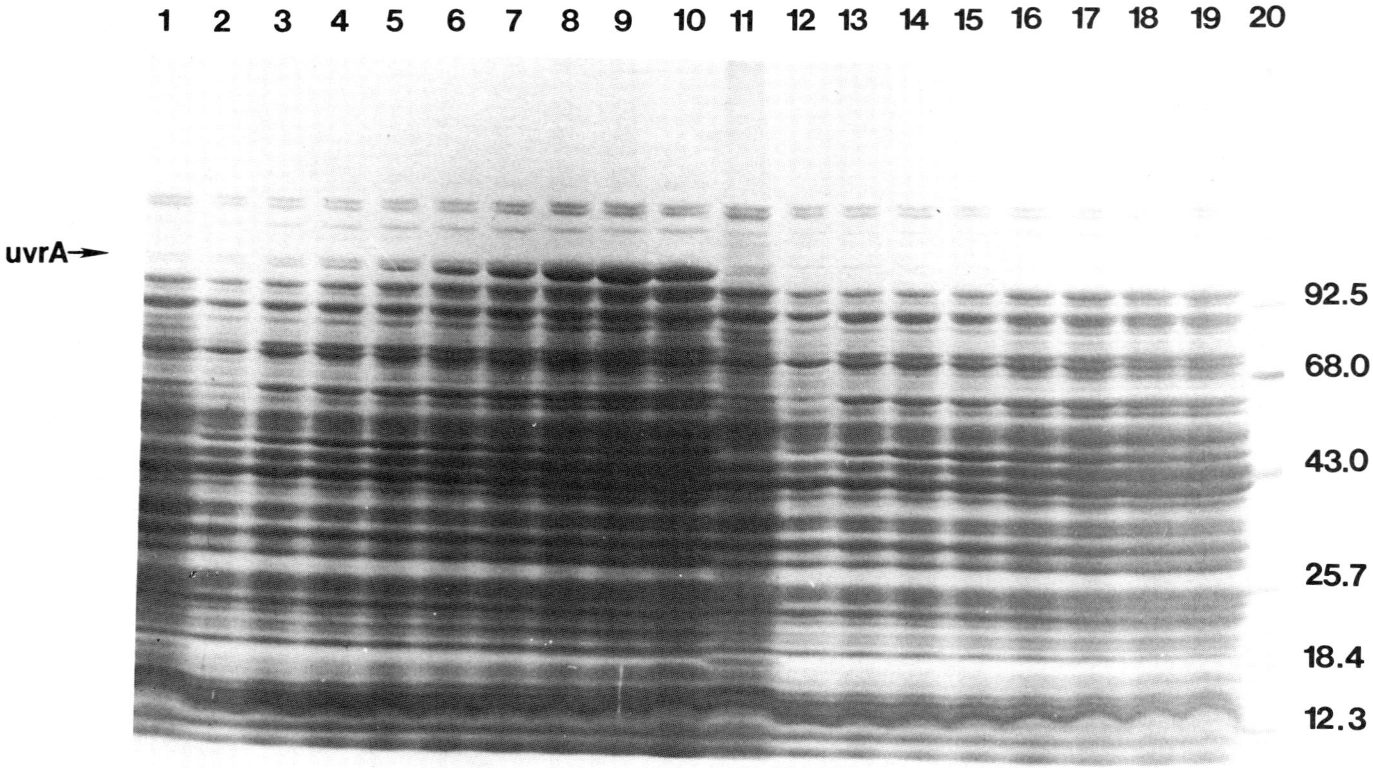

1 2 3 4 5 6 7 8 9 10 11 12 13 14 15 16 17 18 19 20
uvrA→
92.5
68.0
43.0
25.7
18.4
12.3

the *uvrA* polypeptide under P_L-transcription control permitted the rapid selection of hybrid plasmids capable of optimal *uvrA*-polypeptide expression on pkC30 hybrid plasmids. This screening method provides a step to permit selection of hybrid plasmids that amplify the gene product without the need to assay enzymically each potentially useful plasmid. In addition, the approximate levels of polypeptide can be estimated by conventional SDS–polyacrylamide gel analysis to provide some estimates that are useful in determining the scope of problems to be met when purification of the protein is initiated (Fig. 5).

Discussion

The methods described here are intended to facilitate the application of pKC30 and similarly regulated amplification vectors to the general problem of gene product amplification. During the process of devising the procedures described, consideration was given to the problems associated with selecting the optimal hybrid plasmid constructs from recombinant DNA mixtures with many different plasmid molecules. In addition, methods were devised to permit rapid testing of hybrid plasmids for their ability to meet the primary requirement of an amplifiable plasmid: the expression of an inserted gene under regulatory control. The procedures designed for this project depended heavily on previously established methodology and represent a tailoring of these methods for use in a variety of gene-product amplification projects. The utility of the approach described

FIG. 4. Optimal conditions for batch-culture amplification of *uvr*A polypeptide from L-broth cultures of GHY8533/pGHY5003 (*recA*$^+$ N99 lambda lysogen described in the test/pGHY5003), assayed by Coomassie-Blue staining of cell lysates resolved by electrophoresis on 7.5% polyacrylamide–SDS gels.[16] The protein profile from GHY8533/pGHY5003 was compared to GHY8531/pKC30 to provide an estimate of the background bands that would arise from the host strain and vector under the conditions used for *uvrA* amplification.[16] The profiles of total cell protein were indexed against a fluorograph of this gel, and the position of *uvrA* polypeptide labeled by the maxicell procedure in other experiments is indicated at the side of the gel. The [^{35}S]methionine labeled *uvrA* extracts used for fluorography are in lanes 1 and 11. Lanes 2–10 contain cell lysates after 42° induction of GHY8533/pGHY5003: lane 2, 0 min; lane 3, 15 min; lane 4, 30 min; lane 5, 1 hr; lane 6, 1.5 hr; lane 7, 2 hr; lane 8, 3 hr; lane 9, 4 hr; lane 10, 6 hr. Lanes 12 through 19 contain 0–4 hr 42° induction of pKC30 in the same genetic background (GHY8531/pKC30). Lane 20 contains protein standard markers (BRL), and the molecular weight of each is indicated in the margin. Cellular levels of other proteins increase during the 42° induction, and determination of the number of viable cells and the increase in cell mass (OD_{550}) indicate that the cell numbers increase by less than a factor of 2 during the 4-hr period considered optimal for *uvrA*-polypeptide induction.[16] Many of the proteins induced during the 42° incubation may be lambda proteins.

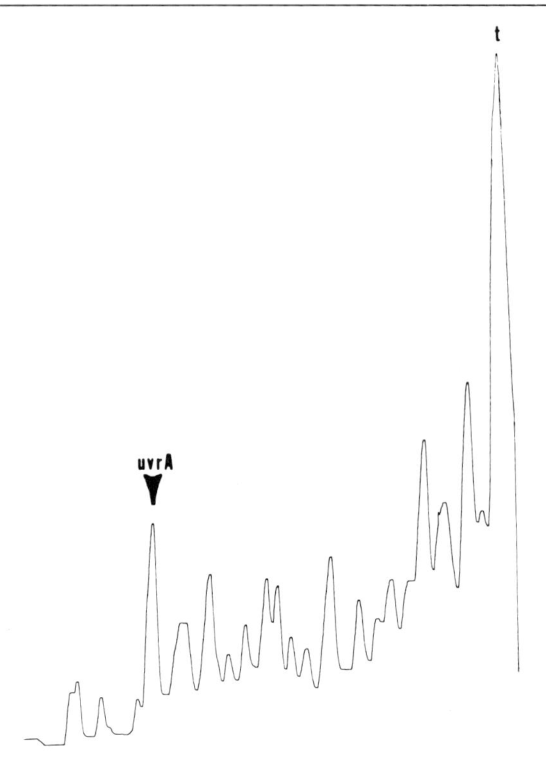

FIG. 5. Densitometric scan of the protein profile of GHY8533 after 4 hr of incubation at 42°, cell lysis in SDS, and electrophoresis on a 7.5% polyacrylamide–SDS gel stained with Coomassie Blue. The arrow indicates the position of the *uvrA* band, and *t* indicates the position of the bromophenol blue tracking dye. Quantitation of the area under the *uvrA* peak indicates that *uvrA* is approximately 7% of total cellular protein.[16]

here include (*a*) reduction in the amount of work required to accomplish goals; (*b*) improved quality of final products, since more potentially useful hybrid plasmid constructs can be easily tested; and (*c*) decreased cost to accomplish the project goal.

The value of such procedures is best proved by the products resulting from their application. Plasmid pGHY5003 is capable of amplification of *uvrA* polypeptide to 7% of cellular protein, which is an approximate amplification of this gene product by 10,000-fold.[16] The problems of protein purification for *uvrA* gene product is, thus, substantially reduced from the range of 10,000- to 30,000-fold purification to approximately 20- to 50-fold purification to obtain homogeneous protein preparations.

Acknowledgments

I thank Larry Grossman, Rebecca Goldfarb, Robert Grafstrom, William Mattes, and Tony Yeung for their help and intellectual stimulation during the course of this work. In addition, I appreciate being provided the N99 lambda lysogen and the pKC30 vector by Martin Rosenberg, and the phage lysates I used for lysogen construction by Roger McMacken.

The experiments were supported by grants from the American Cancer Society (NP-8L), the National Institutes of Health (5 R01 GM22846), and The National Science Foundation (PCM-79-14191) to Larry Grossman.

[8] Plasmids Containing the *trp* Promoters of *Escherichia coli* and *Serratia marcescens* and Their Use in Expressing Cloned Genes

By BRIAN P. NICHOLS AND CHARLES YANOFSKY

Studies on the regulation of the tryptophan (*trp*) operon in enteric bacteria, including *Escherichia coli* and *Serratia marcescens,* have established the use of two transcriptional control mechanisms: repression and attenuation. As bacterial cells become deficient in tryptophan, both repression and attenuation are relieved, generally resulting in a several hundredfold elevation of overall expression of the operon.[1] This regulatory change accompanying tryptophan deprivation has been used as an effective means of producing large amounts of the *E. coli trp* operon polypeptides.[2-4] When the chromosomally located *trp* operon is maximally expressed, the five *trp* operon polypeptides constitute greater than 10% of the total cellular protein. Thus, the *trp* regulatory region directs high level expression of genes in the same operon, and expression is subject to control by tryptophan availability.

In order to achieve even higher levels of expression, *trp* promoters and associated operon segments have been introduced into multicopy plasmids, where they lead to the overproduction of tryptophan biosynthetic enzymes.[5,6] Plasmids containing the *trp* promoter of *E. coli* have also been joined to foreign DNA segments and employed to direct the

[1] C. Yanofsky, T. Platt, I. Crawford, B. Nichols, G. Christie, H. Horowitz, M. van Cleemput, and A. Wu, *Nucleic Acids Res.* **9,** 6647 (1981).

[2] C. Yanofsky, *J. Biol. Chem.* **224,** 783 (1957).

[3] U. Henning, D. Helinski, F. C. Chao, and C. Yanofsky, *J. Biol. Chem.* **237,** 1523 (1962).

[4] C. Yanofsky, *J. Am. Med. Assoc.* **218,** 7 (1971).

[5] V. Hershfield, H. W. Boyer, C. Yanofsky, M. A. Lovett, and D. R. Helinski, *Proc. Natl. Acad. Sci. U.S.A.* **71,** 3455 (1974).

[6] G. F. Miozzari and C. Yanofsky, *J. Bacteriol.* **133,** 1457 (1978).

ISBN 0-12-182001-7

synthesis of a variety of polypeptides.[7-11] In this report we describe expression plasmids containing the *trp* promoters of *E. coli* and *S. marcescens* and their use in the overproduction of *trp* operon polypeptides.

Principles of the Method

Fusion of any operon, gene, or segment thereof, to a *trp* promoter will, under the appropriate conditions, ensure a high level of transcription of the fused sequences. Fusions may be constructed either by inserting DNA sequences into the expression plasmids described in the following section, or by excising the promoter-containing DNA fragments and inserting them into suitable sites preceding a cloned gene of interest. Note that the choice of fusion sites within the *trp* regulatory region can lead to either "transcriptional" or "translational" expression plasmids. Transcriptional fusions are those that ensure a high level of transcription of the fused segment, but depend on the existing translational signals within that segment for protein synthesis. Translational fusions, on the other hand, may be made by joining DNA sequences directly to the *trpL* peptide coding region, and as long as the fused segment is in the proper translational phase, synthesis of protein will be directed by the efficient translational initiation sequence of *trpL*.

If necessary, minor sequence additions, deletions, or alterations may be carried out following a fusion, in order to optimize the proximity of an effective ribosome recognition (Shine–Dalgarno) sequence to the desired initiation codon[7,11] or to remove potentially untranslated polar DNA sequences between the promoter and the translation initiation site.[12,13]

Expression of the DNA segment fused to the *trp* promoter may be accomplished in a variety of ways. By placing the *trp* expression plasmids in *E. coli* strains lacking the *trp* aporepressor gene (*trpR*), constitutive high level expression can be attained. Alternatively, the plasmid may be introduced into a strain that is temperature sensitive for *trp* aporepressor synthesis. In this background, a temperature shift will shut off aporepressor

[7] D. V. Goeddel, H. M. Shepard, E. Yelverton, D. Leung, and R. Crea, *Nucleic Acids Res.* **8,** 4057 (1981).

[8] J. S. Emtage, W. C. A. Tacon, G. H. Catlin, B. Jenkins, A. G. Porter, and N. H. Carey, *Nature (London)* **283,** 171 (1981).

[9] R. A. Hallewell and S. Emtage, *Gene* **9,** 27 (1981).

[10] J. Rose and A. Shafferman, *Proc. Natl. Acad. Sci. U.S.A.* **78,** 6671 (1981).

[11] D. V. Goeddel, E. Yelverton, A. Ullrich, H. L. Heyneker, G. Miozzari, W. Holmes, P. H. Seeburg, T. Dull, L. May, N. Stebbing, R. Crea, S. Maeda, R. McCandliss, A. Sloma, J. M. Tabor, M. Gross, P. C. Familletti, and S. Pestka, *Nature (London)* **287,** 411 (1981).

[12] T. M. Roberts, R. Kacich, and M. Ptashne, *Proc. Natl. Acad. Sci. U.S.A.* **76,** 761 (1979).

[13] S. Adhya and M. Gottesman, *Annu. Rev. Biochem.* **47,** 967 (1978).

production, and escape from repression will occur as the aporepressor is diluted out during cell division. (A strain that produces a temperature-sensitive repressor would be useful; unfortunately, none is available at the present time.) In *E. coli* strains that contain functional aporepressors, expression from the *trp* regulatory region can be induced by creating mild tryptophan starvation conditions. This can be most conveniently accomplished by addition to the medium of a tryptophan analog such as indolyl-3-acrylic acid or indolyl-3-propionic acid.[14] A suitable concentration can usually be found that will cause a mild tryptophan deficiency, thereby fully derepressing transcription from the *trp* promoter, yet inhibiting cell growth only slightly.[15] In addition, certain strains are depressed when grown in minimal medium. These strains have a mutation in one of the *trp* structural genes and produce an altered protein that is inefficient in tryptophan biosynthesis.[16]

Choice of a particular method of turning on a *trp* promoter should be influenced by consideration of the tryptophan content of the protein to be overproduced. Although in our experience mild tryptophan starvation does not appreciably reduce the yield of tryptophan-containing *trp* operon polypeptides, severe starvation does have a marked effect. If the tryptophan content of the protein of interest is high, host strains should be used that are deficient in the repressor protein and may be grown without imposing tryptophan starvation.

Since the chromosomal *trp* promoter can direct the synthesis of an average-size protein to a level approximately 2% of the cellular protein, the 20- to 30-fold copy number amplification afforded by many plasmids should be adequate for high level expression of any gene of interest. If higher rates of expression are desired, plasmids containing temperature-sensitive replication control regions may be employed for additional amplification.[17] Alternatively, two or more tandem copies[7] of the *trp* regulatory region could be used to increase expression. We have found that a single *trp* promoter fused to a gene of interest on a pBR322 derivative can direct the synthesis of a polypeptide to levels that approach 20–30% of the total cellular protein, after 3 hr of growth in the presence of indolyl-3-acrylic acid.

[14] W. F. Doolittle and C. Yanofsky, *J. Bacteriol.* **95,** 1283 (1968).

[15] R. Baker and C. Yanofsky, *J. Mol. Biol.* **69,** 89 (1972).

[16] C. Yanofsky, *in* "De la Physique Théorique à la Biologie," Comptes rendus de la Seconde Conference Internationale de Physique Theorique et Biologie (M. Marois, ed.), p. 191. CNRS, Paris, 1971.

[17] D. H. Gelfand, H. M. Shepard, P. H. O'Farrell, and B. Polisky, *Proc. Natl. Acad. Sci. U.S.A.* **75,** 5869 (1978).

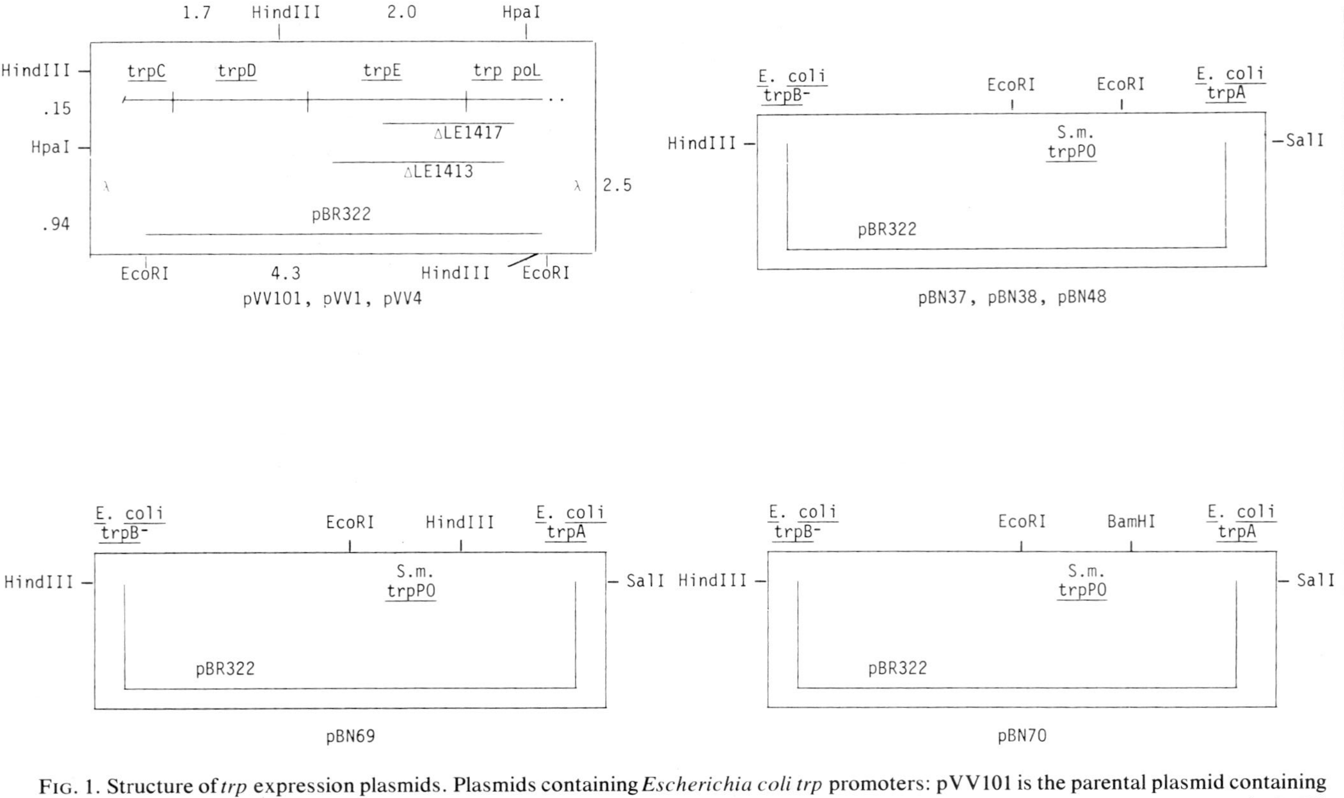

FIG. 1. Structure of *trp* expression plasmids. Plasmids containing *Escherichia coli trp* promoters: pVV101 is the parental plasmid containing intact *trpE* and *trpD*. pVV1 and pVV4 are identical to pVV101 except that they contain deletions Δ*trpLE1413* and Δ*trpLE1417*, respectively.[6,18] Plasmids containing *Serratia marcescens trp* promoters: the fine differences among pBN37, pBN38, and pBN48 are illustrated in Fig. 2.

Methods and Reagents

Plasmids

The *E. coli trp* plasmids, pVV101, pVV1, and pVV4 (Fig. 1) may be used as sources of promoter fragments for operon or gene fusions.[6,18] The sequences of segments of convenient fragments are indicated in Fig. 2. The *TaqI* partial fragment ends between the Shine–Dalgarno region and the leader peptide start codon and may be used in fusions to the AUG of any coding region. The *Bgl*II site of Δ*trpLE1413* or the *Hae*III site of Δ*trpLE1417* may be used for in-phase coding region fusions.

The *S. marcescens trp* expression plasmids diagrammed in Fig. 1 may be used for construction of gene fusions or excision of promoter-containing DNA fragments. The *trp* regulatory region DNA sequences are detailed in Fig. 2.[1,19] pBN37 contains the promoter and attenuation region and would give the greatest range of expression in shifts from tryptophan-containing to tryptophan-poor media. pBN38 contains the leader peptide coding region but lacks the transcription terminator. pBN48 may be used to fuse a foreign coding region to the leader peptide coding region. pBN69 and pBN70 may be used to introduce any genetic region into the *trp* transcription unit at *Hin*dIII or *Bam*HI sites just beyond the *trp* promoter.

Strains

Bacterial strains are available that allow considerable freedom in plasmid construction and *trp* operon expression. Strains *trpR*$^+$ Δ*trpE5* or Δ*trpLD102* can be used to select recombinant plasmids containing a functional *trpE* and/or *trpD* gene, on either side of a cloned gene. Strain CY15042 is temperature sensitive for *trp* aporepressor production. This strain has an amber mutation in *trpR*, and contains a temperature-sensitive amber suppressing tRNA.[20] Strain W3110 *trpA46PR9* contains a functional *trp* aporepressor, but the alteration in *trpA* results in mild tryptophan starvation, and hence relief from tryptophan regulation when the strain is grown in minimal medium.[16] Numerous strains lacking a functional aporepressor are also available.

[18] K. Bertrand, C. Squires, and C. Yanofsky, *J. Mol. Biol.* **113,** 319 (1976).

[19] G. Miozzari and C. Yanofsky, *Nature (London)* **276,** 684 (1978).

[20] M. P. Oeschger and S. L. Woods, *Cell* **7,** 215 (1976).

Experimental

High Level Expression of trpE and trpD Using a Plasmid Containing the Escheridia coli trp Promoter

Strain W3110 Δ*trpE5*/pVV101 was grown at 37° in minimal medium containing 20 μg of L-tryptophan per milliliter and ampicillin to a cell den-

A E. coli promoter sequences

```
ΔtrpLE1413

(-214).....                  -40                     -20                   +1
(TaqI).....                   ¦                      TaqI                   ¦
(TCGA).....GGCAAATATTCTTGAAATGAGCTGTTGACAATTAATCATCGAACTAGTTAACTAGTACGCAAGTTCACG

                  27                                  54  1016
            TaqI   ¦                                    \ /              BglII
TAAAAAGGGTATCGACA ATG AAA GCA ATT TTC GTA CTG AAA GGT TCA CTG GAC AGA GAT CTC
                  Met Lys Ala Ile Phe Val Leu Lys Gly Ser Leu Asp Arg Asp Leu


ΔtrpLE1417

(-214).....                  -40                     -20                   +1
(TaqI).....                   ¦                      TaqI                   ¦
(TCGA).....GGCAAATATTCTTGAAATGAGCTGTTGACAATTAATCATCGAACTAGTTAACTAGTACGCAAGTTCACG

                  27                              50 568
            TaqI   ¦                               ¦  ¦                HaeIII
TAAAAAGGGTATCGACA ATG AAA GCA ATT TTC GTA CTG AAA AGC ACC CGT ATT CAG GCC
                  Met Lys Ala Ile Phe Val Leu Lys Ser Thr Arg Ile Gln
```

B S. marcescens promoter sequences

```
pBN37

             -60                   -40                   -20                    +1
 EcoRI        ¦                     ¦                     ¦                      ¦
GAATTCACGCTGATCGCTAAAACATTGTGCAAAAAGAGGGTTGACTTTGCCTTCGCGAACCAGTTAACTAGTACACAAGT

TCACGGCAACGGCCGTGTCGGATGAGAGTTAACAAAGAGAGTCTGCAA ATG AAC ACA TAC ATT TCT CTT CAC
                                                 Met Asn Thr Tyr Ile Ser Leu His

GGT TGG TGG CGT ACC TCC CTC TTG CGG GCG GTG TAA TCGCGCATAGCTGTCATCTGACAATGCAGATT
Gly Trp Trp Arg Thr Ser Leu Leu Arg Ala Val End

                                        EcoRI
TCCTGAGCCCGCACCTGATGCGGGCTTTTTATGGACAGAATTC
```

FIG. 2. Nucleotide sequences of the *trp* regulatory regions available on various plasmids.[1,19] Numbering is from the site of transcription initiation. (A) Nucleotide sequences of *Escherichia coli* Δ*trpLE1413* and Δ*trpLE1417*. (B) Nucleotide sequences of the *Serratia marcescens trp* regulatory regions.[19]

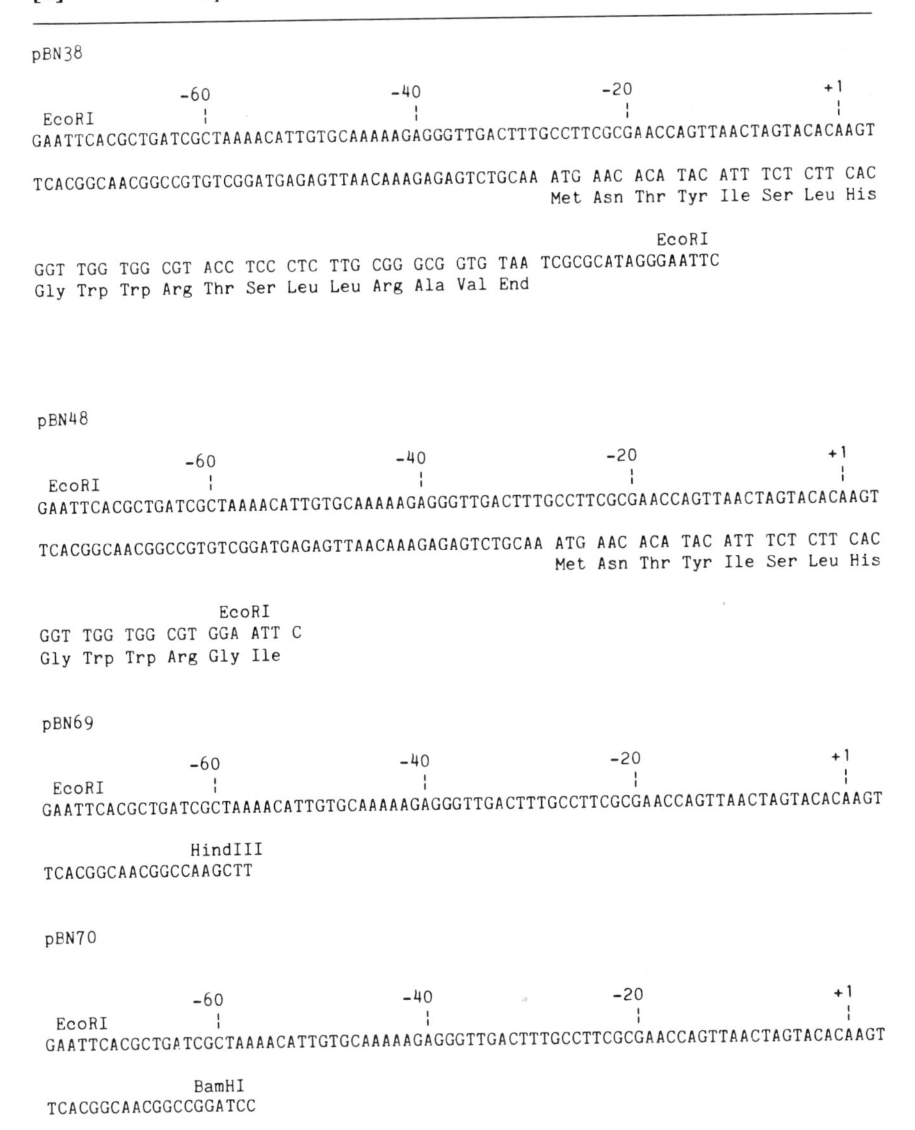

FIG. 2. (*Continued*)

sity of 3 to 5 × 10^8 cells/ml. The cells were sedimented and washed, and then resuspended in minimal medium containing 40 μg of indolyl-3-acrylic acid per milliliter. Samples taken at 2 and 5 hr were assayed for *trpE* and *trpD* encoded enzymic activity. The results indicate that by 2 and 5 hr, 15% and 30%, respectively, of the cellular protein consist of these two polypeptides. An SDS–polyacrylamide gel of a typical crude extract is shown in Fig. 3. Note that there is less *trpD* protein than *trpE* protein at the high indolyl-3-acrylate concentration used in this particular experiment. The *trpD* protein contains one tryptophan residue, whereas the *trpE* protein lacks tryptophan.

High-Level Expression of Plasmid Specified trpA Using Serratia marcescens *trp Promoters*

Escherichia coli strain W3110 Δ(*tonB-trpA905*) was used as the recipient for plasmids containing the *S. marcescens trp* promoter fused near *E. coli trpA*. A strain containing a functional aporepressor was chosen to harbor these plasmids because of plasmid instability in *trpR* strains (see Comments). Plasmid-containing strains were grown in minimal medium containing 0.05% casamino acids, ampicillin, and 20 μg of L-tryptophan

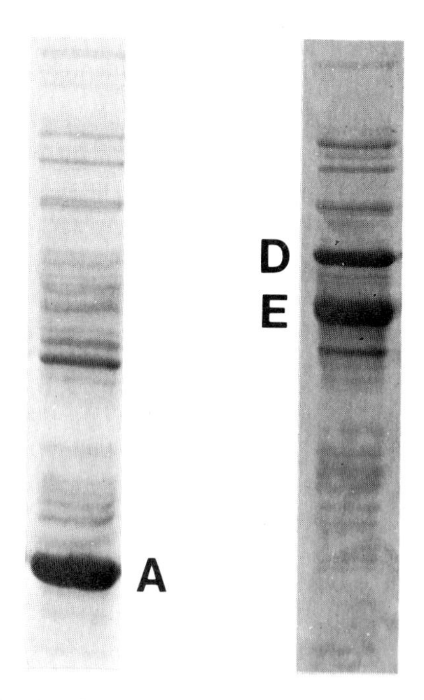

FIG. 3. Sodium dodecyl sulfate–polyacrylamide gels of crude extracts of strains W3110 Δ*trpE5*/pVV101 and W3110 Δ(*tonB-trpA905*)/pBN69. These strains were grown with indolyl-3-acrylate for 5 and 3 hr, respectively. The positions of the *trpE*, *trpD*, and *trpA* polypeptides on a gel stained with Coomassie Blue R are indicated. Forty micrograms of each crude extract were applied to the gel.

per milliliter to a cell density of 3 to 5 × 10^8 cells/ml. The cells were sedimented, washed, and resuspended in the same medium except that 20 μg of indolyl-3-acrylic acid per milliliter was substituted for tryptophan. Samples taken after 3 hr were assayed for tryptophan synthetase α-polypeptide (the *trpA* gene product). The results are shown in the table, and a representative gel is displayed in Fig. 3.

EXPRESSION OF *Escherichia coli trpA* BY THE *Serratia marcescens trp* PROMOTER

Plasmid	Specific activity[a] (units/mg protein)	Estimated percent of cellular protein[b]
pBN31[c]	13	0.2
pBN37	192	4
pBN38	159	3
pBN48[d]	464	9
pBN69	963	19
pBN70	787	16

[a] After 3 hr of growth in the presence of indolyl-3-acrylic acid.
[b] Based on the specific activity of purified α-polypeptide = 5000 units/mg protein.
[c] pBN31 is the *trp* promoter-less parent of pBN37, 38, 48, 69, and 70.
[d] pBN48 is unstable when grown in minimal medium.

The level of *trpA* expression differs among the promoter-containing fragments. pBN48 is unstable when grown in this strain in minimal medium, and the value given in the table represents an underestimate of the true level of synthesis. If the plasmid-bearing strains are grown in a rich medium, and then transferred to minimal medium containing indolyl-3-acrylic acid, pBN48 directs a higher level of α-polypeptide synthesis than either pBN69 or pBN70 (data not shown). The differential expression from these promoters, even when fully induced, may be due to the polarity of the DNA sequences between the initiation sites of transcription and translation. pBN48 is a translational fusion of the *trpL* coding region with the terminal portion of *trpB*. Although the fusion is out of phase relative to the wild-type *trpB* sequence, translation can proceed to within 15 nucleotides of the beginning of *trpA*. The translation of this portion of *trpB* may relieve the polarity that would otherwise result from the presence of an untranslated region between the promoter and *trpA*.

Comments

Bacteria containing *trp* expression plasmids lacking the *trp* leader-attenuator control region are notoriously unstable when the *trp* operon is turned on, and they segregate plasmid-free cells at high frequency. In fact it is difficult, and sometimes impossible, to select or maintain such plasmids in repressor-deficient (*trpR*) strains. The explanation for this instability is not certain, but it is thought that one contributing factor may be that transcription beyond the *trp* operon segment interferes with plasmid replication. If this explanation is correct, then insertion of an efficient transcription termination sequence downstream from the *trp* genes should alleviate the problem. This possibility is being investigated; preliminary indications are that insertion of a transcription terminator beyond the *trp* operon segment does increase the yield of *trp* polypeptides.

Instability is generally avoided by performing all cloning steps in $trpR^+$ strains, either in a rich medium or in a medium containing repressing levels of tryptophan. In some cases (e.g., pBN48), minimal medium containing 20 μg of tryptophan per milliliter is not sufficient to ensure plasmid stability.

For high-level expression, the *trp* operon may be induced by tryptophan starvation. Note that indolyl-3-acrylic acid, a good inducer, is not very effective in the presence of tryptophan. However, we have found that addition of high concentrations of indolyl-3-acrylic acid (100 μg/ml) to a rich medium can slightly induce the *trp* operon, although not to the levels observed in the absence of tryptophan. When the host bacterium is a tryptophan auxotroph, growth on limiting level of indole or tryptophan may also be used as a means of turning on the *trp* operon of the recombinant plasmid.

Acknowledgments

The authors are indebted to Virginia Horn and Magda VanCleemput for assistance and to Anathbandhu Das, Michael Schechtman, and Richard Kelley for help and advice. These studies have been supported by grants from the National Institutes of Health and the National Science Foundation. C. Y. is a Career Investigator of the American Heart Association.

Section II

Cloning of Genes into Yeast Cells

[9] Construction and Use of Gene Fusions to *lacZ* (β-Galactosidase) That Are Expressed in Yeast

By MARK ROSE and DAVID BOTSTEIN

Gene fusion is a powerful tool for the analysis of the expression, regulation, and structure of a gene. Typically the regulatory and/or structural sequences of the gene of interest are covalently linked to the structural sequences of another gene whose protein is easily assayed. Many of the uses for gene and operon fusions in the study of prokaryotic genes are discussed in the excellent review of Bassford *et al.*[1] The use of gene fusions has been extended to the analysis of the genes of the simple eukaryote *Saccharomyces cerevisiae*.[2,3]

The advent of techniques for DNA sequence analysis has raised new problems that are particularly amenable to solution by gene fusion techniques. For example, the discovery of an open translational reading frame raises the question of whether the sequence is functionally expressed *in vivo*. By fusing a known assayable protein to the sequence, it becomes possible to detect its expression. In addition, the hybrid protein provides a means for the purification of the gene product normally encoded by the sequence.[4] Two other uses derive from the novel junction of the DNA segments that are joined in the fusion process. Gene fusions necessarily delete DNA sequences that are internal to the structural gene of interest and therefore can be used to determine the position of any internal sites that might be essential for gene expression. Furthermore, the novel junction can be used to introduce a useful restriction enzyme recognition site, which can in turn be used to accelerate DNA sequence analysis.

The protein most commonly used for gene fusions has been the enzyme β-galactosidase. There are several advantages to this enzyme. The assay for the activity is extremely sensitive and easy both as a plate assay, in which expressing colonies are colored, and as a quantitative enzyme

[1] P. J. Bassford, J. Beckwith, M. Berman, E. Brickman, M. Casadaban, L. Guarente, I. Saint-Girons, A. Sarthy, M. Schwartz, H. Shuman, and T. Silhavy, *in* "The Operon" (J. H. Miller and W. S. Reznikoff, eds.), p. 245. Cold Spring Harbor Laboratory, Cold Spring Harbor, New York, 1978.

[2] M. Rose, M. J. Casadaban, and D. Botstein, *Proc. Natl. Acad. Sci. U. S. A.* **78,** 2460 (1981).

[3] L. Guarente and M. Ptashne, *Proc. Natl. Acad. Sci. U. S. A.* **78,** 2199 (1981).

[4] H. A. Shuman, T. J. Silhavy, and J. R. Beckwith, *J. Biol. Chem.* **255,** 168 (1980).

METHODS IN ENZYMOLOGY, VOL. 101

ISBN 0-12-182001-7

assay in cell extracts.[5] It has been demonstrated that as many as 30 amino acids at the amino terminus of the protein may be replaced with other sequences of varying length without substantially affecting the activity of the enzyme.[6] Various techniques have been devised that select for the presence or the absence of the enzyme in *Escherichia coli*.[5] These facts have led several workers to construct mutations of the *lacZ* gene (which encodes β-galactosidase) in which the protein cannot be expressed unless it is first fused to the beginning of another gene.[7,8] We have made extensive use of one such defective gene (called *'lacZ*) constructed by Casadaban and Cohen,[7] to investigate the expression of the sequences around the yeast *URA3* gene.[9]

Principles

Two general methods for the construction of sets of gene fusions have been used: *in vivo* selection[2] and *in vitro* manipulation of the DNA. In either case the sequence of interest is placed onto a plasmid containing the unexpressed *'lacZ* gene. The DNA fragments are oriented so that the putative coding sequences would be read from the same DNA strand, with the β-galactosidase coding sequence on the 3′ side of the sense strand. *In vivo* selection depends on spontaneous deletions formed in plasmids residing in *E. coli*. If the original construction has a Lac^- phenotype, selection for Lac^+ results in fusion of the *'lacZ* segment to sequences in frame with a functional initiation codon. Since the sequence of interest may not contain a functional *E. coli* promoter, a plasmid promoter must be situated so as to provide transcription of the selected hybrid gene.

The *in vitro* approach relies on the presence of unique restriction sites in the gene of interest and at the beginning of the *'lacZ* segment. Deletion of material between the gene and *'lacZ* is accomplished by means of the double-stranded DNA exonuclease *Bal*31 acting on the ends produced within or near the gene by an appropriate restriction endonuclease. Ligation using DNA linkers containing the restriction site at the beginning of *'lacZ* allows fusion between the two genes. Plasmids recovered in *E. coli* are then transferred into yeast for analysis of the expression of the hybrid protein.

[5] J. H. Miller, "Experiments in Molecular Genetics." Cold Spring Harbor Laboratory, Cold Spring Harbor, New York, 1972.

[6] E. Brickman, T. J. Silhavy, P. J. Bassford, Jr., H. A. Shuman, and J. R. Beckwith, *J. Bacteriol.* **139,** 13 (1979).

[7] M. J. Casadaban, J. Chou, and S. N. Cohen, *J. Bacteriol.* **143,** 971 (1980).

[8] L. Guarente, G. Lauer, T. M. Roberts, and M. Ptashne, *Cell* **20,** 543 (1980).

[9] M. Bach, F. Lacroute, and D. Botstein, *Proc. Natl. Acad. Sci. U. S. A.* **76,** 386 (1979).

Materials and Reagents

Escherichia coli strains. Strains M182, (*lacIPOZYA*) Δ*X74 galU galK strA* (from J. Beckwith) and a derivative containing an insertion of Tn*5* in the *pyrF* gene have been used. Any other extensive deletion of the *lac* operon should serve as well.

Media. *Escherichia coli* strains containing a plasmid conferring ampicillin resistance are propagated on LB[5] supplemented with 100 μg of ampicillin per milliliter. Selection for Lac^+ derivatives of plasmid containing strains is accomplished on M-9 minimal medium[5] containing 0.2% lactose. Strains that produce as few as 10 units of β-galactosidase are able to grow on M-9 lactose plates. Higher levels of β-galactosidase can be distinguished on MacConkey-lactose plates (Difco) containing ampicillin. Intermediate levels of expression can be distinguished by supplementing media with 5-bromo-4-chloro-3-indolyl-β-D-galactoside (X-gal; Bachem Inc., Torrance, California) to 40 μg/ml. The X-gal is dissolved in *N,N*-dimethylformamide at a concentration of 20 mg/ml and can be stored at −20°. The intensity of the blue color varies with the level of β-galactosidase in the cell, the rate of cell growth, and the type of medium. The most intense color is obtained on M-9; less color is obtained on the richer media such as LB or MacConkey.

Yeast strains containing plasmids are propagated on SD minimal medium[10] with constant selection for the presence of the plasmid. SD is composed of 6.7 g per liter of yeast nitrogen base (without amino acids, Difco) supplemented to 2% glucose added after autoclaving. Solid SD medium is made by mixing sterile double-strength SD with an equal volume of molten 4% agar. The expression of β-galactosidase activity is observed on agar plates buffered at pH 7.0 and supplemented with X-gal at 40 μg/ml. The acid media (such as SD) usually employed in culturing yeasts do not allow color development. Previously we have used SD medium buffered by the addition of potassium phosphate. Better results have been obtained using the buffered minimal medium of Clifton *et al.*[11] Buffered minimal medium contains a final concentration of 0.1 *M* KH_2PO_4, 0.015 *M* $(NH_4)_2SO_4$, 0.075 *M* KOH, 0.8 m*M* $MgSO_4$, 2 μ*M* $Fe_2(SO_4)_3$. After autoclaving, sterile glucose is added to 2%, and $\frac{1}{100}$ volume of a vitamin stock is added. The vitamin stock consists, per milliliter, of 40 μg of thiamine, 40 μg of pyridoxine, 40 μg of pantothenic acid, 200 μg of inositol, and 2 μg of biotin. Solid medium is made by making up the salts at double

[10] F. Sherman, G. R. Fink, and C. W. Lawrence, "Methods in Yeast Genetics." Cold Spring Harbor Laboratory, Cold Spring Harbor, New York, 1979.

[11] D. Clifton, S. Weinstock, and D. G. Fraenkel, *Genetics* **88,** 1 (1978).

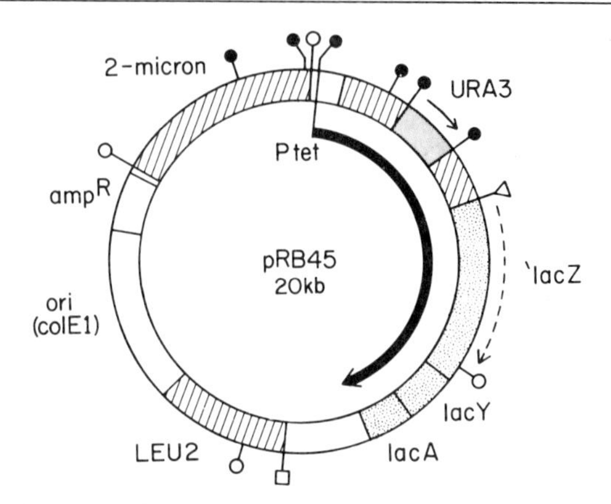

FIG. 1. Structure of plasmid pRB45. Yeast sequences, except for the *URA3* gene, are hatched. The stippled area is derived from the *Escherichia coli lac* operon and was obtained from plasmid pMC874.[7] Other areas are derived from pBR322.[14,15] The tetracycline-resistance gene promoter initiates transcription in a clockwise direction. The *URA3* gene and the *'lacZ* gene are both oriented so that clockwise transcription is from the sense strand. Restriction enzyme sites are as follows: ○, *Eco*RI; ●, *Hin*dIII; ▽, *Bam*HI; □, *Sal*I.

strength and adding to an equal volume of sterile molten 4% agar. X-gal is added just before pouring the plates, as for the bacterial media.

Plasmids. All fusions between the *URA3* gene and *lacZ* were obtained from the plasmid pRB45,[2] shown in Fig. 1. The *'lacZ* fragment is derived from pMC874.[7] Plasmid pRB45 contains a unique *Bam*HI site at the beginning of *'lacZ* and a unique *Sma*I site near the end of the *URA3* gene. (Other derivatives of the *'lacZ* segment with different unique restriction sites at the 5′ end have been constructed by M. Casadaban[11a] and Guarente *et al.*[8]) The other relevant features are a portion of the yeast 2-μm (also called 2-micron) plasmid[12] for autonomous replication in yeast, the yeast *LEU2* gene[13] which serves as a selectable marker in yeast and the pBR322-derived ampicillin resistance gene and *E. coli* origin of replication.[14,15] In *E. coli* the *URA3* gene is transcribed by the tetracycline promoter derived from pBR322 (unpublished results).

Enzymes and Nucleic Acids. All the restriction enzymes and T4 DNA ligase were purchased from New England BioLabs and were used according to the recommendations of the manufacturer. The enzyme *Bal*31 was

[11a] M. Casadaban, this series, Vol. 100 [21].

[12] J. L. Hartley and J. E. Donelson, *Nature* (*London*) **286,** 860 (1980).

[13] B. Ratzkin and J. Carbon, *Proc. Natl. Acad. Sci. U. S. A.* **74,** 487 (1977).

[14] F. Bolivar, R. L. Rodriguez, P. J. Greene, M. C. Betlach, H. L. Heyneker, H. W. Boyer, J. H. Crosa, and S. Falkow, *Gene* **2,** 95 (1977).

[15] JG. Sutcliffe, *Cold Spring Harbor Symp. Quant. Biol.* **43,** 77 (1979).

purchased from Bethesda Research Laboratories and used according to the conditions described by Legerski *et al.*[16] *Bam*HI linkers[16a] were obtained from Collaborative Research Inc. T4 polynucleotide kinase was the generous gift of O. Uhlenbeck.

In Vivo Selection of Fusions

To express β-galactosidase activity, *'lacZ* must be fused to a DNA segment so that an initiation codon is provided upstream in the correct reading frame without any intervening nonsense codons. Starting with a plasmid constructed as described under Principles, the simplest way to obtain such fusions is to select the rare cells in which the plasmid suffered a spontaneous deletion event that allows the *'lacZ* sequence to express β-galactosidase. This can be done by inserting the plasmid into a strain carrying a deletion of the chromosomal *lacZ* gene and then demanding that the strain grow on lactose.

The original unfused construction containing *'lacZ* behind the gene of interest must not express sufficient β-galactosidase for the cell to grow on lactose. It has been our experience that some such constructions are Lac^+. This property will prevent selection of fusions to the sequence of interest. One approach to remedy this problem is to change the reading frame to which the *'lacZ* has been fused by insertion of extra nucleotides at the beginning of the *'lacZ* segment. When this is done, the phenotype usually becomes Lac^-.

Nonreverting point mutations of *lacZ* do not suffice as the method of inactivating the chromosomal source of β-galactosidase. The homologous sequences on the plasmid will result in a high frequency of Lac^+ recombinants. It is likely that most deletion events are catalyzed by the *E. coli* recombination system so that the strain should be Rec^+. The cell must also be $lacY^+$ so that lactose can enter the cell by the permease. Most of the *'lacZ* fragments include the *lacY* gene so that it is expressed by plasmid promoters along with *lacZ* and is not subject to *lacI* repression.

The second requirement for expression of the β-galactosidase activity is that there be transcription of the hybrid gene in *E. coli.* To ensure this, it is convenient to orient the fragments such that they are transcribed by one of the plasmid promoters. A simple test for whether the plasmid transcription will be sufficient is to construct the analogous plasmid using an intact *lacZ* gene instead of *'lacZ*. The expression of β-galactosidase from the control plasmid is a measure of the level of transcription impinging on the *lacZ* gene.

[16] R. J. Legerski, J. L. Hodnett, and H. B. Gray, Jr., *Nucleic Acids Res.* **5,** 1445 (1978).
[16a] C. P. Bahl, K. J. Marians, R. Wu, J. Stawinsky, and S. Narang, *Gene* **1,** 81 (1976).

Prior to selection for fusions many independent clones of the strain carrying the plasmid are grown up to saturation in LB broth containing antibiotics to select for the presence of the plasmid. One milliliter of each culture (around 10^9 cells) is harvested by centrifugation and washed free of the medium by repeated resuspension and centrifugation out of M-9 salts without sugar. The actual selection is begun by spreading the resuspended cells on the surface of agar plates containing M-9 salts and 0.2% lactose and incubated at 37°. After 2 days, Lac^+ colonies appear initially at a frequency of around 10^{-8}, but new colonies continue to appear for several days owing both to slow growers and events that occur on the plate. The colonies that appear later are largely independent isolates, and it is most efficient to pick several of these after marking the colonies that appear early. Leaky growth of the bacterial lawn is often observed and is probably due partly to contamination of lactose with other sugars and partly to residual β-galactosidase activity from internal translation initiation events in the *'lacZ* fragment.

The Lac^+ colonies are purified by restreaking on M-9 lactose plates and then tested semiquantitatively for levels of β-galactosidase by spotting onto a series of plates containing various indicators as described in the materials and reagents section. Quantitative assays of *E. coli* stains containing plasmids are performed on exponentially growing cells in LB broth containing ampicillin. Cells are grown to a density of about 2×10^8 cells per milliliter, chilled on ice, centrifuged to remove the medium, and resuspended in an equal volume of chilled M-9 salts lacking sugar. Assays are then preformed as described by Miller[5] using cells permeabilized with SDS and $CHCl_3$. Plasmid DNA is purified from several independent Lac^+ colonies and the DNA is analyzed by restriction enzyme digestion to determine the extent of the deletion.

In Vitro Fusion Construction

Two different *in vitro* approaches have been taken to construct fusions. If the DNA sequence of the region of interest is known, then it is often possible to find a restriction enzyme cleavage site that allows the direct construction of a fusion.[3] However, for small targets or for genes in which the sequence is not known, this approach will not suffice. We have utilized the double-stranded DNA exonuclease *Bal*31 and DNA linkers to introduce the requisite restriction enzyme sites at random positions in the region of interest.

A plasmid is constructed using the same criteria as for the *in vivo* method. The general structure of such a plasmid and the scheme for *in vitro* fusion construction are diagrammed in Fig. 2. The requisite features

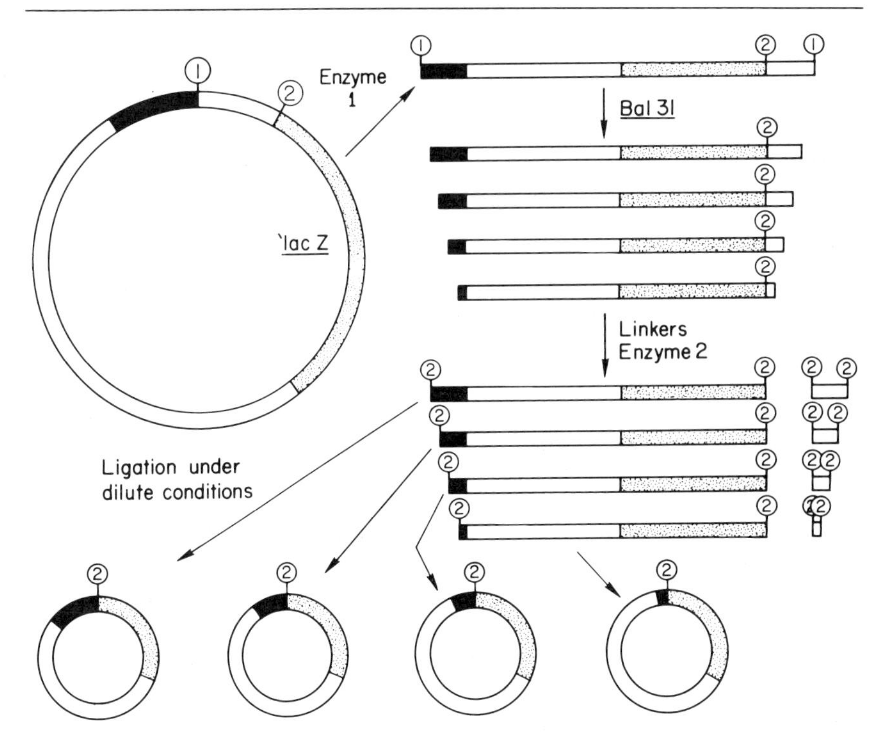

FIG. 2. Scheme for *in vitro* gene fusion construction. The plasmid shown is analogous to pRB45 shown in Fig. 1. The stippled area is the *'lacZ* segment. The black area represents the target gene of interest. See the text for details.

of the plasmid are the two unique restriction sites, 1 and 2. Site 2 is the restriction site at the boundary of the *'lacZ* segment and may be *Bam*HI, *Eco*RI, *Hin*dIII, *Sma*I or *Xho*I. Site 1 can be in or near the end of the gene of interest. In the case of pRB45 the site is *Sma*I. The distance from site 1 to site 2 must not be shorter than the distance that must be deleted on the other side of site 1, or portions of *'lacZ* may get deleted. If necessary, a DNA fragment can be ligated into site 2 to provide a more favorable spacing.

The first step is to cut with enzyme 1 to completion. Since the later steps in the procedure are not completely efficient, it is advisable to remove the small fraction of uncut supercoils from the population, as many of these are likely to be preexisting deletions and will be selected for later on. This can be done conveniently by chromatography on acridine yellow-ED acrylamide gel[17] (Boehringer Mannheim) or by agarose gel elec-

[17] H. Buenemann and W. Mueller, *in* "Affinity Chromatography" (O. Hoffmann-Ostenhof *et al.*, eds.), p. 353. Pergamon, New York, 1978.

trophoresis. After restriction enzyme digestion, the DNA is phenol extracted and ethanol precipitated. To perform the chromatography, the DNA pellet is dissolved at 500 μg/ml in 10 mM sodium citrate, pH 7.0, 0.5 mM EDTA and then brought to 0.5 M NaCl (loading buffer). The dissolved DNA is slowly passed through a Pasteur pipette column containing 0.25 ml of packed gel preequilibated in the loading buffer. The column is washed with 10 column volumes of loading buffer, and fractions that contain DNA are pooled and ethanol precipitated. Usually essentially all of the linear DNA is contained in the first three column volumes. Under these conditions supercoiled DNA is bound to the column and can later be eluted with 1.0 M NaCl. Any relaxed circular DNA elutes with the linear DNA.

Linear DNA is then digested with *Bal*31 using the conditions of Legerski *et al.*[16] Reactions are carried out in 20 mM Tris-HCl, pH 8.0, 0.45 M NaCl, 1 mM Na_2EDTA, 12.5 mM $MgSO_4$, and 12.5 mM $CaCl_2$. For our experiments, DNA was present at a concentration of 100 μg/ml, which corresponds (for our plasmid) to a DNA end concentration of around 15 nM. *Bal*31 was added to a concentration of approximately 35 units/ml, and reaction mixtures were incubated at 28°. Under these conditions the enzyme digested approximately 100 bp from each end per minute. It is important to titrate the enzyme to determine a rate for the reaction under the conditions used for the experiment. This can be done by cutting with a second restriction enzyme to generate a fragment from one end of the linear molecule whose size can be observed to decrease by as little as 100 bp on agarose gels. The *Bal*31 reaction is terminated by addition of Na_2EDTA to 25 mM. To obtain a broad spectrum of fusions, aliquots are removed from the reaction mixture at different times. The distribution of ends is fairly broad, so that close spacing of time points is unnecessary. After the reaction is terminated the DNA is phenol extracted and ethanol precipitated.

Linker DNA molecules are then ligated onto the ends to facilitate making fusions to the enzyme 2 site. Commercially prepared linker molecules containing the enzyme 2 restriction site are first phosphorylated with T4 polynucleotide kinase and then ligated together in the presence of the *Bal*31-treated DNA. We have routinely phosphorylated linkers in 6 mM Tris-Cl, pH 7.5, 10 mM $MgCl_2$, 12.5 mM dithiothreitol, 1 mM spermidine, 1 mM ATP, and 200 μg of bovine serum albumin (BSA) per milliliter with linkers at a concentration of 300 μg/ml (about 50 μM for a 10-mer) and T4 polynucleotide kinase at 2000 units/ml. The phosphorylation reaction mixture is incubated at 37° for 1 hr and then is diluted by the addition of five volumes of *Bal*31-digested DNA dissolved in the phosphorylation buffer. The final linearized plasmid DNA concentration is approximately 200 μg/ml (30 n*M* in ends). T4 DNA ligase is then added to about 20,000 units/ml and incubated at 13° for 2 hr. Ligation of the linkers can be checked by electrophoresis of a portion of the ligation mixture on a 12% polyacrylamide gel and staining the gel with ethidium bromide. A successful ligation is indicated by the appearance of an extensive ladder of linker polymers. Alternatively, the linkers can be phosphorylated with ^{32}P at the kinase step, and incorporation of radioactivity into the plasmid DNA can be followed.

It is necessary both to trim away the extra linkers from the linears and to cut the plasmid at the site next to *'lacZ*. The presence of the linker polymers will inhibit the restriction enzyme by competing with the plasmid DNA. The plasmid DNA is purified away from the linkers by agarose gel electrophoresis. DNA can be purified out of the agarose by various methods. We have used a modification[18] of the hydroxyapatite chromatography method described by Tabak and Flavell,[19] which works well for purification of large DNA fragments from agarose. The DNA is adsorbed onto hydroxyapatite and sequentially washed with two column volumes of fresh running buffer, two volumes of 10 m*M* KH_2PO_4, pH 7.0, two volumes of 100 m*M* potassium phosphate and then eluted with 400 m*M* potassium phosphate. By incorporation of 0.5 μg of ethidium bromide per milliliter into the phosphate buffers, the elution of the DNA can be followed with longwave UV light. The eluted DNA is pooled, 10 μg of carrier tRNA are added and then phenol extracted to remove ethidium bromide. The DNA is dialyzed to remove phosphate and then ethanol precipitated.

The linear plasmid DNA with the linkers ligated onto the ends is then digested with enzyme 2. To remove all of the excess linkers polymerized onto the linear plasmid, it is important to get complete digestion. We have overcut the DNA about 10-fold by adding several aliquots of the restriction enzyme over the course of several hours. The reaction is terminated by the addition of Na_2EDTA to 25 m*M* followed by phenol extraction and ethanol precipitation. In order to circularize the molecules, the redissolved DNA is diluted into ligase buffer to a final concentration of 0.5 μg/ml. Under dilute conditions (0.1 n*M* ends) circulation is favored over polymerization. Ligase buffer is 50 m*M* Tris-Cl, pH 8.0, 10 m*M* $MgCl_2$, 20 m*M* DTT, 1 m*M* ATP, and 50 μg of BSA per milliliter. Ligase is added to between 200 and 2000 units per milliliter, and the reaction is incubated overnight at 14°. The ligation mixtures are then used directly to

[18] R. W. Davis, D. Botstein, and J. R. Roth, "A Manual for Genetic Engineering, Advanced Bacterial Genetics." Cold Spring Harbor Laboratory, Cold Spring Harbor, New York, 1980.

[19] H. F. Tabak and R. A. Flavell, *Nucleic Acids Res.* **5**, 2321 (1978).

transform *E. coli* to drug resistance. By incorporating X-gal into the selection agar, the "in frame fusions" can be easily recognized.

A high proportion of the plasmids recovered in *E. coli* show varying degrees of β-galactosidase expression. In one experiment, as many as 40% showed higher levels of β-galactosidase then the parent plasmid pRB45.

Introduction of Fusions into Yeast and Assay of β-Galactosidase Activity

Interesting fusion plasmids produced by these methods are introduced into appropriate yeast strains by standard transformation procedures.[20] We have taken the approach of using 2-μm plasmid-derived vectors because of the high frequency of transformation and the higher levels of expression from the multicopy plasmid. However, newer techniques of linearizing the vector DNA to obtain increased frequencies of integrative transformation may supersede this approach.[21] Transformants that carry the fusion are easily assayed for their level of β-galactosidase activity by patching the cells onto buffered minimal plates containing X-gal. The neutral pH that is required for efficient color production does not allow regeneration, so that transformants cannot be assayed for β-galactosidase on the transformation selection plates. Cells that express β-galactosidase may turn blue at various rates, from overnight to several weeks, depending on the level of expression. A more quantitative assay (and ultimately quicker for low-level expression), is the direct determination of *o*-nitrophenylgalactoside cleavage by cell extracts.

Yeast cells containing plasmids are grown under selection to about 2×10^7 cells per milliliter. Cells are chilled, concentrated about 20-fold by centrifugation and resuspension in 100 m*M* Tris-Cl, pH 8.0, 1 m*M* dithiothreitol, and 20% (v/v) glycerol and routinely frozen at $-20°$. The frozen cells are thawed on ice, 0.45 mm glass beads (Glasperlen, VWR-Scientific) are added to fill the liquid volume up to the meniscus, and phenylmethylsulfonyl fluoride (Sigma) is added to 1 m*M* from a 40 m*M* stock in 95% ethanol stored at $-20°$. Cells are broken open by vigorous agitation on a Vortex mixer (six times, 15 sec each) at 4°. An equal volume of the breaking buffer is mixed in, and the solution above the beads is clarified by 15-min centrifugation in an Eppendorf centrifuge. β-Galactosidase is then assayed as for a bacterial extract except that the activity is normalized to the protein concentration of the extract. We have measured the

[20] A. Hinnen, J. B. Hicks, and G. R. Fink, *Proc. Natl. Acad. Sci. U. S. A.* **75,** 1929 (1978).

[21] T. L. Orr-Weaver, J. W. Szostak, and R. J. Rothstein, *Proc. Natl. Acad. Sci. U. S. A.* **78,** 6354 (1981).

protein concentration by the method of Bradford,[22] as this method is not sensitive to high Tris concentrations. Activity normalized to protein was found to be a more reliable measure of specific activity than that normalized to the optical density of the culture, particularly when nonisogenic strains are used.

The range of activities of β-galactosidase we have measured in yeast is from 2 units to 20,000 units. Five milliliters of exponentially growing cells is sufficient to assay this range of activity within a few hours of incubation. The reactions have been carried out for up to 24 hr with less than a 20% loss of activity. Autonomously replicating yeast plasmids are rather unstable, so that recombinants that transfer the selectable marker to the chromosome are at a selective advantage. This can result in fast growing strains that may contain the multicopy plasmid in much less than 10% of the cells in the culture. It is therefore best to assay exponentially growing cells grown from fresh overnights made from fresh colonies of the transformants.

Typical Results of the Two Methods

The distribution of the deletion end points produced by the application of the two techniques to pRB45 is shown in Fig. 3. Of the total number of Lac^+ colonies obtained by *in vivo* selection, approximately 80% remained Ura^+ due to deletion hotspots between *URA3* and *'lacZ*. The remaining plasmids were Ura^- and contained large deletions that were broadly distributed over a large region of the plasmid. By contrast, the deletion end points constructed by the *in vitro* method were distributed rather uniformly throughout a small region of the plasmid. The larger deletions produced *in vivo* are useful for determining the location of a structural gene within a large DNA fragment. The *in vitro* method of fusion construction produces more tightly clustered deletion end points, which are useful for determining function over small regions of DNA or for obtaining fusions within a specific DNA segment.

The *in vitro* method allowed the recovery of *'lacZ* fusions to an open reading frame of 84 base pairs that begins with an ATG and overlaps the start of the *URA3* coding sequence. The DNA sequence of one of the fusions to the peptide coding sequence is illustrated in Fig. 4. After transformation of yeast with a derivative of pRB45 containing this fusion, we found that β-galactosidase is expressed at 10% the level of equivalent fusions made to the *URA3* gene. Fusions to the sequences that are not part of yeast genes give no detectable expression of β-galactosidase. As the

[22] M. M. Bradford, *Anal. Biochem.* **72,** 248 (1976).

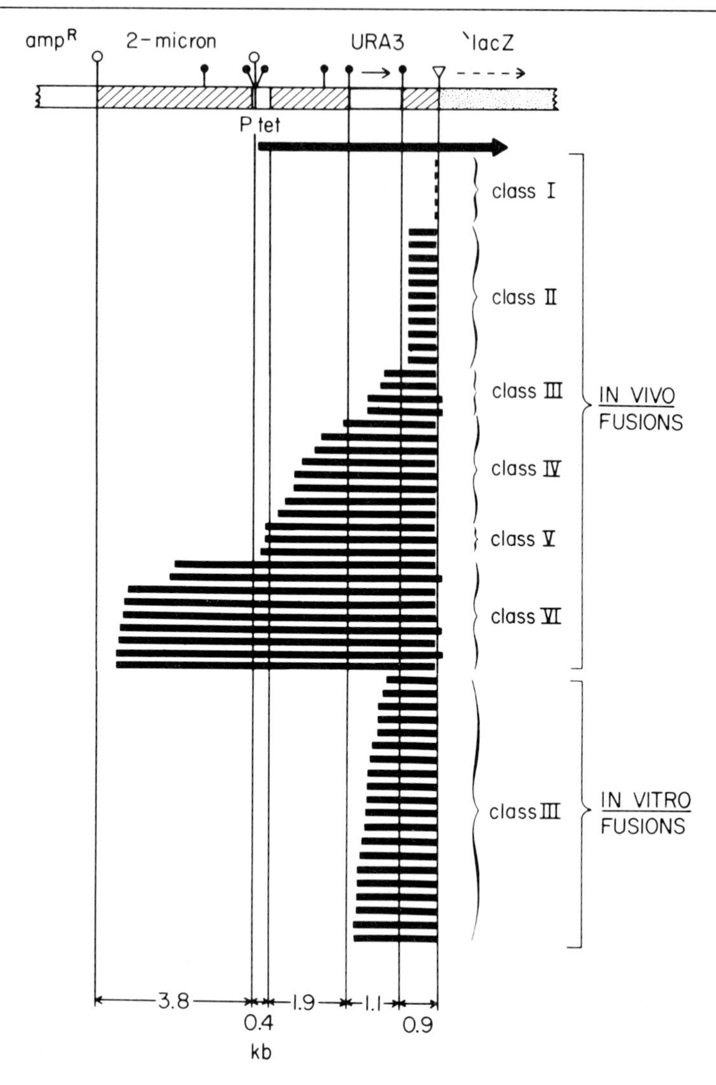

FIG. 3. Deletion map of fusion plasmids. The structure of the relevant portions of pRB45 are shown at the top. Left to right corresponds to clockwise in Fig. 1; the map is drawn approximately to scale. Restriction enzyme site symbols are as in Fig. 1. For the *in vivo* fusions, the deletion end points are assumed to lie close to the *Bam*HI site. Some of the deletions remove the *Bam*HI site and must therefore end in the *'lacZ* sequence. All of the *in vitro* fusions necessarily end at the *Bam*HI site. The various fusion classes are defined by the left-hand deletion end point. In yeast, only class II, class III, and some of class VI fusions express significant β-galactosidase activity. All the fusions shown express β-galactosidase activity in *Escherichia coli*.

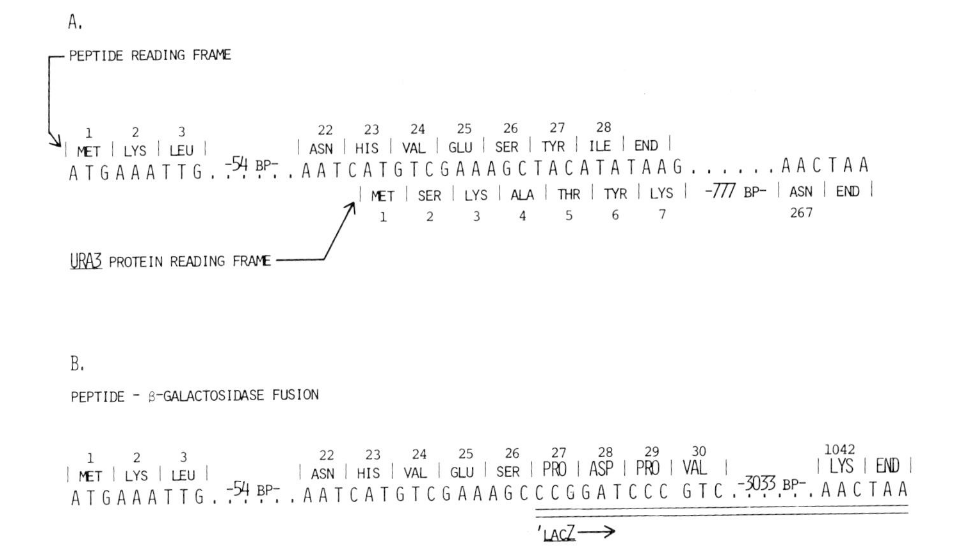

FIG. 4. (A) DNA sequence at the beginning of the *URA3* gene. An open reading frame potentially encoding a peptide of 28 amino acids overlaps with an open reading frame encoding a protein of 267 amino acids. The two reading frames overlap by 17 base pairs. (B) DNA sequence of a fusion of *'lacZ* to the peptide coding sequence. *'lacZ* is fused after the 26th codon of the peptide. The *'lacZ* segment is underlined. This fusion is expressed in yeast and *Escherichia coli*. A derivative of this fusion in which the *'lacZ* sequence is frameshifted so as to be fused to the beginning of the *URA3* protein coding sequence is also expressed in yeast.

hybrid protein is expressed in yeast, the 84 base pair sequence encoding the peptide must be transcribed. Moreover, since the AUG at the start of this sequence is capable of initiating translation of the fusion protein it must also be capable of initiating translation of the peptide itself. Evidence for the expression of this open reading frame would be difficult to obtain by other methods.

Fusions of *'lacZ* to the *URA3* gene show a gradient of β-galactosidase expression in yeast. Fusions obtained near the amino terminus of the gene tend to show higher levels of β-galactosidase (as much as 10-fold higher) than fusions obtained near the carboxyl terminus. The difference is apparently not simply due to variation in the inherent specific activity of the proteins as all of the fusions make similar levels of β-galactosidase in *E. coli*. The reason for the differences in expression are not yet understood but invalidate arguments based on absolute levels of expression.

The presence of the defined linker segment in the *in vitro* constructed fusions is extremely useful in determining the exact position and reading frame of the fusion. The linker is a unique restriction site so that DNA sequencing from this site rapidly yields the exact position of the fusion. This is particularly useful in determining the reading frame of new sequence. If the fusion expresses β-galactosidase then the reading frame of the sequence to the 5′ side of *'lacZ* can be extrapolated by extension of the *lacZ* reading frame. If the deletions are constructed in an unsequenced gene, the set of fusion plasmids provides an ordered set of unique restriction sites from which to sequence. This greatly accelerates the rate at which one can sequence a new gene as it does not require knowledge of a detailed restriction map. We have sequenced most of the *URA3* gene by this method.

The *in vivo* selected deletions end at various points 5′ to *'lacZ* resulting in a set of hybrid genes that are ill defined at the fusion junction. Most of the deletions do not remove material from the *'lacZ* segment, and thus the hybrid gene must also contain some material that was immediately adjacent to the *'lacZ* segment. The specific activity of the hybrid protein may vary owing to differences in this intervening segment. This problem is avoided with the *in vitro* method as all the fusions must contain the same linker segment.

It is worthwhile to point out that many of the fusions that are functional in *E. coli* are not necessarily true fusions to a yeast gene. We have observed several different events that complicate the analysis of fusions in *E. coli*. Fusions to the yeast sequences to the 5′ side of the *URA3* gene are strongly expressed in *E. coli* but are not at all expressed in yeast. Fusions to the peptide sequences are indistinguishable from fusions to *URA3* in *E. coli,* but show much lower levels of expression in yeast and are not regulated by pyrimidine metabolism as is *URA3*. A fusion was obtained to a point within the *URA3* coding segment that made high levels of β-galactosidase activity in *E. coli* undetectable levels in yeast. DNA sequence analysis revealed that the fusion is out of frame with respect to the *URA3* sequence. Whereas *E. coli* is capable of either an internal initiation event, or frameshifting during translation, yeast is unable to express this false hybrid. In the last analysis, the best criterion for a true fusion of *lacZ* to a yeast gene is the observation of expression of the β-galactosidase activity in yeast.

Acknowledgments

We thank Arlene Wyman, David Shortle, and Michael Lichten for helpful comments on this manuscript. We thank Jack Szostak for recommending the buffered minimal medium of Clifton *et al.* This work was supported by NIH Grants R01-GM21253-08 and R01-GM18973-11 and ACS Grant MV 90C. M. R. was supported by an NSF fellowship and by NIH training Grant GM07287.

[10] Yeast Promoters and *lacZ* Fusions Designed to Study Expression of Cloned Genes in Yeast

By LEONARD GUARENTE

Experiments probing gene regulation in bacteria have yielded two pertinent kinds of results. First, they have shed light on what bacterial signals regulate gene expression. Second, they have provided systematic methods to monitor expression, the most notable of which is the *lac* fusion technique (for review, see Bassford *et al.*[1]). This technique involves fusing genes of the *lac* operon (e.g., *lacZ*) to the promoter under study so that regulation can be monitored by changes in levels of *lac* enzymes (e.g., β-galactosidase), which can be assayed with ease. The above results have led to the development of methods that employ bacterial regulatory signals to direct expression in *Escherichis coli* of cloned genes from any source.[2]

It is now possible to apply most of the above methodology to the yeast *Saccharomyces cerevisiae*. Laying the groundwork for this extension was the development of a yeast transformation procedure[3] and shuttle plasmids that can be moved back and forth between *E. coli* and yeast.[4] More recently, the *lacZ* fusion technique has been adapted to yeast.[5,6] In this report, I describe plasmids and methods for constructing fusions of any cloned gene to *lacZ* for study in yeast. One application of this technique to be discussed is in the study of regulation of expression of yeast genes. Our studies to date in this area have given rise to plasmids containing yeast promoter regions and conveniently located restriction enzyme sites that facilitate fusion of the promoters to cloned genes. The potential use of these plasmids and *lacZ* fusions to obtain expression in yeast of cloned (heterologous) genes is discussed.

[1] P. J. Bassford, J. Beckwith, M. Berman, E. Brickman, M. Casadaban, L. Guarente, I. Saint-Girons, A. Sarthy, M. Schwartz, H. A. Shuman, and T. Silhavy, *in* "The Operon" (J. H. Miller and W. S. Reznikoff, eds.), p. 245. Cold Spring Harbor Laboratory, Cold Spring Harbor, New York, 1978.

[2] L. Guarente, G. Lauer, T. M. Roberts, and M. Ptashne, *Cell* **20,** 543 (1980).

[3] A. Hinnen, J. B. Hicks, and G. R. Fink, *Proc. Natl. Acad. Sci. U.S.A.* **75,** 1929 (1978).

[4] D. Botstein, C. Falco, S. Stewart, M. Brennan S. Scherer, D. Stinchcomb, K. Struhl, and R. W. Davis, *Gene* **8,** 17 (1979).

[5] L. Guarente and M. Ptashne, *Proc. Natl. Acad. Sci. U.S.A.* **78,** 2199 (1981).

[6] M. Rose, M. J. Casadaben, and D. Botstein, *Proc. Natl. Acad. Sci. U.S.A.* **78,** 2460 (1981).

ISBN 0-12-182001-7

Materials, Reagents, and Assays

DNA Enzymology

Restriction Digests. These are performed in prescribed buffers (New England BioLabs).

Conversion of Cohesive Ends into Flush Ends. 5′ or 3′ extensions are converted into flush ends in a 50 μl reaction containing about 1 unit[6a] of DNA polymerase I large fragment (New England BioLabs), 6.6 mM $MgCl_2$, 6.6 mM Tris-HCl, pH 7.4, 50 mM NaCl, 6.6 mM 2-mercaptoethanol, 500 μM of each of the four deoxynucleoside triphosphates, and 1–10 pmol of ends. The reaction is carried out at 23° for 1 hr and terminated by heating to 70° for 20 min, or by phenol extraction of the DNA.

Ligations. DNA ligations are performed in a 25-μl reaction volume containing 5×10^3 units[6a] of T4 DNA ligase (New England BioLabs), 50 mM Tris-HCl pH 7.8, 10 mM $MgCl_2$, 20 mM dithiothreitol, 600 mM ATP, and 1–10 pmol of ends to be joined. The reaction is performed at 15° for 2 hr. Linkers to be attached are added to 0.01 A_{260} unit. After ligation, the DNA either is used directly for transformation or gel electrophoresis (see below) or is extracted by phenol if additional enzymic reactions are desired.

Isolation of Component DNA Fragments for Use in Constructions. DNA fragments are typically separated on a 3.5% acrylamide gel (30, 0.8, acrylamide–bis), and the gel is stained in H_2O containing 0.02% methylene blue and destained in H_2O until the DNA bands are visible. Bands are excised from the gel and pulverized in a round-bottom plastic tube with a tight-fitting Teflon plunger; the DNA is eluted overnight at room temperature in a buffer containing 50% saturated phenol, 0.25 M NaCl, 5 mM Tris-HCl, pH 8.0, and 5 mM EDTA. DNA recovered from the aqueous phase is then precipitated in ethanol, dried down, and subsequently used in a ligation reaction.

Transformation

Escherichia coli. One hundred milliliters of a log-phase culture are chilled on ice for 10 min; the cells are spun down, resuspended in 50 ml of

[6a] Units. DNA polymerase large fragment: the amount of enzyme that will convert 10 nmol of deoxyribonucleotides to an acid-insoluble form in 30 min at 37°. T4 DNA ligase: the amount of enzyme required to give 50% ligation of *Hae*III fragments of phage λ DNA in 30 min at 16° with a fragment end concentration of 0.12 μM. β-galactosidase: $1000 \times OD_{420}$/time (min) × volume assayed (in ml) × OD_{600}.[7]

0.1 *M* $CaCl_2$, chilled for 20 min on ice, spun again, and resuspended in 2 ml of 0.1 *M* $CaCl_2$. Cells are then chilled on ice overnight and used in 0.1-ml aliquots per transformation. Cells prepared in this way may be used for 3–4 days.

Yeast. The method of Hinnen *et al.*[3] is used with the following modifications: 10 μg of calf thymus DNA are added along with the DNA to be used, and spheroplasts are incubated at 30° for 20 min in 150 μl of 1 *M* sorbitol, 33% yeast extract peptone dextrose media, 6.5 m*M* $CaCl_2$, 6 μg or uracil per milliliter (for URA^+ selection), and then added to regeneration agar for plating.

The *E. coli* strain LG90 (F^-, Δ*lacpro XIII*)[2] and the yeast strain BWG2-9A (α, *URA3-52*, *Ade*$^-$, *gal4*$^-$, *his4-519*) (B. Weiffenbach, unpublished observations) are routinely used because they can be transformed at high efficiency by the above methods.

β-Galactosidase Assays

Plates. These contain an M63 salt base (per liter, 13.6 g of KH_2PO_4, 2 g of $(NH_4)_2SO_4$, 0.5 mg of $FeSO_4 \cdot 7H_2O$, pH adjusted to 7.0),[7] and, per milliliter, 4 μg of thiamin, 200 μg of biotin, 4 μg of pantothenic acid, 20 μg of inositol, 4 μg of pyridoxine, 40 μg of required amino acids, 2% carbon source, and 40 μg of 5-bromo-4-chloro-3-indolyl-β-D-galactoside (XG). It is important that the pH of the plates be about 7.0.

Liquid Cultures. Cells are grown in minimal medium. Then 10^6 to 10^7 cells are spun down and resuspended in 1 ml of Z buffer (per liter, 16.1 g of $Na_2HPO_4 \cdot 7\ H_2O$, 5.5 g of $NaH_2PO_4 \cdot H_2O$, 0.75 g of KCl, 0.246 g of $MgSO_4 \cdot 7\ H_2O$, and 2.7 ml of 2-mercaptoethanol; pH adjusted to 7.0).[7] Three drops (Pasteur pipette) of $CHCl_3$ and 2 drops of 0.1% SDS are added, and cells are vortexed at high speed for 10 sec. *o*-Nitrophenyl-β-D-galactoside (ONPG) hydrolysis is measured as described.[7] Briefly, 0.2 ml of a 4 mg/ml solution of ONPG (dissolved in H_2O) is added to samples preincubated at 28°. The reaction is stopped by adding 0.5 ml of 1 *M* Na_2CO_3, and the cell debris is spun out. The OD_{420} is measured. Assays are normalized to the OD_{600} of the culture and to the assay time. Alternatively, cells in Z buffer may be broken by adding glass beads and vortexing. The ONPG hydrolyzing activity of the extract is then determined, and activity is normalized to the protein concentration of the extract. Both methods are satisfactory.

[7] J. H. Miller, ed., "Experiments in Molecular Genetics." Cold Spring Harbor Laboratory, Cold Spring Harbor, New York, 1972.

Methods

Construction of lacZ Fusions to Cloned Genes

DNA Construction

Plasmid pLG670-Z, shown in Fig. 1, has been developed for the construction of *lacZ* fusions for study in yeast.[5] It contains markers selectable in *E. coli* and yeast (Amp^R and *URA3*, respectively) and origins of replication that function in the respective organisms (from ColE1 and the yeast 2 μm plasmid). Also carried on the plasmid is a large 3′ fragment of *lacZ* immediately preceded by a *Bam* site. Insertion of DNA encoding the amino terminus of a protein into that site will result in a gene fusion to *lacZ*, as long as the translational reading frame is preserved across the junction. The restriction sites *Xho*, *Sal*, and *Sma* lie upstream of *Bam* and facilitate the insertion of DNA fragments into the plasmid as described below. All these sites occur but once on the plasmid.

Several possible approaches for constructing fusions include insertion of *Sau* 3A fragments, fusion by joining flush ends, and use of *Bam* linkers.

Insertion of Sau3A Fragments The enzyme *Sau*3A recognizes a 4-base sequence and cleaves to generate cohesive ends with the same se-

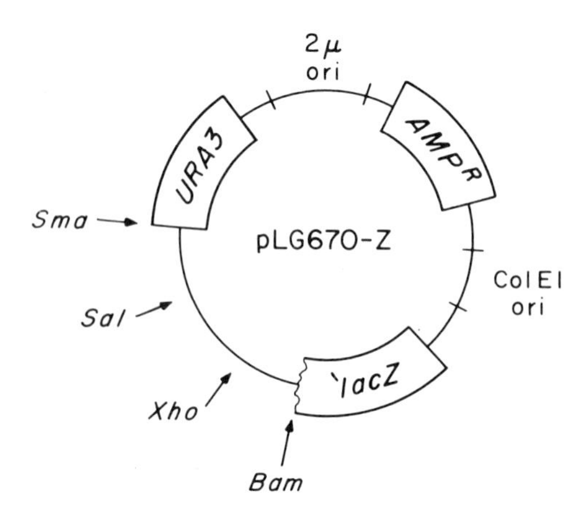

FIG. 1. Depicted is the shuttle vector pLG670-Z used for constructing *lacZ* gene fusions for study in yeast. It contains the markers Amp^R (*Escherichia coli*) and *URA3* (yeast) and origins of replication from ColE1 (*E. coli*) and the 2 μm circle (yeast). Insertion of gene fragments into the *Bam* site can result in fusion to *lacZ*. The *Xho*, *Sal*, and *Sma* sites are also of use in constructing fusions (see text). The *Xho* and *Sal* sites actually lie in a segment of DNA from the *CYC1* leader. This region, however, does not drive significant synthesis of β-galactosidase in yeast because the *CYC1* promoter has been inactivated by a deletion of the UAS.[5]

quence as those generated by *Bam* (GATC). Thus, if a gene contains an internal *Sau*3A site, its fusion to *lacZ* is facilitated. A DNA fragment extending from an internal *Sau*3A site to a *Sau*3A site upstream of the gene can readily be inserted into a pLG670-Z recipient molecule that has been cleaved with *Bam*. If one uses 1 μg of the recipient and insert DNA, about 10^3 to 10^4 transformants of *E. coli* may be obtained. Treatment of the recipient DNA with alkaline phosphatase prior to ligation will increase the percentage of transformants that contain the *Sau*3A insert. *A priori,* the insert will have a 1:2 probability of being correctly oriented; of those correctly oriented, the inserted coding sequence will have a 1:3 probability of being in frame with *lacZ*.

In some cases, for reasons that are not clear, recovered plasmids may show a strong bias in the orientation of their inserts. This problem may complicate the task of constructing in-frame fusions. If *lacZ* is being fused to an *E. coli* or yeast gene, the in-frame fusion may be identified by its encoded β-galactosidase activity in the homologous host organism. However, if *lacZ* is being fused to a gene from some other source, enzyme activity in *E. coli* or yeast resulting from the fused gene is not assured. In this case, it is desirable to insert the gene fragment in a directed orientation.

To direct the gene fragment to insert in the desired orientation only, one must employ a fragment extending from the internal *Sau*3A site to a site recognized by a different enzyme that lies upstream of the gene. If restriction at the upstream site leaves a flush end, or if the end has been rendered flush-ended by DNA polymerase (see Materials, Reagents, and Assays), then the fragment can be inserted into a pLG670-Z backbone that extends from the *Sma* (flush) site to the *Bam* site. Such inserts will all be in the desired orientation. (In addition, the number of background products of ligations, particularly plasmid backbones that self-ligate, is minimized) Alternatively, DNA fragments extending from a *Xho* (or *Sal*) upstream site to an internal *Sau*3A site may be cloned into a pLG670-Z backbone extending from the *Xho* (or *Sal*) site to the *Bam* site.

Fusion by Joining Flush Ends. If the gene to be fused to *lacZ* contains no appropriate *Sau*3A sites, alternative strategies to the above must be employed. One such strategy is to generated a gene fragment that terminates in a flush end. This may be done by cleaving the gene with an enzyme that gives flush ends, or by use of DNA polymerase following cleavage by an enzyme that leaves cohesive ends. Alternatively, exonucleases such as *Bal31*[8] or exonuclease III plus S1,[9] can be used to generate flush

[8] H. B. Gray, D. Ostrander, H. Hodne, R. J. Legerski, and D. Robberson, *Nucleic Acids Res.* **2,** 1459 (1975).

[9] T. M. Roberts and G. D. Lauer this series, Vol. 68, p. 473.

ends after cleavage. In this case, a population of fragments will be generated with different (flush) ends internal to the gene.

If the cloned gene is preceded by a *Sal* (or *Xho*) site, a DNA fragment may be prepared extending from that site to the site internal to the gene that has been rendered flush. This fragment could be inserted in directed fashion into a pLG670-Z recipient molecule extending from the *Sal* (or *Xho*) site to the *Bam* site preceding *lacZ*, which itself has been rendered flush ended. Joining of the flush ends will result in gene fusion if the translational reading frame is preserved across the junction. Alternatively, a DNA fragment bearing two flush ends may be inserted into a pLG670-Z molecule that has been cleaved with *Bam* and rendered flush ended.

Use of Bam Linkers. Oligonucleotide molecules containing the sequence cleaved by *Bam* (*Bam* linkers) may be attached by DNA ligase to gene fragments that terminate with flush ends. Cleavage of the ligated molecules with *Bam* will generate gene fragments that can be fused to the *lacZ* gene of pLG670-Z that itself has been digested with *Bam*. The orientation of such inserts may be directed as described for insertion of *Sau*3A fragments. Methods of constructing fusions using flush ends or linkers provide sufficient flexibility so that virtually any cloned gene can be fused in frame to *lacZ*.

Detection of in-Frame Fusions

If the DNA sequence of the gene to be fused to *lacZ* is known, in-frame fusions can be made simply by choosing the appropriate gene fragment. In the absence of sequence information, confirming that a fusion is in frame must proceed on a case by case basis. In constructing gene fusions between the yeast gene encoding the iso-1-cytochrome *c* (*CYC1*) and *lacZ*, it was noticed that the in-frame fusion displayed significant β-galactosidase activity in *E. coli*.[5] This expression is likely to originate from transcription that is initiated in the *CYC1* leader DNA and translation that is initiated at the ATG at the start of *CYC1* coding DNA (Guarente, unpublished results). Thus, the in-frame fusion could be detected by the appearance of β-galactosidase activity after transformation of a Lac$^-$ *E. coli* indicator strain with DNA containing the fused gene.

Fusions to Yeast Genes

In general, in-frame fusions of yeast genes to *lacZ* should result in β-galactosidase expression in yeast. This expression can be monitored by a plate assay that employs the chromogenic β-galactosidase substrate, XG, or by a quantitative determination of ONPG hydrolyzing activity in a liquid-grown culture (see Materials, Reagents, and Assays). Enzyme expressed from such fusions should respond to the regulatory signals of the yeast genes to which they are fused.

Use of 2 μm Plasmids. We have studied fusions of *lacZ* to regulatory signals of each of two different yeast genes, *CYC1*[9] and *GAL10*.[10] In each case, the fusion resides on 2 μm-containing plasmids and β-galactosidase expression follows the regulation of the promoter to which *lacZ* is fused (see Fusion of Yeast Promoters, below).

In some cases, the high copy number (about 20/cell) of the 2 μm plasmids might interfere with normal gene regulation. Here, the use of vectors that integrate into the yeast genome or that replicate autonomously in low copy is called for.

Effects of Vector Sequences. The vector sequences that precede the yeast DNA is a yeast gene–*lacZ* fusion may affect gene regulation. For example, we have found that a sequence of pBR322 DNA (between the *Hin*dIII and *Sal* sites) is homologous to a portion of the *CYC1* promoter region (unpublished observation). This region of the wild-type *CYC1* promoter lies about 250 base pairs upstream from the transcriptional start (−250) and activates gene expression. If this upstream activation site (UAS) is deleted, expression is reduced about 500-fold. However, if it is replaced with the pBR322 sequence, expression is only slightly reduced.

Fusion of Yeast Promoters to Cloned Genes

The CYC1 Promoter. Transcription of a wild-type *CYC1* gene has been shown to be glucose repressible.[11] The *CYC1–lacZ* fused gene directs synthesis of 100 units (see Materials, Reagents, and Assays) of β-galactosidase in minimal glucose medium, and 1000 units of enzyme in medium supplemented with glycerol and ethanol as carbon sources.[10] The *CYC1* promoter contains two important regions, the UAS mentioned above, and the region around the TATA box and sites of transcription initiation.[10] (*CYC1* mRNA, has been found to contain multiple 5′ ends.[12])

Two shuttle vectors have been constructed that carry the *CYC1* promoter followed by a *Bam* site (Fig. 2). In vector C1 the *Bam* site immediately follows the ATG at the start of *CYC1* coding DNA.[10] In vector C2, this ATG and three nucleotides that precede *CYC1* coding DNA have

[9] L. Guarente, R. Yocum, and P. Gifford, *Proc. Natl. Acad. Sci. U.S.A.* **79,** 7410 (1982).

[10] L. Guarente and T. Mason, *Cell,* in press.

[11] R. Zitomer, D. Montgomery, D. Nichols, and B. Hall, *Proc. Natl. Acad. Sci. U.S.A* **76,** 3627 (1979).

[12] G. Faye, D. Leung, K. Tatchell, B. Hall, and B. Smith (1981). *Proc. Natl. Acad. Sci. U.S.A.* **78,** 2258 (1981).

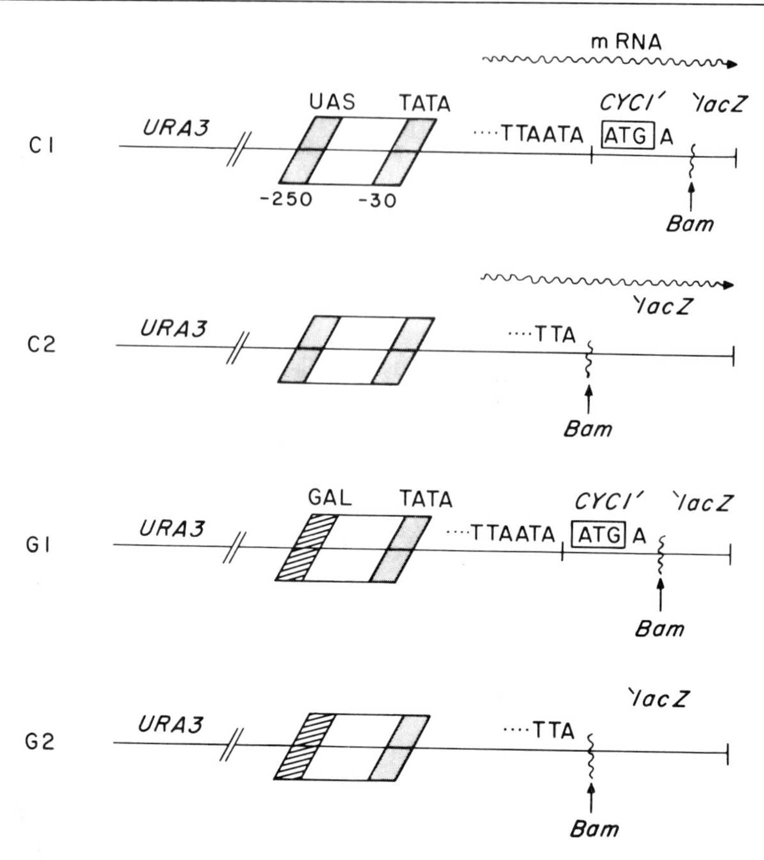

FIG. 2. Yeast expression vectors are shown. These vectors have the same markers and origins of replication as pLG670-Z (Fig. 1). C1 bears the *CYC1* promoter which contains a TATA box and an upstream activation site (UAS) at −250 required for activation of the promoter. The terminal nucleotides of *CYC1* leader DNA and the ATG at the start of *CYC1* coding DNA (boxed) are shown. C2 is identical to C1 except that the sequence ATA[ATG]A (adjacent to the *Bam* site of C1) has been deleted. G1 and G2 are analogous to C1 and C2 except that the UAS has been substituted with a segment of DNA that lies upstream of *GAL10* (hatched box). This segment confers galactose inducible regulation on the promoter. The use of these plasmids for the expression of cloned genes in yeast is discussed in the text.

been deleted (Guarente, unpublished observation). (In both cases, the *Bam* site is immediately followed by the 3′ *lacZ* fragment.) Thus, the C1 DNA upstream of the *Bam* site encodes a mRNA leader and AUG initiator codon, whereas the C2 DNA upstream of the *Bam* site encodes the leader without the initiator codon.

Expression of a cloned gene, X, in yeast may be achieved using either

C1 or C2. If the X coding sequence is inserted into the *Bam* site of C1, it may be expressed as a result of transcription that is directed by the *CYC1* promoter, and translation that initiates at the *CYC1*-encoded AUG. Translation would proceed through any X leader RNA into the X coding sequence. Expression of *lacZ* has been achieved using C1 and analogous plasmids.[5] If the X coding sequence is inserted into the *Bam* site of C2, expression of the unfused X protein may be achieved. In this case, translation would initiate at the start of the X coding sequence. Although we have not yet probed the expression potential of C2, genetic experiments in yeast[13] and biochemical experiments in higher eukaryotic systems[14] suggest that translation initiates at the AUG closest to the 5′ end of the mRNA. If the inserted X leader DNA does not encode any AUG triplets, then the AUG encoded at the start of X coding DNA would be the first AUG of the *CYC1*–X chimeric mRNA. Translation of such an mRNA, therefore, might be expected to initiate at this codon. This point in greater detail in the next section.

A set of vectors, G1 and G2, that are analogous to C1 and C2 have also been constructed[9] (Fig. 2). G1 and G2 differ from C1 and C2 only in the DNA in the −250 region of the promoter. In these plasmids, the UAS has been replaced by a DNA segment that lies upstream of the *GAL10* gene of *S. cerevisiae*. Although this segment does not encode a transcriptional start site, it does encode galactose-inducible expression.[9] Thus, the chimeric promoter, consisting of the *GAL10* segment and the *CYC1* TATA box and initiator region directs synthesis of <1 unit of β-galactosidase in medium with glucose and about 1000 units in galactose medium. Translation of sequences encoded by a gene X cloned into G1 and G2 will proceed in a manner identical to that of C1 and C2. In this case, however, because of the strong galactose inducibility of the promoter, levels of expression of the X product may be varied over a 1000-fold range.

Will Codons at the Start of Cloned Genes Function as Initiators in Yeast?

More detailed experiments designed to troubleshoot problems encountered in expressing cloned genes in yeast are described here. For example, if one does not observe expression of a protein product after cloning a gene X into C2 or G2 as described above, one may wish to pinpoint where the block occurs. The block may be due to an inability of yeast ribosomes to initiate translation at the AUG triplet encoded at the start of

[13] F. Sherman, J. Stewart, and A. Schweingruber, *Cell* **20,** 215 (1980).

[14] M. Kozak, *Cell* **22,** 7 (1980).

X, or to other factors, such as instability of the protein in yeast. Therefore, to determine directly whether the AUG encoded by the start of X functions as an initiator in yeast, I have devised the following scheme. First, a DNA fragment encoding an amino-terminal portion of X is cloned into C2 (or G2) so that an X–*lacZ* fused gene is formed (see Fig. 3B). Detection of β-galactosidase expression from such a fusion would demonstrate that the AUG encoded at the start of X functions as an initiator in yeast. In this case, insertion into C2 of the intact X gene should result in synthesis of the unfused protein product. If the product is not observed, it is likely that it is unstable in yeast.

If no β-galactosidase expression derives from the fused gene in the above case, it is likely that the AUG encoded at the start of X does not

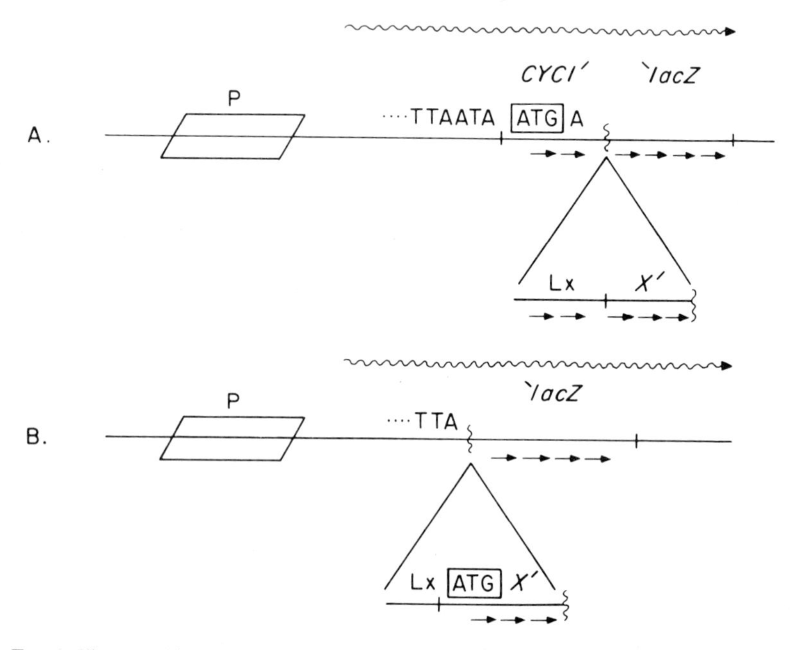

FIG. 3. Illustrated in the utilization of C1 and C2 (Fig. 2) (or G1 and G2) to probe whether AUG triplets encoded at the starts of cloned genes are recognized as initiator codons in yeast. (A) A segment of DNA encoding the leader (L_x) and a portion of the coding sequence of a cloned gene, X, has been inserted into C1 (or G1). Translation (arrows) will initiate at the AUG corresponding to the *CYC1* ATG (boxed), proceed through the leader and gene X sequences, and finally through *lacZ* encoded sequences, providing that the translational reading frame is in register across each junction. (B) The X gene segment has been inserted into C2 (or G2). In this case, translation may initiate at the triplet corresponding to the gene X ATG (boxed) and proceed through gene X and *lacZ* sequences, if the translational reading frame is in register across the X–*lacZ* junction. The use of these manipulations to infer whether the gene X encoded AUG functions as an initiator in yeast is discussed in the text.

function as an initiator in yeast, but other possibilities exist to explain this result. For example, it is possible that internal codons of the X gene are not utilized in yeast or that the X–β-galactosidase hybrid protein is inactive. To distinguish between these possibilities, the X gene fragment may be fused to the *lacZ* fragment in C1 (or G1) (Fig. 3A). In this case, care must be taken so that the AUG initiation codon of *CYC1* is in frame with the inserted X codons and that these in turn are in frame with *lacZ*. Translation, thus, will initiate at the AUG encoded by *CYC1*, proceed through the X leader and coding sequences, and finally into *lacZ* sequences. Failure of such a construct to give rise to β-galactosidase expression would suggest that at least one X codon is not utilized in yeast, or possibly that the hybrid protein is inactive. Moreover, if β-galactosidase is expressed upon insertion of the X gene fragment into C1, but *not* C2, then it is likely that the AUG encoded at the start of X does not function as an initiator in yeast.

With this information, site-specific mutagenesis around the start of X could be performed, and mutations isolated that render the AUG encoded at the start of X as an active initiator. These mutations could be easily recognized, since they would result in β-galactosidase expression, which can be detected on plates containing the dye XG (see Reagents section). After the isolation of such mutations, the X gene could be regenerated intact by recombination *in vitro*. A procedure analogous to this has been developed to express cloned genes in *E. coli*.[2]

Comments

Plasmids described in this chapter may be used to probe the signals that govern the intiation of transcription and translation in *S. cerevisiae*. These methods provide powerful tools in the analysis of the expression of yeast genes. Further, the plasmids should facilitate the expression in yeast of any cloned gene to produce the native, unfused product. If there is a block to the expression of a particular gene, the methods outlined herein should pinpoint the stage at which the block occurs. With this information, it should be possible to isolate mutations that overcome the block, as discussed.

[11] Expression of Genes in Yeast Using the *ADCI* Promoter

By GUSTAV AMMERER

Compared to higher eukaryotic cells, the yeast *Saccharomyces cerevisiae* seems to recognize slightly different signals for gene expression. Therefore it was desirable to develop a vector system capable of promoting high levels of transcription for any coding sequence introduced into yeast cells. Such vectors should be suitable for studying the production of foreign proteins in yeast. They also may be used to enhance the synthesis of yeast gene products, for example, regulatory proteins, which are normally present only at low levels. In this context it is also possible to study the physiological effects of constitutive expression of otherwise highly regulated genes. Because yeast alcohol dehydrogenases and their genes have been well characterized,[1] we chose to use the 5′-flanking sequence of the *ADCI* gene (coding for *ADHI*) as a portable promoter. Although the *ADCI* gene probably does not contain the most powerful *polII* promoter in yeast, the relative abundance of this glycolytic enzyme is reflected on transcriptional level. *ADHI* mRNA is estimated to be 1–2% of poly(A) RNA.

Bennetzen sequenced about 2100 nucleotides of the *ADCI* region including 750 base pairs (bp) flanking the 5′ end and 320 bp flanking the 3′ end.[2] A presumptive Goldberg–Hogness box can be found at position −128 from the initiator codon. The 5′ ends of the mature mRNA have been mapped around nucleotides −37 and −27. These presumptive transcription initiations sites are preserved when the gene is maintained on a plasmid. As with other highly expressed yeast genes, the untranslated leader region of the mRNA is almost completely devoid of G residues. This strong bias is probably not essential for transcription initiation but might be critical for the translational capacity of the mRNAs. In order to ensure high translation rates for hybrid mRNAs containing non-*ADHI* coding information, it would be preferable to leave the untranslated leader of *ADCI* intact. By joining the different genes within the translated regions one normally generates a fusion protein, which might be unacceptable for many purposes. The cleanest approach would be to replace the

[1] T. Young, V. Williamson, A. Taguchi, M. Smith, A. Sledziewski, D. Russell, J. Osterman, C. Denis, D. Cox, and D. Beier, *in* "Genetic Engineering of Microorganisms for Chemicals," (A. Hollaender, D. Demoss, S. Kaplan, J. Konisky, D. Savage, and R. Wolfe, eds.), p. 335. Plenum, New York, 1982.

[2] J. Bennetzen and B. D. Hall, *J. Biol. Chem.* **257,** 3018 (1982).

ISBN 0-12-182001-7

coding region of *ADCI* directly at the initiator *ATG*. This could be accomplished, for example, by the lengthy procedure of chemically synthesizing an oligonucleotide "bridge" fragment to reach from a restriction site 5′ to *ATG* in *ADCI* to a site 3′ to *ATG* in the coding region to be expressed. As a more rapid means of joining the *ADCI* 5′-flanking region to a variety of genes, I introduced convenient restriction endonuclease sites proximal to the initiator *ATG* of the yeast *ADCI* gene. A 1600-bp *Sau*3A fragment served as starting material and was subcloned after *Bam*HI linkers (5′) and *Hin*dIII linkers (3′) had been attached. This piece contained 108 bp of translated *ADCI* sequence and 1500 bp of the 5′-flanking sequence. It was trimmed further on its 3′ end by either a combination of ExoIII and Sl nucleases or later by *Bal*31 nuclease, which proved to be preferable over the first method. The pool of fragments shortened to the desired size was ligated to different molecular linkers, recut with the specific restriction endonucleases, purified by electrophoresis, and ligated into appropriate vectors. After transformation into *E. coli,* individual colonies were screened for inserts, and the position of the linker was determined.

Construction of Promoter Fragments

All enzymes were obtained from Bethesda Research Laboratories (BRL) and normally used as recommended by the supplier. Molecular linkers were purchased from Collaborative Research or BRL.

Step 1. Bal31 Nuclease Digestions. When *Bal*31 nuclease was used in our experiments we started with a promoter fragment that was already shortened and subcloned after a previous *Exo*III–Sl nuclease treatment and had a linker at position +28. Ten micrograms of linearized DNA (about 5 pmol of DNA ends) were dissolved in H_2O and brought to a final concentration of 20 m*M* Tris-HCl, pH 8.1, 200 m*M* NaCl, 12 m*M* $MgCl_2$, 1 m*M* EDTA, and 100 μg of bovine serum albumin (BSA) per milliliter in a volume of 200 μl. The prewarmed sample was digested with 0.5 unit of *Bal*31 enzyme at 30°. After 15 and 30 sec, 100-μl aliquots were mixed with stop solution (100 μl of phenol–chloroform–isoamyl alcohol, 50:50:1 + 5 μl of 250 m*M* EDTA). Under these conditions we found that the enzyme removed 1–2 bp per second per DNA end. To get an even distribution of fragments digested to different length, the reaction mixture was sometimes slowly dripped into the stop solution, using prewarmed pipette tips. After ether extraction and ethanol precipitation of the aqueous phase, the DNA was dissolved in 50 μl of H_2O. Two-microliter aliquots were cut with a convenient restriction enzyme (*Bam*HI or *Sph*I), and the amount of degradation was determined by separating the fragments on agarose gels. Digestions with *Bal*31 nuclease turned out to be quite variable depending on the DNA preparation, DNA concentration, or the preparation and storage of the enzyme. Therefore it is useful to optimize incubation conditions using preparative amounts of DNA.

Step 2. Linker Addition. Although *Bal*31 nuclease should leave a large fraction of DNA molecules with blunt ends, we found that ligation efficiency was much higher after incubation with DNA polymerase I Klenow fragment. Ten microliters of *Bal*31-treated DNA solution (~1 μg) was brought to a final volume of 40 μl in 10 m*M* Tris-HCl, pH 7.4, 6 m*M* $MgCl_2$, 2 m*M* 2-mercaptoethanol, 23 μ*M* concentration of each deoxyribonucleotide, and 100 μg of BSA per milliliter. The sample was incubated with 2 units of DNA polymerase I Klenow fragment for 30 min at room temperature. The reaction was stopped by heating for 10 min at 65°. At the same time molecular linkers were phosphorylated in a 40-μl volume of 10 m*M* Tris-HCl, pH 7.6, 10 m*M* $MgCl_2$, 1 m*M* spermidine, 10 m*M* DTT, and 100 μg of BSA per milliliter with 30 μCi of labeled [^{32}P]ATP and 10 units of T4 polynucleotide kinase. After 10 min at 37°, 1 μl of 20 m*M* ATP was added and the incubation was continued for 30 more minutes. The reaction was stopped by heating for 10 min at 65°. Forty microliters of blunt-end DNA was combined with 20 μl of phosphorylated linkers and 1 μl of 20 m*M* ATP, and 2.5–3 units of T4 ligase were added. The sample was kept at 14° for more than 12 hr. The T4 ligase was inactivated by heat (10 min, 65°). Sodium chloride was added to a final concentration between 60 m*M* (*Hin*dIII linkers) and 150 m*M* (*Xho*I linkers). The sample was recut with an excess of the enzyme specific for the linker (100 units) and the normal amount of enzyme cutting upstream from the promoter (*Bam*HI, 5 units) for 2–3 hr. The reaction was terminated by heating to 65°. Small aliquots were taken before and after digestion with the restriction enzymes and separated on a 8% nondenaturing acrylamide gel and autoradiographed for several hours. In this way kinasing of the linkers, efficient ligation, and complete recutting could be assured.[3] Fragments were purified and separated from the molecular linkers by electrophoresis on agarose or acrylamide gels. The DNA was visualized by staining with ethidium bromide, and the fragments were electroeluted and ligated into plasmids with the appropriate restriction endonuclease sites.

Step 3. Screening for Position of the Molecular Linker. Minipreparations of plasmid DNA were isolated from 5 ml of L-broth culture using the method described by Birnboim and Doly.[4] The DNA was dissolved in 100 μl of H_2O. Five microliters were cut with the restriction enzyme corresponding to the linker in 20-μl reaction volumes. After digestion, the

[3] H. M. Goodman and R. J. MacDonald, this series, Vol. 68, p. 75.

[4] A. C. Birnboim and J. Doly, *Nucleic Acids Res.* **7**, 1513 (1979).

sample was heated for 10 min at 65°. DNA polymerase I Klenow fragment (0.4 unit) and 0.5 μCi of the suitable ^{32}P-labeled deoxyribonucleotide were added and incubated for 30–40 min at room temperature. The reaction was terminated again by heat. The DNA was then recut with *Alu*I, generating small, labeled fragments. The size of these fragments was dependent on the distance from the linker to the *Alu*I site at position −36. After ethanol precipitation, the pellets were dissolved in formamide dye mixture and analyzed on a 20% sequencing gel. Chemical sequencing reactions from a fragment of known size were used as size markers.[5] In this way the position of the linker could normally be calculated with an error of plus or minus one nucleotide.

Principally, the screening for a functional yeast promoter can be facilitated by restoring promotion and function of a selectable or easily detectable gene product in yeast, e.g., β-galactosidase,[6] cytochrome *c*, enzymes of the adenine pathway, or galactokinase.

Vectors Containing the *ADCI* Promoter

By the method described above, a variety of promoter fragments were obtained and characterized. Vectors with *Bam*HI, *Hin*dIII, *Xho*I, and *Eco*RI linkers inserted into the untranslated leader region are available (Figs. 1A and 1B). These vectors are related either to *ars1*-containing vectors such as YRp7[7] or 2 μm DNA vectors such as YEp13.[8] In all cases a functional promoter can be cut out as a 1500-bp fragment, using *Bam*HI

```
A   -1500                                     -410                  -128
                                                                      |
    |-----------------------------//------------+----------------------TATAAATA------------/

    Bam HI                                     SphI


     -40        -30        -20        -10       -1    met ser ile pro glu thr
      |  *       |  *       |          |         |
 /----TCAAGCTATACCAAGCATACAATCAACTATCTCATATACA    ATG TCT ATC CCA GAA ACT-----
       |                        / /  |  /               /    /  \
      AluI                     R6 H5 X2 B8             H9  H21 H10
```

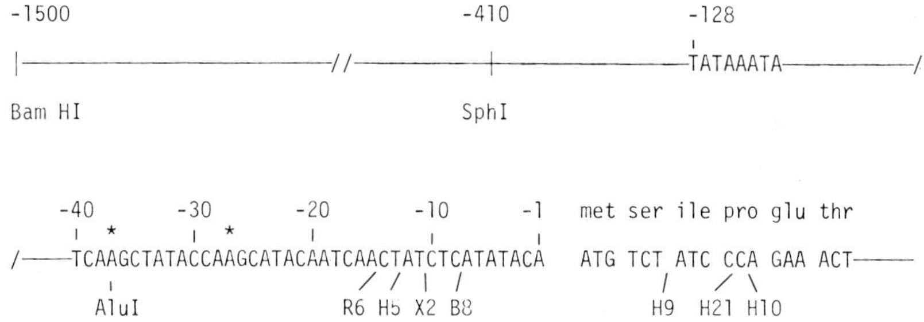

FIG. 1A. Sequence of the N-terminal region of the *ADCI* gene. the stars mark the 5′ ends of the mature *ADHI* mRNAs. The letters indicate the site of the linker attachment to the promoter. H = *Hin*dIII linker CCAAGCTTGG; B = *Bam*HI CCGGATCCGG; X = *Xho*I CCTCGAGG, R = *Eco*RI GGAATTCC.

[5] A. Maxam and W. Gilbert, this series, Vol. 65, p. 497.

[6] M. Rose and D. Botstein, this volume [42].

[7] A. Tschumper and J. Carbon, *Gene* **10,** 157 (1980).

[8] J. R. Broach, J. N. Strathern, and J. B. Hicks, *Gene* **8,** 121 (1979).

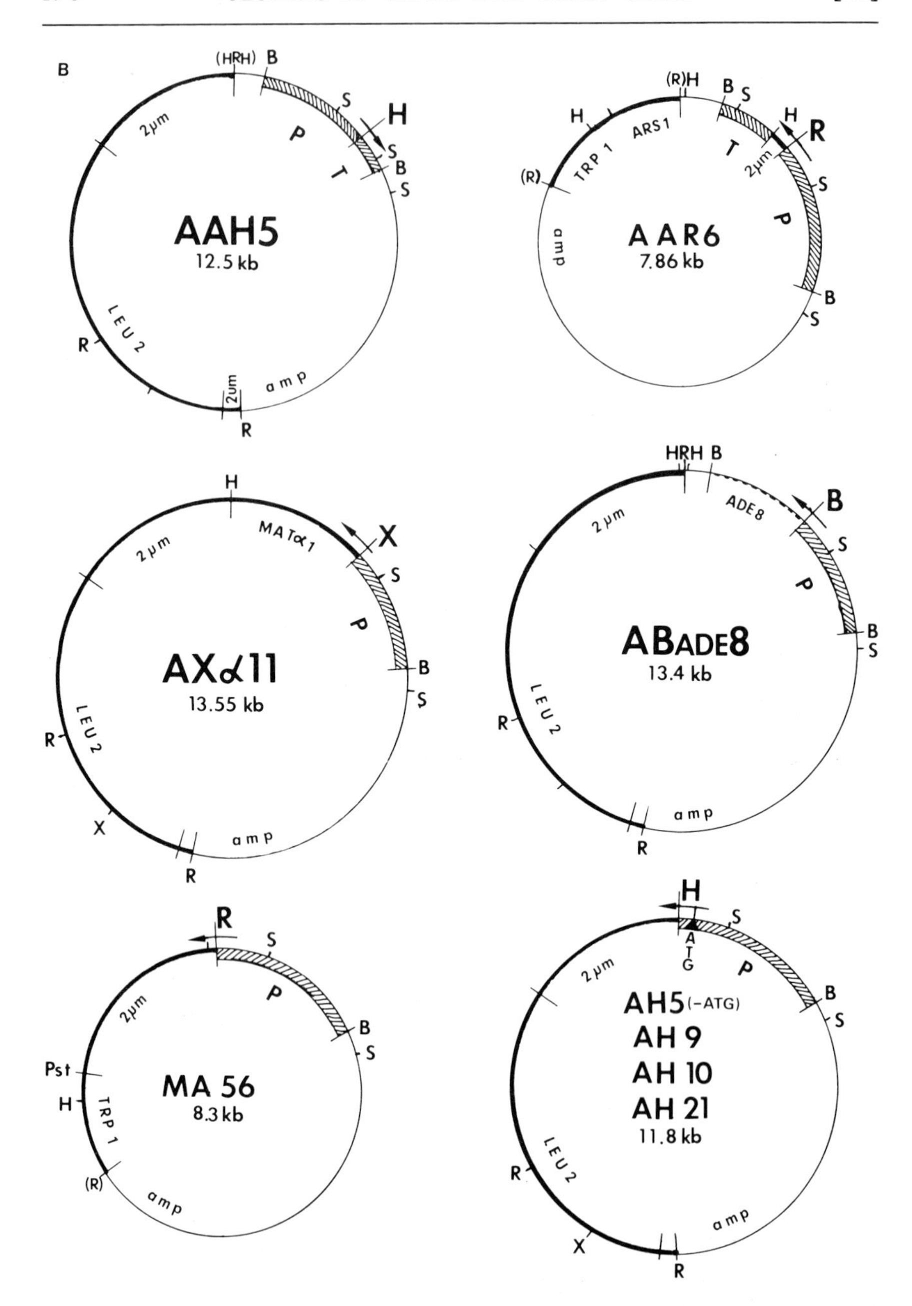

FIG. 1B. Vectors containing the *ADCI* promoter fragment. The promoter with the *Eco*RI linker at position −14 (R6) is used in plasmid pMA56 and in plasmid pAAR6. The fragment with the *Hin*dIII linker at position −12 (H5) is used in pAAH5 and pAH5. Fragments with a *Hin*dIII linker inserted at positions +7, +11, +12, are used in plasmids pAH9, pAH21, and the enzyme specific for the linker, or as a ~410 bp fragment, using *Sph*I endonuclease. Plasmids pAH9, pAH10, and pAH21 retained the initiator methionine of the *ADCI* gene and result in the synthesis of a fusion protein when joined to a coding region. Each plasmid has a single *Hin*dIII site in a different reading frame. Vector pAH5 is similar to the previous plasmids except that the *ADCI* sequence is deleted to position −12 upstream from the *ATG*. In pMA56 the *ars1* sequence of YRp7 was replaced by the *Pst*I–*Eco*RI fragment containing the replication origin of yeast 2 μm DNA. The *Eco*RI site between the *TRP1* gene and the amp^r region is destroyed. The *ADCI* promoter was inserted as a *Bam*HI–*Eco*RI piece into the tet^r region. In respect to stability in yeast, the plasmid behaves similarly to YEp13. Five primed flanking sequences seem to be generally responsible for directing transcription to initiate at specific sites. As expected, we found in Sl nuclease experiments using a chimeric combination of *ADCl* with rat growth hormone cDNA[9] that the mRNA ends were unaltered for the hybrid gene. In addition to faithful transcription initiation, termination and poly(A) tailing of the RNA may be important for gene expression. In our experiments, the mammalian cDNA did not provide functional signals for termination in yeast. Instead, transcription initiating from the *ADCI* promoter fragments terminated within the 2 μm part of the vector. A 2 μm DNA fragment containing the replication origin (bp 105–1998 of the B form)[10] hybridizes specifically to a distinct poly(A) transcript starting from the *ADCI* promoter (Fig. 2). The transcript covers about 600 bp including the carboxy-terminal end of an open-reading frame located on the same DNA strand as the *ADCI* promoter.[11] This reading frame was assigned to the flipping enzyme (Able) encoded by 2 μm plasmid. It seems that this gene provides a signal for poly(A) addition and its RNA usually terminates within the long inverted repeat of 2 μm DNA.[10]

We also constructed vectors with the C-terminal and 3′-flanking region of the *ADCI* gene (450 bp *Hin*dIII–*Bam*HI fragment, referred to as ADH terminator) downstream from the promoter (Fig. 1B). These vectors

[9] P. H. Seeburg, J. Shine, J. Martial, J. D. Baxter, and H. M. Goodman, *Nature (London)* **270,** 486 (1977).

[10] J. R. Broach, J. F. Atkins, J. F. McGill, and L. Chow, *Cell* **16,** 827 (1979).

[11] J. L. Hartley and J. E. Donelson, *Nature (London)* **286,** 860 (1980).

pAH10, respectively. The plasmid pAXα11 contains the promoter fragment with an *Xho* linker at −10 (X2); the plasmid ABade8 contains the fragment with the *Bam*HI linker at position −7 (B8). P = *ADC*I promoter, T = *ADC*I terminator, X = *Xho*I, B = *Bam*HI, R = *Eco*RI, *H* = *Hin*dIII, S = *Sph*I. A letter set in parentheses means that a previous restriction endonuclease site has been destroyed. This was accomplished by cutting the site with the specific enzyme, filling in the recessive ends with DNA polymerase I Klenow fragment, and religating the blunt ends.

showed two to three times higher expression of an inserted human interferon gene when compared to a plasmid terminating in 2 μm DNA. Effects of the 3'-flanking sequence on the mRNA stability are possible but not proved. Plasmids pAAR6 and pAAH5 also offer the advantage that one can transfer from them to other vectors a single integral unit of promoter–inserted sequence–terminator as either *Bam*HI or *Sph*I fragment. Many of the common yeast–*E. coli* vectors contain a single *Bam*HI or *Sph*I site within the *tet*^{r} region of pBR322. In pAAR6, which contains a single *Eco*RI site for cloning, a small 2 μm *Eco*RI–*Hin*dIII (105 bp) fragment served as adapter between the terminator and promoter of *ADC*I. The vector was constructed from YRp7[7] after the two *Eco*RI sites had been eliminated. Plasmid pAAH5 provides a *Hin*dIII site for cloning. The short DNA sequence between the original two *Hin*dIII sites of YEp13 was deleted, and the sites were destroyed.

In the plasmid ABade8 a *Bam*HI–*Bam*HI promoter piece was placed in front of a truncated *Drosophila melanogaster ADE*8 gene.[12] This plasmid can complement *ade*8 mutations in yeast. Plasmid *AX*α11 (*Xho*I linker at position −10) was constructed by ligating a *Xho*I–*Hind*III fragment of the *MAT*α*l* gene[13] together with a *Bam*HI–*Xho*I promoter fragment into *Bam*HI, *Hin*dIII cut YEp13. All the vectors described can confer ampicillin resistance to *E. coli* cells. In yeast plasmids pAH5–pAH21, pAAH5 pABade8, and pAXα11 complement *leu2* mutations. Plasmid pMA56 and pAAR6 contain the *TRP1* gene as a selectable marker.

Regulation of ADHI

Because control of transcription plays a major role in regulating gene expression, one might anticipate that the joining of the 5'-flanking region of *ADCI* to other coding regions would impart *ADCI*-specific regulation on these genes. Yeast *ADHI* was originally considered to be a constitutively produced enzyme. Studies by Denis *et al.*[14] have shown that expression of the *ADCI* gene is regulated to some extent. When yeast cells are shifted from glucose to ethanol-containing medium, the amount of the enzyme, the level of its translatable mRNA, and the amount of RNA detectable by hybridization on Northern blots decreases considerably. A quite similar effect can be observed when cells reach late log or stationary growth phase. The drop in *ADHI* expression is normally accompanied by derepression of the isoenzyme *ADHII*.[1] At the same time a new relatively

[12] S. Henikoff, K. Tatchell, B. D. Hall, and K. A. Nasmyth, *Nature* (*London*) **289,** 33 (1981).
[13] K. Tatchell, K. Nasmyth, B. D. Hall, C. R. Astell, and M. Smith, *Cell* **27,** 26 (1981).
[14] C. Denis, J. Ferguson, and T. Young, *J. Biol. Chem.*, in press (1983).

weakly expressed transcript appears that hybridizes not only to *ADCI* coding region, but also specifically to the 5′-flanking sequence (Fig. 2). It starts about 1000–1100 nucleotides upstream from the start points originally mapped for *ADHI* message. Because of the upstream start and stop signals for translation, it should not be a translatable *ADHI* mRNA. Additionally it should not act as a precursor to *ADHI* mRNA. Beier and Young[15] have shown that deletions of the region upstream from the TATA box or upstream from the *Sph*I site do not abolish *ADHI* activity. Instead, in this case high expression of *ADHI* continues even when cells are grown on ethanol as carbon source. If more than 1400 bp of flanking sequence are present, *ADHI* activity in those cells is regulated in a similar way as in wild-type strains. In all these experiments the *ADCI* gene was maintained on a plasmid.

No systematic studies of transcriptional regulation have been done with fusions using the *ADCI* promoter. However, most of the data indicate that production of foreign proteins behaves similar to *ADHI* expression. For those cases in which yeast cultures are grown to high density, it might therefore be preferable to use a small promoter piece (e.g., between the linker and the *Sph*I). We also found that on a plasmid the expression of the larger transcript seemed considerably enhanced compared to the amount of the same RNA detected from chromosomal *ADCI*. One is tempted to speculate that read-through from the upstream promoter site is responsible for the low expression of *ADHI* in cells grown on ethanol. But generally speaking the significance of this read-through transcript to *ADHI* regulation, as well as the whole nature of the *ADCI* control mechanism, is still obscure.

Conclusion

The *in vivo* expression obtained by joining the yeast *ADC1* promoter to other yeast genes and to heterologous genes must be evaluated at two different levels. For all coding regions tested, attachment to the *ADC1* 5′ flanking sequence promoted active transcription in yeast, as evidenced by a strongly hybridizing band on Northern blots probed with the non-*ADH* coding sequence. However, at the level of stable protein product accumulated in yeast, the results have been much more variable, with high expression in the case of human α-interferon[16] and moderate or undetectable levels for other genes (hepatitis B surface antigen, bovine parathyroid

[15] D. Beier and T. Young, *Nature* (*London*) **300,** 724 (1982).

[16] R. A. Hitzeman, T. E. Hagie, H. L. Levine, D. V. Goeddel, G. Ammerer, and B. D. Hall, *Nature* (*London*) **293,** 717 (1981).

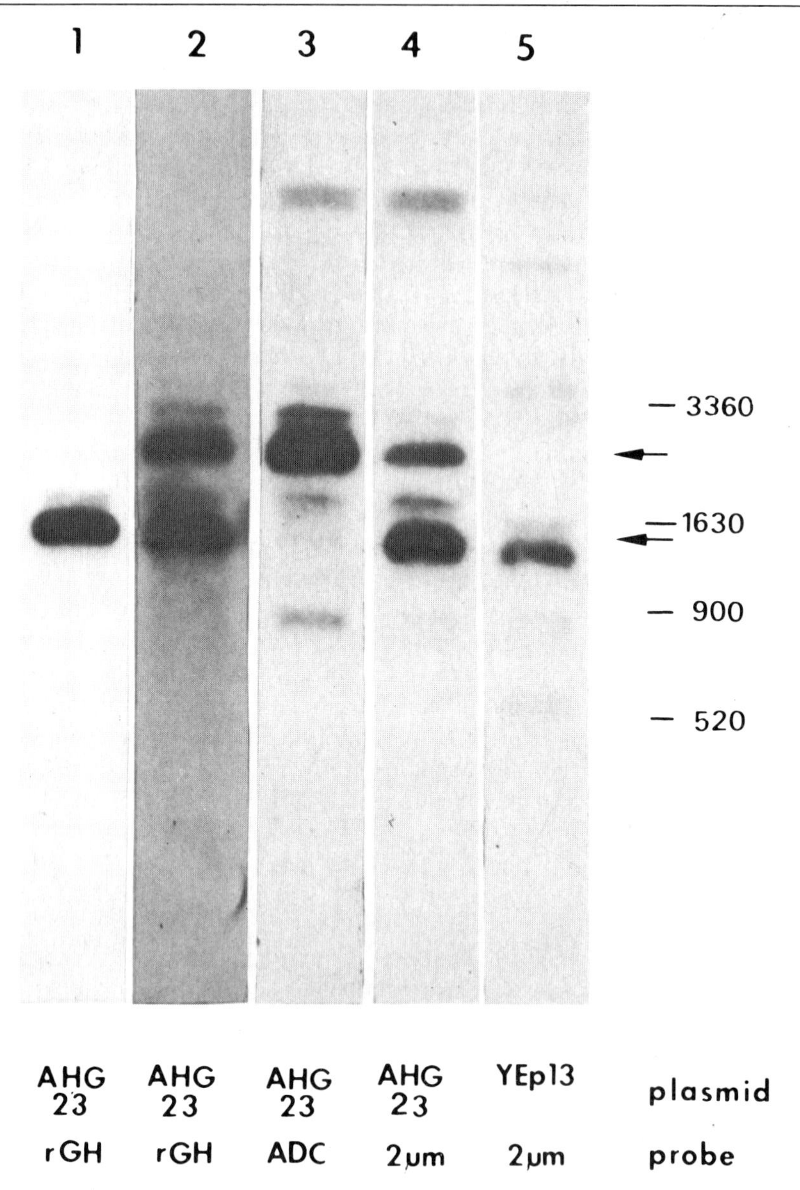

FIG. 2. Characterization of transcripts initiating from the *ADC1* promoter. Yeast strain PS23-6A (α *leu2 trp1*) was transformed with YEp13 or the vector pAHG23 related to pAH5 containing a ~800 bp fragment of rat growth hormone cDNA[9] inserted into the *Hin*dIII site. Glyoxylated samples of poly(A)-containing RNA (10 μg) were fractionated on a 1.2 agarose gel, transferred to nitrocellulose paper, and probed with different DNA fragments ^{32}P-labeled by nick translation (see P. S. Thomas, this series, Vol. 100 [18]). 3 to 5 × 10^5 cpm were used for hybridization. The autoradiograms were exposed with intensifying screen at

hormone, or rat growth hormone, respectively). These differences in expression from one case to another may result from a variety of posttranscriptional and posttranslational effects, e.g., mRNA stability, mRNA processing, translational capacity of the mRNA, stability of the protein, modification or processing of the protein, and more. Low expression results are also explainable by differences in the preferred codon usage between higher eukaryotes and *S. cerevisiae*.[17] But unfortunately no systematic experiments have been done so far to substantiate any of these explanations. Finally it should be mentioned that identical plasmids gave variable expression of a foreign gene in different yeast strains. This, however, is no surprise considering how much the expression of yeast genes themselves depends on the genetic background of the yeast strain.

Acknowledgments

I am greatly indebted to people from the laboratories of Ted Young, Ben Hall, and Mike Smith for providing information and help. During the work I was recipient of a postdoctoral fellowship from the Max Kade Foundation.

[17] J. Bennetzen, and B. D. Hall, *J. Biol. Chem.* **257,** 3026 (1982).

−70° for 2–4 hr. Lanes 1–4 contain RNAs from cells transformed with pAHG23; lane 5 contains RNA from cells transformed with YEp13. In lanes 1 and 5, strains were grown to a density of 1 to 2 × 10^7 cells/ml; in lanes 2–4, to a density of 5 to 6 × 10^7 cells/ml. Lanes 1 and 2 are hybridized to a probe of rat growth hormone cDNA. The size of the specific mRNAs are 1400 ± 50 bp and 2400 ± 50 bp, respectively. The larger mRNA is present only in cells grown to higher density. Lane 3 shows hybridization to the 1500 bp *ADCI* promoter fragment. Only the large mRNA gives a major positive signal. Lanes 4 and 5 are probed with the 2 μm DNA fragment (bp 105–1998 of 2 μm plasmid) flanking the rGH cDNA on the 3′ end. The RNA hybridization pattern detected in lane 5 is consistent with published results concerning transcription from 2 μm plasmid in yeast.[10]

[12] One-Step Gene Disruption in Yeast

By RODNEY J. ROTHSTEIN

The accessibility of the yeast genome to genetic manipulation using plasmid technologies is reviewed in this volume.[1] This chapter describes a relatively simple technique for gene disruption or replacement in yeast that requires a single transformation. The method can be used (*a*) to determine whether a cloned fragment contains a specific gene; (*b*) to determine whether a cloned gene is essential; and (*c*) to alter or completely delete a specific region. The method takes advantage of our previous observations that during yeast transformation free DNA ends are recombinogenic, stimulating recombination by interacting directly with homologous sequences in the genome.[2]

Figure 1 outlines the principle of the procedure. A cloned DNA fragment containing *GENE* Z^+ is digested with a restriction enzyme that cleaves within the *GENE Z* DNA sequence. Another DNA fragment containing a selectable yeast gene ($HIS3^+$ in this example) is cloned into the cleaved *GENE Z*. The outcome of this cloning strategy is to disrupt *GENE Z*. As illustrated in Fig. 1, the cloned *gene Z'* and *'gene Z* (we have adopted the nomenclature for protein fusions from *E. coli*) are separated by the selectable yeast fragment containing the $HIS3^+$ gene. The *in vitro* disrupted gene is liberated from the bacterial plasmid sequences by restriction enzyme digestion. It is important that the linear fragment contain sequences homologous to the chromosomal *GENE Z* region on both sides of the inserted fragment to direct the integration into the *GENE Z* region. The digested DNA is transformed into a $his3^-$, *GENE* Z^+ yeast cell and $HIS3^+$ transformants are selected. Among the transformants are strains that simultaneously become $HIS3^+$ and *gene* Z^-. Thus in these transformants *GENE Z* has been replaced with a disrupted *gene* Z^- in one step.

The method outlined above may be modified to delete an entire gene or a gene fragment simply by using the appropriate cloning strategy (Fig. 2, type II). It can be used to determine whether or not a gene is essential by first disrupting the gene in a diploid (Fig. 2). Subsequent genetic analysis of a disrupted essential gene will reveal linkage of the lethal function to the selectable genetic marker used to disrupt the gene. As has been

[1] T. L. Orr-Weaver, J. W. Szostak, and R. J. Rothstein, this volume [14].

[2] T. L. Orr-Weaver, J. W. Szostak, and R. J. Rothstein, *Proc. Natl. Acad. Sci. U.S.A.* **78,** 6354 (1981).

ISBN 0-12-182001-7

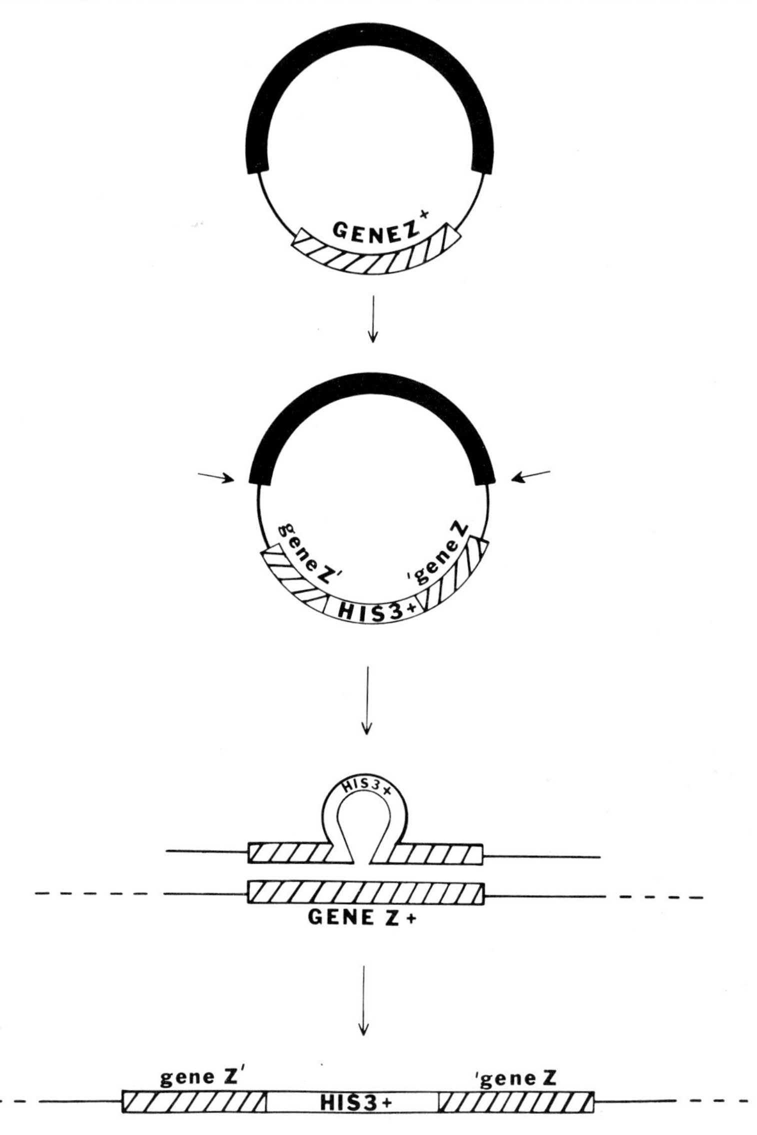

FIG. 1. One-step gene disruption. The cloned fragment containing *GENE* Z^+ is digested with a restriction enzyme that cleaves within the *GENE Z* sequence. A fragment containing a selectable yeast gene ($HIS3^+$ in this example) is cloned into the site. The fragment containing the disrupted *gene Z* is liberated from plasmid sequences, making certain that homology to the *GENE Z* region remains on both sides of the insert. Transformation of yeast cells with the linear fragment results in the substitution of the linear disrupted sequence for the resident chromosomal sequence.

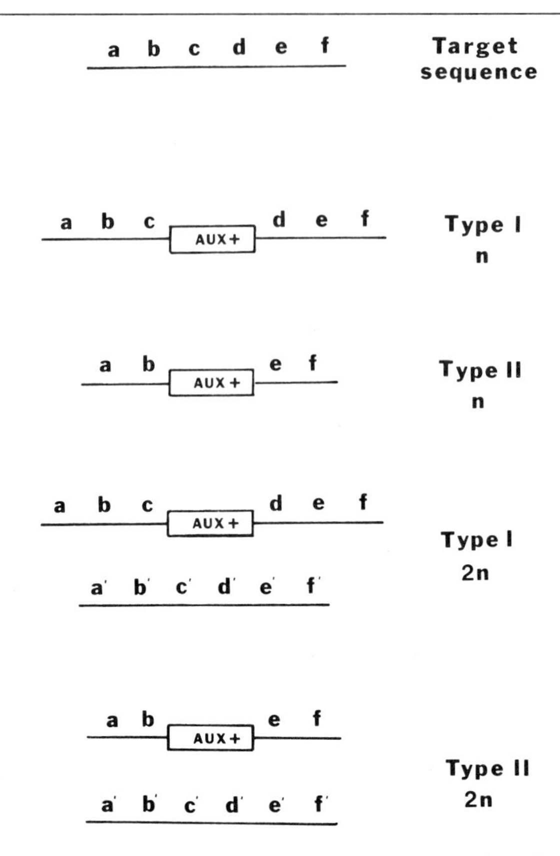

FIG. 2. Various types of one-step gene disruptions. A type I disruption is created after cloning a selectable marker (AUX^+) into a single restriction site within the target sequence. A type II disruption is created when two restriction sites are present and AUX^+ is inserted while c-d is deleted. For diploid type I and II disruptions, a′, b′, etc., represent the undisrupted homologous chromosome. In the case of the disruption of an essential gene in a diploid, spores containing AUX^+ would be inviable.

pointed out before,[3] gene disruption techniques in yeast offer similar advantages as the use of transposon insertions in prokaryote genetic studies.

Results

The methods for the manipulation of plasmids, yeast transformation and genomic blot analysis are described in this volume.[1] In this section, three examples of successful one-step gene replacement experiments will be described (Fig. 2, type II). The first involves the *put2* gene, which

[3] D. Shortle, J. E. Haber, and D. Botstein, *Science* **217**, 371 (1982).

codes for an enzyme needed for proline utilization,[4] cloned by Brandriss.[5] Brandriss identified a 5.6-kb *Sac*I fragment containing *put2* complementing activity in a yeast shuttle vector and undertook a one-step gene disruption experiment to prove that the fragment coded for the *PUT2*$^+$ gene and not for a phenotypic suppressor. Preliminary restriction mapping and subcloning experiments indicated that two *Bgl*II restriction sites flanked a 0.3-kb region necessary but not sufficient for *PUT2*$^+$ function. Standard cloning methods were used to insert a 1.7-kb *Bam*HI fragment containing the *HIS3*$^+$ region[6] into these *Bgl*II sites. Restriction digestion with *Sac*I liberated a linear fragment that contained the *put2* genetic region that had been disrupted by a 1.7-kb *HIS3*$^+$ insert. The *HIS3*$^+$ insert was flanked by 3.1 kb and 2.2 kb of homology to the *put2* region. Approximately 120 ng of fragment were used to transform 2×10^7 cells of W301-18A (*MAT*α *ade2-1 leu2-3,112 his3-11,15 trp1-1 ura3-1*), a *his3*$^-$, *PUT2*$^+$ strain. Among the 14 *HIS3*$^+$ transformants were 3 that simultaneously became *put2*$^-$. Yeast colony hybridization was performed on these colonies using vector sequences as probe.[7] Two of the colonies also contained the vector sequences and were presumably due to contaminating linear molecules that integrated and converted the *put2* region during integration (for discussion of conversion associated with integration, see Orr-Weaver *et al.*[1]) The other transformant did not contain vector sequences, but did contain a *put2* region in which the *HIS3* DNA substituted for part of the *put2* gene. This result was confirmed by a genomic blot[8] of a *Sac*I digest of both the parent strain and the transformant probed with labeled *PUT2* DNA (Fig. 3). Tetrad analysis confirmed tight linkage of *put2*$^-$ and *HIS3*$^+$.

In a second example of gene disruption, Orr-Weaver and I[9] inserted one selectable genetic function into a region and simultaneously deleted another from the same region (Fig. 2, type II). The recipient strain contained, integrated at the *HIS3* genetic region, a plasmid into which the *SUP3*-**a** suppressor fragment was cloned. In a one-step gene replacement experiment, we successfully substituted a plasmid containing the *LEU2*$^+$ genetic region for the *SUP3*-**a** plasmid sequence. A *LEU2*+, *HIS3*$^+$ plasmid was linearized by a restriction cut within the *HIS3* sequence. The free ends of the incoming plasmid molecule stimulate recombination with the resident *HIS3* sequences. Since in this case the recipient strain contained two *HIS3* regions separated by the *SUP3*-**a** plasmid sequence, homolo-

[4] M. C. Brandriss and B. Magasanik, *J. Bacteriol.* **140,** 498 (1979).
[5] M. C. Brandriss, personal communication, 1982.
[6] K. Struhl and R. W. Davis, *J. Mol. Biol.* **136,** 309 (1980).
[7] A. Hinnen, J. B. Hicks, and G. R. Fink, *Proc. Natl. Acad. Sci. U.S.A.* **75,** 1929 (1978).
[8] E. M. Southern, *J. Mol. Biol.* **98,** 503 (1975).
[9] T. L. Orr-Weaver and R. J. Rothstein, unpublished observations, 1982.

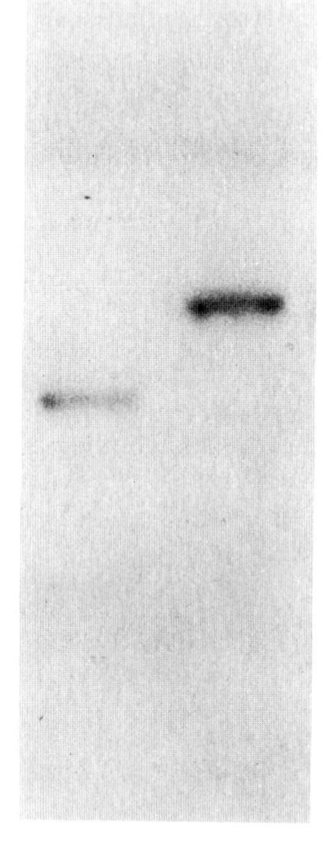

FIG. 3. Genomic blot of the *put2* gene disruption. W301-18A, a *his3*$^-$ *PUT2*$^+$ strain was transformed in one step to a *HIS3*$^+$ *put2*$^-$ strain using a *PUT2* fragment which was disrupted within the *put2* region with *HIS3*$^+$ DNA sequence. DNA was isolated from both the parent W301-18A and the transformant MB1426; 0.5 μg of DNA was digested with *Sac*I, electrophoresed on a 0.7%agarose gel, and transferred to nitrocellulose. Hybridization with radioactively labeled *Sac*I–*PUT2* DNA revealed that the *PUT2* region contained the *HIS3*$^+$ insert.

gous pairing could result in either targeted integration (see Orr-Weaver *et al.*[1]) or a one-step gene replacement. For the pairing configuration that resulted in the one-step replacement, there was 0.3 kb of homology at one end of the *LEU2*$^+$ insert and 1.7 kb of homology at the other. After transforming with 500 ng of DNA, at least two transformants simultaneously inserted the *LEU2*$^+$ plasmid and deleted the *SUP3*-**a** plasmid as confirmed by genomic blots.

In the third example, Tollervey and Guthrie[10] mutated a small structural RNA transcript (~200 bp) by a one-step gene disruption experiment. They cloned and identified a 3.4-kb fragment from yeast that encodes the RNA transcript. They disrupted the gene by cloning a 2.2-kb *Sal*I-*Xho*I *LEU2*$^+$ fragment into the middle of it, replacing 36 bp of DNA. A linear fragment containing the disrupted gene was used to transform a diploid yeast strain. A diploid was used because it was anticipated that the replacement might be lethal. A *LEU2*$^+$ transformant lacking vector sequences was shown by genomic blots to contain a disrupted gene fragment as well as an unaltered wild-type fragment (Fig. 2, type II diploid).

One-step gene disruptions have been successful when the DNA has as little as 0.3 kb of homology in one of the two regions adjacent to the inserted selectable fragment. It is recommended that the length of homology to the desired chromosomal region be maximized to lower the probability of gene conversion of the selectable marker at its own chromosomal locus. It is not necessary to purify the linear fragment from the vector sequences if at least one of the two restriction sites that liberates the fragment from the vector is at the border of the vector and the fragment. It is desirable to purify the linear fragment when homologous chromosomal sequences are left on both ends of the plasmid vector. Without purification, such a plasmid may interact directly with the homologous chromosomal region, resulting in gap repair and integration.[1,2]

As outlined above, the one-step gene disruption technique requires cloning of a selectable gene fragment into the region to be disrupted. A partial list of fragments available for this procedure is given in the table. Included are the flanking restriction sites that can serve to permit the insertion of the functional gene fragment. The genotypes of several commonly used laboratory strains that contain markers suitable for selecting one-step replacements have also been listed in the table.

Discussion

Cloning techniques have permitted the isolation of numerous yeast genes. Many have been recovered from clone banks of total yeast DNA

[10] D. Tollervey and C. Guthrie, personal communication, 1982.

CONVENIENT DNA FRAGMENTS AND YEAST STRAINS FOR USE IN ONE-STEP GENE DISRUPTIONS[a]

Gene	Plasmid	Flanking restriction sites	Size of fragment (kb)	Reference[b]
LEU2$^+$	YEp13	*Bgl*II	3.0	(1)
		*Sal*I, *Xho*I	2.2	
		*Hpa*I, *Sal*I	1.9	
HIS3$^+$	YIp1	*Sal*I, *Eco*RI	6.0	(2)
		*Bam*HI	1.7	
		*Bam*HI, *Xho*I	1.4	
URA3$^+$	YEp24	*Hin*dIII	1.2	(3)
TRP1$^+$	YRp7	*Eco*RI	1.4	(2)
		*Eco*RI, *Bgl*II	0.9	

Strains	Genotype	Reference
LL20	*MAT*α *leu2-3,112 his3-11,15*	L. Lau, (4)
DBY745	*MAT*α *leu2-3,112 ade1-101 ura3-52*	(3)
SR25-1A	*MAT***a** *his4-912 ura3-52*	S. Roeder (5)
RH218	*MAT*α *trp1 gal2*	(6)
W301-18A	*MAT*α *ade2-1 trp1-1 leu2-3,112 his3-11,15 ura3-1*	R. Rothstein (7)

[a] Gene disruptions with the *Sal*I–*Xho*I *LEU2*$^+$ fragment, the *Bam*HI *HIS3*$^+$ fragment, and the *Hin*dIII *URA3*$^+$ fragment have been successful. The other fragment has not been tried.

[b] Key to references:

(1) J. R. Broach, J. N. Strathern and J. B. Hicks, *Gene* **8,** 121 (1979).

(2) K. Struhl, D. T. Stinchcomb, S. Scherer, and R. W. Davis, *Proc. Natl. Acad. Sci. U.S.A.* **76,** 1035 (1979).

(3) D. Botstein, S. C. Falco, S. E. Stewart, M. Breenan, S. Scherer, D. T. Stinchcomb, K. Struhl, and R. W. Davis, *Gene* **8,** 17 (1979).

(4) T. L. Orr-Weaver, J. W. Szostak, and R. J. Rothstein, *Proc. Natl. Acad. Sci. U.S.A.* **78,** 6354 (1981).

(5) Unpublished strain.

(6) G. Miozzari, P. Niederberger, and R. Hütter, *J. Bacteriol.* **134,** 48 (1978); S. Scherer and R. W. Davis, *Proc. Natl. Acad. Sci. U.S.A.* **76,** 4951 (1979).

(7) Unpublished strain.

by selecting for complementation of mutations in yeast.[11-18] Others have been isolated from clone banks by hybridization with pure or partially purified RNA molecules,[19-23] cDNA probes,[24] or probes from highly conserved genes previously cloned in divergent species.[25-27]

For genes cloned by complementation, it is necessary to demonstrate that a fragment codes for the wild-type gene, not for a phenotypic suppressor. Proof that a fragment actually codes for the complemented gene is generally demonstrated by showing that a plasmid containing the fragment integrates at the genetic locus and shows genetic linkage to the gene. This evidence does not rule out the possibility that the fragment codes for a tightly linked phenotypic suppressor. A gene disruption experiment proves in one step that a cloned fragment codes for the gene function in question by (*a*) using homology on the fragment to integrate into the corresponding chromosomal region; (*b*) simultaneously creating a mutation; and (*c*) showing genetic linkage of the insert by tetrad analysis.

Gene disruption experiments may also be used to examine the phenotype of fragments cloned by hybridization procedures. The two existing methods for gene disruption, transplacement[28] and integration disruption,[3] require several manipulations or detailed information about the cloned fragment. In the Scherer and Davis method[28] an altered sequence cloned on a plasmid is introduced by homologous integration into the de-

[11] A. Hinnen, P. J. Farabaugh, C. Ilgen, G. R. Fink, and J. Friesen, *Eucaryotic Gene Regul. ICN–UCLA Symp. Mol. Cell. Biol.* **14,** 43 (1979).

[12] J. R. Broach, J. N. Strathern, and J. B. Hicks, *Gene* **8,** 121 (1979).

[13] V. M. Williamson, J. Bennetzen, E. T. Young, K. Nasmyth, and B. D. Hall, *Nature (London)* **283,** 214 (1980).

[14] J. Hicks, J. Strathern, and A. Klar, *Nature (London)* **282,** 478 (1979).

[15] K. Nasmyth and K. Tatchell, *Cell* **19,** 753 (1979).

[16] K. A. Nasmyth and S. I. Reed, *Proc. Natl. Acad. Sci. U.S.A.* **77,** 2119 (1980).

[17] M. Crabeel, F. Messenguy, F. Lacroute, and N. Glansdorff, *Proc. Natl. Acad. Sci. U.S.A.* **78,** 5026 (1981).

[18] C. L. Dieckmann, L. K. Pape, and A. Tzagoloff, *Proc. Natl. Acad. Sci. U.S.A.* **79,** 1805 (1982).

[19] R. A. Kramer, J. F. Cameron, and R. W. Davis, *Cell* **8,** 227 (1976).

[20] M. V. Olson, B. D. Hall, J. R. Cameron, and R. W. Davis, *J. Mol. Biol.* **127,** 285 (1979).

[21] J. L. Woolford, Jr., L. M. Hereford, and M. Rosbach, *Cell* **18,** 1247 (1979).

[22] T. P. St. John and R. W. Davis, *Cell* **16,** 443 (1979).

[23] R. A. Kramer and N. Andersen, *Proc. Natl. Acad. Sci. U.S.A.* **77,** 6541 (1980).

[24] M. J. Holland, J. P. Holland, and K. A. Jackson, this series, Vol. 68, p. 408.

[25] L. Hereford, K. Fahvner, J. Woolford, Jr., M. Rosbash, and D. B. Kaback, *Cell* **18,** 1261 (1979).

[26] D. Gallwitz and R. Seidel, *Nucleic Acids Res.* **8,** 1043 (1980).

[27] R. Ng and J. Abelson, *Proc. Natl. Acad. Sci. U.S.A.* **77,** 3912 (1980).

[28] S. Scherer and R. W. Davis, *Proc. Natl. Acad. Sci. U.S.A.* **76,** 3912 (1979).

sired region of the chromosome. The integrated plasmid is flanked by a mutant copy and a wild-type copy of the gene. The next step requires that a recombination event between the duplicated regions flanking the plasmid occur such that the wild-type sequence is excised and the mutant sequence is left in the chromosome. The major disadvantage of this method is that several recombinants must be examined in order to detect the correct replacement. A positive selection for loss of a marker on the plasmid enriches for, but still does not ensure, the successful replacement of the mutant gene. The method of Shortle *et al.*[3] utilizes plasmid integration to create a mutant phenotype. A plasmid containing a cloned internal fragment of a gene is integrated into its homologous chromosomal sequence. Since the internal fragment lacks the 5′ and 3′ sequences of the gene, the integrated plasmid results in two mutated copies of the gene—a 5′ deletion as well as a 3′ deletion. The limitations of this technique are that (*a*) it demands the knowledge of restriction sites with respect to the 5′ and 3′ ends of the desired gene; (*b*) the mutated gene can revert by deletion of the plasmid sequences, since the disrupted gene region is flanked by direct repeats as a consequence of the integration event; and (*c*) the technique may not be applicable to all genes, since it has been observed that some small fragments (less than 200 nucleotides in length) fail to integrate into their homologous region.[1] The methods outlined in this chapter permit simple disruptions (Fig. 2, type I) or deletion-substitution events (Fig. 2, type II) in one step. A detailed restriction map of the fragment is not necessary, since a single insertion (Fig. 2, type I) is sufficient to disrupt a gene. The position, within a fragment, of a gene cloned by complementation may be determined by independently disrupting the fragment at several different restriction sites and testing each for its phenotype. Identification of transformants containing the correct configuration is aided by the fact that the insertion simultaneously results in a mutation whose phenotype can be directly scored. For disruptions of essential genes, the recessive lethal phenotype becomes tightly linked to the insertion, which is nonreverting and can be used as a selectable marker for further genetic manipulation.

To date, successful disruptions have been reported with the *Bam*HI *HIS3*$^+$ fragment, the *Sal*I-*Xho*I *LEU2*$^+$ fragment,[29] and the *Hin*dIII *URA3*$^+$ fragment.[30] The other fragment listed in the table has not been tried in one-step disruption experiments. As with any gene disruption

[29] Amar Klar (personal communication) has replaced yeast mating-type information with the *Sal*I–*Xho*I *LEU2*$^+$ fragment using the techniques described in this paper.

[30] N. Abovich and M. Rosbash (personal communication) have disrupted a ribosomal protein gene with the *Hin*dIII *URA3*$^+$ fragment.

method, a successful experiment must be verified by Southern blots on the DNA of the transformants (e.g., Fig. 3). Finally, it is important to consider the possibility that moving a gene into a new environment may inadvertently create a position effect[31] leading to nonexpression of the selectable marker.

Summary

The one-step gene disruption techniques described here are versatile in that a disruption can be made simply by the appropriate cloning experiment. The resultant chromosomal insertion is nonreverting and contains a genetically linked marker. Detailed knowledge of the restriction map of a fragment is not necessary. It is even possible to "probe" a fragment that is unmapped for genetic functions by constructing a series of insertions and testing each one for its phenotype.

Acknowledgments

The author wishes to thank Nadja Abovich, Christine Guthrie, Amar Klar, Terry Orr-Weaver, Michael Rosbash, David Tollervey, and especially Marjorie Brandriss for permission to quote unpublished data. Thanks are also due to Cindy Helms for technical assistance. This work was supported by NIH Grant GM27916, and NSF Grant PCM 8003805, and Foundation of UMDNJ Grant 24-81.

[31] K. Struhl, *J. Mol. Biol.* **152,** 569 (1981).

[13] Eviction and Transplacement of Mutant Genes in Yeast

By FRED WINSTON, FORREST CHUMLEY, and GERALD R. FINK

Background and Principle for the Methods

Molecular genetic studies in the yeast *Saccharomyces cerevisiae* hold great promise for the elucidation of the mechanisms by which a lower eukaryotic organism controls the expression of its genes (for a review, see Petes[1]). Although no single regulatory mechanism is yet understood in detail, it is likely that several aspects of the problem will soon yield to the efforts of workers in the field. This certainty owes much to the ease with which yeast can be manipulated genetically through *in vitro* recombinant

[1] T. D. Petes, *Annu. Rev. Biochem.* **49,** 845 (1980).

ISBN 0-12-182001-7

DNA techniques. In particular, genetic transformation of yeast using purified DNA has provided simple, rapid methods for the molecular cloning of mutant forms of genes ("eviction" of mutant genes) and for the introduction into yeast of mutant genes constructed *in vitro* ("transplacement" of mutant genes). This chapter will place these techniques in a conceptual framework and provide a practical guide for their application to studies of the expression and regulation of yeast genes.

The classical genetic methods of mutant selection and genotypic analysis continue to yield yeast strains that carry mutations leading to interesting alterations in the pattern of gene activity. The experimentalist is usually interested in cloning a segment of DNA containing the mutation of interest because molecular cloning of a mutant gene permits a powerful analysis of the changes in DNA sequence that underlie the mutant phenotype. Eviction of the mutant gene is the method of choice for molecular cloning whenever a clone of the wild-type gene is already available. The strong points of eviction are speed and accuracy: the method circumvents the need for preparing a clone bank from the mutant strain, and it eliminates the possibility of cloning any DNA sequences other than those of the mutant gene.

In vitro manipulation of cloned genes has advanced quickly to the point where a wide variety of well-defined changes in the DNA sequence can be introduced with precision. The obvious goal of experiments of this type is to determine the genetic consequences of such "engineered" mutations. Transplacement permits a single copy of the mutated gene to be integrated into the yeast genome at its normal chromosomal location, replacing completely the wild-type copy. This provides the ideal situation for observing *in vivo* the effects of the alteration in DNA sequence.

Eviction and transplacement of mutant genes rely on the properties of integrative transformation in yeast.[2] Integrative transformation occurs when a segment of DNA is inserted into the chromosome and is replicated as part of that chromosome. In practice, this type of transformation occurs when yeast spheroplasts are incubated with a bacterial plasmid that contains no origin for DNA replication in yeast and carries a cloned segment of yeast DNA. Transformants are selected relying on the expression of genes present on the plasmid, either within the cloned yeast segment or elsewhere in the vector. These transformants are found to contain a copy of the plasmid DNA integrated into the yeast genetic material. The integrated structure arises by a single crossover event involving the cloned yeast segment and a homologous sequence in the genome, as diagrammed in Fig. 1A. This event results in a direct, nontandem duplication

[2] A. Hinnen, J. B. Hicks, and G. R. Fink, *Proc. Natl. Acad. Sci. U.S.A.* **75,** 1929 (1978).

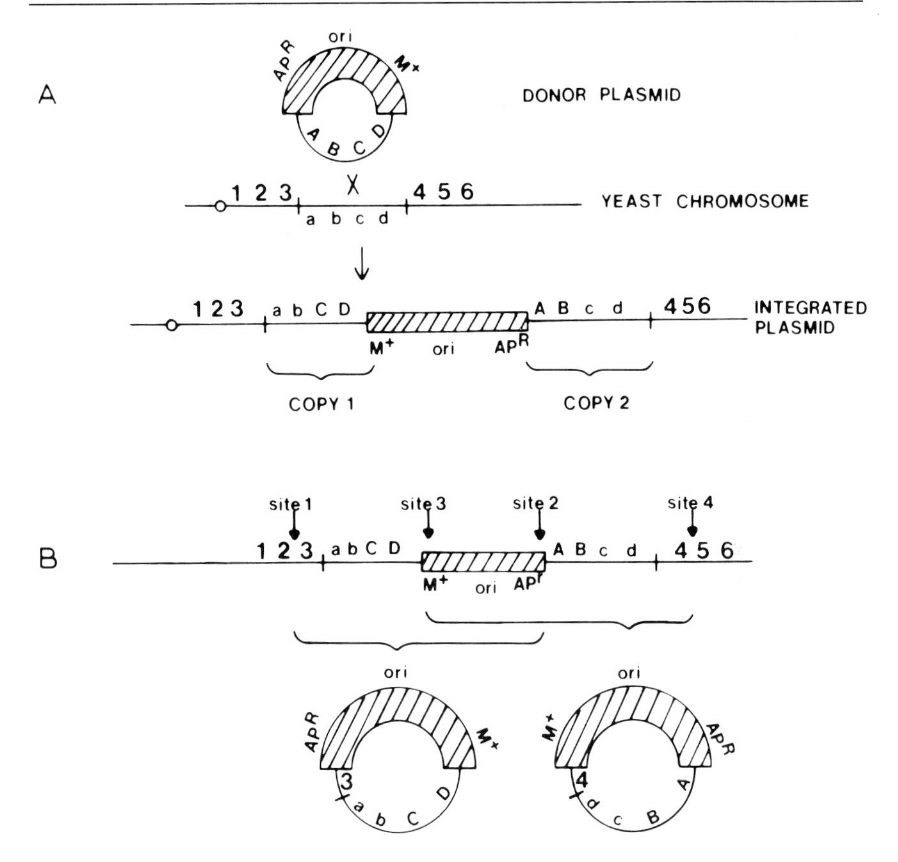

FIG. 1. (A) Generalized diagram of integrative transformation. A circular donor DNA molecule is integrated into a yeast chromosome after a single recombination event between the homologous regions *ABCD,* carried by the plasmid, and *abcd,* resident in the yeast chromosome. The result is a direct nontandem duplication of the *ABCD* interval, with copies bracketing the integrated plasmid DNA. Each copy of the *ABCD* interval is recombinant for the *ABCD* markers, as indicated.

(B) Eviction of a yeast DNA segment. Eviction of the chromosomal markers *ab* is diagrammed. DNA prepared from the duplication strain in Fig. 1A is digested with a restriction enzyme that cuts at sites 1 and 2. The linear fragment that is generated can then be ligated to form the plasmid shown. The restriction enzyme must generate a linear fragment that includes the bacterial origin of replication for the plasmid and a drug resistance marker that can be selected in bacteria. The evicted plasmid differs from the original plasmid in Fig. 1A in two important features. First, from the *ABCD* interval, the plasmid now carries the *ab* sequences that originally resided in the yeast chromosome. Second, the plasmid has acquired the region indicated by the number (3), which flanks *ABCD* in the chromosome. This second new feature demonstrates the utility of eviction as a method for "walking" along the chromosome during successive rounds of molecular cloning. If the DNA of the duplication strain were digested with a restriction enzyme that cuts at sites 3 and 4, ligation of the linear fragment would result in eviction of the chromosomal markers *cd*.

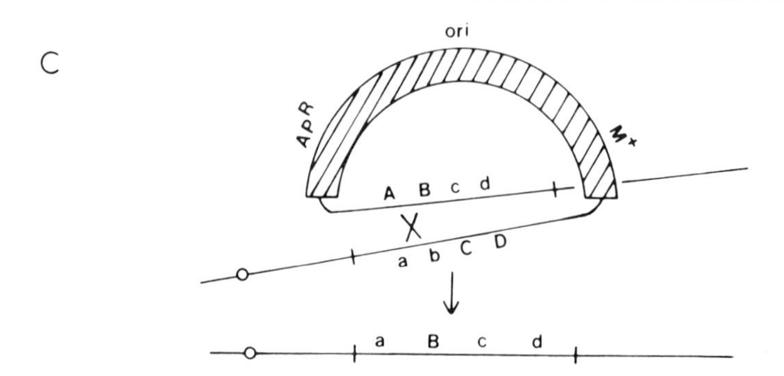

FIG. 1. (C) Transplacement of a yeast DNA segment. Transplacement of the marker *B* is shown in the diagram. Recombination between copies of the duplicated yeast DNA segment reduces the copy number to one and results in the loss of the integrated plasmid. A single crossover in the interval between *A* and *B* results in the transplacement of the sequence *B* originally carried by the donor plasmid in Fig. 1A.

of the cloned segment, with copies bracketing integrated plasmid sequences. This duplication, arising through integrative transformation, provides the basis for both eviction and transplacement, as shown in Figs. 1B and 1C.

In eviction, genomic DNA isolated from the transformant is first cut with a particular restriction enzyme, then ligated and transformed into bacteria so as to re-create a chimeric plasmid. The plasmid contains a copy of sequences that were formerly present in the genome of the integrative transformant. These sequences are said to have been "evicted." Transplacement, on the other hand, relies on the inherent genetic instability of the duplicated structure. A single recombination event between the repeated regions resolves the duplication and excises the integrated plasmid, which is then lost during cell division. Transformants that have undergone such a recombination event can be detected as mitotic segregants that have lost the genetic marker(s) associated with the integrated plasmid. If such a recombination event falls in the appropriate genetic interval, a segment of the yeast genome will have been altered to contain sequence information formerly present in the plasmid used for integrative transformation.

Although methods for the eviction and transplacement of mutant genes were conceived in connection with yeast molecular genetics, the techniques should be applicable to any system where integrative transformation can occur. Furthermore, the basic approach of the methods can be extended to problems other than the manipulation of mutant genes. Eviction, for example, suggests a novel method for "walking" along the chromosome in successive molecular cloning experiments (see Fig. 1B and the discussion below).

Materials and Other Requirements for Application of the Methods

Successful application of the eviction and transplacement methods requires a consideration of the details of the system of interest in view of the paradigms in Fig. 1. The material requirements of the methods can be simply stated, however, and they will be discussed in the following paragraphs.

Molecular Clone of the Gene under Investigation. A clone of the wild-type gene, or a suitable subclone, must be available. The restriction sites of this clone must be mapped, and its orientation in the plasmid vector must be known. Transplacement requires that the cloned segment contain genetic homology on both sides of the mutation to be introduced into the yeast genome. The plasmid to be used must be of a type that does not replicate in yeast, and it should in most cases carry a gene whose expression can be selected in yeast transformation. The vector that is currently most useful for eviction and transplacement is YIP5,[3] which is derived from pBR322 and carries the yeast *URA3*$^+$ gene. The YIP5 plasmid has a number of useful restriction sites for cloning, and expression of the *URA3*$^+$ gene can be selected in any *ura3*$^-$ recipient strain.

Recipient Strain for Integrative Transformation. The recipient strain for the transformation experiment should be chosen with regard for transformability and the suitability of markers in the genetic background. If YIP5 is used as the vector, the genetic background should include a nonreverting *ura3*$^-$ mutation such as *ura3-50* or *ura3-52*, which cause uracil auxotrophy and can be complemented by the *URA3*$^+$ genc on the plasmid. Of course, the strain chosen for transformation must contain genetic markers appropriate for the investigation of the genes of interest. Transplacement of mutant genes is facilitated by the presence in the genetic background of one or more mutations that cause inositol auxotrophy (*ino*$^-$ mutations[4]). Such mutations permit "inositol-less death" enrichment for mitotic segregants that have lost the integrated plasmid.[5] Genetic selections have been devised for the isolation of mutants that have lost the function of certain genes, including *URA3*[6] and *SUP11*.[7,8] If a plasmid vector used for integrative transformation carries one of these genes, mitotic segregants that have excised the plasmid can be selected directly.

[3] D. Botstein, S. C. Falco, S. E. Stewart, M. Brennan, S. Scherer, D. T. Stinchcomb, K. Struhl, and R. W. Davis, *Gene* **8,** 17 (1979).

[4] M. R. Culbertson and S. A. Henry, *Genetics* **80,** 23 (1975).

[5] S. A. Henry, T. F. Donahue, and M. R. Culbertson, *Mol. Gen. Genet.* **143,** 5 (1975).

[6] M. L. Bach and F. Lacroute, *Mol. Gen. Genet.* **115,** 126 (1972).

[7] A. Singh, *Proc. Natl. Acad. Sci. U.S.A.* **77,** 305 (1977).

[8] T. P. St. John, S. Scherer, M. W. McDonell, and R. W. Davis, *J. Mol. Biol.* **152,** 317 (1981).

Yeast Transformation. Genetic transformation of yeast using plasmid DNA has become routine.[2] Spheroplasts of the recipient strain are prepared using glusulase to digest the cell walls, the spheroplasts are treated with DNA, and transformants are selected in pour plates. When $Ura3^+$ is selected, transformants are plated in specially enriched yeast minimal medium containing no uracil, solidified with 3% agar.

Genetic Analysis of Yeast. Standard materials for the maintenance and genetic analysis of yeast are required. These will include replica printing materials, as well as equipment for dissection and analysis of the products of yeast meiosis, in order to verify the genetic structure of the integrative transformants.[9]

Recombinant DNA Manipulations. Standard materials for preparation of *in vitro* recombinant DNA are required, including restriction endonucleases and DNA ligase. For eviction of mutant genes, the restriction endonucleases to be used are dictated by the sites available in the vector and in the segment to be cloned from the chromosome. For experiments involving YIP5, the most useful enzymes are usually *Eco*RI, *Cla*I, *Hin*dIII, *Bam*HI, or *Sal*I. Bacterial transformation using DNA evicted from yeast proceeds well when *E. coli* strain HB101[10] or one of its derivatives is used as a recipient. Southern blot analysis[11] using yeast genomic DNA can be a useful technique for verification of the genetic structure of integrative transformants and transplaced mutant genes.

Methods

The methods for eviction and transplacement have several steps in common. First, we will describe the method for an eviction and then describe those steps that are different for transplacement.

Cloning of Mutations by Eviction

Choice of Plasmid. One should use a plasmid containing a restriction fragment such that integration will give a predicted DNA structure. An eviction will be greatly facilitated if one knows the relative orientation of the mutation and the plasmid sequences in the transformant. Knowledge of the approximate location of the mutation in the gene and proper selection of a restriction fragment can result in this situation. For cloning the

[9] F. Sherman, G. R. Fink, and C. W. Lawrence, "Methods in Yeast Genetics," p. 98. Cold Spring Harbor Laboratory, Cold Spring Harbor, New York, 1979.

[10] H. W. Boyer and D. Roulland-Dussoix, *J. Mol. Biol.* **41,** 459 (1969).

[11] E. M. Southern, *J. Mol. Biol.* **98,** 503 (1975).

his4-912 insertion mutation, Roeder and Fink[12] used a plasmid containing a fragment of wild-type *his4* DNA that contained the site of the mutation to be cloned but not the entire *his4* gene. By selecting for His$^+$ transformants and by knowing the orientation of the *his4* fragment in the plasmid, they were able to obtain transformants with a predicted orientation of the plasmid with respect to the *his4-912* mutation (Fig. 2). In cloning a *cyc1* mutation, Stiles *et al.*[13] used a fragment of *cyc1* that did not cover the mutation. However, they knew that the mutation was 5′ to their fragment sequences, and they knew the orientation of the fragment in the plasmid. Therefore, they also were able to predict the integrated structure in the transformant.

Transformation and Integration at a Specified Site in the Yeast Genome. The transformation procedure used is the same as that described by Hinnen *et al.*[2]

Since integrative transformation is a low frequency event, we recommend using a substantial amount of plasmid DNA, in the range of 20–50 μg. The larger the insert is, the greater the transformation frequency will be. The transformability of different yeast strains can vary over a range of greater than a thousandfold. Therefore, it is advisable to check the transformation efficiency of the recipient strain by first doing a transformation with a vector containing 2μ plasmid DNA that allows autonomous replication in yeast. These plasmids give a much higher frequency of transformation than an integrating plasmid[14,15]; a highly transformable strain will give several thousand (up to 20,000) transformants per microgram of DNA when transformed with a 2μ vector.[15] If the original recipient strain shows poor transformability, a new one that is highly transformable can be constructed by the appropriate cross. The recipient with poor transformability is crossed by a strain that is highly transformable, the diploid is sporulated, the resulting tetrads are dissected, and 5–10 progeny with the proper genotype are checked for their transformability. Some of these progeny should now be highly transformable and can be used for the integrative transformation. Transformability is not determined by any simple Mendelian inheritance (G. S. Roeder, F. Winston, and G. R. Fink, unpublished results); hence, it is difficult to predict the frequency of transformable progeny arising from a cross.

[12] G. S. Roeder and G. R. Fink, *Cell* **21,** 239 (1980).

[13] J. I. Stiles, J. W. Szostak, A. T. Young, R. Wu, S. Consul, and F. Sherman, *Cell* **25,** 277 (1981).

[14] J. Beggs, *Nature (London)* **275,** 104 (1978).

[15] K. Struhl, D. T. Stinchcomb, S. Scherer, and R. W. Davis, *Proc. Natl. Acad. Sci. U.S.A.* **76,** 1035 (1979).

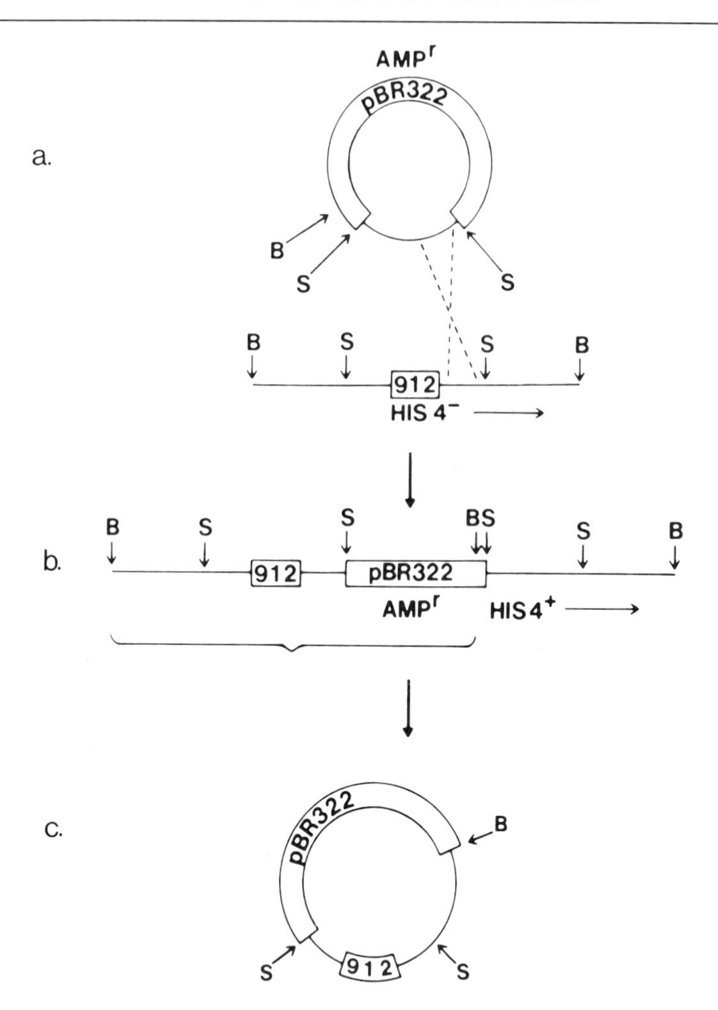

FIG. 2. Eviction of the *his4-912* mutation. The eviction of *his4-912:* (a) A single crossover between plasmid pBR322 containing a *Sal*I restriction fragment of the wild-type *HIS4* gene and the *his4* region of DNA from a yeast strain carrying the *his4-912* mutation. (b) The integrated structure in the transformant. Digestion of the yeast DNA with *Bam*HI releases the fragment indicated by the bracket. This fragment contains pBR322 and the *his4-912* insertion. (c) Ligation of the *Bam*HI fragment results in formation of a plasmid that can be selected in bacteria by transforming for ampicillin resistance.

In an integrative transformation, integration can occur at two different sites: one homologous to the selected marker (usually *ura3, leu2* or *trp1*) and the other homologous to the gene of interest (the desired location of integration). One can discriminate against integration at the region homologous to the selectable marker by different methods.

First, if one uses *URA3*$^+$ as the selectable marker on the plasmid one can enrich for integration at the desired site. For a recipient strain one must use a nonreverting *ura3*$^-$ mutation such as *ura3-52* or *ura3-50; ura3-52* is not a deletion, a common misconception (Mark Rose, personal communication). Integration at *ura3* is low compared to integration at other sites for reasons not totally understood, but at least partly because the *ura3* sequences on the vectors such as YIP5 are only 1.2 kilobases long. Therefore, using a vector with *URA3*$^+$ will generally direct most integrants to the location of choice. However, if the other yeast homology is quite small (much less than 1000 base pairs) one may see a significant proportion of transformants in which the plasmid has integrated at the *ura3* locus.

Second, one can cut the plasmid at a restriction site unique to the cloned yeast DNA sequences. Ilgen *et al.*[16] and Orr-Weaver *et al.*[17] have demonstrated that transformation with a hybrid plasmid that has been cut by a restriction enzyme within a piece of yeast DNA can have three effects: first, the frequency of transformation will be greater than with the uncut plasmid; second, virtually all transformants will have the plasmid integrated in the region homologous to the sequence in which the cut was made; and third, transformants will often but not always have multiple tandem insertions of the plasmid. Therefore, while this is a good technique to direct integration to the desired location, one must screen the transformants by Southern hybridization to find those that have only a single insertion of the plasmid. This technique requires that the plasmid contain a unique restriction site in the cloned yeast segment.

Verification of the Correct Structure of the Transformant. Verification of the location of integration and the correct DNA structure at that site must include both genetic and Southern analysis. One can be sure of the location of integration if, as in the case of Roeder and Fink[12] in the eviction of the *his4-912* insertion mutation, the desired transformant will have a distinct phenotype. For the *his4-912* eviction, the desired integration at *his4* resulted in a His$^+$ phenotype; integration at *ura3* would not have yielded a His$^+$ transformant.

Location of integration of the plasmid can always be shown by a genetic cross even when the map position of the gene is unknown. The general principle is that a plasmid integrated into a particular gene will segregate as an allele of that gene in a cross. The most common situation will be

[16] C. Ilgen, P. J. Farabaugh, A. Hinnen, J. M. Walsh, and G. R. Fink, *in* "Genetic Engineering" (A. Hollaender and J. Setlow, eds.), Vol. 1, p. 117. Plenum, New York, 1979.

[17] T. L. Orr-Weaver, J. W. Szostak, and R. J. Rothstein, *Proc. Natl. Acad. Sci. U.S.A.* **78**, 6354 (1981).

one where, for example, a *ura3⁻ his4⁻* mutant has been transformed to Ura^+ with a plasmid containing part or all of the *HIS4* gene. This plasmid has two regions of homology, *his4* and *ura3*. To determine whether the plasmid is integrated at *ura3* or *his4*, one simply mates the transformant with a *URA3*$^+$ strain, sporulates the diploid, and dissects 10 tetrads to examine the *ura* segregation patterns. Integration at *ura3* will result in segregation patterns of 4 Ura^+ : 0 Ura^-, and integration at *his4* will result in segregation patterns of 4 : 0, 3 : 1, and 2 : 2 for Ura^+ : Ura^-, the 3 : 1 segregation being the majority class. A cross of this type is the definitive method to demonstrate location of integration.

That the structure of the integrated plasmid is correct must also be verified by Southern analysis to show that only a single copy of the plasmid is present. Southern hybridizations are also useful if, as in the case of *his4-912*, the mutant and wild-type alleles can be distinguished, to verify the presence of both.

At this point in the procedure one should be confident of having a transformant with a single copy of the plasmid integrated in the proper location. Complete understanding of the structure at the locus also depends on prior knowledge of the orientation of the cloned fragment within the plasmid.

Preparation of Yeast DNA of the Transformant. Large quantities of pure yeast DNA can be prepared by the procedure of Cryer *et al.*[18] By an abbreviated method, yeast DNA can also be prepared from small cultures (40 ml) in quantity sufficient for an eviction experiment. This "mini prep" method, which greatly facilitates the handling of several different strains, is listed in the protocol below.

Protocol for Rapid Mini Yeast DNA Preps

1. Grow cells in 40 ml of YEPD until saturated.
2. Spin cells at 5000 rpm for 5 min. Use a Sorvall type SS34 rotor for all centrifugations unless otherwise noted. Use 50-ml Nalgene tubes for all steps up to 6.
3. Resuspend the pellet in 3.2 ml of 0.9 *M* sorbitol, 0.1 *M* EDTA, 14 m*M* mercaptoethanol, pH 7.5. Make a stock of 0.9 *M* sorbitol, 0.1 *M* EDTA, pH 7.5, add 3.2 ml of stock to cells and then 3.2 μl of 2-mercaptoethanol to the mix.
4. Add 0.2 ml of 5.0 mg/ml zymolyase (zymolyase 60,000, Kirin Brewery) and incubate at 37° for 60 min. Spheroplast formation is sufficient when over 90% of the cells lyse upon mixing a drop of the zymolyase-treated cell suspension with a drop of 1% SDS.

[18] D. R. Cryer, R. Eccleshall, and J. Marmur, *Methods Cell Biol.* **12,** 39 (1975).

5. Spin at 5000 rpm for 5 min.
6. Add 0.32 ml 0.5 *M* EDTA, pH 8.5; 0.16 ml of 2 *M* Tris base, and 0.16 ml of 10% sodium dodacyl sulfate (SDS) to a polyallomer tube. (Use Beckman polyallomer tubes No. 338818.)
7. Resuspend the cells in 3.2 ml 10 m*M* Tris, pH 7.4, 1 m*M* EDTA (TE buffer) by repeatedly sucking the cells into the pipette and releasing them into the tube. Transfer cells to the centrifuge tube containing ingredients listed in step 6.
8. Add 10 μl of diethylpyrocarbonate to each tube (in a hood) and mix thoroughly by inverting.
9. Incubate (uncapped) at 65° in a hood for 30 min.
10. Add 0.8 ml of 5 *M* potassium acetate and let sit on ice for 60 min. (The tube can be left on ice overnight if convenient.)
11. Spin in type 65 rotor at 20,000 rpm for 30 min.
12. Transfer the supernatant to a 30-ml Corex tube and add 12 ml of 100% ethanol at room temperature. Mix by inverting, and spin at 10,000 rpm for 15 min.
13. Dry the pellet under a vacuum, vortex to break up the pellet, and resuspend in 3.0 ml of 10 m*M* Tris, pH 7.4, 1 m*M* EDTA (TE). This may take several hours on a shaker; it can be left overnight.
14. Spin at 10,000 rpm for 15 min and transfer the supernatant to a 15-ml Corex tube.
15. Add 150 μl of 1 mg/ml pancreatic RNase (Millipore Corp.; prepared in 10 m*M* sodium acetate, pH 7.0; boil for 10 min after it is first made up to inactivate DNase), and incubate at 37° for 30 min.
16. Add 3.0 ml of isopropanol and rock gently to mix. Remove the precipitate with a small glass hook (which can easily be fashioned by melting the tip of a Pasteur pipette). Do not spin.
17. Resuspend the DNA in up to 0.5 ml of TE. The final concentration should be about 200 μg/ml.

Restriction Digestion, Ligation, and Bacterial Transformation. These steps are done as described by Roeder and Fink.[12] Yeast DNA is digested with the enzyme of choice, the enzyme is inactivated either by heating the digestion reaction or by phenol extraction and the DNA is ligated under conditions favoring self-closure of linear molecules. The choice of the correct restriction enzyme is based on the structure of the transformant (see Figs. 1 and 2). One must choose an enzyme such that the resulting fragment will contain both the yeast sequences with the mutation and the plasmid sequences; in this way the only fragment that will be capable of transforming bacteria to drug resistance carries the mutation to be cloned.

Roeder and Fink[12] digested 25 μg of DNA with *Bam*HI. After digestion they inactivated the *Bam*HI at 65° for 5 min and then diluted the reaction mixture to 1 ml with T4 ligase buffer, added ligase, and incubated at 14° for 16 hr. They obtained ampicillin-resistant transformants at a frequency of 40 per microgram of DNA. Stiles *et al.*[13] followed a similar procedure.

The bacterial transformation is done by the standard procedure as described in Davis *et al.*[19] The most commonly used bacterial strain for transformation is HB101.[10]

Final Verification of a Successful Eviction. Demonstration that a plasmid obtained by eviction contains the desired mutation must be done by different methods, depending on the molecular nature of the mutation. In every case, the first step in the analysis is to examine the restriction digest pattern of the plasmid to show that it correlates with that predicted from Southern analysis using the cloned gene as a probe.

In the case of cloning the *his4-912* mutation,[12] analysis was simplified because *his4-912* is an insertion mutation that results in predicted restriction site pattern differences from *HIS4*$^+$. Restriction fragments from the new plasmid comigrated in a pattern consistent with the results of Southern analysis of the chromosomal DNA of a strain containing the *his4-912* mutation. This type of analysis is desirable for mutations resulting in a difference detectable by restriction pattern changes.

The cloned mutation will not always differ from the wild-type allele in its restriction pattern. Stiles *et al.*[13] designed their eviction such that the region containing the mutation to be cloned was not duplicated in the transformant. By evicting using a restriction enzyme that they knew cut beyond the gene, they were confident that the final plasmid would contain the mutation. Their final verification came from DNA sequence analysis.

In some cases, the investigator may not know the relative orientation of the mutation and plasmid in the transformant and, therefore, may not know in which direction to evict the DNA. One then must evict in both directions (see Fig. 1B) and determine which of the two types of evicted plasmids carries the mutation. If the mutation does not alter the restriction pattern, then one must use another method to determine which evicted plasmid carries the mutation.

In conjunction with the predicted restriction pattern, one can genetically verify the presence of the mutation on the plasmid by transforming each plasmid back into the mutant yeast strain used initially for the eviction. If the mutation is present on the plasmid, then transformants will be homozygous for the mutation and should not give rise to wild-type recombinants at a level above the reversion frequency of the mutation in ques-

[19] R. W. Davis, D. Botstein, and J. R. Roth, "Advanced Bacterial Genetics." Cold Spring Harbor Laboratory, Cold Spring Harbor, New York, 1980.

tion. If the wild-type allele is present on the plasmid, one would expect to see wild-type recombinants at a frequency greater than the reversion frequency of the mutation. If the entire wild-type gene is on the plasmid, then one would see the wild-type phenotype in the initial transformant. The ultimate proof of correct eviction in this case must come from DNA sequencing.

Replacement of a Chromosome Segment by Transplacement

To carry out a transplacement one must first obtain the proper integrative transformant by the same procedure of transformation and verification of the transformant as described for evictions. Next, the goal is to find, by some means, a recombinant that has excised the plasmid and original chromosomal sequences in that region, leaving behind the DNA alteration that was initially on the plasmid. The transplaced mutation must be detectable either by phenotype or by Southern analysis. There are several ways to screen, enrich and select for an isolate that has undergone the type of excision described above.

Screen for Spontaneous Loss. Scherer and Davis[20] replaced the wild-type *HIS3* gene with a deletion mutation made *in vitro*. After isolation of the proper transformant, they showed that the plasmid-encoded Ura3$^+$ phenotype was somewhat unstable owing to the duplicated *his3* DNA flanking the plasmid sequences. To isolate a strain that lost the plasmid sequences they grew their transformant permissively (in the presence of uracil) for 10 generations and then screened colonies for a Ura$^-$ phenotype. Seven of 900 colonies had become Ura$^-$, and three of the seven had left behind the desired *his3* deletion as determined by Southern analysis.

Enrichment by Inositol-less Death. One can enrich for a particular auxotrophy by inositol-less death.[5] This technique can greatly enrich for cells that have lost the plasmid vector. To utilize this technique, the recipient yeast strain used in the transformation must contain a stable *ino*$^-$ mutation. T. F. Donahue, C. A. Styles, and G. R. Fink (unpublished results) have used inositol-less death enrichment when doing transplacements of *his4* deletion mutations constructed *in vitro* into the chromosome in place of wild-type *HIS4*. They modified the method of Henry *et al.*,[5] for use directly on agar plates. This method is outlined below. The medium is the same as that described in the paper by Henry *et al.*,[5] and the Cold Spring Harbor Yeast Manual[9] except that it is solidified with 2% agar. In the procedure describing inositol-less death on plates we have used the example of a *ura3*$^-$ strain that has been transformed to Ura$^+$ with a plasmid con-

[20] S. Scherer and R. W. Davis, *Proc. Natl. Acad. Sci. U.S.A.* **76,** 4951 (1979).

taining the *URA3*$^+$ gene. Those segregants that have lost the vector sequences containing *URA3*$^+$ will survive the starvation for inositol.

Inositol-less Death Enrichment on Plates

1. Plate 10^7 cells on a YEPD plate and grow at 30° for approximately 18 hr. This step prepares the strain for the procedure.
2. Replica print to a starvation plate lacking the nutrient for the desired auxotrophy (uracil in our example) but containing all other growth requirements including inositol; grow for 30 hr at 30°. This step allows the Ura$^-$ cells in the population to deplete their uracil pools.
3. Replica print to a plate lacking inositol and uracil and containing all other nutrients required by the strain and incubate for 40 hr at 30°. Ura$^+$ cells are killed on this medium.
4. Replica print to a YEPD plate and screen colonies that arise for the desired phenotype. The Ura$^-$ segregants (and other auxotrophs) that survived will grow into colonies.

Selection for Loss of the Plasmid Marker. To date, two types of yeast vectors have markers that can be selected against, thereby facilitating isolation of recombinants that have excised the plasmid sequences by recombination. Singh *et al.*[7] showed that yeast strains having a SUP mutation are sensitive to high levels of ethylene glycol in the media. St. John *et al.*[8] and S. Scherer (personal communication) have taken advantage of this observation and have used a vector that contains the *SUP11* gene of yeast (as well as the *URA3* gene) to select directly for loss of an integrated plasmid.

A selection for *ura3*$^-$ mutants has been described by Bach and Lacroute.[6] This procedure has been applied successfully in transplacement experiments[21] using a modified procedure developed by M. Rose and D. Botstein (personal communication). The protocol is listed below.

Protocol for Positive Selection of *ura1*, *ura3* and *ura5* Mutants

1. Plates: Use 1.7 g of yeast nitrogen base without amino acids and without ammonium sulfate, 5 g of proline (as nitrogen source), and 2 g of orotic acid (not neutralized; from Sigma No. 02750).
   Dissolve the ingredients listed above in 500 ml of H_2O, and autoclave separately from 20 g of agar in 500 ml of H_2O.
   When the yeast nitrogen base solution is relatively cool, add: (*a*) uracil to 40 mg/liter; (*b*) 200 mg of dry ureidosuccinate; (*c*) 50 ml of 40% glucose. Mix the solutions, pour into petri dishes, and use after 2 days.

[21] M. C. Rykowski, J. W. Wallis, J. Chol, and M. Grunstein, *Cell* **25,** 477 (1981).

2. Protocol
   a. Spread up to 10^5 cells per plate on YEPD plates.
   b. Incubate 1–2 days at 30°.
   c. Place plates at 4° for 5–7 days.
   d. Replica print to the ureidosuccinate plates described above.
   e. Incubate at 30° for 24–72 hr; Ura^- papillae appear proportionally to the number of cells plated.
3. Notes
   a. The refrigeration *does* improve the selection.
   b. *ura1*$^-$, *ura3*$^-$, and *ura5*$^-$ colonies grow on these plates.
   c. *ura1*$^-$ mutants grow very poorly.
   d. All *ura5*$^-$ mutants are leaky Ura^- on minimal plates but are tight Ura^- on the special medium described above.
   e. Never mix together the two solutions for making the ureidosuccinate plates while they are still hot, as the acid will break down the agar.
   f. This procedure does not work for strains that carry an *ade*$^-$ mutation in the genetic background.

Verification of Transplacement

Recombination events in which the plasmid marker is lost can result either in transplacement, where the segment containing the desired mutation is now present in the chromosome, or restoration of the segment present before the transformation. The outcome depends upon the position of the crossover with respect to the mutation. For example, in the case of transplacement of a *his4* deletion constructed *in vitro* (Fig. 3; and T. F. Donahue, unpublished results) a crossover of type 1 will result in restoration of the wild-type *HIS4* gene in the chromosome, while a crossover of type 2 will result in transplacement of the *his4* deletion into the chromosome. The relative frequency of these two types of crossovers will depend upon the amount of homology on either side of the mutation. If the homology is too small (less than around 300 base pairs) a particular crossover may occur at an undetectably low frequency; therefore the amount of homology must be taken into account when planning a transplacement. One should be able to determine by the phenotype or by Southern analysis which type of crossover has occurred. In the case of Scherer and Davis,[20] of seven Ura^- segregants, three were His^- and four were His^+.

Southern analysis can also demonstrate that the predicted structure exists at the locus: no plasmid sequences should remain, and the mutant allele should be present. The latter will be possible only in the case of a mutation that alters the restriction digest pattern, such as a deletion or insertion.

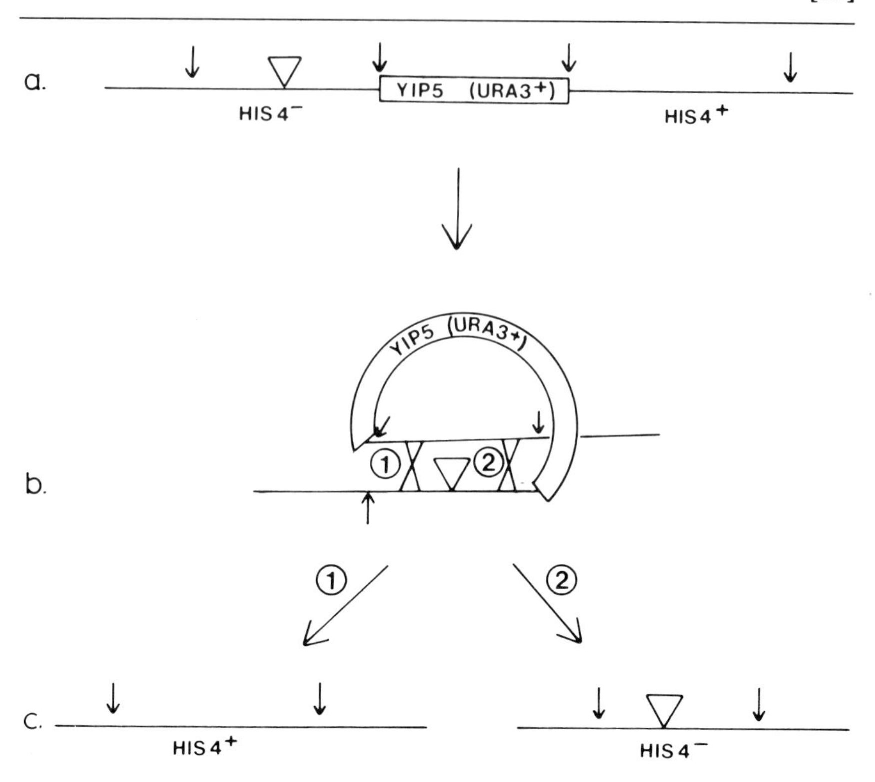

FIG. 3. Transplacement of a *his4* deletion mutation into the yeast chromosome. (a) Structure at the *his4* locus after transformation and integration of a plasmid (YIP5) containing a *his4* fragment with an *in vitro*-constructed deletion. The integration results in a duplication of the *his4* region, one copy carrying the deletion, the other carrying the wild-type gene. (b) The plasmid sequences can excise by a crossover in one of two regions. (c) A crossover of type 1 results in restoration of $HIS4^+$, and a crossover of type 2 results in transplacement of the *his4* deletion into the chromosome. The arrows indicate *Pst* restriction sites.

Comments

With the materials and methods presented here one should be able, given sufficient knowledge about a cloned yeast gene, (*a*) to clone any mutation in that gene by eviction, and (*b*) to replace the wild-type gene with an *in vitro* generated mutation of that gene by transplacement. These are both extremely powerful techniques allowing molecular analysis of mutations.

Pitfalls can best be avoided if the transformants are carefully checked in the ways described, both genetically and physically, to verify their structure.

One additional application of evictions for yeast is to clone easily the neighboring sequences. An eviction will often result in cloning of neighboring sequences not on the original clone. For example, the eviction of *his4-912*[12] included the cloning of DNA beyond the initial *Sal*I site to the next *Bam*HI site. A continued application of this technique could mean a directed way of "walking" from one cloned segment of yeast DNA down the chromosome.

Since the techniques of eviction and transplacement depend upon integrative transformation they would seem to be limited to those organisms known to have a means of taking up DNA and integrating it by homology. Currently those techniques are limited to yeast and bacteria. Integrative transformation into higher cells does not seem to depend upon DNA sequence homology[22]; therefore, it is difficult to imagine direct application of these techniques to other eukaryotic systems for which transformation systems currently exist.

In bacterial systems, techniques to move mutations between plasmids, phages, and the chromosome by *in vivo* recombination have been successfully employed. Some examples are in study of transposition,[23,24] regulation of the histidine operon in *Salmonella typhimurium*,[25] and mutations affecting secretion of proteins, such as the *E. coli lamB* protein.[26] One *in vivo* technique used in bacteria, that may have application for the eviction of mutations in yeast is homogenotization. This technique[27–29] involves a "gene conversion" of the wild-type allele on a plasmid using the mutant chromosomal allele as a template, resulting in a plasmid-carrying strain that now contains the mutant gene both on the chromosome and the plasmid. Homogenotization in yeast would permit the transfer of mutations from the chromosome to an autonomously replicating vector and would avoid the necessity of the integration, excision, and ligation steps in the current eviction procedure.

[22] D. M. Robins, S. Ripley, A. S. Henderson, and R. Axel, *Cell* **23,** 29 (1981).

[23] T. J. Foster, M. A. Davis, D. E. Roberts, K. Takeshita, and N. Kleckner, *Cell* **23,** 201 (1981).

[24] D. E. Berg, A. Weiss, and L. Crossland, *J. Bacteriol.* **142,** 439 (1980).

[25] H. M. Johnston and J. R. Roth, *J. Mol. Biol.* **145,** 713 (1981).

[26] S. D. Emr and T. J. Silhavy, *J. Mol. Biol.* **141,** 63 (1980).

[27] F. Jacob and E. A. Adelberg, *C. R. Hebd. Seances Acad. Sci.* **249,** 189 (1959).

[28] F. Jacob and E. Wollman, "Sexuality and the Genetics of Bacteria," pp. 251–255. Academic Press, New York, 1961.

[29] D. E. Sheppard and E. Englesberg, *J. Mol. Biol.* **25,** 443 (1967).

Acknowledgments

We thank Tom Donahue and Mark Rose for sharing unpublished results. We also thank our colleagues in the Fink laboratory: Tom Donahue, Karen Durbin, Alan Hinnebusch, Jennifer Jackson, Giovanna Lucchini, Franz Meussdoerffer, Francisco del Rey, Sandy Silverman, and Barbara Valent for their critical reading of the manuscript. Additionally, we thank Fran Ormsbee for help in preparing the manuscript.

[14] Genetic Applications of Yeast Transformation with Linear and Gapped Plasmids

By TERRY L. ORR-WEAVER, JACK W. SZOSTAK, and RODNEY J. ROTHSTEIN

Transformation of yeast by plasmids can occur by integration of the plasmid into a homologous sequence on the chromosome, producing a stably transformed strain.[1] The integration event yields an integrated copy of the plasmid vector sequences flanked by direct repeats of the homologous region on the plasmid and the chromosome (Fig. 1). In approximately 20% of stable transformants, the homologous DNA on the chromosome has been replaced by the corresponding plasmid sequence without integration of the plasmid vector DNA. These substitution transformants arise either via a gene conversion or by a double-crossover event. Among transformants containing integrated plasmids, many contain multiple copies of the plasmid arranged in tandem repeats (Fig. 1).[2] These multimers do not arise by ligation or recombination of the plasmid prior to integration.[3] In their initial studies on yeast transformation, Hinnen and co-workers found that the *LEU*$^{+}$ plasmid pYe*leu*10 occasionally integrated at genomic sites other than at the homologous *leu2* region. Subsequently it was found that a subcloned fragment of pYe*leu*10 contains a repeated sequence.[4] This subcloned plasmid integrated at the dispersed copies of the repeated sequence as well as at the *LEU2* locus.[5] Thus all integrations appear to result from recombination between homologous sequences.

[1] A. Hinnen, J. B. Hicks, and G. R. Fink, *Proc. Natl. Acad. Sci. U.S.A.* **75,** 1929 (1978).

[2] J. W. Szostak and R. Wu, *Plasmid* **2,** 536 (1979).

[3] T. L. Orr-Weaver, J. W. Szostak, and R. J. Rothstein, *Proc. Natl. Acad. Sci. U.S.A.* **78,** 6354 (1981).

[4] A. Kingsman, R. Gimlich, L. Clarke, A. Chinault, and J. Carbon, *J. Mol. Biol.* **145,** 619 (1981).

[5] H. Klein and T. D. Petes, personal communication, 1981.

ISBN 0-12-182001-7

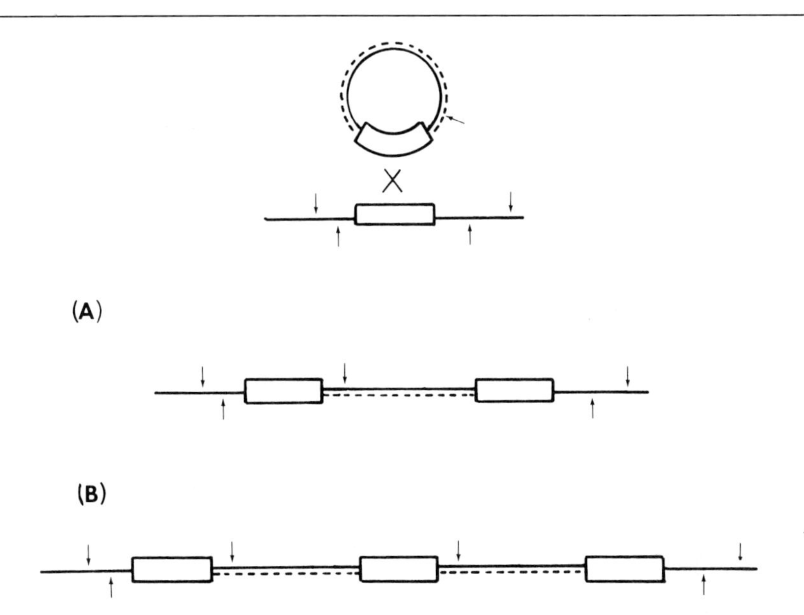

FIG. 1. Plasmid integration. (A) Recombination between homologous regions on the plasmid and the chromosome results in integration of the plasmid vector sequences flanked by direct repeats of the homologous DNA. (B) Multiple, tandem copies of the integrated plasmid DNA are observed in up to 50% of integrated transformants. (↓) A restriction enzyme site in the vector DNA, and in flanking sequences; (↑) a restriction enzyme site present only in flanking chromosomal sequences; ⚌, plasmid vector DNA; ▭; homologous DNA; ——, chromosomal DNA.

A second class of transformants, characterized by high frequencies of transformation and an unstable phenotype for selectable markers on the plasmid, was described by Struhl *et al.*[6] This high-frequency transformation requires the presence of a chromosomal ARS (autonomously replicating sequence) on the plasmid, which permits autonomous replication of the plasmid DNA. The extrachromosomal nature of the plasmids is demonstrated by the instability of the plasmid markers, and by Southern blot analysis which shows the plasmid to be unlinked to chromosomal DNA. ARS-containing plasmids transform at a high frequency because they are maintained autonomously rather than requiring a recombination event with chromosomal DNA.

We have described high-frequency transformation with linear-integrating plasmids in which a double-strand break is introduced in the region homologous to the yeast genome.[3] These double-strand breaks, in-

[6] K. Struhl, D. T. Stinchcomb, S. Scherer, and R. W. Davis, *Proc. Natl. Acad. Sci. U.S.A.* **76,** 1035 (1979).

troduced by restriction enzyme digestion, are recombinogenic and interact with homologous chromosomal DNA. A complex plasmid, containing more than one yeast gene, can be directed to integrate at a single site in the genome by making a restriction enzyme cut *within* the corresponding region on the plasmid. A plasmid cut twice within a yeast fragment to generate a double-strand gap also transforms at high frequencies. Moreover, the gap is always repaired from chromosomal information during the integration event.[3]

Yeast transformation has been an increasingly useful technique not only for isolating genes, but also for reintroducing sequences modified *in vitro*. Numerous genes have been recovered from libraries of total yeast DNA cloned into replicating vectors by selecting for complementation of mutations.[7-11] Integrating plasmids have been used to mutagenize chromosomal loci. In the case of actin, a mutant gene was obtained through gene disruption by transforming with a plasmid that contained a fragment of the actin gene lacking both the 5′ and 3′ ends of the gene. Integration of this plasmid into the chromosomal actin locus resulted in two mutated copies of the actin gene, one with a deletion of the 5′ end and the other with a deletion of the 3′ end.[12] Another approach (transplacement) has been to substitute an *in vitro* generated mutation on a plasmid for a chromosomal gene by transforming and integrating the plasmid into the corresponding chromosomal locus. Colonies are screened for those excisions that leave the mutated gene copy in the chromosome.[13] Linear and gapped plasmids can be employed to expand the uses of transformation. In this chapter we describe the use of linear and gapped plasmids (*a*) to obtain higher transformation frequencies; (*b*) to target complex plasmids to a specific chromosomal site; (*c*) to facilitate integration of replicating plasmids; and (*d*) to isolate rapidly and map chromosomal mutations.

Materials and Methods

Strains

Strain LL20 (α, *canl, his3 -11,15, leu2 -3,112,* [*Psi*$^+$]) was obtained from Lester Lau (Cornell University). D234-3B (α, *his3 -11,15, leu2 -3,-*

[7] A. Hinnen, P. J. Farabaugh, C. Ilgen, G. R. Fink, and J. Friesen, *Eucaryotic Gene Regul., ICN-UCLA Symp. Mol. Cell. Biol.* **14,** 43 (1979).

[8] J. R. Broach, J. N. Strathern, and J. B. Hicks, *Gene* **8,** 121 (1979).

[9] V. M. Williamson, J. Bennetzen, E. T. Young, K. Nasmyth, and B. D. Hall, *Nature (London)* **283,** 214 (1980).

[10] J. B. Hicks, J. N. Strathern, and A. Klar, *Nature (London)* **282,** 478 (1979).

[11] K. Nasmyth and K. Tatchell, *Cell* **19,** 753 (1979).

[12] D. Shortle, J. E. Haber, and D. Botstein, *Science* **217,** 371 (1982).

[13] S. Scherer and R. W. Davis, *Proc. Natl. Acad. Sci. U.S.A.* **76,** 4951 (1979).

112, tcm1, trp1, ura3) was constructed by P. Brown (Dana-Farber Cancer Institute). The *ura3* mutation was generated by ethylmethanesulfonate mutagensis.[14] Bacterial strain SF8 (C600, $rK^- \ mK^- \ recB^-C^- \ lop\text{-}11$) was obtained from Lester Lau.[15]

DNA Preparation

Plasmid DNA was isolated from *Escherichia coli* by the rapid-boiling procedure[16] or by centrifugation in CsCl gradients.[17] Plasmid DNA purified by either procedure is suitable for yeast transformation, but DNA purified on gradients transforms more efficiently. Linear-plasmid DNA was prepared by restriction enzyme (New England BioLabs and Bethesda Research Laboratories) digestion and was then phenol extracted and precipitated with ethanol. *Escherichia coli* carrier DNA was dissolved in 10 m*M* Tris, 1 m*M* EDTA (pH 7.4) at approximately 2 mg/ml, then sonicated 10 times for 30 sec interspersed with 30-sec periods to allow the sample to cool. The DNA concentration was then determined spectrophotometrically.

Yeast Transformation

Media were as described.[14] Yeast transformation was as described by Hinnen *et al.*[1] with slight modifications. Fifty milliliters of a mid-log culture of cells (1×10^7 cells/ml) was used for no more than eight transformation samples. The cells were centrifuged at maximum speed in an IEC clinical centrifuge and washed twice with 10 ml of 1 *M* sorbitol (Aldrich). The cells were resuspended in 5 ml of 1 *M* sorbitol, an aliquot was diluted to 10^{-5} in sterile water, and 100 μl of this were plated on a YPD plate; 150 μl of glusulase (Endo Laboratories, Inc.) and 5 μl of 2-mercaptoethanol were added to the cells, and they were incubated at 30° for 1 hr with slow shaking. An aliquot was diluted to 10^{-3} in sterile water and 100 μl were plated on YPD; this measures the percentage of killing, an indication of the efficiency of spheroplast formation. The spheroplasts were pelleted in the clinical centrifuge for 4 min and resuspended gently in 2 ml of 1 *M* sorbitol by slow rocking. This volume was adjusted to 10 ml with 1 *M* sorbitol, and the spheroplasts again were centrifuged. This wash was repeated, then the spheroplasts were pelleted. The spheroplasts were resuspended in 9 ml of 1 *M* sorbitol; 1 ml of 0.1 *M* Tris (7.4), 0.1 *M* $CaCl_2$ was

[14] F. Sherman, G. R. Fink, and C. Lawrence, "Methods in Yeast Genetics". Cold Spring Harbor Laboratory, Cold Spring Harbor, New York 1977.

[15] J. F. Cameron, S. Panasenko, I. Lehman, R. W. Davis *Proc. Natl. Acad. Sci. U.S.A.* **72,** 3416 (1975).

[16] D. S. Holmes and M. Quigley, *Anal. Biochem.* **114,** 193 (1981).

[17] Y. Kupersztock and D. R. Helinski, *Biochem. Biophys. Res. Commun.* **54,** 1451 (1973).

added, and the spheroplasts were centrifuged as before. The final resuspension was in 1 *M* sorbitol, 10 m*M* Tris (pH 7.4), 10 m*M* $CaCl_2$ at 0.2 ml per DNA sample to be transformed. The plasmid DNA was added and, when used, 10 μg of sonicated *E. coli* carrier DNA were also added. The total volume of DNA added was always less than one-tenth the volume of the spheroplasts. After 10 min at room temperature, 2 ml of 50% (w/v) PEG-4000 (Fisher Lot No. 712265) mixed with 0.2 ml of 0.1 *M* Tris (pH 7.4), 0.1 *M* $CaCl_2$ were added, and the samples were mixed well. We routinely allowed the samples to remain in PEG for 10 min at room temperature before centrifuging the spheroplasts for 10 min in the clinical centrifuge. However, occasionally we directly plated the cells 10 min after the addition of the PEG-4000; the volume of cells and PEG plated must be less than one-tenth of the volume of the regeneration agar. When pelleted, the spheroplasts were resuspended in 0.5 ml of 1 *M* sorbitol, 10 m*M* Tris (pH 7.4), 10 m*M* $CaCl_2$ and plated into regeneration agar, or if necessary, further diluted into the same buffer and then plated.

We have observed large variations in transformation frequencies with different batches of PEG-4000 from both Fisher and Baker. Frequencies obtained with a standard replicating plasmid such as YEp13 can be used to compare lots. As discussed below, different strains vary in their ability to be transformed. This may in part result from differing sensitivities to glusulase. Dilution of cells into water and plating onto YPD before and after glusulase treatment should give approximately a 100-fold decrease in survival. In addition, the number of viable spheroplasts can be estimated by dilution into 1 *M* sorbitol and plating in nonselective regeneration agar. For example, a *leu2* derivative of Y55 (HO a/α, *leu2-1* obtained from J. McCusker, Brandeis University) requires a 10-min glusulase treatment, and LL20 requires a 1-hr incubation, to give the same number of viable spheroplasts (14% of the original cells).[18]

Competent yeast spheroplasts can be frozen and stored at $-70°$ with a 5- to 10-fold loss in transformation ability. The spheroplasts from 50 ml of culture were resuspended in 1.5 ml of 1 *M* sorbitol, 15% DMSO following the 1 *M* sorbitol washes but before the $CaCl_2$ addition. The cells were then slowly frozen and stored at $-70°$. For use, the spheroplasts were slowly thawed, brought to 10 m*M* Tris (pH 7.4), 10 m*M* $CaCl_2$ by the addition of 0.1 *M* Tris (pH 7.4), 0.1 *M* $CaCl_2$, pelleted, and resuspended in 1 *M* sorbitol, 10 m*M* Tris (pH 7.4), 10 m*M* $CaCl_2$ at 0.2 ml per DNA sample.

Transformants were purified by streaking on selective media. The stability of a selectable marker on the plasmid was checked by streaking

[18] A. Murray and J. Szostak, unpublished results, 1982.

transformants onto nonselective media, replica plating to selective media, and monitoring for growth.

Southern Blots

The structure of integrated plasmids should be determined by Southern blot analysis prior to extensive use of the strain. Plasmid integration frequently results in multiple, tandem insertions of the plasmid DNA, and aberrant integration events may occasionally occur. We purified yeast DNA by the method of Davis *et al.*[19] except that diethyl pyrocarbonate was not added during the lysis step. Genomic DNA was digested by restriction enzymes, electrophoresed on agarose gels, and transferred to nitrocellulose.[20] The filters were prehybridized for 12 hr in 4× SSC (1× SSC is 0.15 *M* NaCl, 0.015 *M* sodium citrate), 1× Denhardt's solution[21]; hybridization was at 67° for 12 hr with 3 μCi of nick-translated probe[22] in 4× SSCP (4× SSC with 80 m*M* sodium phosphate, pH 7.0), 1× Denhardt solution, 0.2% SDS, with 0.1 mg/ml sonicated calf-thymus carrier DNA. At the end of the hybridization the Southern blots were washed at room temperature four times, for 15 min each, in 4× SSCP, 1× Denhardt solution 0.1% SDS, and three times for 5 min each in 3 m*M* Tris base. Autoradiography was done with Kodak XR-5 film and intensifying screens at −70°.[23] Multiple, tandem plasmid integrations were detected by digesting the DNA from transformants with a restriction enzyme that cuts the plasmid at a single site in the vector sequences. A Southern blot was prepared and hybridized to nick-translated vector probe. The presence of a band with a size equal to the length of the plasmid on the Southern blot is diagnostic of multiple, tandem plasmid integration (see Fig. 1).

Plasmid Rescue from Yeast

Integrated plasmids in yeast transformants were isolated by digesting approximately 1 μg of yeast DNA with the appropriate restriction enzyme. After phenol extraction and precipitation of the DNa with ethanol, the DNA was ligated with T4 ligase (New England BioLabs) under conditions specified by the supplier and at dilute DNA concentration (1 μg/300 μl). The dilute DNA concentration reduces the ligation of multiple fragments into the plasmid. The DNA was then transformed into fro-

[19] R. W. Davis, M. Thomas, J. R. Cameron, T. P. St. John, S. Scherer, and R. A. Padgett, this series, Vol. 65, p. 404.

[20] E. M. Southern, *J. Mol. Biol.* **98,** 503 (1975).

[21] D. T. Denhardt, *Biochem. Biophys. Res. Commun.* **23,** 641 (1966).

[22] P. W. Rigby, M. Dieckmann, C. Rhodes, and P. Berg, *J. Mol. Biol.* **113,** 237 (1977).

[23] R. A. Laskey and A. D. Mills, *FEBS Lett.* **82,** 314 (1977).

zen competent bacterial cells,[24] and ampicillin-resistant colonies were selected on LB plates[25] supplemented with 100 μg of ampicillin per milliliter. The plasmid DNA was isolated and purified by centrifugation in CsCl gradients and digested with the restriction enzyme *Hin*fI. End-labeling of *Hin*fI restriction fragments was performed by incubating approximately 10 pmol of *Hin*fI-digested ends with 20 pmol of [$\alpha - {}^{32}$P]dATP (2000–3000 Ci/pmol, Amersham) and 2 units of reverse transcriptase (obtained from J. Beard, Life Sciences) for 30 min at 37° in 50 m*M* Tris-HCl (pH 7.4), 50 m*M* KCl, 10 m*M* dithiothreitol, 10 m*M* $MgCl_2$.[26] Care was taken to keep the concentration of dATP greater than 1 μ*M* for efficient end labeling.

Results and Discussion

Transformation Frequencies

The frequency of yeast transformation with either replicating or integrating plasmid is highly dependent on strain background. We have observed up to two orders of magnitude difference in the ability of various strains to be transformed. Therefore strains should be screened for transformation frequencies with a standard replicating plasmid and an integrating plasmid. We routinely use YEp13 as a replicating standard and the *rDNA-LEU2* plasmid pSZ32 as an integrating standard because both transform at high frequency and consequently give measurable numbers of transformants with most strains (Fig. 2). pSZ32 yields more transformants than other integrating plasmids because the repeated copies of the *rDNA* region present in the genome provide a large homologous region for plasmid integration.

Transformation frequencies with integrating plasmids are stimulated if the plasmid contains a double-strand break in the region homologous to the yeast genome. The double-strand break is generated by digesting the plasmid DNA with a restriction enzyme that makes a single cut within the homologous region; the nature of the cut, whether flush or possessing 5′ or 3′ overhanging ends, does not appear to affect this stimulation. However, the increase in frequency varies with the fragment. For example, we consistently observe a 2000- to 3000-fold stimulation in frequency when transforming with a linear plasmid containing a double-strand break in the

[24] D. Morrison, *J. Bacteriol.* **132,** 349 (1977).

[25] J. Miller, "Experiments in Molecular Genetics." Cold Spring Harbor Laboratory, Cold Spring Harbor, New York, 1972.

[26] C. P. Bahl, R. Wu, J. Stawinsky, and S. A. Narang, *Proc. Natl. Acad. Sci. U.S.A.* **74,** 966 (1977).

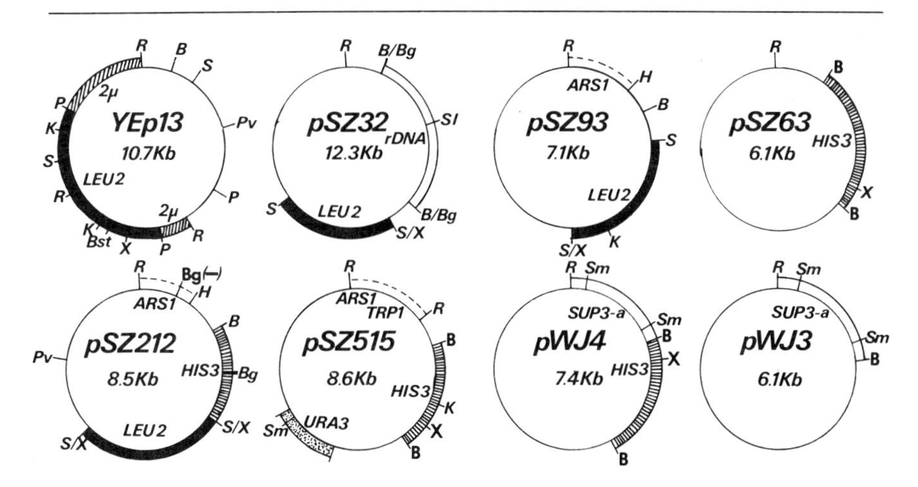

FIG. 2. Plasmids. YEp13 was obtained from J. Hicks; it contains a 4.8 kb *LEU2 Pst*I fragment inserted into a 1.5 kb *Eco*RI fragment containing the 2 μm origin of replication [J. B. Hicks, A. Hinnen, and G. R. Fink, *Cold Spring Harbor Symp. Quant. Biol.* **43,** 1305 (1978)]. pSZ32, pSZ93, and pSZ212 contain the *LEU2* gene on a *Sal*I–*Xho*I fragment from pYe*leu*10 [B. Ratzkin and J. Carbon, *Proc. Natl. Acad. Sci. U.S.A.* **74,** 487 (1977)]. pSZ32 has IS2 on this fragment. It also contains a 4.5 kb *Bgl*II fragment of yeast *rDNA* inserted into the *Bam*HI site of pBR322. Both pSZ212 and pSZ93 have a 0.7 kb *Eco*RI *Hin*dIII fragment with the *ARS1* sequence.[5] pSZ63 contains a 1.7 kb *Bam*HI *HIS3* fragment (obtained from K. Struhl) inserted into the *Bam*HI site of pBR322. The *Bgl*II site in the *ARS1* fragment of pSZ212 was removed by bisulfite mutagenesis [D. Shortle and D. Nathans, *Proc. Natl. Acad. Sci. U.S.A.* **75,** 2170 (1978)], and the 1.4 kb *HIS3 Bam*HI–*Xho*I fragment was inserted between the *Bam*HI and *Sal*I sites. pSZ515 is derived from pJT29 (obtained from J. Thomas, MIT), which has a 1.5 kb *Eco*RI fragment containing both *ARS1* and *TRP1* and a 1.05 kb *Hin*dIII *URA3* fragment tailed into the *Ava*I site of pBR322. The 1.7 kb *HIS3 Bam*HI fragment from pSZ63 was inserted into the *Bam*HI site of pJT29 to generate pSZ515. Both pWJ3 and pWJ4 contain a 1.7 kb *Eco*RI–*Bam*HI fragment with the *SUP3-a* gene ligated into the *Eco*RI–*Bam*HI region of pBR322. pWJ4 also contains the 1.7 kb *HIS3*–*Bam*HI fragment. B, *Bam*HI; Bg, *Bgl*II; Bst, *Bst*EII; H, *Hin*dIII; K, *Kpn*I; P, *Pst*I; Pv, *Pvu*II; R, *Eco*RI; S, *Sal*I; SI, *Sac*I; Sm, *Sma*I; X, *Xho*I. ——— , pBR322; ▬, ▨, ▭, - - -, ▥, ░, yeast DNA, as indicated.

HIS3 fragment, while a plasmid cut in the *LEU2* fragment shows only a 50- to 100-fold effect.[3] Consequently, plasmids containing certain genes are more difficult to integrate than others. The location of the double-strand break relative to the chromosomal mutation(s) may also affect the stimulation observed, because degradation past the sequences homologous to the chromosomal mutation(s) makes it impossible to obtain a wild-type transformant (see below).

The frequency of transformation for both replicating and integrating plasmids is increased by the addition of sonicated *E. coli* carrier DNA

TABLE I
EFFECT OF CARRIER DNA ON TRANSFORMATION FREQUENCY

Plasmid DNA	Plasmid DNA added to spheroplasts[a] (μg)	Spheroplasts plated (%)	*E. coli* carrier DNA[b]	Transformants per microgram of plasmid DNA
pSZ32 uncut	10	5	−	72
	2.5	20	−	48
	0.5	100	−	70
	10	5	+	170
	2.5	20	+	250
	0.5	100	+	260
pSZ32 *Sac*I	10	1	−	1400
	1	10	−	1100
	0.1	100	−	1100
	10	1	+	2600
	1	10	+	12,000
	0.1	100	+	26,000
YEp13 uncut	2	1	−	2500
	0.5	10	−	1400
	0.02	100	−	4500
	0.02	100	+	120,000
YEp13 *Bam*HI	5	10	−	6200
	1	50	−	3400
	0.5	100	−	420
	5	10	+	4200
	0.5	100	+	3200
pSZ63 *Xho*I	20	0.5	−	260
	2	5	−	220
	0.2	50	−	200
	20	0.5	+	2500
	2	5	+	2400
	10[c]	10	−	45
	5[c]	20	−	60
	1[c]	100	−	60
	10[c]	10	+	560
	5[c]	20	+	3900
	1[c]	100	+	2900

[a] Plasmid DNA was added to spheroplasts derived from 10^8 cells of strain LL20.
[b] Ten micrograms of sonicated *Escherichia coli* carrier DNA were added to spheroplasts.
[c] A different transformation experiment.

during the DNA incubation step of the transformation procedure (Table I). Carrier DNA stimulates transformation by both circular and linear plasmids. The effect of the *E. coli* DNA is most dramatic when low concentrations of plasmid DNA are incubated with the cells. The structure of plasmid DNA in both stable and unstable transformants is not affected by the presence of carrier DNA in the transformation (data not shown). We presume the carrier DNA acts to reduce degradation of the transforming plasmid DNA.

Transformation with Linear Plasmids

Complex plasmids that contain more than one fragment of yeast DNA can be directed to integrate at one specific integration site by making a double-strand break in the corresponding region on the plasmid. We employed this targeting technique to direct a plasmid containing the *HIS3* and *sup3*$^+$ genes to either site in the genome[3] and a plasmid with *URA3*, *CYH2*, and *LEU2* to *URA3*.[27] The ability to direct a plasmid to integrate allows placement of an easily scored marker adjacent to a gene that is difficult to map. It also facilitates the use of transplacement[13] and of gene disruption[12] for generating mutants because the plasmid is targeted to the desired gene.

There are several constraints on the general application of this technique. The region on the plasmid that is homologous to the desired chromosomal integration site must be large enough to permit a recombination event and to avoid degradation past the end of the homologous sequences. Plasmid integration cannot occur if the DNA is degraded past the border of targeting and vector DNA because homology must be present on both sides of the double-strand break to obtain a crossover (Fig. 1). For example, targeting did not work when yeast was transformed with a plasmid containing the *URA3* selectable marker and a 183 base-pair (bp) fragment with the *H2B* histone gene.[28] The plasmid was made linear either by digestion with a restriction enzyme that cuts 31 bp from one end of the *H2B* fragment, or by an enzyme that cuts at a site 41 bp from the other end. Thirty-one transformants were analyzed by Southern blots; no integration events at *H2B* were detected. A plasmid with a 6 kb fragment containing the *H2B* gene was readily directed to the *H2B* locus. We have successfully targeted a plasmid with a *sup3*$^+$ 1.7 kb fragment using the *Sma*I site 170 bp from the junction with vector DNA.[3] If the double-strand break is close to the edge of homology, degradation may occur, and substitution rather than integrations will be obtained. Therefore it is preferable to make the restriction enzyme cut in the middle of the fragment. If this is not possible, integration events can be selected if the plasmid contains a additional selectable marker. We transformed D234-3B with the *URA3-HIS3* plasmid, pSZ515 (Fig. 2), which had been made linear by digestion with *Sma*I and thus had a double-strand break 58 bp from the

[27] P. Brown and J. Szostak, unpublished results, 1982.

[28] M. A. Osley and L. Hereford, personal communication, 1981.

pBR322 *URA3* junction. HIS$^+$ selection yielded stable, integrated plasmids (Table II). Similar to the variation seen in the stimulation of transformation frequencies by double-strand breaks, the ability to target a plasmid to integrate varies with the fragment. We have found it difficult to isolate single copy integrants at the *LEU2* gene.[27]

Replicating plasmids can be stimulated to integrate by making a double-strand break in the yeast gene on the plasmid prior to transformation. We obtained integration of ARS1 replicating plasmids containing either *HIS3*, *LEU2*, or *URA3* (Table II). When circular, replicating plasmids are transformed $<1\%$ of the transformants are integrated and stable for the selectable marker. In contrast, transformation with linear plasmids, when the break is in DNA homologous to yeast, results in up to 50% stable

TABLE II
INTEGRATION OF REPLICATING PLASMIDS

Plasmid DNA		Site cut	Selection[a]	Transformants[b]	
				Unstable	Stable
pSZ93	Uncut		LEU$^+$	64	0
	*Bam*HI	pBR322	LEU$^+$	64	0
	*Kpn*I	*LEU2*	LEU$^+$	52	12
pSZ212	Uncut		LEU$^+$	32	0
	*Pvu*II	pBR322	LEU$^+$	21	0
	*Bam*HI	border	LEU$^+$	124	1
	*Bgl*II	*HIS3*	LEU^{+c}	103	27
pSZ511[d]	*Bgl*II	*HIS3*	HIS$^+$	57	63
pSZ515	Uncut		URA$^+$	55	0
	*Kpn*I	*HIS3*	URA^{+c}	24	31
	*Xho*I	*HIS3*	URA^{+c}	81	39
	*Kpn*I + *Xho*I	*HIS3*	URA^{+c}	25	31
	*Sma*I	*URA3*	HIS$^+$	45	3

[a] Transformed spheroplasts were plated into regeneration agar lacking the amino acid indicated.

[b] Transformants were checked for stability of the selected marker by streaking onto nonselective media, replica plating to selective media, and monitoring for growth.

[c] Many of these transformants are *his*$^-$ but retain the restriction enzyme site at which the plasmid was made linear. They may arise by degradation of the linear plasmid DNA past the region homologous to the chromosomal point mutations, followed by repair of the resulting double-strand gap from chromosomal information (see Fig. 4). Alternatively, branch migration to form heteroduplex DNA followed by repair could result in transfer of the chromosomal mutations to the plasmid.

[d] pSZ511 was made from pSZ212 by digestion with *Bgl*II and religation under dilute conditions. Therefore pSZ511 contains a 60 base-pair deletion at the *Bgl*II site, and digestion with *Bgl*II generates a gapped plasmid.

transformants (Table II). Linear plasmids containing a break in vector DNA do not yield integrated plasmids.

Transformation with Gapped Plasmids

When two restriction enzyme cuts are made within the region on the plasmid homologous to chromosomal DNA, a double-strand gap is generated. This gap is faithfully repaired from chromosomal information during plasmid integration.[3] pSZ501 is an integrating plasmid containing a 4.5 kb *Bgl*II *rDNA* fragment and the selectable marker *LEU2* (Fig. 3). The plasmid was digested with *Xba*I and religated to produce pSZ503, which has a 3.5 kb deletion at the *Xba*I site in the *rDNA* fragment (Fig. 3). Digestion of pSZ503 with *Xba*I produces a 3.5 kb double-strand gap. We obtained transformants using circular pSZ501, circular pSZ503, and *Xba*I-cut pSZ503, and their DNA was isolated and analyzed by Southern blots. Digestion of the DNA from circular pSZ501 transformants with *Pst*I and *Bgl*II gives a 10 kb and a 5.6 kb fragment that hybridize to pBR322. A 11.1 kb fragment, the unit length of the plasmid, appears if there are multiple tandem copies of the plasmid (Fig. 3). Transformants from pSZ503 cut with *Xba*I show a 10 kb and 5.6 kb band (Fig. 3); therefore the gap is repaired. Transformants with circular pSZ503 show a 3.5 kb deletion in either the 10 kb or 5.6 kb fragment, depending on which side of the deletion the crossover occurs. The unit length band, diagnostic of multiple insertions, is reduced in size by 3.5 kb to give a 7.6 kb band. Thus free DNA ends allow double-strand gap repair during plasmid integration. We postulate that the DNA ends invade homologous chromosomal DNA, serve as primers for repair synthesis to repair the gap, and yield an intermediate that can be resolved to give an integrated plasmid or a free, repaired plasmid.[3,29]

The gap repair reaction provides a simple method for the cloning of chromosomal alleles of genes for which a cloned fragment and restriction map are available. The information used to repair the gap on the plasmid is from chromosomal DNA. Consequently, transformation with an integrating plasmid containing a double-strand gap that deletes the region homologous to the chromosomal mutation results in an integrated plasmid in which both copies of the homologous region contain the mutation (Fig. 4). Reisolation of the plasmid always results in recovery of the chromosomal allele. The plasmid pWJ4 contains the *SUP3-a* supressor gene and the *HIS3* gene (Fig. 2). Transformation of LL20 with up to 150 μg of circular pWJ4 DNA did not yield transformants, presumably because suppressors

[29] T. L. Orr-Weaver and J. W. Szostak, *Proc. Natl. Acad. Sci. U.S.A.*, in press (1983).

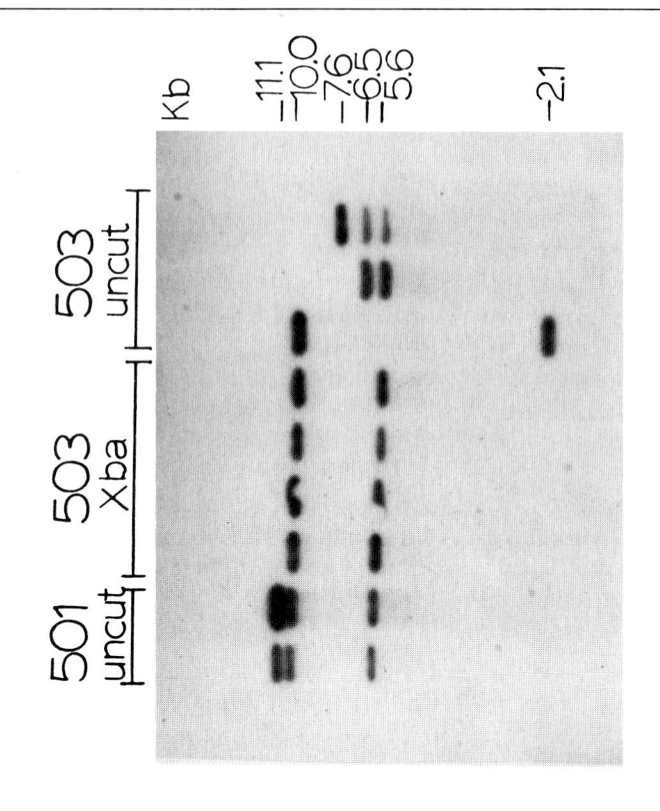
501
uncut
503
Xba
503
uncut
Kb
11.1
10.0
7.6
6.5
5.6
2.1

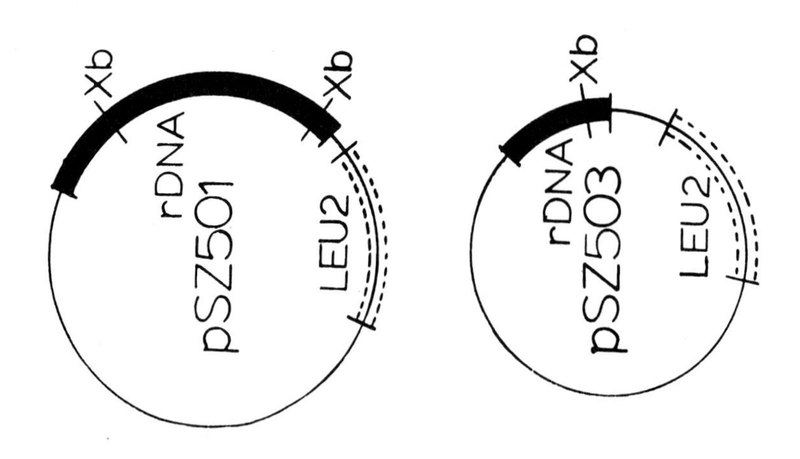
Xb
rDNA
pSZ501
LEU2
Xb
Xb
rDNA
pSZ503
LEU2

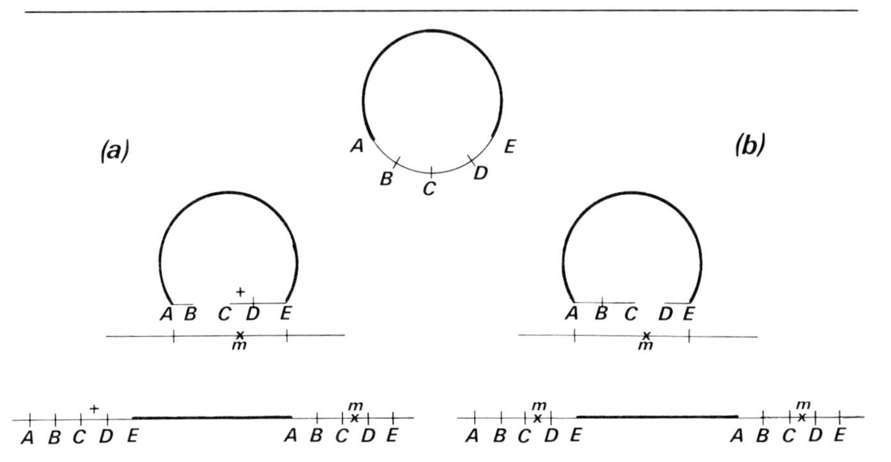

FIG. 4. Repair of plasmid gaps from chromosomal information. (a) When the double-strand gap on the plasmid does not extend past the chromosomal mutation, the gap is repaired during plasmid integration to yield a wild-type and a mutant gene. (b) If the plasmid gap extends past the chromosomal mutation, during gap repair the chromosomal information is transferred to the plasmid, resulting in two copies of the homologous region each containing the mutation. X, mutation; ——— , homologous DNA; ▬▬▬ , plasmid vector sequences.

make cells sensitive to the high osmotic conditions of the regeneration agar. Digestion of pWJ4 with the restriction enzyme *Sma*I creates a 1.2 kb gap in the *SUP3-a* fragment; this gap extends into the *SUP3-a* gene and removes the amber anticodon. When LL20 was transformed with *Sma*I-cut pWJ4, 10–50 transformants per microgram of DNA were obtained. Eight of these transformants were analyzed by Southern blots; all had integrated at the *sup3*$^+$ locus and had the gap repaired (data not shown). By digesting the chromosomal DNA from the transformants with *Eco*RI or *Bam*HI and ligating under dilute conditions, we recovered two

FIG. 3. Repair of gapped plasmids. pSZ501 (11.1 kb) contains a 4.5 kb *Bgl*II *rDNA* fragment inserted into the *Bam*HI site of pBR322 and the 2.3 kb *LEU2 Sal*I–*Xho*I fragment in the *Sal*I site; pSZ503 (7.6 kb) was generated by digesting pSZ501 with *Xba*I and religating under dilute conditions. DNA was isolated from transformants obtained with circular pSZ501, circular pSZ503, and *Xba*I-cut pSZ503. The DNA was digested with *Pst*I and *Bgl*II, Southern blots were made, and they were hybridized to nick-translated pBR322. *Hin*dIII-digested λ DNA was used as size standards. Sizes of the fragments are labeled. Circular pSZ501 transformants have a 10 kb and a 5.6 kb fragment that hybridize to pBR322. A unit length band of 11.1 kb is present if multiple tandem integrations have occurred. The DNA from *Xba*I-cut pSZ503 transformants gives full-length 10 kb and 5.6 kb fragments. However, transformants from circular pSZ503 have a 10 kb and 2.1 kb, or a 6.5 kb and 5.6 kb band, showing that the deletion was not repaired. One circular pSZ503 sample has a deleted multimer fragment of 7.6 kb. ——— , pBR322; ▬▬▬, rDNA; - - - - , *LEU2*.

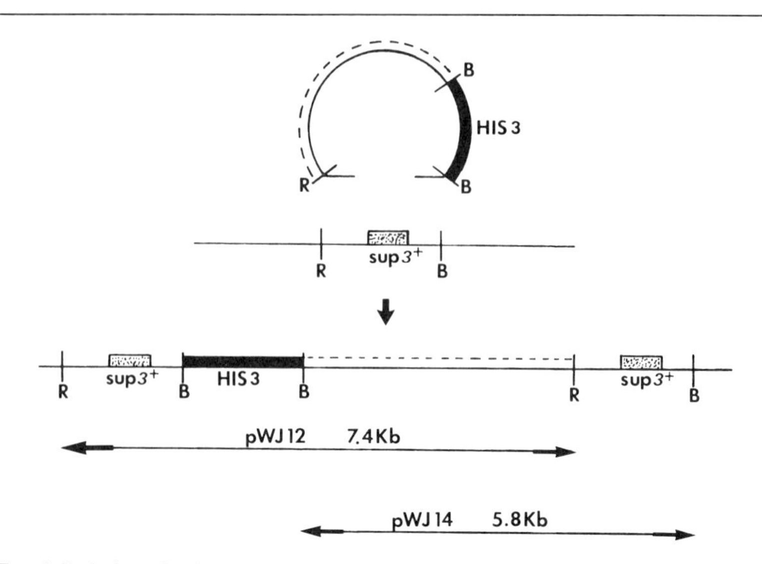

FIG. 5. Isolation of a chromosomal allele. Integration of pWJ4 containing a double-strand gap that extends across the chromosomal *sup3*$^+$ gene produces integrated vector sequences flanked by two repeated copies of the wild-type *sup3*$^+$ gene. The DNA ends of the gap invade the homologous region on the chromosome and serve as primers for repair synthesis. After the gap is repaired, an intermediate is formed that can be resolved to yield an integrated, repaired plasmid. The transformant DNA was digested with *Eco*RI and religated to produce the 7.4 kb plasmid pWJ12 that contains *sup3*$^+$, *HIS3*, and pBR322. Digestion with *Bam*HI and religation produces pWJ14, 5.8 kb, and containing *sup3*$^+$ and pBR322. ------, pBR322 DNA; ■, HIS3; ▒, *sup3*$^+$; ——, homologous DNA on plasmid and chromosome; R, *Eco*RI site; B, *Bam*HI site.

plasmids that contain the right-hand and left-hand copies, respectively, of the duplicated *sup3*$^+$ gene. *Eco*RI digestion of chromosomal DNA generates a 7.4 kb fragment with the *HIS3* gene, amp^r, and one *sup3*$^+$ gene (Fig. 5). A 5.8 kb fragment with amp^r and the other *sup3*$^+$ gene is formed by *Bam*HI cleavage (Fig. 5). The fragments from the digests were ligated separately; the two plasmids were amplified and purified from *E. coli* and are designated pWJ12 and pWJ14. Since the *SUP3-a* anticodon mutation generates a new *Hin*fI restriction site, we verified that both the left side and the right side of the integrated plasmid contain the chromosomal *sup3*$^+$ gene by digesting pWJ12 and pJW14 with *Hin*fI, end-labeling the fragments, and electrophoresing the DNA on a 5% polyacrylamide gel. The *Hin*fI site at the amber anticodon creates a 41 bp fragment, while in the wild-type gene this fragment is fused to the adjacent one to generate a 366 bp fragment. Both plasmids show this 366 bp fragment on *Hin*fI digestion (Fig. 6), and therefore both contain the wild-type chromosomal gene.

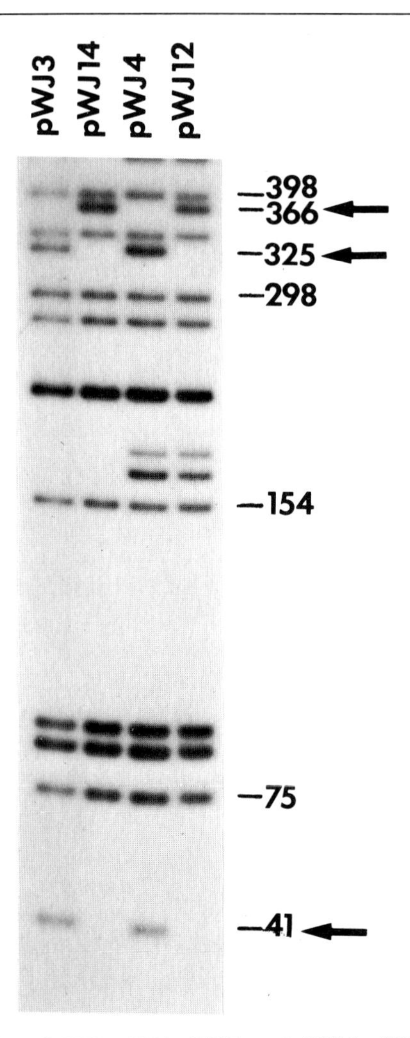

FIG. 6. *Hin*fI digestion of pWJ3, pWJ4, pWJ12, and pWJ14. pWJ3 and pWJ4 contain the *SUP3-a* gene, and pWJ4 contains also the *HIS3-Bam*HI fragment. pWJ12 and pWJ14 were isolated from a transformant of *Sma*I-gapped pWJ4 as described in the text. The plasmid DNA was digested with *Hin*fI and end-labeled with reverse transcriptase. The DNA was electrophoresed on a 5% polyacrylamide gel and autoradiographed. The amber anticodon forms a new *Hin*fI site, producing a 41 and a 325 base-pair fragment. These fragments are fused to produce a 366 base-pair fragment in the wild-type gene. Size standards are pBR322 fragments from the *Hin*fI digestion.

Chromosomal alleles can be directly isolated on plasmids by transforming yeast with a replicating plasmid containing a double-strand gap that covers a chromosomal mutation. We showed that replicating plasmids with a double-strand gap are repaired from chromosomal information to give repaired, replicating plasmids as well as integrated ones. Gap repair to regenerate a replicating plasmid will permit isolation of a chromosomal allele without requiring digestion and religation of chromosomal DNA.[29] Similar observations have been obtained by C. Chan.[30]

Transformation with gapped plasmids can be used for fine-structure mapping of chromosomal mutations within genes. When a strain with a mutation is transformed with a plasmid containing a double-strand gap in that gene, wild-type transformants can arise only if the gap does not include the chromosomal mutation. Our results demonstrate that a gap that deletes a region homologous to a chromosomal allele will be repaired to give an integrated plasmid flanked by two copies of the allele. Therefore if wild-type transformants are obtained, the gap did not extend into the region of the chromosomal allele. In this manner we have mapped the *his3-11,15* mutations to a region between the *Bgl*II site and the 5′ end of the gene.[3] This technique has also been used by Shortle to map a mutation at the actin locus; the location of this mutation is being verified by DNA sequencing.[31] However, gapped plasmids may be degraded past the region homologous to the chromosomal mutation, greatly reducing the recovery of the wild-type transformants. Therefore, when possible, selection should be made directly for the phenotype of the gene of interest.

Transplacement of in Vitro-Generated Mutations

Linear plasmids are useful for directing the plasmid to the desired site for transplacement of a mutated sequence. On occasion we observe gene conversion of mutations on the plasmid accompanying plasmid integration.[32] If the conversion occurs in the direction to convert the chromosomal copy to two mutant copies, it is not necessary to ask for excision of the plasmid to obtain a mutant strain. Consequently transplacement of mutations from plasmids to the chromosome can be simplified if transformants are screened for conversion of the mutation by Southern blots or genetic analysis.

[30] C. Chan and B. Tye, personal communication, 1981.

[31] D. Shortle, personal communication, 1982.

[32] T. L. Orr-Weaver and J. W. Szostak, unpublished results, 1982.

Summary

Techniques for high frequency yeast transformation have been described. A double-strand break introduced by restriction enzyme cleavage can be used to direct a plasmid to integrate into a particular chromosomal locus. Plasmids containing a double-strand gap can be used in a straightforward method for the isolation and mapping of chromosomal alleles. These techniques extend the genetic applications of yeast transformation.

Acknowledgments

We thank Ray Wu, in whose laboratory this work was initiated, and Burt Beames and Randy Holcombe for technical assistance. We thank Clarence Chan, Bik Tye, Hannah Klein, Tom Petes, Mary Ann Osley, Lynna Hereford, David Shortle, David Botstein, Jim Haber, Pat Brown, and Andrew Murray for communication of results prior to publication. T. L. O.-W. was supported by National Institutes of Health Training Grant CA09361. This work was supported by National Institutes of Health Grant GM27862 to J. W. S. and National Science Foundation Grant PCM8003805 to R. J. R.

[15] A Rapid Procedure for the Construction of Linear Yeast Plasmids

By J. W. SZOSTAK

All plasmid cloning vectors in current general use are circular DNA molecules. Only linear restriction fragments of DNA or shear fragments bounded by two cut DNA ends can be inserted into circular vectors. Telomeres, which are pieces of DNA with one protected end, cannot be cloned in such vectors. Most linear cloning vectors, generally modified phage or viruses, are subject to the same limitation, in that they are designed to accept pieces of DNA defined by two restriction sites. The discovery[1] that the ends of the *Tetrahymena* rDNA linear plasmid will function as telomeres in yeast has resulted in the availability of a new class of cloning vectors that are linear molecules. These new linear vectors are potentially of wide-ranging utility. A linear plasmid with two *Tetrahymena* ends was used as a vector to clone yeast telomeres.[1] Given the apparently strong conservation of telomere structure and function through evolution, it is possible that yeast linear plasmids may be useful in the cloning of telomeres from many different organisms. Conversely, linear

[1] J. W. Szostak and E. H. Blackburn, *Cell* **29**, 245 (1982).

ISBN 0-12-182001-7

plasmids with *Tetrahymena* or yeast ends may be useful in the development of cloning vectors that could be maintained as linear episomes in many different organisms, e.g., plants, nematodes, *Drosophila*, and mammalian cells. This possibility is quite attractive, since a linear plasmid, especially if endowed with a centromere, ought to behave as a complete artificial chromosome. There is no intrinsic limit to the size of chromosome that could be built up in yeast; linear plasmids might therefore facilitate the study of large eukaryotic genes containing many introns, or genes embedded in large chromosomal regulatory domains. Linear plasmids must, so far, be grown in yeast; however, the technology of yeast molecular genetics is quite advanced and is sufficient for the rapid manipulation of such artificial minichromosomes. Linear plasmids may also be useful for studies on recombination and chromosome mechanics. Recombination between different linear plasmid molecules does not give rise to plasmid multimers, as it does with circles. Indeed, only linear monomers are detectable in strains carrying linear plasmids. However, recombination between a linear plasmid and a chromosome can break the chromosome, generating a centric and an acentric fragment. The manipulation of such chromosome fragments, and the formation of new chromosomes by the addition of cloned centromeres[2] to chromosome fragments, should be useful in studying the behavior of chromosomes. Finally, linear plasmids can be used to study regulatory chromatin domains in a way that circular molecules cannot, because the ends of a linear molecule can be used to interrupt and destroy the continuity of a domain. Linear episomal vectors might be useful in studying the effects of sequences that act at a distance, such as the SV40 72 base-pair repeats, since polarity can be unambiguously determined only within a linear molecule.

In order to use linear plasmids effectively in experiments such as those described above, it will be necessary to construct many different linear vectors. The first linear plasmid was constructed entirely *in vitro*. The method used was designed to produce a fully characterized molecule before transformation and is not suitable for the construction of large numbers of new plasmids. In this chapter I describe a simplification of this method that is faster and uses less material, and that can be generally applied to the construction of new linear plasmids.

Principle of the Method

Linear DNA molecules with one functional telomere and one restriction enzyme cut end transform yeast very poorly (less than 10^{-3} of the

[2] L. Clarke and J. Carbon, *Nature* (*London*) **287,** 504 (1980).

efficiency of linear molecules with two functional telomeres). Transformation itself, therefore, acts as a strong selection against linear molecules that do not have two functional ends. This suggested that it might be possible to construct new linear molecules with two *Tetrahymena* ribosomal DNA ends without first purifying the desired molecules *in vitro* or screening an inordinate number by Southern blot restriction mapping. The simplified method for the construction of new linear plasmids involves the ligation of an end fragment of *Tetrahymena* rDNA to a circular plasmid that has been linearized by restriction enzyme digestion. The restriction sites are chosen such that after the ligation the DNA can be redigested with the same restriction enzymes, leaving only the desired molecules intact. This is readily accomplished by using pairs of restriction enzymes (such as *Bam*HI and *Bgl*II[3,4] or *Sal*I and *Xho*I[5,6]) that generate identical cohesive ends, but have different flanking nucleotides. For example, in the construction described below, we cut *Tetrahymena* rDNA with *Bam*HI to generate a 1.5 kb terminal fragment, and we cut a circular yeast–*Escherichia coli* shuttle vector with *Bgl*II to generate a 5.0 kb linear molecule. The DNAs were ligated together in the presence of both restriction enzymes. This increases the efficiency of the reaction, since *Bam*HI end fragments that ligate to other *Bam*HI fragments of *Tetrahymena* rDNA are regenerated; similarly, vector molecules that circularize or dimerize are recut to monomeric linear molecules. Only molecules in which *Bgl*II ends are joined to *Bam*HI ends accumulate in the reaction. After the ligation, the DNA is digested to completion with *Bam*HI and *Bgl*II, leaving only joint molecules and unreacted substrates. Among these molecules will be vector molecules with 0, 1, or 2 *Tetrahymena* rDNA ends. Many vector molecules will have nonterminal *Bam*HI fragments of *Tetrahymena* rDNA joined to them, but even if these molecules subsequently had a terminal *Bam*HI fragment ligated on, that end would be removed in the final redigestion with *Bam*HI. After yeast transformation, only circular molecules or molecules with two functional ends are recovered. Linear plasmids derived by the simple addition of two *Tetrahymena* rDNA ends to the vector are the great majority of products, since linear molecules with 0 or 1 functional ends transform at low efficiency. The structure of the linear plasmids that are recovered after transformation is easily verified by Southern blot restriction mapping.

[3] R. J. Roberts, G. A. Wilson, and F. E. Young, *Nature (London)* **265,** 82 (1977).

[4] V. Pirotta, *Nucleic Acids Res.* **3,** 1747 (1976).

[5] J. R. Arrand, P. A. Myers, and R. J. Roberts, *J. Mol. Biol.* **118,** 127 (1978).

[6] T. R. Gingeras, P. A. Myers, J. A. Olson, F. A. Hanberg, and R. J. Roberts, *J. Mol. Biol.* **118,** 113 (1978).

Materials and Reagents

Preparation of DNA

Plasmid DNA was prepared from *E. coli* by the rapid boiling procedure[7] and further purified by phenol extraction, RNase treatment, and ethanol precipitation, if necessary.

Tetrahymena ribosomal DNA was prepared as described[8] and was the generous gift of E. H. Blackburn.

Enzymes

Restriction enzymes and T4 polynucleotide ligase were purchased from New England BioLabs or Bethesda Research Laboratories and were used according to the directions of the supplier.

Strains

Yeast strain A2 (LL20 from Lester Lau) is *leu2-3,112, his3-11,15, can1*. The double point mutations in *leu2* and *his3* ensure that revertants arise at a frequency of less than 10^{-10}. This strain transforms well, with the replicating plasmid CV13 (YEP13) yielding 10^5 transformants per microgram of DNA.[8a]

Escherichia coli 5346 (C600, *leuB*, thi^-, thr^-, r^-, m^-) was from J. Calvo.

Plasmid pSZ218 was constructed by the insertion of a 2.8 kb *Bgl*II fragment of yeast DNA containing the *LEU2* gene into the *Bam*HI site of plasmid pSV10, which was obtained from P. Mellon. The structure of pSZ218 was confirmed by standard restriction mapping methods (see Fig. 1).

Media

Yeast media were as described.[9] YPD was used as a complete, nonselective medium. SD supplemented with all the amino acids but one was used as a selective medium for yeast transformation and for growth of mitotically unstable transformants.

[7] D. S. Holmes and M. Quigley, *Anal. Biochem.* **114,** 193 (1981).

[8] M. A. Wild and J. G. Gall, *Cell* **16,** 565 (1979).

[8a] T. L. Orr-Weaver, J. W. Szostak, and R. J. Rothstein, this volume [14].

[9] F. Sherman, G. R. Fink, and C. Lawrence, "Methods in Yeast Genetics." Cold Spring Harbor Laboratory, Cold Spring Harbor, New York, 1977.

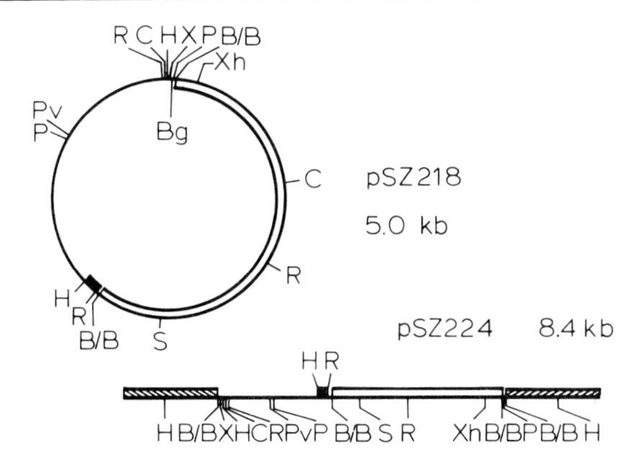

FIG. 1. Plasmid restriction maps. The circular plasmid pSZ218 has a single *Bgl*II restriction site; the linear plasmid pSZ224 was made by the addition of *Tetrahymena* rDNA ends at the *Bgl*II site of pSZ218. Thin line, pBR322; thick line, SV40 origin of replication; double line, the yeast *LEU2* gene; hatched line, *Tetrahymena* end fragment. Restriction sites: R, *Eco*RI; C, *Cla*I; H, *Hin*dIII; X, *Xba*I; Xh, *Xho*I; P, *Pst*I; Pv, *Pvu*I; S, *Sal*I; Bg, *Bgl*II; B/B, *Bam*HI–*Bgl*II junction.

Procedures

The linear plasmid pSZ224 was constructed from pSZ218 and *Tetrahymena* rDNA ends. Purified DNA of plasmid pSZ218 (2 μg) was digested to completion with the restriction enzyme *Bgl*II (5 U) for 2 hr at 37° in a reaction volume of 20 μl of 50 m*M* NaCl, 6 m*M* Tris-HCl (pH 7.4), 6 m*M* $MgCl_2$, 5 m*M* 2-mercaptoethanol, 100 μg of bovine serum albumin per milliliter. Purified *Tetrahymena* ribosomal DNA (5 μg) was digested in the same manner with 5 U of *Bam*HI in an 80-μl reaction volume. One-fifth of each reaction was removed, and the completeness of digestion was assayed by agarose gel electrophoresis. The remainders of each reaction were then mixed together, ATP was added to a final concentration of 1 m*M*, and 100 U (New England BioLabs) of T4 ligase were added. The restriction enzymes were not destroyed by heating, but were left in the reaction, which was incubated at 12.5° overnight. The reaction was then incubated at 75° for 10 min to destroy the ligase, after which fresh *Bam*HI (5 U) and *Bgl*II (5 U) were added. After an additional 2-hr incubation at 37°, the DNA was stored at −20° until further use.

Half of the above reaction was used in a yeast transformation[8a,10] experiment. Sonicated *E. coli* DNA (10 μg) was added as a carrier. Several

[10] A. Hinnen, J. B. Hicks, and G. R. Fink, *Proc. Natl. Acad. Sci. U.S.A.* **75,** 1929 (1978).

hundred Leu$^+$ transformants were obtained. Twelve transformants were picked with toothpicks and streaked out to single colonies on selective medium (complete synthetic minus leucine). One colony of each transformant was then picked and tested for mitotic stability of the Leu$^+$ marker, by streaking to single colonies on complete medium (YPD) and replica plating to selective (minus leucine) medium. Transformant number 1 was mitotically stable, but transformants 2 through 12 were mitotically unstable and therefore appeared to carry autonomously replicating plasmids.[11] The degree of instability observed was comparable to that previously observed for a linear plasmid with two *Tetrahymena* ends, and an *ars1* element.[1] Transformants with circular *ars1* plasmids were much less stable than the transformants described above. Since pSZ218 does not have an *ars* element, it cannot exist as an autonomous plasmid in yeast.

DNA was prepared from 40-ml selective cultures of each of the above transformants, as previously described.[12] An aliquot (1 μg) of each DNA preparation was electrophoresed on an 0.5% agarose gel, which was stained with ethidium bromide and photographed with shortwave UV illumination. Lambda DNA digested with *Hin*dIII was included to provide a set of size markers. Most of the transformants had a faint band visible on the gel at a mobility corresponding to that of an 8.4 kb linear duplex (not shown). This band is not observed in DNA from the parental yeast strain and is not due to any form of the endogenous 2 μm circle plasmid. The 8.5 kb band was not seen in the DNA from transformant No. 1. This is consistent with the transformant being stable and hence due to a chromosomally integrated *Leu*$^+$ gene. Transformant 10 lacked the 8.5 kb band, but had an additional new band at about 9 kb. Transformants 2, 3, and 6 were examined more carefully, by Southern blot[13] restriction mapping. Undigested DNA, and *Sal*I-, *Xho*I-, and *Xba*I-digested DNA samples were subjected to agarose gel electrophoresis, blotted to nitrocellulose paper, and probed with a pBR322 probed labeled with ^{32}P by nick translation. All fragments from the linear plasmids in transformants 3 and 6 were of the predicted sizes (data not shown). Transformant 2 appears to have one *Tetrahymena* end 200 bp longer than expected, and the other end 200 bp shorter than expected. The nature of this rearrangement is unknown. An unequal crossover between the two *Tetrahymena* ends of one molecule is a possible explanation. Transformant 6 was subjected to a

[11] D. T. Stinchcomb, K. Struhl, and R. W. Davis, *Nature* (*London*) **282,** 37 (1979).

[12] R. W. Davis, M. Thomas, J. Cameron, T. P. St. John, S. Scherer, and R. A. Padgett, this series, Vol. 65, p. 404.

[13] E. M. Southern, *J. Mol. Biol.* **98,** 503 (1975).

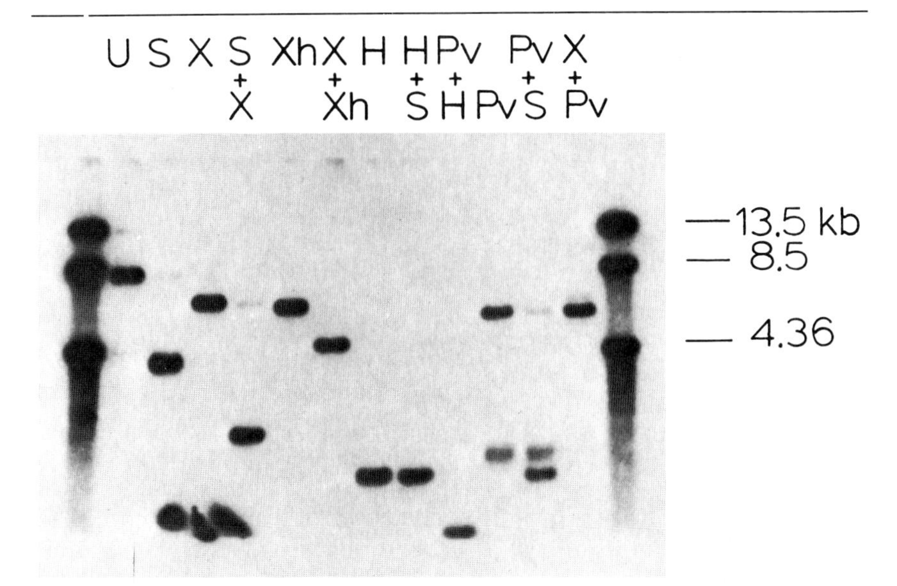

FIG. 2. Southern blot restriction mapping of pSZ224. DNA from transformant T633-6 was digested with the restriction enzymes indicated, electrophoresed on an 0.7% agarose gel, blotted onto nitrocellulose paper, and hybridized with labeled pBR322 DNA. The size markers in the outer lanes are linear duplex DNAs of 13.5, 8.5, and 4.36 kb. The sizes of the fragments detected in this blot were consistent with the map of pSZ224 shown in Fig. 1. Restriction sites: U, uncut DNA; S, *Sal*I; X, *Xba*; Xh, *Xho*I; H, *Hin*dIII; Pv, *Pvu*II.

more thorough Southern blot restriction mapping analysis (see Figs. 1 and 2). All fragments were of the predicted sizes.

Discussion

The method described in this chapter for the construction of new linear plasmids has several advantages over the previously described[1] *in vitro* construction. This procedure uses much less *Tetrahymena* ribosomal DNA (which is tedious to prepare), since there are no losses due to gel elution of the end fragment. Furthermore, less time is required, since gel purification of both the *Tetrahymena* end fragment and the desired linear plasmid molecules are eliminated.

The linear plasmids that are made by this method must be carefully analyzed by Southern blot restriction mapping. This is necessary no matter what the method of linear plasmid construction. In the case of linear plasmids with *Tetrahymena* ends, this mapping is made easier by the moderately high copy number of the plasmids (an average of about 20 copies per cell, or 0.5 to 1% of total genomic DNA). This high copy number is associated with a greater mitotic stability of the linear plasmid compared to similar circular plasmids; thus transformants carrying circular plasmids can be detected on the basis of their extreme mitotic instability. The high copy number also means that the linear plasmid can be seen in an ethidium bromide stained agarose gel of undigested genomic DNA. The sizes of the plasmids in a set of transformants can therefore be determined quickly, and without the need for a Southern blot. This allows the more detailed restriction analysis to be done on a smaller number of transformants.

Other methods for the construction of novel linear plasmids are being developed. There is, in principle, no reason why new linear plasmids cannot be made simply by the integration of a circular plasmid into a linear plasmid. Targeted plasmid integration[14] may be applicable to such constructions. The stability of such plasmids will have to be evaluated, since they could decay by recombination between sequences in direct orientation. Direct repeats are fairly stable in chromosomal structures (e.g., see Brown and Szostak[15]). By extension of the plasmid integration method, it should be possible to add DNA fragments to an existing linear plasmid by transplacement.[16]

Further advances in linear plasmid technology will probably depend upon a greater understanding of telomeric function, and the structural requirements of that function. The creation of linear plasmids containing centromeric DNA is in progress. Such "artificial chromosomes" may be quite useful in understanding the nature of chromosome behaviour.

Acknowledgments

I would like to thank Toby Claus for technical assistance. This work was supported by NIH Grant GM27862.

[14] T. L. Orr-Weaver, J. W. Szostak, and R. J. Rothstein, *Proc. Natl. Acad. Sci. U.S.A.* **78,** 6354 (1981).

[15] P. A. Brown and J. W. Szostak, this volume [18].

[16] S. Scherer and R. W. Davis, *Proc. Natl. Acad. Sci. U.S.A.* **76,** 4951 (1979).

[16] Cloning Regulated Yeast Genes from a Pool of *lacZ* Fusions

By STEPHANIE W. RUBY, JACK W. SZOSTAK, and ANDREW W. MURRAY

Gene fusions have been used during the last decade to study such diverse problems as transcriptional regulation[1,2] translational regulation,[3] and protein localization.[4,5] Their use has been particularly valuable for the study of genes that are autogenously regulated, or whose products are not easily measured. The *Escherichia coli lacZ* gene has been utilized most often in the generation of these fusions for several reasons: the *lac* operon is well characterized[6]; the *lacZ* gene product, β-galactosidase, is easily assayed[7]; the first 27 amino acids of the N terminus of the protein may be replaced by other peptides with little or no effect on enzyme activity[8]; there are several substrates for the selection of Lac^- and Lac^+ mutants, and there are several nonselective, indicator substrates.[6]

In prokaryotes, various genetic methods[1,9,10] have been employed to create operon and protein fusions. Recombinant DNA techniques are now commonly applied to engineer fusions between eukaryotic and prokaryotic genes. Rose *et al.*[11] isolated many different fusions between the yeast *URA3* and the *lacZ* genes by placing the two genes near each other on a plasmid and then selecting for deletions between them by selecting

[1] N. C. Franklin, *Annu. Rev. Genet.* **12,** 193 (1978).
[2] J. R. Beckwith, *Cell* **23,** 307 (1981).
[3] A. Miura, J. H. Krueger, S. Itoh, H. A. de Boer, and M. Nomura, *Cell* **25,** 773 (1981).
[4] T. Silhavy, H. Shuman, J. Beckwith, and M. Schwartz, *Proc. Natl. Acad. Sci. U.S.A.* **74,** 5411 (1977).
[5] K. Ito, P.H. Bassford, and J. Beckwith, *Cell* **24,** 707 (1981).
[6] J. Beckwith, *in* "The Operon" (J. H. Miller and W. S. Reznikoff, eds.), p. 11. Cold Spring Harbor Laboratory, Cold Spring Harbor, New York, 1972.
[7] J. H. Miller, "Experiments in Molecular Genetics," p. 466. Cold Spring Harbor Laboratory, Cold Spring Harbor, New York, 1972.
[8] E. Brickman, T. J. Silhavy, P. J. Bassford, H. A. Shuman, and J. R. Beckwith, *J. Bacteriol.* **139,** 13 (1979).
[9] P. Bassford, J. Beckwith, M. Berman, E. Brickman, M. Casadaban, L. Guarente, I. Saint-Girons, A. Sarthy, M. Schwartz, H. Shuman, and T. Silhavy, *in* "The Operon" (J. H. Miller and W. S. Reznikoff, eds.), p. 245, Cold Spring Harbor Laboratory, Cold Spring Harbor, New York, 1980.
[10] M. J. Casadaban and S. N. Cohen, *Proc. Natl. Acad. Sci. U.S.A.* **76,** 4530 (1979).
[11] M. Rose, M. J. Casadaban, and D. Botstein, *Proc. Natl. Acad. Sci. U.S.A.* **78,** 2460 (1981).

ISBN 0-12-182001-7

for a lac$^+$ phenotype in *E. coli*. Guarente and Ptashne[12] joined the yeast *CYCl* gene to the *lacZ* gene directly *in vitro*. Both groups subsequently showed that the gene fusions could be expressed in yeast as protein fusions, and that the levels of β-galactosidase in yeast were regulated in the appropriate manner. Furthermore, they found that yeast cells, which are lac$^-$, can hydrolyze the chromogenic substrate Xgal when they contain an active fusion, and that the resulting blue color of the colony varied with the cellular enzyme levels.

We have utilized *lacZ* gene fusions in the yeast *Saccharomyces cerevisiae* to clone regulated yeast genes. These genes were identified as regulated protein fusions and detected by screening a library of random yeast gene:*lacZ* fusions. This method allows the cloning of yeast genes based solely on the way in which they are regulated and thus is applicable to genes whose products are not easily assayed or whose functions are unknown. It is particularly useful for cloning genes that are similarly or coordinately regulated, and it complements other cloning techniques such as complementation[13,14] and differential plaque hybridization.[15] In this chapter we describe the procedures that we have used to isolate DNA *d*amage *in*duced (*DIN*) genes as an illustration of the usefulness of the fusion cloning method.

General Outline of the Method

The active, regulated gene fusions constructed by the procedures described below contain a yeast DNA fragment with sequences for the control and initiation of transcription and translation joined to a *lacZ* fragment from which the 5′ portion including the first few N-terminal amino acid codons of the gene have been deleted. Expression of such fusions in yeast results in hybrid proteins with β-galactosidase activity.

The first step of the method is the creation of a library of random yeast gene fusions. A fusion shuttle vector is described that has a unique restriction endonuclease site at the 5′ end of the *lacZ* fragment. Random yeast DNA fragments are inserted at this site to create a pool of ligated DNA molecules, which are then used to transform yeast cells.

The second step is to screen the library of yeast transformants for colonies with regulated β-galactosidase activity. Yeast colonies are replica-plated onto media containing Xgal, and the colonies are monitored for differential expression by differences in color intensities.

[12] L. Guarente and M. Ptashne, *Proc. Natl. Acad. Sci. U.S.A.* **78,** 2199 (1981).
[13] B. Ratzkin and J. Carbon, *Proc. Natl. Acad. Sci. U.S.A.* **74,** 487 (1977).
[14] K. A. Nasmyth and S. I. Reed, *Proc. Natl. Acad. Sci. U.S.A.* **77,** 2119 (1980).
[15] T. P. St. John and R. Davis, *Cell* **16,** 225 (1979).

The third step is to recover the fusion plasmids from the yeast strains of interest and to confirm that the fusion plasmids have the expected regulated expression when reintroduced into yeast.

The final step is to demonstrate by an alternative assay such as Northern blot analysis that the identified genes are regulated in the expected manner.

Materials and Reagents

Strains

Yeast strain A2 (α *his3-11,15 leu2-3,112 canl*) was originally LL20 from L. Lau. Strain A15 (α *gal2, mal*) was originally S288C. The strain DA151 is α *his3-11,15 leu2-3,112, ura3, trpl, tcml*,[*LEU2, his3:lacZ*] × *a leu2-3,112, met10,* which was derived from a transformant of strain D234-3B (α *his3-11,15, leu2-3,112, ura3, trpl, tcml*) from P. Brown. *Escherichia coli* strain 5346 (C600 *leuB6, thr-, thi-, r-, m-*) was obtained from J. Calvo. The plasmid pMC1403 was obtained from M. Rose. The plasmids pSZ62 and pSZ93 were given by T. L. Orr-Weaver.

Media

Yeast synthetic selective medium (lacking leucine), YPD, and regeneration agar were as described.[16] Yeast calcium-free, Xgal medium (lacking leucine) contained amino acids, adenine, and uracil as in synthetic medium, as well as 13.6 g of KH_2PO_4, 2 g of $(NH_4)_2SO_4$, 4.2 g of KOH, 0.2 g of $MgSO_4 \cdot 7\ H_2O$, 0.5 mg of $FeCl_3 \cdot 6\ H_2O$, 400 μg each of pantothenic acid, biotin, and pyridoxine monohydrocholoride, 2 μg of biotin, 2 mg of myoinositol, 20 g of dextrose, 20 g of agar, and 40 mg of Xgal per liter. The KH_2PO_4, $(NH_4)_2SO_4$, and KOH were made up as a 10 × stock, the $MgSO_4$ and $FeCl_3$ were made up as a 1000 × stock, and both unsterile stocks were stored at room temperature. The vitamins were kept as a 100 × frozen, sterile stock, and the amino acids were stored as sterile stock solutions as described.[16] Xgal was prepared as 20 mg/ml in dimethylformamide and stored frozen. To prepare the medium, the amino acids (minus leucine, threonine, aspartic acid, and tryptophan) and salts were brought to a volume of 500 ml with H_2O, the agar and sugar were put into 500 ml of H_2O, and the two solutions were autoclaved separately. After cooling to 55°, the two solutions were mixed, and the vitamins, remaining amino acids (minus leucine), and Xgal were added. Bacterial media were as described.[7]

[16] F. Sherman, G. R. Fink, and C. W. Lawrence, "Methods in Yeast Genetics," p. 98, Cold Spring Harbor Laboratory, Cold Spring Harbor, New York, 1979.

Solutions

Z buffer,[7] pH 7.0, contained, per liter, 16.1 g of $Na_2HPO_4 \cdot 7\ H_2O$, 5.5 g of $NaH_2PO_4 \cdot H_2O$, 0.75 g of KCl, 0.25 g of $MgSO_4 \cdot 7\ H_2O$, and 2.7 ml of 2-mercaptoethanol.

1× RNA gel buffer, pH 7.3, was 20 m*M* morpholinopropanesulfonic acid (MOPS), pH 7.3, 50 m*M* sodium acetate, and 1 m*M* EDTA.

20× Denhardt's solution contained 0.4 g of poly(vinylpyrollidone), 0.4 g of bovine serum albumin, and 0.4 g of Ficoll per 100 ml of H_2O.

4× RNA hybridization buffer, pH 7.5, contained 175.3 g of NaCl, 73.1 g of Tris base, 22.1 g of NaH_2PO_4, 64.3 g of $Na_2HPO_4 \cdot 7\ H_2O$, 4 g of sodium pyrophosphate, and 35 ml of 12 N HCl per liter.

20× SSC was 3.0 *M* NaCl and 0.3 *M* sodium citrate.

Reagents

5-Bromo-4-chloroindolyl-β-D-galactoside (Xgal) and *O*-nitrophenyl-β-D-galactopyranosyl (ONPG) were obtained from Bachem, Marina del Ray, California, and Sigma, respectively. Formaldehyde and formamide were from Mallinckrodt. Restriction endonucleases and T4 DNA ligase were from New England BioLabs, and all enzyme units were those of the supplier.

Procedures and Results

Fusion Vector

The plasmid pSZ211 was used as a fusion shuttle vector (Fig. 1). It is composed of (*a*) the 3.7 kb *Eco*RI–*Sal* fragment of pBR322, encoding ampicillin resistance and a replication origin for selection and replication in *E. coli;* (*b*) the yeast 2.2 kb *Sal*I–*Xho*I *LEU2*[14] fragment and the 0.84 kb *Eco*RI–*Hin*dIII *ars1*[17] fragment for selection and replication in yeast; and (*c*) the *E. coli* 6.3 kb *Bam*HI–*Sal*I *lac* fragment, which contains the *lacZ* gene deleted for the N-terminal codons of β-galactosidase, the *lacY* gene, and a portion of the *lacA* gene. The *lac* fragment was obtained from plasmid pMC1403, which was constructed by M. Casadaban *et al.*[18]; pSZ211 was constructed by inserting the *Bam*HI–*Sal*I *lac* fragment into *Bam*HI, *Sal*I-digested pSZ93 (not shown).[19]

The plasmid pSZ211 does not produce detectable β-galactosidase activity in yeast, but it does produce easily detectable enzyme activity in

[17] D. T. Stinchcomb, K. Struhl, and R. W. Davis, *Nature (London)* **282,** 39 (1979).

[18] M. J. Casadaban, J. Chou, and S. N. Cohen, *J. Bacteriol.* **143,** 971 (1980).

[19] T. L. Orr-Weaver, J. W. Szostak, and R. J. Rothstein, this volume [14].

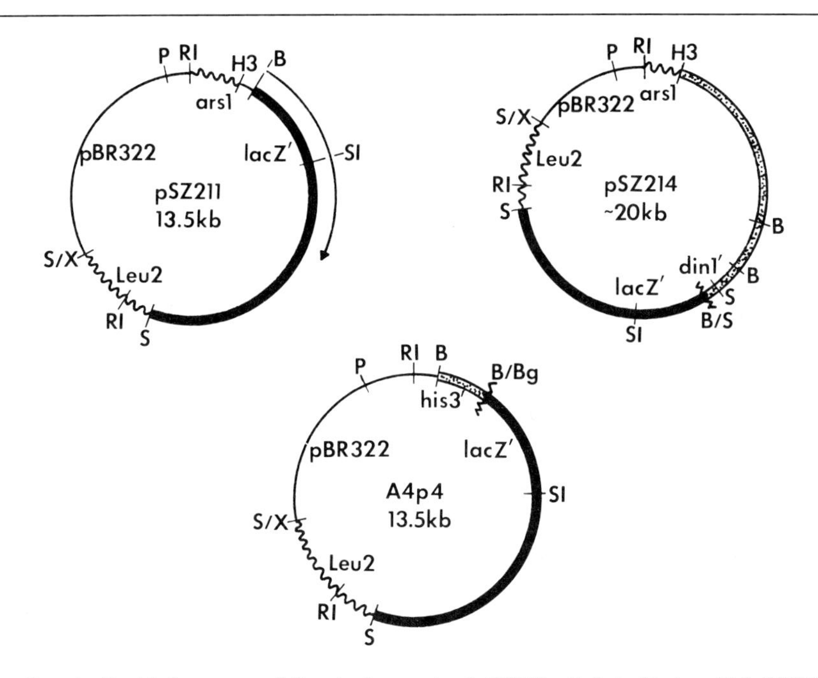

FIG. 1. Restriction maps of the cloning vector (pSZ211), *din1:lacZ* plasmid (pSZ214), and *his3:lacZ* plasmid (A4p4). An arrow indicates the direction of transcription of the *lacZ* gene. Its 3′ end indicates the approximate end of the *lacZ* coding sequences, but it does not indicate the 3′ end of a transcript. Plasmids are not drawn to scale. Restriction endonuclease sites are indicated as follows: B, *Bam*HI; Bg, *Bgl*II; H3, *Hin*dIII; P, *Pst*I; RI, *Eco*RI; SI, *Sac*I; S, *Sal*I; S3, *Sau*3A; and X, *Xho*I.

E. coli. It contains an *ars* sequence and therefore transforms yeast at high frequency[20] and is usually maintained extrachromosomally. The plasmid has a unique *Bam*HI site at the 5′ end of the *lacZ* gene into which random yeast DNA fragments can be inserted to create gene fusions.

Construction of Random Fusions

A DNA pool of random fusions was made by ligating yeast genomic DNA that had been partially cut with *Sau*3A, to *Bam*HI-cleaved vector. Yeast DNA from haploid strain A15 was prepared by the method of Davis *et al.*[21] The amount of *Sau*3A required to achieve the desired partial digestion was estimated from a set of small-scale reactions. One of these was

[20] K. Struhl, D. T. Stinchcomb, S. Scherer, and R. W. Davis, *Proc. Natl. Acad. Sci. U.S.A.* **76,** 1035 (1979).

[21] R. W. Davis, M. Thomas, J. Cameron, T. P. St. John, S. Scherer, and R. Padgett, this series, Vol. 65, p. 404.

subsequently scaled up. About 200 μg of A15 DNA were cleaved in 6 m*M* Tris HCl, pH 7.5, 50 m*M* NaCl, 6 m*M* $MgCl_2$, and 6 m*M* 2-mercaptoethanol in a volume of 4.5 ml with 3.6 units of *Sau*3A for 1 hr at 37°, after which the reaction was stopped by heating at 65° for 10 min. A 40-μl aliquot of this DNA was fractionated by agarose gel electrophoresis and visualized by ethidium bromide staining; the average molecular weight was estimated to be 6 kb. The remainder of the DNA was added to 30 μg of *Bam*HI-cut pSZ211 DNA, and the mixture was brought to a volume of 6 ml and a final concentration of 66 m*M* Tris-HCl, pH 7.5, 0.4 m*M* ATP, 10 m*M* $MgCl_2$, and 10 m*M* dithiothreitol. Three hundred units of T4 DNA ligase were added, and the mixture was incubated overnight at 4°. The DNA was EtOH precipitated, washed once with 70% EtoH, dried, and resuspended in 200 μl of 10 m*M* Tris-HCl, pH 7.5, and 1 m*M* EDTA (TE). The pool of ligated DNA molecules was used to transform yeast directly without prior amplification in *E. coli*. Ligated DNA (150 μl) was added to 1.8 ml of competent yeast cells derived from 90 ml of a log-phase culture of A2 cells.[19] Approximately 10,000 colonies were obtained.

Screening

The number of independent colonies that must be screened to ensure isolation of at least one fusion that is regulated in the desired manner was estimated. Assuming that (*a*) the average yeast gene is 1 kb long; (*b*) the yeast haploid genome is 2×10^7 bp; (*c*) the number of yeast genes is 10^4; and (*d*) one out of six fusions contains yeast and *lacZ* coding sequences in the same reading frame and orientation, then 8% of the colonies will yield productive fusions. From the Poisson distribution, the number of colonies that must be screened to isolate at least one regulated fusion with a 95% probability is 360,000 divided by the number of regulated genes. The fusion cloning method then, is more convenient for identifying fusions to members of a group of genes that are similarly or coordinately regulated than a fusion to a particular gene.

Yeast colonies containing productive fusions were detected by replica plating onto Xgal medium. Since the yeast transformants were embedded in an overlay of regeneration agar, they could not be replica plated directly from the transformation plates. We were not able to detect enzyme activity consistently in colonies within the agar overlay by including Xgal in buffered regeneration agar. We used two simple procedures for extracting cells from the overlay. One method was to use toothpicks or a wire inoculating needle to transfer each independent colony manually onto a master plate. Alternatively, all the colonies from each transformation plate were pooled by breaking up the overlay in water in a blender and filtering out the agarose with cheesecloth. An aliquot of cells from each pool were plated onto selective medium, and the remainder of each cell pool was stored in 50% glycerol at −70°. Each 100-mm master plate was routinely inoculated with 200 colonies or cells.

The screening for productive fusions, or for regulated fusions, worked best when the colonies were actively growing on the master plate just prior to replica plating. Under these conditions a few yeast colonies were blue on Xgal plates within 4 hr of replica plating. Most colonies, however, required 1 to 2 days to develop a detectable blue color, and a few were blue only after 1–2 weeks of incubation at 30°. When master plates were inoculated with cells extracted from the overlay by either method, 6% of the colonies contained productive fusions.

The transformants were screened for the presence of regulated fusions by two protocols. In the first protocol, all transformants were prescreened to detect productive fusions, and those containing productive fusions were screened for regulated β-galactosidase activity. In the second protocol, the prescreen for productive fusions was omitted, and all transformants were screened for regulated *lacZ* expression. About 1% of all the colonies were pale blue on experimental plates but white on control plates in a screen for DNA *d*amage *in*duced (*din*) fusions, and therefore, would have been missed by the first screening protocol.

The results of the screening for *din* fusions illustrate the plate screening technique. Approximately 20,000 colonies from 20 independent yeast transformant pools were replica plated onto Xgal medium. Nearly 6% of the colonies were blue after 2 weeks of incubation at 30°. The blue colonies were transferred to selective medium, grown for 2 days, and replica plated onto 3 sets of Xgal plates that were treated in the following manner: (*a*) UV irradiated with two doses of 8.4 J/m^2 separated by 2 hr; (*b*) given nitroquinoline oxide (NQO),[22] a DNA damaging agent, in the medium at a concentration of 0.05 μg/ml; and (*c*) untreated control. After 1 week of incubation, 4 colonies were detected as darker blue under treatment than under control conditions. The increased color of the colonies was not due to growth of mutants generated by the treatments, because whole colonies, rather than sectors, were darker than control colonies. Subsequent quantitative assays (see below) showed that one of the colonies contained a *din*:*lacZ* fusion. The haploid strain T378 (see below) containing this fusion (*din1*:*lacZ*) was used to further optimize the screening conditions for identifying other *din* fusions. Under these conditions,[23] the induction of the *din1*:*lacZ* fusion can be detected on a screening plate between 16 and 24 hr after the initial exposure of the colony to the DNA damaging agent.

[22] L. Prakash, *Genetics*, **83**, 285 (1975).

[23] S. Ruby and J. Szostak, in preparation.

Quantitative β-Galactosidase Assays

Once potentially interesting colonies were identified, they were streaked on selective medium to obtain pure clones, which were then analyzed quantitatively during growth in liquid medium. Some fusions appeared to be regulated in the desired manner on Xgal plates, but were not so regulated when assayed quantitatively. In one screen for *din* fusions, only $\frac{4}{20}$ of the candidates identified on Xgal plates behaved as *din* fusions in subsequent quantitative assays.

β-Galactosidase assays using permeabilized cells gave good reproductibility and speed. Guarente and Ptashne[13] have shown for a *cyc1:lacZ* fusion that the same changes in β-galactosidase specific activity were detected whether permeabilized cells or cell extracts were assayed. If the enzyme activity was high enough, we assayed cells directly in liquid, calcium-free medium (minus Xgal). It was necessary to use a calcium-free medium as a precipitate formed when Z buffer was added to selective medium. Cells having low enzyme activity were concentrated by centrifugation and resuspended in 50 m*M* potassium phosphate buffer, pH 7.0. The enzyme assay was similar to that used for bacteria.[7] Z buffer (0.4 ml) containing 0.025% SDS and 1 m*M* phenylmethylsulfonyl fluoride (PMSF) was added to 0.4 ml of cell suspension, and the mixture was vortexed vigorously for 5 sec. After exactly 10 min at 23°, 2 drops of $CHCl_3$ were added to the mixture, which was immediately vortexed vigorously for 5–8 sec and placed in a 28° water bath. After exactly 10 min again, 0.2 ml of ONPG (4 mg/ml in 0.1 *M* potassium phosphate, pH 7.0) was added, and the sample was vortexed and incubated at 28°. The reaction was terminated by the addition of 0.5 ml of 1 *M* Na_2CO_3. The samples were spun briefly at 3000 rpm in a Sorvall GLC-4 centrifuge before measuring the OD_{420} of the supernatants. The units were calculated as follows: units = $OD_{420} \times 1000/t(\text{min}) \times V(\text{ml}) \times OD_{600}$; where v was the volume of the cell sample assayed and OD_{600} was the cell density of the culture as measured spectrophotometrically.

An example of the type of induced response that is seen for the haploid yeast strain T378, which contains the *din1:lacZ* fusion stably integrated in genomic DNA, is seen in Fig. 2. Cells were grown at 30° to mid-log phase in selective medium. At 0 hr, 0.05 μg of NQO per milliliter was added to half of the culture. During subsequent incubation at 30°, cell samples were removed at the times indicated and assayed for cell density and β-galactosidase activity. The NQO-treated culture reached 86% of the control culture cell density at stationary phase, which they both entered at the same time, between 4 and 6 hr. Enzyme activity increased 100-fold above that of control levels at 3–4 hr after NQO addition. After 4 hr the induced levels declined. The reason for this decline is not known,

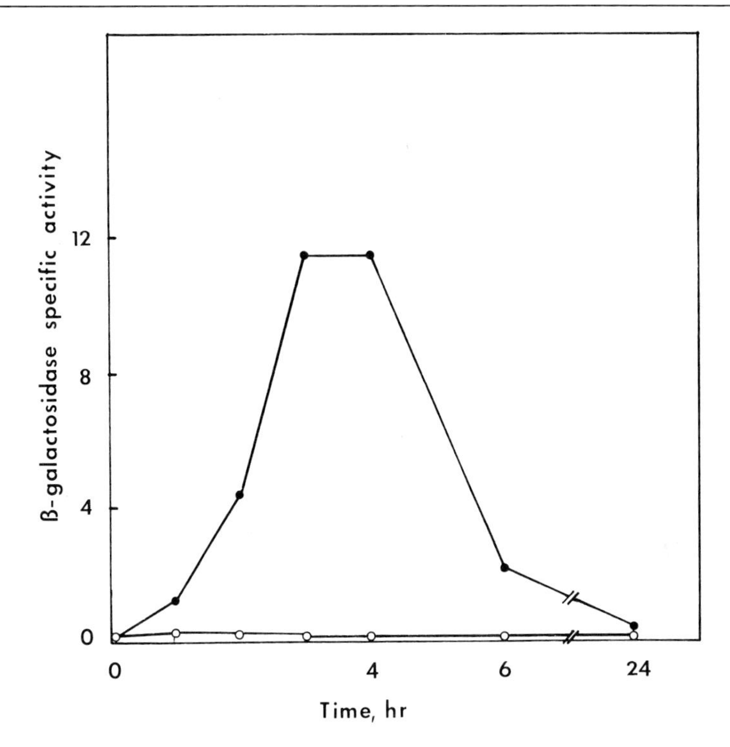

FIG. 2. Kinetics of NQO induction of β-galactosidase in strain T378 containing the *din1:lacZ* fusion. Mid-log phase cultures received 0.05 μg of NQO per milliliter (●) or no treatment (○) at 0 hr. Cells were assayed for enzyme activity during subsequent growth at 30° at the times indicated as described in text. All samples were assayed in duplicate.

although it is not due to inactivation of NQO. A similar response is observed in T378 cells treated with other DNA damaging agents and in A2 cells containing other *din:lacZ* fusions and other fusions.[24] The decline in β-galactosidase activity occurs when the culture is entering stationary phase. It is also noteworthy that increased levels of β-galactosidase are detectable much more rapidly in quantitative assays than on plates.

Recovery of Plasmids from Yeast Transformants

All *lacZ*-fusion strains of interest were tested for the mitotic stability of the *LEU2* marker of their fusion plasmids. In general, a stable strain contains an integrated plasmid, and an unstable strain contains a free plasmid. The strains were streaked onto nonselective YPD medium and incu-

[24] S. Ruby, unpublished data, 1982.

bated for 2 days at 30°. The colonies were replica plated onto selective medium (minus leucine) and monitored for growth after 24 hr.

To recover plasmids from strains that were mitotically unstable, the extracted yeast DNA was used directly to transform competent bacterial cells. Miniprep yeast DNA was prepared by the method of Davis *et al.*[21] except that the diethylpyrocarbonate treatment was omitted and the total supernatant was precipitated with isopropanol. DNA from 50 ml of cells was resuspended in 0.5 ml of TE; the concentration was about 100 μg/ml. Competent *E. coli* 5346 cells were prepared by a modification of the method of Morrison.[25] Cells in 1 liter of LB medium were grown to $OD_{550} = 0.3$, after which they were chilled, pelleted in a centrifuge, washed once with 0.1 *M* $CaCl_2$, and resuspended in 15 ml of 0.1 *M* $CaCl_2$ and 15% glycerol. After 16 hr at 0°, the cells were slowly frozen at −70° in 1.0 ml aliquots. From 5 to 20 μl of yeast miniprep DNA was used to transform 0.5 ml of cells, which were then plated onto LB medium containing 100 μg of ampicillin per milliliter. The number of colonies obtained from different fusion plasmids ranged from 0 to 500. These were screened for β-galactosidase production and for complementation of the *leuB* mutation by replica plating onto buffered minimal medium containing 0.04 mg of Xgal per milliliter and 0.5% glucose. It was not necessary to use Lac^- bacteria to detect plasmid encoded β-galactosidase, as the activity of the repressed, endogenous *lacZ* gene was very low.

To recover integrated plasmids, the yeast miniprep DNA was cut by a restriction endocnuclease and then ligated before being used for bacterial transformation. As there are no *Sma*I, *Xba*I, and *Xho*I sites within the vector sequences, and the *Hin*dIII site is between the *ars1* and pBR322 sequences, these four endonucleases can be used to cleave genomic DNA while leaving the vector sequences intact. DNA (80 μl) was brought to a volume of 100 μl by the addition of 20 μl of 5× restriction enzyme buffer, 5 units of restriction enzyme were added, and the sample was incubated at 37° for 2 hr. The reaction was terminated by heating at 65° for 10 min, and a 20-μl aliquot was removed for analysis of the completeness of the reaction by agarose gel electrophoresis. The remaining 80 μl were brought to a volume of 0.8 ml and a final concentration of 66 m*M* Tris HCl, pH 7.5, 10 m*M* dithiothreitol, 6.6 m*M* $MgCl_2$, and 0.4 m*M* ATP. Thirty units of T4 DNA ligase were added, and the reaction was incubated from 4 hr to overnight at 4°. The solution was brought to 0.2 *M* NaCl, and the DNA was EtOH precipitated, washed once with 70% EtOH, and dried. The DNA was resuspended in 60 μl of 0.01 *M* Tris HCl, pH 7.4, 0.1 *M* $CaCl_2$, and 0.01 *M* $MgCl_2$, and 10–40 μl of the solution was added to 0.5 ml of competent 5346 cells. The transformants were plated and tested as above.

[25] D. Morrison, *J. Bacteriol.* **132,** 349 (1977).

DNA of each fusion plasmid was prepared[26] and then analyzed by restriction enzyme mapping and Southern blots[27] to determine the insert size and to detect any insert rearrangements that would result in spurious expression. The *din1:lacZ* fusion plasmid, pSZ214, (Fig. 1) for example, was recovered after the yeast genomic DNA was cut with *Hin*dIII, ligated to circularize the plasmid, and then used to transform bacteria. It has a 7 kb insert that is larger than the average insert size (4 kb) of the eight other fusion plasmids examined. The 1.3 kb fragment between the *Bam*HI/*Sau*3A junction and the adjacent *Bam*HI site on pSZ214 is found in genomic DNA of wild-type yeast strain A2. Yeast sequences upstream of the *Bam*HI site, however, are not contiguous in A2 DNA.[24] These sequences were probably joined together in pSZ214 during the ligation of genomic DNA fragments to the fusion vector.

To confirm that the isolated fusion plasmids contained the regulated gene sequences, the circular fusion plasmids were used to transform A2 yeast cells.[19] Strain T378 was obtained by transformation with the plasmid, pSZ214. The regulation of β-galactosidase in T378 is not detectably different from that of the original *din1* fusion transformant, T283.

Detection of Regulated Wild-Type Gene Activity by Northern Blot Analysis

The changes in β-galactosidase specific activity in NQO-induced T378 cells reflect changes in wild-type *DIN1* gene activity. Levels of *DIN1* gene transcripts in NQO-induced and uninduced A2 (wild type, untransformed) cells were examined by Northern blot analysis.[28] A2 cells were grown in YPD at 30° to mid-log phase. At 0 hr, 0.05 μg of NQO per milliliter was added to one half of the culture. During subsequent incubation at 30°, cell samples were removed at the times indicated in Fig. 3, and the total RNA was extracted according to Hereford *et al.*[29] Total cell RNA (20 μg per sample in 50% formamide, 2.2 *M* formaldehyde, and 1 × RNA gel buffer) was heated at 60° for 10 min, cooled rapidly, and then size-fractionated on a 1% agarose gel in 1 × RNA gel buffer containing 2.2 *M* formaldehyde.[30] The gel was treated with 50 m*M* NaOH[31] and 150 m*M* NaCl for 30 min, neutralized with 1 *M* Tris-HCl, pH 7.6, and 3 *M* NaCl for 30 min, and washed for 10 min in 10 × SSC. The RNA was transferred

[26] D. S. Holmes and M. Quigley, *Anal. Biochem.* **114,** 193 (1981).
[27] E. M. Southern, *J. Mol. Biol.* **98,** 503 (1975).
[28] P. S. Thomas, *Proc. Natl. Acad. Sci. U.S.A.* **77,** 5201 (1980).
[29] L. M. Hereford, M. A. Osley, J. R. Ludwig, and C. S. McLaughlin, *Cell* **24,** 367 (1981).
[30] H. Lehrach, D. Diamond, J. M. Wozney, and H. Boedtker, *Biochemistry* **16,** 4743 (1977).
[31] B. Seed and D. Goldberg, in preparation, 1982.

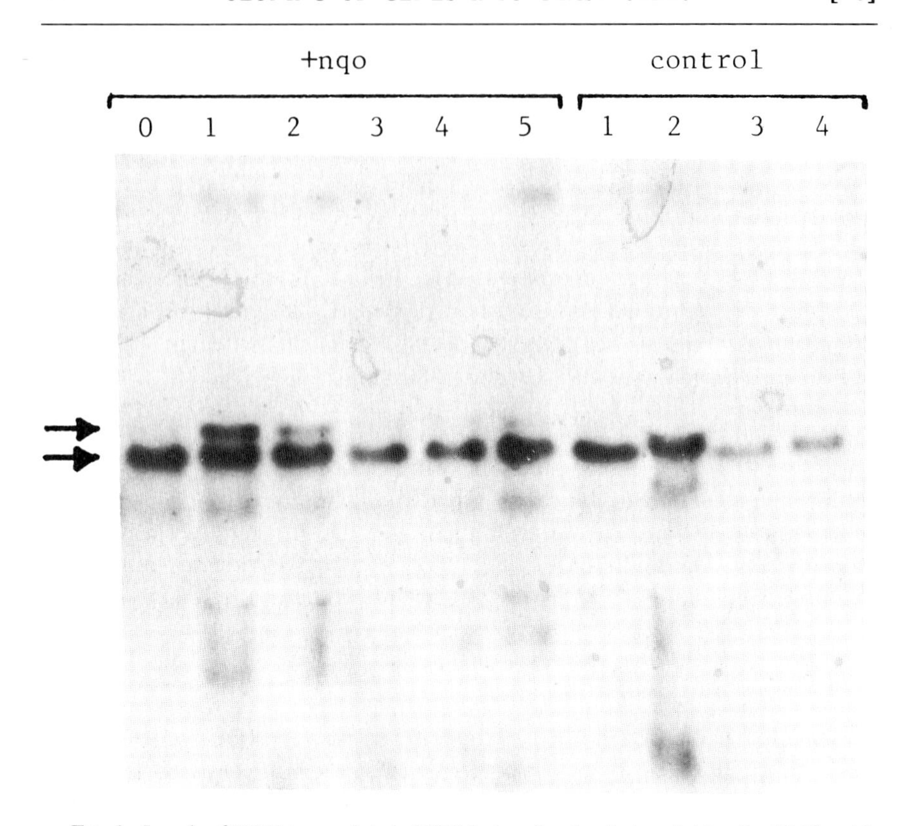

FIG. 3. Levels of *DIN1* transcripts in NQO-induced and uninduced A2 cells. Mid-log A2 cultures received 0.05 μg of NQO per milliliter or no treatment (control) at 0 hr. During subsequent growth, cell samples were removed at the hours indicated, and their RNA was extracted and assayed by Northern blot analysis as described in text. The autoradiogram of a Northern blot probed for *DIN1* (upper arrow) and control gene (lower arrow) transcripts is shown.

to nitrocellulose in 10× SSC for 4 hr. The blot was washed once briefly in 5× SSC, and then baked under vacuum at 80° for 2 hr.

The blot was prehybridized in 20 ml of 50% formamide, 0.1% SDS, 4× Denhardt's solution, 1× RNA hybridization buffer, 4 m*M* vanadyl adenosine complex,[32] and 100 μg of boiled calf thymus DNA per milliliter for 30 min at 42°. It was then hybridized in 10 ml of fresh solution to which 10^7 cpm of boiled ^{32}P-labeled DNA probe were added. Incubation of the blot was at 42° overnight. The blot was washed four times at 23° with 2× SSC and 0.1% SDS for 15 min each, followed by four times with

[32] S. L. Bezger and C. S. Birkenmeir, *Biochemistry* **18,** 5143 (1979).

3 m*M* Tris base for 5 min each at 23°, and exposed to X-ray film in the presence of an intensifying screen for 4 days. Two ^{32}P-labeled probes specific for *DIN1* and control transcripts were prepared by nick translation.[33] The *DIN1* probe (7×10^7 cpm/μg) was the 2.3 kb *Sal*I–*Sac*I fragment of pSZ214, which contained 0.5 kb of *din1* and 1.8 kb of *lacZ* sequences (Fig. 1). It was obtained by cutting pSZ214 DNA with *Sal*I and *Sac*I, fractionating the DNA by agarose gel electrophoresis, electroeluting the fragment from the gel onto DEAE paper, and eluting it from the paper with 1 *M* NaCl, 10 m*M* Tris HCl, pH 7.5, and 1 m*M* EDTA. The control probe (10^8 cpm/μg) was plasmid pSZ62[34] (not shown), which contains pBR322 and the 1.7 kb *Bam*HI-*HIS*3 fragment.[35]

As shown in Fig. 3, there was a single size class of *DIN1* transcript, which increased within 1 hr to detectable levels in NQO-treated cells, remained present for an additional hour, and then declined. These changes correlate with changes in β-galactosidase levels in T378 cells. There were no detectable *DIN1* transcripts in untreated cells. The 2.3 kb control transcript[35] from a gene adjacent to the *HIS3* gene reflects the amount of RNA present in each lane. The *lacZ* sequence alone does not hybridize detectably to yeast RNA under these conditions.

Sensitivity of the Fusion Cloning Method

The *lacZ* fusion method can be used to detect regulated genes whose products are at very low concentrations within the cell because the development of a blue color depends on cumulative β-galactosidase activity. An estimation of the sensitivity of fusion screening was made by examining the β-galactosidase levels of a yeast strain containing a fusion to a gene whose product levels are known. The diploid yeast strain, DA151, contains a single copy of a *his3 : lacZ* fusion integrated at the *HIS3* locus.[36] The *his3 : lacZ* fusion plasmid A4p4 is illustrated in Fig. 1. During log-phase growth in histidine-free medium, the β-galactosidase specific activity was 0.67 unit as measured in permeabilized cells. The level of *HIS3* mRNA in wild-type, haploid cells growing in liquid under similar conditions has been estimated to be 1 or 2 copies per cell.[35] When this diploid was grown on histidine-free medium and replica plated onto Xgal medium lacking histidine, a light blue color was detectable after 2 days. Fusions expressed at one-tenth of this level can be reproducibly detected.

[33] P. Rigby, D. Deichmann, C. Rhodes, and P. Berg, *J. Mol. Biol.* **113,** 237 (1977).

[34] T. L. Orr-Weaver, J. W. Szostak, and R. J. Rothstein, *Proc. Natl. Acad. Sci. U.S.A.* **78,** 6354 (1981).

[35] K. Struhl and R. Davis, *J. Mol. Biol.* **152,** 535 (1981).

[36] A. Murray, unpublished data, 1981.

Discussion

Genes Not Detectable by This Method

There are probably regulated yeast genes that are not identifiable by the fusion cloning method. A gene fusion product could disrupt normal cellular functions and cause lethality either because of its abundance or location. A fusion on a multicopy replicating plasmid might be improperly regulated. One solution to this problem might be to include a centromere fragment[37] on the fusion vector. Finally, the fusion itself might disrupt regulation if sequences adjacent to the 3′ end of the gene are required for proper control.

Not all yeast genes will be represented in a fusion library as constructed here. The *lac* fragment in pMC1403[18] has three restriction endonuclease sites (*Eco*RI, *Sma*I, and *Bam*HI) adjacent to each other and occurring just within the N-terminus codons of the *lacZ* gene. Any one of these sites can be used for making fusions. We chose the *Bam*HI site because *Sau*3A-cut DNA can be ligated into the unique *Bam*HI site of pSZ211. The *Sau*3A recognition sequence occurs frequently enough to be within most yeast genes. A partial *Sau*3A digest would have DNA fragments large enough to contain all the necessary regulatory, and transcription and translation initiation sequences. At least some of the fusions made with *Sau*3A fragments would have a *Bam*HI site regenerated at the fusion junction. Clearly, fusions will not be obtained to genes lacking *Sau*3A sites in the correct reading frame in their coding regions. A similar strategy could be used to obtain complementary fusion libraries by using the other two restriction sites on the *lac* fragment from pMC1403. A single, complete fusion library could be achieved by using sheared DNA fragments since the breakpoints would be sequence independent; however, this is not necessary for the isolation of a few representatives of a large class of genes.

A library of fusions could also be constructed and amplified in *E. coli*.[38] This would enrich the DNA pool for intact replicating plasmids, and would be useful if it was necessary to screen for fusions in several different strains. It would though, reduce the probability that an independent yeast transformant contained a unique species of fusion plasmid. Using the ligated pool of DNA to transform yeast directly as described here saves time and possibly increases the proportion of integrated plasmids among the transformants. Two out of 6 *din* fusions for example, were obtained as mitotically stable transformants.[23] In contrast, for

[37] L. Clarke and J. Carbon, *Nature* (*London*) **287,** 504 (1980).
[38] L. Clarke and J. Carbon, this series, Vol. 68, p. 396.

closed, circular, replicating plasmids containing the *ars1* sequence and 1.7 kb to 2.3 kb yeast DNA inserts, less than 1% of the yeast transformants are mitotically stable.[19]

Problems in Screening lacZ Fusions

We have encountered certain problems in screening *lacZ* fusions and in subsequently working with specific fusions. To obtain consistent results among screenings of a pool of fusions, the colonies should be of the same size and at the same stages of growth in different experiments. It is important to use optimal time intervals consistently between (*a*) inoculation of the master plate and replica plating and (*b*) replica plating and exposure of the colonies to treatment conditions.

The genetic background of a strain may affect the expression of a fusion, and the development of blue color on Xgal media. The blue color varied greatly among nonisogenic strains containing the same fusion integrated at the same locus. When a diploid formed from two nonisogenic haploid strains was sporulated, the colonies from the four spores of a tetrad had different induced and uninduced color intensities. A significant reduction in this variability often required more than three backcrosses.[24] This variation may give rise to a large number of false positives when screening for differential fusion expression in haploid and diploid cells unless isogenic strains are used. It may also complicate the use of Xgal plates in the study of the effects of mutations on fusion expression, when the mutation is crossed in from a nonisogenic strain.

The biochemical and physiological characteristics of chimeric transcripts and proteins in yeast are possible sources of artifacts in the use of *lacZ* fusions to isolate and study regulated yeast genes. Active β-galactosidase is a tetramer,[39] so that any foreign peptide in the amino terminus of the monomer that would interfere with subunit interactions could reduce or destroy β-galactosidase activity. Oliver and Beckwith[40] have found a significant reduction in β-galactosidase activity of a hybrid protein in *E. coli* when it is membrane-associated compared with when it is in the cytoplasm as the result of a signal sequence mutation in the *malE:lacZ* gene fusion. It is interesting that for different yeast *cyc1:lacZ*[41] and *ura3:lacZ* fusions,[42] in general, the more of the yeast peptide in a fusion, the lower the specific activity of β-galactosidase. We know little about the

[39] I. Zabin and A. V. Fowler, *in* "The Operon" (J. H. Miller and W. S. Reznikoff, eds.), p. 89. Cold Spring Harbor Laboratory, Cold Spring Harbor, New York, 1980.

[40] D. B. Oliver and J. Beckwith, *Cell* **25,** 765 (1981).

[41] L. Guarente, personal communication.

[42] M. Rose, personal communication.

processing and degradation of chimeric molecules or their distribution within the yeast cell. It is conceivable that in some cases the blue color developed on Xgal plates might not reflect enzyme levels measured in permeabilized cells or in cell extracts. A formal possibility for some gene fusions is that changes in protein–protein interactions rather than gene activity could lead to changes in β-galactosidase activity. Therefore, the identification of a regulated gene by the use of a gene fusion should be confirmed by independent evidence such as Northern blot analysis of mRNA levels.

Rearrangements in Fusion Plasmids

In the isolation of fusion plasmids we routinely selected *E. coli* transformants that were blue on minimal medium with Xgal in the cases where there were both blue and white colonies on the transformation plates. The restriction map analyses of four different fusions examined thus far have shown that these plasmids had not undergone rearrangements to allow expression in *E. coli*. Since *E. coli* is not capable of expressing all eukaryotic genes, it is expected that not all fusions isolated will be expressed in bacteria. In fact, a *lacZ* fusion to the yeast histone H2A gene is not productive in *E. coli* but is in yeast, while a fusion to the H2B gene is expressed in both organisms.[43] Thus, it is important to examine plasmids for rearrangements if blue bacterial transformants are used for plasmid isolations.

Conclusions

We have described the use of *lacZ* gene fusions for the identification and isolation of regulated yeast genes. A library of random yeast gene:*lacZ* fusions in yeast transformants was constructed and screened for regulated gene fusions. The screen detected differential β-galactosidase activity in fusion-containing yeast colonies on different Xgal media. We have used this method to clone *DIN1* sequences as a *din1:lacZ* fusion that is induced in yeast by DNA damaging agents. The method is particularly useful for the cloning of coordinately regulated genes.

Once isolated, gene fusions can be applied to the study of gene structure, function, and regulation. The *lacZ* fusion would be useful for fine-structure mapping and sequencing since the 5′ end of the *lacZ* gene has been sequenced.[44] Gene products could be identified by mRNA hybrid selection and *in vitro* translation.[45] Techniques already exist by which fu-

[43] M. A. Osley and L. Hereford, personal communication.

[44] A. M. Maxam, personal communication.

[45] M. L. Goldberg, R. P. Lifton, G. R. Stark, and J. G. Williams, this series, Vol. 68, p. 206.

sions can be mutagenized *in vitro*,[46] and the effects of the mutations examined *in vivo*.[47] *lacZ* fusions could be used to create insertion and deletion mutations in the wild-type gene.[47] Gene fusions could also be employed to isolate unlinked, regulatory mutations. This last application would be aided by the development of yeast able to take up lactose and of selectable substrates for fusion-containing yeast.

Acknowledgments

We thank Burt Beames for excellent technical assistance at the beginning of this work and Sandra Phillips for typing. We also thank T. L. Orr-Weaver and C. Potrickus for helpful comments on the manuscript. This work was supported by NIH Grant GM 27862 to J. W. S. S. W. R. was supported by NIH Training Grant 5T32GM07196, and A. W. M. by NIH Training Grant CA 09361.

[46] D. Shortle and D. Nathans, *Proc. Natl. Acad. Sci. U.S.A.* **75,** 2170 (1978).
[47] S. Scherer and R. W. Davis, *Proc. Natl. Acad. Sci. U.S.A.* **76,** 4951 (1979).

[17] Construction of Specific Chromosomal Rearrangements in Yeast

By NEAL SUGAWARA and JACK W. SZOSTAK

Investigations have shown that repeated DNA sequences located at nonhomologous chromosomal locations in the yeast *Saccharomyces cerevisiae* are able to recombine during mitotic growth. The type of recombination most frequently observed is gene conversion, which does not result in any gross change in chromosomal structure.[1-4] Reciprocal recombination between dispersed sequences is observed less frequently but results in the generation of chromosomal rearrangements. For example, two ring derivatives of chromosome III have been identified that arose by homologous, reciprocal recombination between the mating type sequences *HML, MAT,* and *HMR*. One was generated by selecting for recombination between the homologous sequences of *HML* and *MAT*.[5]

[1] S. Scherer and R. W. Davis, *Science* **209,** 1380 (1980).
[2] J. F. Ernst, J. W. Stewart, and F. Sherman, *Proc. Natl. Acad. Sci. U.S.A.* **78,** 6334 (1981).
[3] N. Sugawara and J. W. Szostak, in preparation, 1982.
[4] M. Mikus and T. D. Petes. Genetics **101,** 369 (1982).
[5] J. N. Strathern, C. S. Newlon, I. Herskowitz, and J. B. Hicks, *Cell* **18,** 309 (1979).

ISBN 0-12-182001-7

The other was the product of recombination between *HML* and *HMR*.[6,7] A deletion in the right arm of chromosome III was similarly generated by recombination between *MAT* and *HMR* sequences.[8]

Specific chromosomal rearrangements also have been generated by employing recombinant DNA techniques to place defective, auxotrophic heteroalleles at two dispersed loci and then selecting for prototrophy. Several reciprocal translocations have been generated using recombinant DNA methods. In these cases recombination was selected to occur between the *his3*,[3] *leu2*, and *ura3*[4] genes, and their respective heteroallelic sequences were inserted into the rDNA cluster on chromosome XII. In *Saccharomyces*, chromosomal rearrangements arise infrequently; the use of recombinant DNA techniques allows the generation of many types of rearrangements at defined sites. This chapter describes and illustrates this general method for the construction of specific chromosomal rearrangements in *Saccharomyces cerevisiae*.

Materials

Plasmids Restriction maps of the plasmids pSZ80 and pSZ81 are shown in Fig. 1. The 1.4 kb *Bam*HI fragment containing a 5′ deletion of the *his3* gene was derived from YRp7-Sc2712, which was obtained from K. Struhl.[9] The rDNA segment was cloned as described,[10] and *LEU2* was taken from pYeleu10[11] obtained from A. Hinnen.

Strains. The genotype of D154-13A is α, *his3-11,15, leu2-2,112, ade1, trp1, tcm1*. D153-13C is *a*, *his3-11,15, leu2-2,112, asp5, ura4, can1*. Both resulted from backcrosses to strain LL20 obtained from L. Lau. The markers *asp5, ura4, his3-11,15, leu2-2,112,* and *can1* were obtained from LL20 or 6657-4D (G. Fink). *Tcm1* came from CLP-1 (J. Davies). *Trp1* and *ade1* came from X4001-39A (M. Olson).

Media. Nonselective (YPD), selective, and sporulation media are described in the Cold Spring Harbor Yeast Manual.[12] Trichodermin (the generous gift of W. Gotfredson) was dissolved in 50% ethanol to 20 mg/ml

[6] A. Klar, personal communication, 1981.
[7] J. E. Haber, personal communication, 1981.
[8] D. C. Hawthorne, *Genetics* **48,** 1727 (1963).
[9] K. Struhl, *J. Mol. Biol.* **152,** 553 (1981).
[10] J. W. Szostak and R. Wu, *Plasmid* **2,** 536 (1979).
[11] B. Ratzkin and J. Carbon, *Proc. Natl. Acad. Sci. U.S.A.* **74,** 487 (1977).
[12] F. Sherman, G. R. Fink, and C. Lawrence, "Methods in Yeast Genetics." Cold Spring Harbor Laboratory, Cold Spring Harbor, New York, 1979.

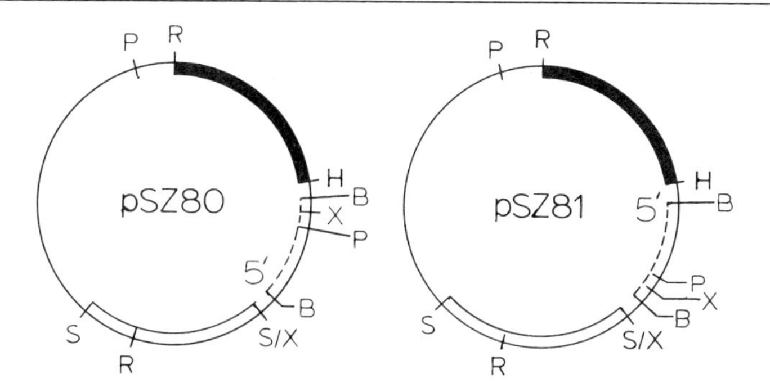

FIG. 1. Plasmids. pSZ80 and pSZ81 were constructed by inserting the rDNA (dark lines), *his3*-Δ (broad filled lines), and *LEU2* (open lines) segments into pBR322 (thin lines). The 5′ indicates the 5′ end of the *his3*-Δ gene. Both plasmids are 10.3 kb in size. Restriction sites: R, *Eco*RI; H, *Hin*dIII; B, *BamHI;* P, *Pst*I; X, *Xho*I; and S, *Sal*I.

and added with a final concentration of 20 mg/liter. Canavanine plates were prepared with a final concentration of 60 mg/liter. Amino acids, adenine sulfate, uracil, and 1-canavanine sulfate were obtained from Sigma; salts and dextrose from Fisher; and yeast extract, peptone, agar, and yeast nitrogen base from Difco.

Enzymes. Enzymes were obtained from New England BioLabs or Bethesda Research Laboratories. Manufacturer's reaction conditions were followed.

Transformation and Blots. Reagents and methods used for yeast transformation were as described by Hinnen *et al.*[13] and modified by Orr-Weaver *et al.*[14] Southern blots were as described.[15]

Principle of the Method

In order to construct a specific chromosomal rearrangement, auxotrophic heteroalleles must be placed at the sites where the crossover is desired and in the correct orientations. To do this the regions surrounding these sites must first be cloned. This is not difficult to do in yeast, where any gene with a scorable mutant phenotype can be readily cloned by transformation-complementation.[13] The next step is to introduce the auxotrophic heteroalleles at each site. One of the mutant alleles may reside at

[13] A. Hinnen, J. B. Hicks, and G. R. Fink, *Proc. Natl. Acad. Sci. U.S.A.* **75,** 1929 (1978).
[14] T. L. Orr-Weaver, J. W. Szostak, and R. J. Rothstein, this volume [14].
[15] E. M. Southern, *J. Mol. Biol.* **98,** 503 (1975).

the usual chromosomal location of the gene. The second mutant allele must be inserted at the second site by transformation and plasmid integration. This is done by constructing a plasmid containing DNA from the second site, the heteroallele, and a selectable marker for transformation. This plasmid can integrate into a chromosome by a single, homologous recombination event with any of these three yeast segments. It is possible, however, to target the plasmid to the correct site by cutting the plasmid in the segment containing that site and then transforming.[16]

The ideal mutations to use in this scheme are deletions that extend into the 5′ or 3′ end of a gene. Such deletions cannot revert and are not subject to gene conversion if there is homology to the heteroallele on only one side of the deletion. These can be constructed *in vitro* using standard techniques and then introduced into the cell at the new site by plasmid integration as described above. Transplacement[17] may be used to substitute a deletion allele for the wild-type gene. In the latter case the plasmid is integrated into the wild-type site by homologous recombination and then excised by the reverse process in such a way as to substitute the new allele for the one on the chromosome. Transplacement can also be used to delete the wild-type site if a crossover is desired between heteroalleles placed at two other sites. If the orientations of the two sites with respect to the centromeres are not known, it may be necessary to integrate one of the deletion alleles in both orientations to ensure recovery of the desired rearrangement. Once it is confirmed that the deletion alleles are integrated at the proper sites, recombinants between the two alleles are isolated by selecting prototrophs. Since a large number of prototrophs may have to be examined genetically to obtain a specific rearrangement, it is easiest to isolate prototrophs in a diploid that may be sporulated directly and then examined by tetrad analysis.

Construction of a Reciprocal Translocation with a Breakpoint in the rDNA Locus

Construction of Strains

We wished to construct a reciprocal translocation with a crossover point located in the rDNA locus of yeast on chromosome XII for the purpose of studying copy number regulation in this repeated gene cluster.[18]

[16] T. L. Orr-Weaver, J. W. Szostak, and R. J. Rothstein, *Proc. Natl. Acad. Sci. U. S. A.* **78,** 6354 (1981).

[17] S. Scherer and R. W. Davis, *Proc. Natl. Acad. Sci. U.S.A.* **76,** 4951 (1979).

[18] G. P. Smith, *Cold Spring Harbor Symp. Quant. Biol.* **38,** 507 (1974).

The entire rDNA repeat unit has been cloned in several laboratories, and many useful fragments have been subcloned.[10,19–21]

We selected *HIS3* prototrophs from a strain that contained two defective *his3* alleles at dispersed loci. One defective allele was a nonreverting double point mutation, *his3-11,15,* at the *his3* locus on chromosome XV. For the second *his3* allele, we used a deletion constructed by K. Struhl, which extends into the 5′ end of the structural gene and was available on a cloned 1.4 kb *Bam*HI fragment.

The deletion allele of *his3* was integrated into the rDNA cluster as follows. A hybrid plasmid, pSZ81 (Fig. 1), that contained rDNA sequences, the *his3* deletion allele, and the *LEU2* gene, was constructed *in vitro.*[3] This plasmid was transformed into the haploid yeast strain D153-13C (*a, his3-11,15, leu2-2,112, asp5, ura4, can1*) by selecting leu$^+$ transformants. This plasmid integrated into the rDNA on chromosome XII by homologous recombination. A Southern blot of genomic DNA from our transformant when probed with pBR322 established that only a single copy of the plasmid integrated into the rDNA (Fig. 2). This transformant was mated with strain D154-13A (α, *his3-11,15, leu2-2,112, ade1, trp1, tcm1*) to yield the strain D214 with the genotypic configuration shown in Fig. 3. Since we did not know the orientation of the *his3* gene with respect to centromere XV, we also made a second, plasmid, pSZ80, with the *his3* segment oriented in the opposite direction (Fig. 1). The plasmid was integrated into the rDNA of D153-13C, and one transformant with a single-copy plasmid was mated to D154-13A to yield strain D231. Our results show that D214 produced the reciprocal translocation.

Selection and Genetic Analysis of His$^+$ Recombinants

Strain D214 (Fig. 3) is a diploid that is homozygous for the double point mutation *his3-11,15* at the *his3* locus on chromosome XV, and heterozygous for an integrated plasmid that carries the *his3* 5′ deletion in the rDNA locus on chromosome XII. A single reciprocal recombination event between these two heteroalleles of *his3* would result in the desired reciprocal translocation. Independent his$^+$ recombinants were selected by plating a culture of D214 in stationary phase on rich medium (YPD) at 100–200 cells per plate. After 2 days at 30° the colonies were replica-plated to complete medium lacking histidine. All colonies were his$^-$, but after 3–6 days his$^+$ papillae appeared on some of the colony prints ($<5\%$).

[19] G. I. Bell, L. J. DeGennaro, D. H. Gelfand, R. V. Bishop, P. Valenzuela, and R. V. Rutter, *J. Biol. Chem.* **252,** 8118 (1977).

[20] J. M. Cramer, R. W. Farrelly, J. T. Barnitz, and R. M. Rownd, *Mol. Gen. Genet.* **151,** 229 (1977).

[21] P. Philippsen, R. A. Kramer, and R. W. Davis, *J. Mol. Biol.* **123,** 371 (1978).

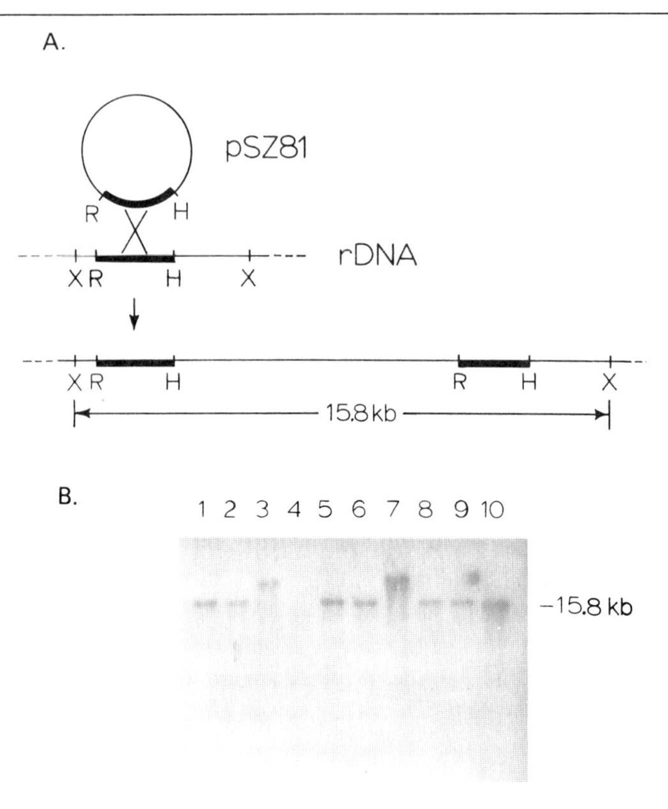

FIG. 2. Integration of plasmids. (A) pSZ81 or pSZ80 is transformed into yeast and integrated into the rDNA by a single recombination event between homologous rDNA sequences on the *Eco*RI–*Hin*dIII segment. Plasmids can integrate as single or as tandem, multiple copies. A single-copy insertion of pSZ80 or pSZ81 has a 15.8 kb *Xba*I fragment. Restriction sites are R, *Eco*RI; H, *Hin*dIII; and X, *Xba*I. (B) A Southern blot of *Xba*I-cut DNA from transformants with pSZ80 (lanes 1–3 and 5) and pSZ81 (lanes 6–10) probed with pBR322. Lanes 1, 2, 5, 6, 8, 9, and 10 are single-copy inserts. Lanes 3 and 7 are multiple-copy inserts. Two transformants with single-copy inserts were used in this study (lanes 2 and 8).

Sixty such colonies were picked and purified by streaking out to single colonies on YPD plates and then picking a single his$^+$ colony from each isolate. Each his$^+$ recombinant is independent because it was derived from a separate his$^-$ colony.

Each of these isolates was grown as a 2 cm^2 patch on YPD for 1–2 days and then replica-plated to sporulation medium. Ten of the 60 isolates did not sporulate well enough to examine; 5–10 tetrads were initially dissected from each of the remaining strains and analyzed by conventional genetic procedures.[12]

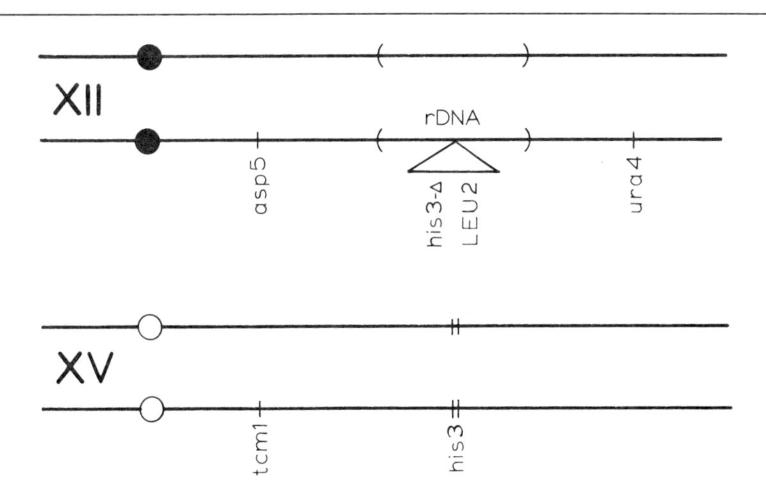

FIG. 3. Genotypic configuration. Strain D214 is heterozygous for pSZ81 (*LEU2, his3-*Δ) integrated into the rDNA locus on chromosome XII. Strain D231 is similarly heterozygous for pSZ80, which contains the *his3-*Δ fragment in the opposite orientation. Both strains are homozygous for the double point mutation *his3-11,15* on chromosome XV. These *his3* alleles are flanked by the markers *asp5, ura4,* and *tcm1*.

Strains with a wild-type chromosomal configuration normally produce four viable spores per tetrad. A heterozygous reciprocal translocation, however, will produce tetrads with predominantly 0, 2, or 4 viable spores per tetrad if the crossover points are meiotically unlinked to the centromeres,[22] as are the rDNA and *his3* loci. Such inviability results from the segregation of a normal chromatid and a translocation chromatid to the same spore so that all the information distal to one on the crossover points on one of the chromosomes is missing. Of the 60 his^+ recombinants isolated, four were tentatively identified as reciprocal translocations based on their spore viability patterns. One was subcloned and examined more closely. This recombinant, designated D214R3S3, had a spore viability pattern of 57:9:166:12:49 (4:3:2:1:0 viable spores per tetrad). *HIS3* segregates $1^+:1^-$ or $2^+:2^-$ in 97% of the tetrads, this indicates that the crossover point maps to the *HIS3* gene. The remaining 3% probably result from random spore inviability, gene conversion, or errors in dissecting tetrads.

Owing to the small size of yeast chromosomes, cytological studies cannot be carried out. This makes it necessary to establish the presence of a reciprocal translocation by genetic procedures. Conventional meiotic mapping can be carried out in a translocation homozygote constructed by

[22] D. D. Perkins and E. G. Barry, *Adv. Genet.* **19,** 133 (1977).

crossing two spores containing the translocation and which upon sporulation yields four viable spores per tetrad. However, meiotic mapping cannot be used because there are no distal markers that are meiotically linked to the rDNA locus. We therefore used mitotic mapping to show that *ura4*, which normally resides on chromosome XII (Fig. 3), has become mitotically linked to *tcm1*, trichodermin resistance, on chromosome XV. To show this, we constructed a translocation homozygote with the following genotypic configuration: *tcm1*^{r} *ura4*/*TCM1*^{s} *URA4*. Subclones of this diploid were grown on YPD plates in 1 cm^2 patches and replica-plated to YPD + trichodermin plates. After 2–3 days at 30°, trichodermin-resistant colonies appeared; 71% of these were *ura*$^-$. These resistant colonies (*tcm1*^{r}/*tcm1*^{r}) presumably arose from a crossover between *tcm1* and centromere XV, which caused homozygosis of *tcm1* and all markers distal to it. The fact that most colonies homozygous for *tcm1* were also homozygous for *ura4*$^-$ implies linkage of the two genes. This procedure was repeated for the parental strain D214 in which *URA4* resides on chromosome XII. In this strain, less than 1% of the tcm-resistant papillae were *ura*$^-$.

C. Other Classes of His$^+$ *Recombinants*

Reciprocal translocations comprised only a minor fraction of the recombinants isolated from D214. The remaining recombinants consisted of several classes.[3] Class I: Gene conversions of the *his3-11,15* to *HIS3*$^+$ comprised the largest class (29/60) and were identified by independent segregation of *HIS3* and *LEU2*. Gene conversion of the *his3* 5′ deletion did not occur because of the lack of homology to the *his3* sequences on the 5′ side of the deletion. Class II: One transposition of the plasmid to the *his3* locus was observed and was identified by complete linkage of *HIS3* and *LEU2*. Both *HIS3* and *LEU2* mapped to the *his3* locus on chromosome XV. Class III: A substantial fraction did not sporulate (10/60) and were not analyzed. Class IV: 9/60 became homozygous for *HIS3*, or for the *LEU2* gene on the plasmid integrated in the rDNA and were not analyzed further. Class V: Several exhibited aberrant spore viabilities and remain to be examined more closely.

Discussion

We have used recombinant DNA methods to place homologous sequences on nonhomologous chromosomes. After selection for recombination between these sequences, we were able to identify genetically the expected chromosomal rearrangement. This method has been demonstrated for the construction of reciprocal translocations[3,4] and should work equally well for the construction of other rearrangements (e.g., duplications, deletions, and inversions).

Certain variations on the method used here may facilitate the isolation of chromosomal rearrangements. In this study we used a deletion allele and a double point mutation to minimize the reversion frequency of the mutations. This is important since recombinants arose at a rate comparable to the spontaneous mutation rate. While we found it useful to use a deletion allele on the plasmid, it is also possible to use undefined mutations that are nonsuppressible and have a low reversion rate.[4] As long as one allele has a region of homology to the other allele on both sides of the mutation, the dominant class of recombinants will be comprised of gene convertants. The frequency of this class could be reduced by the use of two deletion alleles, one covering the 5′ end of the gene and one covering the 3′ end. Another possible way to reduce the proportion of gene conversions would be to isolate the prototrophs in a *rad52* strain, since *rad52* is known to reduce mitotic gene conversion.[23–25] Unfortunately *rad52*/*rad52* diploids do not yield viable spores. This can be circumvented by selecting recombinants in a *rad52* haploid and then mating to a wild-type *RAD52* tester strain. Spores from *rad52*$^-$/*RAD52*$^+$ diploids are viable.

The methods described above should allow the construction of a variety of chromosomal rearrangements with breakpoints at any site that has been cloned. These rearrangements may be used for many purposes, e.g., the study of chromosome behavior (disjunction, pairing), recombination, sister chromatid exchange, breakage-fusion-bridge cycles, and various mapping problems.

Chromosomal aberrations such as translocations can be constructed by this method in order to investigate aspects of chromosomal behavior such as centromere disjunction and chromosome pairing. Translocating varying lengths or different regions of chromosomes to new locations allows one to examine sequence and homology requirements for correct disjunction and pairing. It may also be possible to use the prototroph selection scheme outlined in this paper to study the constraints on pairing and recombination between homologous sequences on nonhomologous chromosomes.

Ring chromosomes are particularly useful in the study of sister chromatid exchanges, because SCEs within a ring chromosome lead to the production of easily identifiable dicentric chromosomes. A ring derivative of chromosome XII would be useful in determining the relative contributions of unequal crossing over and gene conversion toward sequence rec-

[23] J. A. Jackson and G. R. Fink *Nature* (*London*) **292,** 306 (1981).

[24] S. Prakash, L. Prakash, W. Burke, and B. Montelone, *Genetics* **91,** 31 (1980).

[25] R. Malone and R. Esposito, *Proc. Natl. Acad. Sci. U.S.A.* **77,** 503 (1980).

tification in the rDNA locus. Ring chromosomes may provide a valuable source of DNA for cloning fragments from individual chromosomes, since a CsC1–ethidium bromide gradient can be used to enrich for circular DNA molecules.[5] Ring chromosomes that are viable as haploids, i.e., with no essential genes deleted, can be used to map specific telomeric restriction fragments to individual chromosomes.[26] Since ring chromosomes lack telomeres, a Southern blot using telomeric DNA to probe a restriction digest of genomic DNA from such a strain should show how two bands less than genomic DNA of a wild-type strain.

Other problems may also be approached with this method of constructing chromosomal aberrations. For example, the generation of a reciprocal translocation by homologous recombination can be used to determine the orientation of a gene with respect to its centromere. A reciprocal translocation can occur only if both homologous sequences are in the same orientation relative to their respective centromeres. Thus, if the orientation of the region at one of the two sites is known, the orientation of the second can be determined. Even if neither orientation is known, the orientations can be deduced by genetic analysis of the breakpoint regions.[3] Finally, our goal in constructing a reciprocal translocation with one breakpoint in the rDNA locus was to be able to construct strains carrying various amounts of ribosomal DNA, so that we could study the regulation of rDNA copy number.

The method described here may be applicable to other organisms where efficient transformation is possible. These include fungi, such as *Neurospora*[27,28] and *Schizosaccharomyces pombe*,[29] and perhaps higher organisms, such as *Drosophila* and mammalian cells.

[26] J. W. Szostak and E. H. Blackburn, *Cell* **29,** 245 (1982).

[27] M. E. Case, M. Schweizer, S. R. Kushner, and N. H. Giles, *Proc. Natl. Acad. Sci. U.S.A.* **76,** 5259 (1979).

[28] R. Radford, S. Pope, A. Sazci, M. J. Fraser, and J. H. Parish, *Mol. Gen. Genet.* **184,** 567 (1981).

[29] D. Beach and P. Nurse, *Nature (London)* **290,** 140 (1981).

[18] Yeast Vectors with Negative Selection

By PATRICIA A. BROWN and JACK W. SZOSTAK

It is often desirable to isolate derivatives of a stably transformed strain of yeast (*Saccharomyces cerevisiae*) that no longer contain integrated plasmid sequences. We have devised a method of isolating such deriva-

ISBN 0-12-182001-7

tives in which the loss of a plasmid results in a drug-resistant phenotype; thus we are able to select for drug resistance and simultaneously for plasmid loss. A number of drug-resistance loci in yeast have been characterized in which the wild-type (sensitive) marker is dominant to the resistant allele. These dominant sensitivity markers include *CYH*^{s} (cycloheximide sensitivity),[1] *TCM*^{s} (trichodermin sensitivity),[2] *CAN*^{s} (canavanine sensitivity),[3] and *LYS2* (a lysine biosynthetic gene that confers sensitivity to aminoadipic acid).[4] A diploid heterozygous for any of these markers will not grow on media containing the appropriate drug. Similarly, a resistant haploid transformed with a plasmid containing the dominant sensitive allele will exhibit the sensitive phenotype. It is easy to obtain segregants of such a transformant that have lost the integrated plasmid sequences by selecting for the drug-resistant phenotype.

We describe DNA fragments carrying the *TCM*^{s} and *CYH*^{s} genes and their use in the construction of plasmids with negative selection. We have characterized transformants containing integrated copies of such plasmids and have demonstrated the efficiency of negative selection in the isolation of recombinants that have lost the integrated plasmid. We also discuss the application of negative selection systems to transplacement,[5] genetic mapping, the isolation of recessive lethal mutations, and the study of recombination.

Methods

Strains. D234-3B (*α leu2-2,112, his3-11,15, trp1, ura3, tcm1*) was constructed using standard genetic methods. The *leu2-2,112* and *his3-11,15* alleles were derived from the strain LL20 obtained from G. Fink. *Tcm1* was derived from CLP-1 (J. Davies), and *trp1* from X4001-39A (R. K. Mortimer). The *ura3* marker was obtained by EMS mutagenesis of an LL20 derivative. D578-7D (*a leu2-2,112, his3-11,15 ura3, tcm1, cyh2/HIS3, TCM*^{s}) was constructed by backcrossing T400 (Table I) with a spontaneous cyh^{r} derivative of LL20. Except for *cyh2*, all markers are the same as those found in D234-3B. Transformants are described in Table I.

[1] J. E. Middlekauff, S. Hino, S. P. Yang, G. Lindegren, and C. C. Lindegren, *Genetics* **42,** 66 (1957)

[2] P. G. Grant, D. Schindler, and J. E. Davies, *Genetics* **83,** 667 (1976).

[3] M. Grenson, M. Mousset, J. M. Wiame, and J. Bechet, *Biochim. Biophys. Acta* **127,** 325 (1966).

[4] B. B. Chattoo, F. Sherman, D. A. Azubalis, T. A. Fjellstedt, D. Mehnert, and M. Ogur, *Genetics* **93,** 51 (1979).

[5] S. Scherer and R. W. Davis, *Proc. Natl. Acad. Sci. U.S.A.* **76,** 4951 (1979).

TABLE I
CHARACTERISTICS OF TRANSFORMANTS

Transformant	Strain	Plasmid	Description
T400	D234-3B (α *leu2-2,112, his3-11,15, ura3, trpl, tcml*)	pSZ414 (uncut)	Single copy of pSZ414 at rDNA
K5	D234-3B	pSZ414 (*Kpn*1 cut)	Single copy of pSZ414 at *HIS3*
T972	D578-7D (*a leu2-2,112, his3-11,15 ura3, tcml, cyh2/HIS3 TCM*^{S} at rDNA)	pSZ430 (*Sma*1 cut)	Single copy of pSZ430 at *URA3*/single copy of pSZ414 at rDNA
T948	D578-7D	pSZ430	pSZ430 at *CYH2* (copy number not known)/ single copy pSZ414 at rDNA

Media. Nonselective (YPD) and drop-out media (complete medium minus amino acid, CM − AA) are described in the Cold Spring Harbor Yeast Manual.[5a]

Drugs. Trichodermin is synthesized by the organism *Trichoderman viridae.* Our trichodermin stock was a gift from W. Gotfredsen. The drug is not commercially available. Trichodermin is stored as a 20 mg/ml solution in 50% ethanol at −20°. Selective medium contains 20 μg of trichodermin (TCM) per milliliter. TCM plates may be stored for several weeks at room temperature.

Cycloheximide was purchased from Sigma; stock solutions of 10 mg/ml in water may be stored frozen for several weeks. Selective plates contain 10 μg of cycloheximide per milliliter.

Drug Markers. TCM^s, the trichodermin-sensitivity marker, was isolated from a Yep13 yeast library.[6] This library contains a partial *Sau*3A digest of yeast genomic DNA ligated into the *Bam*HI site of the vector pYe13. The vector replicates autonomously and carries a *LEU2* selectable marker. Leu$^+$ transformants of J3, a spontaneous tcm resistant mutant of LL20, were screened for trichodermin sensitivity. The plasmid pSZ206 isolated in this way contains an insert of approximately 7 kb containing the TCM^s determinant. We have subcloned a 3.2 kb *Bam*HI–*Ava*I fragment conferring TCM sensitivity. This fragment has been modified to possess *Bam* HI restriction sites at both ends (Fig. 1).

[5a] F. Sherman, G. Fink, and C. Lawrence "Methods in Yeast Genetics." Cold Spring Harbor Laboratory, Cold Spring Harbor, New York, 1977.

[6] K. A. Nasmyth and S. I. Reed, *Proc. Natl. Acad. Sci. U.S.A.* **77,** 2119 (1980).

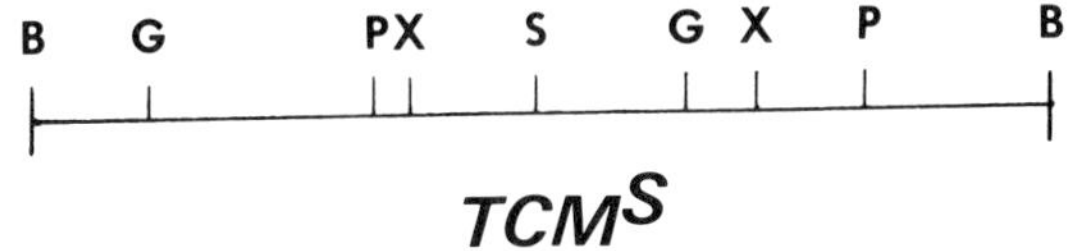

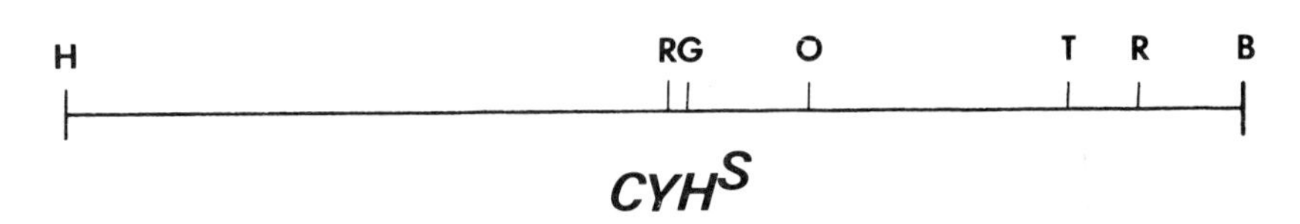

FIG. 1. Restriction maps of the *TCM*^{s} and *CYH*^{s} fragments. The derivation of these fragments is described in the text. The 3.2 kb *TCM*^{s} fragment contains no *Hin*dIII, *Sma*I, *Kpn*I, or *Xho*I sites. It has two unmapped *Bst*EII sites in addition to the sites shown here. The 5.0 kb *CYH*^{s} fragment contains no *Sal*I or *Kpn*I sites; it may contain other *Bst*EII sites. Restriction enzyme cleavage sites are designated as follows: B, *Bam*HI; G, *Bgl*II; H, *Hin*dIII; O, *Xho*I; P, *Pvu*II; R, *Eco*RI; S, *Sal*I; T, *Bst*EII; X, *Xba*I.

The *CYH2* fragment was obtained from pCYH2, the generous gift of H. Fried. We have subcloned a 5kb *Hin*dIII–*Bam*HI fragment carrying the CYH^s determinant (Fig. 1).

Transformation. Yeast transformations were done by the procedure of Hinnen *et al.*[7] as modified by Orr-Weaver *et al.*[8]

Determination of Recombination Frequency. Each transformant was grown overnight in 3 ml of a positively selective medium (CM − HIS for T400 and K5; CM − HIS − LEU for T972 and T948). Recombinants that have lost the integrated plasmid can grow only to a limited extent (a few divisions) in selective medium, and thus accumulate to a constant level. Dilutions of these cultures were plated on YPD and on YPD + TCM or YPD + CYH plates, and colonies were counted 3 days after plating. Frequencies are expressed as drug-resistant segregants (colonies on drug plates) per viable cell (determined from efficiency of plating on YPD). These numbers are not true frequencies (events per generation) but are useful for purposes of comparison.

[7] A. Hinnen, J. B. Hicks, and G. R. Fink, *Proc. Natl. Acad. Sci. U.S.A.* **75,** 1929 (1978).
[8] T. L. Orr-Weaver, J. W. Szostak, and R. J. Rothstein, this volume, [14].

Results

Negative Selection Systems

We have constructed two independent negative-selection systems with which to monitor the loss of an integrated plasmid in yeast. For each system we use a haploid strain containing the recessive allele of a drug-resistance marker. This strain is transformed with a plasmid consisting of three components: the dominant drug-sensitivity marker, a positively selectable marker for the introduction and maintenance of the plasmid in the transformed strain, and a chromosomal sequence to which plasmid integration can be directed. We have constructed plasmids pSZ414 and pSZ430 (Fig. 2) for use in such selections.

pSZ414 consists of the trichodermin-sensitivity marker *TCM*^{s}, the selectable marker *HIS3*, and a fragment of rDNA. The rDNA fragment directs the plasmid to integrate at the rDNA locus where the repeated rDNA sequences form a large region of homology for the integration event. PBR322 sequences provide a replication origin and the ampicillin-

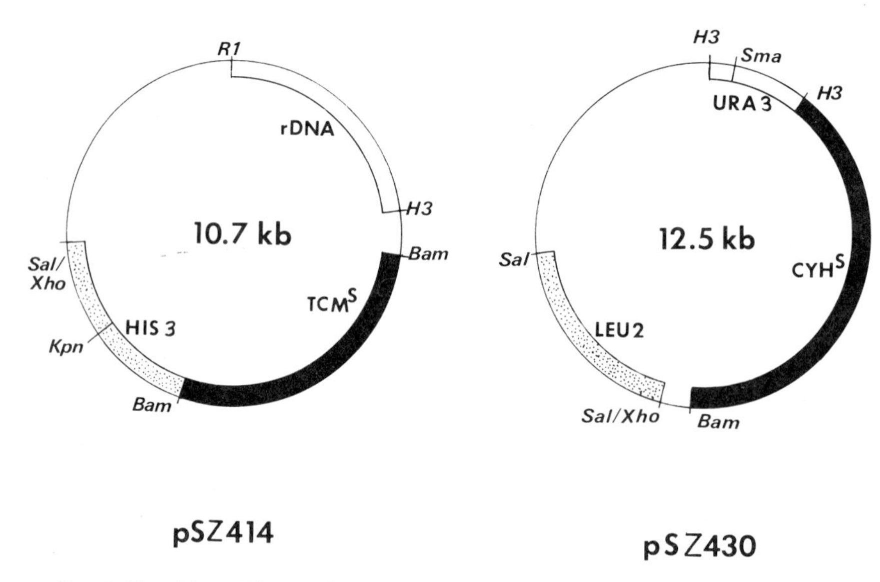

FIG. 2. Plasmids used in negative selections. The plasmids pSZ414 and pSZ430 were constructed by standard techniques. Each plasmid contains a DNA fragment homologous to the desired region of integration in the yeast genome (open bar), a dominant drug-sensitivity marker (filled bar), a selectable marker (stippled bar), and sequences derived from the plasmid pBR322 (thin line). Selected restriction enzyme cleavage sites are shown.

resistance marker for replication and selection of the plasmid in *Escherichia coli*.

The TCM^s marker is located on a 3.2 kb *Bam*HI fragment (Fig. 1). Resistance to trichodermin, a protein synthesis inhibitor, is caused by mutations in the ribosomal protein L3[9]; the coding sequence for this protein has been mapped to chromosome XV.[2] Negative selection using the plasmid pSZ414 is performed in strains carrying a *tcm1* mutation (recessive drug resistance). A *tcm1* haploid grows normally on media containing 20 μg of trichodermin per milliliter, whereas the same haploid strain with a single inserted copy of pSZ414 fails to grow under these conditions.

We have tested a second negative-selection procedure with pSZ430 (Fig. 2), which contains the *CYH2* sensitivity marker, the *LEU2* gene, and the *URA3* gene. We have used the *URA3* fragment only as a site for integration.

The *CYH2* marker is carried on a 5 kb *Hin*dIII-*Bam*HI fragment (Fig. 1). Resistance to the protein synthesis inhibitor cycloheximide (*cyh2*) is the result of an altered ribosomal protein encoded on chromosome VII.[10] The *CYH2* allele is genetically dominant; a resistant haploid can grow on media containing 10 μg of cycloheximide per milliliter, whereas a CYH^s transformant will not grow under these conditions.

We have constructed haploid strains suitable for transformation with pSZ414 and pSZ430 using standard genetic methods (Table I). The plasmids were introduced into these strains by the yeast transformation technique developed by Hinnen *et al.*[7] Complex plasmids such as pSZ414 or pSZ430, which contain multiple regions of homology with the yeast genome, may be targeted to integrate at a specific site by cutting the transforming plasmid DNA within the desired integration site.[8,11] We have used this technique in directing plasmid integration in our transformants.

Characterization of Transformants

It is necessary thoroughly to characterize transformed strains before proceeding with an experiment. We have found that both genetic and physical analyses of transformants are necessary when using vectors with negative selection.

An integrated plasmid confers a stably transformed phenotype; that is, a transformant streaked on nonselective medium almost always retains the selectable marker. The presence of the drug-sensitivity marker on the

[9] H. Fried and J. R. Warner, *Proc. Natl. Acad. Sci. U.S.A.* **78,** 238 (1981).

[10] R. K. Mortimer and D. C. Hawthorne, *Genetics* **74,** 33 (1973).

[11] T. L. Orr-Weaver, J. W. Szostak, and R. J. Rothstein, *Proc. Natl. Acad. Sci. U.S.A.* **78,** 6354 (1981).

integrated plasmid can be verified by replica plating to drug-containing plates. No growth occurs on the first day following replica plating, but drug-resistant papillae appear on succeeding days. If the plasmid has integrated at the targeted site, these papillae occur because of loss of the plasmid (and the dominant sensitivity marker it carries) by plasmid excision or unequal sister chromatid exchange[12–14] (Fig. 3). The loss of the selectable marker occurs concomitantly with the loss of the drug-sensitivity marker. This implies both that the drug-resistant recombinants should be auxotrophic for the selectable marker and that recombinants that are drug resistant but retain the selectable marker should be quite rare. The trans-

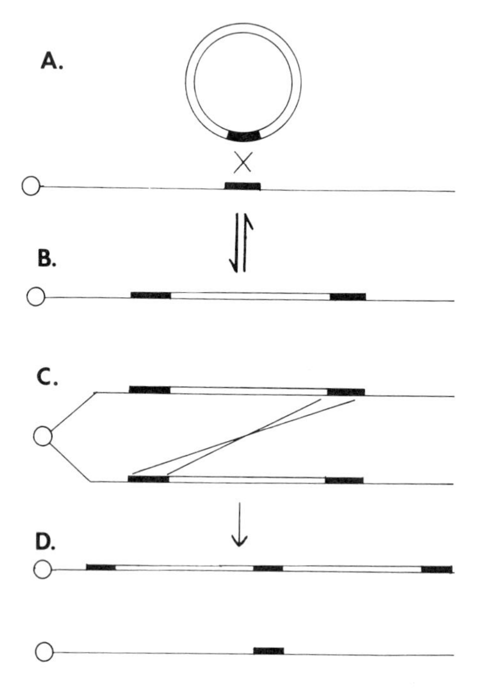

FIG. 3. Plasmid integration and loss in yeast. Plasmid integration by homologous recombination with the yeast genome (A) creates a duplication of the region of homology flanking the inserted plasmid sequences (B). Region of homology, filled bar; other plasmid sequences, open bar; yeast chromosome, thin line; centromere: small circle. Plasmid loss may occur by excision (B to A) or by unequal sister chromatid exchange (C to D). An unequal exchange (C) results in the loss of the inserted plasmid sequences in one segregant and their duplication in the other (D).

[12] T. D. Petes, *Cell* **19,** 765 (1980).

[13] J. W. Szostak and R. Wu, *Nature* (*London*) **284,** 426 (1980).

[14] J. A. Jackson and G. R. Fink, *Nature* (*London*) **292,** 306 (1981).

TABLE II
FREQUENCY OF DRUG-RESISTANT RECOMBINANTS

Transformant	Frequency of recombinants[a]			
	tcm^r	*his*$^+$*tcm*^{r}	*cyh*^{r}	*leu*$^+$*cyh*^{r}
T400	4.8×10^{-3}	9.2×10^{-6}	—	—
K5	1.8×10^{-3}	7.2×10^{-4}	—	—
T972	1.8×10^{-3}	9.2×10^{-8}	6.9×10^{-5}	5.1×10^{-6}
T948	3.5×10^{-3}	2.6×10^{-6}	2.6×10^{-4}	$\sim 2 \times 10^{-4}$

[a] Frequencies were determined as described in Methods.

formant T400 (Table I), which contains a single copy of pSZ414 integrated at the rDNA locus, exhibits just such behavior (see Table II). Trichodermin-resistant (*tcm*^{r}) segregants of T400 arise at a frequency of 5×10^{-3}, but *his*$^+$*tcm*^{r} segregants occur at less than 10^{-5}.

The frequency of drug-resistant recombinants is affected by the plasmid integration site. T400 and K5 are a pair of isogenic strains in which plasmid pSZ414 is integrated at the rDNA locus and at the *HIS3* locus, respectively. Plasmid loss from the rDNA locus is reproducibly three times more frequent than from the *HIS3* locus. T972 contains pSZ414 integrated at the rDNA locus and pSZ430 integrated at *URA3;* in this strain, pSZ414 is lost 20 times more frequently than pSZ430. Strain differences in excision frequency also occur; e.g., there is almost a 3-fold difference in the frequency of loss of pSZ414 from the rDNA locus in T400 compared to T972.

When a plasmid suitable for negative selection is integrated at the drug-resistance locus, a high frequency of prototrophic drug-resistant recombinants arise, probably by gene conversion between the direct repeats generated by plasmid integration. T948 contains plasmid pSZ430 integrated at the *cyh2* locus; the majority of cycloheximide resistant recombinants are leu$^+$ and have not lost the plasmid. Prototrophic drug-resistant recombinants can also arise at high frequency when a negative selection plasmid is integrated at the site of the positively selected gene. Transformant K5 contains pSZ414 integrated at *HIS3;* 40% of the tcm^r recombinants are his$^+$. These recombinants probably arise by loss of the plasmid in such a way that the *HIS*$^+$ allele of the plasmid is retained. Transformants in which the plasmid has integrated at the desired locus rather than at sites homologous to selective or drug-sensitivity markers are easily distinguished by the low frequency of prototrophic drug-resistant recombinants.

Additional genetic analysis is necessary to verify that the transformed strain is still haploid and to confirm the location of the integrated plasmid. During polyethylene glycol-mediated transformation,[7] cell fusions can occur that result in a/a or α/α diploids. These fusion diploids, which constitute approximately 10% of the tranformants, mate normally to a or α haploids, but the resultant triploids show poor spore viability (<10%). The genetic analysis of crosses with the transformant also confirms that the integrated plasmid carries functional and linked genes for the selectable and dominant sensitivity markers. If the transformant is mated to a strain with an identifiable marker at the locus in which the plasmid was integrated, the plasmid markers will segregate in opposition to that marker.

Southern blot analysis[15] of DNA from yeast transformants is required to determine the copy number of an integrated plasmid. A single-copy plasmid insertion contains integrated plasmid sequences flanked by a duplication of the sequences at which integration occurred (Fig. 3). Digestion of genomic DNA from a transformant using an enzyme that cuts once within the plasmid sequences generates two fragments with homology to the plasmid; each fragment spans one copy of the repeated sequences. Transformants containing multimeric inserts possess, in addition to these junction fragments, a characteristic third fragment the size of the linear form of the integrated plasmid. Southern blot analysis may also be used to determine the site of plasmid integration. If a detailed restriction map of the region of insertion is available, the location of the integrated plasmid may be inferred from the sizes of the junction fragments observed on the blot.

Discussion

The Construction of Other Negative Selection Systems

We have discussed two independent systems that can be used to select for the loss of an integrated plasmid. We have described two plasmids, pSZ414 and pSZ430, that we have used in our investigations of plasmid loss. Similar plasmids suitable for negative selections in other strains may be readily constructed. A generalized plasmid for negative selection includes a selectable marker, a site for integration, and a dominant drug sensitivity marker (Fig. 2). We have isolated the *TCM*^{s} and *CYH*^{s} markers on DNA fragments suitable for recloning in other plasmids (Fig. 1). Haploids in which these markers may be used for negative selection are easily

[15] E. M. Southern, *J. Mol. Biol.* **98**, 503 (1975).

isolated in almost any strain background by selecting for spontaneous drug-resistant mutants. All mutations that we have examined that confer resistance to 20 μg of trichodermin or to 10 μg of cycloheximide per milliliter fall within the *tcm1* or *cyh2* loci, respectively. Diploids suitable for use in negative selection schemes are constructed by mating independently isolated drug-resistant haploids. The principle of negative selection may thus be adapted to a variety of problems.

The frequency with which drug-resistant segregrants are obtained from a newly constructed negative selection system will depend not only on the site at which the plasmid is integrated, but also on the strain background and the composition of the plasmid itself. The frequency of drug-resistant segregants may vary quite widely: *tcm*^{r} segregants from K5 are about 100 times as frequent as *cyh*^{r} segregants from T972 (Table II). Both of these transformants contain single-copy plasmids inserted at unique chromosomal DNA sequences. Interchromosomal gene conversions of the plasmid drug sensitivity marker from the chromosomal drug-resistance alkele result in drug-resistant recombinants that retain the selectable marker. The frequency of interchromosomal gene conversion (*his*$^+$*tcm*^{r} segregants of T400 or *leu*$^+$*cyh*^{r} segregants of T972) is less than 10^{-5} (Table II). Although the frequency of drug-resistance recombinants varies greatly between these two transformants, the frequency of interchromosomal gene conversion is approximately the same. It is advisable to characterize any newly constructed transformant to be used in a negative selection system to determine the proportion of drug-resistant segregants that is due to plasmid loss.

Other Dominant Sensitivity Markers

Other dominant drug-sensitivity markers such as *CAN*^{s} or *LYS2* could also be used in vectors for negative selection.[4,16] In principle, tRNA suppressors may also be used. Suppressors, or *SUP* markers, confer dominant sensitivity to high osmotic pressure, inhibiting growth on media containing 2.5 *M* ethylene glycol. However, there are several problems with the use of *SUP* markers for negative selections. First, many yeast strains are osmotically sensitive even in the absence of suppressors; this sensitivity can be eliminated only by conventional crosses to truly resistant strains. Second, many yeast strains contain ψ (psi), a cytoplasmic factor that enhances the effectiveness of suppressors.[17,18] The strain must be

[16] J. R. Broach, J. N. Strathern, and J. B. Hicks, *Gene* **8,** 121 (1979).

[17] B. S. Cox, *Heredity* **20,** 505 (1965).

[18] B. S. Cox, *Heredity* **26,** 211 (1971).

cured of $\psi^{19,20}$ for transformation with a suppressor to work. Even in ψ^- strains, it is difficult to recover transformants with a strong suppressor such as *SUP3*.[21] Weaker suppressors, such as *SUP11*, are less deleterious; however transformation with *SUP11* increases the proportion of fusion diploids among the transformants to greater than 90%.[22] We have found drug-sensitivity markers more convenient to use in negative-selection systems.

Applications

Negative selection systems such as those we have described can be used to facilitate any technique in which loss of an integrated plasmid is desirable. Almost any haploid strain suitable for yeast transformation can be adapted for a negative-selection system; spontaneous drug-resistant mutants are easily isolated from patches replica-plated to drug-containing plates. Plasmids containing suitable markers can be constructed using standard techniques. We discuss four applications of the negative selection system: gene transplacement,[5] genetic mapping, the isolation of mutations, and the analysis of recombination.

Transplacement. The gene transplacement technique developed by Scherer and Davis[5] substitutes a cloned DNA sequence for the corresponding sequence in the yeast genome. This allows the *in vivo* analysis of the effects of *in vitro* alterations in the DNA sequence. Transplacement requires the integration of a plasmid containing an altered sequence at the homologous site in the yeast genome. Subsequent plasmid loss leaves either the altered or the wild-type sequence in the genome. Strains that have undergone plasmid loss must be tested to discover whether they retain the altered sequence. The isolation of strains that have lost the integrated plasmid is frequently accomplished by screening for the loss of a positively selectable marker. The low frequency of plasmid loss, especially when the plasmid is inserted at a unique gene (see T972, Table II), renders this process quite tedious. Use of the negative-selection technique simplifies the isolation of such strains. Negative selection is particularly useful when the use of UV irradiation or other methods to promote plasmid loss is undesirable, as would be the case with radiation-sensitive (rad) strains.

Mapping. The discovery that extensive aneuploidy is tolerated in diploid yeast has led to the use of mitotic chromosome loss for genetic map-

[19] A. Singh, C. Helms, and J. Sherman, *Proc. Natl. Acad. Sci. U.S.A.* **76,** 1952 (1979).

[20] M. F. Tuite, C. R. Mundy, and B. S. Cox, *Genetics* **98,** 691 (1981).

[21] R. J. Rothstein, personal communication, 1981.

[22] K. Struhl, personal communication, 1981.

ping.[23] Loss of a chromosome carrying a dominant trait in a heterozygous diploid permits the exposed recessive marker to be expressed. This observation has been used to map genetic loci to specific chromosomes after chromosome loss. *Rad52* mutants,[24] *cdc6* and *cdc14* mutants,[25] or methylbenzimidazole carbamate[26] may be used to induce chromosome loss for mapping. If the gene to be mapped has been cloned, then a modification of the negative selection technique can be used in conjunction with any of these chromosome loss schemes to simplify the mapping.

The plasmid constructed for this selection must include the gene to be mapped and the other positively and negatively selectable elements described in Fig. 2. This plasmid is introduced into a suitable drug-resistant haploid by yeast transformation, with integration targeted to the unmapped locus.[8,11] The transformant should be characterized as we have described to ensure that the plasmid has not integrated at some other site. A transformant carrying the integrated plasmid at the unknown site is then mated to appropriately marked tester strains carrying recessive markers on each chromosome for mapping and the recessive drug-resistant marker required for the negative selection. Loss of the chromosome containing the plasmid integrated at the unmapped locus in the diploid results in the expression of drug resistance. The recessive markers consistently unmasked in the drug-resistant segregants reside on the same chromosome as the unmapped locus.

Isolation of Mutations. The plasmids we have constructed for use in negative selections may be useful in the isolation of severely detrimental or recessive lethal mutations. A haploid strain with an integrated plasmid carries two copies of any gene on the plasmid. Mutations in the chromosomal copy of such a gene will be complemented by the plasmid copy of the gene; loss of the plasmid, however, will result in expression of the mutant phenotype. A transformant containing a recessive lethal mutation in a chromosomal gene that is complemented by the plasmid will appear to be unable to lose the plasmid; a detrimental chromosomal mutation will result in the appearance of slow growing recombinants. Plasmids with negative selection can therefore be used to screen for mutations in any cloned region of DNA.

Analysis of Recombination. The negative-selection system allows the isolation of mutants altered in the ability to lose the integrated plasmid.

[23] E. M. Parry, and B. S. Cox, *Genet Res.* **16,** 333 (1970).

[24] R. K. Mortimer, R. Contopoulou, and D. Schild, *Proc. Natl. Acad. Sci. U.S.A.* **78,** 5778 (1981).

[25] G. Kawasaki, personal communication, 1979.

[26] J. Wood, personal communication, 1981.

We are studying such mutants in an attempt to elucidate the mechanics of homologous recombination and unequal sister chromatid exchange.

There exist many variations on the negative selection for plasmid loss that may be applied in other situations. The method promises to be a powerful tool for genetic manipulation.

Acknowledgments

We wish to thank R. Fishel for assistance in plasmid construction and S. Phillips for help in preparing the manuscript. P. A. B. was supported by postdoctoral fellowship 5F32 GM07283 from the National Institutes of Health. This work was supported by ACS Grant NP-357 to J. W. S.

[19] Use of Integrative Transformation of Yeast in the Cloning of Mutant Genes and Large Segments of Contiguous Chromosomal Sequences

By JOHN I. STILES

A number of convenient methods for cloning yeast genes are now available. In this chapter I describe a method that has been developed for rapidly cloning a defined region of the yeast genome by retrieval of a site-specific integrated vector. The system to be described has been developed for a region of chromosome that includes the iso-1-cytochrome *c* gene, the *CYC1* genetic locus, and adjacent sequences.[1,2] Analogous methods have been developed in other laboratories, indicating that this approach is general and can be applied to other regions of the yeast genome.[3]

The basis for this approach is the ability to transform yeast by integration of the transforming vector into the yeast chromosomal DNA at sites of homology between the yeast chromosome and the vector. This allows the placement of a vector adjacent to a particular gene or region of the genome by including the appropriate homologous sequence on the vector. If the vector also includes sequences for selection and replication in *E. coli*, the region adjacent to the integration site can be easily cloned by digesting total genomic DNA from the transformed yeast with a restriction endonuclease that does not cut the vector or between the vector and

[1] J. I. Stiles, J. W. Szostak, A. T. Young, R. Wu, S. Consaul, and F. Sherman, *Cell* **25,** 277 (1981).

[2] J. F. Ernst, J. W. Stewart and F. Sherman, *Proc. Natl. Acad. Sci. U.S.A.* **78,** 6334 (1981).

[3] G. S. Roeder and G. R. Fink, *Cell* **21,** 239 (1980).

ISBN 0-12-182001-7

the chromosomal sequence of interest. The DNA fragments are then ligated under conditions that favor intramolecular events. This produces circles from the linear genomic restriction fragments that are used to transform *Escherichia coli*. Transformants are selected using the marker present on the original vector. Since the vector has integrated into the yeast genome specifically at a sequence of interest, the plasmids recovered into *E. coli* should all contain the desired yeast sequences. In addition, large segments of genome flanking the integration site can be recovered by partial restriction endonuclease digestion of the transformed yeast genomic DNA. While this method requires the construction of a vector specific for the region of the yeast genome of interest, it eliminates the need for construction and screening of a random gene bank and simplifies the subsequent verification of clones. This approach is especially convenient if the same chromosomal region is to be cloned repeatedly, as might be the case in the analysis of a number of mutations of a particular locus. This method also simplifies the cloning and characterization of large regions of adjacent chromosomal DNA, since it eliminates the need for the construction and repeated screening of overlapping banks. It also alleviates the associated problems of small fragments and small overlaps giving weak hybridization signals and of reiterated sequences giving numerous false-positive clones.

Principle of the Method

It is most convenient if the vector can be made to integrate outside of, but immediately adjacent to, the sequence of interest. Since the vector integrates by homologous recombination, which results in the production of a duplication of the homologous sequence region, it is preferable that this structure occur outside of the region of interest. An additional point that should be considered is maximizing for integration at the target site. Several factors can affect the probability of integration at a specific site. The presence of other yeast sequences, such as the yeast transformation marker, on the transformation vector can result in recombination at nontarget areas of the genome. The effect of these sequences can be limited by using strains containing deletions or other types of mutations that suppress integration. Also, it appears that the frequency of integration at a particular site is at least somewhat proportional to the length of the homologous sequence so that minipulation of sequence lengths can give some direction to the integration process. In the example shown in Fig. 1, the chromosomal target sequence is about 1.8 kb in length and the *URA3* gene sequence about 1 kb in length. In transformation of strains containing the *ura3-52* mutation, about 80% of the transformants contain the vec-

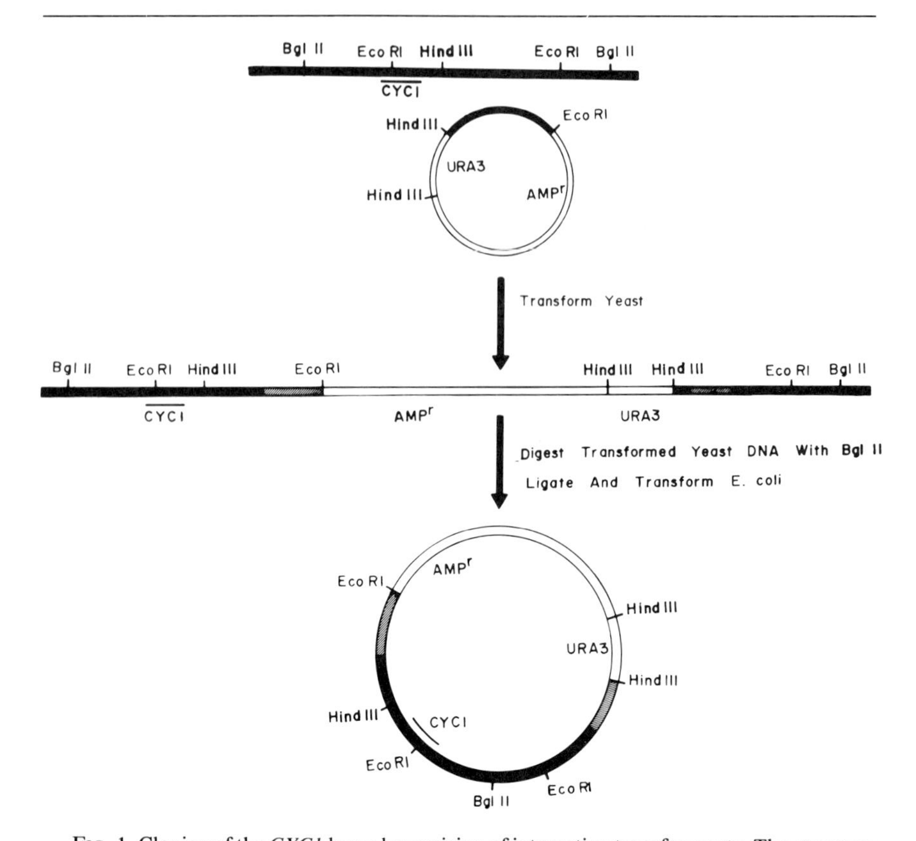

FIG. 1. Cloning of the *CYC1* locus by excision of integrative transformants. The structure of the DNA in the region of the *CYC1* locus is shown at the top of the figure. The cross-hatched area of the plasmid, bounded by a *Hin*dIII site and an *Eco*RI site, is homologous to the chromosomal DNA immediately above the plasmid. Integrative transformation resulting from homologous recombination results in the chromosomal structure shown in the center of the figure. Digestion and intramolecular ligation of chromosomal DNA from the transformed yeast results in the plasmid shown at the bottom.

tor integrated at the target site on the chromosome. Increased direction of integration can also be achieved by cleavage of the vector in the sequence homologous to the target site. The free ends that are generated are increased in recombinagenic potential.[4,5] This approach is most easily achieved by designing the vector with a unique restriction site in the homologous sequence area.

[4] T. L. Orr-Weaver, J. W. Szostak, and R. J. Rothstein, *Proc. Natl. Acad. Sci. U.S.A.* **78,** 6354 (1981).

[5] T. L. Orr-Weaver, J. W. Szostak, and J. R. Rothstein, this volume [14].

For repeated cloning of a small region, such as a single gene, it is most convenient if a vector can be constructed that can be recovered by complete digestion of the transformed genomic DNA using a restriction endonuclease that produces cohesive ends. A specific example of this arrangement is shown in Fig. 1. Here the vector, the gene of interest (*CYC1*) and the sequences between are free of recognition sites for *Bgl*II, so that digestion of genomic DNA from the transformed yeast strain with this restriction endonuclease will produce a single fragment containing both the vector and the *CYC1* gene. And, this fragment will have cohesive ends. This specific fragment can then be recovered from the pool of genomic fragments by ligation using conditions that favor intramolecular ligation events followed by transformation of *E. coli* with selection for the marker present on the original vector.

Larger regions adjacent to the integration site can be recovered by partial restriction endonuclease digestion. Using this approach regions of over 30 kb have been obtained. In addition, if clones are recovered using

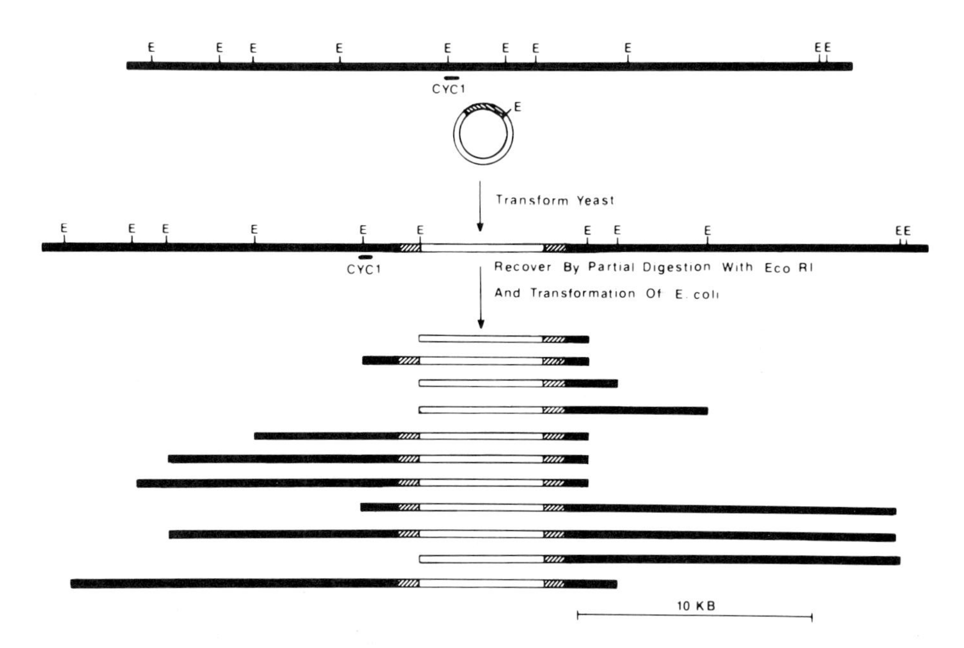

FIG. 2. Cloning of extended regions encompassing the *CYC1* locus. The *Eco*RI restriction sites from the chromosomal region surrounding the *CYC1* locus are shown at the top of the figure. Transformation occurs as shown in Fig. 1. The chromosomal structure of the *CYC1* region in the transformed strain is shown in the center of the figure. The sequences present in various plasmids recovered by partial *Eco*RI digestion are shown at the bottom.

varying degrees of digestion, a series of plasmids carrying genomic DNA of increasing size can be isolated. If representatives of each size class of plasmid are subjected to complete digestion and analyzed for the presence of specific fragments, the order of each fragment along the chromosome can be easily deduced. An example of this approach is shown in Fig. 2.

Example of the Approach

Excision of integrative transformants has been used extensively in cloning alleles of the *CYC1* locus as well as flanking regions.[1,2,6] A segment of DNA that lies just distal to the *CYC1* locus was inserted into the *Escherichia coli* plasmid pBR322 as a 1.8 kb *Hin*dIII–*Eco*RI fragment. Subsequently a *Hin*dIII fragment containing the yeast *URA3* gene was recovered from YIp31 and inserted into the *Hin*dIII site to create the vector shown in Fig. 1.[7] This vector is capable of growth and selection in *E. coli*, since it contains the origin of replication and the ampicillin-resistance gene from pBR322. In addition, the yeast *URA3* gene serves as a selectable marker for transformation of yeast.

Using standard yeast genetic techniques, the *CYC1* allele of interest is crossed with a strain containing the *ura3-52* allele. The *ura3-52* mutant allele has been chosen because it does not readily revert and the recovery of integrative transformants at this locus is quite low. Haploid progeny are selected that contain both the *CYC1* allele of interest and the *ura3-52* allele and then transformed with the vector. Transformants are selected on the basis of uracil prototrophy. The transformants are then screened genetically to test for linkage between the *URA3* gene and the *CYC1* locus. In normal strains the *URA3* and *CYC1* loci are not genetically linked, however if the plasmid containing the *URA3* gene has integrated adjacent to the *CYC1* gene, then they will show tight genetic linkage. Alternatively, restriction mapping of the genomic DNA from the transformed yeast strain can confirm that integration is at the expected site. This is especially useful if a detailed restriction map of the region is available or if genetic analysis is not convenient.

Clones containing the *CYC1* locus are recovered by digestion with *Bgl*II. Genomic DNA from the transformed strain is digested with *Bgl*II and ligated under dilute conditions that favor intramolecular ligation events. *Escherichia coli* are then transformed with this DNA. Colonies containing the vector sequence and additional flanking material to the

[6] J. I. Stiles, L. R. Friedman, and F. Sherman, *Cold Spring Harbor Symp. Quant. Biol.* **45,** 602 (1981).

[7] D. Botstein, S. C. Falco, S. E. Stewart, M. Brennan, S. Scherer, D. T. Stinchcomb, K. Struhl, and R. W. Davis, *Gene* **8,** 17 (1979).

nearest *Bgl*II restriction sites in the yeast chromosomal DNA, which in this case includes the *CYC1* gene, are then recovered by selecting for ampicillin-resistant transformants. These steps are summarized in Fig. 1.

Extended regions adjacent to the integration site for the vector have been recovered by partial digestion with *Eco*RI.[6] DNA from a transformed yeast strain is digested for various periods of time using a limiting amount of *Eco*RI. After ligation, plasmids are recovered as previously described. These transformants consist of a collection of plasmids of discrete sizes. Analysis of representatives of each size class by complete *Eco*RI digestion shows that each different class contains a particular number of specific *Eco*RI fragments. By observing the order of appearance of a specific fragment as a function of the total number of fragments present in the plasmids, it is possible to order the fragments in respect to the chromosomal integration site of the vector. For example, two classes of plasmids containing one *Eco*RI fragment in addition to the complete digest vector fragment, one extending in each direction from the integration site (second and third plasmids shown at the bottom of Fig. 2) are found. Plasmids having more *Eco* RI fragments exist as extensions in one or both directions from the insertion site such that the order of appearance of a specific fragment defines its position on the chromosome. An example of this type of analysis is shown at the bottom of Fig. 2.

Experimental Procedures

Transformation of Yeast

Transformation of yeast is carried out by a slight modification of the procedure of Hinnen *et al.*[8] The yeast culture is grown to a density of 1×10^7 to 2×10^7 cells/ml in YPD. YPD contains 1% yeast extract, 2% peptone, and 2% dextrose. The cells are collected by centrifugation at 4000 *g* for 5 min, washed in 25 ml of 1 *M* sorbitol, and again collected by centrifugation. The pellet is resuspended in 25 ml of 1 *M* sorbitol, and 0.25 ml of 1 mg/ml Zymolyase 60,000 (Miles Laboratories) is added. This suspension is incubated at 30° with gentle shaking until spheroplasts have formed. This generally takes from 30 to 60 min and can be checked by diluting a drop of cells into 10% sodium dodecyl sulfate (SDS). When formation of spheroplasts has occurred, the SDS will cause lysis of the cells and the suspension will clear. Alternatively, the cells can be observed under the microscope while a drop of SDS is added. A quantitative estimate of spheroplast formation can be obtained by diluting an aliquot of

[8] A. Hinnen, J. B. Hicks, and G. R. Fink, *Proc. Natl. Acad. Sci. U.S.A.* **75,** 1929 (1978).

the cells before and after spheroplasting into sterile water and then plating on YPD. The low osmotic strength of the dilution will lyse the spheroplasted cells but will not lyse the intact cells.

When spheroplast formation is complete the cells are collected by centrifugation at 1000 *g* for 5 min. The spheroplasts are washed by gently suspending in 20 ml of 1 *M* sorbitol and then in 20 ml of 1 *M* sorbitol, 10 m*M* Tris (pH 7.5), and 10 m*M* $CaCl_2$ (STC). After each wash the cells are collected by centrifugation at 1000 *g* for 5 min.

The cells are suspended in 1 ml of STC and divided into two equal aliquots, each of which is placed in a sterile 15-ml centrifuge tube. Up to 50 μl of vector DNA (1–10 μg) is added to each tube. The tubes are gently rotated to mix the cells and the DNA and then incubated at room temperature for 10 min. Next, 4.5 ml of a solution containing 40% polyethylene glycol 6000 (Sigma Chemical Company) and 10 m*M* $CaCl_2$ is added. After gently mixing, the tubes are incubated for 10 min at room temperature. The spheroplasts are pelleted by centrifugation at 1000 *g* for 5 min, resuspended in 5 ml of STC, and then added to 500 ml of regeneration agar that has been melted and kept liquid by incubation at 45°. Regeneration agar consist, per liter, of 182 g of sorbitol, 30 g of agar, 6.7 g of yeast nitrogen base without amino acids (Difco), and any necessary nutrients except that used for the selection of the transformed cells. After mixing, the cells are rapidly poured into petri dishes. The cells should remain at 45° for as brief of a time as possible, as high temperature decreases the regeneration frequency. It is sometimes advantageous to cool the regeneration agar slightly just before addition of the cells to limit the damage due to the high temperature. An aliquot of the transformation mix is also plated in complete regeneration media to test for regeneration of the spheroplasts. The plates containing the transformed spheroplasts are incubated at 30° for 3–5 days. If the relative humidity is low, it may be necessary to enclose the plates in a plastic bag to prevent excessive dehydration during this time. Transformed colonies should be subcloned on omission media specific for the transformation marker, uracil in the example shown in Fig. 1 and 2.[9]

Extraction of DNA from Transformed Yeast

Yeast DNA is extracted using a slight modification of the procedure of Roeder (personal communications). A 50-ml culture of the transformed strain is grown in YPD to a density of 1×10^8 to 2×10^8 cells/ml. The cells are collected by centrifugation at 4000 *g* for 5 min and resuspended in 3.5 ml of a solution containing 1 *M* sorbitol, 0.1 *M* EDTA (pH 8.0), and

[9] F. Sherman, G. R. Fink, and C. W. Lawrence, "Methods in Yeast Genetics." Cold Spring Harbor Laboratory, Cold Spring Harbor, New York, 1974.

14 m*M* 2-mercaptoethanol. The cells are spheroplasted by the addition of 0.1 ml of 2.5 mg/ml Zymolyase 60,000 and incubation at 37°. Generally 1 hr is sufficient to achieve a high level of spheroplast formation. The spheroplasts are collected by centrifugation at 4000 *g* for 5 min and then suspended in 3.2 ml of a solution of 10 m*M* Tris and 1 m*M* EDTA (TE). When the cells are suspended, 0.32 ml of 0.5 *M* EDTA (pH 8.0), 0.16 ml of 2 *M* Tris base, 0.16 ml of 10% SDS, and 10 μl of diethyl pyrocarbonate are added. This solution is mixed by inverting the tube and is then incubated at 65° for 30 min. After the addition of 0.8 ml of 5 *M* potassium acetate, the solution is placed on ice for 1 hr and then centrifuged at 45,000 *g* for 10 min at 0°. The supernatant is poured off into a 30-ml centrifuge tube and the DNA is precipitated by the addition of 12 ml of ethanol at room temperature. The DNA is collected by centrifugation at 15,000 *g* for 15 min. The DNA is resuspended in 5 ml of TE and 7.74 g of CsCl are added. After the CsCl has dissolved, the volume is adjusted to 9.8 ml with water, and 0.2 ml of 10 mg of ethidium bromide per milliliter is added. The sample is centrifuged at 50,000 rpm in a Beckman type 70-Ti rotor for at least 18 hr. The band containing the DNA is visualized by UV illumination and recovered by puncturing the side of the centrifuge tube with a 16-gauge needle attached to a 1-ml syringe. The sample is diluted with 1.5 volumes of water and extracted with equal volumes of isoamyl alcohol until the organic phase is clear and then twice more. The DNA is precipitated by the addition of 2.5 volumes of ethanol with incubation at −20° overnight. It should be noted that incubation at a lower temperature will result in the precipitation of CsCl, which will interfere with recovery of the DNA. The DNA is recovered by centrifugation at 45,000 *g* for 15 min, dried, and redissolved in 0.3 ml of TE. This solution is made 0.25 *M* in NaCl and reprecipitated by the addition of 2.5 volumes of ethanol. The purity of this DNA is sufficient for restriction endonuclease digestion and ligation.

Recovery of Integrative Transformants

Isolation of Mutant Genes by Complete Digestion One to 2 μg of the total genomic DNA from the transformed yeast strain is digested to completion with the appropriate restriction endonuclease in about 50 μl. *Bgl*II was used in the example shown in Fig. 1. The reaction is stopped by heating at 65° for 10 min. Water is added to give a final volume of 0.44 ml, and then 50 μl of 10× ligation cofactor mix and 10 μl of 25 m*M* ATP are added. The 10× ligation cofactor mix consists of 0.3 *M* Tris (pH 8.0), 40 m*M* $MgCl_2$, 12 m*M* EDTA, 0.1 *M* dithiothreitol, and 0.5 mg of bovine serum albumin (BSA) per milliliter. After the addition of 1.2 units of T4 DNA ligase, the reaction is incubated at 14° for at least 5 hr. This ligated

DNA is then used to transform *E. coli* cells made competent by Ca^{2+} treatment.[10] Approximately 30–60 transformants are obtained per microgram of genomic DNA from the transformed yeast strain.

Recovery of Extended Regions. Genomic DNA (20 μg) from the transformed yeast strain is digested with 5 units of restriction endonuclease in 100 μl. Ten-microliter aliquots are removed at various times from 0.5 min to 60 min and placed in tubes that have been preheated to 65°. Each sample is heated at 65° for 10 min to inactivate the restriction endonuclease and then ligated and recovered by transformation of *E. coli* as described above. Digestion times of 5–20 min give reasonable levels of plasmid containing multiple segments. Times shorter than 5 min generally result in few recovered plasmids, and times longer than 20 min give high levels of plasmids that are complete digest products.

Analysis of Clones

Plasmids from Complete Digests. Plasmids recovered from complete digests would be expected to be of a single size class. The size of the plasmid in several of the clones that are recovered is determined by agarose gel electrophoresis. Plasmid DNA is obtained from individual colonies using a rapid plasmid isolation technique such as that described by Davis *et al.*[11] Typically, 80% to 90% of the plasmids will be of one size, the size predicted for the complete digest product. Occasionally, larger plasmids are observed that generally result from partial digest products. Less frequently, small plasmids of unknown origin are found. A colony containing a plasmid of the majority size class is selected for further restriction analysis to confirm the identity of the clone.

Plasmids from Partial Digests. Plasmids recovered from the different restriction endonuclease digestion times are analyzed by the rapid plasmid screening method described above. Results from one experiment are shown in Fig. 3. The plasmids that are recovered fall into various discrete size classes depending on the number and composition of the fragments present. Plasmid DNA from one member of each size class is isolated and analyzed by complete restriction endonuclease digestion using the same enzyme that was used in the cloning step. The fragments are then separated by agarose gel electrophoresis. The order of appearance of a particular fragment is indicative of its order in the chromosome. Typical results of this type or analysis are shown at the bottom in Fig. 2.

[10] M. Mandel and A. Higa, *J. Mol. Biol.* **53,** 159 (1970).

[11] R. W. Davis, M. Thomas, J. Cameron, T. P. St. John, S. Scherer, and R. A. Padgett this series, Vol. 65 [49].

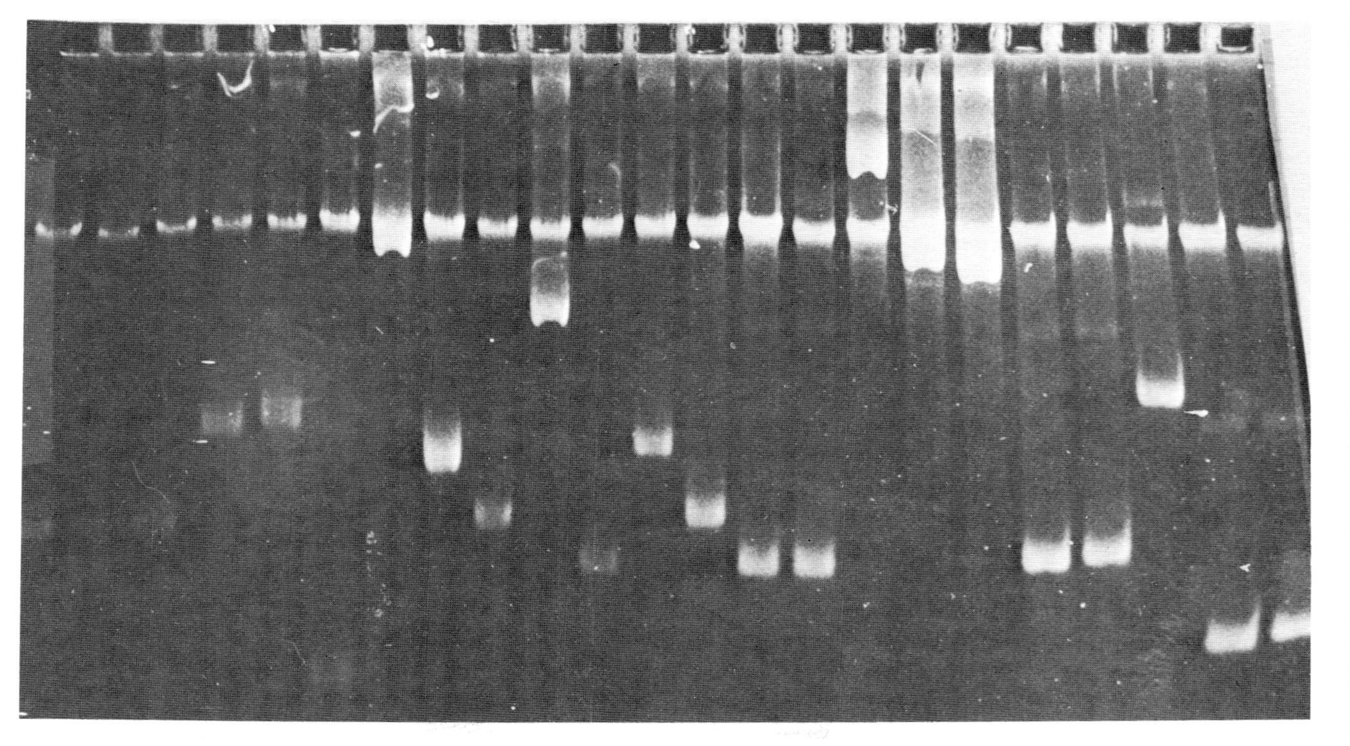

FIG. 3. Analysis of partial-digest plasmids. Plasmids recovered after partial *Eco*RI digestion have been analyzed by electrophoresis in 0.75% agarose gels. The first and second lanes from the left are plasmids recovered from a 0.5-min digestion. The third, fourth, and fifth lanes from the left are from a 10-min digestion, and the remaining lanes are from a 20-min digestion. The common band present in all lanes is the *Escherichia coli* chromosomal DNA.

Discussion

The procedures that have been discussed provide a rapid and reliable method for the repeated cloning of a specific segment of the yeast genome. It is also possible to easily recover extended regions of contiguous DNA. This approach is particularly useful if a large number of mutants of a particular locus are to be analyzed or if specific chromosomal regions from different strains are to be compared.

One problem which is often encountered is the variability in transformation frequency of laboratory strains of yeast. Certain strains transform at much lower frequencies than others. This problem can usually be overcome by increasing the amount of DNA used in the yeast transformation step. Another approach is to cut the transformation vector in the homologous region. The transformation frequency is increased 5- to 10-fold by cutting the vectors to generate free ends.[4,5] In addition, the specificity of the integration site is also increased.[4,5] An alternative approach is to include an origin of replication for yeast on the vector. This will increase the transformation frequency approximately 100-fold. These vectors will replicate extrachromosomally but can also integrate into a specific chromosomal site by homologous recombination. The episomally replicating plasmids are much less stable than those that have integrated, so that it is possible to select for integrated vectors by several rounds of growth on complete media followed by growth on selective media.

[20] Selection Procedure for Isolation of Centromere DNAs from *Saccharomyces cerevisiae*

By LOUISE CLARKE, CHU-LAI HSIAO, and JOHN CARBON

Hybrid plasmid derivatives of pBR322 that contain yeast (*Saccharomyces cerevisiae*) centromere DNA in combination with a yeast DNA replicator (*ars1* or *ars2*[1,2]) and a suitable genetic marker are stably maintained in yeast through both mitotic and meiotic cell divisions and exhibit typical Mendelian segregation (2+ : 2−) through meiosis.

The centromere DNA from yeast chromosome III (*CEN3*) was originally isolated by cloning the centromere III-linked genes *LEU2*[3] and

[1] D. T. Stinchcomb, K. Struhl, and R. W. Davis, *Nature* (*London*) **282,** 39 (1979).
[2] C.-L. Hsiao and J. Carbon, *Proc. Natl. Acad. Sci. U.S.A.* **76,** 3829 (1979).
[3] B. Ratzkin and J. Carbon, *Proc. Natl. Acad. Sci. U.S.A.* **74,** 487 (1977).

ISBN 0-12-182001-7

CDC10,[4] obtaining flanking regions by overlap hybridization,[5] and testing individual DNA segments from the *LEU2-CDC10* region that were carried on hybrid *ars*-containing plasmids for proper mitotic and meiotic segregation in yeast.[6] Similarly, the centromere from yeast chromosome XI (CEN11) has been isolated by cloning the centromere XI-linked gene *MET14* and assaying flanking regions for centromere function *in vivo*.[7] Obviously these methods are most successful for isolation of those centromeres that are closely adjacent to a known genetic marker.

Taking advantage of the unique mitotic stability of *CEN* hybrid plasmids (minichromosomes) in yeast, however, and the characteristic instability of other *ars* plasmids lacking centromere sequences,[6,8] we have developed[9] and present here a procedure that should permit the isolation of centromere DNA sequences of any yeast chromosome directly from a yeast recombinant DNA library.

Principle of the Method

Escherichia coli–S. cerevisiae shuttle vectors, such as YRp7, that contain a presumptive chromosomal replicator (*ars1*[1]) are unstably maintained in budding yeast cultures[1,8] and are almost completely lost after 10–20 generations of cell growth in nonselective media. However, when the vector contains functional yeast centromere DNA (*CEN3* or *CEN11*), the hybrid plasmid is maintained in up to 90% of the cells after 20 generations of nonselective growth.[6,7] The mitotic stabilization of plasmids in yeast induced by centromere DNAs is sufficient to permit their direct selection from yeast cultures transformed with genomic library DNA constructed in the vector YRp7.

Three alternative routes can lead to mitotic stabilization of plasmids in yeast. The first is integration of the plasmid into host chromosomal DNA; the second is enhanced stability of *ars* plasmids in diploid versus haploid yeast; and the third is the presence of cloned 2 μm (sometimes termed "2 μ") plasmid DNA sequences. These may readily be distinguished, however, by the methods described here.

[4] L. Clarke and J. Carbon, *Proc. Natl. Acad. Sci. U.S.A.* **77,** 2173 (1980).

[5] A. C. Chinault and J. Carbon, *Gene* **5,** 111 (1979).

[6] L. Clarke and J. Carbon, *Nature (London)* **287,** 504 (1980).

[7] M. Fitzgerald-Hayes, J.-M. Buhler, T. G. Cooper, and J. Carbon, *Mol. Cell. Biol.* **2,** 82 (1982).

[8] A. J. Kingsman, L. Clarke, R. K. Mortimer, and J. Carbon, *Gene* **7,** 141 (1979).

[9] C.-L. Hsiao and J. Carbon, *Proc. Natl. Acad. Sci. U.S.A.* **78,** 3760 (1981).

Materials and Techniques

Commonly employed techniques for construction of genomic libraries, isolation and purification of DNAs, yeast and *E. coli* transformations, genetic manipulations of yeast, and preparation of culture media have all been described in this volume and in (Vol. 68) and in references cited in footnotes 2–11.

Method

Selection of Mitotic Stabilizing Sequences

Five micrograms of hybrid plasmid DNA from a yeast genomic library consisting of partial *Sau*3A restriction fragments of total yeast DNA ligated into the single *Bam*H1 site of the *E. coli*–yeast shuttle vector Yrp7 (*Trp1 ars1 ampicillin*^{R})[10] are used to transform yeast strain Z136-1-13C (a *trp1* strain) to Trp^{+}, selecting for the *TRP1*$^{+}$ marker on the vector. With sterile toothpicks, 500 of the fastest growing transformants are transferred to YEP (1% yeast extract, 2% peptone, 2% glucose; nonselective) agar plates in a grid array and grown overnight at 32°. The colonies are replica-plated onto fresh YEP plates and grown overnight for a total of at least five successive transfers. Finally the transformants are replica-plated onto yeast minimal (0.65% Difco yeast nitrogen base, 2% glucose) agar plates lacking tryptophan, and those colonies that are still strongly Trp^{+} after prolonged nonselective growth (5–10% of the original 500 clones) are colony purified and tested for mitotic stability of the *TRP1* marker.

Assay for Mitotic Stabilization

Individual putative stably transformed clones are grown nonselectively on YEP agar or in 10 ml of liquid YEP medium for 24 hr at 32°. Each culture is then streaked or spread for single colonies on YEP agar plates, and the resulting colonies are replica plated onto minimal plates with and without tryptophan. The relative numbers of Trp^{+} and Trp^{-} colonies are scored after overnight incubation at 32°.

For nearly all the cultures selected from the original 500, the fraction of cells remaining Trp^{+} after overnight growth on nonselective media should be roughly 60–100%, whereas the parent vector YRp7 is usually completely segregated from strain Z136-1-13C under identical conditions.

[10] K. A. Nasmyth and S. I. Reed, *Proc. Natl. Acad. Sci. U.S.A.* **77**, 2119 (1980).

Recovery of Mitotically Stable Hybrid Plasmids from Individual Yeast Clones

A crude DNA preparation (containing both chromosomal and plasmid DNA)[9] is isolated from each stably transformed yeast clone and used to transform *E. coli* strain JA228(W3110 $hsdM_K^+$ $hsdR_K^-$ *argH* Str^R), selecting for ampicillin resistance, a marker carried on the pBR322-derived Yrp7 shuttle vector. With approximately 85% of these preparations, large numbers of ampicillin-resistant transformants are obtained from which recombinant plasmids may be isolated and purified. The yeast clones whose DNAs yield no bacterial transformants presumably either contain an integrated or recombined *TRP1* gene or are Trp^+ revertants.

The set of hybrid plasmids now contained in *E. coli* may be screened directly by colony hybridization[5] for the presence of all or a portion of the 2 μm yeast plasmid sequences. Thus those plasmids stabilized in yeast by 2 μm DNA are identified and may be eliminated from the set.

Purified plasmid DNAs from the remainder readily transform Trp^- yeast to Trp^+ with high frequency ($\sim 10^4$ transformants per microgram), and all the transformants should be mitotically stable. These plasmids, therefore, contain cloned DNA segments capable in some way of stabilizing plasmid replication and maintenance in yeast during mitotic cell division.

Identification of CEN-Containing Plasmids by their Meiotic Segregation Pattern

Hybrid plasmids carrying cloned yeast centromere DNA sequences (minichromosomes) segregate as ordinary yeast chromosomes (2+ : 2−) through meiosis.[6] On the other hand, multiple copy number plasmids, specifically, 2 μm-derived vectors, would be expected to segregate 4+ : 0−. Therefore, plasmids carrying functional centromere sequences can be distinguished from 2 μm-derived vectors, which segregate predominately 4+ : 0−, and *ars* vectors lacking centromere sequences, which are unstable and are normally lost in the process of meiosis, by the following genetic manipulations.

The plasmid to be tested is contained in a yeast strain with a mutation in the gene for which there is a wild-type copy on the plasmid (for example, $trp1^-$ in the host genome and $TRP1^+$ on the plasmid). The strain is crossed with another strain of the opposite mating type that also contains a mutation in the gene whose wild-type allele is carried by the plasmid. In this way the plasmid may be followed easily through the cross by the presence of the wild-type marker. Either one or both strains in the cross should also contain at least one centromere-linked marker so that sister spores, the products of the second meiotic division, may be identified.

After mating, diploids are purified and sporulated; resulting asci are dissected on YEP agar plates for genetic analysis.[11] At least 10–15 tetrads having four viable spores are scored on appropriate minimal plates for all pertinent markers in the cross. If a plasmid carries functional centromere sequences, it will segregate as a chromosome in the first meiotic division and, as a consequence, is found predominantly in the two sister spores, which are the products of the second meiotic division. Centromere-containing plasmids are not completely stable, however, and segregate 0+ : 4− in 10–15% of the asci. The marker on the plasmid should behave as a centromere-linked gene, so that when the plasmid marker is scored versus another centromere-linked gene (such as *met14* on chromosome XI), principally parental ditype and nonparental ditype (but very few tetratype) asci are obtained in roughly equal numbers in the cross.

The table lists a number of stable plasmids isolated by the direct selection procedure.[9] Plasmids pCH4 and pCH25 have been shown by hybridization to contain *CEN3* DNA.[9] Plasmids pCH3, pCH9, and pCH10 are also presumably centromere-containing plasmids, since their behavior in genetic crosses is typical of that observed with minichromosomes carrying the previously identified centromere DNAs, *CEN3*[6] and *CEN11*.[7]

A hybrid plasmid would also exhibit 2+ : 2− segregation if it were integrated or recombined into one of the host chromosomes. If this were the case, however, the small proportion of 4+ : 0− and 0+ : 4− tetrads would not be obtained in crosses, nor would all integration events be expected to occur near a centromere. Although the reason for the small percentage of 4+ : 0− and 0+ : 4− tetrads seen in crosses involving CEN plasmids is still unclear, the pattern is characteristic of plasmids carrying *CEN3*,[6] *CEN11*,[7] and presumably other yeast centromeres.[9]

The table gives two examples of the behavior of 2 μm-derived plasmids (pCH21 and pCH27; both isolated by the direct selection procedure) in genetic crosses. As expected, these high copy number plasmids are predominantly distributed to all four progeny of the tetrads analyzed.

Biochemical Confirmation of Autonomous Replication of CEN Plasmids

Confirmation that prospective *CEN*-containing plasmids are replicating autonomously in yeast may be obtained by a routine Southern blot hy-

[11] F. Sherman, G. R. Fink, and J. B. Hicks, "Methods in Yeast Genetics;" (Laboratory Manual). Cold Spring Harbor Laboratory, Cold Spring Harbor, New York, 1979.

[12] E. M. Southern, *J. Mol. Biol.* **98**, 503 (1975).

MEIOTIC SEGREGATION OF STABLE YEAST PLASMIDS ISOLATED BY THE DIRECT-SELECTION PROCEDURE[a]

Plasmid in cross	Stabilizing DNA	Distribution in tetrads of *TRP1*$^+$ marker on plasmid[b]					Test for centromere linkage of marker on plasmid			Reference centromere marker
		4+ : 0−	3+ : 1−	2+ : 2−	1+ : 3−	0+ : 4−	PD	NPD	T	
pCH3	(*CEN*)	2 (13)	0	13 (87)	0	0	6	7	0	*met14*
pCH4	(*CEN3*)	7 (27)	0	17 (65)	0	2 (8)	4	13	0	*met14*
pCH9	(*CEN*)	2 (9)	1 (4)	13 (57)	0	7 (30)	4	9	0	*met14*
pCH10	(*CEN*)	2 (10)	0	17 (85)	0	1 (5)	8	7	2	*ura3*
pCH25	(*CEN3*)	0	1 (7)	12 (86)	0	1 (7)	5	7	0	*met14*
pCH21	(2 μm)	16 (100)	0	0	0	0				
pCH27	(2 μm)	19 (76)	0	1 (4)	0	5 (20)				

[a] Transformants in strain Z136-1-13C [*a trp1 ade1 leu1 gal1 arg4*/*pCH*(*TRP1*$^+$)] were crossed with strain X2928-3D-1C (*α trp1 leu1 his2 ura3 ade1 met14 gal1*). PD, parental ditype; NPD, nonparental ditype; T, tetratype.

[b] Values in parentheses represent percentage of total.

bridization experiment.[6,9,12] The presence of unintegrated plasmid is shown by hybridization of ^{32}P-labeled pBR322 DNA to Southern blots of fractionated restriction digests of total cell DNA isolated from plasmid-bearing progeny of selected crosses. For example, the cell DNA is predigested with a restriction enzyme that cleaves the plasmid once in the pBR322 vector region. A single radioactive band on the Southern autoradiogram corresponding in mobility to control plasmid DNA digested with the same enzyme indicates the presence of unintegrated plasmid in the cell and the absence of integrated copies of pBR322 DNA.

Similar experiments with 2 μm-derived plasmids yield a single band of much greater intensity, reflecting the high copy number of these autonomously replicating plasmids.[9]

Screening Clones for Haploidy Using Resistance to Canavanine

Standard procedures for the transformation of yeast in the presence of polyethylene glycol result in a significant proportion of diploid (a/a or α/α) transformants. Plasmid YRp7 (*TRP1 ars1*) and its derivatives are mitotically appreciably more stable in these diploid cells[8] and may be confused with stable *CEN* or 2μm-containing plasmids in the original selection. In order to avoid this problem, a simple screening procedure may be carried out on the yeast transformants to distinguish haploid from diploid colonies in the collection. For example, the fast growing clones from the original transformation are distributed in small patches on YEP plates, grown to confluency, and replica-plated onto minimal plates containing 60 μg of the amino acid antagonist canavanine per milliliter. (The plates cannot contain arginine, an inhibitor of canavanine action.) Mutations to canavanine resistance in yeast are recessive, thus haploid cells may readily be distinguished from diploids by a brief exposure to ultraviolet light, resulting in canavanine resistance at a much higher frequency in haploids. Using control haploid and diploid strains, UV exposure conditions are sought that yield good growth in patches of haploid clones and little or none in diploid patches. This test is also useful for identification of haploid (versus diploid) transformants prior to performing genetic crosses or studies of plasmid stability and copy number.

Although centromere-carrying plasmids have been identified with relative ease by the direct selection procedure without introducing the canavanine- resistance screen, the addition of this step does eliminate at the onset further analysis of apparently stable plasmids that turn out to be mitotically unstable when introduced into haploid cell.

Comments

The direct-selection procedure for isolation of functional centromere DNA by mitotic stabilization of *ars* plasmids has been used successfully in *S. cerevisiae* with cloned yeast DNA segments. The applicability of the method to other eukaryotic DNAs or host cells is untested and ultimately depends on the proper functioning and the mitotic behavior of *ars*-carrying plasmids in these systems.

[21] Construction of High Copy Yeast Vectors Using 2-μm Circle Sequences

By JAMES R. BROACH

This chapter addresses one aspect of the problem of expressing any cloned gene at very high levels in yeast. Since, as a first approximation, it is reasonable to assume a gene dosage effect for the expression of cloned sequences in yeast, maximizing expression will involve maximizing the number of copies of that gene in the cell. For reasons elaborated below, this in turn will undoubtably require propagating that gene on a vector derived from the yeast plasmid 2-μm circle. Thus in this chapter I discuss various aspects of the construction and use of 2-μm circle vectors for propagating cloned genes in yeast.

The yeast plasmid 2-μm circle is a 6318 bp double-stranded DNA species present at 60–100 copies per cell in most *Saccharomyces cerevisiae* strains.[1-3] Although the replication of this plasmid during normal exponential growth is strictly under cell cycle control, it can escape from this control in certain situations to increase its copy number from as low as a single copy per cell to its normal high level.[4,5] As discussed below, this capability is a property of the 2-μm circle origin of replication functioning in conjunction with several proteins encoded by the plasmid itself. As a consequence of this amplification potential, many hybrid plasmids constructed from 2-μm circle sequences can establish and maintain high copy number in yeast strains, following their introduction by transformation.[5-7]

[1] J. L. Hartley and J. E. Donelson, *Nature (London)* **286,** 860 (1980).

[2] C. P. Hollenberg, P. Borst, and E. F. J. Van Bruggen, *Biochim. Biophys. Acta* **209,** 1 (1970).

[3] G. D. Clark-Walker and G. L. G. Miklos, *Eur. J. Biochem.* **41,** 359 (1974).

[4] D. C. Sigurdson, M. E. Gaarder, and D. M. Livingston, *Mol. Gen. Genet.* **183,** 59 (1981).

[5] J. B. Hicks, A. H. Hinnen, and G. R. Fink, *Cold Spring Harbor Symp. Quant. Biol.* **43,** 1305 (1978).

[6] K. Struhl, D. T. Stinchcomb, S. Scherer, and R. W. Davis, *Proc. Natl. Acad. Sci. U.S.A.* **76,** 1035 (1979).

[7] C. Gerbaud, P. Fournier, H. Blanc, M. Aigle, H. Heslot, and M. Guerineau, *Gene* **5,** 233 (1979).

ISBN 0-12-182001-7

In addition, these hybrid plasmids containing 2-μm circle sequences are relatively stable during mitotic growth and meiosis. Although a number of specific yeast chromosomal sequences are also capable of promoting autonomous replication in yeast of plasmids in which they are incorporated[8-10]—a property that is presumably a reflection of their normal function in the cell as chromosomal origins of replication—the ability to establish and maintain high copy number and to propagate stably during mitotic growth is apparently unique to plasmids derived from 2-μm circle sequences.

At this point an unequivocal recommendation of the best possible 2-μm circle vector for achieving high copy number in yeast is not possible. First, our knowledge of the mechanism of 2-μm circle copy control is not complete. In addition, the influence on plasmid copy number—whether that plasmid is 2-μm circle or a hybrid cloning vector—of such factors as genetic background of the host yeast strain, the size of the vector, the presence on the plasmid of extra yeast or bacterial sequences, or the presence of other plasmids in the cell, has not been rigorously examined. Nonetheless, the components of the copy control system of 2-μm circle are reasonably well defined. On the basis of this information we can develop various strategies for using 2-μm circle sequences to maximize the copy number of a cloned gene in yeast. Thus, in this chapter, I discuss various approaches to the construction and use of 2-μm circle vectors that possess the potential for stable, high-copy propagation in yeast. By pursuing several of these avenues and assessing the copy number attained with each, one can optimize the synthesis of a cloned gene in yeast. Other aspects of the problem of maximizing expression of a particular gene in yeast—such as attachment to appropriate promoter and terminator sequences—are addressed in other chapters in this volume.

In the first section of this chapter I describe the structure of 2-μm circle and the features of the plasmid replication apparatus salient to a discussion of vector construction. I then describe hybrid 2-μm circle vectors suitable for use with yeast strains containing endogenous 2-μm circles (designated [cir^+] strains). There are some reasons to suspect, however, that one can achieve higher copy numbers by propagating appropriate hybrid 2-μm circle vectors in strains that lack 2-μm circles ([cir^0] strains). Thus, in the third section I provide a list of available [cir^0] strains, present a procedure for isolating [cir^0] derivatives of [cir^+] strains, and describe 2-μm circle vectors suitable for use in [cir^0] strains. Then I describe sev-

[8] D. T. Stinchcomb, K. Struhl, and R. W. Davis, *Nature* (*London*) **282,** 39 (1979).
[9] C. Chan and B.-K. Tye, *Proc. Natl. Acad. Sci. U.S.A.* **77,** 6329 (1980).
[10] D. Beach, M. Piper, and S. Shall, *Nature* (*London*) **284,** 188 (1980).

eral unusual vectors and suggest several strategies for possible additional improvements in copy levels. Finally, in the last section I describe several rapid procedures for determining copy number of hybrid plasmids in yeast.

In addition to using 2-μm circle vectors to optimize the copy number of cloned sequences, there are other contexts in the molecular biology of yeast in which hybrid 2-μm circle vectors play or can play a significant role. These uses of 2-μm circle vectors are described briefly in the following, and, for each procedure, a reference to the chapter in this volume in which it is discussed more fully is provided. First, 2-μm circle vectors have been useful for cloning yeast genes by complementation. That is, several hybrid plasmids that will be described in this chapter have been used as vectors for constructing libraries of random genomic sequences from yeast as well as from other organisms. Specific genes have been readily recovered from these libraries by transforming an appropriate yeast strain with the pooled plasmids and selecting for complementation of specific markers.[11,12] This technique was made possible by the high transformation frequencies obtainable with 2-μm circle vectors, the relative stabilities of such plasmids in yeast, and the ease with which such plasmids can be recovered from yeast into *E. coli*. This procedure for cloning genes by complementation is mentioned in the chapter by MacKay [22]. Second, various 2-μm circle vectors are useful in recovering specific alleles of a cloned gene from the yeast genome.[13] This procedure—which involves transformation by a 2-μm circle vector carrying sequences spanning the gene of interest, after linearizing the plasmid by restriction enzyme digestion at a site within the gene—is addressed in the chapters by Stiles [19] and Rothstein [14]. Finally, 2-μm circle plasmids can be used in a quite novel fashion to determine the chromosomal location of a cloned yeast fragment. This involves integrating 2-μm circle sequences at the chromosomal site homologous to the cloned fragment. High-frequency chromosomal breakage at this site, which results from the specialized recombination system encoded by 2-μm circle, permits a rapid genetic determination of the site of integration.[14]

The Copy Control System of 2-μm Circle

A diagram of the structure of 2-μm circle is presented in Fig. 1. The entire nucleotide sequence of the plasmid has been determined[1] and the

[11] J. R. Broach, J. N. Strathern, and J. B. Hicks, *Gene* **8,** 121 (1979).

[12] K. A. Nasmyth and S. I. Reed, *Proc. Natl. Acad. Sci. U.S.A.* **77,** 2119 (1980).

[13] J. N. Strathern, personal communication.

[14] C. Falco, Y.-Y. Li, J. R. Broach, and D. Botstein, *Cell* **29,** 573 (1982).

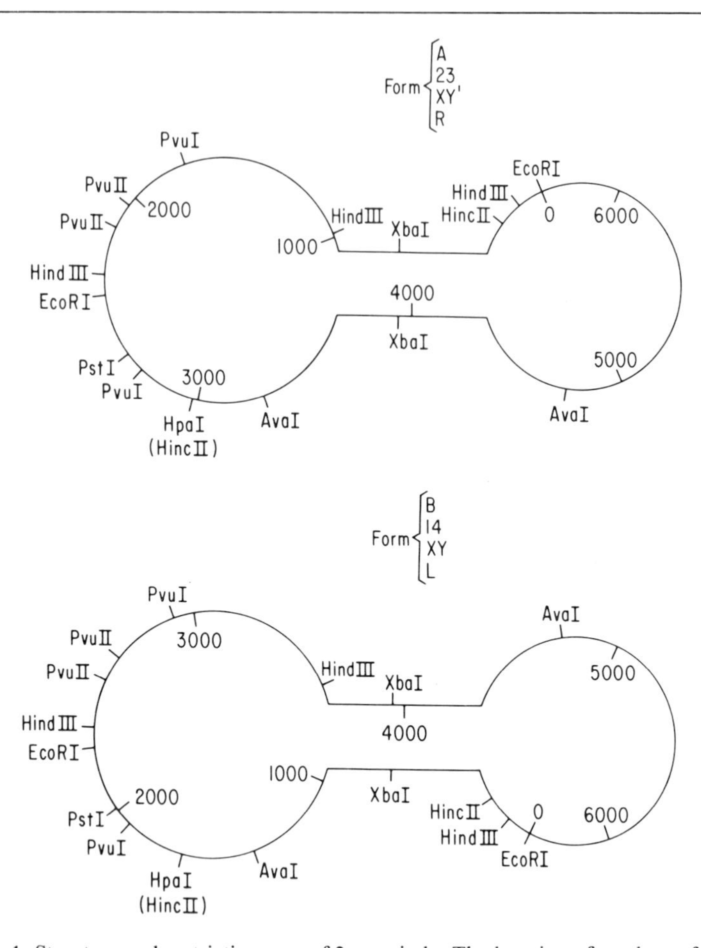

FIG. 1. Structure and restriction map of 2-μm circle. The location of a subset of restriction sites as determined from the published sequence of the plasmid are indicated on schematic diagrams of the two forms of 2-μm circle. Those terms variously used to designate the two forms are listed above each. The circular portion of each figure represents unique sequences, whereas the linear portion of the figures represents the repeated regions. The base-pair numbering system, shown inside each figure, is consistent with that used by Hartley and Donelson.[1]

cleavage sites for a subset of restriction enzymes as determined from the sequence are indicated in the figure. In confirmation of earlier observations on the structure of the plasmid, analysis of the sequence demonstrated that the plasmid contains two regions of 599 bp each, which are precise inverted repeats of each other and divide the molecule approximately into halves. Since recombination readily occurs between these repeated sequences, 2-μm circles isolated from yeast actually consists of a mixed population of two plasmids that differ only in the orientation of one unique region with respect to the other.[15] Both forms of the plasmid—which have been variously designated as A and B, R and L, XY′ and XY, or 23 and 14—are diagrammed in Fig. 1.

The salient features of the 2-μm circle replication system are indicated in Fig. 2. The origin of replication spans a region extending from the middle of one of the inverted repeats 100 bp into the adjacent unique sequences (approximately nucleotides 3650 to 4000 in the A form or nucleo-

FORM A

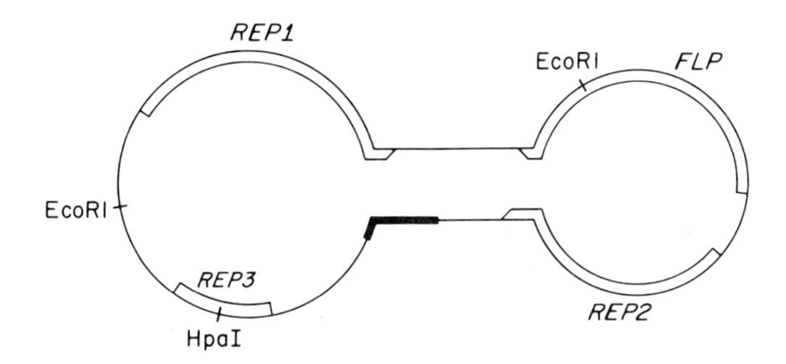

FORM B

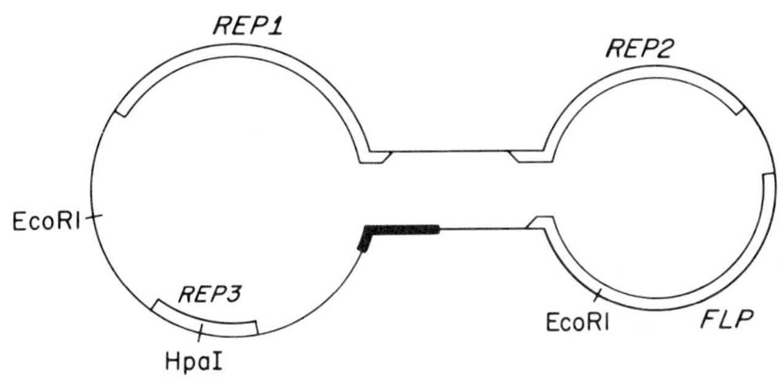

FIG. 2. Components of the 2-μm circle replication system. The locations of the origin of replication (filled line), the three known 2-μm circle genes (open double lines), and the cis-acting replication region (*REP3*) are shown on schematic diagrams of the two forms of the plasmid.

[15] J. D. Beggs, *Nature (London)* **275,** 104 (1978).

tides 650 to 1000 in the B form). This site was initially identified as the sole sequence within the 2-μm circle that would promote autonomous replication in yeast of hybrid plasmids into which it was inserted.[16] That this region actually functions *in vivo* as the primary 2-μm circle origin of replication was subsequently confirmed.[17,18]

Although autonomous replication is conferred by the 2-μm circle origin of replication, propagation at high copy number requires, in addition, two proteins encoded by the plasmid itself. Hybrid plasmids containing 2-μm circle sequences spanning only the origin of replication are present at low copy number in [cir^0] yeast strains and are lost rapidly during nonselective growth. However, in a [cir$^+$] strain, the same plasmids are present stably and at high copy number.[16] Similarly, hybrid plasmids containing the entire 2-μm circle genome are present stably and at high copy number even in [cir^0] strains. Thus, some trans-acting function encoded in a site or sites away from the origin of replication is necessary for high copy propagation of the plasmid in yeast. The genes encoding this function have been identified by genetic analysis of hybrid plasmids containing the entire 2-μm circle genome. Mutations that interrupt either of two plasmid genes, designated *REP1* and *REP2* in Fig. 2, abolish efficient, high copy maintenance of the plasmid.[19,20] We have demonstrated that high copy number and stability also require sequences between the *Pst*I and *Ava*I sites in the large unique region. This region is active only in cis with respect to the 2-μm circle origin and probably defines the site through which *REP* proteins act to promote amplification. We have designated this region *REP3*.[20] Thus, the requisite components for high-copy, stable propagation of a hybrid plasmid in yeast are the presence on the plasmid of a fragment from 2-μm circle spanning both the origin of replication and *REP3* and the presence in the cell of the *REP1* and *REP2* gene products. These proteins can be provided either by the endogenous 2-μm circles in the cell or by the hybrid plasmid itself.

Vectors for Use with [cir$^+$] Strains

Since [cir$^+$] strains contain a sufficient complement of 2-μm circle *REP* proteins, all that is required for high copy number propagation of a particular hybrid plasmid is the presence on that plasmid of sequences

[16] J. R. Broach and J. B. Hicks, *Cell* **21,** 501 (1980).

[17] C. S. Newlon, R. J. Devenish, P. A. Suci, and C. J. Roffis, *Initiation of DNA Replication, ICN–UCLA Symp. Mol. Cell. Biol.* **22,** 501, (1981).

[18] H. Kojo, B. D. Greenberg, and A. Sugino, *Proc. Natl. Acad. Sci. U.S.A.* **78,** 7261 (1981).

[19] J. R. Broach, V. R. Guarascio, M. H. Misiewicz, and J. L. Campbell, *Alfred Benzon Symp.* **16,** 227 (1981).

[20] M. Jayaram, Y.-Y. Li, and J. R. Broach, *Cell,* in press (1983).

spanning the 2-μm circle origin of replication and the *REP3* locus. In order to introduce the plasmid into yeast by transformation and assure its maintenance, the plasmid must also contain a gene that can be selected in an appropriate yeast strain. There are essentially two strategies for constructing such a high copy, selectable plasmid that also contains a particular gene of interest. Either the gene can be inserted into a preexistent plasmid vector, or sequences for selection and high copy number propagation can be inserted into a plasmid containing the gene of interest.

Preexisting Vectors. A number of plasmid vectors currently available are capable of selection and high copy number propagation in yeast.[6,11] The relevant properties of a number of these plasmids are identified in Table I, and their structures are diagrammed in Fig. 3. All these plasmids are constructed from the bacterial plasmid pBR322 and thus can be maintained in and purified, in reasonable quantities, from appropriate *E. coli* strains. Each of these vectors carries the 2-μm circle origin of replication and *REP3*, most often in the form of the 2240 bp *Eco* RI fragment from the B form of the plasmid. The vectors differ though in the particular yeast gene used as the selectable marker for transformation and maintenance in yeast and in the availability of restriction sites for inserting a gene of interest. Nonetheless, each of these plasmids is maintained at approximately 30 copies per cell in [cir$^+$] yeast strains. In addition, all these plasmids are

TABLE I
USEFUL 2-μm CIRCLE VECTORS

Plasmid	Selectable markers: Yeast	Selectable markers: *E. coli*	Available cloning sites	Comments
CV7	*LEU2*[a]	*Amp*^{R}, *Leu*B	*Bam*HI, *Sal*I, *Hin*dIII	—
YEp13	*LEU2*[a]	*Amp*^{R}, *Tet*^{R}, *Leu*B	*Bam*HI, *Hind*III	Inserts in *Bam*HI can be detected as *Tet*^{S} clones
YEp13S	*LEU2*[a]	*Amp*^{R}, *Tet*^{R}, *Leu*B	*Bam*HI, *Sal*I, *Hin*dIII	Inserts in *Bam*HI or *Sal*I can be detected as *Tet*^{S} clones
YEp6	*HIS3*[a]	*Amp*^{R}, *His*B	*Bam*HI	—
YEp24	*URA3*[a]	*Amp*^{R}, *PyrF*	*Eco*RI, *Bam*HI	—
YEp16	*URA3*[a]	*Amp*^{R}, *Tet*^{R}, *Pyr*F	*Bam*HI, *Sal*I, *Hin*dIII	Inserts into any of these sites can be detected as *Tet*^{S} clones
CV19	*LEU2*[b]	*Amp*^{R}, *Leu*B	*Bam*HI, *Sal*I	—
CV21	*LEU2*[b]	*Amp*^{R}, *Leu*B	*Bam*HI, *Sal*I	—
pSI4	*LEU2*[b]	*Amp*^{R}, *Tet*^{R}, *Leu*B	*Bam*HI, *Sal*I	Derived from pJDB219; insertions detectable as *Tet*^{S} clones

[a] For use with [cir$^+$] strains only.
[b] Can be used with both [cir^0] and [cir$^+$] strains.

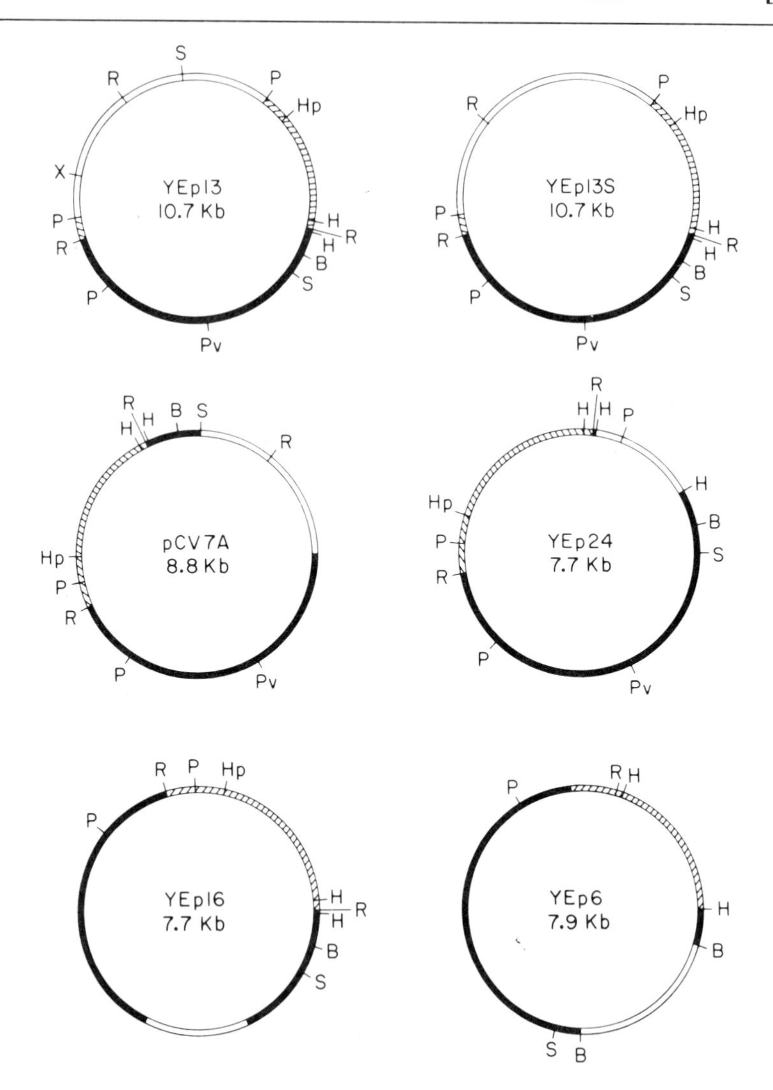

FIG. 3. Hybrid 2μm circle vectors. In each vector diagram pBR322 sequences are represented by filled double lines, 2-μm circle sequences by hatched lines, and yeast chromosomal sequences by open double lines. For YEp13, YEp13S, and CV7, the chromosomal sequences span the *LEU2* gene; for YEp24 and YEp16, these sequences span *URA3;* and for YEp6 these sequences span *HIS3*. The locations of a number of restriction enzyme sites are indicated on the diagrams. These sites are abbreviated as follows: B, *Bam*HI; H, *Hin*dIII; Hp, *Hpa*I; P, *Pst*I; Pv, *Pvu*II; R, *Eco*RI; S, *Sal*I; and X, *Xho*I.

reasonably stable during mitotic growth, being lost in less than 20% of the cells after 10 generations of growth in nonselective media.[6,16] Since there is apparently little difference among the plasmids in terms of stability or copy number, the choice of a particular plasmid would be dictated by the genetic markers of the yeast strain to be used and by the particular restriction sites bracketing the gene to be inserted.

A hybrid plasmid lacking the 2-μm circle origin of replication, but containing an inverted repeat, can also propagate at moderately high copy number in yeast.[5] This is possible since such a plasmid can recombine very efficiently with an endogenous 2-μm circle in the cell and thus replicate by virtue of its physical attachment to an autonomously replicating molecule.[16] Since the plasmid by itself is not able to replicate, and since the reverse recombination event also occurs frequently, the plasmid is lost at a reasonably high rate. Thus such a plasmid is of little use in the context of maximizing expression, but can be of value in other situations.

Contructing New Vectors. If a gene of interest resides on a plasmid containing a selectable yeast marker or if the gene itself can be selected in yeast, then propagating this gene at high copy number can be accomplished by inserting a DNA fragment spanning the 2-μm circle origin of replication and *REP3* into the plasmid and transforming the resultant plasmid into a [cir$^+$] strain. From the B form of 2-μm circle one can recover several different restriction fragments spanning the origin and *REP3* which could be inserted readily into a plasmid at a single restriction site. These include the 2242 bp *Eco*RI fragment, the 2214 bp *Hin*dIII fragment, and the 1576 *Sau*3A fragment (which can be inserted into either a *Bam*HI, a *Bgl*II, or a *Bcl*I site). A ready source of any of these fragments is plasmid pMJ5 (cf Fig. 4), which consists of the entire B form of 2-μm circle cloned at the *Eco*RI site in the small unique region into the *Eco*RI site of pBR322.

For those situations in which the gene of interest resides on a plasmid lacking a suitable selectable marker, there are several plasmids available from which a single restriction fragment can be obtained that carries the 2-μm circle origin of replication as well as a gene that can be selected in yeast. Thus, by inserting such a fragment into a plasmid containing a gene of interest, it is possible to generate in one step a vector by which the gene can be introduced into yeast and propagated at high copy number. For instance, the 3.2 kb *Sal*I to *Hin*dIII fragment from plasmid pC4 [or the equivalent *Bam*HI to *Hin*dIII fragment of the derivative plasmid pC4(B); see Fig. 4] carries both the *LEU2* gene of yeast and the 2-μm circle origin.[21] Similarly, the 3.6 kb *Hin*dIII fragment from pJDB219 carries the origin and the *LEU2* gene[15] (see Fig. 6; the special properties of plasmid

[21] J. R. Broach, unpublished results.

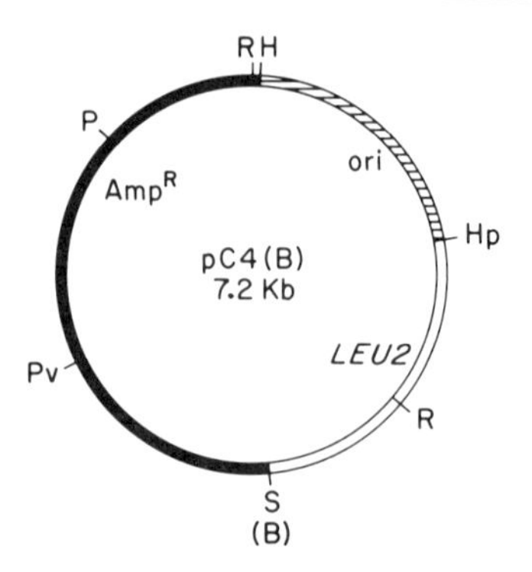

FIG. 4. Plasmid pC4. The representation of the sources of the various portions of the plasmid and the abbreviations for restriction sites on the plasmid are the same as those used in Fig. 3. In plasmid pC4B the *Sal*I site has been converted into a *Bam*HI site. As described in the text, the right-hand *Hin*dIII to *Sal*I restriction fragment of pC4 (or the equivalent *Hin*dIII to *Bam*HI fragment of pC4B) contains all the sequences required for selection and high copy propagation of a plasmid in a *leu2* [cir$^+$] yeast strain.

pJDB219 are discussed below). Insertion of any of these fragments into a bacterial plasmid is readily monitered since the *LEU2* gene complements the *E. coli leu*B mutation (present in the commonly used *E. coli* strain C600). There are certainly other plasmids from which one can obtain such dual-function fragments, but an exhaustive list is not provided here. As above, the primary criteria in choosing a particular fragment are the constraints of available restriction sites in the recipient plasmid and of available genetic markers in the recipient yeast strain.

Propagating Cloned Genes in [cir^0] Strains

If two distinct bacterial plasmids share a common copy control system, then they can exhibit a phenomenon termed incompatibility. One manifestation of incompatibility is that the number of copies of either plasmid in a cell when both plasmids cohabit the cell is less than the number of copies of either plasmid when either is the sole occupant of the cell. There is some indication that 2-μm circles and hybrid 2-μm circle plasmids display incompatibility. That is, the presence of endogenous 2-μm circle in a yeast strain may reduce the potential copy number of a hybrid 2-μm circle plasmid introduced into the strain.[22] Thus, in seeking to optimize copy number of a cloned gene it is worth considering using strains devoid of endogenous circle ([cir^0] strains) in conjunction with appropriate 2-μm circle-derived vectors.

[22] C. Gerbaud and M. Guerineau, *Curr. Genet.* **1**, 219 (1980).

[*cir*^{0}] *Strains*. There are a number of [cir^0] strains currently available that contain nonreverting alleles (or alleles that revert at low frequency) of loci complemented by the cloned yeast genes most commonly used as selectable markers for transformation. A partial list of such strains is given in Table II. These strains either arose as spontaneous [cir^0] derivatives of [cir$^+$] strains or were induced by the technique described below.[4, 7, 23–25] None of the strains, though, displays any obvious barrier to propagation of 2-μm circles when they are reintroduced by cytoduction or of hybrid 2-μm circle plasmids introduced by transformation.

In addition to these preexisting [cir^0] strains, [cir^0] derivatives can be recovered from almost any *leu2* [cir$^+$] strain after transformation with the hybrid 2-μm circle-*LEU2*-pMB9 plasmid, pJDB219.[24,25] Since pJDB219 propagates in yeast transformants at unusually high levels (discussed below), it is possible that elimination of endogenous 2-μm circles from such transformants is a manifestation of incompatibility; that is, pJDB219 molecules outcompete for a replication apparatus specific for 2-μm circle origins to the exclusion of the endogenous circles. However, the fact that this incompatibility is observed following introduction of pJDB219 by transformation, but is not observed if pJDB219 is introduced by cytoduction or mating, suggests that the actual explanation of the phenomenon is more complex.[25] Nonetheless, by using the following procedure one can obtain [cir^0] derivatives of almost any *leu2* strain.

The *leu2* strain is transformed to Leu$^+$ with pJDB219 DNA using the standard transformation protocol.[15] Cells from a single transformant are resuspended in selective media (ca. 2 ml of synthetic complete medium minus leucine[11]) and grown overnight at 30° with agitation. A sample of this culture is then diluted 1:1000 in YEPD (1% yeast extract, 2% Bacto peptone, and 2% glucose) and grown overnight at 30°. An appropriate dilution of the culture is spread on solid YEPD medium to give ca. 200 colonies per plate, and colonies auxotrophic for leucine are identified by replica plating to synthetic complete minus leucine medium. Approximately

TABLE II
AVAILABLE [cir^0] STRAINS

Designation	Genotype	Reference
AH22	**a** *leu2-3,112 his4-519 can1-11*	5, 25
SC3	α *trp1 ura3-52 his3* ∇ *gal2 gal10*	4
DC04	**a** *leu2-04 ade1*	16
SB1-3B	α *leu2-04 ade1 ade6*	14

[23] D. M. Livingston, *Genetics* **86,** 73 (1977).

[24] M. J. Dobson, A. B. Futcher, and B. S. Cox, *Curr. Genet.* **2,** 193 (1980).

[25] A. Toh-e and R. I. Wickner, *J. Bacteriol.* **145,** 1421 (1981).

20 Leu^- colonies are recovered and tested for the presence of 2-μm circle sequences by colony hybridization[5,7] using ^{32}P-labeled pMJ5 DNA (or any readily available DNA containing 2-μm circle sequences). Leu^- colonies which lack 2-μm circle sequences by this criterion can be further tested by Southern analysis of genomic DNA isolated from a small culture of the strain.[26]

There is, in addition, a genetic test for the absence of 2-μm circles that one can use as an alternative procedure for identifying the [cir^0] clones arising from the pJBD219-transformed strain. This procedure is based upon the observation that the presence of 2-μm circles in a diploid strain in which an inverted repeat from 2-μm circle has been inserted into a chromosome causes a marked instability of that chromosome.[14] The instability can be tested by assessing whether there is an unusually high number of cells, in a culture of the diploid strain, that possess a phenotype associated with the loss of a dominant marker on the insertion-containing chromosome for which the equivalent locus on the homolog chromosome carries a recessive allele. Specifically, for identifying the [cir^0] clones obtained by the above procedure, the Leu^- segregants are mated to strain DC-04::LCV7 (**a** *leu2-04*::CV7 *ade1 ade6* [cir^0]) or strain SB1-3B (α *leu2-04*::CV7 *ade1 ade6* [cir^0]) and diploids are obtained by selecting for Ade^+ Leu^+ colonies (both DC-04::CV7 and SB1-3B are Ade^- but Leu^+, since the CV7 plasmid integrated at *leu2* contains an intact *LEU2* gene).

After the diploid has been colony purified, it is streaked on nonselective media (YEPD, for example) for single colonies so that at least 100 individual colonies are distinguishable. (Alternatively, the colony-purified diploid can be grown in liquid YEPD and appropriate dilutions can be spread on YEPD plants to give 100 or so colonies.) These colonies are then replica-plated onto SD minus leucine media. One of two possible outcomes should be obtained: either all the colonies will be Leu^+ or a reasonable proportion (5% to 25%) of the colonies will be Leu^-. If all the colonies are Leu^+, then the strain tested was [cir^0]; if a reasonable proportion are Leu^-, then the strain tested was [cir^+]. This is explained in the following manner. If the diploid strain does not contain 2-μm circles—that is, if the haploid strain being tested is [cir^0]—then the chromosome in which CV7 is inserted will be stable, the *LEU2* will be retained in all the cells, and all the colonies will be Leu^+. If, on the other hand, the strain does contain 2-μm circles, then frequently the chromosome in which CV7 is inserted will be lost. Since this chromosome contains the only functional copy of *LEU2*, cells that lose this chromosome will give rise to Leu^- colo-

[26] R. W. Davis, M. Thomas, J. R. Cameron, T. P. St. John, S. Scherer, and R. A. Padgett, this series, Vol. 65, p. 404.

nies. In this protocol this chromosome loss occurs in approximately 5% to 25% of the cells.

Vectors for Use with [*cir*^{0}] *Strains*. Since [cir^0] strains lack the *REP* gene products, both *REP* genes must be carried by a hybrid 2-μm circle plasmid if that plasmid is to propagate at high copy number in the strain. Owing to the genetic organization of 2-μm circles, such vectors have been most readily constructed by inclusion of the entire 2-μm circle genome. Two vectors appropriate for use in [cir^0] strains are diagrammed in Fig. 5 and listed in Table I. They contain pBR322 plus *LEU2* sequences inserted in the *Eco*RI site either in the small unique region (pCV20) or in the large unique region (pCV21). For pCV20 this insertion interrupts the 2-μm circle *FLP* gene. However, this insertion does not appear to impair significantly the replication potential of the plasmid. Similarly, although the pBR322-*LEU2* sequences in pCV21 interrupt an open coding region within the 2-μm circle genome, the replication potential of the plasmid is apparently unimpaired. Both of these plasmids are maintained at approximately 40 copies per cell in [cir^0] strains and are lost from less than 20% of the cells following growth in nonselective media for 10 generations. Plasmids pMJ5 and pMJ6 are equivalent to pCV20 and pCV21, respectively, except that they lack *LEU2* sequences and therefore possess intact *Tet*^{R} and *Amp*^{R} genes within the pBR322 moiety. Thus sequences containing other selectable yeast genes can be easily cloned onto these plasmids to construct vectors suitable for use with other [cir^0] strains.

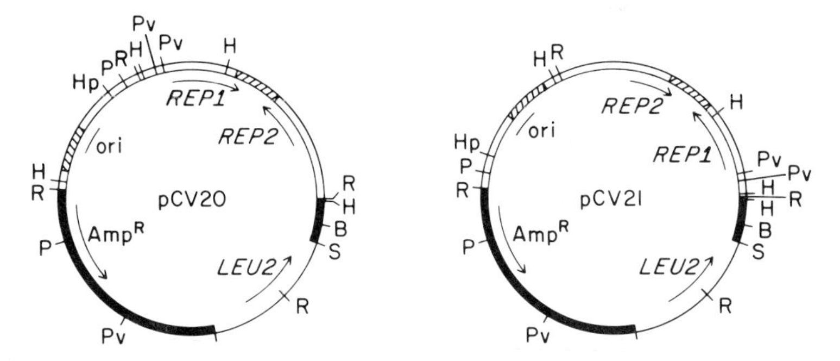

FIG. 5. Hybrid 2-μm circle vectors suitable for use with [cir^0] yeast strains. pBR322 sequences are shown as filled double lines and 2-μm circle sequences as open double lines with the inverted repeat regions indicated by hatching. The location of chromosomal sequences spanning the *LEU2* gene is indicated by the single line. The directions of transcription of *REP1*, *REP2*, *Bla*, and *LEU2* are shown, as is the approximate location of the 2-μm circle origin of replication (ori). Restriction site abbreviations are the same as those used in Fig. 3.

Other Approaches to Vector Construction

pJDB219 and Derivative Plasmids. The structure of plasmid pJDB219 is diagrammed in Figure 6. It consists of the entire 2-μm circle genome cloned, at the *Eco*RI site in the small unique region, into the *Eco*RI site of the bacterial plasmid pMB9. The plasmid contains a fragment of yeast DNA spanning the *LEU2* gene—isolated from randomly sheared, total genomic DNA—inserted at the *Pst*I site in the 2-μm circle moiety through complementary homopolymer extensions[15] (since the homopolymers in this case are dA and dT, *Pst*I sites on either side of the insert are not reconstructed). This fragment of yeast DNA is 1400 bp or so in length, which is not much larger than the *LEU2* gene itself. Several derivative plasmids constructed from pJDB219 are also diagrammed in Fig. 6. Plasmid pSI4 differs from pJDB219 by the substitution of pBR322 sequences for the pMB9 moiety. Plasmid pJDB207 consists of that region of pJDB219 spanning the *LEU2* gene and the origin of replication cloned

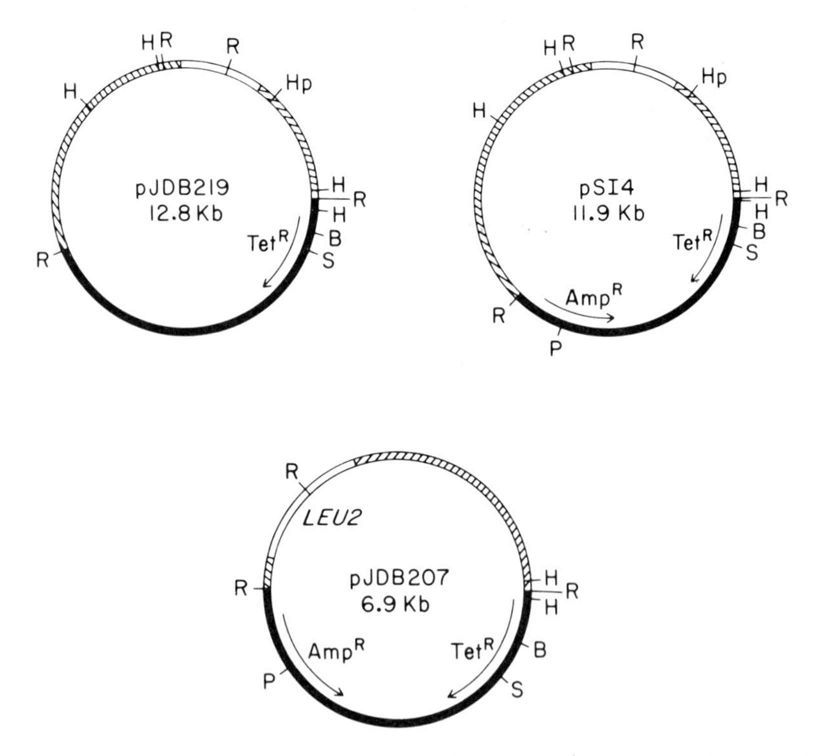

Fig. 6. Plasmid pJDB219 are its derivatives. Sequence source and restriction site abbreviations are the same as those in Fig. 3.

onto the bacterial vector pAT153, a smaller derivative of pBR322.[27] All three of these plasmids can be propagated in *E. coli* and can transform *leu2*$^-$ yeast strains to Leu$^+$ at high frequency. In addition, the latter two plasmids contain intact *Amp*^{R} and *Tet*^{R} genes, with the *Bam*HI and *Sal*I sites in the *Tet*^{R} gene available for insertion of exogenous restriction fragments.

The property of pJDB219 and its derivatives that warrants their separate discussion in this chapter is that they display unusually high copy number in yeast.[27] All three of the plasmids shown in Fig. 6 propagate at an average of 200 copies per cell in yeast (cf. Fig. 7). The reason for this high copy number is unknown. A plasmid essentially identical to pJDB219, except that the *LEU2* insert is somewhat larger, does not display an elevated copy number. At this point the high copy number is best ascribed to an as yet unexplained activation of the 2-μm circle replication origin fortuitously induced by the singular nature of the *LEU2* insertion in the plasmid. As is evident in Fig. 7, the high copy number and stability of pJDB219 derivatives is dependent in part on 2-μm circle *REP* functions. That is, plasmid pSI5, a plasmid essentially equivalent to pJDB207, is present at 200 copies per cell in a [cir$^+$] strain, but is present at only 15 copies per cell in a [cir^0] strain. The *REP* functions can also be provided by the hybrid plasmid itself. This is evident since pJDB219 and pSI4 display high copy number in [cir^0] as well as [cir$^+$] strains.[21]

Vectors Free of Bacterial Sequences. The presence of certain bacterial sequences on hybrid plasmids has been shown to inhibit the replication of these hybrid plasmids in cultured mammalian cells.[28] Whether the presence of extraneous bacterial sequences on 2-μm circle vectors inhibit their replication potential as well has not been established. It is certainly true that, except for pJDB219 and its derivatives, all hybrid vectors—even those containing the entire 2-μm circle genome—are present in the cell at lower copy number than authentic 2-μm circles. Therefore, in attempting to maximize the copy number of a gene, it may be worth considering the use of vectors free of bacterial sequences.

This approach has not been appreciably explored and is presented here primarily as an indication of a potential future direction for the development of yeast expression vectors. In the following, a specific construction is described as an illustration of the way in which this technique can be applied.

A plasmid consisting of pBR322 and the entire 2-μm circle genome, joined through their respective *Pst*I sites (cf. Fig. 1), is digested with an

[27] J. D. Beggs, *Alfred Benzon Symp.* **16,** 383 (1981).

[28] M. Lusky and M. Botchan, *Nature* (*London*) **293,** 79 (1981).

enzyme that makes a single cut in the plasmid, at a site in the 2-μm circle moiety that is nonessential for replication. The *Bcl*I site, which lies at the 3′ end of the *FLP* gene, should be one such site (after isolation of the plasmid from a *dam*$^-$ *E. coli* strain) as may be the *Hpa*I site in the large unique region. Sequences to be used for selection in yeast as well as sequences that one wishes to propagate in yeast are then inserted by ligation into this site and the resultant plasmid is recovered in *E. coli*. After purification of this plasmid from *E. coli*, it is digested with *Pst*I and used to transform an appropriate [cir$^+$] yeast strain, or it is digested, religated at low DNA concentration, and used to transform an appropriate [cir^0] yeast strain. In the former case, the recombinogenic nature of the ends of the 2-μm circle fragment leads to restoration of a covalently closed circle through recombination with endogenous plasmids.[29,30] Thus by this technique one can readily insert and propagate sequences of interest on a 2-μm circle vector that is free of bacterial DNA.

Dobson *et al.* have described a high-copy 2-μm circle hybrid plasmid that should be useful as a yeast vector free of bacterial sequences.[24] The plasmid, designated pYX, arose as a recombinant in yeast between pJDB219 and an endogenous 2-μm circle and consists of the entire 2-μm circle genome containing the pJDB219 *LEU2* insertion in the large unique region. Since this species is the sole extrachromosomal plasmid in their strain MC16 L$^+$1 (α *ade2-1 leu2 his4-712 SUF2* [pYX$^+$, cir^0]) and appears to be present at approximately 150–200 copies per cell in this strain (thus constituting ca 5% of the total DNA), it should be readily purifiable in reasonable amounts by isopycnic centrifugation of total genomic DNA in CsCl plus ethidium bromide. Sequences of interest can then be inserted at any one of a number of sites within the purified plasmid (as mentioned above, either the *Bcl*I or the *Hpa*I site should be suitable) and then introduced into yeast by transformation of an appropriate *leu2*$^-$ [cir^0] strain.

Determination of Plasmid Copy Number

Since there is presently no definitive vector of choice for maximizing copy number of a sequence of interest, the approach to achieving high-level propagation that I have recommended in this chapter is one of trial and error. That is, several different strategies have been outlined with the anticipation that one or more of these can be pursued for any particular gene of interest. The plasmid that provides the highest copy number is

[29] T. L. Orr-Weaver, J. W. Szostak, and R. J. Rothstein, *Proc. Natl. Acad. Sci. U.S.A.* **78,** 6354 (1981).

[30] M. Jayaram, S. Sumida, and J. R. Broach, unpublished results.

then retained. It is obvious, then, that the ability to assess readily the copy numbers of the various constructions must play in integral part of this approach.

Measurements of the cellular copy number of 2-μm circles or of various hybrid yeast vectors have been accomplished through a variety of techniques. All require isolating total DNA from the appropriate yeast strain and then determining the ratio of plasmid-specific DNA to total genomic DNA. This has been done either by direct measurement of circular versus linear DNA in electron micrographs.[2,3] or by determining the amount of plasmid DNA in a given amount of total DNA by various hybridization or reassociation protocols.[22,31,32] None of these techniques includes a suitable control for possible errors arising from differential extraction of chromosomal versus plasmid DNA. In addition, each of these methods possesses one or more specific potential sources of error, which could lead to an inaccurate estimate of the plasmid copy number. Nonetheless, the similarity of the values obtained by each of these procedures for the copy number of 2-μm circles lends greater credence to each procedure than theoretical considerations might warrant.

Although any of the above referenced methods can be used, the easiest method for determining the copy number in yeast of a high copy number plasmid is to compare the intensity of ethidium bromide staining of a plasmid specific restriction fragment with respect to a fragment corresponding to the repeated rDNA sequences in restriction digests of total DNA isolated from an appropriate transformed strain. As can be seen in Fig. 7, both rDNA restriction fragments and plasmid specific restriction fragments stand out sharply against the background of staining of random genomic DNA fragments, after fractionation of the digested DNA by electrophoresis on agarose gels. The relative amount of plasmid DNA versus ribosomal DNA can be quantitiated from densitometer scans of photographic negatives of such a gel, as shown in Fig. 7. Assuming 100 repeats of rDNA per haploid genome[33] and normalizing the staining intensities for the sizes of the fragments (a restriction map of the ribosomal DNA repeat unit is given by Bell *et al.*[34]), one can determine an absolute value for the copy number of the hybrid plasmid. With lower copy number, pBR322-containing plasmids for which the plasmid specific restriction fragments are not so prominent, the relative copy number can be determined by

[31] D. T. Stinchcomb, C. Mann, E. Selker, and R. W. Davis, *Initiation of DNA Replication, ICN–UCLA Symp. Mol. Biol.* **22,** p. 473 (1981).

[32] V. L. Seligy, D. Y. Thomas, and B. L. A. Miki, *Nucleic Acids Res.* **8,** 3371 (1980).

[33] T. J. Zamb and T. D. Petes, *Cell* **28,** 355 (1982).

[34] G. I. Bell, L. J. DeGennaro, D. H. Gelfand, R. J. Bishop, P. Valenzuela, and W. J. Rutter, *J. Biol. Chem.* **252,** 8118 (1977).

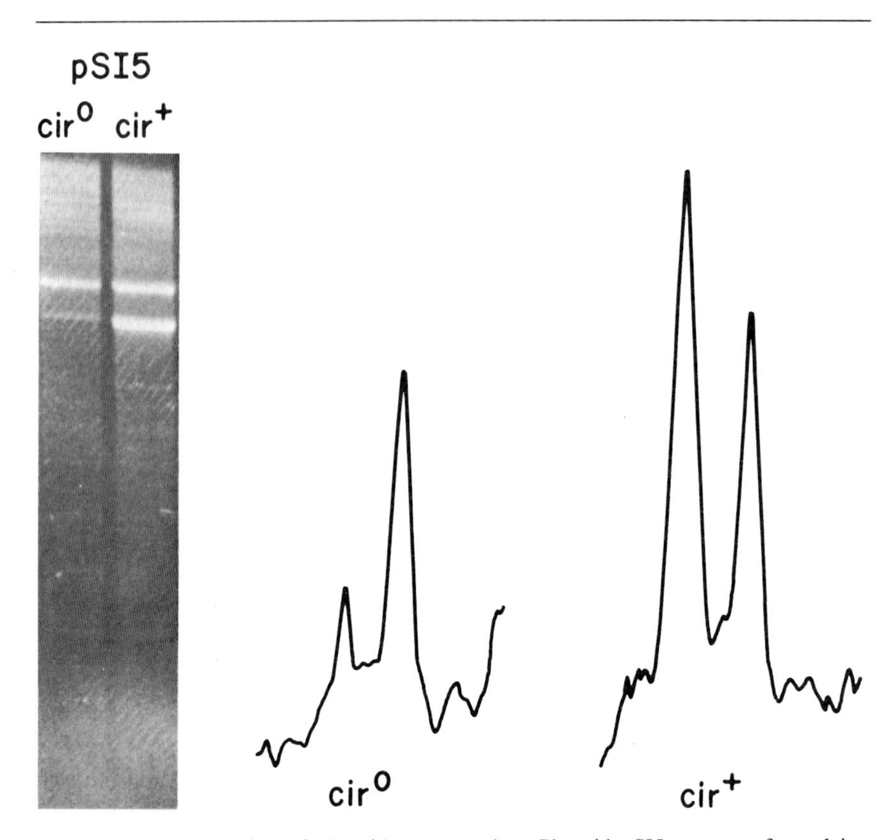

FIG. 7. Determination of plasmid copy number. Plasmid pSI5 was transformed into strains DCO4 [cir⁰] and DCO4 [cir⁺], and total DNA was isolated from a representative transformant from each. Plasmid pSI5 consists of the *Hin*dIII B fragment of pJDB219, which spans *LEU2*, *REP3*, and the 2-μm circle origin of replication, inserted into the *Hin*dIII site of pBR322. Samples (2 μg) of each DNA preparation were digested with *Kpn*I and fractionated by electrophoresis on a 0.8% agarose. The gel was photographed on a UV transilluminator with Polaroid type 665 Pos/Neg film. The positive photograph of the two tracks is shown on the left. The upper band visible against the background is the 9060 bp rDNA repeat fragment. The lower band is the 8080 bp plasmid fragment. Each of these tracks was scanned on the photographic negative using a Joyce-Loebl microdensitometer, the tracings of the relevant portions of which are shown on the right. The areas under the peaks corresponding to plasmid and rDNA bands were determined using a Numonics Corporation digitizer. The copy number of the plasmid in each of the two strains could then be determined using the formula

$$\text{Copy number} = \frac{(\text{area under plasmid peak})(\text{length of rDNA fragment})}{(\text{area under rDNA peak})(\text{length of plasmid fragment})} \times 100$$

The final term is the rDNA repeat value. The data shown in the figure yield values of 15 and 190 for the copy number of pSI5 in the [cir⁰] and [cir⁺] strain, respectively.

Southern analysis, hybridizing the transferred genomic DNA with a pBR322-rDNA probe. As with ethidium staining, an absolute value for the copy number can be determined from densitometer tracings of the autoradiogram, with due consideration given to the relative lengths of the fragments and the extent to which the fragments are covered by the labeled probe. Care should be taken in selecting a plasmid specific fragment upon which to make the measurements. Any 2-μm circle plasmid containing two inverted repeat sequences is capable of undergoing inversion and thus may exist in the yeast in two forms. Thus some of the restriction fragments from such a plasmid may represent only half of the total cellular population of that plasmid. Care should also be taken in assuring oneself that autoradiographic exposures yield a linear response with respect to amount of label hybridized. Using preflashed film usually avoids any nonlinearity of response.

This procedure is an unsophisticated but rapid technique for determining copy number. Besides speed, the principal advantage of the technique is that it incorporates an internal standard to permit ready extraction of absolute values for copy number. The method obviously goes no further in addressing the problem of differential extraction of plasmid versus chromosomal DNA than those previously mentioned. Indeed, this procedure adds the additional complication of potential differential extraction of rDNA, which serves as the normalizing component. Nonetheless, this procedure has provided reproducible values for copy levels of various hybrid plasmids we have examined and has given values of 2-μm circle that are certainly comparable to those obtained by more rigorous methods.

Acknowledgments

I would like to thank M. Jayaram and D. Shortle for critical evaluation of this manuscript.

[22] Cloning of Yeast *STE* Genes in 2 μm Vectors

By VIVIAN L. MACKAY

The development of genetic transformation and molecular cloning methods in the yeast *Saccharomyces cerevisiae,* coupled with its well-established genetics and ease of manipulation, allows detailed investigations at the molecular level of complex cellular processes involving numerous genes and regulatory factors. One such function is sexual conjugation and

ISBN 0-12-182001-7

mating-type determination, which is one of the few cases in eukaryotes where a single genetic locus (i.e., *MAT*) has been demonstrated to regulate the expression of a number of unlinked genes denoted *STE*. Several of the *STE* genes have now been cloned, leading to useful and in some cases unexpected results, as described below.

In heterothallic strains of *S. cerevisiae*, mating occurs between haploid cells of opposite mating type (**a** or α). The mating type of a cell is determined by which allele, either *MAT***a** or *MAT*α, is present at the mating-type locus. (In homothallic strains of this yeast, DNA transposition events lead to frequent interconversions between *MAT***a** and *MAT*α by transposition from "silent copies" elsewhere in the genome.[1]) Cells with the *MAT*α locus secrete the peptide hormone α-factor and respond to **a**-factor, while **a** cells produce **a**-factor and are sensitive to α-factor.[2,3] The hormones arrest cells of the opposite mating type in the G_1 phase of the cell cycle and induce other physiological changes that are preparatory to conjugation. The product of mating between **a** and α haploids is a diploid that is heterozygous (**a**/α) at *MAT* and is completely defective in mating, hormone secretion, and hormone response.[1]

Conjugation requires the products of a number of genes that have been identified genetically and designated *STE* for their *ste*rile mutant phenotype.[4–9] Since these are not linked to *MAT*, it has been proposed that their expression could be regulated by the products of the *MAT* genes.[5,6] The *STE* genes fall into several classes (Table I): Class I, those that are apparently required only in **a** cells (**a**-specific genes), such as *STE2*, *STE6*, and *STE14*; class II, α-specific genes, such as *STE3* and *STE13*; class III, nonspecific genes essential for conjugation in both **a** and α cells. As yet, there is no evidence of genetic linkage between any of the *STE* genes, although not all of them have been mapped. It should be emphasized that none of these mutations seems to affect cellular functions other than those involved in mating; they are therefore probably conjugation-specific, al-

[1] I. Herskowitz and Y. Oshima, *in* "The Molecular Biology of the Yeast Saccharomyces" (J. N. Strathern, E. W. Jones, and J. R. Broach, eds.), p. 181. Cold Spring Harbor Laboratory, Cold Spring Harbor, New York, 1981.

[2] T. R. Manney, W. Duntze, and R. Betz, *in* "Sexual Interactions in Eukaryoptic Microbes," p. 21. Academic Press, New York, 1981.

[3] J. Thorner, *in* "The Molecular Biology of the Yeast Saccharomyces" (J. N. Strathern, E. W. Jones, and J. R. Broach, eds.), p. 145. Cold Spring Harbor Laboratory, Cold Spring Harbor, New York, 1981.

[4] V. L. MacKay and T. R. Manney, *Genetics* **76,** 255 (1974).

[5] V. L. MacKay and T. R. Manney, *Genetics* **76,** 273 (1974).

[6] L. H. Hartwell, *J. Cell Biol.* **80,** 811 (1980).

[7] J. D. Rine, Ph.D. Thesis, University of Oregon, Eugene, 1979.

[8] L. C. Blair, Ph.D. Thesis, University of Oregon, Eugene, 1979.

[9] G. F. Sprague, Jr., J. Rine, and I. Herskowitz, *J. Mol. Biol.* **153,** 323 (1981).

TABLE I
CLASSES AND PHENOTYPES OF *STE* MUTATIONS

Gene	Hormone secretion	Hormone response	References[a]
I. **a**-Specific			
ste2	+	−	4–6
ste6	−	+	7
ste14	−	+	8
II. α-Specific			
ste3	+	−	4, 5
ste13	−	+	9
III. Nonspecific			
ste4	−	−	4–6
ste5	−	−	4–6
ste7	−	−	6
ste8[b]	−	−	6
ste9[b]	−	−	6
ste11	−	−	6
ste12	−	−	6
IV. α-Specific pleiotropic			
kex2	−	(+)[c]	12
tup1	−	(−)[d]	13, 14
V. *MAT* mutations (proposed to be regulatory)			
*mat*α1	−	−	4, 5
*mat*α2	−	(−)[e]	1, 4, 5,
*mat***a**1[f]	+	+	1[g]

[a] Numbers refer to text footnotes.

[b] *STE8* and *STE9* were shown to be the same genes as *SIR3* and *SIR4*, respectively, and therefore probably do not code for conjugation functions, but act to repress expression of *HML* and *HMR*.[6] *HML* and *HMR* contain the same genetic information as *MAT*α and *MAT***a**, but these silent genes are not transcribed in wild-type cells.[1]

[c] Mutations in *KEX2* prevent expression of the killer plasmid, a double-stranded RNA. The α *kex2* mutants probably respond normally to **a**-factor.

[d] α *tup1* strains exhibit decreased mating with **a** cells but increased mating with α cells.

[e] Mutations in the *MAT*α2 gene cause reduced mating with **a** cells but increased mating with α cells.

[f] Mutations in the *MAT***a**1 gene do not block mating, **a**-factor secretion, or response to α-factor, although the mating efficiency may be slightly reduced.

[g] Y. Yassir and G. Simchen, *Genetics* **82,** 187 (1976).

though their products and roles in conjugation have not been elucidated. Interestingly, mutations in the structural genes for the hormones α-factor and **a**-factor have not yet been identified. The hormones are not encoded by the *MAT*α and *MAT***a** alleles which have been sequenced.[10,11] There-

[10] K. A. Nasmyth, K. Tatchell, B. D. Hall, C. Astell, and M. Smith, *Cold Spring Harb. Symp. Quant. Biol.* **45,** 961 (1980).

[11] C. R. Astell, L. Ahlstrom-Jonasson, M. Smith, K. Tatchell, K. A. Nasmyth, and B. D. Hall, *Cell* **27,** 15 (1981).

fore, more conjugation-specific genes, perhaps including those for the hormones, may yet be identified genetically. In addition to the *STE* genes, two other genes are required for conjugation specifically in α cells. The products of these (*KEX2* and *TUP1*) are clearly not conjugation-specific, but have a broader role in the cell.[12-14]

The general hypothesis for mating-type regulation holds that the products of the *MAT* loci regulate expression of some or perhaps all of the unlinked *STE* genes[1,10] (Table I and Fig. 1). Specifically, in α cells, the product of the *MAT*α1 gene is proposed to turn on the expression of α-specific genes, while the *MAT*α2 product blocks expression of **a**-specific genes. In

[12] M. J. Leibowitz and R. B. Wickner, *Proc. Natl. Acad. Sci. U.S.A.* **73,** 2061 (1976).
[13] J. F. Lemontt, D. R. Fugit, and V. L. MacKay, *Genetics* **94,** 899 (1980).
[14] D. R. Fugit, Ph.D. Thesis, Rutgers University, New Brunswick, New Jersey, 1981.

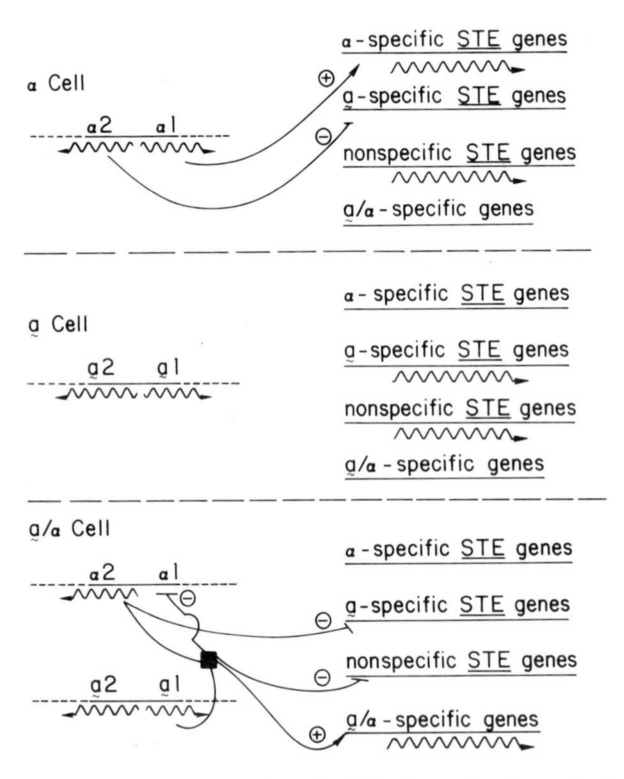

FIG. 1. Schematic model for regulation of unlinked genes by the products of the *MAT* locus. See text for explanation. Wavy lines indicate transcription; ⊕ and ⊖ indicate positive and negative regulation, respectively; ■ represents the joint action of the *MAT*α2 and *MAT***a**1 products in **a**/α cells. Modified from Herskowitz and Oshima.[1]

a cells, the **a**-specific phenotype is constitutive, since there is no *MAT*α*2* product; i.e., the *MAT***a** locus may have no function in **a** haploids. However, in **a**/α diploids, the *MAT***a**1 and *MAT*α*2* products combine to establish the **a**/α phenotype and to repress transciption of *MAT*α*1*. Thus, in **a**/α cells, both **a**-specific and α-specific gene expression would be blocked, but **a**/α-specific genes, such as those for meiosis, could be expressed under appropriate physiological conditions. If the nonspecific *STE* genes are conjugation-specific, then one or more of them might not be expressed in **a**/α cells.

The model described above makes several predictions for the presence or the absence, in different cell types or mutants, of gene products from certain *STE* or other conjugation-related genes. Some of these predictions can be tested by cloning the genes and using them to detect RNA transcripts or to identify protein products. Hence, in addition to the *MAT* loci themselves,[15,16] at this writing the following genes have been cloned and positively identified: *STE3*,[17] *STE5*,[18,19] *STE8*,[20] *STE9*,[20] *STE13*,[17] *TUP1*,[21] and the structural gene for α-factor (which contains a fourfold tandem repeat of the coding sequence).[22] Several other *STE* genes may also have been cloned, although their identification has not yet been rigorously confirmed. As described in detail later, except for the α-factor structural gene the cloned genes were identified biologically by their complementation in yeast of chromosomal mutations. For the α-factor gene, secretion of the hormone in an appropriate regulatory mutant was used to detect transformants containing a plasmid with the structural gene.[22]

Use of the cloned *STE3* and *STE5* genes as probes has demonstrated that these are regulated transcriptionally; transcription of *STE3* (an α-specific gene) requires a functional *MAT*α*1* product[17] while transcripts homologous to *STE5* (a nonspecific gene) are present in both **a** and α cells, but are not detected in **a**/α cells.[18,19] Thus, these results are consistent with the model described above.

All the genes were cloned in yeast from a library of yeast DNA fragments inserted into the plasmid vector YEp13[16] that was derived from the *E. coli* plasmid pBR322 with the addition of (*a*) a yeast chromosomal fragment, carrying the *LEU2* gene as a selectable marker; and (*b*) a fragment

[15] J. B. Hicks, J. N. Strathern, and A. J. S. Klar, *Nature* (*London*) **282,** 478 (1979).

[16] K. A. Nasmyth and K. Tatchell, *Cell* **19,** 753 (1980).

[17] G. F. Sprague, Jr. and I. Herskowitz, personal communication, 1981.

[18] V. L. MacKay and K. A. Nasmyth, submitted (1982).

[19] A. Brake, K. A. Nasmyth, and J. Thorner, personal communication (1981).

[20] J. B. Hicks and K. A. Nasmyth, personal communication, 1981.

[21] M. N. Jagadish and V. L. MacKay, unpublished observations, 1982.

[22] J. Kurjan and I. Herskowitz, *Cell* **30,** 933 (1982).

containing the origin of replication from the endogenous yeast 2 μm plasmid. In general, YEp-type plasmids are maintained moderately stably at a high copy number, variously estimated at 10–50 copies per cell.[23] This high copy number of the vectors is responsible for unexpected results that arose during attempts to clone *STE4* and *STE13*. An insert that complemented two different *ste4* mutant alleles was subsequently shown to carry *STE5*, not *STE4*. Subcloning the *STE5* insert into low copy number plasmids significantly decreased its complementation (suppression) of *ste4* and *ste4* was not suppressed when the *STE5* gene was integrated.[18] Similarly, in the cloning of *STE13*, two different DNA fragments in YEp13 complemented the *ste13* mutation, one of which was shown to contain the *STE13* gene. In an integrated (i.e., low copy number) state, the second insert failed to complement *ste13*.[17]

In summary, the use of 2 μm-derived, high copy number plasmid vectors to clone *STE* and other conjugation-related genes has permitted not only the isolation of many of these genes, but has also provided new information about possible interactions between genes or gene products or the possible existence of other regulatory factors.

Principle of the Method

Two general biological approaches have been used to clone *STE* and other conjugation-related genes in yeast: (*a*) complementation of chromosomal mutations; and (*b*) escape synthesis of a gene product (i.e., α-factor) in a regulatory mutant.[22] In the first, appropriate strains carrying a chromosomal mutation that blocks conjugation, such as *ste5* or *tup1*, are transformed with a plasmid library of total yeast DNA fragments and transformants are selected; these are then screened for those in which mating ability has been restored. Although each of the *ste* mutations confers other phenotypic properties (e.g., the absence of hormone response) that should also be complemented by its wild-type gene, the mating screen is feasible for large numbers and is quite sensitive. A colony containing cells that can mate at only 1% the wild-type efficiency can easily be distinguished among as many as 2000 colonies per plate that mate at $\leq$0.01% the wild-type frequency. An attempt to establish a single-step selection for transformants that had acquired mating ability[24] indicated that this technique would be more cumbersome and less reliable than the two-step selection/screen that has been used.

In general, it is advantageous to use a library constructed in 2 μm-

[23] J. R. Broach, this volume [21].
[24] V. L. MacKay, unpublished observations, 1980–1982.

derived plasmid vectors such as YEp13, since these plasmids are moderately stable and maintained at high copy number in the cell.[23] As discussed later, plasmid stability is particularly important in either of the two approaches to be described. Although a yeast library constructed in a centromere-bearing plasmid that would be quite stable could also be used, the copy number of such a plasmid (1–2 per cell)[25] would be much lower than that of YEp-type vectors. Unexpected, potentially informative results, such as the suppression of *ste4* by the *STE5* gene, might have been missed if *STE5* had been cloned initially in a centromere vector.[18]

The initial mating-capable transformants are subsequently tested to eliminate false-positives and revertants of the chromosomal mutation. However, even those isolates in which a plasmid-borne gene clearly complements the *ste* mutation for mating ability should not be assumed to contain the desired cloned *STE* gene. As discussed above, the limited results to date from studies of conjugation genes indicate that the phenotype conferred by one mutant gene can often be suppressed by a different gene carried in a high copy number plasmid. Furthermore, some suppression of *ste4* mutations by *STE5* was still detectable even when *STE5* was inserted into a low copy number plasmid.[18] Positive identification of the cloned gene should be obtained by methods such as integration at the site of the mutation or by cosegregation of the mutation with restriction-site polymorphisms in the cloned gene.

A description of the methodology of the second approach, used to clone the α-factor gene, is beyond the scope of this chapter; however, an explanation of the logic behind it seems appropriate. Since structural gene mutations for the mating hormones have not been isolated and/or identified, an alternative approach that exploits the properties of *MAT* regulation was devised to isolate the α-factor gene[22] and possibly also the **a**-factor gene.[26] Strains with mutations in the *MAT*α*2* gene produce both **a**-factor and α-factor (which is under the positive control of the *MAT*α*1* product[1]; see Fig. 1). The secretion of α-factor cannot be detected, however, because *mat*α*2* mutants also make an **a**-specific product, called Bar, that inactivates α-factor.[27,28] It was reasoned that a high level of α-factor synthesis would result if the structural gene was cloned on a high copy number plasmid. The increased synthesis was predicted to overcome the Bar activity in a *mat*α*2* mutant, so that a halo of secreted α-factor could be detected around the transformant colony. (The secretion of α-factor by a colony results in the cell division arrest of nearby **a** cells in a

[25] L. Clarke and J. Carbon, *Nature* (*London*) **287,** 504 (1980).
[26] T. R. Manney and V. L. MacKay, unpublished observations, 1982.
[27] G. R. Sprague, Jr. and I. Herskowitz, *J. Mol. Biol.* **153,** 305 (1981).
[28] R. K. Chan and C. A. Otte, *Mol. Cell. Biol.* **2,** 21 (1982).

lawn spread over the surface of the plate.) A DNA fragment conferring this phenotype on the *mat*α2 recipient strain was shown to contain the α-factor gene, since the DNA sequence matched the published amino acid sequence of the hormone.[29] (As mentioned above, there are actually four copies of the α-factor coding sequence on this fragment.[22])

We have used a similar approach in an effort to clone the **a**-factor gene(s),[26] which might also be repeated.[30] A temperature-sensitive *mat*α2 mutation confers temperature-dependent mating and hormone characteristics on the cell.[31] At 25°, the mutant secretes α-factor, but not **a**-factor, whereas at 36°, it secretes **a**-factor, but not α-factor. Presumably, the α-factor is inactivated by the Bar activity, which is also made at 36°. At the semipermissive temperature 33.8°, **a**-factor secretion is just slightly detectable by a halo assay; apparently, **a**-factor synthesis is neither fully inhibited nor fully expressed at this threshold temperature. The *mat*α2 mutant was transformed with the yeast library in YEp13, and transformants were screened at 33.8° for **a**-factor secretion. Although several transformants that secrete the hormone have been isolated, we have not yet determined whether any of the plasmids obtained from them have the **a**-factor gene. One of the plasmids confers a complex phenotype that may be derived in part from elevated expression of one or more genes on the insert.[26]

For the cloning of both hormone genes, a high copy number vector that allows increased expression of the cloned gene is necessary to overcome the inactivating protein or negative regulatory protein in the cell. Furthermore, a stable plasmid that is carried in most of the cells in the colony leads to greater hormone secretion and larger halos around the colony.

Materials and Reagents

Yeast DNA Library

In all cases, the library that has been used to clone *STE* genes is the one constructed by Nasmyth and Tatchell,[16] in which total yeast DNA from strain AB320 (see below) was partially digested with the restriction endonuclease *Sau*3A to yield fragments averaging 5–20 kb. These were ligated into the *Bam*HI site of YEp13 (Fig. 2), which carries the yeast

[29] D. Stötzler, H. H. Kiltz, and W. Duntze, *Eur. J. Biochem.* **69**, 397 (1976).
[30] R. Betz and W. Duntze, *Eur. J. Biochem.* **95**, 469 (1979).
[31] T. R. Manney, personal communication, 1981.

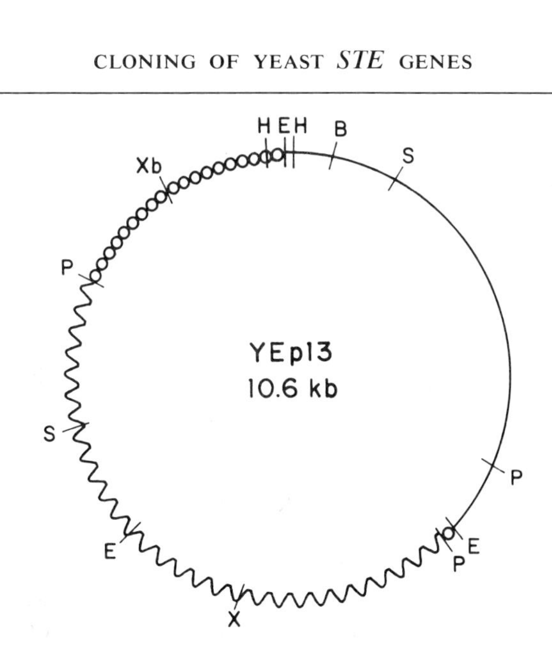

FIG. 2. Schematic representation of vector YEp13. The solid line indicates pBR322, the wavy line represents the yeast chromosomal fragment containing the *LEU2* gene, and the circles indicate sequences derived from the yeast 2 μm plasmid. Restriction sites: *Eco*RI (E), *Hin*dIII (H), *Bam*HI (B), *Sal*I (S), *Pst*I (P), *Xho*I (X), *Xba*I (Xb).

LEU2 gene as a selection marker. Enrichment for ampicillin-resistant, tetracycline-sensitive *Escherichia coli* transformants yielded a library in which most of the plasmids contain yeast inserts. The experimental methodology described in the rest of the chapter will be based on the use of a YEp13 library. However, other vectors with different selection markers are also suitable, although the stability and copy number of YEp-type vectors provide some advantages.

Yeast Strains

Strain AB320, used to construct the library, has the following genotype[16]

$$\frac{\mathbf{a}}{\alpha}\ \frac{HO}{HO}\ \frac{ade2}{ade2}\ \frac{lys2}{lys2}\ \frac{trp5}{trp5}\ \frac{leu2}{leu2}\ \frac{can1}{can1}\ \frac{met4}{met4}\ \frac{ura3}{ura3}\ \text{and/or}\ \frac{ura1}{ura1}$$

Mutants with defects in the *STE* genes listed in Table I have been isolated by several research groups and are generally available from the Yeast Genetics Stock Center (University of California, Berkeley) or from the individual investigators. To reduce the background interference, mutant alleles with insignificant residual activity (leakiness) and low reversion rates are preferable.

Since the mutant strain to be transformed must also have a mutation in the *LEU2* gene, which is the selection marker on the plasmid, it is necessary to construct the double *leu2 steX* mutant by crossing the *steX* mutant with a *leu2* strain of the opposite mating type. Although the *ste* mutations block mating (which is a necessary step in strain construction), many of the mutations are temperature-sensitive or leaky enough so that diploids can be selected.[4,6] For the screening methods described below, the transformed strain must have at least one additional, and preferably more, auxotrophic mutations, such as *trp1*, *his4*.

In addition to the strain to be transformed, a test strain of the opposite mating type is needed to detect mating-competent transformants. This should carry one or more auxotrophic mutations that complement those in the transformed strain and may also have a *leu2* mutation, depending on which variation of the screening method is employed (see below).

Yeast Media

YEPD (1% yeast extract, 2% peptone, 2% glucose, 2% agar) is a rich complex medium used for routine culturing and rapid nonselective growth of cultures. For chemically defined medium, a basic minimal medium (MV) of 0.67% Difco-yeast nitrogen base without amino acids, 2% glucose, and 2% agar can then be supplemented with adenine, uracil, and a number of amino acids, as described previously.[4,32] This synthetic complete (SC) medium contains all the nutritional supplements any of our mutants might require. – Leu is SC lacking leucine and is therefore selective for the growth of cells containing YEp13.

Enzymes

Glusulase, the lytic enzyme preparation used in yeast transformation, is a crude mixture of digestive enzymes from the snail *Helix pomatia* and is available from Endo Laboratories (Garden City, New York). Some investigators prefer Zymolyase (Miles), a lytic enzyme preparation from *Arthrobacter luteus*. Restriction enzymes for plasmid analysis are purchased from standard suppliers and used in digests as recommended.

[32] R. K. Mortimer and D. C. Hawthorne, *in* "The Yeasts" (A. H. Rose and J. S. Harrison, eds.), Vol. 1, p. 385. Academic Press, New York, 1969.

Other Materials

Escherichia coli strain RR1[33] or an equivalent strain is used as host for the amplification and purification of plasmids. All other chemicals and reagents are available from standard sources.

Methods

The general outline of the experimental approach is diagrammed in Fig. 3 and described in detail below. Table II summarizes the results at each stage in attempts to clone several *STE* genes and *TUP1*.

Transformation of a leu2 steX *Mutant with the YEp13 Library*

Detailed descriptions of methods for the genetic transformation of yeast have been published.[34,35,35a] We use Beggs' procedure[34] with two modifications.[24]

1. Cells are incubated at 30° in buffer containing 1% (v/v) glusulase only until the majority of cells in the population have become osmotically sensitive and lyse in water or a detergent-containing buffer. With most of our strains, 50–95% of the cells are osmotically sensitive after 20 min of Glusulase treatment. This limited exposure, in which most of the treated cells retain their morphology, leads to rapid regeneration and colony formation by transformed spheroplasts and high yields of transformants, probably because a higher percentage of the spheroplasts regenerate.
2. All but one of the washes has been eliminated; the only one retained is immediately after Glusulase treatment when the cell–spheroplast mixture is washed with 1 *M* sorbitol, 0.01 *M* $CaCl_2$, 0.01 *M* Tris-HCl, pH 7.5, before resuspension in a small volume of this buffer and the addition of DNA. Omitting the five other washes of the Beggs procedure significantly reduces the time and quantities of materials required, particularly when several strains are to be transformed at the same time, although the overall yield of transformants in one experiment was estimated to be reduced about two- to three-fold. Since even with the shortened method most of our strains transform at sufficiently high frequency (generally 10^3 to 10^4 transformants per microgram of DNA per 10^8 initial cells), the decreased yield is acceptable. However, transformation efficiencies can vary widely between strains.

[33] K. A. Nasmyth and S. I. Reed, *Proc. Natl. Acad. Sci. U.S.A.* **77,** 2119 (1980).
[34] J. D. Beggs, *Nature* (*London*) **275,** 104 (1978).
[35] C. Ilgen, P. J. Farabaugh, A. Hinnen, J. M. Walsh, and G. R. Fink, *in* "Genetic Engineering" (J. K. Setlow and A. Hollaender, eds.), p. 117. Plenum, New York, 1979.
[35a] H. Ito, Y. Fukuda, K. Murata, and A. Kimura, *J. Bacteriol.* **153,** 163 (1983).

Transformation of leu2 steX with the YEp13 library

↓

Resuspension of Leu$^+$ transformants and replating on -Leu medium

↓

Screening of Leu$^+$ colonies for mating ability

↓

Isolation of and E. coli transformation with plasmids from Ste$^+$ transformants

↓

Rapid isolations and screening of plasmids from E. coli

Retransformation of leu2 steX with isolated plasmids

Positive identification of cloned STEX gene

FIG. 3. Flow diagram for the cloning of *STE* genes.

TABLE II

SUMMARY OF RESULTS AT STAGES IN ATTEMPTS TO CLONE *STE* GENES AND *TUP1* FROM THE LIBRARY OF *Saccharomyces cerevisiae* DNA FRAGMENTS IN YEp13

	TUP1[a]	*STE2*[b]	*STE6*[b]	*STE7*[c]	*STE11*[c]	*STE12*[c]
Total transformants	~18,000	~70,000	~170,000	~11,000	~45,000	~28,000
Replated colonies tested for mating	5,500	~21,000	~61,000	3,500	~84,000	~79,000
Initial mating-positive colonies	36	~60	~130	19	96	137
Mating-positive colonies after retesting	1	8	4	3	0	0
Plasmids/transformants	3/1	8/6	2/4	2/1	—	—
Plasmids that complement the mutation	1	5	2	INC[d]	—	—
Positively identified gene	1	1	1	INC	—	—

[a] M. N. Jagadish and V. L. MacKay, unpublished, 1982.
[b] A. Hartig and V. L. MacKay, unpublished, 1982.
[c] M. Cortelyou and V. L. MacKay, unpublished, 1982.
[d] INC = incomplete analysis.

Selection of YEp13-transformed cells is done by mixing the transformed cell–spheroplast suspension with −Leu top agar (−Leu medium containing 2.5% agar and 1 *M* sorbitol) that has been melted and held at 52°, and plating them on −Leu agar medium that contains 1 *M* sorbitol. Regenerated colonies are frequently just visible after 24 hr at 30° and usually are grown enough for further handling after an additional day.

Resuspension and Replating of Regenerated Colonies

Because the transformant colonies are embedded within the top agar, this layer must be removed and homogenized to obtain resuspended cells from each colony. The top agar can be homogenized in a blender with enough −Leu liquid medium or buffer to break up visibly all the agar pieces (final volume of approximately 100 ml when top agar from five plates was homogenized). Microscopic examination in a hemacytometer of a sonicated dilution from this suspension showed little debris and only single cells at a frequency within the expected range for the number of transformants. This dilution can be further diluted, if necessary, for plating on −Leu to yield approximately 500 colonies per plate. In the first screening of a transformant population, we usually test approximately 10,000 colonies; the cell suspension can be stored at 4° and plated again for additional colonies.

An alternative approach that avoids resuspension and replating has been developed by Garvik and Hartwell[36]; the top agar contains 3% Sea-Plaque (FMC) low gelling-temperature agarose instead of agar. Transformants are mixed with 2.5 ml of the top agar (melted and held at 40°) and plated onto 40° prewarmed plates of selective medium containing sorbitol. These are incubated until the embedded colonies have grown onto the agar surface (approximately 90% efficient), which can then be screened directly for mating ability by replica plating as described below.

Screening for Mating Ability

After the transformed cells have formed colonies on −Leu, they are replica-plated to fresh lawns of a test strain of the opposite mating type having one or more genetic markers that complement those in the transformed strain. At least two variations of this step are possible (Fig. 4).

Method a. If the test strain does not require leucine for growth, then the transformant colonies are replica-plated (and marked for orientation of the plates) to lawns of this strain on −Leu, so that the selective pressure for plasmid-bearing cells is maintained. The plates are incubated at

[36] B. Garvik and L. H. Hartwell, personal communication, 1982.

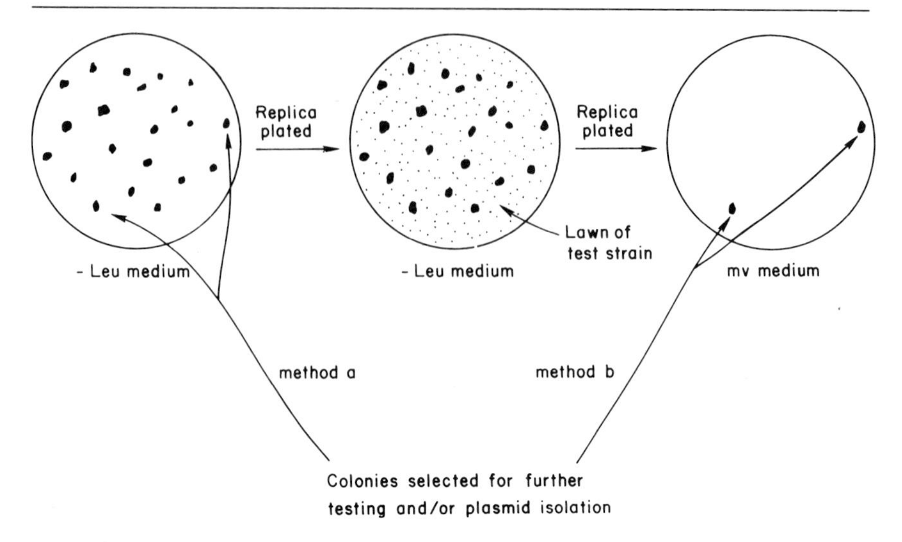

FIG. 4. Screening methods for Ste^+ transformants. See text for explanation.

30°, or at the restrictive temperature for a temperature-sensitive *ste* mutant, for approximately 24 hr to allow mating, then replica-plated to a medium, usually MV, that selects for the diploid progeny of a mating and reincubated. Progeny from colonies with mating efficiencies $\geq 1\%$ that of wild-type cells can be detected on the MV plate after 24 hr. After an additional day of incubation, however, the background on the MV plate can begin to obscure these results, particularly when the *ste* mutation is somewhat leaky. Therefore, mating is usually scored after 24 hr. The MV plate is compared to the −Leu master to mark the colony or area on the master that gave rise to high-frequency mating. Cells from the master plate are transferred to fresh −Leu and retested against test strains of both mating types to eliminate false-positives, contaminants, etc.

To ensure that the acquired mating ability in the positive transformants is due to a plasmid-borne gene, we routinely test for co-loss of both phenotypes, i.e., leucine prototrophy and mating ability. Transformants are grown overnight (10–12 generations) under nonselective conditions (YEPD or SC) in liquid medium or on plates. The culture is then plated or streaked out for single colonies on YEPD or SC, which develop in about 2 days. These are replica-plated to −Leu medium to identify Leu^- colonies that arose from segregants that had lost the plasmid. Representative Leu^- and Leu^+ colonies are transferred from the master plate to SC and −Leu, respectively, and tested for mating. If the chromosomal *ste* mutation is complemented by a plasmid-borne gene, then all Leu^- segregants should be nonmaters. However, some Leu$^+$ single colonies may also now fail to mate, since under our transformation conditions most cells take up and maintain more than one plasmid. During mitotic growth, each daughter cell may not inherit each of the plasmids in the original transformant.

Method b. If the *ste* mutation in the transformed strain allows little or no residual mating (e.g., the amber allele *ste4-4*[6,24]), then a direct selection for mating may be more expeditious and allows screening of many more colonies per plate. In this case, the test strain of opposite mating type would carry a *leu2* mutation in addition to markers that complement auxotrophic mutations in the transformed strain. The transformant colonies on −Leu are replicated to a fresh lawn of the test strain on −Leu containing a low level of leucine (approximately 10–20% of our standard concentration which is 80 mg/liter). The low level of leucine permits in the test strain a limited period of macromolecular synthesis that is required for mating. Once the leucine is exhausted, growth of this strain and mating will cease, although the transformed cells and any diploid progeny bearing a plasmid will continue to grow. After 24 hr, the plates are replica-plated to MV, or some other appropriate medium, to select for diploid progeny. Since these must contain plasmids in order to grow in the absence of leucine, this method provides a direct selection for cells with plasmids that complement the *ste* mutation. However, because the **a**/α diploid progeny will be nonmaters and cannot be retested for mating ability, false-positives, revertants, etc., cannot be eliminated at this stage. Unless the *ste* mutation completely blocks residual mating, only large diploid colonies appearing on MV at 24 hr should be selected for further analysis. Moreover, as discussed above, during transformation most cells take up and maintain more than one plasmid. After mating, there is no further selection for a plasmid carrying the *STE* gene, but only for the *LEU2* gene on all the plasmids. The *STE*-bearing plasmid, therefore, may be lost during subsequent mitotic divisions of the diploids. It may be necessary then to isolate plasmids from a number of initial positives, amplify them in *E. coli,* and retransform a *steX* recipient with one of each type to find one that complements the *steX* mutation.

Isolation of Plasmid from Ste$^+$ *Transformants and Retransformation*

Several methods have been published for the rapid isolation of DNA from yeast;[33,34,37–40] we usually use a minor variation[41] of the procedure of

[37] D. R. Cryer, R. Eccleshall, and J. Marmur, *Methods Cell Biol.* **12,** 39 (1975).

[38] F. Sherman, G. Fink, and J. Hicks, "Laboratory Manual for Methods in Yeast Genetics." Cold Spring Harbor Laboratory, New York, 1979.

Nasmyth and Reed.[33] For each 5 ml of yeast culture grown in – Leu liquid medium, the cells are resuspended in 0.4 ml of 1 *M* sorbitol, 0.1 *M* sodium citrate, pH 5.8, 10 m*M* EDTA, 0.1 *M* 2-mercaptoethanol, 0.1 mg of Zymolyase 60,000 per milliliter. After 30 min at 37°, 0.2 ml of 50 m*M* Tris-HCl, pH 8, 50 m*M* EDTA, 0.5% SDS is added to lyse the cells. The mixture is incubated with 2 μl of diethyl pyrocarbonate at 65° for 30 min, cooled on ice, and kept on ice for 60 min after the addition of 0.1 ml of 5 *M* potassium acetate. After centrifugation, the nucleic acid is precipitated with 2 volumes of 95% ethanol for 30–60 min at – 70°. The pellet is resuspended in 10 m*M* Tris-HCl, pH 8, 1 m*M* EDTA, adjusted to 0.2 *M* NaCl, and reprecipitated with ethanol. The final pellet is resuspended in 25 μl of 10 m*M* Tris-HCl, pH 8, 1 m*M* EDTA.

The plasmid preparations from yeast are used to transform a suitable *E. coli* strain[42]; with the YEp13 vector, transformants are selected on ampicillin-containing medium. Plasmids from individual *E. coli* transformants are then isolated, usually by a rapid, small-scale method,[43,44] and compared by agarose gel electrophoresis of restriction fragments. Since in our experiments most of the original yeast transformants have contained more than one plasmid, a number of *E. coli* transformants should be screened to ensure that at least one carrying the putative *STEX* gene is represented; usually it is sufficient to screen 12 *E. coli* transformants for each initial yeast transformant. One representative of each plasmid type is then used to transform the *leu2 steX* yeast recipient to identify which if any carries a fragment that complements the *steX* mutation. The yeast transformants are selected for Leu$^+$ and screened for mating ability as described above.

It is also possible to transform the *leu2 steX* yeast strain directly with the plasmid preparations from the original yeast transformant without passage and purification through *E. coli*.[24] While this approach allows more rapid confirmation that one of the plasmids contains a gene that complements the *steX* mutation, subsequent screening of plasmids from *E. coli* transformants is still necessary to determine which plasmid is responsible for the complementation.

[39] A. Zamir, C. V. Maina, G. R. Fink, and A. A. Szalay, *Proc. Natl. Acad. Sci. U.S.A.* **78,** 3496 (1981).

[40] K. Struhl, D. T. Stinchcomb, S. Scherer, and R. W. Davis, *Proc. Natl. Acad. Sci. U.S.A.* **73,** 1035 (1979).

[41] K. Tatchell, personal communication, 1980.

[42] F. Bolivar, R. L. Rodriguez, P. J. Greene, M. C. Betlach, H. L. Heyneker, M. W. Boyer, J. H. Crosa, and S. Falkow, *Gene* **2,** 95 (1977).

[43] D. Ish-Horowicz and J. F. Burke, *Nucleic Acids Res.* **9,** 2989 (1981).

[44] D. S. Holmes and M. Quigley, *Anal. Biochem.* **114,** 193 (1981).

Confirmation of the Cloned Gene as STEX

Isolation of a plasmid carrying a gene that complements the *steX* mutation is not sufficient evidence that the cloned gene is *STEX*, since there are now several examples of genes cloned in YEp13 that suppress mutations in other genes. For the suppression of *ste4* mutations by the cloned *STE5* gene, this effect was allele specific; two different temperature-sensitive, presumably missense, *ste4* mutations were each suppressed by the *STE5* gene in YEp13, whereas the nonsense mutation *ste4-4* was not.[18] Therefore, if a second, independent mutation in the *STEX* gene is available (preferably a nonsense mutation), it may be possible to ascertain that the cloned gene is not *STEX*.

Positive identification of the cloned gene as *STEX* requires further evidence, such as integration analysis; integration of the plasmid by sequence homology at the *steX* locus establishes that the cloned fragment has that sequence. However, it is generally difficult to obtain integration directly with YEp13 vectors for at least two reasons. The plasmids are usually rather stable, so that integrants with increased mitotic stability are difficult to detect. Also, the *Pst*I fragment containing the *LEU2* gene also has part of a Tyl element, a sequence that is moderately repeated throughout the yeast genome,[45] so integration by sequence homology can occur at these sites or at *LEU2*. However, YEp-type plasmids can be readily integrated at the chromosomal site homologous to the insert if the transforming plasmid is first converted to a linear form by a restriction enzyme that cuts within the insert.[46] The entire linear molecule is integrated, including the 2 μm sequences, which could subsequently recombine with endogenous 2 μm plasmids, leading to instability, rearrangements, etc., that might complicate genetic analysis of the integrants. However, we have used this approach to establish that the cloned gene that complements the *tup1* mutation is *TUP1*.[21]

Integration can also be achieved by subcloning the gene into an integrative vector, the YIp type, that is incapable of autonomous replication.[47] The plasmid must be an integrated state in the transformants[48] and, as above, integration can be directed to the chromosomal site homologous to the insert by transforming with linear molecules cut within the insert.[46] Alternatively, in the identification of *STE5*, the gene was subcloned into

[45] A. J. Kingsman, R. L. Gimlich, L. Clarke, A. C. Chinault, and J. Carbon, *J. Mol. Biol.* **145,** 619 (1981).

[46] T. L. Orr-Weaver, J. W. Szostak, and R. J. Rothstein, *Proc. Natl. Acad. Sci. U.S.A.* **78,** 6354 (1981).

[47] D. Botstein, S. C. Falco, S. E. Stewart, M. Brennan, S. Scherer, D. T. Stinchcomb, K. Struhl, and R. W. Davis, *Gene* **8,** 17 (1979).

[48] A. Hinnen, J. B. Hicks, and G. R. Fink, *Proc. Natl. Acad. Sci. U.S.A.* **75,** 1929 (1978).

YRp7, an autonomously replicating plasmid (due to a chromosomal *ars* sequence on the plasmid[40]) that is mitotically unstable.[49,50] The selecting marker on YRp7 is the yeast *TRP1* gene and integrants of YRp7-derived plasmids are easily isolated as those yeast transformants that retain the Trp^+ phenotype after nonselective growth. In the four independent *STE5* integrants that were analyzed, the plasmid had integrated at or very close to the chromosomal *ste5* site (see below).

The site of integration can be readily determined by classical genetic analysis of the integrants.[32] For example, in the analysis of *STE5*, an α *ste5 trp1* strain was transformed with a YRp7 plasmid carrying the subcloned *STE5* fragment and stable Trp^+ integrants were selected from the transformants. These were crossed with an **a** *trp1* strain with the wild-type *STE5* gene. Meiotic segregations from all four diploids were consistent with the predicted results if the plasmid integrated by sequence homology at the *ste5* locus; i.e., the tetrads segregated 2 Trp^+ : 2 Trp^- and 4 maters : 0 nonmaters. Genetic analysis of the integrants can be complicated, however, by the occurrence of spheroplast fusion during the transformation procedure. Usually, haploid strains are transformed, but frequently the transformant colonies that arise are found to be diploid,[24] completely homozygous at the mating-type locus and for the genetic information in the original haploid strain. If such diploid cells are mated with a haploid, viability of the meiotic progeny derived from the resulting triploid will be extremely low.[32] Spheroplast fusion during transformation can perhaps be avoided by using lower cell densities after the spheroplasting step.

The identity of the cloned gene can also be confirmed by the segregation of restriction site polymorphisms for the cloned fragment in meiotic products. This approach was used to identify *MAT, HML,* and *HMR,*[12] and *STE5*.[19]

Comments

The experimental data in Table II show the substantial variability among attempts to clone different *STE* or mating-related genes such as *TUP1*. These differences probably reflect random chance to some extent, but may also suggest that certain DNA fragments and/or specific genes are underrepresented in the YEp13 library or that these genes might be deleterious in yeast when present in high copy number plasmids. Use of a different library in a low copy number vector, such as a centromeric plas-

[49] D. T. Stinchcomb, K. Struhl, and R. W. Davis, *Nature* (*London*) **282,** 39 (1979).
[50] A. J. Kingsman, L. Clarke, R. K. Mortimer, and J. Carbon, *Gene* **7,** 141 (1979).

mid, could circumvent these problems, although with this vector the potential suppression of a mutation in one gene by a different cloned gene probably would not be detectable.

Technical problems, including those stemming from leaky mutations, plasmid instability, multiple plasmids in the initial yeast transformants, and the formation of diploids by spheroplast fusion during transformation, have been discussed in previous sections.

In the mating-type system, there are now several examples of suppression of a mutation by another gene in a high copy number state. Such interactions probably could occur for other complex cellular functions as well and could provide new insights about the process. It was fortuitous that we were easily able to establish *STE5* as the gene that suppressed *ste4* mutations,[18] but it should be possible in other cases to identify the cloned gene. After integration of a plasmid carrying the cloned gene at the chromosomal site homologous to the insert, the integration site can be mapped by classical genetic methods.[32] Although in the integrated state the cloned gene may no longer suppress the *steX* mutation, the selecting marker on the vector can be readily scored to determine whether integration occurred at or near a known *ste* mutation.

Acknowledgments

I wish to thank the many investigators who allowed the citation of their unpublished results and also Mike Leibowitz and Bill McAllister, who provided useful criticism of the manuscript. This work was supported by grants from the National Institutes of Health, the National Science Foundation, and the Charles and Johanna Busch Memorial Fund.

Section III

Systems for Monitoring Cloned Gene Expression

A. Intact Cell Systems
Articles 23 through 25

B. Introduction of Genes into Mammalian Cells
Articles 26 through 33

C. Cell-Free Systems; Transcription
Articles 34 through 36

D. Cell-Free Systems; Translation
Articles 37 through 45

[23] Analysis of Recombinant DNA Using *Escherichia coli* Minicells

By JOSEPHINE E. CLARK-CURTISS and ROY CURTISS III

Minicells are a product of an aberrant cell division due to frequent septations occurring at the polar ends of cells.[1,2] Minicells, which are produced continually during the growth of the bacterial culture, contain RNA and protein but little or no chromosomal DNA. Minicells produced by plasmid-containing strains usually, but not always, contain plasmid DNA. These plasmid-containing minicells have been particularly useful for studies on the replication, recombination, and repair of plasmid DNA and for studies on the synthesis of plasmid-specified gene products.[3]

The most widely studied minicell-producing strains are derivatives of *Escherichia coli* K12. However, minicell-producing mutants from strains of *Salmonella, Shigella, Erwinia, Vibrio, Bacillus,* and *Haemophilus* species have also been isolated and studied.[3,4]

Since a diversity of cosmid[5] and plasmid[6] cloning vectors segregate into minicells produced by strains harboring these vectors, minicells constitute a useful system for studying the synthesis and stability of gene products specified by fragments of "foreign" DNA inserted into these vectors. Thus, it is possible to study the size, rate of synthesis, and stability of individual messenger RNA transcripts, and to measure the size, rate of synthesis, and stability of individual translation products. In addition, it is possible to fractionate minicells containing radioactively labeled plasmid-specified proteins to determine whether these proteins are located in the cytoplasm, the cytoplasmic membrane, the periplasm, the outer membrane or are secreted.[7–11] By analyzing the various fractions, it is also

[1] H. I. Adler, W. D. Fisher, and G. E. Stapleton, *Science* **154,** 417 (1966).
[2] H. I. Adler, W. D. Fisher, A. Cohen, and A. A. Hardigree, *Proc. Natl. Acad. Sci. U.S.A.* **57,** 321 (1967).
[3] A. C. Frazer and R. Curtiss III, *Curr. Top. Microbiol. Immunol.* **69,** 1 (1975).
[4] P. Gemski and D. E. Griffin, *Infect. Immun.* **30,** 297 (1980).
[5] J. Collins and B. Hohn, *Proc. Natl. Acad. Sci. U.S.A.* **75,** 4242 (1978).
[6] F. Bolivar and K. Backman, this series, Vol. 68, p. 245.
[7] R. C. Gayda and A. Markovitz, *J. Bacteriol.* **136,** 369 (1978).
[8] B. R. Glick, J. Zeisler, A. M. Banaszuk, J. D. Friesen, and W. G. Martin, *Gene* **15,** 201 (1981).
[9] J. B. Hansen, Y. Abiko, and R. Curtiss III, *Infect. Immun.* **31,** 1034 (1981).
[10] T. J. Larson, V. A. Lightner, P. R. Green, P. Modrich, and R. M. Bell, *J. Biol. Chem.* **255,** 9421 (1980).
[11] K. Motojima, I. Yamato, Y. Anraku, A. Nishimura, and Y. Hirota, *Proc. Natl. Acad. Sci. U.S.A.* **76,** 6255 (1979).

METHODS IN ENZYMOLOGY, VOL. 101

ISBN 0-12-182001-7

possible to study the types of processing and/or modification involved in the movement of a protein from the cytoplasm to some extracytoplasmic location.[12] The experimental use of minicells is often facilitated by employing drugs that inhibit specific classes of macromolecular synthesis or of cellular function.[12,13] The use of minicell-producing strains with specified gene defects can also be advantageous in increasing the stability of macromolecules or in the study of protein processing mechanism.

Material and Methods

Media

Minimal Media. Several minimal media have been used to obtain essentially equivalent yields of minicells from minicell-producing strains. M9 minimal medium contains 60 g of Na_2HPO_4, 30 g of KH_2PO_4, 5 g of NaCl, 10 g of NH_4Cl per liter of deionized water to yield a 10× concentrated stock solution.[14] ML medium contains 50 g of NH_4Cl, 10 g of NH_4NO_3, 20 g of Na_2SO_4 (anhydrous), 90 g of K_2HPO_4 (anhydrous), and 30 g of KH_2PO_4 (anhydrous) per liter to yield a 10× stock.[15] Each salt added should be completely dissolved before adding the next salt. The pH of both media should be about 6.8–6.9 prior to autoclaving for 20–30 min. Prior to use, both media are diluted 10-fold with sterile deionized water and a carbon source at 0.5% (final concentration), and nutrient supplements appropriate for the bacterial strain are added. Addition of 1 ml of 1 *M* $MgSO_4 \cdot 7\ H_2O$ per liter is necessary for 1 × M9 medium and 0.4 ml of 1 *M* $MgSO_4 \cdot 7\ H_2O$ for 1 × ML medium. Improved yields of minicells can usually be achieved by addition of casamino acids to a 0.5–1.5% final concentration. The addition of casamino acids can sometimes be deleterious if protein synthesis is to be measured in plasmid-containing minicells or in instances in which expression of the genetic information of interest is regulated by free amino acids.[3]

3XD Medium.[16] The 3XD medium is a rich minimal medium that allows growth of bacteria to a very high density. It is prepared by adding 150 ml of a phosphate concentrate [30 g of KH_2PO_4 (anhydrous) and 70 g of Na_2HPO_4 (anhydrous) per liter of water], 10 ml of a 10% NH_4Cl solu-

[12] R. C. Gayda, G. W. Henderson and A. Markovitz, *Proc. Natl. Acad. Sci. U.S.A.* **76,** 2138 (1979).

[13] K. J. Roozen, R. G. Fenwick, Jr., and R. Curtiss III, *J. Bacteriol.* **107,** 21 (1971).

[14] R. W. Davis, D. Botstein, and J. R. Roth, *in* "Advanced Bacterial Genetics," p. 203. Cold Spring Harbor Laboratories, Cold Spring Harbor, New York, 1980.

[15] R. Curtiss III, *J. Bacteriol.* **89,** 28 (1965).

[16] D. Fraser and E. A. Jerrell, *J. Biol. Chem.* **205,** 291 (1953).

tion, 6 ml of a 10% $MgSO_4 \cdot 7 H_2O$ solution, 10 ml of a 0.1% gelatin solution, 24.0 ml of glycerol (30 g), 3.0 ml of a 0.05 *M* $CaCl_2$ solution (add last and slowly with stirring), 797 ml of H_2O, and 15 g of casamino acids (Difco).

Buffered Saline with Gelatin (BSG).[15] BSG is used as a diluent for suspending minicells during their purification and as the basic buffer to which sucrose or glycerol are added in constructing linear sucrose or glycerol gradients. BSG contains 8.5 g of NaCl, 0.30 g of KH_2PO_4 (anhydrous), and 0.60 g of Na_2HPO_4 (anhydrous) in 990 ml of H_2O to which 10 ml of 1% gelatin are added.

Isolation of Minicell-Producing Mutant Strains of Bacteria

Minicell-producing mutants of various bacterial species have been isolated following mutagenic treatment with triethylenemelamine, UV irradiation, and nitrosoguanidine.[3] Some minicell-producing mutants differ in colony morphology from the parental strains, but most minicell-producing mutants have been isolated by screening cells in surviving colonies by phase-contrast microscopy. Minicell-producing strains of *Escherichia* and *Salmonella* have approximately two times greater resistance to ionizing radiation than do the parental strains. Selection for resistance to X-rays has been successfully employed to obtain minicell-producing mutants.[1,2,17]

Minicell-Producing Strains of Escherichia coli

The table lists some of the available minicell-producing strains of *E. coli* that have potential utility in recombinant DNA research.[1-3,9,18,19] χ1488, χ1849, χ1776, χ2001, χ2341, and χ2363 all have the *hsdR2* mutation, which abolishes the K12 restriction enzyme activity but allows for the modification methylase to be fully functional. This permits subsequent transfer of recombinant DNA introduced into these strains into a diversity of other K12 strains that are proficient at both restriction and modification. Some of the strains, such as χ984, χ1274, χ1411, χ2001, χ2341, and χ2363, are sensitive to bacteriophage λ and thus are suitable for use with cosmid or lambda cloning vectors. χ1274, χ2001, and χ2363 have *recA* mutations that facilitate stability of some recombinant plasmids. χ2341, χ2363, and χ2651 lack nonsense suppressors, whereas all the other

[17] N. A. Epps and E. S. Idziak, *Appl. Microbiol.* **19,** 338 (1970).

[18] R. Curtiss III, D. A. Pereira, J. C. Hsu, S. C. Hull, J. E. Clark, L. J. Maturin, Sr., R. Goldschmidt, R. Moody, M. Inoue, and L. Alexander, *Recomb. Mol. Impact Sci. Soc., Miles Int. Symp. Ser.* **10,** 64 (1977).

[19] B. J. Bachmann and K. B. Low, *Microbiol. Rev.* **44,** 1 (1980).

MINICELL-PRODUCING STRAINS OF *Escherichia coli* K12

Strain designation	Genotype[a]	Reference
χ925	F⁻ *thr-1 ara-13 leu-6 azi-8 tonA2 lacY1 minA1 glnU44 gal-6* λ⁻ *minB2 rpsL135 malA1 xyl-7 mtl-2 thi-1*	Single colony isolate of P678-54; Adler *et al.*[1,2]
χ984	F⁻ *minA1 tsx-63 purE41 glnU42* λ⁻ *pdxC3 minB2 his-53 metC65 rpsL97 tte-1 xyl-14 ilv-277 cycB2 cycA1*	Frazer and Curtiss[3]
χ1274	F⁻ *minA1 tsx-63 purE41 glnU42* λ⁻ *pdxC3 minB2 recA1 metC65 rpsL97 tte-1 xyl-14 ilv-277 cycB2 cycA1*	Frazer and Curtiss[3]
χ1411	F⁻ prototroph *minA1 glnU42* λ⁻ *minB2*	Frazer and Curtiss[3]
χ1488	F⁻ *minA1 purE41 glnU42* λ⁻ *pdxC3 minB2 his-53 metC65 rpsL97 tte-1 xyl-14 ilv-277 cycB2 cycA1 hsdR2*	This text
χ1776	F⁻ *tonA53 dapD8 minA1 glnU42* Δ40[*gal-uvrB*] λ⁻ *minB2 rfb-2 nalA25 thyA142 oms-2 metC65 oms-1 tte-1* Δ29[*bioH-asd*] *cycB2 cycA1 hsdR2*	Curtiss *et al.*[18]
χ1849	F⁻ *tonA53 dapD8 minA1 purE41 glnU42* Δ40[*gal-uvrB*] λ⁻ *minB2 his-53 nalA25 metC65 oms-1 tte-1* Δ29[*bioH-asd*] *ilv-277 cycB2 cycA1 hsdR2*	Hansen *et al.*[9]
χ2001	F⁻ Δ*araC766 tonA53 dapD8 proA370* Δ*lacZ39 minA1* Δ69[*gal-chlD*] λ⁻ *tyrT58* Δ*galU183* Δ*trpE5 minB2 rfb-2 recA56 relA1* Δ*thyA57 endA1 oms-1* Δ*asd-4 rpoB402 cycB2 cycA1 hsdR2*	This text
χ2341	F⁻ Δ*araC766 tonA53 dapD8 proA318 minA1* Δ69[*gal-chlD*] λ⁻ Δ*trpBC13 minB2 rfb-2 nalA25* Δ*thyA57 endA1 oms-1* Δ*asd-4 cycB2 cycA1 hsdR2*	This text
χ2363	F⁻ Δ*araC766 tonA53 dapD8 proA70* Δ*lacZ39 minA1* Δ69[*gal-chlD*] λ⁻ Δ*galU183* Δ*trpE5 minB2 rfb-2 recA56 relA1* Δ*thyA57 endA1 oms-1* Δ*asd-4 rpoB401 cycB2 cycA1 hsdR2*	This text
χ2651	F⁻ *minA1 tsx-63* λ⁻ *pdxC3 minB2 his-53 metC65 rpsL97 tte-1 xyl-14 ilv-277 cycB2 cycA1*	From χ984 by R. Goldschmidt

[a] Genotype symbols as defined by Bachmann and Low.[19]

strains have *glnU* (*supE*) or *tyrT* (amber suppressor) mutations. χ1849, χ1776, χ2001, χ2341, and χ2363 all require diaminopimelic acid (DAP) and the omission of DAP from the minimal medium (containing lysine), which is used to suspend purified minicells results in DAP-less death of contaminating parental cells and their ultimate lysis. Because of alterations in their cell surfaces, χ1776, χ2001, χ2341, and χ2363 are sensitive to macrolide antibiotics such as erythromycin. This permits direct selection of plasmids that carry genes for macrolide resistance, especially those from gram-positive microorganisms. This property also allows the use of erythromycin (at 25–100 μg/ml) for amplification of copy number of plasmid vectors that express resistance to chloramphenicol.

Protein products expressed by recombinant molecules are often unstable in *E. coli* owing to proteolytic degradation.[20] Mutant derivatives of minicell-producing strains that have *lon* (*capR, deg*) mutations are often quite useful in such instances. When the "foreign" genetic information expressed in *E. coli* complements a gene defect in the minicell-producing strain, we have frequently isolated spontaneous mutants for their better growth under conditions that select for maximal expression of the genetic information on the recombinant molecule. Very often these mutants will possess mutations in the *lon* locus, which decreases the rate or extent of proteolytic degradation of the "foreign" gene products (J. B. Hansen, personal communication).

Transfection and Transformation

Introducing Recombinant Cosmid Vectors Packaged in λ Particles into Minicell-Producing Strains

The λ-sensitive, minicell-producing strains are grown in medium containing 1% tryptone, 0.5% yeast extract, 0.5% NaCl, and 0.4% maltose (DAP must be added for strains requiring this amino acid). When the cells reach a density of about 5×10^8/ml, they are sedimented by centrifugation and suspended in 10 mM $MgSO_4$. Preadsorption of the packaged cosmid recombinants is allowed to occur for about 20 min at 37°, after which suitably diluted samples are plated on a complex agar medium containing an antibiotic to which the cosmid vector confers resistance.

Introduction of Plasmid Recombinants by Transformation

Numerous protocols exist for the transformation of bacterial strains with recombinant plasmid DNA.[9,21–23] Two methods, developed in our laboratory by Gurnam Gill and Laura Alexander, that work quite well

[20] R. B. Meagher, R. C. Tait, M. Betlach, and H. W. Boyer, *Cell* **10,** 521 (1977).

[21] M. Mandel and A. Higa, *J. Mol. Biol.* **53,** 159 (1970).

[22] R. Curtiss III, *in* "Manual of Methods for General Bacteriology" (P. Gerhardt, R. G. E. Murray, R. N. Costilow, E. W. Nester, W. A. Wood, N. R. Krieg, and G. B. Phillips, eds.), p. 248. American Society for Microbiology, Washington, D.C., 1981.

[23] S. R. Kushner, *in* "Genetic Engineering" (H. W. Boyer and S. Nicosia, eds.), p. 17. Elsevier/North-Holland, Amsterdam, 1978.

with various strains are described below. For a given strain, one method sometimes works better than the other, but the low pH-$MnCl_2$ method has the advantage of working with nonenteric gram-negative bacterial species.

$MgCl_2$ Method. After growth overnight as a standing culture, the bacterial strain is diluted 1:100 into 20 ml of L broth (10 g of tryptone, 5 g of yeast extract, 5 g of NaCl, 0.1 g of glucose per liter,[24] and grown with aeration at 37° to an optical density of about 0.07 at 600 nm. The culture is chilled in an ice-water bath, and cold $MgCl_2$ is added to a final concentration of 100 m*M*. After 20 min on ice, cells are sedimented by centrifugation at 4° and suspended in 0.4 ml of transformation buffer (50 m*M* $MgCl_2$, 75 m*M* KCl, 100 m*M* $CaCl_2$, 10 m*M* Tris-HCl, pH 8.0). Plasmid DNA (10–500 ng in 1–10 μl) and 0.2 ml of cells are mixed, held at room temperature for 25 min, and then heat-shocked for 5 min in a 30° water bath. After 60 min of incubation without aeration at 37° for phenotypic expression of drug resistance, suitably diluted samples are plated on complex medium selective for the antibiotic to which the vector confers resistance.

Low pH–$MnCl_2$ Method. A standing overnight culture of the recipient strain is diluted 1:100 into 40 ml of L broth and incubated with aeration (either by bubbling or shaking) at 37° until the cells reach early log phase (about 1 to 2 $\times$ 10^8 cells/ml, which is equivalent to an optical density of 0.1–0.15 at 600 nm). The culture is chilled on ice, and the cells are sedimented by centrifugation at 7000 *g* at 4°. The cell pellet is gently suspended in 10 ml of ice-cold 10 m*M* NaCl–50 m*M* $MnCl_2$–10 m*M* sodium acetate buffer (pH 5.6). The cell suspension is held on ice for 20 min, and the cells are sedimented by centrifugation. The cell pellet is gently suspended in 1–2 ml of ice-cold 75 m*M* $CaCl_2$–100 m*M* $MnCl_2$–10 m*M* sodium acetate buffer (pH 5.6). (It should be noted that the $CaCl_2$ concentration can be varied from 25 to 100 m*M*, and the $MnCl_2$ concentration from 50 to 200 m*M* so as to optimize the method for specific strains of *E. coli*.) Competent cells (200 μl) are added to chilled 13 $\times$ 100 mm test tubes, and 1–10 ng of plasmid DNA (in 1–10 μl) are added to each sample. If the plasmid DNA is in a volume of 10 μl or less, the buffer used to suspend the DNA will have little effect on the result. However, if the DNA sample is larger than 10 μl, the DNA should be diluted into the $CaCl_2$–$MnCl_2$-sodium acetate buffer (pH 5.6) used to suspend the competent cells. The DNA:cell mixture is held on ice in the cold room for approximately 30 min. Then the mixture is heat-shocked at 30° for 2.5 min. The rapidity of the temperature shift is important, but the actual temperature for the heat

[24] E. S. Lennox, *Virology* **1,** 190 (1955).

shock and the duration will vary depending upon the properties of the recipient strain. After heat shock, the mixture is neutralized to pH 7.0 by addition of 3 μl of 2 *M* Tris buffer (pH 7.4). The mixture is diluted 1:10 into a growth medium and incubated without aeration at 37° for 60–90 min to allow phenotypic expression of drug-resistance markers. Recombinant transformants are selected on appropriate complex medium containing the antibiotic to which the vector confers resistance.

Selection of Strains That Give High Yields of Minicells That Are Easily Separable from Parent Cells[3]

Microscopic Examination. The expression of the minicell-producing phenotype in newly constructed strains is variable. Similarly, in minicell-producing strains passaged on slants or subcultured, there is often a decrease in yield and purity of minicells as a function of increasing numbers of strain transfers. It is thus necessary initially to isolate transfectants or transformants that produce abundant minicells of uniform size and have parental cells whose size and length facilitate purification of minicells. Such initial and periodic screening of isolates is best accomplished by phase-contrast microscopic examination of suspensions of cells propagated from individual colonies. Derivatives that have uniform small minicells of an average diameter of about 0.8 μm are best. We have also found it beneficial to select derivatives that have rather long parental cells, as this facilitates minicell purification.

Quantitative Determination of Minicell Production. A more quantitative measure of yield of minicells and the ease with which they can be separated from the parental cells that produce them is achieved by layering 100–200 μl of a concentrated suspension of a minicell-producing culture in BSG on a 5-ml linear 5 to 20% sucrose in BSG gradient. Numerous such gradients can be conveniently centrifuged at top speed in a clinical centrifuge for 20–45 min, depending on the centrifuge model. Gradients can be examined visually to determine the relative yield of minicells and degree of separation between the minicell and parental cell bands. Alternatively, these gradients can be fractionated from the top using a Buchler density flow pump with an optical density profile monitor and a flow cell to obtain a more quantitative measure of minicell yield and degree of purification from cells. Moreover, the degree of cell contamination in the minicell band should be quantitated by plating for viable cells. A good minicell-producing strain will give no more than one contaminating parental cell for every 10^3 to 10^4 minicells after one sucrose gradient centrifugation. It should be noted that the minicell-producing cultures should be checked at the stage of growth that is likely to be used in subsequent experiments with minicells. Thus, cultures in the late log to early stationary phase are best scrutinized by these methods.

Growth of Minicell-Producing Strains

Minicells from log-phase cultures contain stable messenger RNAs that specify production of several membrane-associated proteins.[25] Translation of these messages disappears because of message instability in minicells harvested from late log-, early stationary-, or stationary-phase cultures. A number of studies have also demonstrated that synthesis of plasmid-directed mRNA and protein is highest in minicells from early stationary-phase cultures.[3,13] Thus, minicell-producing strains are grown in M9 or ML media supplemented with 0.5% casamino acids, 0.5% carbon source, and additional required supplements such as DAP, tryptophan, purines, pyrimidines, and vitamins. The strains are grown with vigorous aeration by shaking at 37° until total densities of 1 to 2 $\times$ 10^9 cells/ml are achieved. Supplemented 3XD medium is also excellent for achieving high yields of minicells. By appropriate choice of inoculum dilution (10^{-4}), one can inoculate cultures in the late afternoon to achieve early stationary phase by the next morning.

Labeling of Plasmid DNA in Minicell-Producing Strains

The amount of plasmid DNA segregating into minicells is not a constant. Indeed, some plasmids, such as prophage P1, are not segregated into minicells, and other plasmids, such as the F fertility plasmid, segregate only into about 1% of the minicells produced.[3,13,26] Thus, the fraction of minicells produced that contain plasmid DNA and the mean number of plasmid copies per minicell vary depending upon the plasmid replicon being employed as a cloning vector. The size and type of DNA insert can also influence plasmid segregation into minicells.

Because of these considerations, it is wise when comparing minicells containing two or more different recombinant plasmids with different inserts to establish the relative amounts of plasmid DNA contained in the different minicell preparations from the different strains. This is best done by allowing for the incorporation of [^{3}H]thymidine during growth of the minicell-producing culture. Addition of 1–10 μCi of [^{3}H]thymidine per milliliter is usually adequate for these purposes. The presence of *thyA* in the minicell-producing strains greatly facilitates labeling, but *thy*$^+$ strains can also be labeled if 200 μg of deoxyadenosine per milliliter are added

[25] S. B. Levy, *Proc. Natl. Acad. Sci. U.S.A.* **72,** 2900 (1974).

[26] L. R. Kass and M. B. Yarmolinsky, *Proc. Natl. Acad. Sci. U.S.A.* **66,** 815 (1970).

concurrently to block conversion of thymidine to thymine and deoxyribose 1-phosphate.[27] The amount of plasmid DNA in the minicells can be established after purification of minicells by comparing the acid-precipitable counts per optical density unit of purified minicells. Using this information, one can correct for observed differences in the amounts of plasmid-directed RNA or protein synthesis in the purified minicells depending on differences in the amounts of plasmid DNA.

Purification of Minicells

Many methods have been used to purify minicells.[3] For highly purified preparations, a combination of purification procedures is often necessary. The level of parental cell contamination in purified minicell preparations should be determined by plating appropriate dilutions for viable cell contaminants. Such cell contamination varies considerably from strain to strain and from experiment to experiment. The number of minicells per milliliter in a purified preparation is determined using a Petroff–Hauser counting chamber with a phase-contrast microscope, and the number can be converted to optical density units, which are more readily measured in subsequent experiments.

Differential Centrifugation. Differential low speed centrifugation of cultures of *E. coli* minicell-producing strains is quite effective in reducing the number of cells as much as 100-fold while leaving 50–70% of the minicells in the supernatant fluid to be subsequently collected by high-speed centrifugation. It is best to conduct such differential centrifugation at room temperature, since chilling to 4° can induce physiological changes that are detrimental in subsequent uses of the purified minicells.[3] Differential centrifugation is performed at 500–2000 *g* for 5 min. However, the choice of rotors, centrifugal force, and time of centrifugation must be determined empirically for different minicell-producing strains because of differences among strains with respect to cell length and minicell size. When the cultures are at high density, lower speeds for longer times lead to less cosedimentation of minicells with cells than higher speeds for shorter centrifugation times. Although one can achieve relatively high-purity minicell preparations by performing several successive cycles of differential centrifugation, differential rate sedimentation on sources or glycerol gradients is usually necessary to achieve high purity preparations.[3]

Differential Rate Sedimentation in Sucrose or Glycerol Gradients. High-purity minicell preparations (1 vegetative cell per 10^6 to 10^7 minicells) can be obtained by two cycles of differential rate sedimentation in linear gradients of glycerol (10 to 30%) or sucrose (5 to 20%), especially if

[27] D. R. Budman and A. B. Pardee, *J. Bacteriol.* **94,** 1546 (1967).

these gradient centrifugations follow a low speed differential centrifugation to remove many of the vegetative cells. In making the linear gradients, the glycerol or sucrose should be made up in BSG, since we have found that the minicells harvested in such gradients are more physiologically active than minicells harvested from gradients where the sucrose or glycerol is made up in Tris or phosphate buffers.[3] Since cold shock has often been found to be deleterious to later physiological function of plasmid-containing minicells, most purification steps for minicells are best done at room temperature.

Cultures containing cells and minicells are concentrated 20-fold (late log phase) to 10-fold (stationary phase) in BSG. The suspension is layered onto the surface of one or more 35-ml gradients with a Pasteur pipette in a volume equal to 4–8% of the volume of each gradient. If large quantities are required, minicells should be partially purified by differential centrifugation and then concentrated 100-fold (late log phase) to 50-fold (stationary phase) before layering on the gradients. The concentrated suspension should always be vortexed vigorously for about 2 min prior to layering on the gradients to eliminate aggregates and thus reduce loss of minicells by cosedimentation with cells. The gravitational force and duration of centrifugation will depend upon the minicell-producing strain, but a sedimentation regimen of 5000 rpm for 15 min in a Beckman SW27 swinging-bucket rotor works well for purifying minicells from many strains.

After centrifugation, the minicell band is withdrawn from the gradient by inserting a syringe with the needle bent at a right angle through the top of the tube. A 10-ml syringe is adequate for gradients of 35 ml. The minicell suspension, which is now in 13–15% glycerol or 7–8% sucrose, is slowly diluted with an equal volume of BSG at room temperature and then centrifuged at 10,000 *g* for 10 min to pellet the minicells. The slow dilution is important so as not to subject the minicells to osmotic shock, which could adversely affect their physiological activity. (When the cells that produce minicells are relatively short, it is sometimes necessary to take the minicells purified on the first linear gradient and allow a 30–45-min period of growth in nutrient medium at 37° to increase the size of the contaminating parental cells prior to performing the second gradient purification.) Very often the minicells obtained from two of the first gradients can be combined and layered on a second linear sucrose or glycerol gradient.

Rate-Zonal Centrifugation. For a large-scale purification of minicells, rate-zonal centifugation in linear sucrose or glycerol gradients in BSG can be successfully used.[3,28]

[28] G. B. Barker, C. S. Cordery, D. Jackson, and S. F. J. LeGrice, *J. Gen. Microbiol.* **111**, 387 (1979).

Decreasing Viable Cells in Preparations of Minicells. Minicells purified either by differential rate sedimentation on linear sucrose or glycerol gradients or by rate-zonal centrifugation should have no more than one contaminating parental cell for every 1×10^6 minicells and often have less than that if appropriate care has been taken with regard to selection of the minicell-producing strain for uniformity of minicells and ease of separation from the parental cells that produced them. However, it may be desirable in some experiments to completely eliminate viable contaminating parental cells, especially when minicells are to be incubated under growth conditions for a length of time that would be sufficient for contaminating parental cells to regrow and contribute significantly to the types of biosynthetic activities being observed. A number of methods have been devised to eliminate contaminating parental cells, but the following methods are the most applicable to the strains listed in the table.

1. With minicell-producing strains that have *recA* mutations, one can use low doses of ultraviolet light followed by a 20-min incubation period at 37°. Because of its large target size, chromosomal DNA will be degraded, whereas the plasmid DNA within contaminating bacterial cells or minicells will be comparatively much more resistant to degradation.[29]
2. Another approach is to inhibit cell wall synthesis by the addition of 25 μg of ampicillin per milliliter or 25 μg of cycloserine per milliliter or by starving diaminopimelic acid-requiring strains for DAP in medium that contains lysine. In order for sufficient killing by ampicillin and/or cycloserine, the contaminating cells must first commence to grow, and this generally requires 30 min at 37° after recovery from sucrose or glycerol gradients, which induce a lag in initiation of cell growth and division. Since lysis of cells by DAP-less death occurs at a more rapid rate by the addition of low levels (1–5 μg/ml) of ampicillin or cycloserine, one can commence DAP starvation of DAP-requiring strains immediately after suspension of minicell bands from sucrose or glycerol gradients and after 30 min at 37° add cycloserine and/or ampicillin and continue incubation for another 60–90 min. After this time, viable cells should be impossible to recover.

Labeling of Macromolecules Synthesized in Plasmid-Containing Minicells[3,7,13,30]

Purified minicells can be incubated in a minimal medium supplemented with 0.5% glucose or 3% glycerol and containing appropriate

[29] A. Sancar, A. M. Hack, and W. D. Rupp, *J. Bacteriol.* **137,** 692 (1979).

[30] P. J. M. van den Elzen, R. N. H. Konings, E. Veltkamp, and H. J. J. Nijkamp, *J. Bacteriol.* **144,** 579 (1980).

purine, pyrimidine, vitamin, and amino acid supplements. Under such circumstances, newly synthesized proteins can be labeled by incorporation of [^{35}S]methionine, [^{3}H]leucine, or a mixture of [^{14}C]amino acids during a 1–2-hr incubation at 37°. Lysine must be added for DAP-requiring strains. One can also label proteins by addition of [^{14}C]tryptophan in medium supplemented with 0.5% casamino acids. Newly synthesized RNA can be labeled in either of the above-described media by addition of [^{3}H]uridine or [^{14}C]uracil and incubation at 37° for 30–60 min. When measuring total protein synthesis rather than just analyzing the proteins by gel electrophoresis, one should determine the amount of hot acid-precipitable counts since charging of tRNAs with amino acids can account for 10–20% of the incorporated counts during short periods of incubation.

Synthesis and Stability of Plasmid-Specified RNA

Minicells from plasmid-containing minicell-producing strains can be used to study the rates of synthesis of different size classes of RNAs as a function of environmental conditions. The quantity of any given RNA species synthesized per unit time can be determined directly following fractionation of minicell lysates if the size class is unique or by several types of hybridization analyses using DNA complementary to the RNA species of interest.[13,31,32] Minicells can also be used to determine the stability of given RNA species quantitatively by addition of rifampin following synthesis of radioactively labeled RNA. Using pulse-chase methodology, it is possible to follow processing of RNA species in plasmid-containing minicells.

Lysis of Minicells and Isolation of RNA. To lyse minicells, a minicell suspension (1 to 5 $\times$ 10^9 minicells per milliliter in the growth medium used for labeling RNA) is added to a Tris-NaCl buffer containing SDS and diethyl pyrocarbonate at room temperature. The final concentrations of the components should be 0.02 *M* Tris (pH 7.6)–0.1 *M* NaCl–1% SDS and 2.8% diethyl pyrocarbonate. After shaking at room temperature for 15 min, an equal volume of liquid phenol equilibrated with 0.1 *M* Tris-HCl (pH 7.2) is added and the mixture is shaken for at least 30 min at room temperature before centrifugation at 12,000 *g* for 30 min. The aqueous phase is removed, and 2.5 volumes of ethanol are added to precipitate the RNA. After overnight storage at −20°, the precipitate is collected by centrifugation and dissolved in 0.1 *M* NaCl. This solution is dialyzed at 4° against 1 *M* NaCl for several hours and then against 0.01 *M* NaCl overnight.[13]

The above purification has been used and modified by others. These

[31] D. Gillespie and S. Spiegelman, *J. Mol. Biol.* **12,** 829 (1965).

[32] J. C. Alwine, D. J. Kemp, and G. R. Stark, *Proc. Natl. Acad. Sci. U.S.A.* **74,** 5350 (1977).

modifications include additions of 0.05 *M* EDTA and the use of proteinase K and purified bentonite to destroy or inhibit ribonucleases.[33]

Determination of RNA Species Synthesized. Polyacrylamide gel electrophoresis can be used to determine the sizes and amounts of each size class of RNA synthesized in plasmid-containing minicells. Very often an excess of unlabeled ribosomal RNA is added to provide size standards when analyzing gels at 260 nm. Tube gels can then be frozen at their original length over Dry Ice, slab gels dried, and the frozen or dried gels sliced into 1-mm transverse sections for quantitation of the radioactivity.

Various RNA–DNA hybridization procedures can also be used to quantitate the amounts of specific RNA species. In these studies, one would employ a DNA fragment that is only complementary to the RNA species of interest. One can use the RNA synthesized by the plasmid-containing minicells for hybridization to the DNA probe contained on a nitrocellulose filter.[13,31] One can also electrophorese the RNA in methylmercuric hydroxide agarose gels followed by transfer of the RNA species to diazobenzyloxymethyl paper for hybridization to the DNA probe.[30,32]

Stability and Processing of RNA Species Synthesized. After labeling of RNA with [^{3}H]uridine or [^{3}H]uracil in plasmid-containing minicells, rifampicin can be added at 25–200 μg/ml to prevent further RNA synthesis. (Rifampicin at 1 μg/ml can block RNA synthesis in minicell-producing strains such as χ1776.) As a function of time thereafter, polyacrylamide gel electrophoresis and/or hybridization can be used to quantitate changes in the amounts of specific RNA species. In addition, one can couple this technique with short pulses of radioactive label so as to follow potential processing of given RNA species.

Synthesis and Stability of Plasmid-Specified Proteins

Minicells from plasmid-containing minicell-producing strains are frequently used to identify the size of proteins specified by plasmid genetic information.[20] By placing the cloned DNA fragment in both orientations in the vector and evaluating the protein synthesized, one can deduce whether a promoter sequence controlling expression of the cloned gene exists on the cloned gene fragment or not. Determining the size of proteins specified by a cloned fragment facilitates devising a strategy for purification of a protein of interest. It should be noted that not all protein-synthesizing systems function in minicells and that some systems function less well in minicells than in vegetative cells.[3,34] It has been hypothesized that these deficiencies may be due to nuclease or ribonuclease activity in

[33] A. J. Kool, M. S. van Zeben, and H. J. J. Nijkamp, *J. Bacteriol.* **118,** 213 (1974).

[34] E. K. Jagusztyn-Krynicka, M. Smorawinska, and R. Curtiss III, *J. Gen. Microbiol.* **128,** 1135 (1982).

minicells[3] or to competition between promoter sequences for limiting amounts of RNA polymerase present in minicells.[3,34]

Lysis of Minicells. Minicells at 1 to 5 × 10^9/ml, suspended in the medium used to label proteins, are chilled on ice, sedimented at 10,000 *g* for 10 min at 4°, and suspended in a small volume of Laemmli's[35] sample buffer [62.5 m*M* Tris-HCl (pH 6.8), 2% sodium dodecyl sulfate, 10% glycerol, and 5% 2-mercaptoethanol (final concentrations)] and then boiled for 5 min. Lysates can be dialyzed overnight at 4° prior to analyzing the polypeptides synthesized by SDS–polyacrylamide gel electrophoresis.[35]

When proteins synthesized in plasmid-containing minicells are to be examined under nondenaturing conditions, lysis of minicells is best accomplished by the use of T4 lysozyme, which is more active on the *E. coli* murine layer than is hen egg white lysozyme.[36] Zylicz and Taylor[36] have taken minicells with labeled proteins, chilled the minicells, sedimented and washed them, and then suspended them in a lysis buffer containing 50 m*M* Tris (pH 7.3), 5 m*M* dithiothreitol, 1 m*M* EDTA, 40 μ*M* spermine, and 10% sucrose. T4 lysozyme is added to a final concentration of about 2 μg/ml. After gentle mixing the suspension is incubated at 0° for 30 min. Lysis is initiated by adding an equal volume of cold lysis buffer (lacking sucrose), and the unlysed minicells are removed by sedimentation prior to analysis of proteins by polyacrylamide gel electrophoresis with or without SDS or a reducing environment.

A number of studies have employed hen egg white lysozyme, added to 1–2.5 mg/ml final concentration.[3,26,37–39] We have found that preparing the suspension of the hen egg white lysozyme immediately before its addition to the minicell preparation increases the activity and reduces the concentration of lysozyme needed to lyse minicells in the absence of detergents.[3]

Determination of Protein Species Synthesized. The molecular weights of polypeptides synthesized in plasmid-containing minicells is best analyzed by electrophoresis on SDS–polyacrylamide gels,[35] using molecular weight standards. One can also fractionate proteins on nondenaturing gels,[36,40,41] although separation under these conditions is based on both size and charge. The latter method permits one to measure enzymic activ-

[35] U. K. Laemmli, *Nature* (*London*) **227,** 680 (1970).

[36] M. Zylicz and K. Taylor, *Eur. J. Biochem.* **113,** 303 (1981).

[37] F. W. Shull, Jr., J. A. Fralick, L. P. Stratton, and W. D. Fisher, *J. Bacteriol.* **106,** 626 (1971).

[38] M. C. Paterson and R. B. Setlow, *Proc. Natl. Acad. Sci. U.S.A.* **69,** 2927 (1972).

[39] R. J. Sheehy, D. P. Allison, and R. Curtiss III, *J. Bacteriol.* **114,** 439 (1973).

[40] U. K. Laemmli and M. Favre, *J. Mol. Biol.* **80,** 575 (1973).

[41] S. W. Kessler, *J. Immunol.* **115,** 1617 (1975).

ity of proteins within gels if a method is available for the analysis of the proteins of interest.

Stability of Proteins Synthesized. Many proteins specified by "foreign" DNA in *E. coli* are unstable, being subject to degradation by various proteolytic enzymes. Plasmid-containing minicells can be conveniently used to investigate the occurrence, rate, and extent of such instability. After a period of labeling proteins synthesized in plasmid-containing minicells, chloramphenicol can be added to 25–100 μg/ml. As a function of time thereafter, samples of minicells can be lysed and the proteins analyzed on SDS–polyacrylamide gels.

Localization and Processing of Proteins in Minicells

Minicells from plasmid-containing minicell-producing strains can be conveniently used to examine whether plasmid-specified proteins are located in the cytoplasmic membrane, the periplasm, or the outer membrane, or whether they are excreted to the surrounding medium, and to examine the changes in protein structure that accompany such translocations.[7–12]

Lysis of Minicells. After labeling of proteins, minicell suspensions are chilled on ice, sedimented, and suspended in BSG. To avoid using detergents to lyse the minicells, they are disrupted by two passages through a French pressure cell at 16,000–17,000 psi.[3,7] Alternatively, the minicells can be lysed with T4 lysozyme.[36]

Prior to lysis of minicells by use of French pressure cells or T4 lysozyme, one can release periplasmic proteins by subjecting the minicells to cold osmotic shock by the method of Hazelbauer and Harayama.[42] After release of periplasmic proteins, minicells can be sedimented and subjected to one of the methods of lysis described above. Efficiency of release of periplasmic proteins without concomitant release of cytoplasmic protein should be monitored. β-Lactamase and β-galactosidase constitute two readily assayable control proteins. Furthermore, release of periplasmic proteins can be monitored by SDS–polyacrylamide gel electrophoresis to determine the molecular weights of the β-lactamase from the periplasmic extract as opposed to the cell extract, since the periplasmic protein has a mass 2000 daltons smaller than the cytoplasmic form.[43–45]

Fractionation of Minicell Lysates. Large membrane fragments and unlysed minicells are pelleted by centrifugation at 27,000 *g* for 15 min at 4°,

[42] G. L. Hazelbauer and S. Harayama, *Cell* **16,** 617 (1979).
[43] R. B. Ambler and G. K. Scott, *Proc. Natl. Acad. Sci. U.S.A.* **75,** 3732 (1978).
[44] J. G. Sutcliffe, *Proc. Natl. Acad. Sci. U.S.A.* **75,** 3737 (1978).
[45] J. G. Sutcliffe, *Cold Spring Harbor Symp. Quant. Biol.* **43,** 77 (1978).

and the small membrane fragments that remain in the supernatant fluid are pelleted by centrifugation at 200,000 *g* for 2 hr at 4°. If lysis of minicells is not complete, it may be necessary to separate the large membrane fragments from unlysed minicells by sucrose gradient centrifugation.[7] The cytoplasmic membrane is solubilized by 1% sodium lauryl sarcosinate (Sarkosyl) in the presence of 7 m*M* EDTA. The outer membrane can then be sedimented, and the proteins in the supernatant fluid can be analyzed on SDS–polyacrylamide gels. The outer membrane can then be solubilized in SDS, and the proteins can be analyzed by SDS–polyacrylamide gel electrophoresis.

Processing of Proteins. Changes in the molecular weight of proteins depending on location can be determined by analyzing fractions of proteins obtained in the cytoplasm, cytoplasmic membrane, periplasm, outer membrane, or supernatant fluid on SDS–polyacrylamide gels. Proof of relationship between proteins having different migrations due to the presence or the absence of modification require identification by immunoprecipitation with monospecific antibodies and *Staphylococcus aureus* protein A.[41]

Acknowledgments

Methods and observations from the authors' laboratories arose during research supported by the National Institutes of Health (AI-11456, AI-62508, AI-72533, CA-13148, and DE-02679), the National Science Foundation (PCM-7726477), the Damien Foundation, and the Chemotherapy of Leprosy (THELEP) and Immunology of Leprosy (IMMLEP) components of the UNDP/World Bank/World Health Organization Special Programme for Research and Training in Tropical Diseases.

[24] Uses of the Transposon $\gamma\delta$ in the Analysis of Cloned Genes

By MARK S. GUYER

This chapter describes a technique for isolating insertion mutations in DNA sequences that have been cloned in the widely used cloning vector pBR322 and certain other plasmids. The discussion is presented in terms of the use of the transposon $\gamma\delta$ (Tn*1000*). $\gamma\delta$ is a 5.7-kilobase (kb) transposable element that is naturally found on the *Escherichia coli* chromosome[1] and on the conjugative plasmid F[2]; the experimental use of $\gamma\delta$ to be

[1] M. S. Guyer, R. R. Reed, J. A. Steitz, and K. B. Low, *Cold Spring Harbor Symp. Quant. Biol.* **45,** 135 (1981).

[2] M. S. Guyer, *J. Mol. Biol.* **126,** 347 (1978).

ISBN 0-12-182001-7

described uses F as the transposon donor. $\gamma\delta$ is a member of the class of transposons related to Tn3.[3] It is flanked by 35-base pair (bp) perfect inverted repeats,[4] encodes transposase (R. Reed, personal communication) and resolvase[5,6] functions, and generates a 5 bp duplication at the target site upon insertion.[4] In addition to $\gamma\delta$, however, a number of other transposable elements have been used for generating insertions in cloned DNA sequences.

One use for the insertion mutations in cloned sequences is to identify and map the region within the cloned DNA segment that encodes the function of interest. In theory, any transposon would be suitable for this purpose. However, it is frequently important to isolate and characterize several insertions in the gene of interest. For example, a reasonable number of insertions[7–12] are necessary to ensure that the extent of the DNA coding sequence is defined as fully as possible. Similarly, in order to determine the direction of transcription of the gene by this technique, it is necessary correlate the positions of each of several insertions with the size of the truncated protein produced by each mutant plasmid. Thus, the specificity of insertion with respect to choice of target site[3] is an important consideration in the choice of the transposable element to be used in such experiments. The insertion of $\gamma\delta$ appears to be random enough to be generally useful. In addition, $\gamma\delta$ has other properties that make it a particularly good choice for use as a plasmid mutagen.

Transposition of $\gamma\delta$ during F-Mediated Conjugal Mobilization

This mutagenesis technique is based upon the observation that conjugal transmission (mobilization) of pBR322 by the conjugative plasmid F is dependent upon, or at least completely correlated with, the transposition of $\gamma\delta$ from F to pBR322[2]. This conclusion is based upon the results of three experiments. First, selection for transfer of the Tc^R and Ap^R mark-

[3] N. Kleckner, *Annu. Rev. Genet.* **15,** 341 (1981).

[4] R. R. Reed, R. A. Young, J. A. Steitz, N. D. Grindley, and M. Guyer, *Proc. Natl. Acad. Sci. U.S.A.* **76,** 4882 (1979).

[5] R. R. Reed, *Cell* **25,** 713 (1981).

[6] R. R. Reed, and N. D. F. Grindley, *Cell* **25,** 721 (1981).

[7] F. Heffron, P. Bedinger, J. J. Champoux, and S. Falkow, *Proc. Natl. Acad. Sci. U.S.A.* **74,** 702 (1977).

[8] R. Gill, F. Heffron, G. Dougan, and S. Falkow, *J. Bacteriol.* **136,** 742 (1978).

[9] A. Arthur, and D. J. Sherratt, *Mol. Gen. Genet.* **175,** 267 (1979).

[10] P. A. Kitts, A. Lamond, and D. Sherratt, (1982) *Nature (London)* **295,** 626 (1982).

[11] N. D. Grindley, and D. J. Sherratt, *Cold Spring Harbor Symp. Quant. Biol.* **43,** 1257 (1978).

[12] J. A. Shapiro, *Proc. Natl. Acad. Sci. U.S.A.* **76,** 1933 (1979).

ers of pBR322 in a conjugal mating in which the donor strain carries both pBR322 and F (which carries a single copy of $\gamma\delta$), invariably results in the recovery of pBR322::$\gamma\delta$ plasmids from the $Tc^R Ap^R$ transconjugants.[2] The presence of pBR322 itself in a $Tc^R Ap^R$ transconjugant has never been observed in such an experiment. Second, two derivatives of F from which $\gamma\delta$ (and adjacent sequences) have been deleted, FΔ (0–15)[13] and pOX38,[1] can each mobilize a pBR322::$\gamma\delta$ plasmid at the same frequency with which F mobilizes pBR322 (M. Guyer and R. Reed, unpublished results). In these experiments, $\gamma\delta$ is invariably transposed back to the conjugative plasmid. The frequency of conjugative transfer of $Tc^R Ap^R$ is the same in these two experiments. Finally, if neither plasmid carries a copy of $\gamma\delta$ (i.e., in an experiment in which the donor strain carries pOX38 and pBR322), the frequency of transfer of pBR322 is reduced by at least three orders of magnitude. However, the few $Tc^R Ap^R$ transconjugants obtained in such a mating carry both pOX38::$\gamma\delta$ and pBR322::$\gamma\delta$ (M. Guyer, unpublished results). The source of the transposon in this last case must be the chromosome of the donor strain, which happened to be a strain with three chromosomal copies of $\gamma\delta$ (*E. coli* K12 strains are known to carry from 0 to 3 chromosomal copies of $\gamma\delta$[1]).

The dependence of the conjugal transmission of pBR322 on $\gamma\delta$ transposition can be accounted for in terms of the current understanding of the mechanism of transposition. $\gamma\delta$ is a member of a group of closely related transposons known collectively as the Tn*3*-like elements.[3] Transposition of these elements involves an intermediate structure known as a cointegrate, in which the donor and recipient molecules are covalently linked by copies of the transposable element at the two junctions[7–10] (Fig. 1a). The mechanism by which the cointegrate intermediate is generated is not yet established, although a number of hypotheses have been offered.[3,11,12,14] In the cointegrate, the two copies of the transposon are oriented as direct repetitions. Subsequent recombination between the two copies of the transposon has the effect of regenerating the donor molecule and a derivative of the recipient molecule, which now contains a transposed copy of the Tn*3*-like element.[8,9,15–18] Such recombination is termed "cointegrate resolution" and is mediated by an element-specified site-specific recombination system[5,6,8,9,15,16] (see also R. R. Reed, this series, Vol. 100 [13]).

[13] P. A. Sharp, M. T. Hsu, E. Ohtsubo, and N. Davidson, *J. Mol. Biol.* **71,** 471 (1972).

[14] R. M. Harshey, and A. I. Bukhari, *Proc. Natl. Acad. Sci. U.S.A.* **78,** 1090 (1981).

[15] R. R. Reed, *Proc. Natl. Acad. Sci. U.S.A.* **78,** 3428 (1981).

[16] R. Kostriken, C. Morita, and F. Heffron, *Proc. Natl. Acad. Sci. U.S.A.* **78,** 4041 (1981).

[17] P. Kitts, R. Reed, L. Symington, M. Burke, and D. Sherratt, *Proc. Natl. Acad. Sci. U.S.A.* **79,** 46 (1981).

[18] M. McCormick, W. Wishart, H. Ohtsubo, F. Heffron, and E. Ohtsubo, *Gene* **15,** 103 (1981).

(a) Structure of Transposon - mediated plasmid cointegrate

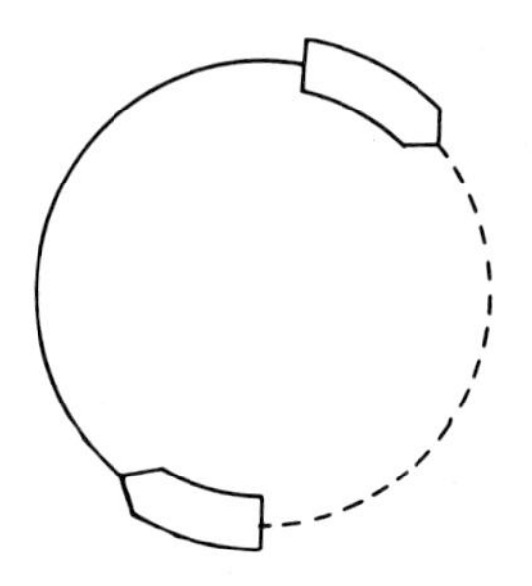

(b) Structure of $\gamma\delta$ - mediated F-pBR322 cointegrate

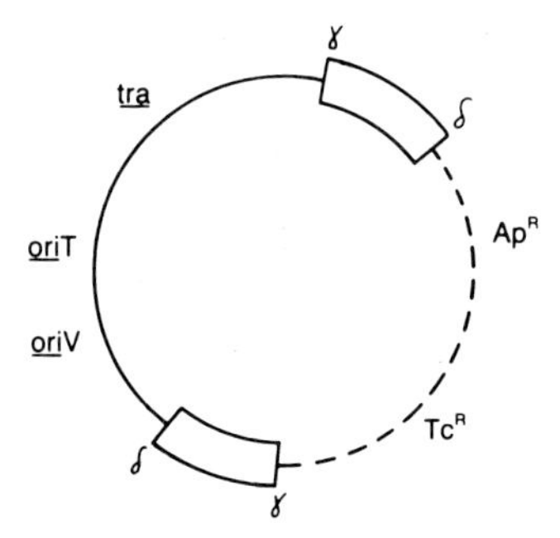

FIG. 1. (a) Diagrammatic structure of a plasmid cointegrate transpositional intermediate. The donor replicon is symbolized by a solid line, the recipient replicon is symbolized by a dashed line, and the transposon is symbolized by an arrow. The direct orientation of the transposon duplication is indicated. (b) Diagrammatic structure of the hypothetical F-pBR322 cointegrate. F is symbolized by a solid line, pBR322 is symbolized by a dashed line, and $\gamma\delta$ is symbolized by a rectangle. Representative markers are shown.

Transposition of $\gamma\delta$ from F to pBR322, then, would involve a cointegrate in which the pBR322 sequence is covalently linked to F (Fig. 1b). In effect, this structure is an F′ element. Its structure is entirely analogous to the well-known F′ elements F14[19] and F105.[20] Conjugal transmission of the cointegrate would then serve to transfer the pBR322 sequence. Resolution of the cointegrate in the recipient would result in the regeneration of F and the generation of pBR322::$\gamma\delta$. Although the hypothetical

[19] E. Ohtsubo, R. C. Deonier, H. J. Lee, and N. Davidson, *J. Mol. Biol.* **89,** 565 (1974).
[20] S. Palchaudhuri, W. K. Maas, and E. Ohtsubo, *Mol. Gen. Genet.* **146,** 215 (1976).

F-pBR322 cointegrate has not yet been observed (Guyer, unpublished results), an analogous cointegrate has been observed in the case of the Tn*3*-mediated mobilization of pSC101 by R100-1.[21]

Plasmids with γδ insertions can be generated by F-mediated mobilization of plasmids other than pBR322. These include pSC101 (M. Guyer, unpublished results, P. Prentki, personal communication), pBR325 (P. Prentki, personal communication), and pNR5311, a derivative of *Rts*1.[22] Reed and Grindley (personal communication) have obtained a significant increase in the frequency of pBR322 mobilization by the use of pOX200, a derivative of pOX38 carrying a *tnpR*$^-$ (derepressed) mutant of γδ, as the conjugative plasmid.

Insertion of γδ into Cloned Genes

γδ can be inserted into chimeric plasmids comprised of one of the nonconjugative vectors listed above and an inserted DNA sequence as a result of conjugal mobilization by a γδ-containing conjugative plasmid. In such experiments, γδ is frequently inserted into the cloned gene. The first experiments of this sort were done by Sancar and Rupp.[23] Their experiments involved F-mediated transfer of pDR2000, a recombinant plasmid carrying the *uvrA* and *ssb* genes of *E. coli* on a 13.7 kb sequence cloned in pBR322. Ten of 480 pDR2000::γδ isolates obtained in these experiments failed to complement mutations in the *uvrA* gene. Sancar *et al.*[24] subsequently used these mutant plasmids to map the *uvrA* gene on pDR2000, to determine the direction of transcription of *uvrA*, and to identify the protein specified by *uvrA*. The same technique has been used by Sancar *et al.*[25] to isolate γδ insertions in the *uvrB* cistron of *E. coli* (46 of 384 isolates were *uvrB*::γδ mutations in this case). Other genes that have been analyzed by this approach include the *uvrC*,[26] *uvrD* (V. Maples and S. Kushner, personal communication), and *umuC* (S. Elledge and G. Walker, personal communication) genes of *E. coli* and the *muc* gene of plasmid pKM101 (K. Percy and G. Walker, personal communication).

In two cases, previously unrecognized genes have been identified by

[21] N. J. Crisona, J. A. Nowak, H. Nagaishi, and A. J. Clark, *J. Bacteriol.* **142,** 701 (1980).

[22] N. Goto, S. Horiuchi, A. Shoji, and R. Nakaya, *in* "Microbial Drug Resistance" (S. Mitsuhashi, ed.), Vol. 3. Japan Scientific Societies Press, Tokyo, in press.

[23] A. Sancar and W. D. Rupp, *Biochem. Biophys. Res. Commun.* **90,** 123 (1979).

[24] A. Sancar, R. P. Wharton, S. Seltzer, B. M. Kacinski, N. D. Clarke, and W. D. Rupp, *J. Mol. Biol.* **148,** 45 (1981).

[25] A. Sancar, N. D. Clarke, J. Griswold, W. J. Kennedy, and W. D. Rupp, *J. Mol. Biol.* **148,** 63 (1981).

[26] G. H. Yoakum, and L. Grossman, *Nature (London)* **292,** 171 (1981).

means of $\gamma\delta$ insertion. Grindey (personal communication) has analyzed a cloned DNA sequence derived from *E. coli* that encodes PolI. The cloned DNA sequence also contains an open reading frame capable of encoding a 20,000-dalton protein. A protein of that size was found in *in vitro* coupled transcription–translation experiments. Insertions of $\gamma\delta$ into the open-reading frame led to the loss of the ability to specify the 20,000-dalton protein, demonstrating that this open-reading frame was being utilized. Devlin *et al.*[27] isolated $\gamma\delta$ insertions into a chimera containing a fragment of phage P1 DNA. One class of $\gamma\delta$-containing derivatives determined a P1-superimmune phenotype, thus defining a new P1 gene, *sim*.

The $\gamma\delta$ system has promise for introducing the techniques of transposon genetics to DNA of non-*E. coli* origin. S. Johnson (personal communication) has introduced into *E. coli* two 6.6 kb R factors (pGR4744 and pGR9091) derived from *Neisseria gonorrhea*. Mobilization of these plasmids by an F′*lac*$^+$ plasmid resulted in conjugal transfer of Pen^R, the phenotype specified by each R factor. Preliminary evidence indicates that most of the Pen^R transconjugants carry plasmids that have incurred $\gamma\delta$ insertions.

It should also be possible to insert $\gamma\delta$ into non-*E. coli* DNA sequences that have been cloned in pBR322. Reintroduction of the derivatized DNA into the original host, either on a plasmid or into the chromosome, would then allow analysis of the effects of the insertion mutation. This should prove useful for those strains in which natural transposon systems are unavailable.

Other Uses for $\gamma\delta$ Insertions

The insertion of $\gamma\delta$ can also be useful for purposes beyond the physical characterization of a cloned gene and the identification of its protein product. To this end, the detailed structural properties of the $\gamma\delta$ sequence have been exploited in a variety of experiments done in the laboratory of Lucien Caro. One application has been the addition of convenient restriction sites to DNA sequences that lack such sites. A number of restriction enzymes have only one or two recognition sites within $\gamma\delta$. A restriction map detailing a number of these is available.[2] Sequence data have demonstrated the presence of additional single or double recognition sites for restriction endonucleases (R. Reed, personal communication).

Prentki and co-workers (personal communication) have isolated a plasmid (op71) in which the *dnaA* gene of *E. coli* on a 3.5 kb DNA fragment has been cloned in the vector pBR325. Analysis of the cloned *dnaA*

[27] B. D. Devlin, B. Baumstack, and J. Scott, *Virology* **120,** 360 (1982).

gene was hampered by the lack of convenient restriction sites to use for subcloning. Plasmids containing additional sites in the gene were obtained after F-mediated mobilization of op71 and identification of those plasmids that contained insertions of $\gamma\delta$ into *dnaA*. Subclones containing portions of *dnaA* (these are effectively deletion mutations) were then isolated after digestion of the op71::$\gamma\delta$ DNA with two restriction enzymes, one that cut within $\gamma\delta$ and one that cut in the plasmid outside of $\gamma\delta$. By using a recombination assay to detect the wild-type allele of a *dnaA* mutation, Prentki was able to map the location of that allele relative to the position of the $\gamma\delta$ insertion. Several *dnaA* mutations have been mapped (with accuracies of about 0.2 kb) in this way.

Linder *et al.*[28] have used a similar approach to map an incompatibility function of pSC101 and to identify a protein required for pSC101 replication. A variation of this use of $\gamma\delta$ is to isolate a series of insertions within a gene of interest and then to subclone the region of the gene that is located to the same side of the insertion from each isolate. This generates a series of plasmids containing increasingly longer regions of the gene of interest, starting from a common point at the start of the gene. Such a series can be used for deletion mapping. Finally, it is important to note that this use of $\gamma\delta$ is not limited to the analysis of DNA sequences encoding proteins, but can also be used to interrupt or generate deletions of the functional sites in DNA.

A series of $\gamma\delta$ inserts in a cloned gene can also be used for rapid DNA sequence determination. DNA sequence analysis of $\gamma\delta$ has shown that there is an *Sst*I recognition site 78bp from the δ end (R. Reed, personal communication) and *Rsa*I recognition sites 30 bp from each end,[4] in the inverted repeat. Using $\gamma\delta$ then as a "vector" for introducing convenient cleavage sites (G. Churchward and P. Prentki, personal communications), long stretches of DNA in which $\gamma\delta$ insertions are located at appropriate intervals can be rapidly sequenced.

Rapid sequencing could also be done by using short oligonucleotides homologous to sequences near the end of $\gamma\delta$ (but not within the inverted repetitions[4]) as primers to sequence out from $\gamma\delta$ into the adjacent region by the dideoxy method of DNA sequence determination.[29] As this technique can be applied (R. McCandliss, personal communication) to plasmid DNA prepared from small volumes of culture,[30,31] it should be possible to determine very rapidly the sequence of a large region of DNA

[28] P. Linder, G. Churchward, and L. Caro, In preparation.

[29] A. J. H. Smith, this series, Vol. 65, p. 560.

[30] H. C. Birnboim and J. Doly, *Nucleic Acids Res.* **7,** 1513 (1979).

[31] D. S. Holmes and M. Quigley, *Anal. Biochem* **114,** 193 (1981).

without having to obtain any restriction site information at all, again by using $\gamma\delta$ insertions at different sites within the region of interest.

Isolation of $\gamma\delta$ Insertions in Chimeric Plasmid DNA

A suitable donor strain is constructed either by transforming a *recA* F^+ donor strain with the chimeric plasmid of interest or by crossing F into a strain containing the desired chimera by conjugation. A population of plasmids carrying $\gamma\delta$ insertions is then isolated by using such a strain as a donor in a mating with a *recA* recipient.

For the mating, cultures of the donor and recipient strains are grown in L broth to a density of 1 to 3 $\times$ 10^8 cells/ml (A_{600} = 0.7–1.0, or 40–60 Klett units, depending upon the particular strains being used). Equal volumes of the two cultures are then mixed. (The frequency of transfer of the chimeric plasmid is low enough in this mating so that the rather high donor : recipient cell ratio of 1 : 1 is useful for obtaining a sufficient number of transconjugants.) It appears that the frequency of transfer varies somewhat with the donor strain (J. L. Rosner, personal communication). Thus, the volume of mating mixture that should be used can range from 1 to 100 ml. The cell mixtures are then incubated at 37°, typically for 60 min without shaking. The matings are then briefly agitated in a Vortex mixer, after concentration in the case of large volume matings. Samples are plated on medium selective for recipient cells to which the R plasmid has been transferred (e.g., in the case of a pBR322 derivative containing passenger DNA cloned at the *Eco*RI site, selection would be for transfer of both Ap^R and Tc^R from the donor and perhaps Sm^R or Nal^R from the recipient). Selection for every available donor plasmid marker is useful because it reduces the number of transconjugants recovered having $\gamma\delta$ insertions in regions of the plasmid that are not of interest, thus reducing the number of isolates that must be screened to find those of interest.

Individual colonies can then be screened for the mutant phenotype. The plasmids present in isolates displaying the desired phenotype can then be analyzed or manipulated according to the needs of the particular experiment.

[25] The Use of *Xenopus* Oocytes for the Expression of Cloned Genes

By J. B. GURDON and M. P. WICKENS

General Properties of Oocytes Compared to Other Recipient Cells

The injection of amphibian oocytes was one of the first systems in which purified DNA was correctly transcribed and expressed as protein.[1] Since then, cell-free systems have been developed which initiate transcription accurately, and the expression of eukaryotic genes has been obtained by infecting cells with genetically manipulated viruses, by transfecting Ca-precipitated DNA into cultured cells, and by the direct injection of DNA into cultured cells or mouse eggs. Compared to these other systems, amphibian oocytes have three characteristics. First, a very small amount of DNA needs to be injected into a single oocyte to obtain recognizable transcription and expression (see sections Transcription and Expression as Protein). Second, the expression of DNA can be monitored within a few hours of injection, during which time it is not replicated or integrated into host cell chromosomes, but is assembled with nucleosomes into an apparently normal chromatin structure. In most other expression systems, DNA is integrated and replicated, as a result of which it may undergo genetic changes and its expression may be influenced by the properties of adjacent host DNA. Third, oocyte injection makes it possible to introduce any cell components, such as RNA, chromatin, or nuclear proteins; this is likely to be particularly valuable for analyzing the regulation of gene expression.

An amphibian oocyte is a single large cell, surrounded by several thousand small follicle cells. It is in meiotic prophase, and active in RNA and protein synthesis, but totally inactive in DNA synthesis. Its composition and general properties are summarized in Table I.

[1] Many of the conclusions stated in this chapter have been documented in recent reviews which should be consulted for all unreferenced statements [see J. B. Gurdon and D. A. Melton, *Annu. Rev. Genet.* **15,** 189 (1981); A. Kressman and M. L. Birnstiel, *in* "Transfer of Cell Constituents into Eukaryotic Cells" (J. Celis, ed.) pp. 383–407 Plenum, New York, 1980; M. P. Wickens and R. A. Laskey, *in* "Genetic Engineering" (R. Williamson, ed.), Vol. 1, pp. 103–167. Academic Press, New York, 1981]. We supply original references in this article only when they provide a source of technical information.

METHODS IN ENZYMOLOGY, VOL. 101

ISBN 0-12-182001-7

Methods

Oocytes

The collection and culture of oocytes, as well as the instruments and techniques needed for injection, have all been described.[2] This source supplements the information given below.

In a typical experiment, 25 nl of DNA at 200 μg/ml (i.e., 5 ng) is injected into an oocyte. The best medium for culturing oocytes is modified Barth solution (MBS) (see Table II).

Labeling of RNA and Protein

For labeling RNA it is generally best to inject 1 μCi of [^{32}P]GTP or [^{32}P]CTP per oocyte (mixed with DNA if desired). ATP or UTP may be used, but the pool sizes are larger (see Table I) and UTP is efficiently converted in oocytes into non-RNA molecules. Nucleosides and amino acids, but not nucleotides, are taken up from the medium. Thus RNA may also be labeled by adding [^{3}H]guanosine to the medium. In this case, transcripts synthesized in the large number of follicle cells account for more than 90% of the total ^{3}H-labeled RNA (see below for follicle cell removal).

The most generally useful procedure for labeling proteins is to incubate an oocyte in 10 μl MBS containing 5 μCi of [^{35}S]methionine for 18–24 hr. The amount of incorporation is not greatly affected by the volume of medium in which the oocytes are incubated, at least between 3 and 15 μl per oocyte.

With isotopes of high specific activity, the amount of incorporation is directly proportional to the amount of radioactive precursor used, at least up to 10 μCi per oocyte.

Injection Technique

The transcription of injected DNA takes place only in the nucleus or germinal vesicle of an oocyte, which is not normally visible. DNA can be deposited in the germinal vesicle with about an 80% success rate by penetrating the oocyte in the center of the black pigmented hemisphere, until the pipette tip is judged to be one-third of the way from this point to the opposite pole of the oocyte.[3] This technique may be readily learned by practising the injection of a concentrated trypan blue solution. Opening the oocyte with forceps just after injection will show whether the dye was

[2] J. B. Gurdon, *Methods Cell Biol.* **16,** 125, 139 (1977).

[3] J. B. Gurdon, *J. Embryol. Exp. Morphol.* **36,** 523, 540 (1976).

TABLE I
COMPOSITION AND GENERAL PROPERTIES OF A FULL SIZED (1250 μm DIAMETER) OOCYTE OF *Xenopus laevis*[a]

	Cytoplasm	Nucleus	Follicle cells
Volume (% of total)[b]	0.5 μl (90%)	40 nl (~10%)	—
DNA content			
Chromosomal[c]	None	12 pg	30,000 pg
Nucleolar[c]	None	25 pg	30 pg
Mitochondrial[c]	4000 pg	None	150 pg
DNA synthesis per day[d]	None	None	Significant
RNA content			
Ribosomal (28 + 18 S)[e]	5 μg (10^{12} ribosomes)	—	
5 S[f]	60 ng	—	
4 S[f]	60 ng	—	
Poly(A)-containing[g]	70[h] ng (10% polysomal)	10 ng	
RNA accumulation[i] (total per day)	Whole oocyte: 20 ng (1 ng mitochondrial)		
Protein content			
Yolk	250 μg	None	
Nonyolk	25 μg	2.5 μg	
Histones[j]	70 ng	70 ng	
Nucleoplasmin[k]	5 ng	250 ng	
RNA polymerase[l]	—	10^5 × somatic cell	
Protein synthesis (total accumulated/day)[m]	400 ng	None	

	Content per oocyte, ignoring follicle cells (pmol)
rATP[n]	1700
rUTP	1200
rCTP	500

rGTP	250
dTTP	7
Methionine[o]	44
Glutamic, aspartic acids	2900, 1600
Other amino acids	30–300

[a] Much of the information in this table on synthesis is discussed in detail in Ref. 28. —, not known.

[b] Total volume is 1 μl, of which half is yolk and is therefore metabolically unavailable space.

[c] Tetraploid nucleus with 1000-fold amplified ribosomal genes. The haploid genome consists of about 3×10^9 base pairs or 3 pg DNA. See Refs. 34–37 for genomic and mitochondrial DNA values. About 5000 follicle cells surround each oocyte.

[d] No replication of double-stranded DNA; single-stranded DNA becomes double-stranded (see Ref. 16). About 5% of follicle cells take up [^{3}H]thymidine.

[e] See Refs. 38 and 39.

[f] See Ref. 40.

[g] See Refs. 41, 44, 49, and 51; 90% is untranslated, not polysome-associated.

[h] Partly mitochondrial, see Refs. 36, 37, and 46.

[i] The typical transcription rate for an oocyte is 15 nucleotides per second, with RNA polymerases spaced 100–200 bp apart (Ref. 28). See also Refs. 11, 36, 42, and 45. At this rate of RNA accumulation (<20 ng/day), it would take 250 days for a full-sized oocyte (5 μg RNA) to be formed.

[j] See Ref. 43.

[k] See Ref. 32.

[l] An oocyte has equal activities of polymerases I, II, and III. Its activity exceeds that of a cultured cell by 60,000 for polymerases I and II, and by 500,000 for polymerase III. Almost all oocyte activity except for IIA is in its nucleus. See Ref. 33.

[m] See Ref. 47, p. 136 and Ref. 48. The rate given may increase by a few fold under some physiological conditions. The typical translation rate for an oocyte is one codon (3 nucleotides) per second, with ribosomes spaced 100 nucleotides apart. Protein synthesis does not include yolk, which is synthesized in the liver and transported to growing oocytes.

[n] For measurement of nucleotide pools, see Ref. 50.

[o] For measurement of amino acid pools, see Ref. 29.

TABLE II
MODIFIED BARTH SOLUTION (MBS) AND ITS PREPARATION

	Concentration in medium (m*M*)	10 × stock[a] (g/liter)
NaCl	88	51.3
KCl	1.0	0.75
$NaHCO_3$	2.4	2.0
Hepes, pH 7.5	10.0	23.8
$MgSO_4 \cdot 7H_2O$	0.82	2.0
$Ca(NO_3)_2 \cdot 4H_2O$	0.33	0.78
$CaCl_2 \cdot 6H_2O$	0.41	0.90

[a] The 10 × stock solution should be prepared by adding reagents in the above order to 900 ml H_2O, finally making up to 1 liter. After filter sterilization, this stock may be stored for months at +4°. After dilution for use, the pH should not need adjustment, but penicillin and streptomycin may be added to give a final concentration of 10 mg/liter.

deposited in the oocyte's nucleus. Some workers prefer to centrifuge oocytes lightly so as to bring the germinal vesicle to the surface where it can be seen,[4] but this can reduce viability. Usually a manually controlled syringe is used to control the volume of fluid injected,[2] but a more sophisticated apparatus has been described[5] which avoids problems of the pipette becoming blocked by backflow.

A single injected oocyte is often adequate for the detection of RNA or protein. However, individual oocytes vary somewhat in the amount of RNA or protein which they synthesize from an injected template. Such individual differences are easily averaged out by injecting 10 or more oocytes from the same ovary with each sample.

Follicle Cell Removal

Each oocyte is closely surrounded by about 5000 follicle cells, which greatly affect the composition and synthesis of ovarian material (Table I) unless removed. This can be done by gently swirling small clusters of oocytes (20 per cluster) overnight in MBS containing 2 mg/ml collagenase.[6] This procedure removes all follicle cells as judged by scanning electron

[4] A. Kressman, S. G. Clarkson, V. Pirrotta, and M. L. Birnstiel, *Proc. Natl. Acad. Sci. U.S.A.* **75,** 1176 (1978).

[5] D. L. Stephens *et al., Anal. Biochem.* **114,** 299 (1981).

[6] T. J. Mohun, C. D. Lane, A. Colman, and C. C. Wylie, *J. Embryol. Exp. Morphol.* **61,** 367 (1981).

microscopy, and leaves oocytes with unimpaired viability.[6] It seems preferable to the use of pronase.[7] In many cases, the presence of follicle cells can be ignored and not removed, e.g., after labeling RNA by the injection of ^{32}P-labeled nucleotides which do not penetrate the follicle cells from oocyte cytoplasm.

RNA Extraction

For the analysis of transcripts, RNA may be extracted from oocytes with good recovery and minimal risk of degradation by the following procedure (other procedures have also been reported[8–11] and may be used successfully). Homogenize 5–15 oocytes in 0.5 ml of 0.3 *M* NaCl, 2% SDS, 50 m*M* Tris, pH 7.5, 1 m*M* EDTA (room temperature). This is conveniently done in a 1.0-ml glass homogenizer. Quickly transfer the homogenate to a 1.5-ml microfuge tube containing 0.5 ml of phenol: chloroform (1:1), and vortex immediately. Centrifuge in a microfuge for 5–10 min. Remove the aqueous phase. Add 0.5 ml of the same homogenization buffer to the phenol: chloroform phase, vortex, and centrifuge again. Remove the aqueous phase, combine with the first, and add 2 volumes of ethanol. A large flocculent white precipitate of carbohydrate (see below for removal) will form immediately upon addition of the alcohol. Ethanol precipitate as desired (−20° for 15 min usually is adequate); then recover the precipitate by centrifugation and wash it with 70% ethanol. Again drain off the alcohol, and dry briefly. If the precipitate is contaminated with brown material, as is often the case when extracting 15 or more oocytes per 0.5 ml, redissolve the precipitate in 50 m*M* Tris, pH 7.5 and extract once more with an equal volume of phenol: chloroform (1:1). Remove the aqueous phase, adjusting to 0.3 *M* NaCl, and precipitate with 3 volumes of ethanol.

Proteinase K may also be included in the homogenization buffer (at 1 mg/ml) when isolated germinal vesicles and cytoplasms are being analyzed.[12] It is simplest to transfer the nucleus or cytoplasm directly into homogenization buffer. Using proteinase K-supplemented buffer, cytoplasm and nuclei may be accumulated for at least 15 min without the RNA suffering any degradation.

When preparing RNA from single oocytes, homogenize each oocyte in 0.2–0.4 ml as above. It is not necessary to add carrier, since each oocyte

[7] L. D. Smith and R. E. Ecker, *Dev. Biol.* **19,** 281 (1969).

[8] J. E. Mertz and J. B. Gurdon, *Proc. Natl. Acad. Sci. U.S.A.* **74,** 1502 (1977).

[9] D. D. Brown and E. Littna, *J. Mol. Biol.* **8,** 669 (1964).

[10] A. Kressmann *et al., Cold Spring Harbor Symp. Quant. Biol.* **42,** 1077 (1978).

[11] D. M. Anderson and L. D. Smith, *Cell* **11,** 663 (1977).

[12] E. Probst *et al., J. Mol. Biol.* **135,** 709 (1979).

contains 5 μg of rRNA and a considerable amount of carbohydrate. However, in preparing RNA from single germinal vesicles which contain little nucleic acid or carbohydrate, it is prudent to add exogenous RNA or DNA (e.g., tRNA, to a final concentration of 50 μg/ml) to the above homogenization buffer, so as to minimize losses during extraction and precipitation.

Analysis of Labeled RNA

Any standard method used to analyze labeled or unlabeled RNA can, in principle, be applied to transcripts from oocytes. Each of the techniques which follow is discussed in detail elsewhere in this series. Here we discuss only those points which are particularly important for work with oocytes.

1. Direct gel electrophoresis. For injected genes transcribed by RNA polymerase III (tRNA and 5 S RNA), this clearly is the method of choice. Enough radioactive RNA is synthesized in a few hours to be detected by gel electrophoresis of total RNA from less than one oocyte. Although oocytes synthesize tRNA and 5 S RNA from their own genes, this endogenous background is inconsequential in most experiments, since the transcripts of injected genes are usually at least 20 times more abundant. Genes encoding mRNA, transcribed by polymerase II, cannot always be assayed by direct gel electrophoresis of total labeled RNA. This is because (a) such genes initiate 100 to 1000 times less frequently than those transcribed by polymerase III,[1,12] (b) precise initiation and termination are required for the production of a detectable RNA band, and neither process may be efficient on a particular DNA template, and (c) polyadenylation and multiple splicing events may complicate the pattern observed. In addition, several endogenous transcripts are prominent in those regions of an agarose gel in which mRNAs generally lie, and so contribute to the background (see Ref. 46 for a detailed description). In spite of these difficulties, mRNA transcripts from some genes, notably the sea urchin histone genes and the SV40 late genes, are of discrete size, and can be readily detected in total RNA preparations from less than one oocyte.

2. Purification of specific transcripts by hybridization to DNA bound to paper. DNA immobilized on DBM-paper filters may be used to purify template-specific transcripts which are only a small fraction of total oocyte RNA. This reduces the background by two to three orders of magnitude relative to direct gel analysis and provides a reasonable recovery of specific transcripts (10–50% of template-specific transcripts in a 4-hr hybridization). In such experiments as much as 30 oocytes' worth of RNA is redissolved in 100 μl of 50% formamide (deionized), 0.4 *M* NaCl, 0.2% SDS, 20 m*M* Pipes, pH 6.4, 5 m*M* EDTA. Redissolving the RNA at this concentration may require repeated pipetting and some patience. This so-

lution is then hybridized to paper-bound DNA which is then washed and eluted. The large amount of carbohydrate does not increase the background or interfere with the hybridization.

3. Southern filter hybridization. To determine which regions of injected DNA direct the synthesis of labeled transcripts, 50,000–500,000 cpm of ^{32}P-labeled RNA from injected oocytes can be hybridized to a filter bearing DNA restriction fragments transferred after gel electrophoresis.

Analysis of Unlabeled RNA

1. Northern analysis. Sufficient material is generally synthesized from injected genes (Table III) to permit rapid detection. However, it is essential to remove the DNA which was injected from the RNA sample prior to electrophoresis, since it can otherwise contribute a very high background. This may be achieved by standard biochemical techniques, e.g., digestion with RNase-free DNase. It must also be noted that each oocyte contains 5 μg of ribosomal RNA which may prevent the detection of transcripts of similar size.

2. S1 nuclease digestion of RNA:DNA hybrids. The versatility of this method and the large amount of RNA synthesized in oocytes make this technique very useful for analyzing transcripts synthesized by polymerase II. Again, standard techniques may be used without modification. Less than one oocyte's worth of RNA is often adequate for an overnight exposure using an end-labeled DNA fragment probe of only 10^6 cpm/μg specific activity.

A contaminant of oocyte RNA (possibly the carbohydrate) can distort the migration of protected DNA fragments on thin sequencing gels, such that lanes narrow toward the bottom. Results are interpretable, though it may be difficult to deduce precise lengths by comparison with undistorted marker lanes. This problem can be circumvented either by using only a little oocyte RNA in each lane (less than one-tenth oocyte per 5-mm-wide slot), by mixing markers with the protected fragments, or by purifying the RNA free of carbohydrate prior to hybridization [see DBM-paper technique (2) above].

Methods of Protein Analysis

The methods used to prepare homogenates for protein analysis must, of course, vary with the conditions required for the stability of the particular protein examined. A generalized technique useful for direct gel analysis of radioactive proteins is as follows. It includes extraction of the homogenate with freon (see below); this selectively removes yolk proteins, which distort the high-molecular-weight region of SDS polyacrylamide gels, but does not appear to selectively remove any other proteins from an oocyte homogenate (as judged by SDS–polyacrylamide gel electrophoresis). In contrast, the removal of yolk by direct centrifugation results in highly specific losses of basic proteins (R.A. Laskey, personal communication).

Homogenize 10–30 oocytes in 1 ml of ice-cold 15 m*M* Tris, pH 6.8, and 150 μg/ml PMSF. Add an equal or greater volume of freon (1,1,2-trichlorotrifluoroethane) and vortex. Separate the upper, aqueous phase from the lower, freon phase by centrifugation for 10 min in a microfuge. A large dark interphase containing yolk protein and pigment granules will be obvious. Remove the aqueous phase and, if necessary, centrifuge for 10 min to clarify. This homogenate may then be analyzed directly by SDS–polyacrylamide gel electrophoresis.

Immunological techniques may also be used to recognize proteins synthesized in the oocyte from injected DNA templates. It is advantageous that the oocyte extensively and accurately modifies protein posttranslationally (see section Posttranslational Events). Furthermore the selective secretion by oocytes of only those proteins which are normally secreted results in a considerable purification, since all the endogenous nonsecreted proteins are effectively removed.

To detect materials secreted into the medium,[6,13] oocytes should be isolated individually from the loosely attached follicular tissue. It is essential to remove any dead oocytes, which may release proteolytic enzymes. Incubations should be carried out with one or two oocytes in a 5 or 10 μl drop of MBS (Table II) in a water-saturated atmosphere. Increasing the volume in which the oocyte is incubated does not much decrease the amount of [^{35}S]methionine which it takes up from the medium. The medium may be collected after 1–2 days.

Isolation of Germinal Vesicles and Cytoplasms

For some experiments, it is useful to separate the germinal vesicle and cytoplasm of injected oocytes. When this is anticipated, it is desirable to inject only about 15 nl into the germinal vesicle, so that it is not damaged by inflation. The germinal vesicle and cytoplasm can be separated by opening the oocyte with forceps in MBS (Table II), and removing adhering yolk by passing the germinal vesicle into and out of a pipette. Alternatively, an incision can be made with a syringe needle (26 G) in the oocyte's animal pole (center of pigmented region) and the germinal vesicle gently squeezed out with forceps.[2] Since the germinal vesicles are small, they are best collected in homogenization medium containing proteinase K and carrier DNA, RNA, or protein (see above). Isolated germinal

[13] A. Colman and J. Morser, *Cell* **17,** 517 (1979).

vesicles of DNA-injected oocytes may be used for the electron microscope examination of active transcription complexes. To avoid spending time on the analysis of germinal vesicles which happened to miss an injection of DNA, a trace of any iodinated large protein such as serum albumin, may be added to the DNA which is injected.[14] Individual germinal vesicles are then counted in a drop in a gamma counter before making nuclear spreads.

Transcription

The configuration of DNA injected into oocytes has a substantial effect on the efficiency of its transcription. A small linear molecule (5000 base pairs long) yields 10–20 times less RNA in oocytes than the same kind of DNA in circular form.[1,12] On the other hand all forms of circular molecule are equally well transcribed. Single-stranded molecules, such as M13 phage DNA, are copied into a double-stranded form,[15] and nicked circles are ligated. These are then converted into double-stranded supercoiled molecules and assembled with nucleosomes.[16]

A generally appropriate amount of DNA to inject is 5 ng (25 nl at 200 μg/ml). This amount seems to saturate the transcriptional capacity of an oocyte, at least with genes transcribed by polymerase II and III,[8,17] though an oocyte has sufficient histones to assemble much more than 5 ng DNA (Table I). The efficiency with which DNA is transcribed (transcripts/gene/hour) increases when less DNA is injected, but the absolute amount of gene product is less.

The amount of RNA typically obtained from injected DNA, and its specific activity, can be calculated as follows, assuming circular molecules containing one copy of a gene are injected: 1 μCi (2×10^6 dpm) of [^{32}P]nucleoside triphosphate (either GTP or CTP) is injected. Since the oocyte contains about 250 pmol of GTP (Table I), this corresponds to 8000 dpm/pmol GTP, or 5 dpm/pg RNA, which therefore has a specific activity of about 5×10^6 dpm/μg.

Oocytes not injected with DNA generally incorporate about 5% of the GTP pool in 24 hr, corresponding to 12.5 pmol GTP, or 17 ng RNA. Most of the stable RNA seen after a 24-hr incorporation period is rRNA; in shorter labeling periods, pre-rRNAs predominate. In addition to their endogenous transcription, oocytes injected with DNA transcribed by polymerase II generally incorporate 1% of the GTP pool into RNA com-

[14] M. F. Trendelenburg and J. B. Gurdon, *Nature* (London) **276,** 292 (1978).

[15] R. Cortese, R. Harland, and D. A. Melton, *Proc. Natl. Acad. Sci. U.S.A.* **77,** 4147 (1980).

[16] A. H. Wyllie, *et al., Dev. Biol.* **64,** 178 (1978).

[17] J. B. Gurdon and D. D. Brown, *Dev. Biol.* **67,** 346 (1978).

plementary to injected DNA (or about 5% with DNA transcribed by polymerase III). These values correspond to 3 ng (polymerase II) and 8 ng (polymerase III) of transcript accumulated per day in each oocyte (Table III).

The amounts of accumulated transcript per day given above relate to *stable* RNAs and are presented as an aid in experimental design. They are not equivalent to *rates* of synthesis, since as much as 90% of the newly synthesized RNA from some templates—SV40, for instance—may be rapidly degraded.[17a]

α-Amanitin can be used to determine the type of oocyte polymerase which transcribes injected genes (Fig. 1). The injection of α-amanitin (mixed with DNA) at 5 pg per oocyte eliminates polymerase II transcription without decreasing the activity of polymerases III or I; 500 pg α-amanitin per oocyte greatly reduces polymerase III transcription, but even 5 ng per oocyte has no substantial effect on transcription by polymerase I (Fig. 1). It has been found so far that eukaryotic genes are transcribed in injected oocytes by the same type of polymerase as is used in the cells where these genes are normally expressed.

TABLE III
RNA SYNTHESIS IN DNA-INJECTED *Xenopus* OOCYTES[a]

DNA injected (No. of genes)[b]	Synthesis of complementary (5 S or SV40) RNA[b]		
	Total RNA dpm (% of [^{32}P]GTP injected)	dpm in RNA (% of total labeled RNA)	Amount of 5 S or SV40 RNA[c]
None	100,000 (5%)	[5 S: 0.25% (endogenous); SV40: none]	(5 S: 40 pg)
Xenopus 5 S in plasmid[b] (5 ng; 5 × 10^8 mol)	200,000 (10%)	50,000[d] (25%)	8 ng; 10^{11} mol
SV40 DNA[b] (5 ng; 5 × 10^8 mol)	120,000 (6%)	20,000 (16%)	3 ng; 10^9 mol[e]

[a] Labeled for 24 hr with 1 μCi [^{32}P]GTP per oocyte. These results apply only to newly synthesized RNA which is stable for several hours.

[b] See Ref. 17 for quantitation of results with 5 S genes in a plasmid, and Refs. 8 and 30 for SV40 transcription.

[c] Values calculated assuming that an uninjected oocyte synthesizes 20 ng RNA per day (Table I).

[d] The remaining 50,000 dpm RNA synthesized from injected DNA is complementary to the plasmid region of the injected DNA.

[e] Assuming an average transcript length of 3000 nucleotides.

[17a] A. A. Miller *et al. Molec. Cell Biol.* (1982) (in press).

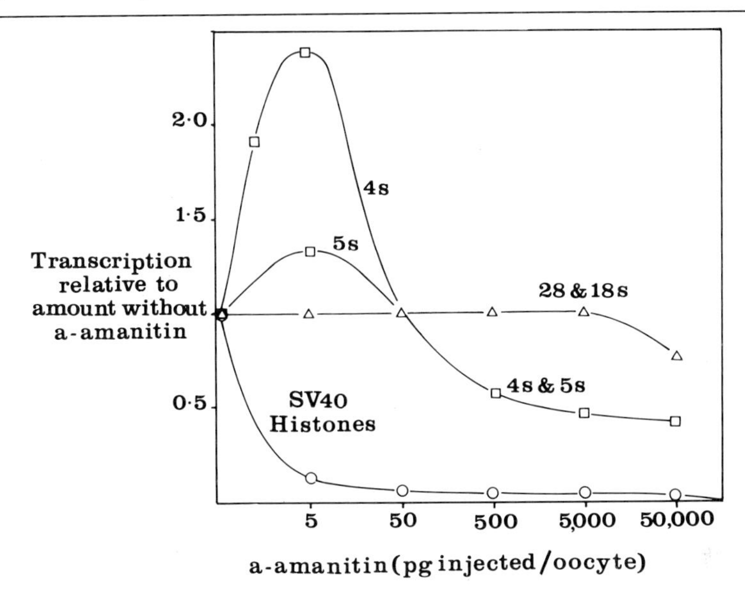

FIG. 1. α-Amanitin sensitivity of transcription in DNA-injected oocytes.

Injected prokaryotic plasmids (pBR322, ColE1, etc.) are transcribed by polymerase II, and produce approximately the same amount of transcript per day per oocyte as do eukaryotic genes which direct accurate initiation and termination, like SV40. Not surprisingly then, read-through transcription from vector DNA into eukaryotic genes inserted into recombinant clones can complicate studies of transcription initiation. Consequently, eukaryotic DNAs are often excised from the vector using a restriction enzyme and are then religated into a circle free of vector DNA prior to injection.[12]

The fidelity with which injected genes are transcribed by oocytes varies considerably according to the type of gene used.[1] Genes transcribed by polymerase III (e.g., 5 S genes and tRNA genes) generally show good strand selectivity, accurate initiation, and good but not perfect termination. Some genes transcribed by polymerase II, such as herpes virus thymidine kinase, certain sea urchin histone genes, and the SV40 late genes also show accurate initiation and termination, while others, such as ovalbumin (in a plasmid or on its own), do not. The cause of this difference in the transcription of these genes is not yet clear.

In those cases in which the oocyte has been used to map promoter regions, it appears that oocytes provide more *in vivo*-like conditions than cell-free systems, in that regions of DNA other than the TATA box are required. In this context the fact that injected DNA is assembled into nucleosomes may be relevant.

Posttranscriptional Processing

RNA polymerase III transcribes tRNA genes injected into oocytes into a primary transcript which is then matured in the nucleus into tRNA by a series of modifications. These include the removal of 5′ and 3′ sequences, base modification, and the precise excision of an intervening sequence. Fully matured tRNA is produced with high efficiency and is transported to the cytoplasm.

Oocytes also carry out several posttranscriptional modifications of mRNA precursors synthesized by polymerase II from injected genes. For example, the proper 3′ end of SV40 late transcripts is synthesized efficiently in oocytes; in infected cells, it has been shown that this requires posttranscriptional cleavage of an RNA precursor.[18] Furthermore, roughly 80% of those transcripts which terminate at the proper position are polyadenylated, as judged by their binding to poly(U)-Sepharose and their electrophoretic mobility.[18] Other genes tested in these respects include histones and herpes virus thymidine kinase. Some of the histone genes direct efficient and accurate termination, while other histone genes[19] and the thymidine kinase gene do not.[20] Neither histone nor thymidine kinase transcripts are efficiently polyadenylated. The presence of a 5′ cap has not been investigated.

The splicing of mRNA precursors has been inferred when proteins are synthesized from genes in which intervening sequences interrupt the protein-coding region. This is true for T antigen of SV40 and for ovalbumin (see the following section). The direct demonstration of RNA splicing has so far been presented only for tRNA genes transcribed by RNA polymerase III and for SV40 late genes transcribed by polymerase II.[18] At present it is difficult to predict whether a particular type of injected gene will produce abundant correctly spliced transcripts.

The passage of transcripts from an occyte nucleus to the cytoplasm has been followed in some detail because it is easy to separate manually the nucleus from the cytoplasm. The most significant result of such investigations is that incorrect or incomplete transcripts generally fail to reach the cytoplasm whereas mature correctly processed transcripts clearly do so. This has been documented for tRNA genes,[1] histone genes,[12] and SV40[18]

[18] M. P. Wickens and J. B. Gurdon, *J. Mol. Biol.* **163,** 1 (1983); J. P. Ford and M.-T. Hsu, *J. Virol.* **28,** 795 (1978).

[19] C. C. Hentschel and M. L. Birnstiel, *Cell* **25,** 301 (1981).

[20] S. L. McKnight and E. R. Gavis, 1980. *Nucleic Acids Res.* **8,** 5931 (1980), and personal communication.

TABLE IV

PROTEIN SYNTHESIS IN INJECTED OOCYTES[a]

Material injected[b]	Complementary RNA synthesized[c]	Functional mRNA	Protein synthesis due to injection	
			dpm in specific protein (% of total)[d]	Amount[e]
SV40 DNA[b] (1 ng; 10^8 mol)	3 ng; 10^9 mol	10^6 mol[f]	5000 dpm VP1 (0.5%)	1.25 ng; 3×10^{10} mol
Chick ovalbumin[b] DNA in plasmid (5 ng; 3×10^8 mol)	3 ng; 3×10^8 mol	10^4 mol[f]	100 dpm (0.01%)	25 pg; 5×10^8 mol
Rabbit β-globin[b] mRNA (1 ng)	—	3×10^9 mol	500,000 dpm (50%)	120 ng; 5×10^{12} mol

[a] Labeled for 24 hr, with 5 μCi [^{35}S]methionine per oocyte.

[b] Results based on Ref. 31 (for SV40 DNA), Ref. 30 (for chick ovalbumin DNA), and Ref. 21 (for rabbit globin mRNA).

[c] Based on values in Table III for SV40, and in Ref. 30 for ovalbumin.

[d] One oocyte incubated in 5 μCi [^{35}S]methionine in 10 μl MBS for 24 hr synthesizes a total of 10^6 dpm protein.

[e] Calculated from percentage values in last column assuming that an oocyte synthesizes 250 ng protein per day (Table I).

[f] Estimated assuming that each molecule of mRNA makes 30 proteins per minute (Ref. 21).

Expression as Protein

DNA injected into oocytes has been shown to be expressed as protein for SV40 and polyoma virus (T antigens and virion proteins), *Drosophila* and sea urchin histones, chick ovalbumin, and thymidine kinase.[1] Protein will possibly be found whenever sufficiently sensitive methods of detection are used. The amount of protein synthesized can be estimated as follows. Under standard conditions (see Methods section) an oocyte incorporates about 10^6 dpm of [^{35}S]methionine into protein per day, during which time it synthesizes 250 ng of total protein. Virion protein 1 of SV40 is synthesized in DNA-injected oocytes in greater amounts than most other proteins coded for by injected genes (Table IV). It constitutes 0.5% of total synthesis, corresponding to 5000 cpm (1 ng) after standard labeling conditions for 1 day. A similar value is obtained if a calculation is based on the specific activity of an oocyte's methionine pool (44 pmol) and the methionine content of the protein.

There is considerable variation in the yield of protein synthesized by different types of injected DNA. For example 50 times less ovalbumin is synthesized than SV40 VP1 from similar amounts of DNA (Table IV). The efficiency with which oocytes translate mRNA is known from the injection of many different kinds of purified mRNA. A pure preparation of rabbit β-globin mRNA is translated 30 times per minute.[21] If we assume that fully processed cytoplasmic mRNA for SV40 proteins and ovalbumin are all translated at the same rate, we can deduce the amounts of these mRNAs in DNA-injected oocytes. It is clear from Table IV that only a small fraction of the total RNA transcribed from injected DNA becomes translatable mRNA.

Posttranslational Events

Many proteins are accurately modified in oocytes after translation. These posttranslational steps include the modification of amino acids, proteolytic cleavage of a primary protein chain, and the transport and secretion of selected proteins within and from a cell.[22] The ability of oocytes to carry out these activities has been tested in mRNA rather than DNA-injection experiments. In general oocytes perform correctly most post-translational events characteristic of wholly unrelated cells of different types and different species. For example,[23] the N-terminal methionine is

[21] J. B. Gurdon, "The Control of Gene Expression," p. 59. Harvard Univ. Press, Cambridge, 1974.

[22] C. D. Lane, *Cell* **24,** 281 (1981).

[23] C. D. Lane and J. S. Knowland, *in* "Biochemistry of Development" (R. Weber, ed.), Vol. 3. Academic Press, New York, 1974.

acetylated on calf lens crystalline, a terminal 15 amino acids are removed from mouse light chain immunoglobulin, ovalbumin is glycosylated, and the primary polypeptide of mouse encephalomyocarditis virus is cleaved into virion proteins.

The selective secretion of proteins also takes place in injected oocytes.[13] For example, mammalian interferon and the milk protein casein are secreted from mRNA-injected oocytes, as they are from cells in which they are normally synthesized. In contrast, globin and other proteins not normally secreted are also not secreted from oocytes. The recovery of materials secreted by oocytes is described in the Methods Section.

It seems, from these examples, that the cellular mechanisms responsible for posttranslational events are fairly universal in their occurrence in cells, and particularly in oocytes. Proteins synthesized from injected eukaryotic genes of nonamphibian species are therefore very likely to undergo their normal modifications even though synthesized in oocytes. The selective secretion of proteins can greatly help their identification. The recovery of proteins from the oocyte culture medium rather than from a crude oocyte homogenate already eliminates most of the high background of normal oocyte proteins, and antibody precipitations can be further used to recognize minute amounts of material.

Other Uses of Oocytes

Assay for mRNA Purification. Gene isolation usually requires at least partially pure mRNA for the preparation of cDNA or for screening a genomic DNA library directly. Since oocytes were first used for translating mRNA, cell-free systems have been greatly improved and generally have a lower background than oocytes. However, the efficiency of mRNA translation in oocytes is much greater than cell-free systems. Thus if only small amounts of mRNA are available, oocytes provide an especially sensitive assay.[24] As far as is known, all types of eukaryotic mRNA [including some that do not normally carry a poly(A) tail] are translated in oocytes, whether they come from mammals, nonamphibian vertebrates, or invertebrates. Large amounts of contaminating rRNA or tRNA can be tolerated (up to 1 mg/ml or 50 ng per oocyte) but, since all mRNAs are in competition for a limited translational capacity, contaminating mRNAs reduce the efficiency of translation of the one being assayed.

DNA Expression in Somatic Cells. DNA can be conveniently injected into the cytoplasm of fertilized eggs. More than 250 pg per egg usually causes abnormal development,[25] and least damage is sustained if about

[24] S. Nagata *et al. Nature (London)* **284,** 316 (1980).
[25] J. B. Gurdon, *Nature (London)* **248,** 772 (1974).

30 nl containing 200 pg (or 2×10^7 molecules of 5 kb DNA) of DNA is injected into the vegetal pole of eggs undergoing cleavage into the two-cell stage. The injected DNA replicates, and by the late blastula stage has increased the injected amount by 10 times or more.[26,27] Some of the injected DNA is expressed and is probably integrated into the host chromosomes. The large amount of yolk in amphibian eggs makes it impossible to see the egg pronuclei (as can be done in mouse eggs), but DNA deposited in the cytoplasm has a good chance of becoming included in nuclei as they undergo over 10 rounds of rapid division during the 12 hr that follow fertilization.

Acknowledgments

We gratefully acknowledge the comments of L. Dennis Smith, E. J. Ackerman and P. Farrell.

[26] M. M. Bendig, *Nature* (*London*) **292,** 65 (1981).
[27] S. Rusconi and W. Schaffner, *Proc. Natl. Acad. Sci. U.S.A.* **78,** 5051 (1981).
[28] E. H. Davidson, "Gene Activity in Early Development." Academic Press, New York, 1976 (contains original references).
[29] J. J. Eppig, Jr., and J. N. Dumont, *Dev. Biol.* **28,** 531 (1972).
[30] M. P. Wickens *et al., Nature* (*London*) **285,** 628 (1980).
[31] D. Rungger and H. Turler, *Proc. Natl. Acad. Sci. U.S.A.* **75,** 6073 (1978).
[32] A. D. Mills *et al., J. Mol. Biol.* **139,** 561 (1980).
[33] R. G. Roeder, *J. Biol. Chem.* **249,** 249 (1974).
[34] I. B. Dawid, *J. Mol. Biol.* **12,** 581 (1965).
[35] D. D. Brown and I. B. Dawid, *Science* **160,** 272 (1968).
[36] J. W. Chase and I. B. Dawid, *Dev. Biol.* **27,** 504 (1972).
[37] A. Webb and L. D. Smith *Dev. Biol.* **56,** 219 (1977).
[38] D. D. Brown and E. Littna, *J. Mol. Biol.* **20,** 95 (1966).
[39] L. Golden, U. Schafer, and M. Rosbash, *Cell* **22,** 835 (1980).
[40] M. Mairy and H. Denis, *Dev. Biol.* **24,** 143 (1971).
[41] M. Rosbash and P. J. Ford, *J. Mol. Biol.* **85,** 87 (1974).
[42] I. B. Dawid, *J. Mol. Biol.* **63,** 201 (1972).
[43] H. R. Woodland and E. D. Adamson, *Dev. Biol.* **57,** 118 (1977).
[44] M. O. Cabada *et al., Dev. Biol.* **57,** 427 (1977).
[45] A. C. Webb, M. J. La Marca, and L. D. Smith, *Dev. Biol.,* **45,** 44 (1975).
[46] E. Rastl and I. B. Dawid, *Cell* **18,** 501 (1979).
[47] E. D. Adamson and H. R. Woodland, *Dev. Biol.* **57,** 136 (1977).
[48] W. J. Wasserman, J. D. Richter, and L. D. Smith, *Dev. Biol.* **89,** 152 (1982).
[49] G. J. Dolecki and L. D. Smith, *Dev. Biol.* **69,** 217 (1979).
[50] J. Maller, M. Wu, and J. C. Gerhart, *Dev. Biol.* **58,** 295 (1977).
[51] D. G. Capco and W. R. Jeffrey, *Dev. Biol.* **89,** 1 (1982).

[26] Eukaryotic Cloning Vectors Derived from Bovine Papillomavirus DNA

By PETER M. HOWLEY, NAVA SARVER, and MING-FAN LAW

SV40, the human adenoviruses, and the human herpesviruses are currently being exploited and developed as viral vector systems for the transient expression of foreign genes in permissive cells. Similarly, the LTR segments of cloned retrovirus DNAs are being used as components of viral vectors to enhance transformation efficiency and integration into the host chromosome. The finding that bovine papillomavirus (BPV) genome remains exclusively extrachromosomal as a circular plasmid in transformed mouse cells[1] suggested the potential utility of the papillomavirus DNAs as plasmid cloning vectors for introducing foreign DNA fragments into cells susceptible to papillomavirus-mediated transformation. We demonstrated the utility of the subgenomic transforming segment of the BPV genome as a eukaryotic cloning vector using the 1.62 kb fragment of the rat preproinsulin gene (rI_1).[2] We describe here the use of BPV as a cloning vector and recent experiments that permit the shuttling of plasmids between eukaryotic and bacterial cells.

Materials and Reagents

BPV DNA. The BPV-1 genome was cloned from a cutaneous bovine fibropapilloma (isolate 307).[3] The genome consists of 7944 base pairs, and the complete nucleotide sequence is now known.[4] Only a specific 69% subgenomic segment of the genome mapping between the *Bam*HI and *Hin*dIII sites is required for transformation of mammalian cells[5] and au-

[1] M.-F. Law, D. R. Lowy, I. Dvoretzky, and P. M. Howley, *Proc. Natl. Acad. Sci. U. S. A.* **78,** 2727 (1981).

[2] N. Sarver, P. Gruss, M.-F. Law, G. Khoury, and P. M. Howley, *Mol. Cell. Biol.* **1,** 486 (1981).

[3] P. M. Howley, M.-F. Law, C. A. Heilman, L. W. Engel, M. C. Alonso, W. D. Lancaster, M. A. Israel, and D. R. Lowy, *in* "Viruses in Naturally Occurring Cancers" (M. Essex, G. Todaro, and H. zur Hausen, eds.), p. 233. Cold Spring Harbor Laboratory, Cold Spring Harbor, New York, 1980.

[4] E. Y. Chen, P. M. Howley, A. Levinson, and P. H. Seeburg, *Nature (London)* **299,** 529 (1982).

[5] D. R. Lowy, I. Dvoretzky, R. Shober, M.-F. Law, L. Engel, and P. M. Howley, *Nature (London)* **287,** 72 (1980).

ISBN 0-12-182001-7

tonomous extrachromosomal replication.[1] A detailed restriction endonuclease map of the BPV-1 genome indicating the specific subgenomic transforming segment is shown in Fig. 1.

Cells. We have used the mouse C127 cells exclusively in our transformation studies owing to the distinctive characteristic of the BPV-induced

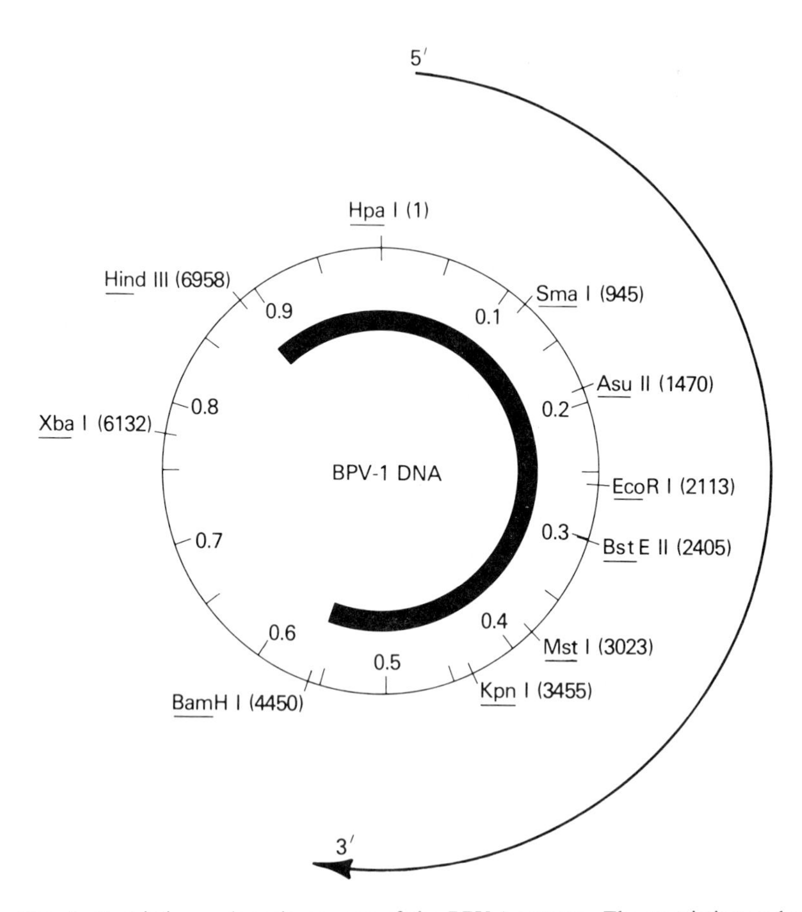

FIG. 1. Restriction endonuclease map of the BPV-1 genome. The restriction endonuclease sites present once in the genome are indicated. The map units and base numbering are derived from the sequence using the *Hpa*I site as the 0/1.0 map unit.[4] The subgenomic (*Bam*HI/*Hin*dIII) transforming segment is indicated by the filled bar.[5] The transcriptional direction and the extent of the genome represented by polyadenylated transcripts found in BPV-1 transformed mouse cells are indicated by the outer arrow.[8] Restriction endonucleases that do not recognize sites in the BPV-1 genome include *Bal*I, *Sst*I, *Sst*II, *Pvu*I, *Sal*I, and *Xho*I.

transformed phenotype.[6] NIH-3T3 cells[6] and rat 3T3 cells[7] are also susceptible to BPV-mediated transformation.

Methods

Construction of Hybrid DNAs

The BPV DNA used as a vector in all our experiments to date has been the 69% subgenomic transforming segment of the BPV-1 genome, extending from the *Hin*dIII site to the *Bam*Hi site.[5,8] Modification of the termini of this segment with synthetic linkers does not reduce the efficiency of transformation. Earlier studies in this laboratory have indicated the pBR322 sequences covalently linked to the subgenomic transforming segment of the bovine papillomavirus is inhibitory to cellular transformation. For this reason, hybrid DNAs have been constructed in such a way as to allow for the physical separation of the pBR322 sequences from the remainder of the BPV-hybrid molecule. This is accomplished by modifying the termini between the BPV-hybrid and the pBR322 portions of the DNA to homologous restriction endonuclease recognition sites that are not present elsewhere in the molecule. Cleavage with the appropriate restriction endonuclease, therefore, separates the molecule into its pBR322 and BPV-hybrid moieties. The inhibition of transformation occurs only when the pBR322 sequences are covalently linked to the BPV-hybrid portion, thus it is not necessary to purify the viral DNA sequences away from the pBR322 sequences prior to transfection.

The steps involved in a typical construction of a hybrid DNA containing pBR322 and BPV sequences is diagrammed in Fig. 2.[2] The source of the DNA to be cloned is the 5.3 kb fragment of the rat preproinsulin gene (rI_1)[9] cloned in pBR322 at the *Bam*HI site. A 1.62 kilobase (kb) segment containing the coding sequences of the rI_1 gene, its intervening sequence, and the regulatory signals at the 5′ and 3′ ends was generated from the cloned 5.3 kb DNA by a *Bam*HI and *Hin*cII restriction endonuclease digestion. After two purification steps through agarose gels, the DNA was electroeluted from the gel, extracted with phenol, and precipitated with

[6] I. Dvoretzky, R. Shober, and D. R. Lowy, *Virology* **103,** 369 (1980).

[7] B. Binétruy, G. Rautmann, G. Menequzzi, R. Breathnach, and F. Cuzin, *in* "Eukaryotic Viral Vectors" (Y. Gluzman, ed.), p. 87. Cold Spring Habor Laboratory, Cold Spring Harbor, New York (1982).

[8] C. A. Heilman, L. Engel, D. R. Lowy, and P. M. Howley, *Virology* **119,** 22 (1982).

[9] P. Lomedico, N. Rosenthal, A. Efstratiadis, W. Gilbert, R. Kolodner, and R. Tizard, *Cell* **18,** 545 (1979).

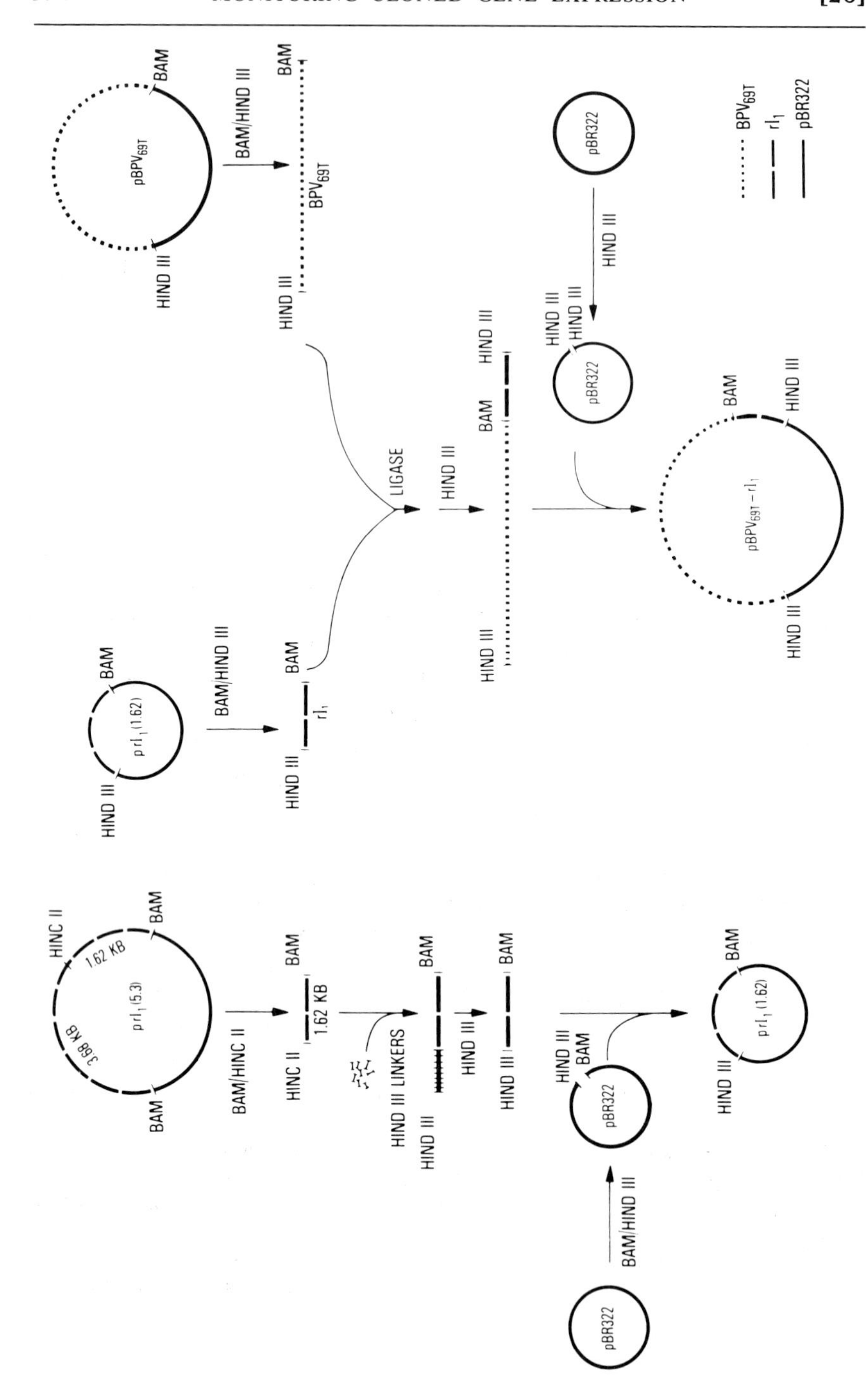
pBPV$_{69T}$
BAM
HIND III
BAM/HIND III
BPV$_{69T}$
pBR322
HIND III
rI$_1$
pBR322
LIGASE
pBPV$_{69T}$ – rI$_1$
prI$_1$ (1.62)
HINC II
1.62 KB
prI$_1$ (5.3)
3.68 KB
BAM/HINC II
HIND III LINKERS
BAM/HIND III

ethanol. Synthetic *Hin*dIII linkers (Collaborative Research) were joined to the *Hin*cII site, and the products were digested with *Hin*dIII to generate ends with monomeric linkers.[10] The modified 1.62 kb fragment was then ligated to the 4.0 kb fragment of the *Bam*HI–*Hin*dIII cleaved pBR322 DNA, and the ligation mixture was used to transform *Escherichia coli* K12 (strain HB101).[11] Plasmid DNA from ampicillin-resistant, tetracycline-sensitive colonies was isolated after an amplification step with chloramphenicol[12] and analyzed with restriction enzymes for the presence of the 1.62 kb fragment insert. One such plasmid, prI_1 (1.62 kb) was selected for further study.

The 69% transforming region of BPV-1 was excised from $pBPV_{69T}$ by *Bam*HI and *Hin*dIII codigestion, purified on an agarose gel, and ligated to gel-purified 1.62 kb fragment of prI_1 (1.62 kb). After digestion with *Hin*dIII, the products were ligated to *Hin*dIII-cleaved pBR322, and a ligation mixture was used to transform susceptible *E. coli* HB101 cells. DNAs from ampicillin-resistant, tetracycline-sensitive colonies were analyzed for the presence of the 1.62 kb insert, and the positive colony thus identified ($pBPV_{69T}$-rI_1) was amplified in *E. coli* HB101.

A number of different BPV_{69T} cloning vectors have been generated in

[10] T. Maniatis, R. C. Hardison, E. Lacy, J. Lauer, C. O'Connel, D. Quon, G. K. Sim, and A. Efstratiadis, *Cell* **15,** 687 (1978).

[11] K. W. Hutchinson and H. O. Halvorson, *Gene* **8,** 267 (1980).

[12] D. B. Clewell, *J. Bacteriol.* **110,** 667 (1972).

FIG. 2. Construction of $pBPV_{69T}$-rI_1. A 1.62-kb segment containing the coding sequences of the rI_1 gene, its intervening sequence, and the regulatory signals at the 5′ and 3′ termini was generated from the cloned 5.3-kb DNA by a *Bam*HI + *Hin*cII digest. After two purification steps through agarose gels the DNA was electroeluted from the gel, extracted with phenol, and precipitated with ethanol. ^{32}P-labeled synthetic *Hin*dIII linkers (Collaborative Research) were joined to the *Hin*cII site, and the products were digested with *Hin*dIII to generate tails with monomeric linkers. Modified 1.62-kb fragments were then ligated to the 4.0-kb fragment of *Bam*HI + *Hin*dIII-cleaved pBR322, and the ligation mixture was used to transform *Escherichia coli* K12 strain HB101. Plasmid DNA from ampicillin-resistant, tetracycline-sensitive colonies was isolated after an amplification step with chloramphenicol and analyzed with restriction enzymes for the presence of the 1.62-kb fragment. A recombinant plasmid, $pBPV_{69T}$, containing the 69% transforming region of bovine papillomavirus 1 DNA, has been described.[3] Viral DNA was excised from this recombinant by *Bam*HI + *Hin*dIII digestion, purified on an agarose gel, and ligated to the gel-purified, 1.62-kb fragment of prI_1. After digestion with *Hin*dIII, the products were ligated to *Hin*dIII-cleaved pBR322, and the resulting DNA was used to transform *E. coli* strain HB101. DNA from ampicillin-resistant, tetracycline-sensitive colonies was isolated and analyzed with restriction endonucleases. One of the colonies containing the recombinant DNA ($pBPV_{69T}$-rI_1) was isolated, amplified, and used for subsequent studies. Reprinted from Sarver *et al.*,[2] with permission.

TABLE I
BPV_{69T} CLONING VECTORS

Vector[a]	Plasmid segment	BPV-1 insert	pBR322/BPV junction[b]	Clonable sites[c]
$pBPV_{69T}$ (52-1)	pBR322	69% transforming segment	*Xho*I	*Bam*HI
$pBPV_{69T}$ (69-6)	pBR322	69% transforming segment	*Sal*I	*Bam*HI
				*Xba*I
				*Xho*I
$pBPV_{69T}$ (70-20)	pBR322	69% transforming segment	*Sal*I	*Bam*HI
				*Hin*dIII
				*Xho*I

[a] Unpublished data of M.-F. Law and P. Howley.
[b] Restriction endonuclease sites generated by *Escherichia coli* DNA polymerase I (Klenow fragment) fill-in reaction and synthetic linkers at junction between pBR322 and BPV-1 DNA sequences. Cleavage with this enzyme separates the molecule into hybrid DNA and pBR322 moieties.
[c] Unique sites present on BPV-1 side of pBR322/BPV-1 DNA junction suitable for the insertion of foreign DNA segments.

pBR322 or into pML2, a deletion derivative of pBR322.[13] These hybrid DNAs have been constructed using fill-in reactions and synthetic linkers to create new restriction endonuclease sites at the junction of the BPV and pBR322 sequences. Unique clonable restriction endonuclease sites are located in these plasmid vectors for the insertion of foreign DNA segments to generate hybrids, which can then be separated from the pBR322-derived sequences by a single restriction endonuclease cleavage.[14] A listing of some of these vectors are presented in Table I.

Transfection of Eukaryotic Cells

Several methods are currently available for delivering naked DNA into eukaryotic cells. These include the calcium phosphate method,[15] the DEAE-dextran method,[16,17] and more recently the protoplast fusion technique.[18] We have tested these three methods, and they are discussed.

As mentioned above, a variety of cell types are susceptible to BPV-

[13] M. Lusky and M. Botchan, *Nature (London)* **293,** 79 (1981).
[14] M.-F. Law, B. Howard, N. Sarver, and P. M. Howley, *in* "Eukaryotic Viral Vectors" (Y. Gluzman, ed.), p. 79. Cold Spring Harbor Laboratory, Cold Spring Harbor, New York, 1982.
[15] F. L. Graham and A. J. van der Eb, *Virology* **52,** 456 (1973).
[16] J. H. McCutchan and J. S. Pagano, *J. Natl. Cancer Inst.* **41,** 351 (1968).
[17] L. M. Sompayrac and K. J. Danna, *Proc. Natl. Acad. Sci. U.S.A.* **78,** 7575 (1981).
[18] W. Schaffner, *Proc. Natl. Acad. Sci. U.S.A.* **77,** 2163 (1980).

mediated transformation. These include the C127 cells derived from a mammary tumor of an RIII mouse, the NIH-3T3 cells,[6] and the Fisher rat 3T3 cells.[7] In our laboratory, the bulk of our experience has been with the C127 cells. To ensure reproducibility from experiment to experiment, we routinely use early-passage cells that have been frozen in parallel. We carry the cells from 1 vial for approximately 8 passages before returning to an early-passage freeze-down. One day prior to transformation, the cells are split and seeded at approximately 5×10^5 cells per 60-mm tissue culture dish. This initial cell density gives a 50–75% confluency level after 24 hr. The medium is changed 3–4 hr prior to transfection.

Calcium Precipitation. The procedure is a modification of that first described by Graham and van der Eb.[15] A sterile 2× $CaCl_2$ solution is dispensed into a sterile tube, and carrier DNA (i.e., calf thymus DNA or salmon sperm DNA) and the desired amount of recombinant DNA are added such that the final DNA concentration in the transfection cocktail will be 20 μg/ml. A second tube contains an equal volume of sterile 2× HeBS solution [280 m*M* NaCl, 50 m*M* HEPES (Sigma), 1.5 m*M* Na_2HPO_4, pH 7.10 ± 0.05]. This buffer may be stored in aliquots at −70° in polyethylene tubes until use. The sterile 2 M calcium chloride solution should either be made fresh or be stored at −20 in plastic tubes until ready for use.

The coprecipitate is formed by adding the 2× $CaCl_2$–DNA solution to the tube containing the equal volume of the 2× HeBS solution. This is accomplished by inserting a 1-ml sterile cotton-plugged pipette into the mixing tube containing the 2× HeBS solution and gently agitating the mixture by blowing bubbles while the 2× DNA-calcium chloride solution is slowly added. The calcium phosphate DNA precipitate is allowed to form without further agitation for 45 min at room temperature. It is then mixed gently and applied (0.5 ml per plate), at an equivalent concentration of approximately 1μg of recombinant DNA per plate, directly into the medium (5 ml) of the 60-mm plate. The precipitate is allowed to remain on the cells for 4 hr at 37°.

The efficiency of DNA-mediated transformation with calcium precipitate is increased by treatment with DMSO or glycerol 4 hr after transfection.[19] We therefore have used this method routinely in our transfection studies. Since the toxicity levels of these agents may vary with different cell types, the toxicity should be titrated with the desired agent and a subtoxic dose applied to the cells. The shock treatment is as follows. Four hours after transfection, the medium is removed, the cells are washed once with fresh medium containing 10% fetal calf serum and drained.

[19] N. D. Stowe and N. M. Wilkie, *J. Gen. Virol.* **33**, 447 (1976).

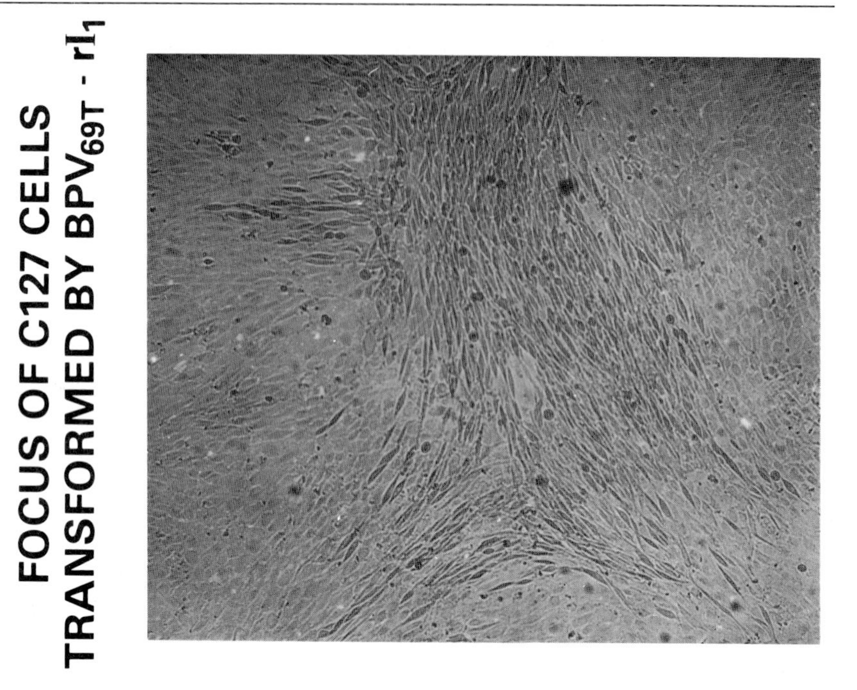
FOCUS OF C127 CELLS
TRANSFORMED BY BPV_{69T} - rI_1

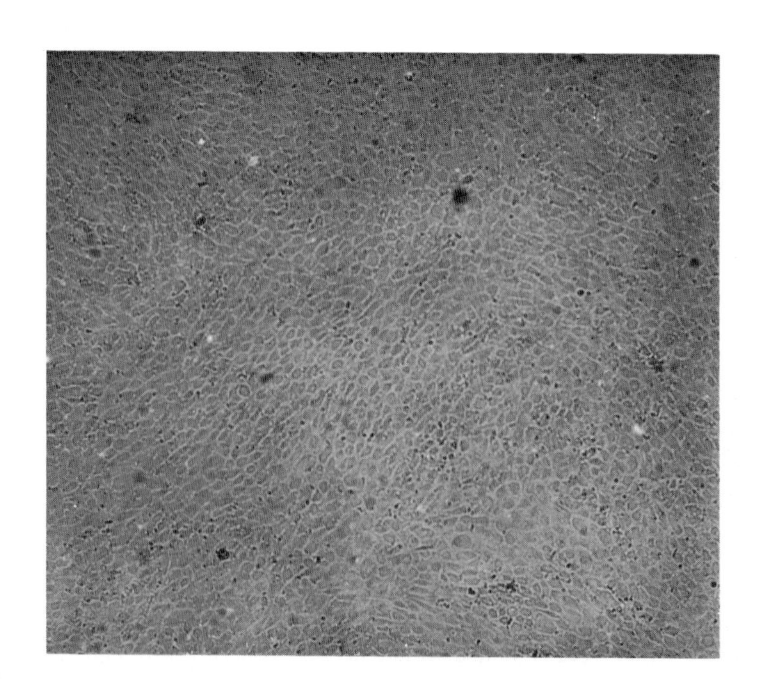
NORMAL C127 CELLS

Sterile DMSO (25% in 1 × HBS) is prepared, and a 1-ml aliquot is applied to the monolayer for 4 min at room temperature with frequent swirling of plates. It is then removed and the cells are washed with medium and refed with medium containing 10% fetal calf serum. Alternatively, the shock may be accomplished using 20–25% glycerol in DME/10% fetal calf serum overlaid on the cells for 1 min. Either technique routinely results in a 3 to 5 fold enhancement in the transformation frequency.

DEAE-Dextran. We have not been able successfully to transform C127 cells by introducing the cloned BPV-1 DNA using DEAE-dextran as a facilitator. We have attempted to transfect mouse cells using DNA at a high concentration (1 mg/ml) of DEAE-dextran for 15 min as originally described by McCutchan and Pagano,[16] as well as at a lower concentration (200 μg/ml) for a prolonged time (8 hr).[17] With neither protocol were BPV transformants induced.

Protoplast fusion. Polyethylene glycol (PEG)-mediated fusion of cells with bacterial protoplasts[18] is an alternative way of effectively delivering BPV-pBR322 hybrid DNAs to C127 cells and rat 3T3 cells.[7] The efficiency of transformation was comparable with *E. coli* containing BPV molecules cloned either into pBR322 or into a derivative pML-2.[13] Further modifications of the protoplast fusion technique that would result in higher levels of transformation would make this an extremely useful approach in the future.

Selection of Transformants

Foci of transformed cells can be recognized 6–7 days after transfection and are of sufficient size to be isolated by day 14 or 16. An illustration of an early transformed focus induced by the BPV_{69T}-rI_1 hybrid DNA on C127 cells is shown in Fig. 3. Individual foci are trypsinized, and subsequent cell lines are established by single cell cloning.

Analysis of Gene Expression

A typical analysis of the lines established from single cell clones involves an analysis of the state of the hybrid DNA within the transformed cell line, an analysis of the RNA transcripts derived from the foreign DNA inserted into the cells, and finally an analysis of any protein products.

FIG. 3. Morphological appearance of untransformed C127 cells (left panel) and a BPV_{69T}-rI_1-induced focus in the same cell line (right panel). The photomicrographs were taken 12 days after transfection (original magnification ×56).

Analysis of DNA in Transformed Cell Lines. Total cellular DNA is extracted from the cell lines by the method of Gross-Bellard[20] and analyzed by the Southern blotting technique.[21] The DNA is either treated with a restriction endonuclease or lightly sheared by passing 10 times through a 25-gauge needle and electrophoresed through a 0.6% agarose gel. After depurination and denaturation *in situ,* the DNA is transferred onto a nitrocellulose membrane (BA85, Schleicher & Schuell Co). The nitrocellulose filters are then hybridized under standard conditions with a nick-translated ^{32}P-labeled probe prepared according to the technique of Rigby *et al.*[22] After hybridization, the filters are washed and exposed to X-ray films as previously described.[1]

Transcriptional Analysis. Polyadenylated cytoplasmic RNA can be isolated from the transformed cells and the RNA analyzed for the presence of transcripts complementary to the cloned foreign DNA segment using the S1 endonuclease and exonuclease VII mapping methods of Berk and Sharp.[23] An example showing the S1 endonuclease and exonuclease VII analysis of rat preproinsulin RNA in mouse cells transformed by the BPV-rI$_1$ hybrid DNA is shown in Fig. 4.[24] In this analysis, polyadenylated RNA is hybridized to a uniformly labeled ^{32}P-labeled probe under conditions that will favor the formation of RNA–DNA duplexes. Hybrids thus formed are treated with either S1 nuclease or exonuclease VII. The former hydrolyzes the single-stranded DNA tails at the 3′ and 5′ termini of the duplex molecule as well as the unhybridized intervening sequences within the gene. The resulting products, analyzed on alkaline agarose gels, thus represent the exons present in specific messenger RNAs (mRNAs). Exonuclease VII, on the other hand, digests only the single-stranded termini, but not internal single-stranded loops. The size of the fragment, therefore, corresponds to the size of the intron plus the exons of the gene.

Protein Analysis. The protein analysis can be done in a variety of ways. If the inserted foreign DNA encodes a biologically active protein, a biological assay can be utilized to assess the synthesis of the protein. Alternatively, a radioimmunoassay may be utilized to detect small levels of the protein either in cellular extracts of the transformed cells or in the media. Finally, the identify of the protein can be established by specific

[20] M. Gross-Bellard, P. Oudet, and P. Chambon, *Eur. J. Biochem.* **36,** 32 (1973).

[21] E. M. Southern, *J. Mol. Biol.* **98,** 503 (1975).

[22] P. Rigby, D. Rhodes, M. Dieckmann, and P. Berg, *J. Mol. Biol.* **113,** 237 (1977).

[23] A. J. Berk and P. A. Sharp, *Proc. Natl. Acad. Sci. U.S.A.* **75,** 1274 (1978).

[24] N. Sarver, P. Gruss, M.-F. Law, G. Khoury, and P. M. Howley, *in* "Developmental Biology Using Purified Genes" (D. Brown and C. R. Fox, eds.), p. 547. Academic Press, New York, 1981.

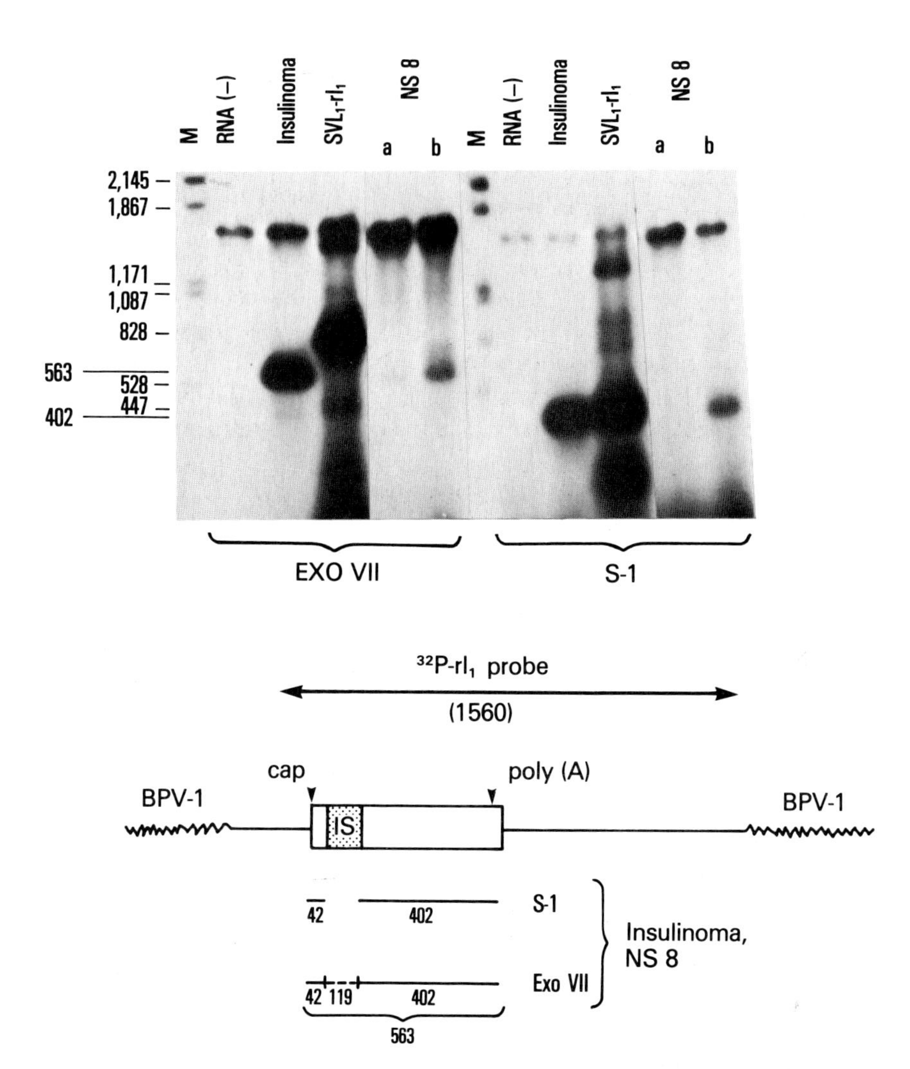

FIG. 4. Poly(A)-selected RNA from 2×10^6 and 5×10^6 cells (lanes a and b, respectively) were mixed with 10,000 cpm of ^{32}p-labeled rat preproinsulin DNA (2×10^6 cpm/μg) purified from recombinant SVL1-rI$_1$ virus by *Hae*II–*Bam*HI digestion.[2] The mixture was precipitated with ethanol, resuspended in 20 μl of formamide buffer, and hybridized for 3 hr at 50°. RNA–DNA duplexes were treated with endonuclease S1 or exonuclease VII,[23] and the digests were analyzed by electrophoresis through a 1.4% alkaline agarose gel. The gel was exposed for 72 hr at −70°. Numbers to the left of the gel indicate the size in base pairs of SV40 DNA segments. The 1560 base species represents the self-annealed ^{32}P-labeled DNA probe. DNA protected by authentic preproinsulin mRNA is 402 nucleotides in the case of endonuclease S1 analysis and 563 nucleotides for exonuclease VII analysis. RNA(−): probe contained no added RNA; insulinoma: RNA from rat insulinoma cells; SVL1-rI$_1$: RNA from AGMK cells infected with SV40-insulin recombinant DNA; NS8: RNA from BPV$_{69T}$-rI$_1$ transformed cells. The diagram depicts the classes of DNA fragments expected if only insulin regulatory signals are involved in transcription. Reprinted, with permission, from Sarver *et al.*[24]

radiolabeling and immunoprecipitation. The identity of the protein was established as that of rat proinsulin by competitive immunoprecipitation studies followed by analysis on SDS–polyacrylamide gels. Cells in 100-ml plates were washed for 3 hr before labeling with Earle's balanced salts containing 5% normal medium and 2% dialyzed fetal bovine serum and labeled in the same medium with 200 μCi of L-[^{35}S]cysteine per milliliter (New England Nuclear Corp.) for 4 hr at 37°. Lysis of cells was performed in 1 ml of Tris-buffered saline (pH 7.6) containing 1% Nonidet P-40, 1 m*M* dithiothreitol, 2 m*M* phenylmethylsulfonyl fluoride, and 2 m*M* *N*-tosylphenylalanine chloromethyl ketone. The labeled immunoreactive proteins were then immunoprecipitated with antibovine insuline serum. For competitive binding studies, the antiserum was first neutralized with bovine insulin (2 μg of bovine insulin per 6 μl of antiserum for 30 min at 4°), after which the mixture was added to the samples. Examples of the immunoprecipitated material from cell extracts (Fig. 5A) and from the medium of the transformed cells (Fig. 5B) are shown.

Comments

There appear to be two limitations in using BPV as eukaryotic cloning vector: (*a*) the limited host range due to the dependence of the BPV-induced transformed phenotype as the selective marker; and (*b*) the cis-inhibition by pBR322 sequence on BPV-induced transformation. These two shortcomings, however, may now be circumvented.

The host range of this BPV vector system may be extended by the use of a dominant selective marker in conjunction with the BPV transforming region. Experiments have been performed to evaluate the utility of the *E. coli* xanthine-quanine phosphoribosyltransferase (XGPRT) gene (*gpt*) as the dominant selective marker for the BPV-induced transformation of C127 cells.[14] A 2.2 kb *Bam*HI fragment containing the *E. coli gpt* harbored in a modified transcriptional unit of the SV40 early region[25] was inserted into the unique *Bam*HI of $pBPV_{69T}$ (52-1), and the resulting recombinant molecules, after the removal of pBR322 sequences were transfected onto C127 cells. Transformants were selected for the expression of the BPV transforming region, which induced phenotypically transformed foci in regular medium, or for the expression of the *E. coli gpt*, which enabled transformed colonies to grow in HAT medium containing xanthine and mycophenolic acid.[14] Cell lines established from individual foci selected for one phenotype were then assayed for the second phenotype. A tight linkage between the two selective markers was established in that 30 of 36 lines transformed with the recombinant DNA molecules

[25] R. C. Mulligan and P. Berg, *Proc. Natl. Acad. Sci. U.S.A.* **78,** 2072 (1981).

TABLE II
TRANSFORMATION OF MOUSE CELLS BY BPV HYBRID PLASMIDS

Plasmid	Bacterial segment	BPV DNA insert	Transformants/μg/10^6 cells	
			Cleaved[a]	Uncleaved
$pBPV_{69T}$ (17-6)	pBR322	69% transforming segment[b]	81, 57	2, 5
$pBPV_{69T}$ (54-2)	pML2	69% transforming segment[b]	56, 46	1, 3
pBPV (8-2)	pBR322	*Bam*HI linear	>200	2, 5
pBPV (142-6)	pML2	*Bam*HI linear	>200	>200

[a] The hybrid recombinant DNA has been cleaved at the restriction endonuclease sites forming the BPV DNA/prokaryotic plasmid junctions.
[b] The subgenomic transforming segment is the large *Hin*dIII–*Bam*HI fragment.

exhibited co-expression of the two phenotypic traits. The orientation of the *E. coli gpt* insert relative to the direction of BPV-1 transcription had no effect on expression. When DNA of these cell lines were examined by blot hybridization, the majority of these lines contained the hybrid DNAs in a plasmid state. It should be noted that a high proportion of the cell lines selected exhibited DNA rearrangement in these molecules. The reasons for these rearrangements are not known. This instability could be a result of the expression of the *E. coli gpt* in eukaryotic cells; alternatively, it could be the result of specific SV40 DNA sequences present in the BPV_{69T}-SV2gpt hybrid molecules, or finally, this instability may be inherent to the use of the 69% subgenomic transforming segment as a vector (see below). Experiments are underway utilizing other dominant selective markers in conjunction with other transcriptional regulatory signals.

The cis-inhibition of pBR322 on BPV DNA-mediated transformation has limited the effective use of BPV DNA as a plasmid vector for shuttling genes between eukaryotic cells and bacteria. More recent results, however, demonstrate that this limitation can be circumvented. It appears that the sequences in pBR322 that are inhibitory to the replication of SV40 sequences in monkey cells[13] may also be responsible for the inhibition of BPV plasmid replication in mouse cells, and hence BPV DNA-mediated transformation of mouse cells. Whereas the complete BPV-1 genome cloned in pBR322 at the *Bam*HI site can efficiently transform C127 cells only after cleavage from the pBR322 sequences, the same full-length BPV-1 DNA cloned into pML2, the "poison sequence" minus derivative of pBR322 can efficiently transform mouse cells with or without such cleavage (Table II).[26] Surprisingly, the 69% subgenomic transforming seg-

[26] N. Sarver, J. C. Byrne, and P. M. Howley, *Proc. Natl. Acad. Sci. U.S.A.* **79,** 7147 (1982).

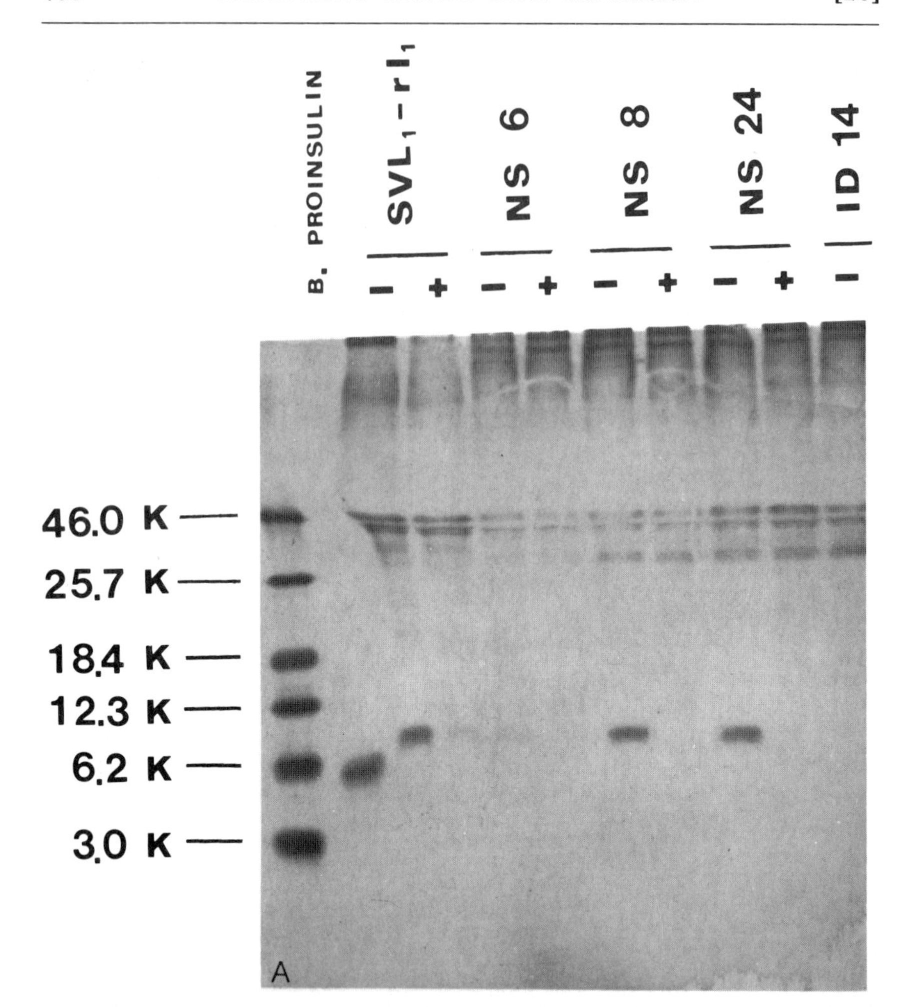

FIG. 5. Cells in 100-mm plates were labeled with 715 μCi (200 μCi/ml) of [^{35}S]cysteine for 4 hr at 37°, and the proteins were recovered as described.[2] Samples equivalent to medium from 10^6 cells or extracts from 1.5×10^6 cells were immunoprecipitated with hamster antibovine serum for 16 hr at 4°. In competitive binding studies the antiserum was first neutralized with 2 μg of bovine insulin (30 min on ice) before it was added to the samples. Immunoprecipitated proteins were analyzed on sodium dodecyl sulfate–polyacrylamide gels. NS6, NS8, and NS24 are cells transformed by BPV_{69T}-rI_1 DNA. ID14 are cells transformed by bovine papillomavirus. (−) No competition; (+) with competition. (A) Analysis of cellular extracts on a 10 to 17% polyacrylamide linear gradient gel. (B) Analysis of proteins secreted into the medium on a linear 16% polyacrylamide gel. Migration of ^{14}C-labeled markers and ^{14}C-labeled bovine proinsulin is indicated. Reprinted, with permission, from Sarver *et al.*[2]

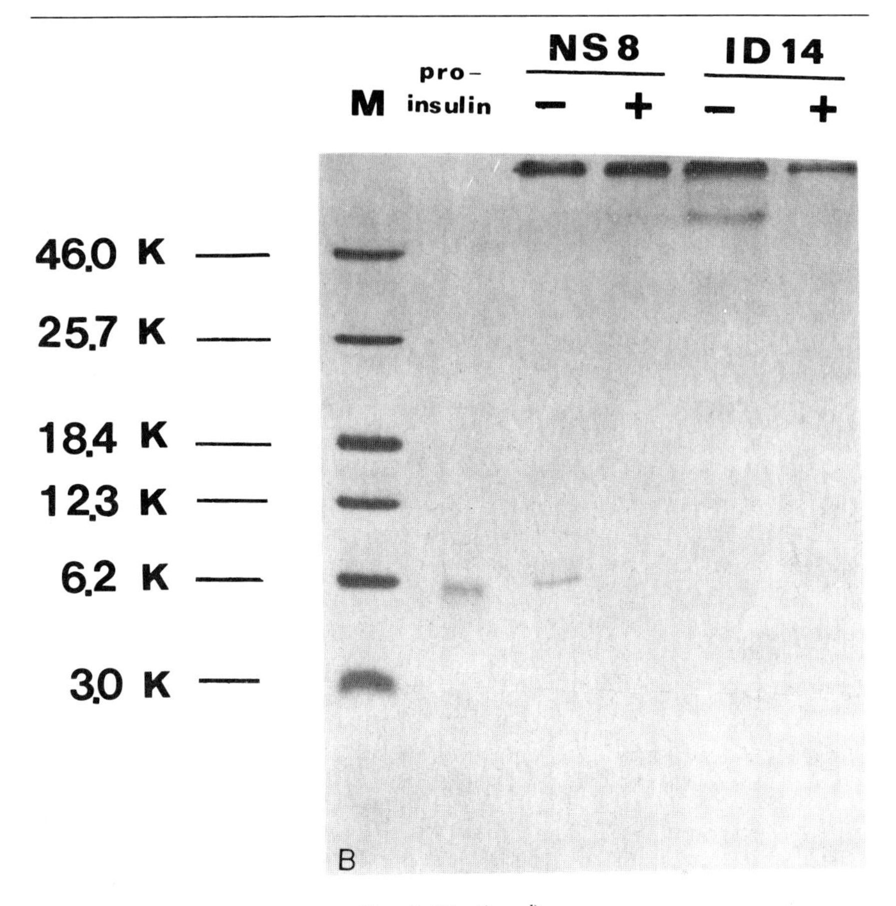

FIG. 5. (*Continued*)

ment cloned into pML-2 will transform mouse cells only after cleavage from the prokaryotic sequences. This suggests that some regulatory region may be present in remaining 31% of the BPV-1 genome that can affect the expression of the transforming segment of BPV-1. Interestingly, a number of eukaryotic gene segments have been shown to have a similar effect as this 31% BPV DNA segment when cloned into a BPV_{69T}−pML2 recombinant. A 7.6 kb segment of the human β-globin gene inserted into a BPV_{69T}−pML2 hybrid can be used to transform mouse cells efficiently.[27] The BPV−pML2−β-globin hybrids are maintained in the transformed cells as plasmids and can be recovered from the transformed cells

[27] D. DiMaio, R. Treisman, and T. Maniatis, *Proc. Natl. Acad. Sci. U.S.A.* **79**, 4030 (1982).

by transformation of susceptible bacteria, thus demonstrating the ability to use BPV–pML2 hybrids as shuttle vectors.[27] In our laboratory we have demonstrated a similar activity in a human growth hormone DNA segment, a 5.3 kb rat preproinsulin DNA segment, and a rat α-fibrinogen DNA segment.[28] Each of these DNA segments when cloned into a BPV_{69T}–pML2 hybrid stimulate the transformation efficiency of mouse C127 cells. The nature of the sequences that result in this stimulation and the molecular basis for this biological effect are not known and are under investigation.

Acknowledgments

We are grateful to Ms. S. Hostler for excellent editorial assistance in the preparation of this manuscript.

[28] N. Sarver, S. Mitrani-Rosenbaum, M.-F. Law, J. C. Byrne, and P. M. Howley, *in* "Gene Transfer and Cancer" (N. Steinberg and M. Pearson, eds.), in press. Raven, New York, 1982.

[27] High-Efficiency Transfer of DNA into Eukaryotic Cells by Protoplast Fusion

By ROZANNE M. SANDRI-GOLDIN, ALAN L. GOLDIN, MYRON LEVINE, and JOSEPH GLORIOSO

Protoplast fusion is a method for directly transferring cloned DNA from bacteria to mammalian cells at high frequency. This technique, unlike calcium phosphate precipitation,[1–5] DEAE-dextran precipitation,[6,7] direct microinjection,[8–11] or the use of liposomes as carrier vehicles,[12–14]

[1] F. L. Graham and A. J. van der Eb, *Virology* **52,** 456 (1973).
[2] N. J. Maitland and J. K. McDougall, *Cell* **11,** 233 (1977).
[3] M. Wigler, S. Silverstein, L.-S. Lee, A. Pellicer, Y.-C. Cheng, and R. Axel, *Cell* **11,** 223 (1977).
[4] M. Wigler, A. Pellicer, S. Silverstein, and R. Axel, *Cell* **14,** 725 (1978).
[5] K. M. Huttner, G. A. Scangos, and F. H. Ruddle, *Proc. Natl. Acad. Sci. U.S.A.* **76,** 5820 (1979).
[6] J. H. McCutchan and J. S. Pagano, *J. Natl. Cancer Inst.* **41,** 351 (1968).
[7] G. Milman and M. Herzberg, *Somatic Cell Genet.* **7,** 161 (1981).
[8] E. G. Diacumakos, *Methods Cell Biol.* **7,** 287 (1973).
[9] C.-P. Lui, D. L. Slate, R. Gravel, and F. H. Ruddle, *Proc. Natl. Acad. Sci. U.S.A.* **76,** 4503 (1979).
[10] M. R. Capecchi, *Cell* **22,** 479 (1980).

ISBN 0-12-182001-7

does not require the isolation or purification of the cloned DNA sequences to be transferred. As first reported by Schaffner,[15] cloned SV40 DNA was transferred directly from bacteria to mammalian cells by converting the bacteria to protoplasts and then fusing with mammalian cells using polyethylene glycol. We have modified the protoplast fusion procedure of Schaffner[15] and have shown that cloned herpes simplex virus type 1 (HSV-1) sequences can be transferred to mammalian cells and expressed at high frequency.[16] Furthermore, stable transformation of Ltk^- cells[17] to a tk^+ phenotype by the HSV-1 *tk* (thymidine kinase) gene also occurred at high frequency in these studies.[16] Here we describe our procedure for protoplast fusion (see Fig. 1).

Principle of the Method

The direct transfer of cloned DNA sequences from bacteria to eukaryotic cells involves only two steps: the conversion of the bacteria to protoplasts or spheroplasts by digestion of the cell wall with lysozyme, and the fusion of the bacterial protoplasts to the animal cells with polyethylene glycol. The second step presumably results in membrane fusion and hybrid cell formation, although this has not been demonstrated in the case of bacterial protoplasts and animal cells. In fact, ultrastructural analysis of plant cell protoplasts fused to bacterial spheroplasts indicated that the spheroplasts were taken into the plant cell cytoplasm by endocytosis, fusion being observed only at low frequency.[18] Whether the transfer occurs by membrane fusion or by endocytosis, genetic material carried by the bacteria is released into the eukaryotic cells and expressed at high frequency.[15,16,18–20]

[11] A. Graessmann, H. Wolf, and G. W. Bornkamm, *Proc. Natl. Acad. Sci. U.S.A.* **77,** 433 (1980).

[12] R. Fraley, S. Subramani, P. Berg, and D. Papahadjopoulos, *J. Biol. Chem.* **255,** 10431 (1980).

[13] T.-K. Wong, C. Nicolau, and P. H. Hofschneider, *Gene* **10,** 87 (1980).

[14] M. Schaefer-Ridder, Y. Wang, and P. H. Hofschneider, *Science* **215,** 166 (1982).

[15] W. Schaffner, *Proc. Natl. Acad. Sci. U.S.A.* **77,** 2163 (1980).

[16] R. M. Sandri-Goldin, A. L. Goldin, M. Levine, and J. C. Glorioso, *Mol. Cell. Biol.* **1,** 743 (1981).

[17] S. Kit, D. R. Dubbs, L. J. Piekarski, and T. C. Hsu, *Exp. Cell. Res.* **31,** 297 (1963).

[18] S. Hasezawa, T. Nagata, and K. Syono, *Mol. Gen. Genet.* **182,** 206 (1981).

[19] B. R. de Saint Vincent, S. Delbrück, W. Eckhart, J. Meinkoth, L. Vitto, and G. Wahl, *Cell* **27,** 267 (1981).

[20] M. Rassoulzadegan, B. Binetruy, and F. Cuzin, *Nature (London)* **295,** 257 (1982).

Materials and Methods

Eukaryotic Cell Lines

High-frequency transfer and expression of genetic sequences following protoplast fusion has been demonstrated with a variety of cell lines. We have used mouse Ltk$^-$ cells and Vero cells (an African green monkey kidney cell line).[16] In addition, two other African green monkey kidney cell lines, CV-1[15,20] and BSC-1,[15] have been used as well as HeLa cells (a human cervix carcinoma line),[15] WI-38 (a diploid human fibroblast line),[15] FR3T3 (a rat cell line)[20] 3T6 (a mouse fibroblast line),[15] and two Chinese hamster ovary cell lines (CHO-K1 and Urd$^-$A).[19] In addition to mammalian cell lines, plant cells (*Vinca rosea*) have been used as recipients in fusion experiments.[18]

Bacterial Strains

The *Escherichia coli* strains that have been used include DH-1,[16] HB101,[15,19] and 1106.[20] *Agrobacterium tumefaciens* was used in fusions with plant cells.[18] The choice of bacterial strain may be an important consideration. While Schaffner[15] and de Saint Vincent *et al.*[19] have reported successful transfer using HB101, we found that this strain required a longer lysozyme treatment than DH-1 for conversion to protoplasts and once protoplasts were formed they lysed more readily than DH-1 protoplasts.[21] Rassoulzadegan *et al.*[20] found strain 1106 to be quite efficient in producing stable protoplasts.

Chimeric Plasmids

Cloned DNA sequences shown to be expressed following transfer by protoplast fusion include the HSV-1 *tk* gene,[16] HSV-1 sequences within an *Eco*RI fragment from the long region of the HSV-1 genome,[16] SV40 viral DNA,[15,20] polyoma early genes,[20] a CAD gene from Syrian hamster cells,[19] an *E. coli* GPT (guanine–xanthine phosphoribosyltransferase) gene linked to eukaryotic control signals,[19] and an octopine-type Ti plasmid.[18]

Formation of Protoplasts

Escherichia coli K12 strain 1100 derivative DH-1 (*rec A1 hsd R hsd M$^+$ Nal A96^R thi-1 end A1 supE 44*)[16] was used as the donor for our protoplast fusion experiments. DH-1 bacteria carrying the appropriate plas-

[21] R. Sandri-Goldin and A. L. Goldin, unpublished observations, 1982.

mids were grown at 37° in 50 ml of M9 salts[22] containing 0.5% casamino acids, 0.4% glucose, 0.012% $MgSO_4$, 5 μg of thiamine per milliliter, and 50 μg of ampicillin per milliliter to an absorbance at 600 nm of 0.7 to 0.8 (about 5×10^8 cells per milliliter). This generally took 5.5–6 hr. Chloramphenicol or spectinomycin (when the plasmid carried a chloramphenicol-resistance gene) was added to 250 μg/ml, and the culture was incubated at 37° for an additional 12–16 hr to amplify the plasmid copy number.[23] We and others[20] have found the chloramphenicol amplification step to be essential for successful transformation following protoplast fusion. No tk^+ transformants were observed in experiments in which the amplification was omitted.[21] Similarly, Rassoulzadegan *et al.*[20] found no polyoma transformants when this step was eliminated. In contrast, Schaffner[15] observed only a twofold enhancement in the percentage of cells expressing SV40 T antigen when amplification was used. This difference in results may be due to the assays used to measure expression. In our experiments and those of Rassoulzadegan *et al.*[20] transformant colonies were scored, whereas Schaffner[15] observed antigen expression shortly after fusion. It is possible that more copies of chimeric plasmid DNA must be introduced into the cell to ensure that stable transformation occurs while even a small number of copies might result in transient gene expression.

After incubation with chloramphenicol, the bacteria were transferred to 25-ml Corex centrifuge tubes, centrifuged at 3000 *g* for 10 min at 4°, and suspended in a total volume of 2.5 ml of chilled 20% sucrose in 0.05 *M* Tris-HCl (pH 8.0). Lysozyme (Millipore Corporation, Freehold, New Jersey) was added (0.5 ml of a freshly prepared solution of 5 mg of lysozyme per milliliter in 0.25 *M* Tris-HCl, pH 8.0), and the mixture was held on ice for 5 min. One milliliter of 0.25 *M* EDTA (pH 8.0) was added to the suspension, which was held on ice for an additional 5 min, after which 1.0 ml of 0.05 *M* Tris-HCl (pH 8.0) was added. The suspension was incubated at 37° until at least 90% of the bacteria were converted to protoplasts as monitored by phase-contrast microscopy. For strain DH-1 this required about 10 min, although careful monitoring was necessary. Low transfer frequencies resulted from either too long an incubation at 37°, causing substantial lysis of the protoplasts, or too short an incubation with lysozyme, so that too few of the bacteria were converted to protoplasts. The protoplast suspension was then slowly and carefully diluted with 20 ml of prewarmed MEM (Eagle minimal essential medium supplemented with

[22] J. H. Miller, "Experiments in Molecular Genetics," p. 431. Cold Spring Harbor Laboratory, Cold Spring Harbor, New York, 1972.

[23] V. Hershfield, H. W. Boyer, C. Yanofsky, M. A. Lovett, and D. R. Helinski, *Proc. Natl. Acad. Sci. U.S.A.* **71,** 3455 (1974).

nonessential amino acids, 100 μg of streptomycin per ml and 100 U of penicillin per milliliter) containing 10% sucrose and 10 mM $MgCl_2$. Lysis of the protoplasts occurred if the MEM was added too quickly. MEM was best added dropwise while gently agitating the tube to mix the solution. After the addition of 7.5–10 ml of medium in this manner, the remainder was added somewhat more quickly. At this point the suspension was kept at ambient temperature for the remaining steps. The final suspension contained about 10^9 bacteria per milliliter and was added directly to the cell monolayers.

Fusion of Protoplasts to Cell Monolayers (see Fig. 1)

Tissue culture cells were seeded in 24-well or 6-well dishes at 12–24 hr before the fusion so that cell monolayers were subconfluent (about 2 to 4×10^4 cells/well of a 24-well dish and about 10^5 cells/well of a 6-well dish). Subconfluent monolayers were used to minimize cell-to-cell fusion, which results in polykaryocyte formation and reduced survival of fused cells. Medium was removed from the cells and either 1.0 ml (24-well dish) or 2.0–4.0 ml (6-well dish) of protoplast suspension was added to each well. The ratio of protoplasts to cells was about 2 to 4×10^4: 1. The protoplasts were pelleted onto the cells by centrifuging at 3000 rpm for 15 min in a swinging Microtiter dish TH-4 rotor of a Beckman TJ-6 centrifuge. As this rotor was designed for Microtiter dishes, not 24-well or 6-well dishes, the latter may be difficult to fit into the rotor. The dishes we used are made by Costar and fit snugly into the rotor, although cracking of the top of the dish occurs occasionally. Spare sterile dishes are kept nearby to replace tops as necessary. After centrifugation, the supernatant was removed by aspiration and the dish was drained by gently tilting. Care must be taken to prevent sloughing the pelleted protoplasts from the cell monolayer. Two milliliters of polyethylene glycol solution (PEG) was added to each well of a 6-well dish or 1.0 ml was added to each well of a 24-well dish. The PEG solution consisted of 50 g of PEG-1000 (Sigma Chemical Co.) in 50 ml of MEM. This solution is most easily made by melting the PEG-1000 at 65°, pouring 50 g into a 100-ml bottle, autoclaving the PEG in the bottle, and then adding 50 ml of MEM after the molten PEG has cooled somewhat but has not yet solidified (about 65°). The PEG solution was left on the cells for different times in various experiments. In general, a 2-minute fusion time has worked well with Ltk^- cells as well as Vero cells, although Vero cells survive well after fusions of up to 4 min. The fusion time may have to be adjusted depending on the cell line used. Some cell lines, such as the HeLa derivative BU-25 or HEF, survived poorly after even short fusion times, so that the PEG concentration may have to be lowered with these cells. After the chosen time, the PEG solution was

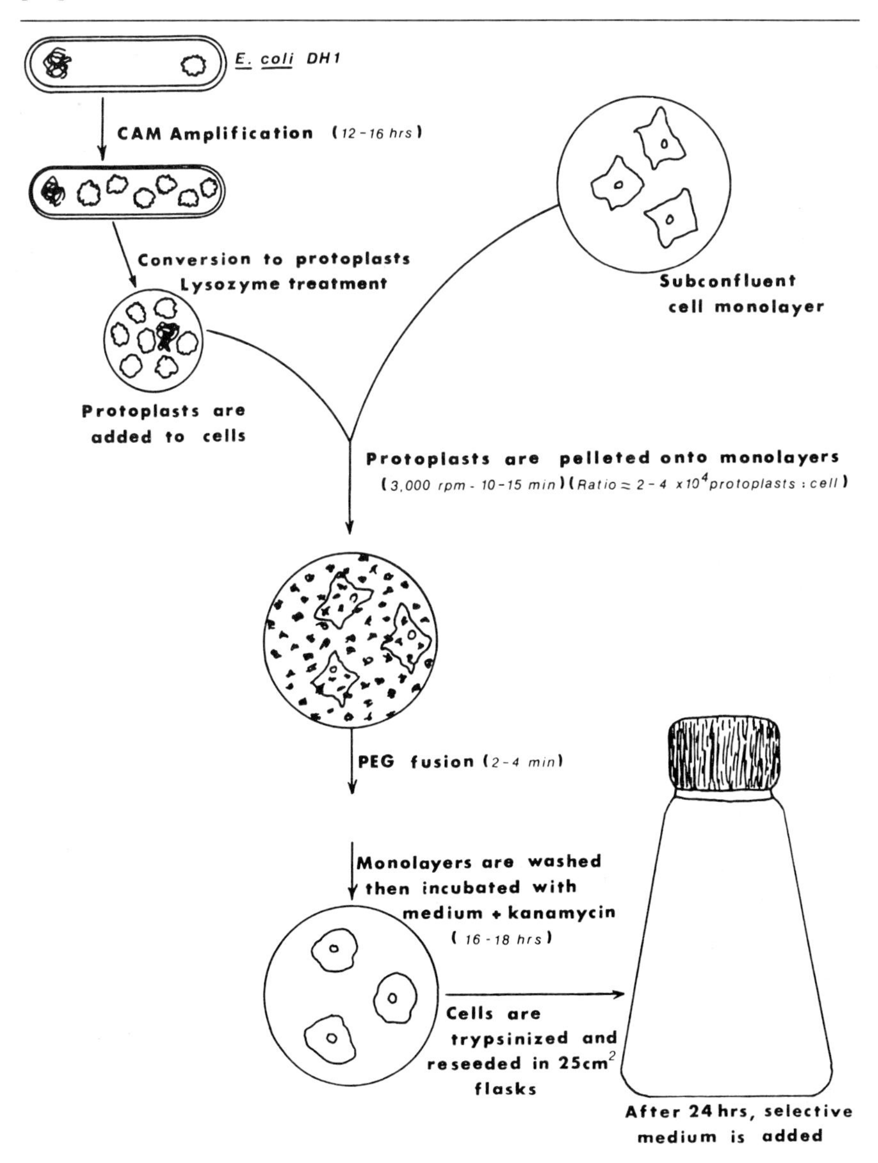

FIG. 1. Schematic representation of the procedure for DNA transfer by protoplast fusion.

removed and the plates were rinsed rapidly with 4 ml of MEM (1.5 ml for 24-well dishes) three successive times, and fresh MEM containing 10% fetal calf serum and 100 μg/ml kanamycin was added to each well. The kanamycin was included to prevent the growth of any bacteria which escaped conversion to protoplasts. Fused monolayers were incubated at 37°.

Fused cells to be analyzed by immunofluorescence were seeded initially onto glass coverslips at the bottom of the wells of 24-well dishes. At various times after fusion (usually 12–48 hr), the cells were fixed in acetone and then treated with the appropriate antibodies. When tk^+ transformant colonies were to be isolated, the fused cells were incubated at 37° for about 16–18 hr after fusion, at which time the medium was removed and fresh MEM containing 10% fetal calf serum was added. Incubation at 37° was continued for an additional 8–12 hr, at which time cells were trypsinized, serially diluted, and reseeded in 25 cm^2 flasks. After 24 hr, medium containing HAT (15 μg of hypoxanthine, 1 μg of aminopterin, 5 μg of thymidine, and 15 μg of glycine per milliliter[3]) was added to the flasks to select for tk^+ transformants. Medium without HAT was added to additional flasks to determine the total number of cells surviving the fusion procedure. Cells were refed every 3 days. Unselected colonies were counted after 1 week, and tk^+ colonies were scored after 2 weeks. Tk^+ colonies were picked with sterile toothpicks into 24-well dishes after 3 weeks and were routinely grown in MEM with 10% fetal calf serum containing HAT.

Results and Discussion

The transfer of material from protoplasts to animal cells was initially monitored by labeling DH-1 protoplasts with the fluorescent dye fluorescein isothiocyanate (FITC).[16] One milligram of FITC was added to the protoplast suspension after the addition of 0.05 *M* Tris-HCl. The labeled protoplasts were fused to Vero cells grown on coverslips for 1, 2, 3, 4, or 5 min. One hour after the fusion, cells were fixed in acetone and observed under a microscope with epifluorescent illumination. In early experiments, around 1–5% of the cells fluoresced after 4- or 5-min fusions, whereas in some later experiments up to 50% of the cells fluoresced after 2- to 4-min fusions. Fluorescence was not observed when protoplasts were pelleted onto monolayers, but no PEG treatment was used.

That DNA was transferred to the cells as a consequence of protoplast fusion was shown by *in situ* hybridization.[16] Protoplasts containing the plasmid pBR325[24] were fused to Vero cells (grown on coverslips) for 90 sec. Twenty-one hours after the fusion, cells were fixed, the DNA was denatured by boiling in 0.1 × SSC, and ^{32}P-labeled pBR325 DNA was hybridized to the cells. In this experiment it was estimated that about 1–2% of the cells showed hybridization to plasmid DNA.

To determine whether transferred sequences were expressed, bacteria

[24] F. Bolivar, *Gene* **4**, 121 (1978).

carrying the plasmid pSG18[25] containing a 16.5 kb HSV-1 *Eco*RI fragment inserted into pBR325 were fused to Vero cells, which were subsequently analyzed by immunofluorescence with antibody made against purified HSV-1 virions.[16] About 5% of the cells fused with these protoplasts fluoresced, whereas no fluorescence was observed when the fusion was done with protoplasts carrying pBR325 but no HSV-1 seqences. This level of expression was consistent with the level of SV40 T antigen expression observed by Schaffner,[15] who fused CV-1 cells with strain HB101 protoplasts carrying the plasmid pBSV-3X which contains three tandem copies of the SV40 genome. More recently, Rassoulzadegan *et al.*[20] reported that as many as 100% of CV-1 cells and around 50% of FR3T3 cells were positive for T antigen expression when these cells were fused with strain 1106 bacteria carrying the plasmid pSV-1, which contains the SV40 early region. This high percentage of transfer was observed at bacteria:recipient cell ratios of about 10^4. Transfer was less efficient at lower ratios. As these authors have pointed out, differences in expression may reflect differences either in the efficiency of fusion or the level of expression of various genes, for example HSV-1 genes as compared to SV40 early genes.

In addition to expression, transformation was found at high frequency following gene transfer by protoplast fusion. We transferred the plasmid pX1,[26] which contains the HSV-1 *tk* gene in pBR322 to Ltk^- cells.[16] Tk^+ colonies (assayed by counting colonies after 2 weeks in HAT selective medium) were found at frequencies of up to one cell in 300–500, though frequencies of about 1 in 10^3 are more usual. No tk^+ colonies were found when pBR322 alone was transferred. High frequencies of stable transformation have also been observed by others. Rassoulzadegan *et al.*[20] found the efficiency of focus formation following transfer of either polyoma or SV40 early genes in pBR322 plasmids was about 40 transformants per 2×10^5 cells at an input ratio of 10^4 protoplasts per cell. These values are comparable to those observed after infection of the same cells with high multiplicities of viral particles.[20,27] Transformation frequencies as high as 0.1 to 0.5% of the recipient cells were also observed by de Saint Vincent *et al.*,[19] who transferred three chimeric plasmids: the Syrian hamster CAD gene cloned in a cosmid vector, the cloned *E. coli GPT* gene,[28,29] and a functional cDNA clone of the mouse dihydrofolate reductase gene.[30]

[25] A. L. Goldin, R. M. Sandri-Goldin, M. Levine, and J. C. Glorioso, *J. Virol.* **38,** 50 (1981).

[26] L. W. Enquist, G. F. Vande Woude, M. Wagner, J. R. Smiley, and W. C. Summers, *Gene* **7,** 335 (1979).

[27] R. Seif and F. Cuzin, *J. Virol.* **24,** 721 (1977).

[28] R. Mulligan and P. Berg, *Science* **209,** 1422 (1980).

[29] R. Mulligan and P. Berg, *Proc. Natl. Acad. Sci. U.S.A.* **78,** 2072 (1981).

[30] S. Subramani, R. C. Mulligan, and P. Berg, *Mol. Cell. Biol.* **1,** 854 (1981).

The transformed cell lines resulting from protoplast fusion gene transfer were stable in nonselective medium. We found that 7 cell lines grown from individual tk$^+$ colonies plated with equal efficiency in selective and nonselective media as soon as they could be tested after transformation and after 5 weeks in nonselective medium.[21] de Saint Vincent *et al.*[19] investigated the stability of five different CAD transformants and found equal plating efficiencies in selective and nonselective media early and after propagation of the lines in nonselective medium for over 100 generations. This stability is comparable to that seen in the stable transformants resulting from microinjection[10] or transfection by calcium phosphate precipitation,[3,31] although numerous instances of unstable transformants have been seen with transfection.[3,31,32] While it has not been reported with transfection or microinjection, we found that protoplast fusion resulted in about 20—30% abortively transformed colonies, which generally did not grow beyond 50–60 cells in selective medium.[21]

The stability of the successfully transformed cell lines may be a direct result of the integration of the transforming DNA into the chromosomal DNA of the recipient cell. Using Southern transfer analysis,[33] we have found that transforming tk DNA was integrated into high molecular weight DNA,[34] and de Saint Vincent *et al.*[19] reported that CAD gene DNA was similarly integrated into high molecular weight DNA. de Saint Vincent *et al.*[19] also used *in situ* hybridization to chromosomal metaphase spreads to demonstrate that most, possibly all, of the CAD gene DNA was associated with a single chromosomal region in each transformed cell line. This suggests integration into cellular genomic DNA.

We have analyzed the structure of the integrated DNA in three tk$^+$ (pX1) transformed cell lines by Southern transfers using four restriction endonucleases that cleave the pX1 plasmid in 0 to 3 sites, respectively. The resulting filters were hybridized to the purified HSV-1 tk transforming fragment of pX1, to the pBR322 vector fragment of pX1, and to *E. coli* DH-1 bacterial DNA. The following preliminary results were obtained.

1. The pX1 plasmid DNA was integrated primarily (if not exclusively) in one site in each of the recipient cell lines.
2. The entire plasmid, including the pBR322 sequences, was present in each cell line.
3. No rearrangements of the plasmid DNA were detectable.

[31] G. A. Scangos, K. M. Huttner, D. J. Juricek, and F. H. Ruddle, *Mol. Cell. Biol.* **1,** 111 (1981).

[32] L. H. Graf, G. Urlaub, and L. Chasin, *Somatic Cell Genet.* **5,** 1031 (1979).

[33] E. M. Southern, *J. Mol. Biol.* **98,** 503 (1975).

[34] R. M. Sandri-Goldin, A. L. Goldin, M. Levine, and J. C. Glorioso, in preparation.

4. The plasmid was integrated in a multimeric configuration in all of the cell lines.

No hybridization of *E. coli* genomic DNA was found with any of the pX1 fragments in the transformed cell lines. This does not rule out the presence of bacterial genomic DNA in the transformants, however, because the highly complex bacterial DNA is an insensitive hybridization probe. All the data are consistent with the integration of an intact multimeric plasmid into a single chromosomal site resulting in transformation of the recipient cell in these three cell lines.[34]

Summary

Protoplast fusion is a highly efficient method for effecting gene transfer to cells in culture resulting in stable transformation at high frequency. A number of cell lines have been used successfully as recipients. There is no need to isolate and purify the DNA, which not only saves time and effort but eliminates steps that cause nicking or breaking of large cloned inserts. The high-frequency transformation achievable by protoplast fusion should make this procedure useful for studies on gene expression and for screening cloned genomic libraries for genes that can be expressed in recipient cells.

Acknowledgments

This work was supported by Grants 5P40-RR00200, AI18228, and AI17900 from the National Institutes of Health.

[28] Gene Transfer into Mouse Embryos: Production of Transgenic Mice by Pronuclear Injection

By Jon W. Gordon and Frank H. Ruddle

Transgenic mice are mice into which have been transferred cloned genetic material. Techniques for production of such mice have only recently been developed.[1] Consequently, exploitation of transgenic mice for studies of mammalian gene regulation has only just begun. These studies have thus far shown that transferred sequences can be retained throughout embryonic development until birth,[1] that they can become integrated into the host genome and transmitted through the germ line to succeeding genera-

[1] J. W. Gordon, G. A. Scangos, D. J. Plotkin, J. A. Barbosa, and F. H. Ruddle, *Proc. Natl. Acad. Sci. U.S.A.* **77,** 7380 (1980).

ISBN 0-12-182001-7

tions,[2-4] and that in at least a subset of instances they can be expressed.[4-6] Preliminary analyses have also indicated that cloned material can be transferred into mice and retained throughout development without becoming integrated into a recipient mouse chromosome.[1,5] These results indicate a significant potential for transgenic mice as experimental tools for the study of mammalian development.

Transgenic mice are produced by microinjecting[7] cloned genetic material into the pronuclei of one-celled mouse embryos. The embryos are then reimplanted into pseudopregnant females and allowed to continue development. DNA is extracted from the resultant animals[8] and subjected to Southern blotting[9,10] in order to identify those animals that have retained the transferred material. This latter technique has been described in detail elsewhere in this series,[11,12] and can be used essentially without modification to analyze DNA from transgenic mice. The present discussion will therefore be confined solely to the actual production of the animals.

Transgenic mouse production can be divided into four main phases: (*a*) preparation of mice; (*b*) recovery and maintenance *in vitro* of one-celled embryos at the pronuclear stage of development; (*c*) microinjection of the embryos; and (*d*) reimplantation of embryos into pseudopregnant females. The entire procedure for producing transgenic mice is outlined in Fig. 1.

Preparation of Mice

Preparing Vasectomized Males

After microinjection, embryos must be reimplanted into pseudopregnant foster mothers for continued development. Foster mothers are pre-

[2] F. Constantini and E. Lacy, *Nature* (*London*) **294,** 92 (1981).

[3] J. W. Gordon, and F. H. Ruddle, *Science* **214,** 1244 (1981).

[4] T. E. Wagner, P. C. Hoppe, J. D. Jollick, D. R. Scholl, R. L. Hodinka, and J. B. Gault, *Proc. Natl. Acad. Sci. U.S.A.* **78,** 6376 (1981).

[5] E. F. Wagner, T. A. Stewart, and B. Mintz, *Proc. Natl. Acad. Sci. U.S.A.* **78,** 5016 (1981).

[6] R. L. Brinster, H. Y. Chen, M. Trumbauer, A. W. Senear, R. Warren, and R. D. Palmiter, *Cell* **27,** 223 (1981).

[7] E. Diacumakos, *Methods Cell Biol.* **7,** 287 (1973).

[8] N. Blin and D. W. Stafford, *Nucleic Acids Res.* **3,** 2303 (1976).

[9] E. M. Southern, *J. Mol. Biol.* **98,** 503 (1975).

[10] G. M. Wahl, M. Stern, and G. R. Stark, *Proc. Natl. Acad. Sci. U.S.A.* **76,** 3683 (1979).

[11] E. Southern, this series, Vol. 68, p. 152.

[12] J. C. Alwine, D. J. Kemp, B. A. Parker, J. Reiser, J. Renart, G. R. Stark, and G. M. Wahl, this series, Vol. 68, p. 220.

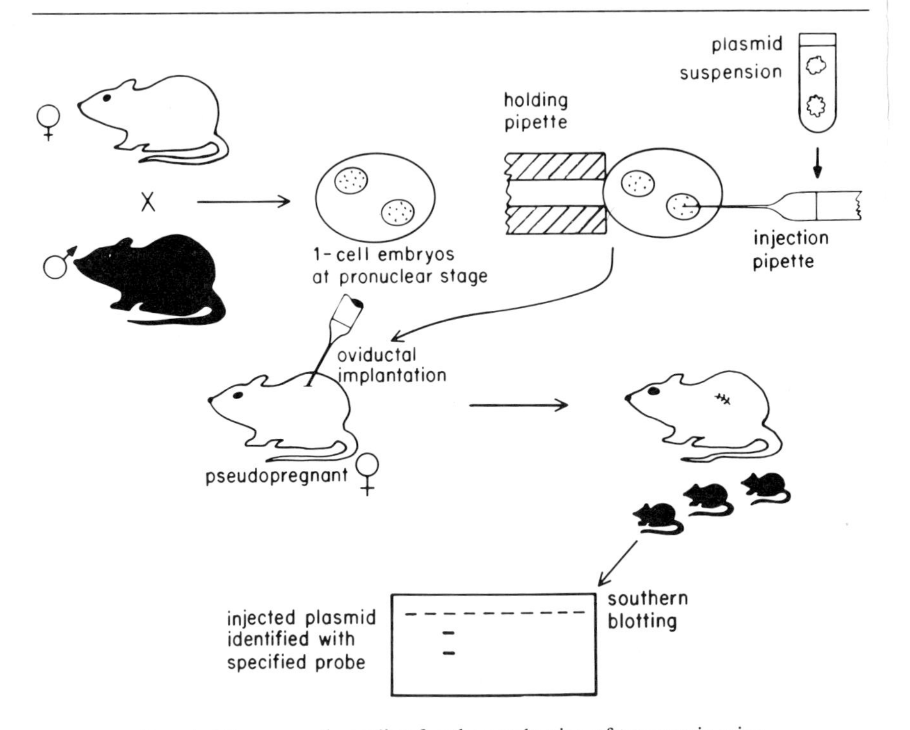

FIG. 1. Diagrammatic outline for the production of transgenic mice.

pared by mating mature females with vasectomized males. Because the males are sterile, the endogenously ovulated eggs are not fertilized and cannot compete with the microinjected embryos that are surgically implanted. It is therefore important that the males be effectively vasectomized. Because even the most carefully performed vasectomies occasionally fail, we take the added precaution of genetically marking our animals so that microinjected embryos carry different pigment alleles than those resulting from fertilization by the vasectomized male. Typically, we arrange our crosses so that fertilization by vasectomized males produces albino mice, whereas microinjected embryos develop normal pigmentation.

Vasectomy is a simple surgical procedure requiring a few minutes for each mouse and is performed as follows.

Preparation of Anesthesia

Reagents
Sodium pentobarbital (Nembutal), 60 mg/ml
Propylene glycol

Ethanol, 95% or 100%
Distilled H_2O
Combine 10 ml of ethanol, 20 ml of propylene glycol, and 70 ml of H_2O to yield 100 ml of diluent for the Nembutal. Dilute Nembutal stock solution 1:10 to yield a final concentration of 6 mg/ml.

Surgery

Equipment
Dumont watchmaker's forceps, No. 5; 2 pair
Stainless steel iris scissors, 1 pair
Hemostat
Surgical suture
Wound clipping device with wound clips (optional)
Syringe, 1 ml, with a 27 gauge, $\frac{1}{2}$ inch needle

Anesthetize an adult (>6 weeks old) male mouse by intraperitoneal injection of 0.10 ml of anesthesia per 10 g body weight. Make a 1-cm longitudinal midline incision in the abdomen just above the preputial gland, which can be identified as a bulge under the skin in the lower abdomen. Gently squeeze the scrotal sac, pushing the testes toward the incision. Identify the testicular fat pads just lateral to the incision and pull them through the wound until the testes are exposed. Expose both testes before vasectomizing either side. Pull one testis farther out of the wound until the red, tubular vas deferens is visible. Using both forceps, control as large a loop of the vas as possible, then clamp the loop in the hemostat. Cut the clamped loop of the vas away from the hemostat, thus resecting a large segment of the vas deferens. Repeat this maneuver on the other side. Replace the testes, and close the peritoneal and skin layers with a single suture. If a wound clip is used, first place a single suture through both skin and peritoneal layers. Then grasp the wound with a forceps underneath the suture loop so that, when the suture is slipped out, both skin and peritoneum are left clamped in the forceps blades. Apply a single wound clip to secure both tissue layers.

After recovering for 1 week, vasectomized males can be test mated to fertile females. Six-week-old males are often sterile at the time of the first mating. Older animals may have enough sperm in the distal vas to be fertile for one mating. As a precautionary measure, all vasectomized males should be mated once before use.

Superovulation and Scheduling of the Experiment

Hormonal superovulation of immature female mice assures that a large number of healthy embryos are available for microinjection (superovulation of 10 immature females yields 50–200 embryos), and allows the experimenter to control the timing of experiments. The estrous cycle of mice is normally 4–5 days, and estrus, ovulation, and mating are related to a diurnal cycle. Estrus normally occurs 4–6 hr after the onset of darkness, and mating requires an additional 2–3 hr. Fertilization is usually complete by 6 hr after mating.[13] Although pronuclei may be visible as early as 6 hr after mating, waiting a few more hours assures that they will be suitably enlarged for microinjection. Because immature females have not yet begun to cycle, superovulation can be used to synchronize the cycles of the mice and to induce ovulation of large numbers of eggs. The timing of ovulation can also be controlled by altering the light: dark cycle of the animal rooms. The following schedule assures that embryos at the pronuclear stage will be available for microinjection in the early afternoon. The day on which microinjection is performed is designated day 0. Biological events that coincide with the procedures listed below are indented.

Schedule for Producing Transgenic Mice

Light: dark cycle: lights on at 8 a.m., off at 10 p.m.

4 p.m., day −3: To 10 immature female mice, administer 5 IU of pregnant mares' serum (PMS) by intraperitoneal injection.

4 p.m., day −1: administer 2.5 IU of human chorionic gonadotropin (HCG) in 0.9% saline by intraperitoneal injection. Immediately place females with males for mating (1 female per male); also place mature (30–35 g) females with vasectomized males (2 females per male).

10 p.m., day −1: lights off.

  4 a.m., day 0: estrus occurs.

  6 a.m., day 0: ovulation and mating completed.

8–10 a.m., day 0: Examine immature and mature females for the presence of vaginal plugs.

  12 p.m., day 0: pronuclei become visible.

1–3 p.m., day 0: Sacrifice immature females and recover one-celled embryos.

3–4 p.m., day 0: Microinject DNA.

4–6 p.m., day 0: Reimplant surviving embryos into pseudopregnant females.

  Days 18–21: Transgenic mice are born.

Note that mature females that serve as foster mothers for the microinjected embryos are not superovulated. While superovulation increases the frequency of mating, many superovulated mice to not carry pregnan-

[13] R. Rugh, "The Mouse, Its Reproduction and Development." Burgess, Minneapolis, Minnesota, 1968.

cies. It is therefore our practice to place 20–30 randomly cycling mature females with 10–15 vasectomized males (2 females per male) on the evening before each experiment. Since the estrous cycle is 4–5 days in length, this procedure assures that a sufficient number of females will mate and be available for implantation.

Examination of Females for Vaginal Plugs

A plug composed of coagulated semen is present in the vaginal orifice of females the morning after mating. This plug can be visualized by immobilizing the female on a flat surface (a cage top is best), lifting the tail, and exploring the vaginal opening with a blunt probe. Because some strains of mice produce less prominent plugs than others, a preferable method of examination is to grasp the female by the scruff of the neck with the thumb and forefinger, pin the tail against the base of the palm, and hold the female with its head toward the floor.

Recovery and Maintenance of One-Celled Embryos

Mouse Embryo Medium MEM

There are several media suitable for maintenance of cleavage stage mouse embryos *in vitro*.[14] The medium we use, modified sightly from that of Mullen and Whitten,[15] is composed of the following reagents (1 liter).

NaCl, 5.15 g
KCl, 0.36 g
KH_2PO_4, 0.16 g
$MgSO_4 \cdot 7\ H_2O$, 0.29 g
$NaHCO_3$, 2.11 g
Sodium pyruvate, 0.04 g
Glucose, 1.00 g
Calcium lactate–5 H_2O, 0.53 g
Penicillin-Streptomycin (10,000 units/ml, 10,000 μg/ml), 10 ml
Sodium lactate, 60%, 3.68 ml
Phenol red, 1% 1.00 ml
Bovine serum albumin (BSA) fraction V, 3 g
Distilled H_2O, 985 ml

Dissolve all solid components except the sodium bicarbonate and BSA, then add the remaining components and stir until dissolved. Filter-

[14] A. McClaren, "Mammalian Chimeras." Cambridge Univ. Press, London and New York, 1976.

[15] R. J. Mullen and W. K. Whitten, *J. Exp. Zool.* **178,** 165 (1971).

sterilize the medium through a 0.22 μm filter (use a prefilter to prevent BSA from clogging the system), and store in sterile, disposable, 50-ml tubes.

Recovery of Mouse Embryos

The following equipment and reagents are needed for recovery and preparation of embryos for microinjection.

Equipment

37° incubator with gas supply of 5% CO_2 in air or 90% N_2, 5% CO_2, 5% O_2. Humidity should be 95–100%.
Dissecting microscope
Tissue culture dishes, 35 mm; five
Tissue culture dishes, 100 mm; two
Optically ground depression slide with glass cover.
Embryo transfer pipette (transfer pipette)
Dumont watchmaker's forceps, No. 5; 2 pair
Iris scissors, 1 pair
Regular surgical scissors, 1 pair
Thin rubber tubing with mouthpiece for transfer pipette
Regular bunsen burner
Small bunsen burner (microflame)
Diamond pen

The microflame is made by cutting off the tip of an 18-gauge hypodermic needle and securing its base into rubber tubing, which in turn is connected to the gas outlet. The needle is embedded in a cork stopper and taped to the counter for support. A screw clamp is attached to the rubber tubing and is used to regulate gas flow.

The transfer pipette is made by drawing the tip of a 9 inch Pasteur pipette in a bunsen flame to an inside diameter of 150–200 μm (1.5–2 times greater than that of an embryo). The tip is cut with a diamond pen to give a clean break. The shaft of the pipette is heated just behind where the taper begins and is drawn out to a thickness compatible with the lumen of the rubber tubing. A clean break is made with the diamond pen, and the base is fire polished. If desired, the tip can also be fire polished in the microflame, but care must be taken to avoid having the lumen of the pipette closed off by melted glass. To prevent this, the pipette can be placed in the rubber tubing and air expelled briskly through the pipette during the fire polishing. The stream of air cools the inside wall of the pipette and prevents melting. Production of the transfer pipette is illustrated diagrammatically in Fig. 2.

Reagents

MEM, 10 ml
MEM, 0.5 ml containing 1 mg/ml hyaluronidase

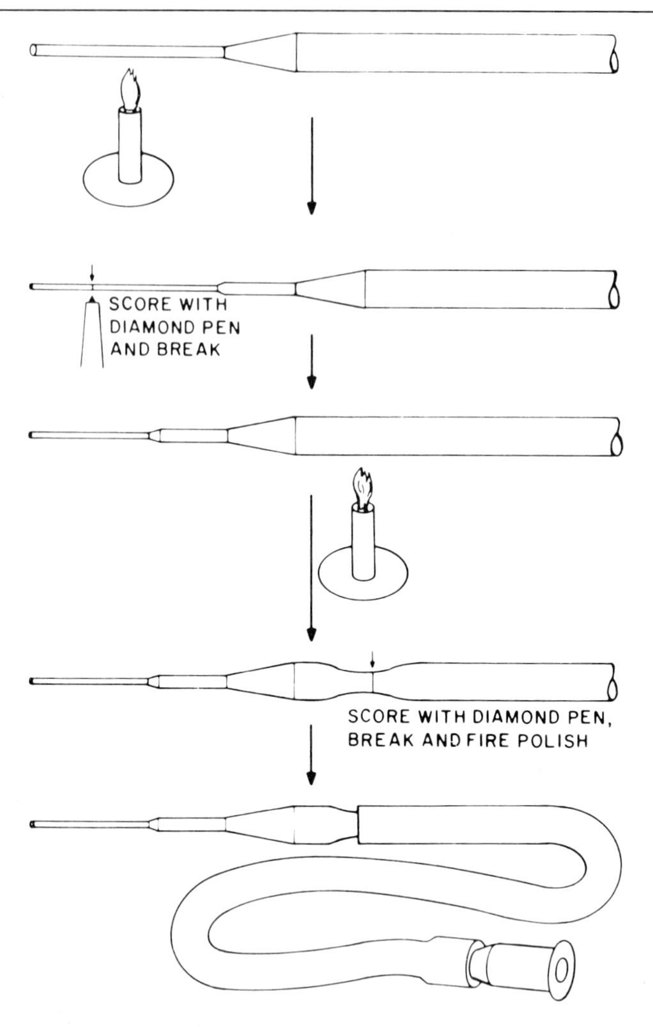

FIG. 2. Diagrammatic outline for the production of an embryo transfer pipette from a 9-inch Pasteur pipette.

Light paraffin oil or mineral oil (for example, Mallinckrodt mineral oil, light, white, No. 6358).

Place 2 ml of MEM in each of five 35-mm tissue culture dishes. Place 0.35–0.5 ml of MEM + hyaluronidase in the depression slide and cover with the glass cover. Fill a transfer pipette with MEM from one of the small dishes and make a 2 × 0.5 cm rectangular microdrop in each of the two 100-mm dishes. The microdrops should be equidistant from the right and left sides of the dishes, about 3 cm from one wall. The long edge of the microdrop should be parallel to an imaginary line drawn from the edge of the circular dish to its center. Place all dishes in the incubator, and do not use them until the medium has been warmed and gas equilibrated.

Sacrifice the immature females by cervical dislocation. Make a single horizontal incision in the mid-abdomen with the regular surgical scissors. Firmly grasp the edges of the wound and pull the skin off the abdomen. Open the peritoneum with the iris scissors. Using both forceps, locate each kidney along the dorsolateral wall. The ovaries are juxtaposed to the kidneys and surrounded by fat. Separate the ovaries from the kidneys with the iris scissors. Remove the ovaries, oviducts, and about 0.5 cm of the attached uterine horns and place them all in one of the 35-mm dishes.

Place the depression slide containing the MEM + hyaluronidase on the dissecting microscope. Place one of the oviducts in the depression slide and identify the loops of the oviduct between the uterine muscle and the ovary. Explore the loops of the oviduct carefully with both forceps until the ampulla is located. The ampulla is the site of fertilization, and can be identified as a swollen portion of the oviduct whose walls are translucent. Longitudinal striations can often be seen along the ampulla as well. Open the ampulla with the forceps. The embryos will be pushed out spontaneously by the intraoviductal fluid pressure, and will appear as a cloud of follicle cells surrounding the large, one-celled zygotes. The number of embryos in each oviduct varies from 0 to more than 30. Use the forceps to express the last of the embryos from the ampulla. Repeat this procedure with each oviduct until all embryos are collected in the depression slide. Allow the embryos to sit for several minutes until the follicle cells have been removed by the hyaluronidase.

Collect the embryos in a transfer pipette and transfer them to one of the four remaining 35-mm dishes. Wash the embryos and transfer them in a minimum volume to another 35-mm dish. Continue washings until 4 of the five 35-mm dishes have been used. Whenever embryos are collected for such transfers, the transfer pipette should first be filled with a small amount of MEM. Thus, when the embryos are expelled into the next dish, air bubbles will not be created. Air bubbles result in dispersion of embryos around the dish and complicate the washing process. Increase the magnification on the dissecting microscope to maximum, and adjust the angle of illumination until the pronuclei become visible. The anatomical landmarks of the embryo include the clear zona pellucida on the outside, one or two polar bodies in the perivitelline space, and the pronuclei. It is frequently possible to see only one of the pronuclei. The cytoplasm of fertilized eggs also has a distinctive granular appearance. Select the embryos with prominent pronuclei and load 20–40 into each microdrop. Return them to the incubator for several minutes prior to microinjection to allow equilibration.

Microinjection of Mouse Embryos

The following equipment and reagents are required for microinjection. The brand names listed are those used in our laboratory, but comparable equipment is available from several other manufacturers.

Equipment

Microscope equipped with two micromanipulators and two injection syringes with micrometers
Micromanipulator instrument collars, three
Vertical pipette puller (David Kopf Instruments, Model 700C)
Microforge (Sensaur)
Glass tubing, 1 mm
Omega dot tubing
Polyethylene tubing equivalent in size to intramedic PE-190
Polyethylene tubing equivalent in size to intramedic PE-90
Wooden-handled metal dissecting probe
Microburner
Syringe, 3 ml, with a long-tipped 30-gauge needle
Syringe, 5 ml, with an 18-gauge needle onto which has been attached 8–10 cm of PE-190 tubing
Plasticene, 1 pound

Reagents

Fluorinert (3M, FC77) or mineral oil
Acetone, 20 ml, in a small vial (e.g., a scintillation vial)

Making the Holding Pipette

The holding pipette is used to fix the position of the embryo for microinjection. It is made by heating a 10-cm piece of 1-mm tubing in the microburner and drawing it by hand to an inside diameter of 50–80 μm. When drawn properly, the thinned portion of the tubing will maintain a consistent diameter over several centimeters of its length. The thinned shaft is broken by hand to a length of 6–8 cm. The point at which the tubing begins to taper is then held over the microflame, and the dissecting probe is held under the drawn out portion of the tubing. Gentle upward pressure is applied with the dissecting probe, and the tubing is bent to a 90-degree angle with respect to the shaft of glass tubing. The tubing is then rotated so that the thinned portion is pointing directly downward, and the dissecting probe is held under the angle of the 90-degree bend.

The shaft is heated at a point approximately 3 cm from the first bend, and pressure is again applied upward with the dissecting probe to put a second 90-degree bend in the tubing. The result is an *S* shaped piece of tubing. The pipette should be heated at the point of the second bend and adjusted so that the shaft and the tip are in the same plane. The shaft is scored with the diamond pen, cut about 3 cm from the second bend, and fire polished in the microflame. These steps are shown diagrammatically in Fig. 3.

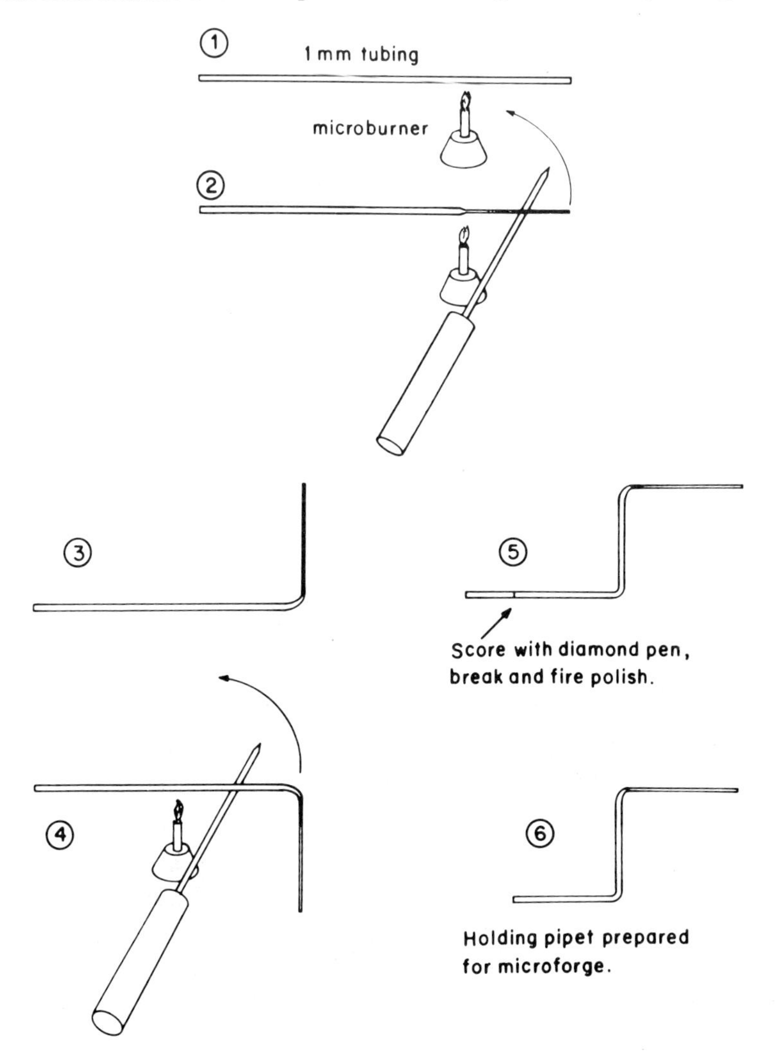

FIG. 3. Outline for the production of the holding pipette. When these steps are completed, the pipette is ready for the microforge (see Fig. 4).

The fire-polished end of the pipette is inserted into an instrument collar and clamped into place on the microforge (Fig. 4). At low magnification (4×), the thinned portion is brought down onto the heating filament so that the glass touches the filament at a point about 1–3 cm from the broken end of the glass. The filament is heated until the glass fuses to it but is not melted to the point where the lumen is narrowed. After the filament has cooled for several seconds, it is pulled briskly away from the holding pipette. This maneuver breaks the glass cleanly at the point of fusion to the filament. The holding pipette and the filament are repositioned in the center of the field so that the tip of the holding pipette is close to the filament and its broken face is parallel to a tangent drawn to the curve of the filament at the point where the filament is closest to the holding pipette. This position is readjusted on high magnification (160×) so that the pipette and the filament are on opposite sides of the circular microscope field. Adjust the position of the pipette to assure that the filament and pipette are in the same focal plane.

Turn on the filament and allow it to reach maximum temperature. (The filament expands as it warms up, so it must be watched to guarantee that it does not touch the holding pipette.) Bring the pipette toward the filament until the glass begins to melt. Allow the glass to melt until the lumen is

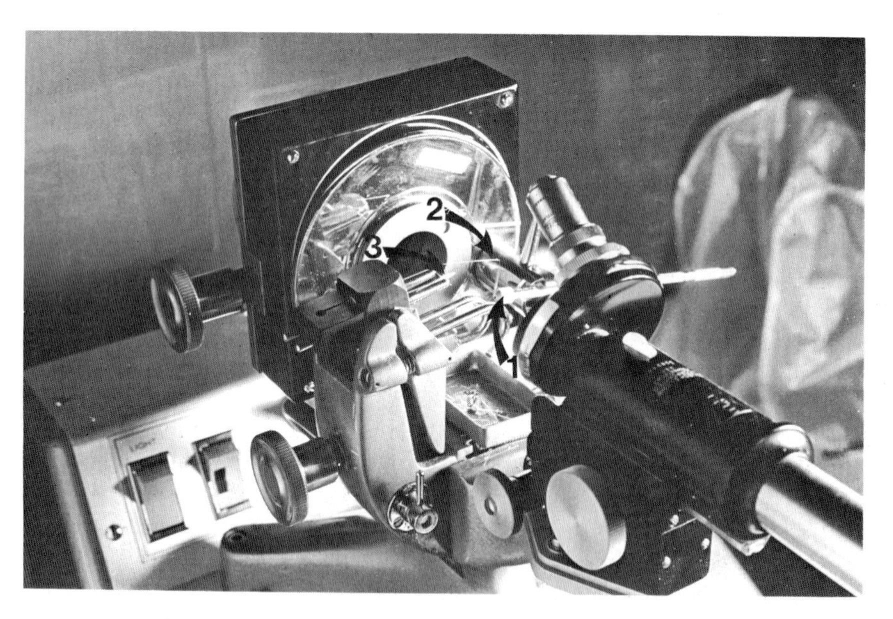

FIG. 4. The holding pipette, positioned on the microforge. 1, Instrument collar; 2, holding pipette; 3, filament.

reduced by 50–75%. Turn off the filament, and store the holding pipette for later use.

Making the Microneedles

Position a piece of omega dot tubing in the pipette puller so that the lower end of the glass is level with the top of the lower clamp. Clamp the tubing in place by gently tightening the upper clamp. Bring up the lower clamp and tighten it sufficiently to hold it in position. Turn on the filament and allow the pipette to be pulled. Remove the lower shaft and insert its base into the PE-190 tubing attached to the 5-ml syringe. Place the tip into the acetone and force air through the microneedle. If air bubbles are not expelled, the tip is not patent and the microneedle is unusable. Large bubbles that rise quickly to the surface indicate that the opening is too broad and that the tip will be too blunt for microinjection. Experience must be gained with pipettes of various calibers in order to define a size range suited to each operator.

Place the base of the pipette into the DNA suspension. The solution is drawn by capillary action to the tip of the pipette. When the suspension has filled 50–75% of the tip, remove the pipette, blot excess DNA from the base, and fire polish the base in the microflame. Fill the 3-ml syringe with Fluorinert, and use the long-tipped 30-gauge needle to fill the barrel of the microneedle. Carefully examine the tip of the microneedle to determine whether air bubbles are present in the DNA suspension. Air bubbles are removed by embedding the shaft of the microneedle in plasticene that has been fixed to the side of the counter top. The microneedle is placed so that the tip is pointing directly downward. After several minutes, the bubbles will rise spontaneously through the DNA and Fluorinet. Five or 6 microneedles should be prepared for each day's work and should be prepared on the day of the experiment.

Preparing the Micromanipulator

Figure 5 shows one micromanipulator with all essential parts labeled. The terms used in the figure will be employed throughout the following discussion. The microscope with micromanipulators should be set up with two syringe micrometers. Our micromanipulator is arranged so that the syringe micrometer on the right side of the stand is connected to the instrument collar on the left micromanipulator. The polyethylene tubing must be purged of all air prior to use. This is done by extending the distal end of the polyethylene tubing through the instrument collar into a large reservoir of whatever medium has been selected (we use Fluorinert because mineral oil is messy and distilled water can become contaminated).

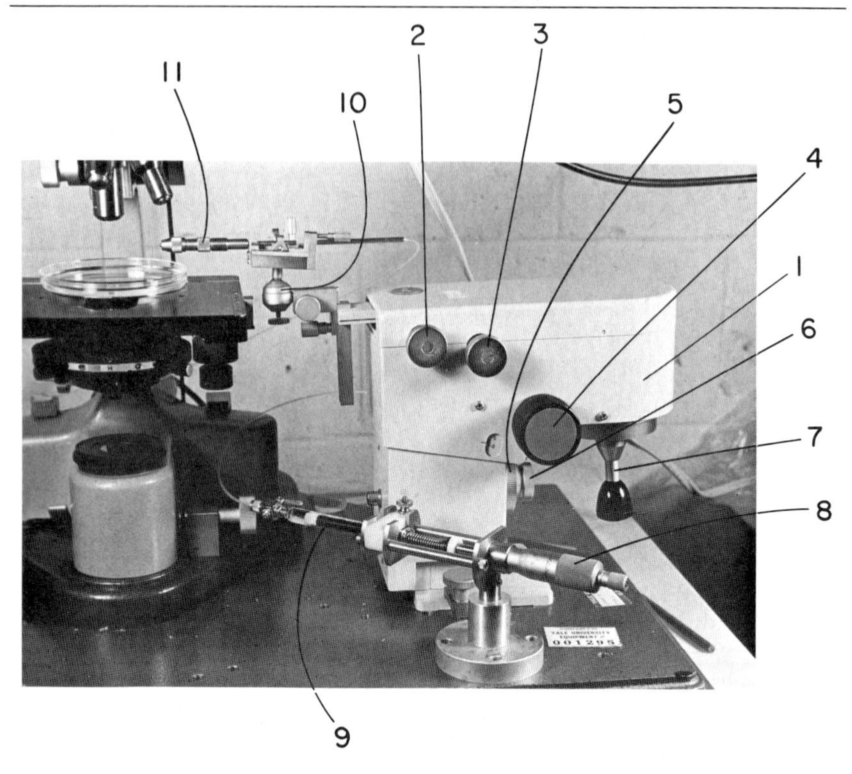

FIG. 5. The right-hand micromanipulator, with essential parts labeled. The syringe micrometer and syringe have been placed in the foreground for easier viewing. 1, Right-hand micromanipulator; 2, knob controlling X-axis motion of the micromanipulator; 3, knob controlling Y-axis motion of the micromanipulator; 4, knob for tilting the micromanipulator; 5, knob controlling coarse vertical motion of the micromanipulator; 6, knob controlling fine vertical motion of the micromanipulator; 7, universal guide lever; 8, syringe micrometer; 9, syringe; 10, ball joint for instrument holder; 11, instrument collar.

The polyethylene tubing and the syringe to which it is attached is completely filled by drawing fluid into the system with the syringe micrometer. The syringe with attached tubing is removed from the micrometer, and the tubing is detached from the syringe. The tubing is lowered below the level of the benchtop, and fluid is allowed to siphon into the system from the reservoir. At the same time, air is carefully expelled from the syringe. When enough fluid has siphoned through the tubing to purge all air, the proximal end is reattached to the syringe, taking care not to introduce air bubbles. The syringe is then reattached to the micrometer. The distal ends of the tubing are allowed to remain in the reservoir until the holding pipette and microneedle are to be attached.

When the holding pipette is attached, care must be taken not to introduce small air bubbles into the system. The vertical shaft of the holding pipette is then filled by advancing the syringe micrometer. The micrometer is given rapid small turns in both directions. If the desired airtight seal has been achieved during attachment of the holding pipette, the fluid level in the pipette should respond quickly to adjustments in the micrometer (this responsiveness will be slowed if mineral oil is used). Before the microneedle is attached, the shaft is rechecked to make sure it is completely filled with fluid. Any air that has entered the distal end of the polyethylene tubing is also removed by advancing the micrometer. The microneedle is then inserted into the tubing and clamped tightly into the instrument collar. Leaks in the holding pipette system manifest as delayed pressure changes in response to changes in the micrometer setting; leaks in the microneedle system almost always appear as fluid leaking from the instrument collar onto the outside of the microneedle shaft.

Microinjection

All microinjection can be carried out under low magnification (160–200 ×). Before the embryos are placed on the stage for microinjection, they are first positioned in the end of the microdrop closest to the edge of the dish. This is done with a transfer pipette at the dissecting microscope. The dish is then carefully placed on the stage for microinjection, with the embryos closest to the operator. The condenser is brought up close to the underside of the dish so that the microscope field can be seen as a circle of light shining through the dish. The Y axis of the stage is then adjusted so that the stage is as close to the operator as possible, and the microdrop is positioned so that the end farthest from the operator is over the condenser. The Y axis is then readjusted so that the embryos are in the microscope field. At this point it is possible to see the embryos in the bright light created by the condenser. The long edge of the microdrop should be parallel to the Y axis of the stage so that movement of the stage along this axis does not move the microscope field beyond the margins of the microdrop. The holding pipette and microneedle, appropriately positioned for lowering into the microdrop, are shown in Fig. 6.

Before the holding pipette is lowered into the microdrop, it is positioned so that its tip is well within the microscope field. Its angle should also be adjusted with the knob for tilting the micromanipulator so that when the pipette is lowered the tip will be the first part to touch the floor of the dish. The holding pipette must also be parallel to the X axis of the stage. When all adjustments are correctly made, the tip of the holding pipette should approach the rectangular microdrop at an angle of 90 de-

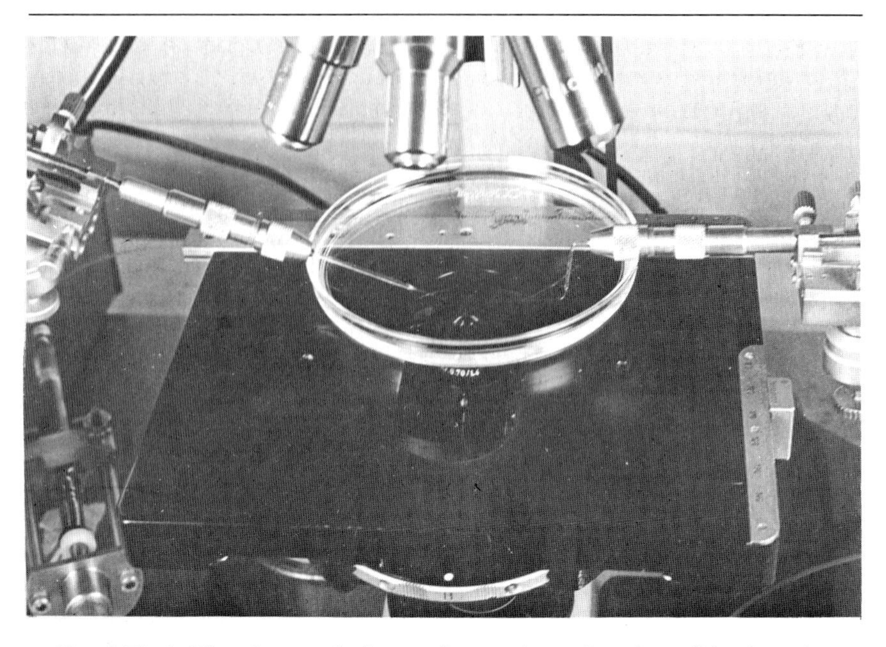

FIG. 6. The holding pipette and microneedle, posed over the culture dish prior to lowering. Note the angle of the microneedle.

grees, and the tip should bisect the circular microscope field without extending beyond the edge of the microdrop. The syringe micrometer is then advanced until fluid completely fills the holding pipette. As the holding pipette is lowered, the syringe micrometer is slowly advanced; this action prevents the pipette from drawing in mineral oil as it is lowered into the microdrop. Using the coarse vertical drive, the pipette is lowered quickly through the mineral oil until the tip touches the floor of the dish. The pipette can then be located under the microscope. If the holding pipette has been properly adjusted prior to lowering, the tip will not be visible until the pipette is drawn back with the micromanipulator knob controlling movement along the X axis. The embryos should be readily visible. When the tip of the holding pipette is brought into the field, an embryo will be drawn to it by capillary action. The function of the holding pipette can then be tested. The tip aperture should not be so large that the embryo is drawn into the pipette, but should be large enough so that the suction holds the embryo firmly. The responsiveness of the holding pipette to the micrometer can then be checked by repeatedly expelling the embryo and drawing it back to the tip. Pressure changes in the holding pipette should be very sensitive to changes in the micrometer. Lack of sensitivity indicates that the holding pipette has probably picked up mineral oil as it was lowered. Such pipettes are usable, but are more difficult to handle. When these adjustments are completed, the holding pipette is left with an embryo attached and the objective is raised so that the microneedle can be lowered.

Before the microneedle is lowered, the coarse vertical drive of the micromanipulator is adjusted to raise the microneedle high above the stage. The heel of the instrument collar is raised, and the knob for tilting the manipulator is also adjusted until the tip of the microneedle is pointing down toward the dish at an angle of 30–60 degrees (Fig. 6). The ball joint for the instrument holder is adjusted so that the microneedle is exactly parallel to the holding pipette (the X axis of the stage). The knob controlling X-axis motion of the instrument holder is used to bring the microneedle over the microscope field, and the coarse knob controlling vertical motion of the micromanipulator is then used to lower the microneedle until its tip touches the surface of the mineral oil. The objective is lowered until the microneedle comes into focus. The microneedle is adjusted with the knob controlling X-axis motion until its tip is in the field. The objective is lowered until the tip of the holding pipette is in focus, and the microneedle is progressively lowered until it begins to come into focus. The microneedle should *not* be lowered until perfect focus is achieved, as the tip might touch the floor of the dish, but it should be lowered until the operator is certain that the tip is in the microdrop. Knobs controlling the X- and Y-axis motion of the microneedle are then used to bring the microneedle onto the same axis as the holding pipette, with the tip of the microneedle far enough away from the holding pipette to assure that the two instruments do not collide when the holding pipette is raised from the floor of the dish.

The holding pipette with the embryo held in place by suction is then raised. The stage is moved along the Y axis toward the operator. When the stage can no longer be moved, the holding pipette and microneedle will be suspended at end of the microdrop farthest from the uninjected embryos. The holding pipette is lowered until the embryo touches the dish. The focus is then adjusted until the pronuclei are located. The prominent nucleoli are readily visible, but the fine focus must be used to visualize the pronuclear membrane. The pronucleus is often more difficult to see when it is in perfect focus. If the objective is focused to a plane above the pronucleus, the pronucleus is seen as a dark, circular region of the cytoplasm. Focusing to a plane below the pronucleus makes the cytoplasm appear dark and the pronucleus relatively light. Because the nucleoli are very large, they may be mistaken for the pronucleus itself. This mistake can be avoided if it is remembered that the pronucleus is very large: its diameter is 20–25% that of the entire embryo. One pronucleus is usually found in the cytoplasm just beneath the second polar body, with a second near the plasma membrane some distance away. If the circle defined by the plasma membrane is considered to be the face of a clock, the second pronucleus should be positioned at 3 o'clock. The embryo can be rotated by expelling it from the holding pipette, adjusting the position of the holding pipette relative to the embryo, and applying suction until the embryo again becomes fixed onto the holding pipette.

With the embryo held by suction and buttressed against the floor of the tissue culture dish, the microneedle is lowered. Using the X and Y axis controls for the instrument holder and the universal guide lever, the microneedle is then placed just outside the zona pellucida at the 3 o'clock position. Using the universal guide lever, the zona is gently probed with the microneedle to determine whether the microneedle and pronucleus are in the same vertical plane (i.e., the same height above the surface of the culture dish). If the microneedle is in the appropriate plane, such probing will result in an immediate recoil of the pronuclear membrane; if not, the pronucleus will not be affected by the probing maneuver. The microneedle is adjusted with the fine vertical drive until it is in the same plane as the pronucleus. The universal guide is then used to pierce the zona, plasma membrane, and pronuclear membrane. The syringe micrometer is turned to inject the DNA solution. If the injection is successful, the pronucleus will swell noticeably.[1] The microneedle is withdrawn as soon as such swelling is seen. The embryo is released from the holding pipette, the microneedle and holding pipette are raised, and the stage is moved along the Y axis until the group of uninjected embryos is seen. The injection process is then repeated until all the embryos have been injected. An embryo, held in position with the microneedle in the pronucleus, is shown in Fig. 7.

Although the microneedles are prechecked for patency by the expulsion of air into acetone, difficulty is frequently encountered in obtaining flow. If such problems occur, the microneedle may be lowered until it touches the culture dish, then gently rubbed on the dish in a circular motion. If this procedure successfully clears the tip of the microneedle, flow will immediately be detected when the microneedle is reinserted into the perivitelline space: the zona will swell and the embryo will be compressed by the buildup of fluid pressure in this space. This phenomenon occurs because of the release of pressure accumulated in the microneedle system during previous attempts to inject with the microneedle blocked. If these efforts fail to clear the microneedle, the tip may be rubbed against the shaft of the holding pipette. This action is repeated with intermittent testing of the microneedle until flow is obtained. These maneuvers may chip the microneedle, thereby rendering it too dull for atraumatic microinjec-

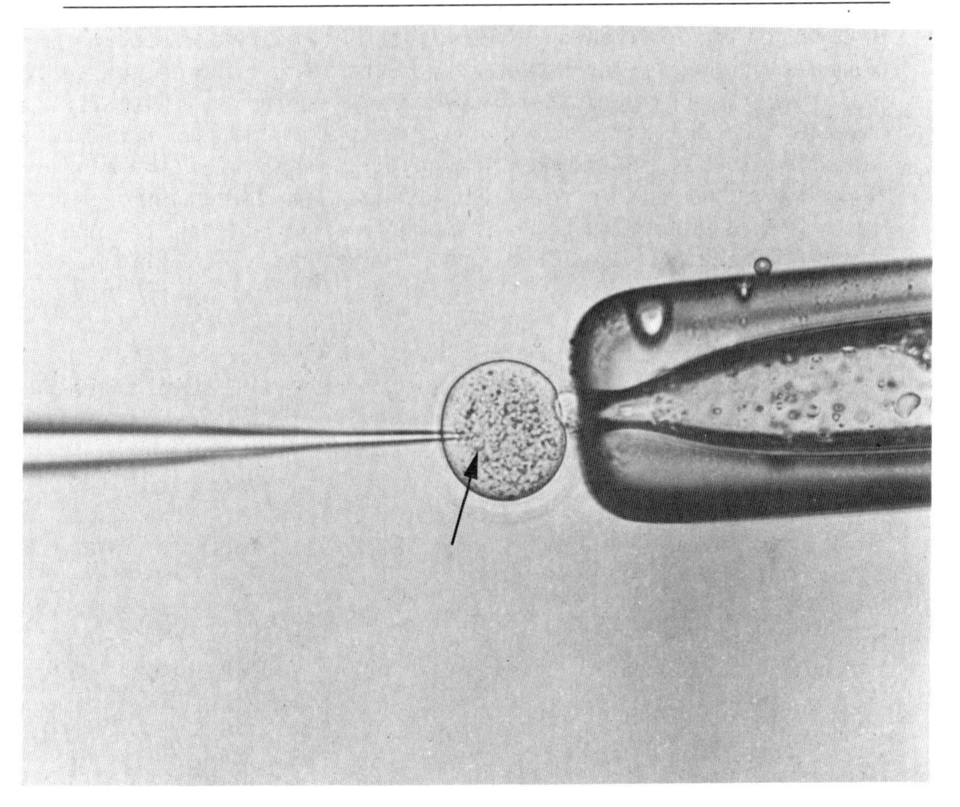

FIG. 7. An embryo held in place by the holding pipette, with the microneedle inside the pronucleus. A large nucleolus can be seen near the tip of the microneedle (arrow).

tion. If this happens, the microneedle must be discarded and replaced. Before the microneedle is discarded, it is important to make certain that the embryo has actually been pierced prior to the attempted injection. The plasma membrane can frequently recoil a significant amount before it actually gives way to the microneedle. In this circumstance, the membrane will be pushed against the microneedle and flow will be obstructed.

Procedures designed to obtain dependable flow from the microneedle frequently require that the first few embryos be sacrificed. Superovulation of 10 females assures, however, that, if the DNA preparation is sufficiently clean, enough embryos will be available for implantation at the end of the experiment.

When all the embryos in the microdrop have been injected, the dish is removed to the dissecting scope. The surviving eggs are collected in a transfer pipette and placed in the last of the five 35-mm tissue culture dishes. Surviving embryos can be distinguished by their well defined plasma membrane, dense granular cytoplasm, and readily visualized pronuclei. Changes associated with death of the embryos can be studied in detail during microinjection in order that they become familiar to the experimenter. When all microinjections are completed, the embryos are allowed to incubate 30–60 min. After this time any additional embryos that have died can be discarded. Any embryo whose status is uncertain should be implanted.

Reimplantation of Embryos

The equipment required for oviductal implantation[16] is the same as that needed for vasectomy, except for the additional requirement of a transfer pipette and an implantation pipette. This latter item is made as follows.

Making the Implantation Pipette

Draw the tip of a 9-inch Pasteur pipette in the same way as described for the transfer pipette. Examine the drawn portion of the glass until a region is found whose diameter is comparable to that of the tip of a transfer pipette. Score with a diamond pen and break cleanly at this point. Make a second break by hand at a position 6–10 cm from the first break: this yields a 6–10-cm piece of glass whose tip is similar to that of a transfer pipette. Heat, taper, break, and fire polish the shaft of the Pasteur pipette in an identical manner as the transfer pipette. Break the tapered tip of this pipette sufficiently proximal to the shaft to give an opening of 1–3 mm Slide the 6–10-cm piece of tapered glass into this opening so that the tip extends 3–4 cm beyond the end of the shaft, with the end that was scored with the diamond pen still exposed. Heat a small amount of sealing wax and seal the 1–3-mm opening in the shaft; this produces an airtight seal that allows medium to be drawn into the tip by applying suction to the shaft of the Pasteur pipette. Production of the implantation pipette is shown diagrammatically in Fig. 8.

Remove the microinjected embryos from the incubator and collect them with a transfer pipette into a small group in the center of the 35-mm dish. Fill 1–3 cm of the pipette with MEM, then draw a 1–2-mm air bubble into the pipette, and then draw another 1–2 cm of MEM. Collect 5–10

[16] K. A. Rafferty, Jr., "Methods in Experimental Embryology of the Mouse." Johns Hopkins Press, Baltimore, Maryland, 1970.

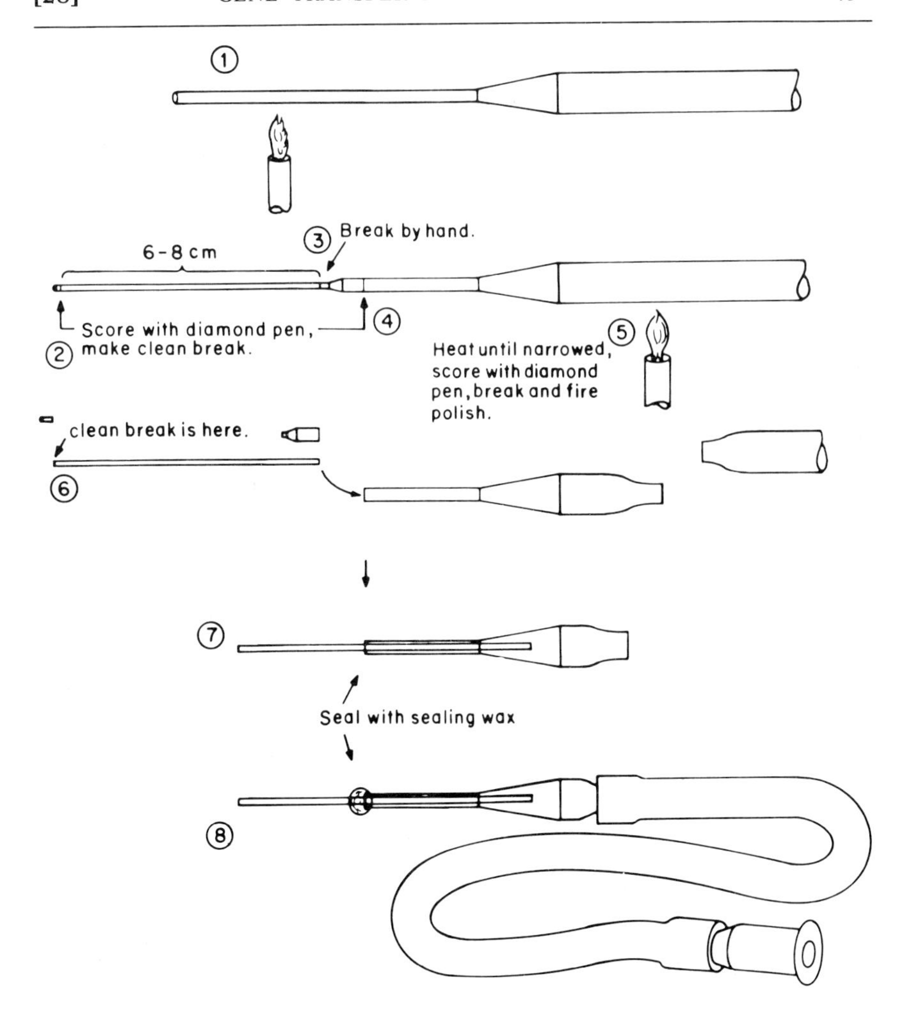

FIG. 8. Diagrammatic outline for the production of the implantation pipette. The steps should be followed numerically.

embryos into the front of the implantation pipette. The embryos should be as closely crowded as possible at the very tip of the pipette. The implantation pipette and rubber tubing are then placed carefully aside on the table.

Anesthetize the pseudopregnant female. With the mouse in the prone position, examine the lumbar region to locate the position where the abdomen bulges maximally from the midline. At this point, make a 1-cm longitudinal incision approximately 1 cm lateral to the dorsal midline. Explore the peritoneal contents until the periovarian fat pad is located. Grasp the fat pad and bring the ovary, oviduct, and a small length of the uterus through the incision. Bring the ovary toward the dorsal midline. Place the female on the dissecting microscope. The opening of the oviduct (the ostium) always faces toward the tail of the mouse. Thus, if the implantation pipette is to be held in the right hand, the nose of the mouse should be oriented toward the left side of the microscope stage. Examine the ovary under the dissecting microscope and locate the ovarian bursa, a thin layer of tissue that encases the ovary. Tear the bursa with two pairs of forceps, taking care not to violate any major bursal blood vessels, and slide the bursa around and under the ovary so that the ovary is exposed. Explore the crevice between the ovary and the loops of the oviduct by inserting the tips of the forceps in this space and then spreading the forceps blades slightly. The ostium will appear in this space. Exert downward pressure with the forceps, thus pinning the oviduct against the back of the mouse and obtaining stable exposure of the ostium. Grasp the implantation pipette with rubber tubing and insert the pipette into the ostium. Do not attempt to force the pipette past the first sharp curve of the oviduct. Slowly apply positive pressure through the mouthpiece until the embryos begin to move into the oviduct. A gentle back and forth motion of the implantation pipette may be necessary to obtain flow. Slowly expel the embryos into the oviduct until the air bubble marker is seen just outside the ostium. Withdraw the implantation pipette, replace the ovary, and close the incision with a single suture or wound clip is described for the vasectomy.

If fewer than 10 embryos survive the microinjection, they may all be implanted into one oviduct; more than 10 should be divided evenly among both oviducts. As many as 30 embryos can be implanted into a single mouse.

The rate of success in producing transgenic mice depends on several factors. Plasmid solutions should be as clean as possible, as dirty solutions interfere with the function of the microneedle and increase mortality rates. Plasmid isolates often contain fine precipitates. In our laboratory, all plasmid stocks are centrifuged in a microfuge for 10 min, and then diluted by aliquoting carefully off the surface of the solution into a filtered diluent. Pregnancy rates can be increased by choosing pseudopregnant females whose weights are not less than 30 g or more than 35 g. Implantation rates can also be improved by inserting the microinjected embryos in a minimum amount of culture medium and by avoiding the introduction of air bubbles into the oviduct. The most important determinant of success is the skill of the experimenter in handling the micromanipulation equipment. Once sufficient skill has been acquired, one can reasonably expect 10–20% of microinjected embryos to survive until birth, and 15–30% of those born to carry transferred DNA sequences.

Acknowledgments

This work was supported by NIH Grants GM09966 to F. H. R., and GM07959-02 to J. W. G.

[29] Introduction of Exogenous DNA into Cotton Embryos

By GUANG-YU ZHOU, JIAN WENG, YISHEN ZENG, JUNGI HUANG, SIYING QIAN, and GUILING LIU

The work of introducing exogenous DNA into cotton embryos was started in 1978. To our surprise, we obtained a great number of mutated offspring. Some of these are now in their fourth generation. The probability of mutation could be up to several per 100.[1]

The method used was a combination of injection and transformation. DNA ($\sim 10^7$ daltons) was partially protected from shearing and hydrolysis and was injected into the axil placenta about a day after self-pollination. The DNA transformed the embryos, probably by entering the ovule, following the path along which the pollen tube grows.

Extraction and Purification of Cotton Seed DNA[2]

Reagents

All reagents used were analytical reagent grade products from the Chinese Chemical Reagent Company. Sodium dodecyl sulfate (SDS) was recrystallized from a chemically pure product; trypsin crystals and nucleic acid were obtained from the Tong Feng Biochemical Reagent Factory, Shanghai Institute of Biochemistry, Academia Sinica. Sepharose 4B was obtained from Pharmacia Fine Chemicals.

[1] J. Huang, S. Qian, G. Liu, J. Weng, Y. Zeng, and G. Zhou, *I Chuan Hsuch Pao* (*Acta Genet. Sini.*) **8,** 56 (1981).

[2] J. Weng and J. Zhao, *Acta Phytophysiologia Sinica* **5,** 363 (1979).

ISBN 0-12-182001-7

Materials

Donors and recipients were grown in the field; they were self pollinated.
Upland cotton (*Gossypium hirsutum*); glandless; Jiangsu I; Jiangsu III; 52-128 (*Fusarium* resistant)
Sea island cotton (*G. barbadense*) 416
Asian cotton (*G. arboreum*)
Wild cotton (*G. thurberi*)
Abutilon avicennae

DNA Extraction and Purification

1. Pretreatment of seeds: Seeds that had been treated with sulfuric acid to remove lint were soaked in water for 24 hr. After peeling off the seed coat, the kernels were used for DNA extraction.
2. Extraction of pigments: Most cotton seeds contain large quantities of gossypol and other pigments. It is absolutely necessary to eliminate them before preparation of DNA.
3. Extraction with SDS solution (5 g of SDS dissolved in 100 ml of 45% ethanol) to obtain a crude nucleic acid–protein product.
4. Further purification, mainly according to the method of Zeng.[3]
5. Experimental procedures (see Scheme 1).

ANALYTICAL DATA FOR ONE DNA PREPARATION FROM SEA ISLAND COTTON 416 (*Gossypium barbadense*) SEEDS

Spectral data		Chemical analysis					
260/230	260/280	RNA	Protein	T_m (°C)	Hyperchromicity at 260 nm	Yield (μg/g wet wt.)	Molecular weight
2.28	1.85	—	—	81.9	40%	200–250	10^6–10^7

6. Methods of analysis (see the table):
   Protein determination according to Lowry *et al.*[4]
   DNA determination according to Burton[5]
   RNA determination according to Schjeide[6]

[3] Y. Zeng (I. Zen), *Sheng Wu Hua Hsueh Yu Sheng Wu Wu Li Hsueh Pao* (*Acta Biochim. Biophys. Sin.*) **10,** 391 (1978).
[4] O. H. Lowry, N. J. Rosebrough, A. L. Farr, and R. J. Randall, *J. Biol. Chem.* **193,** 265 (1951).
[5] K. Burton, *Biochem. J.* **62,** 315 (1965).
[6] O. A. Schjeide, *Anal. Biochem.* **27,** 473 (1969).

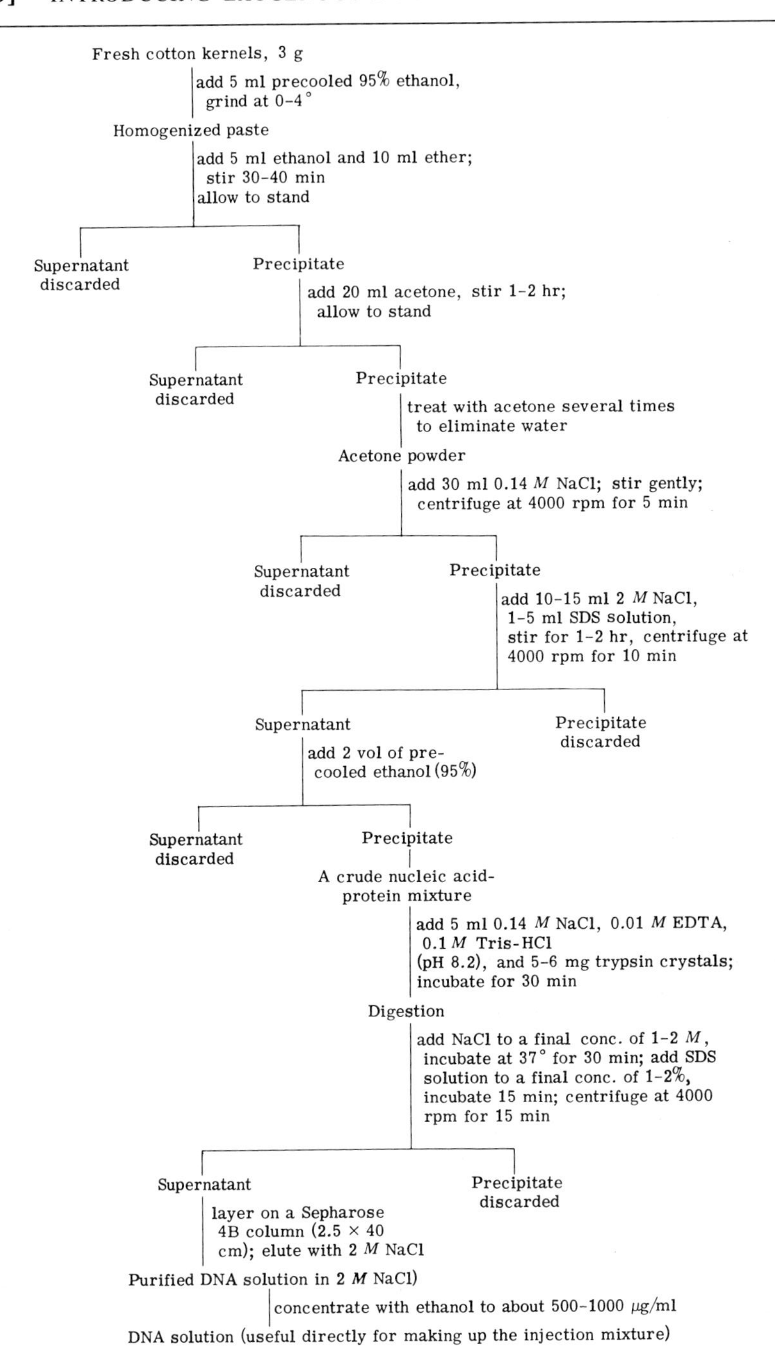
Fresh cotton kernels, 3 g
add 5 ml precooled 95% ethanol, grind at 0-4°
Homogenized paste
add 5 ml ethanol and 10 ml ether; stir 30-40 min
allow to stand
Supernatant discarded
Precipitate
add 20 ml acetone, stir 1-2 hr; allow to stand
Supernatant discarded
Precipitate
treat with acetone several times to eliminate water
Acetone powder
add 30 ml 0.14 M NaCl; stir gently; centrifuge at 4000 rpm for 5 min
Supernatant discarded
Precipitate
add 10-15 ml 2 M NaCl, 1-5 ml SDS solution, stir for 1-2 hr, centrifuge at 4000 rpm for 10 min
Supernatant
add 2 vol of pre-cooled ethanol (95%)
Precipitate discarded
Supernatant discarded
Precipitate
A crude nucleic acid-protein mixture
add 5 ml 0.14 M NaCl, 0.01 M EDTA, 0.1 M Tris-HCl (pH 8.2), and 5-6 mg trypsin crystals; incubate for 30 min
Digestion
add NaCl to a final conc. of 1-2 M, incubate at 37° for 30 min; add SDS solution to a final conc. of 1-2%, incubate 15 min; centrifuge at 4000 rpm for 15 min
Supernatant
layer on a Sepharose 4B column (2.5 × 40 cm); elute with 2 M NaCl
Precipitate discarded
Purified DNA solution in 2 M NaCl)
concentrate with ethanol to about 500-1000 μg/ml
DNA solution (useful directly for making up the injection mixture) in 0.1 × SSC

Scheme 1

Molecular weight of DNA was measured by electron microscopy[7] and agarose gel electrophoresis.

The material for preparation of DNA is more easily extracted from cotton kernels than from leaves and seedlings, which give products low in quality and yield. Any method that can provide DNA segments long enough and of good quality is suitable for DNA injection.

Partial Recombination of DNA with Histone

The method utilized is based on the principle used by the Shanghai Institute of Cell Biology.[8,9]

Bovine thymus histone mixture was a gift from Shanghai Institute of Cell Biology, Academia Sinica.

Histone and DNA are each dissolved in a solution containing 2 *M* NaCl, 5 *M* urea, and 0.01 *M* Tris-HCl (pH 8.3). The histone and DNA solutions are mixed in the mass ratio of 0.5–0.7 : 1, and dialyzed against each of the following solutions in the order mentioned:

2 *M* NaCl, 5 *M* urea, and 0.01 *M* Tris-HCl for 6 hr
0.6 *M* NaCl, 5 *M* urea, and 0.01 *M* Tris-HCl overnight
0.4 *M* NaCl, 5 *M* urea, and 0.01 *M* Tris-HCl for 4 hr
0.15 *M* NaCl, 5 *M* urea, and 0.01 *M* Tris-HCl for 4 hr
0.15 *M* NaCl and 0.01 *M* Tris-HCl overnight

On partial recombination of the DNA with histone, not only the length of the DNA segments is shortened, but also the resistance to mechanical shearing or enzymic hydrolysis is increased. A solution of histone and DNA in a 1 : 1 ratio was found to precipitate easily and was not suitable for transformation of the plant cell.

DNA Solution for Injection

Naked DNA or DNA that has been partially recombined with histones is dialyzed against a solution of 0.1 × SSC, 10 m*M* Ca^{2+},[10] and 10 mM Zn^{2+},[11] overnight with two changes of the solution. It is then adjusted to a final concentration of 3 μg of DNA per 10 μl.

[7] Y. Xu, P. Dai, and Z. Gong, *Sheng Wu Hua Hsuch Yu Sheng Wu Wu Li Jinzhan* (*J. Biochem. Biophys.*) **2,** 9 (1975).

[8] The Cell Research Group of Shanghi Institute of Cell Biology, *Sheng Wu Hua Hsuch Yu Sheng Wu Wu Li Jinzhan* (*Biochem. Biophys.*) **5,** 11 (1977).

[9] T. Y. Shih and J. Bonner, *Biochem. Biophys. Acta* **182,** 30 (1969).

[10] K. Ohyama, O. L. Gamborg, and R. A. Miller, *Can. J. Bot.* **50,** 2077 (1972).

[11] M. Suzuki and I. Takabe, *Z. Pflanzenphysiol.* **78,** 421 (1976).

DNA Injection (see Figs. 1–12)

1. Before injection
   a. The recipient plants are grown in a greenhouse or in the field for about 3 months before the DNA injection; 50–100 healthy plants are required for one experiment.
   b. Fill and seal 0.5-ml portions of sample DNA (donor), control DNA (recipient), and blank (the same solution for injection without DNA) separately in ampoules or tubes with stoppers. Prepare some tubes or ampoules with 5 ml of 0.1 × SSC solution. All operations are conducted under sterile conditions. Keep the solutions in a refrigerator before use.
   c. Prepare receptive flowers a day before injection. Select the first floral bud (white in color in upland cotton) of a branch about to flower the next day. Tie the top to allow self-pollination. About 150 buds are required for one experiment.
   d. Have ready a basket containing several insulated cups and another suitable-sized insulator for ice, a tool for opening ampoules, a few 50-μl micropipettes, pencils, and markers.

2. Injection
   a. Take out the ampoules of DNA and blank solutions and place them in the insulated cups with ice. Fill the other insulator with ice.
   b. The corolla of the floral bud that was tied a day before becomes red and is self-pollinated. Carefully remove the corolla, the stamen tube, and the style of the flower in order to expose the ovary.
   c. Drop gibberellin (20 ppm in H_2O) on the pedicel to prevent the boll from shedding (a common occurrence in cotton, especially after an injection).
   d. Prune away the rest of the branch to ensure that the treated ovary will be well nurtured.
   e. Take out the ampoules, open them carefully, and place them back into the insulated cups. Wash the micropipettes with SSC solution, checking to make sure that they are not plugged.
   f. Pipette the solution from the ampoules. If there is a small piece of tissue remaining on top of the ovary, wipe it away.
   g. Hold the ovary in one hand, puncture vertically with the micropipette about 0.5 cm deep from the top of the ovary into the axil placenta (the middle of the ovary.) Then draw back about 0.2 cm and slowly deliver 10 μl of solution (DNA concentration

is 3 μg/10 μl). Remove the pipette carefully to minimize any possible injury to the ovary. Finally, fix on the tags.

h. A plastic overcap may be used to cover up the ovary, but it is not obligatory.
i. Keep the seeds dry after they have matured. For the next generation, grow them from each single boll in separate rows, side by side, with markers in place. Donor and recipient are to be grown at the same time.
j. Always collect the self-pollinated seeds by tying the floral bud prior to flowering.
k. For specific resistance selection, the seeds should be planted with the controls in inoculated soil or under specified conditions. For some diseases, one may use a pair of scissors smeared with the microorganisms and inoculate the leaves simply by cutting.

Results

Introduction of Sea Island Cotton (G. barbadense) 416 DNA into Glandless Upland Cotton (G. hirsutum) Embryos[1] *(See Figs. 13–37)*

Sixty ovaries of upland cotton plants that had been raised in the field were selected to be injected with sea island cotton DNA in the summer of 1978. During the fall, 52 mature seeds taken from 5 bolls were planted in a greenhouse. These seeds came from bolls that were much smaller and shorter fibered (in some cases, without lint) as compared with the recipient plants. These seeds were designated the zero generation of DNA introduction (D_0). Seventy-eight percent of the seedlings (D_1 or first generation) were nonpubescent, like the donor. When they grew up, 47% had smaller corolla than either the donor or the recipient. Their anthers were golden yellow, similar to the color displayed by the donor. However, some of the D_1 generation had smaller bolls and degenerated anthers. The other 53% possessed white milky anthers and were similar to the recipient.

Ninety-nine seeds from 9 bolls of D_1 were developed in the field to the second generation (D_2) in the summer of 1979. Surprisingly, 90% of the D_2 generation exhibited more visible phenotypic alterations than D_1. Examples of these were samples 99-4 and 102-4. Both samples showed heterosis. Sample 102-4 resembled the recipients except for light golden yellow anthers like those of the donors, while sample 99-4 closely resembled the DNA donor. D_3 and D_4 for sample 102-4 obtained in 1980 and 1981 were nearly the same as D_2.[12] However, sample 99-4 gave only six plants in the D_3 generation from 270 D_2 seeds. These six plants failed to flower during the season. They segregated into two equal categories, one of which eventually resembled the recipient. Both categories were moved into the greenhouse, where five of them gradually flowered and gave mature seeds. These seeds were planted in the field in 1981. The resulting plants (D_4) flowered normally, but their bolls matured later. A high frequency of mutations appeared in D_4. Not only did the shapes and sizes of leaves and plants segregate differently, but also the color of the corolla, pubescence, and gossypol glands, etc., segregated. Some of them eventually resembled the recipient. Others mutated even further (Figs. 13–37).

Introduction of DNA from Sea Island DNA into Upland Cotton Jiangsu I[12]

Figure 38 gives an example of the alterations exhibited in this composition. This is a mosaic offspring that has already passed its third generation. It is half normal and heritable.

Introduction of Sea Island Cotton 416 DNA into Asian Cotton (G. arboreum)[12] *(Figs. 39 and 40)*

In the first generation there were some offspring bearing obvious alterations. A high percentage were sterile.

Introduction of the DNA from Wild Cotton (G. thurberi) into Upland Cotton Xuzhou 142[12]

Some offspring were obtained with better fiber strength, a greater number of bolls, and a shorter period of maturation.

Introduction of DNA from Abutilon avicennae into Sea Island Cotton 416[12]

Abutilon avicennae and sea island cotton belong to different genera of the Malvaceae family. In the first generation the lower leaves on some plants looked like the donor, but this resemblance disappeared in the next generation. However, the plants are always much taller and more thriving than the recipient (See Figs. 41 and 42).

[12] Results in detail will be published elsewhere.

Introduction of the DNA from Fusarium-resistant Upland Cotton 52-128 into Sensitive Species of Upland Cotton Jiangsu I and III[12]

D_0 seeds were planted in the field in soil inoculated with *Fusarium oxysporum*. The resistance of first generation offspring varied widely. Some were as highly susceptible as the recipient; some were as resistant as the DNA donor; and others behaved in varying degrees between these two extremes. The second generation from high-resistance D_1 plants were replanted in an inoculated field. High resistance appeared rather uniformly in this generation. Experiments with Jiangsu III as recipient provided better results than those with Jiangsu I. (See Figs. 43–45.)

Blanks and Controls

In each experiment, blank injections without DNA into the same number of ovaries were made. Alterations never occurred in their offspring.

Since 1980, four control injections have been made with DNA from their own recipient. They were sea island cotton 416, upland cotton glandless, red-leaved upland cotton, and Jiangsu I upland cotton. No noticeable alterations were observed in the first and second (1982) generation grown on a large scale. However, the manifestations of the cotton injected with its own DNA will be followed for several generations, in light of the announcement of Shi *et al.*,[13] that self-DNA produced a balance bar in frogs.

Discussion

A suitable recipient is very important for the introduction of exogenous DNA plants. Cells, pollens, tissues or organs from culture as well as from whole plants have been tried or suggested.[14–16] For any recipient used, the exogenous DNA should finally enter into cells where it is expressed and becomes inherited. From the high mutation frequency observed with the cotton plant ovary into which the exogenous DNA was injected, it is evident that segments of DNA may transform the egg cell or zygote during or after the period of fertilization.

As is well known, the sizes and structure of flowers and ovaries of angiosperms vary greatly. The number of ovules contained in each ovary

[13] L. Shi, Y. Yan, J. Zhang, H. Mo, and Y. Lu, *Sci. Sin* (*Engl. Ed.*) **24,** 402 (1981).

[14] C. I. Kado, *in* Genetic Engineering, (J. K. Setlow and A. Hollaender, eds.) Vol. 1, p. 223. Plenum, New York, 1979.

[15] S. C. Maheshwari, N. Maheshwari, and K. Malhotra, *Sci. Prog.* (*Oxford*) **66,** 435 (1980).

[16] E. C. Cocking, M. R. Davey, D. Pental, and J. B. Power, *Nature* (*London*) **293,** 265 (1981).

also differ. The flowers and ovaries of the cotton plant are of large size. Each ovary has 3–5 loculi compartmented by 3–5 carpels, which are joined at one side to form the axil placenta on which the ovules are seated. In general, there are 30–40 ovules per ovary. The size and structure of the cotton ovary are well suited for the introduction of exogenous DNA.

In nature, when a pollen falls on the stigma, it germinates and a pollen tube grows through the style tissue, arriving ultimately at the ovary. It passes the microphyle, enters the blastophore, and passes through one of the two synergids, which have begun to degenerate. The tip of the pollen tube opens on one side allowing the two nuclei to fertilize the egg cell.[17,18] Direct injection of DNA into the egg cell in the ovule will easily damage it. The process is very difficult, if not impossible, because the egg cell is too small and deeply hidden in the ovule to be readily injected. However, it is crucial to inject DNA into the plant cells, especially when working with crops.

The time from pollen germination to fertilization differs between cotton species. Usually it takes between 24 and 48 hr. In our experiments, flowers were tied for self-pollination. The corolla, stamen, and style were peeled off on the following day to expose the ovary. Finally, the exogenous DNA was injected directly into the axil placenta, while fertilization was still going on or was just over. The DNA should have a chance to leak through the path of the pollen tube to reach the egg cell or zygote. The time for injection is critical because callose will be rapidly produced from the inside to plug the tube.[19]

Figure 46 diagrams the process of pollination.

At present, it appears that animal cells are more amenable to transformation than plant cells. In addition to the lack of good plant cell transformation systems, the process is further complicated by cell walls. Plant protoplasts, especially those from cells of cereals, have yet to show the ability to regenerate cell walls. Embryonic cells, however, lack cell walls or have very thin ones. This should be of advantage in transformation.

The first division of a zygote can be observed after 1–3 days. If the exogenous DNA could be integrated into the cell genome before cell differentiation, it would probably be more easily kept in reproductive cells.

Within 60 days we got seeds from the treated ovaries. These were grown in the same or the next year, from which mutated offspring could be collected. The time necessary for the transformation of a cell to regeneration of a plant through exogenous DNA directly introduced into the

[17] Z. Li, ed., "Cotton Morphology." Academic Press, Beijing, 1979.

[18] W. A. Jensen and D. B. Fisher, *Planta* **78,** 158 (1968).

[19] W. A. Jensen and D. B. Fisher, *Protoplasma* **69,** 215 (1970).

embryos of a living plant should be much shorter than the transformation of the cells. Cells and tissues in culture are attractive indeed in many respects. For instance one can work with them in the laboratory more easily. However, without a specific cell-screening system, it is very difficult and tedious to pick out mutated cells. When the mutated cells have been properly selected, the next thing to do is to grow them into callus, regenerate the plants from the callus, and finally select the plants retaining the exogenous DNA. Working with agricultural breeders and using the live plants, one could effect many combinations between different donor DNA samples and recipients in one season. It is convenient to breed the disease-resistant plant by growing the D_0 seeds directly in the field, or to select resistances to environmental factors by growing the seeds under specified conditions.

Conclusion

The transformation of exogenous DNA into plants could be effected by the injection technique described in this chapter, without vectors, and could be inherited.

The essential points of this technique are the following:

1. Plants with big ovaries and a suitable number of seeds (less than 50), such as cottons, are selected for experiments.
2. For DNA transformation, the time for injection should preferably fall within the period between fertilization and zygote division and the location should preferably be along the path of the pollen tube.
3. The DNA should be protected from being sheared and hydrolyzed.

FIG. 1. A white floral bud near the main stem of glandless upland cotton.

FIG. 2. A piece of string on the pedicel.

FIG. 3. Make a small ring from the other end of the string.

FIG. 4. The ring is placed on top of the floral bud and tied up.

FIG. 5. The flower turns red on the next day.

FIG. 6. An untreated flower at the same stage of development as in Fig. 5.

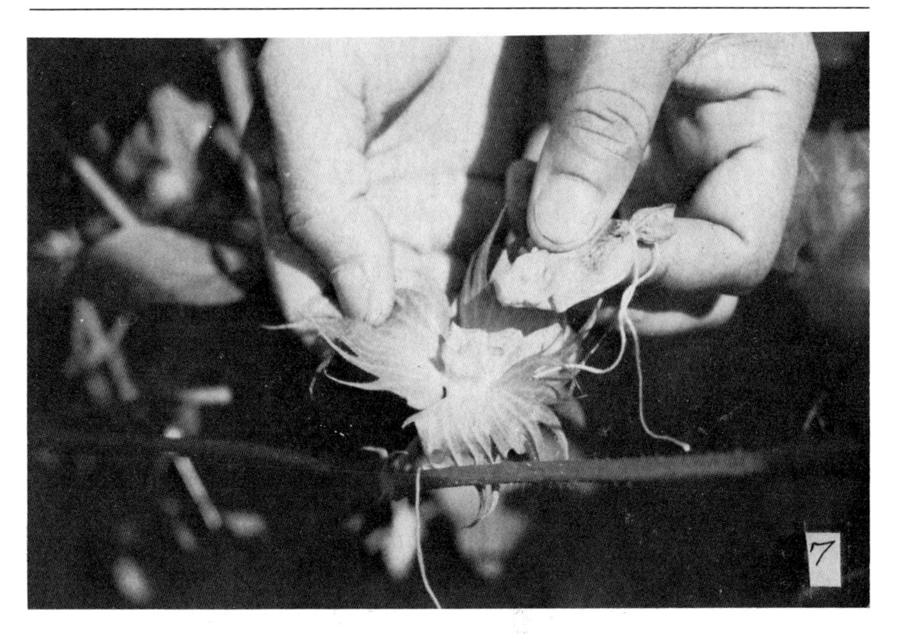

FIG. 7. Using a fingernail, peel off the corolla, the stamen, and the style.

FIG. 8. Getting ready for the injection.

FIG. 9. Injection of DNA into the axil placenta.

FIG. 10. Tie marker on the pedicel and branch.

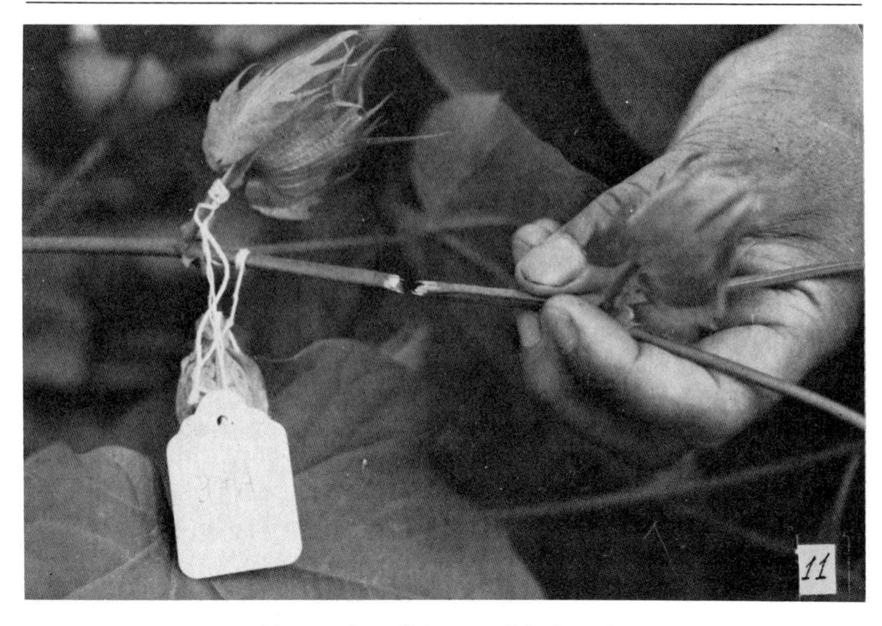

FIG. 11. Cut off the rest of the branch.

FIG. 12. Basket carrying all necessary utensils.

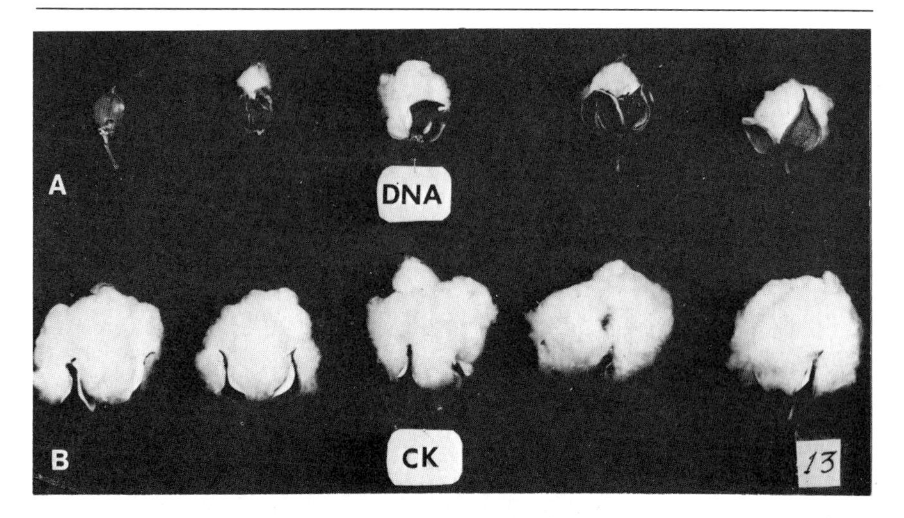

FIGS. 13–37. Upland cotton (glandless) + sea island DNA (D_0).

FIG. 13. Mature cotton bolls. (A) Bolls were injected with sea island 416 DNA (D_0). (B) Recipient normal bolls.

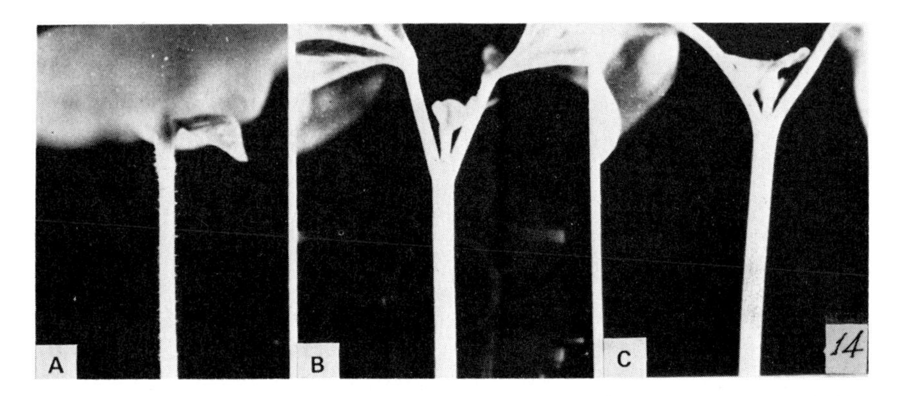

FIG. 14. Seedlings (D_1). (A) Recipient seedlings with pubescence and no gossypol gland. (B) Donor seedlings nonpubescent and with gossypol glands. (C) D_1 seedlings nonpubescent or scantily pubescent with few gossypol glands.

FIG. 15. Recipient plant (control of D_2).

FIG. 16. Donor plant (control of D_2).

FIG. 17. Sample 102-4 plant (D_2).

FIG. 18. Sample 99-4 plant (D_2).

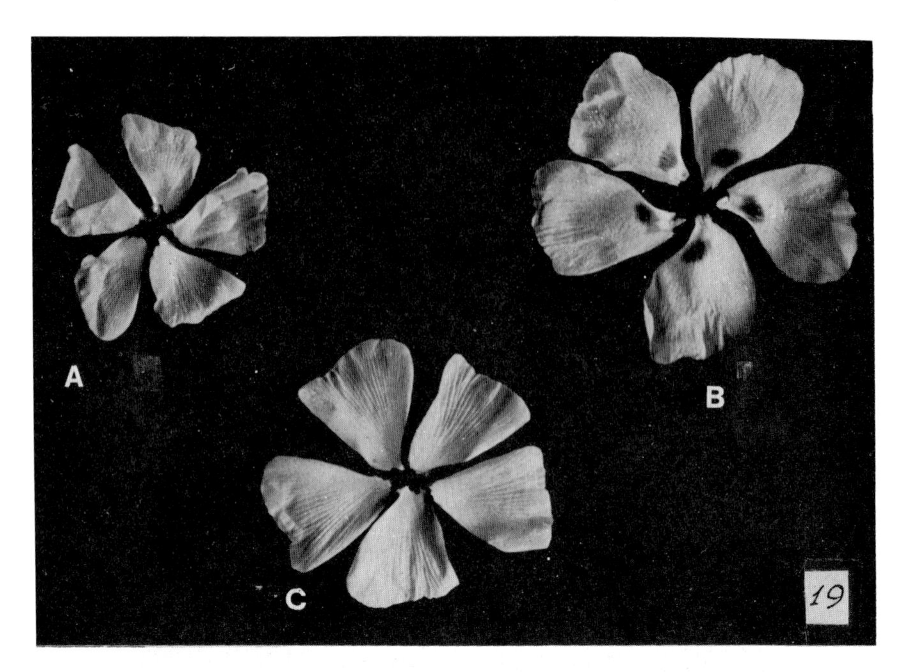

FIG. 19. Flowers. (A) Recipient; (B) donor; (C) sample 102-4 (D_2).

FIG. 20. Flowers. (A) Recipient; (B) donor; (C) sample 99-4 (D_2).

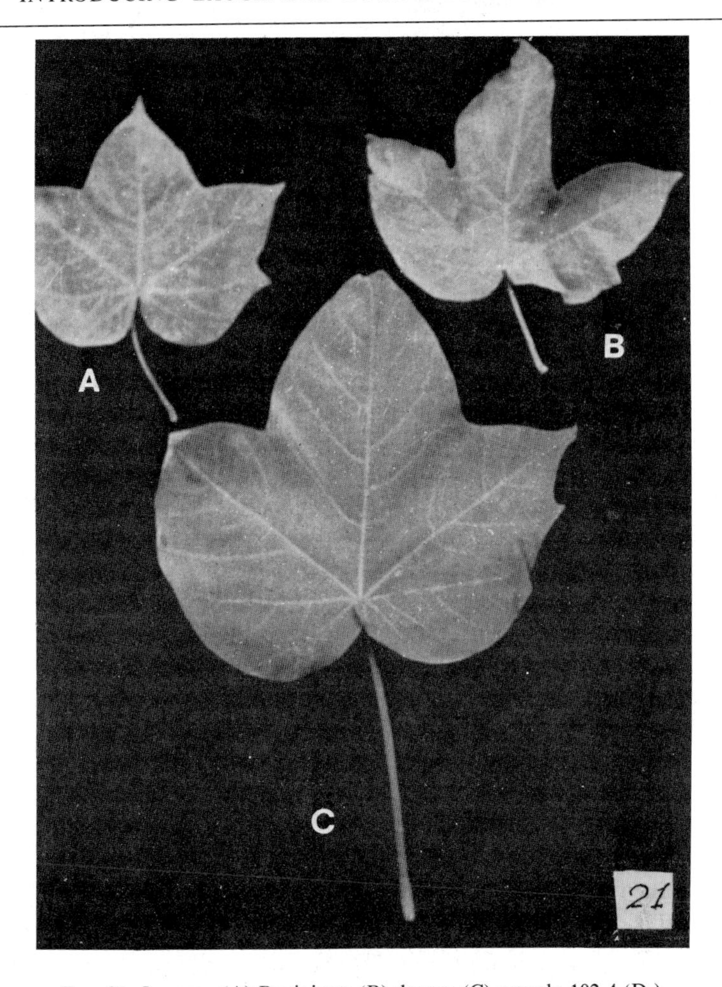

FIG. 21. Leaves. (A) Recipient; (B) donor; (C) sample 102-4 (D_2).

FIG. 22. Leaves. (A) Recipient; (B) donor; (C) sample 99-4 (D_2).

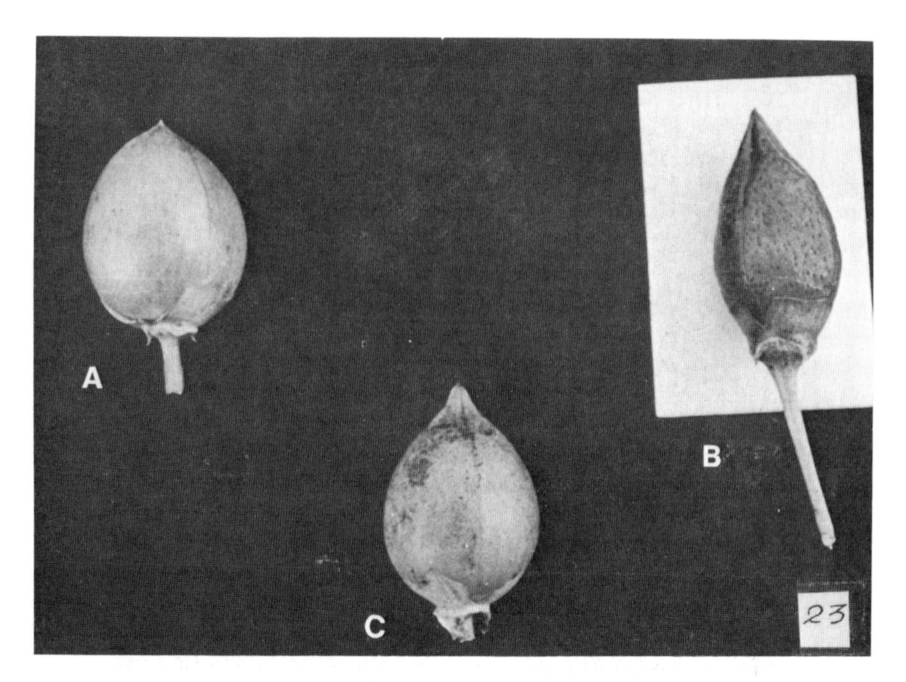

FIG. 23. Bolls. (A) Recipient; (B) donor; (C) sample 102-4 (D_2).

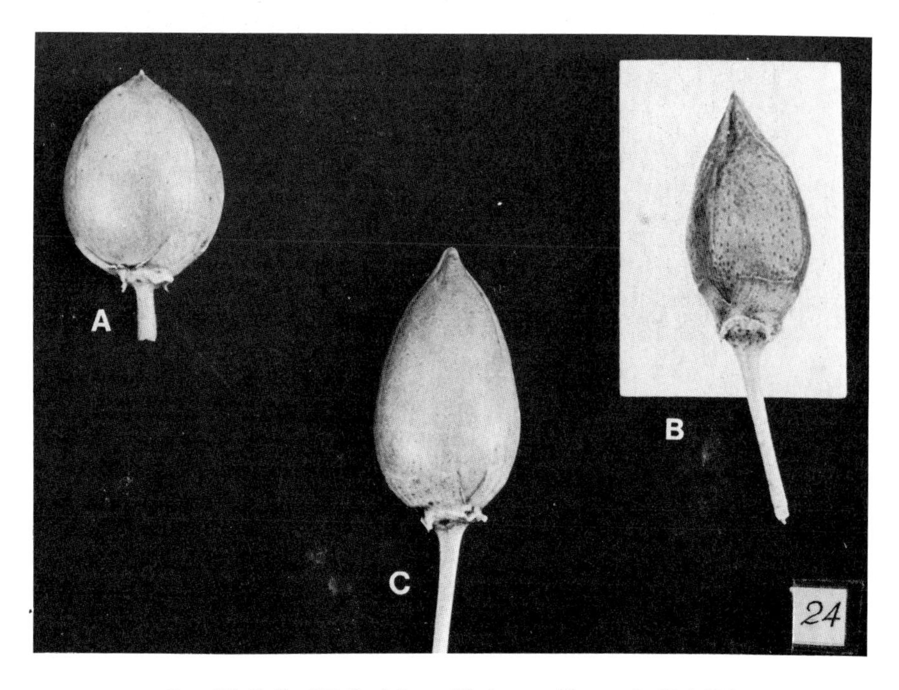

FIG. 24. Bolls. (A) Recipient; (B) donor; (C) sample 99-4 (D_2).

FIG. 25. Sample 102-4 (D_3), flower has lost golden yellow anther color.

FIG. 26. Sample 99-4 (D_3); plant remained principally as in sample 99-4 (D_2).

FIG. 27. Sample 99-4 (D_3); plant turned to resemble the recipient.

FIG. 28. A donor plant (sea island, control of D_4).

FIGS. 29–34. D_4 offspring plant (D_4).

FIG. 30.

Fig. 31.

FIG. 32.

FIG. 33.

FIG. 34.

FIG. 35. A recipient plant (upland cotton glandless, control of D_4).

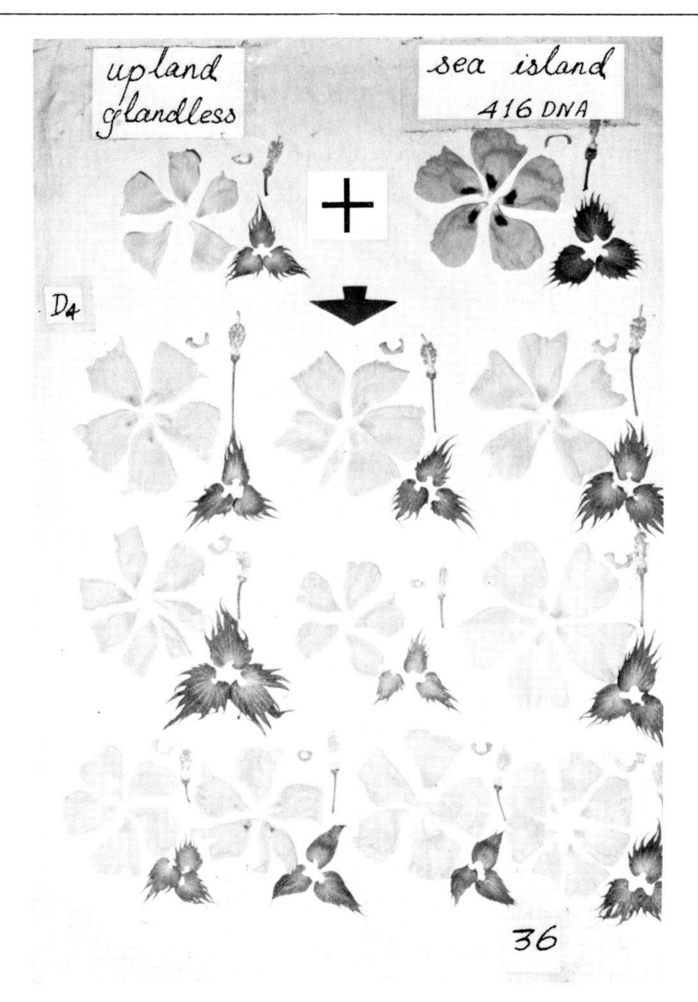

FIG. 36. Flowers collected from different offspring (D_4).

FIG. 37. Leaves collected from different offspring (D_4). It appeared that the offspring from D_4 would continue to alter in their various properties.

FIG. 38. Mosaic plant in D_3.

FIG. 39. Plants. (A) Recipient; (B) D_1 mutant.

FIG. 40. Flowers. (A) Normal recipient flower; (B) D_1 degenerated flower.

FIG. 41. *Abutilon avicennae* plants.

FIG. 42. D_2 mutated offspring.

FIG. 43. Donor 52-128 and recipient Jiangsu III growing on field inoculated with *Fusarium oxysporum*. Right: 52-128; left; Jiangsu III.

FIG. 44. D_1 offspring.

FIG. 45. D_2 of the strong D_1-resistant plants. Middle: 52-128; right: Jiangsu III; left: D_2 plants.

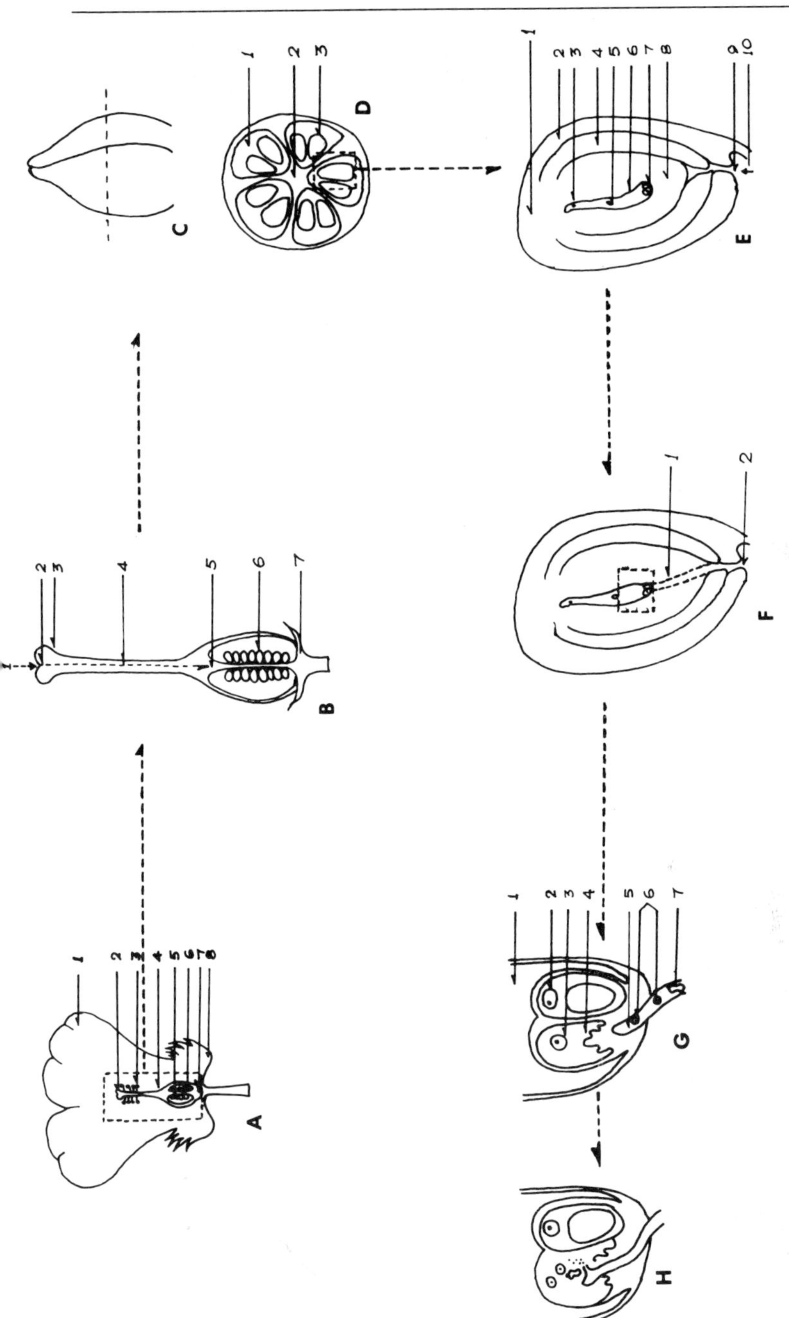

FIG. 46. The pollination process.[17] (A) Cotton flower: 1, corolla; 2, stigma; 3, stamen; 4, style; 5, axil placenta; 6, ovule; 7, ovary; 8, bract. (B) Pistil: 1, pollen dropped on and germinated; 2, the path of a growing pollen tube; 3, stigma; 4, style; 5, axil placenta; 6, ovule; 7, sepal. (C) Ovary. (D) Cross section of an ovary: 1, ovary room; 2, axil placenta; 3, ovule. (E) Vertical section of an ovary: 1, joint end; 2, outer integument; 3, antipodal cells (degenerated); 4, inner integument; 5, polar nucleus; 6, embryo sac; 7, egg cell and synergids; 8, nucleus; 9, microphyle end; 10, path of the pollen tube. (F) Degenerated nucellus cells form a column to facilitate passage of the pollen tube: 1, column cells; 2, microphyle. (G) The pollen tube enters the degenerated synergid: 1, embryo sac; 2, egg cell nucleus; 3, synergid nucleus; 4, degenerated synergid; 5, pollen tube; 6, sperm; 7, nutritive nucleus. (H) Liberation of the sperm from the opened pollen tube.

[30] Microinjection of Tissue Culture Cells

By M. GRAESSMANN and A. GRAESSMANN

During the 1960s and 1970s, different methods were developed for the transfer of macromolecules into tissue cells. These include (*a*) DNA-mediated gene transfer (Ca^{2+}-, DEAE-dextran technique); (*b*) vehicle-mediated transfer (erythrocyte ghost, liposomes, reconstituted Sendai virus, (*c*) the direct microinjection technique.[1]

The microinjection technique is based on the use of small glass capillaries; this allows the transfer of almost any type of molecules into either the cytoplasma or nuclei of the recipient cells. This technique has no cell type restriction, and also suspension culture cells (e.g., lymphocytes) are accessible for microinjection after they are bound to a substrate by suitable linkers.[2] Among these techniques, microinjection is by far the most efficient. Up to 100% of the recipient cells allow expression of the transferred material, and stable transformed cell lines can be isolated with a frequency of 20–30% after intranuclear injection of DNA.[3] Biochemical studies can be performed on 50 injected cells, and the transferred material can be reisolated and further analyzed from only 100–200 recipient cells (Figs. 1 and 2).

Procedure

General. Microinjection is performed under a phase-contrast microscope (Fig. 3): a glass microcapillary, prefilled from the tip with the biological sample, is directed into the cells to be injected with the aid of a micromanipulator, and an appropriate sample volume is transferred by gentle air pressure exerted by a syringe connected to the capillary. Recipient cells are grown on glass slides imprinted with numbered squares for convenient localization of the cells injected.

Preparation of Glass Microcapillaries. Glass tubes of 1.2 mm inner and 1.5 mm outer diameter (SGA Scientific, Inc., Bloomfield, New Jersey) are broken into pieces of about 30 cm in length and cleaned by treatment with a mixture of one-third concentrated H_2SO_4 and two-thirds con-

[1] J. E. Celis, A. Graessmann, and A. Loyter, eds. "Transfer of Cell Constituents into Eukaryotic Cells." Plenum, New York, 1980.

[2] A. Graessmann, H. Wolf, and G. W. Bornkamm, *Proc. Natl. Acad. Sci. U.S.A.* **77**, 433 (1980).

[3] A. Graessmann, M. Graessmann, and C. Mueller, (1981) *Adv. Cancer Res.* **35**, 111 (1981).

ISBN 0-12-182001-7

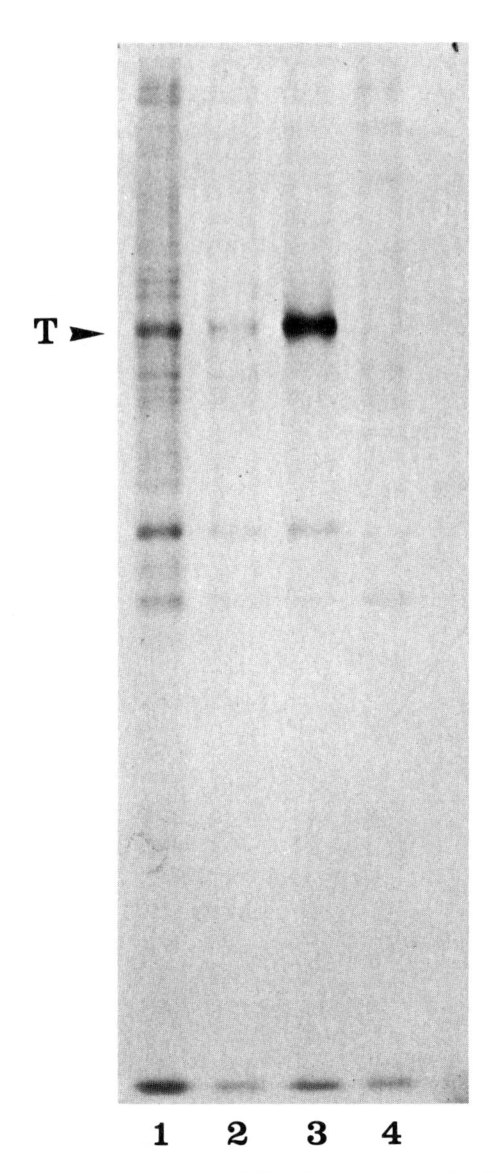

FIG. 1. SV40 T antigen extracted from SV40 cRNA-injected cells. Fluorogram sodium dodecyl sulfate-polyacrylamide gel (10%) loaded with immunoprecipitate from: track 1, 300 cRNA-injected TC7 cells; track 2. 50 poly(A)-cRNA injected TC7 cells; track 3, T-antigen extracted from SV40 virus-infected cells; track 4, mock-infected TC7 cells. In these experiments about 400 TC7 cells were grown on small glass slides (4 × 4 mm). After microinjection and starvation for 1 hr, cells were labeled with 20 μl of [^{35}S]methionine in methionine-free Dulbecco's medium (final concentration 1 mCi/ml). In order to keep the medium as a drop on the cells during the 2-hr labeling period, slides with cells were put on top of smaller glass pieces (2 × 2 mm) in a culture dish (60 mm). Evaporation of the medium was prevented by a water-saturated filter paper placed in the cover of the dish.

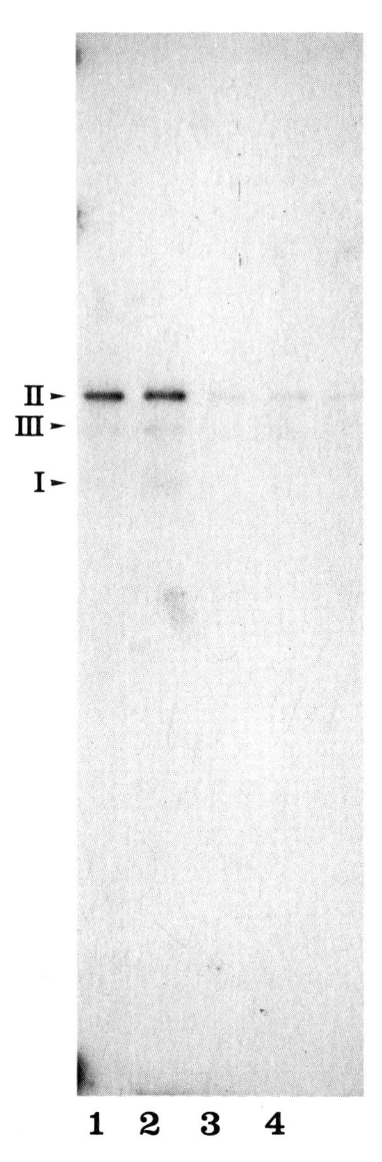

FIG. 2. Reisolation and detection of SV40 DNA from injected monkey cells. TC7 cells grown on small glass slides (2×2 mm) were microinjected with SV40 DNA (concentration, 1 mg/ml). For DNA extraction, glass slides with the recipient cells were directly transferred into Eppendorf test tubes filled with 100 μl of lysis buffer (10 m*M* Tris, pH 7.6, 1 m*M* EDTA, 0.6% sodium dodecyl sulfate), and DNA was extracted by the Hirt method [B. Hirt, *J. Mol. Biol.* **26**, 365 (1967)]. Samples were electrophoresed through 1.2% agarose gels, and blotted to nitrocellulose filters by the technique of Southern *et al.* [E. M. Southern, *J. Mol. Biol.* **98**, 503 (1975)], hybridized to nick-translated virion-derived SV40 DNA (specific activity ~ 1 to 2×10^8 cpm/μg). Track 1; marker DNA I, II, III; track 2; SV40 DNA extracted from 100 injected cells; track 3; SV40 DNA extracted from 25 injected cells; track 4; SV40 DNA extracted from 25 injected cells.

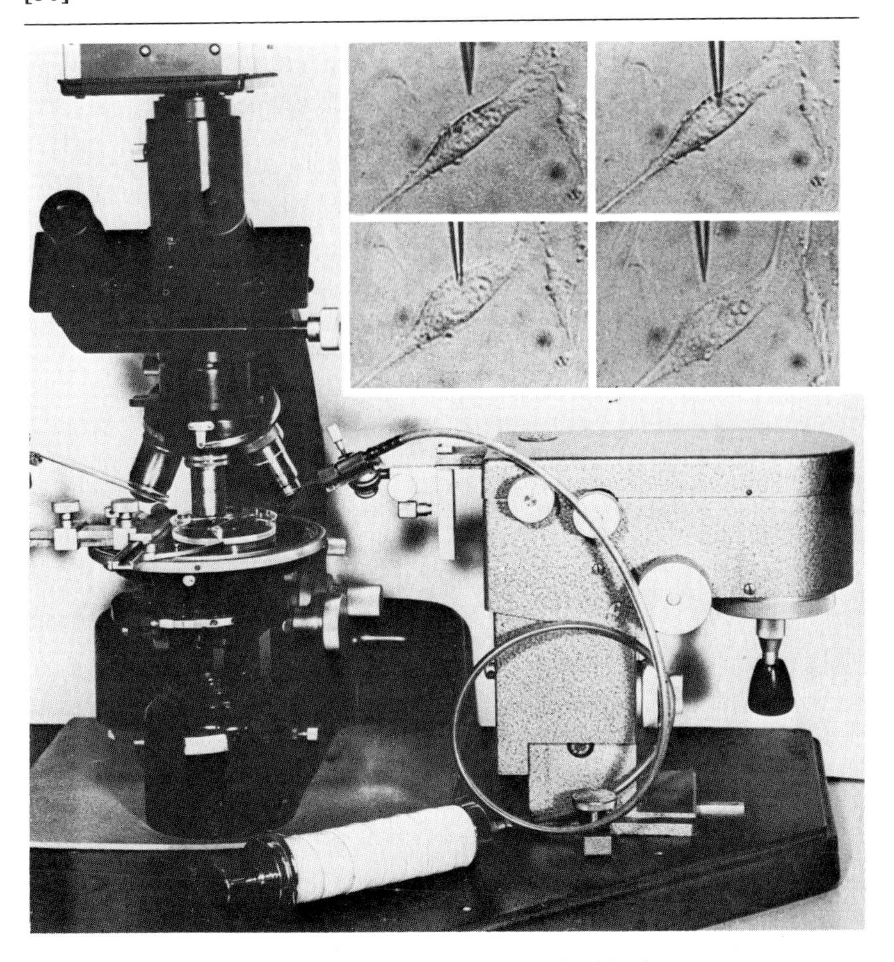

FIG. 3. Assembled instruments for microinjection.

centrated HNO_3 (v/v) for 24 hr and by extensive washings with tap water, double-distilled water, and ethanol (p.a.). Glass tubes pretreated in this way and air-dried at 120° for 24 hr were kept in a closed container until use.

Before pulling the capillary, a constriction is introduced at the middle of the glass tube. This is done by hand, using the small flame of a bunsen burner: the glass is softened in the flame, and the ends are pulled apart outside, resulting in a constriction 8–15 mm in length and 0.3–0.5 mm in diameter (Fig. 4A). The glass tube is then clamped into a capillary puller, the middle of the constriction being surrounded by the heating wire (Fig. 4B). The shape of the capillary to be obtained is shown in Fig. 4C; the tip

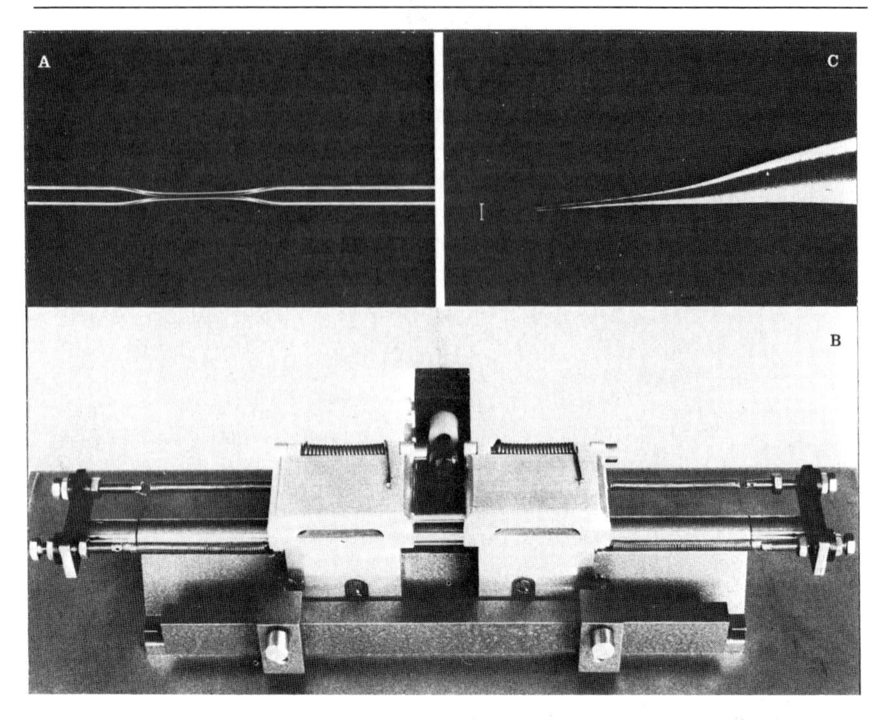

FIG. 4. Pulling the glass microcapillaries. (A) Glass tube with constriction; (B) partial view of the drawing apparatus; (C) microcapillary tip. Bar: 10 μm.

should be rigid and open, with an outer diameter of about 0.5 μm. Conditions for the mechanical pulling must be optimized by varying the temperature of the heating wire and the pulling forces at the carriage. The puller shown in Fig. 4B was built in our workshop.

The capillary tips are further treated by connecting the capillary to a 50-ml syringe, dipping the tip into 50% HF for 1 sec, and then serially washing it by suction and pressure exerted by the syringe with doubly distilled water, ethanol (p.a.), tetrahydrofuran (p.a.), 0.5% dichlorodimethylsilane in tetrahydrofuran (v/v), and again with tetrahydrofuran and ethanol. This treatment produces a smooth capillary tip, but for most purposes it may be omitted without affecting the microinjection procedure. Capillaries are conveniently stored upright (tip end up) by inserting them into fitting holes pierced into a Perspex block. The block is placed into a petri dish and covered with a beaker. Capillaries are then air-dried at 120° for 4 hr.

Preparation of Glass Slides. Glass slides (5 × 1 cm) are coated on both sides with a melted mixture of one-third beeswax and two-thirds stearine (w/w) using a cotton plug, resulting in a thin, supple film. This film is then scored (squares of 1 or 4 mm^2) and numbered (Fig. 5) with a steel needle; the glass is etched with a paste made from CaF_2 (precipitated) and 40% HF, which is spread over the slide and left there for 10–15 min. The slides are kept under running tap water overnight to split off the wax. Finally the slides are treated with a mixture of one-third concentrated H_2SO_4 and two-thirds concentrated HNO_3 (v/v), washed with running tap water for 24 hr, followed by washings with doubly distilled water and ethanol (p.a.). They are air-dried, wrapped in aluminum foil, sterilized in an oven at 200° for 24 hr, and stored until use.

Cells. Cells are grown on plastic dishes or slides under standard conditions for cell culture. Cells growing in suspension culture become accessible to microinjection after they are bound to a substrate by suitable linkers[2] (e.g., concanavalin A, phytohemagglutinin P, poly(L-lysine), pockweed mitogen, IgG).

Figure 6 shows Friend leukemia cells unattached (A) and attached to the plastic surface of a petri dish via concanavalin A (B). For this, a 60-mm petri dish was incubated for 2 hr with 2.5 ml of 2.5% glutaraldehyde at room temperature, washed four times with sterile water, covered with 200 μl of concanavalin A (100 μg/ml), and kept for 1–2 hr at 37° in an incubator. After drying, the dish was washed with PBS without Mg^{2+} and Ca^{2+}, and cells were plated.

Sample. Whenever feasible, the material to be injected is dissolved in injection buffer (0.048 *M* K_2HPO_4, 0.014 *M* NaH_2PO_4, 0.0045 *M* KH_2PO_4, pH 7.2), but other compositions, such as 0.01–0.1 *M* Tris-HCl, pH 7.2, will be tolerated by the cells. Concentrations up to 1 mg of DNA, 5–10 mg of RNA, or 10–20 mg of protein per milliliter of injection solution can be handled. The sample is centrifuged at 10,000–15,000 *g* for 10 min directly before microinjection and then transferred as a small drop to a 60-mm plastic petri dish furnished with a moistened filter paper. The

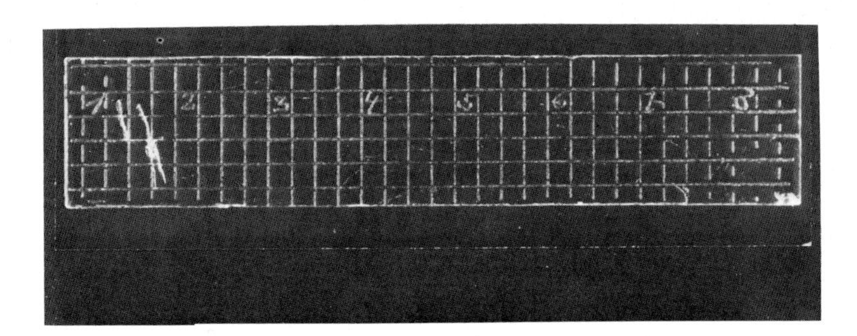

FIG. 5. Scored and numbered glass slide.

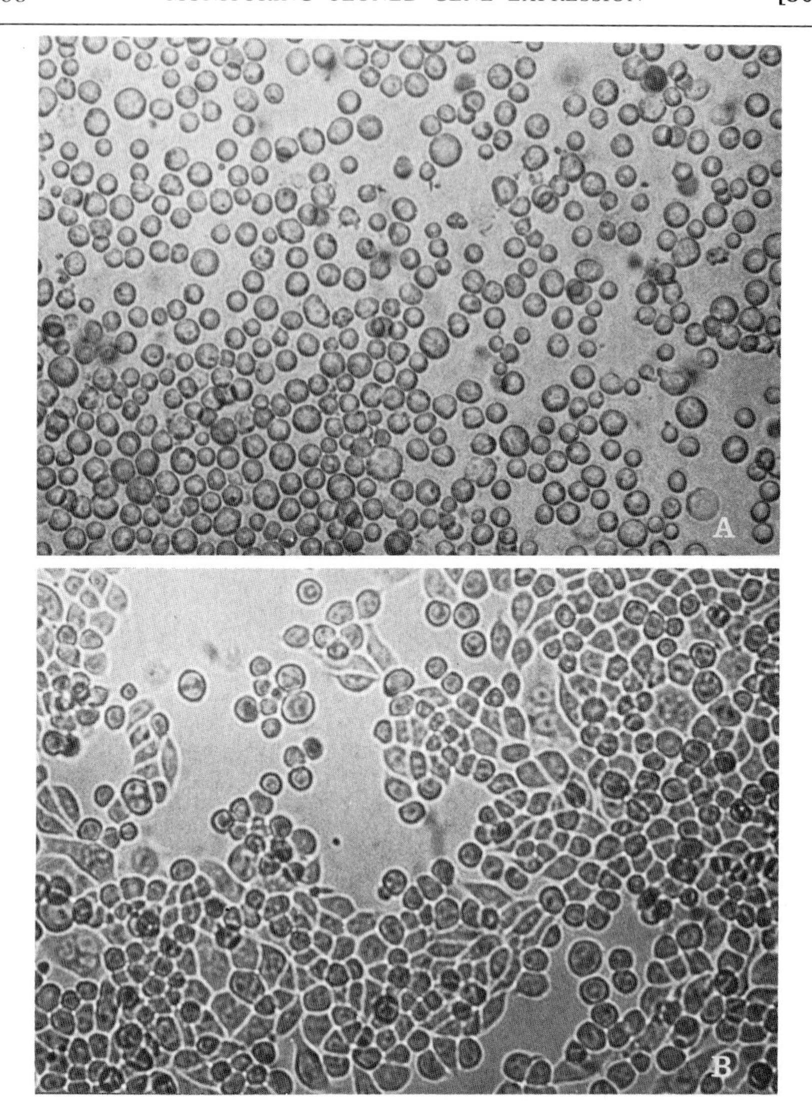

FIG. 6. Friend leukemia cells unattached (A) and attached to the petri dish (B).

drop may be placed either directly on the plastic surface or on a small sheet of Parafilm, a volume of 2 μl being sufficient. The sample dish is kept on ice.

Microinjection. A Leitz Ortholux microscope with phase-contrast equipment (Phaco 10/0.25 objective lens, Periplan GF 16× oculars) and a Leitz micromanipulator (Fig. 3) are employed for microinjection in our laboratory. These instruments are placed vibration-free in a room reserved for the purpose, which can be UV-sterilized for long-term experiments. The microcapillary is fixed to the instrument holder of the manipulator, and the tip is focused under the microscope. Next, the sample drop is brought into plane; and the needle is filled by capillary attraction forces supported by negative pressure exerted by the syringe. The capillary tip is slightly raised by turning the vertical adjustment knob of the manipulator, and the cells (immersed in medium in an open 60-mm petri dish) are placed on the microscope stage. A gentle stream of CO_2 maintains the pH optimum. Cells are brought into focus, and a distinct field is chosen for injection. The capillary is now lowered until it is nearly in focus, and an individual cell is approached by operating the manipulator lever for horizontal movements. The cell is injected by further lowering the capillary tip once it is directly above the cell. Both movements of the capillary are controlled by one hand (the right one), while the other hand exerts a gentle pressure on the syringe. A dent is seen on the cell surface when the capillary touches the cell. Microinjection itself is marked by a slight enlargement (swelling) of the cell. Moreover, the nucleus gains in contrast. Either the nucleus or the cytoplasm can be injected. After transfer of the sample, the capillary tip is brought just above the plane of the cell layer and moved to the next cell. About 150–200 cells can be injected within 10 min with some practice. After microinjection, cells are maintained as usual, checked at appropriate intervals for possible cytotoxic effects of the sample, and finally processed for evaluation.

The volume injected per cell can be estimated by measuring the distance covered by the meniscus within the pipette after microinjection of a certain number of cells and by determining the inner diameter of the capillary. By this method, we determined a mean injection volume of 1×10^{-11} to 2×10^{-11} ml per fibroblast culture cell, but up to 10^{-10} ml can be transferred. Another approach is to microinject a radiolabeled compound of high activity.

The injection volume per cell can be enlarged by a factor of 10^5 using as recipients multinucleated giant cells generated by fusion of tissue culture cells with appropriate agents.[4]

Optimal fusion conditions have to be tested for each individual cell line. Variables are (*a*) the choice of the fusion agent, e.g., polyethylene glycol 1000 or 6000, inactivated Sendai virus; (*b*) concentration; (*c*) time of action of the fusion agent; (*d*) density of cells before fusion; and (*e*) mode of washing the cells after fusion (Fig. 7).

For a comparison with related techniques developed for the transfer of

[4] A. Graessman, M. Graessmann, and C. Mueller, *Biophys. Res. Commun.* **88,** 428 (1979).

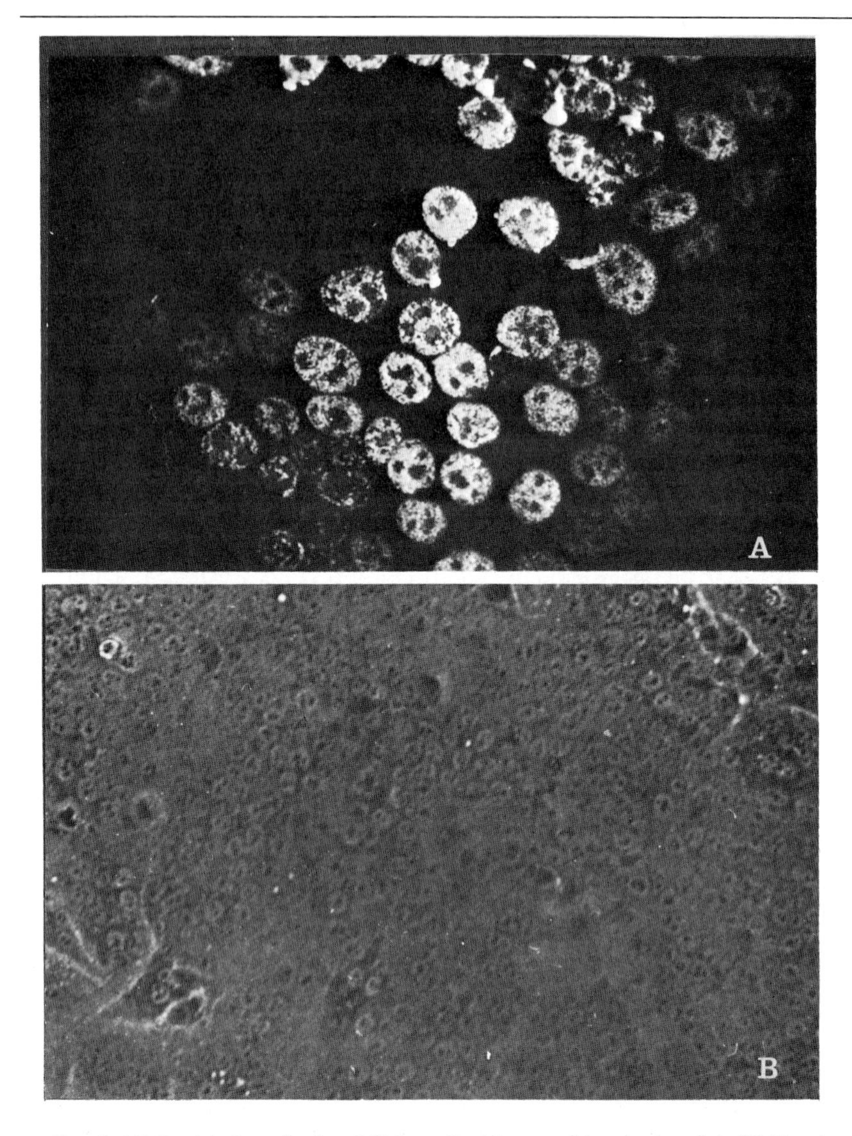

FIG. 7. (A) Partial view of a fused HeLa cell with several hundred nuclei. (B) T-antigen positive multinucleated HeLa cell (partial view), fixed and stained 15 hr after SV40 DNA injection. The nuclei are located in different planes; therefore some are out of focus. Magnification: A, 1:400; B, 1:125.

biomaterials into eukaryotic cells, the following list of general features of the microinjection system (see the table) may be helpful.

GENERAL FEATURES OF THE MICROINJECTION TECHNIQUE

	Single adherent tissue culture cells	Suspension culture cells, fixed to dish	Fused cells; PEG-1000 as fusion agent
Recipient cells	Primary or secondary cells; all permanent cell lines	Lymphocytes	Permanent cell lines; lymphocytes
Material transferable	Cell organelles, viruses, DNA, RNA, proteins, etc.	Lymphocytes	Permanent cell lines; lymphocytes
Site of application	Nucleus/cytoplasm	Nucleus/cytoplasm	Cytoplasm
Injection volume per cell (ml)	10^{-10} to 10^{-11}	10^{-11}	10^{-6} to 10^{-5}
Method of analysis	Biological/ biochemical	Biological/ biochemical	Biochemical/ biological
Efficiency (%)	90–100	90–100	70–90

1. Every tissue culture cell line tested so far has proved to be suitable for microinjection; suspension culture cells, e.g., lymphocytes, are prepared for microinjection by binding them to a substrate via suitable linkers.
2. Virtually no limitations exist regarding the material to be transferred. Intact virions, DNAs, RNAs, and proteins as well as small metabolites or substances unrelated to cellular metabolism can be introduced in purified form without the involvement of helper macromolecules or chemical treatment. The number of molecules transferred is directly correlated to the concentration of the injection solution and to the volume transferred per cell. Even intact cell organelles, e.g., cell nuclei, can be implanted into culture cells by microinjection.[5]
3. The site of inoculation within the recipient cell—nucleus or cytoplasm—can be chosen by the investigator. This allows studies on intracellular transport and compartmentalization, compartment-dependent modification steps, etc.
4. A sample volume of about 2 μl is sufficient for microinjection; most of this material is retained for other experiments.
5. The number and localization of recipient cells is known.

[5] A. Graessmann, *Exp. Cell Res.* **60,** 373 (1970).

6. Injected cells respond efficiently and remain as viable as their uninjected neighbors on condition that the material transferred has no cytopathic effects.

The microinjection technique is restricted, however, to the use of cultured cells as recipients; it is hardly applicable to intact organisms. Yet it seems conceivable to isolate cells from individuals (e.g., bone marrow cells), to passage them *in vitro,* and to clone out lines from injected cells. The donor individual could then be the recipient for these cells.

Acknowledgments

This work was supported by the Deutsche Forschungsgemeinschaft Grants 384/5-9 and 599-3 and by the Verband der Chemischen Industrie eV 1428.

[31] Fusogenic Reconstituted Sendai Virus Envelopes as a Vehicle for Introducing DNA into Viable Mammalian Cells

By A. VAINSTEIN, A. RAZIN, A. GRAESSMANN, and A. LOYTER

It is well established that, under certain conditions, eukaryotic cells can take up DNA molecules from the medium and transport them into the cell nucleus.[1,2] The method commonly used for gene transfer in mammalian cells is the addition of DNA molecules to the cell culture in the presence of facilitators, such as polyornithine[3] or DEAE-dextran,[4] or precipitation of the DNA molecules on the cell surface by calcium phosphate.[5] Coprecipitation of DNA with calcium phosphate has proved, so far, to be the most useful method for transformation of eukaryotic cells by selectable genes.[5,6] However, using this method the reported frequencies of gene transfer have been relatively low.[5,7] This has limited the method for

[1] E. H. Szybalska and W. Szybalski, *Proc. Natl. Acad. Sci. U.S.A.* **48,** 2026 (1962).
[2] P. M. Bhargava and G. Shannugan, *Nucleic. Acids. Res.* **11,** 103 (1971).
[3] F. Farber, J. L. Melnick, and J. S. A. Butel, *Biochim. Biophys. Acta* **390,** 298 (1975).
[4] J. S. Pagano, *Prog. Med. Virol.* **12,** 1 (1970).
[5] F. L. Graham, S. Bacchetti, and R. McKinon, *in* "Introduction of Macromolecules into Viable Mammalian Cells" (R. Baserga, C. Croce, and G. Rovera, eds.), p. 3. Liss, New York, 1979.
[6] M. Wigler, A. Pellicer, S. Silverstein, and R. Axel, *Cell* **14,** 725 (1978).
[7] W. H. Lewis, P. R. Srinivasan, N. Stokes, and L. Siminovitch, *Somatic Cell Genet.* **6,** 333 (1980).

ISBN 0-12-182001-7

those genes for which there exists a good selective system.[6] In addition, the method necessitates, in most cases, the use of specific cell lines, such as the mouse L cell, as recipient.[7] Attempts to transfer specific genes into other cells, such as hamster or human cell lines, either failed or the frequencies of gene transfer obtained were several orders of magnitude lower than with mouse L cells.[7]

Using DNA molecules stained with the fluorescent dye 4, 6-diamino-2-phylindole-dihydrochloride (DAPI), it has been shown that the calcium phosphate–DNA complexes enter mammalian cells by endocyte-like processes.[8] Most of the added DNA was found trapped within phagocytic vacuoles.[8,9] It can be assumed that the low frequencies observed in DNA-mediated gene transfer by the use of calcium phosphate as facilitator could be attributed, at least in part, to the fact that intracellularly these DNA molecules are not free but are enclosed within the phagocytic vacuoles.[8] From these observations it is clear why the method is applicable mostly to cells of high endocytic activity, such as mouse L cells.[7–9]

In light of the above observations and from all the information accumulated during the past few years, it appears that, for an efficient method for the introduction of DNA molecules into as many different cell lines as possible,[6–8,10] it may be necessary that (*a*) the DNA molecules be protected from hydrolysis by serum or cell secreted nucleases[8,9]; (*b*) the method be based on a principle that can be shared by most of the cell lines; (*c*) a high number of DNA molecules, namely, of gene copies, be introduced directly into the cytoplasm of the recipient cells.

It seems that resealed, loaded membranous vesicles may serve as an efficient carrier for the introduction of DNA molecules into mammalian cells. DNA molecules can be enclosed within such vesicles, and a strategy for injecting large numbers of cells simultaneously is to fuse them with the DNA-loaded vesicles. Three different approaches have been used to design such a membranous carrier: the use of vesicles made of pure phospholipids (liposomes) as carriers to transfer macromolecules such as protein or DNA into cells[11,12]; fusion between bacterial protoplasts, carrying plasmids and specific genes, and eukaryotic cells[13]; the

[8] A. Loyter, G. A. Scangos and F. H. Ruddle, *Proc. Natl. Acad. Sci. U.S.A.*, **79,** 422 (1982).

[9] A. Loyter, G. Scangos, D. Juricek, D. Keene, and F. H. Ruddle, *Exp. Cell Res.* **139,** 223 (1982).

[10] G. Scangos and F. H. Ruddle, *Gene* **14,** 1 (1981).

[11] G. Poste, D. Papahadjopoulos, and W. J. Vail, Methods Cell Biol. **14,** 33, (1976).

[12] R. Farley, S. Subramani, P. Berg, and D. Papahadjopoulos, *J. Biol. Chem.* **255,** 10431 (1980).

[13] M. Rassoulzadegan, B. Binetruy, and F. Cuzin, *Nature (London)* **295,** 257 (1982).

use of reconstituted envelopes obtained from fusogenic animal viruses,[14] e.g., envelopes of Sendai virus particles.

Infection of cells by Sendai virus, an enveloped virus belonging to the paramyxovirus group, comprises two steps[15]: (*a*) binding of virus particles to cell surface receptors, mainly sialic acid residues of glycoproteins and glycolipids[15]; and (*b*) fusion of the viral envelopes with the plasma membrane of recipient cells, with the concomitant injection of the viral nucleocapsid. Sendai virus particles contain six different polypeptides,[15] of which two are located within the viral envelope: the HN protein with hemagglutinin and neuraminidase activity, and the F protein, which is required for vius–cell fusion and virus infection.[15] Evidently, the virus–cell binding and the virus–cell fusion activities of Sendai virus particles are located in its envelope. Therefore, it appears that an ideal vehicle for the delivery of DNA molecules directly to the cytoplasm of living cells should be a membranous vesicle that will resemble membrane-enclosed viruses or, more specifically, the envelope of Sendai virus. DNA enclosed within such a vesicle will be protected from digestion by any nucleases present during the transformation process. These vesicles, like intact virus particles, should attach efficiently to specific receptors on cell surfaces and then fuse with their plasma membranes.[15]

Principle of the Method

The envelope of Sendai virus particles, like other biological membranes, can be subjected to solubilization by either nonionic or ionic detergents. Since the viral nucleocapsid is not part of the viral membrane, it remains detergent insoluble when intact viral particles are dissolved by detergents.[16,17]

Fusogenic reconstituted Sendai virus envelopes (RSVE) can be obtained after solubilization of Sendai virus particles either by Triton X-100[16] or by Nonidet P-40 (NP-40).[17] Sendai virus particles can be dissolved also by other detergents such as octylglucoside, cholate, deoxycholate, Lubrol, or lysolectin. However, these detergents cause irreversible inactivation of the virus' fusogenic activity.[18]

[14] A. Vainstein, J. Atidia, and A. Loyter, *in* "Liposomes in Study of Drug Activity and Immunocompetent Cell Function" (A. Paraf and C. Nicolau, eds.), p. 95. Academic Press, New York, 1981.

[15] G. Poste and A. Pasternak, *in* "Membrane Fusion" (G. Poste and G. L. Nicolson, eds.), p. 305. Elsevier/North-Holland, Amsterdam, 1978.

[16] D. J. Volsky and A. Loyter, *FEBS Lett.* **92,** 190 (1978).

[17] Y. Hosaka and K. Shimizu, *Virology* **49,** 627 (1972).

[18] D. J. Volsky, A. Loyter, E. Ferber, and H. U. Weltzien, unpublished results.

A. Intact Sendai virus particles solubilized

in Triton X-100

(3 mg of viral proteins + 6 mg of Triton X-

-100, 10^{-4} M PMSF in 200 μl of Buffer I)

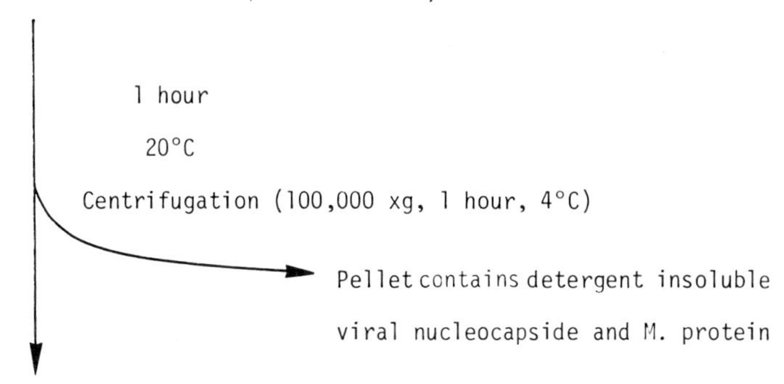

B. Supernatant contains the viral F and

HN glycoproteins

(0.8 mg in 200 μl of 3% Triton X-100)

+

DNA (50-150 μg in 100 μl of Buffer I)

Dialysis against Buffer II containing

SM2 Bio-Beads to remove Triton X-100

for 48-72 hours at 4°C

Centrifugation 100,000 xg, 1 hour

C. After washing and DNase treatment, the

DNA-loaded RSVE (200-300 μg protein) pellet

is suspended in 200 μl of Solution Na to give

protein concentration of 1 mg/ml, and stored

either at 4°C or -70°C

FIG. 1. Schematic summary of the method for preparation of DNA-loaded fusogenic Sendai virus envelopes. For details, see text.

The main features of the method for preparation of fusogenic reconstituted Sendai virus envelopes are summarized in Fig. 1. The viral particles are dissolved by Triton X-100, and the detergent-insoluble viral nucleocapsids are removed by centrifugation.[16] The clear supernatant contains the viral phospholipids and the two viral envelope glycoproteins (the HN protein, M_r 67,000, and the F protein, M_r 54,000) (see also Fig. 2). After removal of the detergent, resealed membranous vesicles that contain the two viral glycoproteins are formed (Fig. 2). Owing to the presence of these two glycoproteins which, as in intact virus particles, form spikes in the viral envelope, the RSVE are fusogenic[16] (see Fig. 2).

Figure 1 also summarizes schematically the way in which RSVE can be used as carriers of DNA. When a water-soluble macromolecule such as DNA (or protein) is added to the reconstitution system, namely, to the mixture of detergent-soluble viral phospholipids and glycoproteins, it will be trapped within the membrane vesicles formed after removal of the detergent (see also scheme in Fig. 4). It should be noted, however, that if a water-insoluble membrane component is added to the detergent-solubilized mixture, it will be inserted into the reconstituted envelope itself, resulting in the formation of "hybrid" virus vesicles.[19]

Fusion of loaded RSVE with mammalian cells will result in fusion-mediated injection of the RSVE content into the cytoplasm of the recipient cells. Uchida *et al.*[20] showed that the protein CRM-45, a nontoxic mutant protein related to diphtheria toxin,[20] can be trapped within RSVE. This protein is not toxic to cells, unless it is injected directly into the cell cytoplasm, because it lacks an amino acid sequence necessary for binding to the surface of susceptible cells.[20] RSVE can be used for the introduction of CRM-45 to cells in culture, which are killed after a short exposure to CRM-45-containing RSVE.[20] In our laboratory in Jerusalem (Israel) we have shown that RSVE can be used to introduce ferritin particles or ^{125}I-labeled IgG into cells in culture.[14] In addition, we have shown that DNA molecules, such as SV40-DNA or PTKx1 (*Herpes simplex* virus, type I, thymidine kinase gene cloned in pBR322),[21] can be trapped within RSVE and transferred to the appropriate living cells with relatively high efficiency.[22]

[19] D. J. Volsky, Z. I. Cabantchik, M. Beigel, and A. Loyter, *Proc. Natl. Acad. Sci. U.S.A.* **76,** 5440 (1979).

[20] T. Uchida, M. Yamaizumi, E. Mekada, and Y. Okada, *in* "Introduction of Macromolecules into Viable Mammalian Cells" (R. Baserga, C. Croce, and G. Rovera, eds.), p. 169. Liss, New York, 1980.

[21] L. W. Enquist, G. F. Van de Woude, M. Wagner, R. Smiley, and W. C. Summers, *Gene* **7,** 342 (1979).

[22] A. Loyter, A. Vainstein, M. Graessman, and A. Graessmann, *Exp. Cell Res.*, in press.

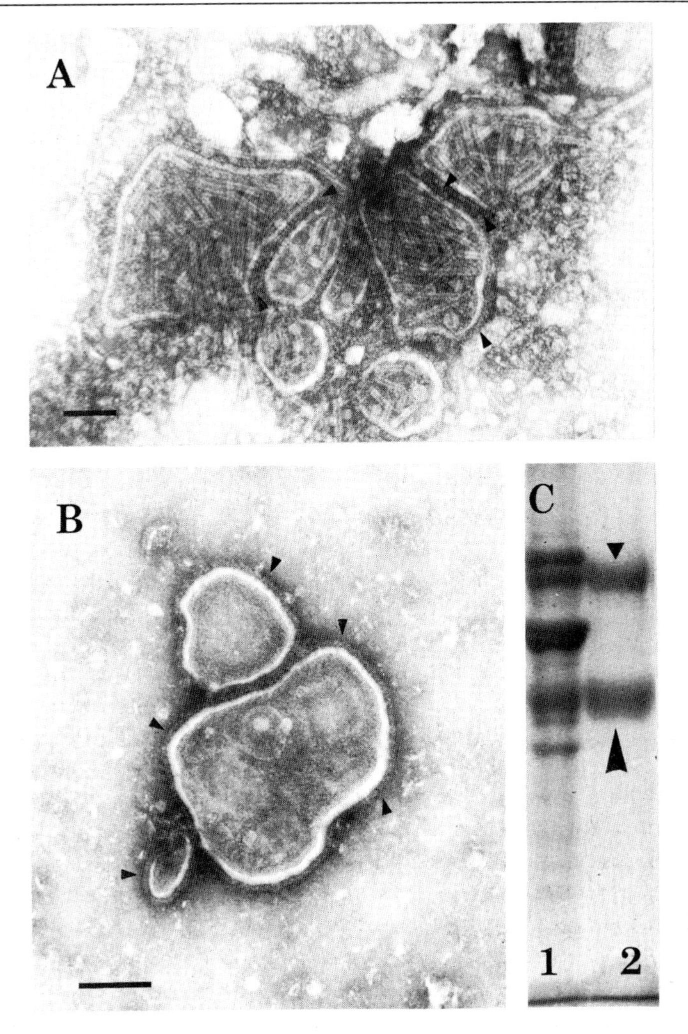

FIG. 2. Electron micrographs (A, B) and gel electrophoresis pattern (C) of intact Sendai virus particles and reconstituted Sendai virus envelopes (RSVE). Intact Sendai virus particles and RSVE were prepared for electron microscopy and observed by the negative staining technique as previously described.[16] (A.) Intact virus particles. (B) RSVE. Arrows show spikes extending from the envelopes of both intact virus and RSVE. (C) Electrophoretic pattern of intact virus (1) or RSVE (2) polypeptides (HN, small arrowhead; F, large arrowhead) (7.5% acrylamide).[16] The major Sendai virus polypeptides were classified as described before.[16] Bars in (A) and (B), 80 nm.

Materials and Reagents

Buffers and Reagents Used for the Preparation of Intact Sendai Virus Particles and RSVE. For injecting into fertilized eggs, Sendai virus particles are suspended in phosphate-buffered saline (PBS) with 50 μg of streptomycin per milliliter and 100 units of penicillin G per milliliter. Intact Sendai virus particles in Solution Na (160 mM NaCl, Tricine-NaOH, pH 7.4) are stored at −70°. Buffer I is 100 mM NaCl, 50 mM Tris-HCl, pH 7.4. Buffer II is 10 mM Tris-HCl, pH 7.4, 2 mM $MgCl_2$, 2 mM $CaCl_2$, and 1 mM NaN_3. Chicken erythrocytes are obtained from blood that was collected from the necks of decapitated chickens into an Erlenmeyer flask containing heparin (100 units/ml).

Buffers and Reagents Used for Preparation of DNA-Loaded RSVE and Their Fusion with Cultured Cells. Triton X-100 (Kotch & Light) is removed by dialysis in semimicro cellulose dialysis tubing (Spectra/por, M_r cutoff 12,000–14,000, Spectrum Medical Industries, Inc.; 15.9 mm in diameter). SM2 Bio-Bead (Bio-Rad) (10 g) is swollen in methanol (100 ml) overnight and then washed on filter (glass fiber paper, Whatman GF/A) with methanol (200 ml) followed by washing with water (2000 ml), and stored in water solution at 4°. PMSF was dissolved in methanol to a concentration of 10^{-2} M. DNase I (Worthington) was dissolved in 5 mM Tris, pH 7.4.

SV40-DNA, unlabeled or labeled with [^{3}H]thymidine, was isolated 48 hr after infection with SV40 strain (777) of TC7 by the method of Hirt.[23] Plasmids pMB9, ϕX174RFI-DNA, pBR322, and pBR322 containing the 3.2-kilobase TK-DNA fragment of *Herpes simplex* virus, type I (PTKx1) were purified by the method of Clewell.[24] ϕX174RFI-DNA was nick-translated with [α-^{32}P]dATP and dCTP, as described before.[25]

Hepatoma tissue culture (HTC) cells were grown in S77 medium supplemented with 10% newborn calf serum (BioLab).

CV1, an African green monkey kidney cell line,[26] TC7, a subline of CV1 cells,[26] F 1′ 1-4, an established rat cell line,[26] and mouse LM(TK$^-$), also an established cell line, were grown in Dulbecco's modified Eagle medium (DMEM) containing 10% (v/v) calf serum, with the following exceptions: (*a*) dTKinase, deficient mouse LM(TK$^-$) cells, which were propagated in medium containing 25 μg of BUdR except for the passage immediately preceding an experiment; and (*b*) a cell line converted to the

[23] B. Hirt, *J. Mol. Biol.* **26,** 365 (1967).

[24] D. B. Clewell, *J. Bacteriol.* **110,** 667 (1972).

[25] R. Weinstock, R. Sweet, M. Weiss, H. Cedar, and R. Axel, *Proc. Natl. Acad. Sci. U.S.A.* **75,** 1299 (1978).

[26] A. Graessmann, M. Graessmann, W. C. Topp, and M. Botchan, *J. Virol.* **32,** 989 (1979).

dTKinase-positive phenotype by fusion-mediated injection of kinase-deficient cells with PTKx1-loaded RSVE, which was transferred to growth medium (36 hr after infection) supplemented with 10^{-4} *M*, hypoxanthine *M* aminopterin, 4×10^{-7} and 4×10^{-5} *M* thymidine (HAT medium), which was changed every 4–5 days for about 3–4 weeks until TK^+ clones developed.

Methods and Results

Preparation of Intact Sendai Virus Particles and Its Quantitative Estimation

Particles of Sendai virus, one of the parainfluenza group I, are estimated quantitatively by their hemagglutinin titer (Salk's pattern method), namely, by their ability to agglutinate chicken erythrocytes at room temperature.[27] In principle, serial dilutions of Sendai virus preparation are incubated with 0.5% (v/v) of chicken red blood cells in PBS, pH 7.0, at room temperature. Hemagglutinating units (HAU) of the Sendai virus preparation are estimated as described before.[27] Generally about 10^4 HAU are present per milligram of viral protein.

High amounts of intact Sendai virus particles can be obtained by propagation in the allantoic cavity of 10-day-old chicken erythrocytes.[27] For this, about 0.2 ml of sterile Sendai virus preparation, containing 2 HAU, is injected into each fertilized egg, after which the injected eggs are incubated in a rotary incubator for 48–72 hr, as previously described.[27] At the end of the incubation period, the allantoic fluid is collected and its hemagglutinating units are estimated. Under optimal growth conditions, about 1500 HAU of Sendai virus are obtained per milliliter of allantoic fluid. About 5 ml of allantoic fluid are obtained from each fertilized egg.

Purification of Viral Particles

The procedure given is described for 100 ml of allantoic fluid, essentially as described before.[27,28] The allantoic fluid (containing about 1.5×10^5 HAU of Sendai virus particles/100 ml) is centrifuged at low speed (1400 *g* for 10 min); the supernatant is collected and recentrifuged at 37,000 *g* for 60 min in the cold. The pellet, containing virus particles, is suspended in 5–10 ml of Solution Na, homogenized to obtain a homogeneous suspension, and then centrifuged at 600 *g* for 5 min to remove large clumps of virus aggregates. The supernatant is loaded on a 15% (w/v) su-

[27] Z. Toister and A. Loyter, *J. Biol. Chem.* **248,** 422 (1973).
[28] Y. Hosaka, H. Kitano, and S. Ikeguchi, *Virology* **29,** 205 (1966).

crose cushion and centrifuged at 100,000 *g* for 1 hr. The pellet obtained contains purified viral particles that are suspended in Solution Na to give about 1×10^4 HAU/mg of protein per milliliter. Approximately 1×10^5 HAU of Sendai virus can be obtained from each 100 ml of allantoic fluid. Other purification methods are described elsewhere.[28,29]

Reconstitution of Sendai Virus Envelopes

Fusogenic envelopes of Sendai virus particles are obtained essentially as described before[16] and summarized in Fig. 1. A pellet of Sendai virus particles, usually containing 3–20 mg of viral protein, is solubilized by Triton X-100 (Triton–viral protein, 2:1, w/v), in a final volume of 0.2–1.3 ml (final concentration of Triton X-100, 3%). It is recommended to solubilize the viral particles in the presence of 10^{-4} *M* PMSF to inhibit degradation of detergent-solubilized proteins by virus-associated proteases.[30]

The turbid suspension obtained contains detergent-soluble viral envelopes and detergent-insoluble viral nucleocapsids. The detergent-insoluble material is removed by centrifugation (100,000 *g*, 60 min), and the clear supernatant contains the viral envelope phospholipids and glycoproteins (HN and F proteins, see Fig. 2), all dissolved in Triton X-100. The Triton X-100 is removed by dialysis for 48–72 hr in the cold against buffer II, which contains SM2 Bio-Beads, (1.2 g) or bovine serum albumin (1 mg/ml).

Efficient removal of Triton is achieved when either Neflex or Spectrophore-2 tubing is used during the dialysis.[16] By the use of ^{3}H-labeled Triton, it was demonstrated that the temperature of the dialysis period greatly affects the rate at which Triton is removed. About 0.01–0.02% of the Triton X-100 is left after 20–30 hr of dialysis at room temperature, whereas 0.015–0.03% remains after 60 hr of dialysis at 4°.[16,31]

The turbid suspension obtained after the end of the dialysis period contains resealed membrane vesicles that are very heterogeneous in size, ranging from 30 to 300 nm. Most of the vesicles resemble viral envelopes, since they reveal a high concentration of spikes extending from their external surface (Fig. 2). In some vesicles the spikes are attached to both sides of the reconstituted envelope.

[29] R. Rott and W. Schaefer, *Virology* **14,** 298 (1961).

[30] A. Loyter, D. Ginzberg, D. J. Volsky, and N. Zakai, *in* "Receptors for Neurotransmitters and Peptide Hormones" (G. Pepeu and M. J. Kuhar, eds.), p. 33. Raven, New York, 1980.

[31] A. Loyter and D. J. Volsky, *in* "Cell Surface Reviews" (G. Poste and G. L. Nicolson, eds.) Vol. 8, p. 215. Elsevier, Amsterdam, 1982.

From 3 mg of protein of intact virus particles, about 0.2–0.4 mg of RSVE are obtained. After centrifugation (100,000 *g*, 60 min), the RSVE are suspended in Solution Na to give about 1 mg/ml, they are kept at 4° for immediate use, or otherwise stored at −70°. Frequent freezing and thawing promotes fusion between the RSVE themselves, resulting in the formation of huge vesicles.

Enclosure of DNA Molecules in RSVE

For enclosure of DNA molecules within the RSVE (see Fig. 1) about 50–150 μg of DNA (dissolved usually in buffer I) are added to the clear supernatant of the detergent-solubilized virus envelopes, to give about 6–20 μg of DNA and 100 μg of detergent-dissolved viral envelope glycoproteins. Subsequent steps are as described above for the reconstition of "empty" unloaded viral envelopes. For removal of externally adsorbed DNA molecules, the RSVE obtained (1 mg of protein per milliliter) are incubated with pancreatic DNase I (200 μg of viral glycoproteins are incubated with 20 μg DNase in 1 ml of Solution Na that contains 5 m*M* $MgCl_2$) for 25 min at room temperature. At the end of the incubation period the DNA-loaded RSVE are collected by centrifugation.

"Biological" Assays (Induction of Hemolysis and Membrane Fusion) of Intact Sendai Virus Particles and RSVE

The amount of intact Sendai virus particles and of RSVE can be estimated, as described above, by their hemagglutinin titer. However, agglutination of chicken erythrocytes represents only binding activity of the viral particles, not necessarily their fusogenic ability.

The quickest and easiest way to check the fusogenic ability of Sendai virus particles or RSVE is by testing their ability to hemolyse and fuse aged human erythrocytes.[32] Fusion of Sendai virus particles with mammalian cells lead to transient leakage of molecules (lysis), a process from which the cells can recover.[33] Fusion with human erythrocytes, on the other hand, induced permanent lysis, concomitantly with the promotion of cell–cell fusion. Eventually giant polyghosts are formed.[32] Therefore, estimation of the extent of hemolysis—which is subjected to inhibition of antiviral antibody—may give a quantitative measurement of the virus–cell fusion process. Hemolysis is estimated essentially as previously described[27,32] and as follows.

Intact Sendai virus particles or RSVE (400 HAU in 10 ml of Solution

ɪ. Peretz, Z. Toister, Y. Laster, and A. Loyter, *J. Cell Biol.* **63,** 1 (1974).
Wasserman, A. Loyter, and R. G. Kulka, *J. Cell Sci.* **27,** 157 (1977).

Na) are added at 4° to a volume of 0.5 ml of washed, aged human erythrocytes (2.5% v/v in Solution Na). Agglutination in the cold reflects virus–cell binding. No agglutination is observed when virus particles (or RSVE) are added to neuraminidase-treated erythrocytes.[34] Treatment with neuraminidase removes the cell membranes' sialic acid residues, which serve as virus receptor.[34] After 10 min of incubation in the cold, the cell agglutinates are incubated at 37° with gentle shaking. Cell–cell fusion can be followed by phase microscopy. At the end of the incubation period (30 min at 37°), the cells or the erythrocyte ghosts are centrifuged (26,000 *g*, 10 min), and the amount of hemoglobin in the supernantant is estimated at 540 nm. The extent of the hemolysis (reflected by the amount of hemoglobin in the supernatant) is linearly correlated to the amount of viral particles present and the duration of incubation at 37°. A highly active virus preparation is considered as one whose 150–200 HAU will induce 50% hemolysis during a 10-min incubation at 37°. A preparation of either intact virus particles or RSVE, which is able to promote cell agglutination but does not induce hemolysis of human erythrocytes at 37°, is considered to be an inactive, nonfusogenic preparation.

Fusion-Mediated Microinjection of RSVE Content into Living Animal Cells

Fusion of RSVE with Cells Grown in Monolayers. Most of our experiments were performed with monolayers of cells growing in plates containing 16-mm wells (Nunelon). Thus, the experimental procedure described here will be related to these conditions. However, the described procedure can be adapted, with slight modifications, to cells grown in petri dishes of various diameters. The description given is for mouse L cells.

Mouse L cells were grown in 16-mm wells (DMEM + 10% calf serum) to give a nonconfluent layer of cells about 4 to 5 × 10^4 cells per 16-mm well). After two washings with cold Solution Na, a volume of 100 μl of Solution Na, containing 10 m*M* $CaCl_2$ and 10 μg of loaded or unloaded "empty" RSVE (about 500 HAU), is added to each well. The wells are then incubated for 10 min in the cold to allow binding of the RSVE to the surface of cells (Table I). To promote RSVE-cell fusion, the wells are transferred at the end of the cold incubation period to CO_2-incubator and incubated at 37° for 20–30 min. During this period some of the RSVE are fused with the cell plasma membranes, thus injecting their content directly into the cell cytoplasm (Table I). Experiments with radiolabeled RSVE showed that about 15–50% of the RSVE remain associated with

[34] R. Rott and H. O. Klenk, *in* "Cell Surface Reviews" (G. Poste and G. L. Nicolson, eds.), Vol. 2, p. 47. North-Holland and Publ. Amsterdam, 1977.

TABLE I
FUSION-MEDIATED INJECTION OF DNA INTO CULTURED CELLS IN MONOLAYERS

Step	Procedure
Binding	DNA-loaded RSVE (10 μg of viral proteins) are added to cells (10^4 to 5×10^4 cells/well) grown in monolayers or to cells in suspension (10^6 to 10^7 cells/ml) for 10–15 min at 4°.
Fusion-mediated injection	Cells are transferred to 37° for 20–30 min of incubation.
Culturing of cells and selection of transformed cells	Cells are washed twice with growth medium and left to grow in a CO_2-incubator.

the cells at the end of this period.[19] Since the cells in the wells are at subconfluent density, the RSVE promote very little, if any, cell–cell fusion. At the end of the incubation period, unbound RSVE are removed by washing with 2 ml of growth medium (DMEM + 10% calf serum); 2 ml of a warm medium are added, and the cells are left for recovery from the fusion event in the CO_2-incubator (Table I).

Fusion of RSVE with Cells Grown in Suspension. Most of these experiments were performed by fusion of RSVE with hepatoma tissue-cultured (HTC) cells, and the procedure described here is based on experiments with these cells.

Fusion-mediated microinjection into cells grown in suspension is somewhat more complicated than microinjection into cells in monolayer. Obviously, in order to ensure high efficiency of injection with high cell viability, virus-cell fusion should be induced without much promotion of cell–cell fusion. This can be achieved, to a large extent, by gentle vortexing of the large cell agglutinates formed after addition of the RSVE in the cold.

A volume of 20–30 μl (20–30 μg of RSVE, containing 1000–2000 HAU) is added to 10^7 cells suspended in 10 ml of Solution Na, containing 0.15 m*M* La^{3+};[33] for other cell lines, Mn^{2+} or Ca^{2+} could be added.[33] To ensure RSVE-cell binding, the suspension is incubated for 10 min at 4° (Table I). At the end of the cold incubation period, the agglutinates formed are gently vortexed to receive as many single cells as possible. This prevents massive cell–cell fusion when the suspension is subsequently incubated at 37° (Table I). The vortexed suspension is incubated at 37° for 20–30 minutes. The cells are then washed twice with warm growth medium (DMEM), and the final pellet of the cells is suspended in 10 ml of growth medium and further incubated at 37° in a CO_2 incubator (Table I).

TABLE II
ENCLOSURE OF VARIOUS DNA MOLECULES WITHIN RECONSTITUTED SENDAI VIRUS ENVELOPES (RSVE)[a]

DNA	DNA added to the reconstitution system (μg/mg of viral protein)	DNA associated with RSVE, after washing (μg)[b]	DNA enclosed within RSVE (after DNase treatment)	
			μg[b]	% of total added
^{3}H-pBR322-DNA (M_r 2.7 × 10^6)	2	0.8	0.08	4.0
	4	2.2	0.26	6.5
	10	5.5	0.55	5.5
	40	5.2	0.40	1.0
	80	1.0	0.30	0.4
	160	1.2	0.30	0.2
^{3}H-pMB9 (M_r 3.5 × 10^6)	10	3.0	0.20	2.0
^{32}P-ϕX174 RFI-DNA (M_r 3.5 × 10^6)	43	3.0	0.30	0.7
^{3}H-SV40-DNA (M_r 3.5 × 10^6)	40	8.0	1.20	3.0

[a] The different species of radiolabeled DNA were trapped within RSVE and the DNA-loaded RSVE were treated with DNase, as described in Methods and Results and elsewhere.[22] Protein was determined by the method of Lowry *et al.* [O. H. Lowry, N. J. Rosebrough, A. L. Farr, and R. J. Randall, *J. Biol. Chem.* **193,** 265 (1951)] with 0.1% sodium dodecyl sulfate in the reaction mixture using bovine serum albumin as standard.

[b] The values are expressed as micrograms of DNA associated with or trapped within RSVE obtained from 1 mg of protein of intact virus particles.

In order to follow quantitatively both RSVE-cell binding and RSVE-cell fusion, ^{125}I-radiolabeled RSVE can be prepared from ^{125}I-labeled viral particles. A procedure for the preparation of ^{125}I-radiolabeled fusogenic Sendai virus particles was previously developed in our laboratory.[35] ^{125}I-labeled viral particles will yield RSVE containing ^{125}I-labeled HN and ^{125}I-labeled F glycoproteins with high specific activity.

Trapping of Radiolabeled (^{3}H or ^{32}P) DNA in RSVE

When DNA molecules are added to the Sendai virus envelopes' reconstitution system, they are adsorbed to and trapped within the RSVE formed upon removal of the detergent. The results (summarized in Table II) are as follows.

[35] D. Walf, I. Kahana, S. Nir, and A. Loyter, *Exp. Cell Res.* **130,** 361 (1980).

1. Regardless of the origin of the DNA, about the same amount remains associated with the viral envelopes after addition of 10–40 μg of DNA to 1 mg of viral protein. As can be seen, about 3–8% of the total DNA remain associated with the RSVE, under the conditions used, at the end of the reconstitution process when four different species of DNA molecules were employed.

2. The amount of the DNA remaining associated with the RSVE was dependent, in a nonlinear fashion, on the amount of the DNA added. With both pBR322 and SV40-DNA (not shown), it was found that there exists an optimal concentration of DNA at which maximum trapping was observed. The reduction in trapping efficiency which was found at higher concentrations of DNA, may indicate that at this concentration of 100–160 μg of pBR322-DNA per milligram of viral protein, for example, the DNA molecules interfere with the reconstitution process.

Table II also shows that about 60–90% of the RSVE-associated DNA are susceptible to hydrolysis by external DNase. DNA molecules associated with RSVE, but not susceptible to hydrolysis by added DNase, are considered to be "trapped" DNA, i.e., DNA molecules that are enclosed within the viral envelope vesicles. Mixing DNA molecules with resealed-reconstituted viral envelopes does not render the DNA molecules resistant to hydrolysis, and they remain susceptible to degradation by added DNases, as do free DNA molecules in solution (not shown).

3. Table III shows that about 55% of the RSVE-associated DNase-resistant DNA molecules can be precipitated by antivirus antiserum. (At

TABLE III
PRESENCE OF ^{32}P-φX174RFI-DNA IN RECONSTITUTED SENDAI VIRUS ENVELOPES (RSVE) PRECIPITATED BY ANTISERUM AGAINST THE VIRUS[a]

System	^{32}P-DNA precipitated (% of total added)
I. ^{32}P-labeled DNA enclosed within RSVE (after DNase treatment)	55.0
II. ^{32}P-labeled DNA	0.3
III. ^{32}P-labeled DNA mixed with RSVE	0.6

[a] DNA was enclosed within RSVE, and the DNA-loaded vesicles were treated with DNase, as described in Methods and Results. Incubation with antiviral antiserum was as follows: 50 μl of antiviral antiserum were incubated with DNA-containing vesicles [50 μg of protein (I)], ^{32}P-labeled DNA [1μg, 3 × 10^5 cpm/μg (II)], or a mixture of ^{32}P-labeled DNA (1 μg) and unloaded RSVE (50 μg of protein) (III) for 1 hr at room temperature and then overnight at 4°. After precipitation (700 *g*), the pellet obtained was washed twice in Solution Na, and the radioactivity was determined after solubilization of the pellet in scintillation liquid (Insta-Gel II, Packard).

TABLE IV
ASSOCIATION OF ^{32}P-LABELED ϕX174 RFI-DNA-CONTAINING RECONSTITUTED SENDAI VIRUS ENVELOPES (RSVE) WITH HTC CELLS AT 4° AND AT 37°[a]

System	^{32}P-DNA associated with HTC cells (% of total label)		
	4° (I)	37° (II)	Cultivation in growth medium (III)
^{32}P-DNA-loaded RSVE	15.0	14.4	5.5
^{32}P-DNA	1.0	1.0	0.6
^{32}P-DNA mixed with RSVE	1.0	1.0	0.3

[a] ^{32}P-labeled ϕX174 RFI-DNA (130 μg of 3×10^5 cpm/μg) was added to the viral envelope reconstitution system (3 mg of viral glycoprotein in 3% Triton X-100). The ^{32}P-labeled DNA-loaded RSVE obtained were treated with DNase, as described in Methods and Results. DNA-loaded RSVE (15 μg of protein), ^{32}P-labeled DNA (1 μg), and a mixture of ^{32}P-labeled DNA (1 μg) and RSVE (15 μg of protein) were incubated with hepatoma tissue culture (HTC) cells. Briefly, various samples of HTC cells, each containing 2×10^6 cells per milliliter of Solution Na with 0.15 m*M* La^{3+}[33] were incubated with RSVE for 15 min at 4°. At the end of each incubation, a sample was withdrawn and, after washing in Solution Na (100 *g*, 5 min), solubilized in solubilization liquid (Insta-Gel II, Packard), and the radioactivity associated with the cells was determined by scintillation counter (Packard) (I). The same was performed after incubation at 37° for 20 min (fusion-mediated injection step II; see Table I) and after culturing in growth medium for 150 min (III).

higher concentrations of antiserum, up to 85% of the RSVE-enclosed ^{32}P-labeled molecules can be precipitated.) As can be seen in Table III, only negligible amounts of the ^{32}P-labeled DNA molecules are precipitated under the same conditions when either free DNA molecules or a mixture of free DNA molecules and unloaded RSVE are incubated with the antivirus antiserum. These observations (resistance to DNase and precipitation by antivirus antiserum) strongly indicate that the DNA molecules are indeed trapped within the viral envelopes.

About 15% of ^{32}P-labeled-DNA-loaded RSVE were found to be adsorbed to HTC cells at 4° when ^{32}P-labeled-ϕX174 RFI-DNA was used as a marker (Table IV). In other experiments, up to 50% of either intact virus particles or added RSVE remained attached to the recipient cells at 4°.[19,36] When either free ^{32}P-labeled DNA or a mixture of free ^{32}P-labeled DNA and unloaded RSVE were incubated with the cells under the same conditions, only 1% of the radioactivity remained associated with the cells (Table IV). After incubation at 37° (fusion-mediated microinjection step)

[36] M. Beigel, G. Eytan, and A. Loyter, *in* "Targeting of Drugs" (G. Gregoriadis, J. Senior, and A. Trowet, eds.), p. 125. Plenum, New York, 1982.

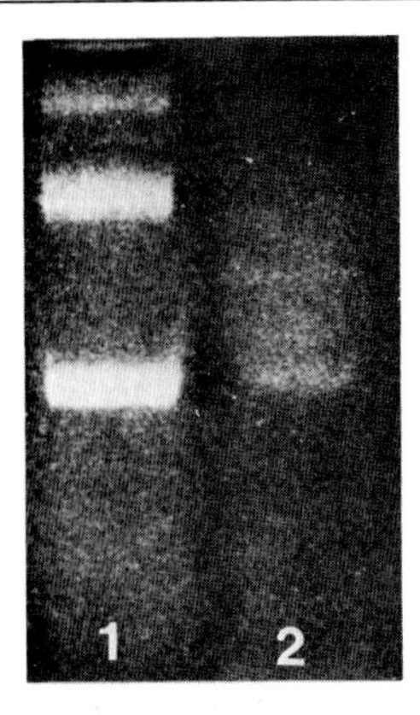

FIG. 3. Gel electrophoresis analysis of pTKx1 trapped within the reconstituted Sendai virus envelopes (RSVE). pTKx1 (150 μg) was trapped within Sendai virus envelopes, as described in Methods and Results. The pellet containing the RSVE was washed twice with Solution Na, suspended in 2 ml of Solution Na (of which 20 μl were withdrawn), and incubated with 4 μg of DNase, in a final volume of 100 μl as described before[22] and above. DNase-treated RSVE were washed once with 2 ml of Solution Na containing 2 m*M* EDTA, and the final pellet was dissolved in 1.5% Triton X-100 and loaded on agarose gel (2). In (1) the gel pattern of the plasmid pTKx-1 (3 μg) used for trapping in RSVE is shown.

and cultivation for 2.5 hr, about 14% and 5% of the added DNA-loaded RSVE remained associated with the cells, respectively (Table IV). This represents about a 10 to 15-fold increase over the amount of DNA that remained associated with the cells after addition of free DNA (Table IV).

Gel electrophoresis studies, both with SV40-DNA[22] and with pTKx1 (Fig. 3), revealed that (*a*) during the reconstitution process, the trapped DNA molecules remained intact and were not hydrolyzed or substantially sheared during the long dialysis period required for efficient removal of the detergent; and (*b*) with pTKx1 only the supercoiled molecules were trapped, but not hydrolyzed by external DNases (Fig. 3). The same was observed with SV40-DNA.[22]

Use of RSVE for Microinjection of SV40-DNA and HSV-TK Genes into Living Cells

SV40-DNA-loaded RSVE can be used as carrier to introduce the trapped DNA either to SV40-susceptible or SV40-resistant cell lines.[22] Epstein-Barr virus (EBV)-DNA was also transferred into various cell lines by using EBV-DNA-loaded RSVE.[37]

RSVE were loaded with naked SV40-DNA as described above in

[37] I. Shapiro, G. Klein, and D. J. Volsky, submitted for publication (1982).

Fig. 1 and Table I. The RSVE were fused with monolayers of CV1 cells (an African mouse kidney cell line permissive to infection by intact SV40 virus) or TC7 cells (subline of CV1). Expression of the inserted SV40-DNA was studied by the appearance of SV40-T antigen in the nucleus of the recipient cells (Table V).[22] Detection of nuclei-associated specific SV40-T antigen is done by the use of fluorescently labeled anti-T antigen antiserum, as previously described.[26] Table V shows that 3% and 20% of TC7 and CV1 cells, respectively expressed SV40-T antigen after fusion-mediated microinjection of SV40-DNA. This high efficiency of transformation was obtained only when SV40-DNA-loaded RSVE were used (Table V). Incubation of the above SV40-permissive cell lines either with naked SV40-DNA or with a mixture of naked SV40-DNA and resealed-unloaded RSVE resulted in the appearance of specific SV40-T antigen in none or only a few of the recipient cells (Table V). This low fluorescence

TABLE V
APPEARANCE OF SV40-T ANTIGEN AND LTK$^+$ CELLS AFTER FUSION-MEDIATED INJECTION OF SV40-DNA AND pTKx1, RESPECTIVELY[a]

System	T antigen-positive cells (% of total)		
	F 1′ 1-4	TC7	CV1
Experiment I			
SV40-DNA-loaded RSVE	1	3	20
Naked SV40-DNA mixed with RSVE	0.15	0.20	0.25
Unloaded RSVE	0.10	0.10	0.10
Naked SV40-DNA	0.10	0.10	0.10
Intact SV40	0	Not determined	Not determined
		LTK$^+$ colonies (% of total cells)	
Experiment II			
pTKx1-loaded RSVE		0.08–0.1	
pTKx1 mixed with RSVE		0	
Unloaded RSVE		0	
pTKx1		0	

[a] Fusion of cells with loaded RSVE was performed as described in Methods and Results. Briefly, 10–20 μl of RSVE (0.1–0.3 μg of DNA) were incubated with 10^4 to 5×10^4 cells for 15 min at 4° and then for 20–30 min at 37°. After this incubation period, cells were washed twice with DMEM supplemented with 10% newborn calf serum or fetal calf serum, and allowed to grow for different periods of time. In case of SV40-DNA loaded RSVE, after growth for 24–48 hr, cells were air dried, fixed with acetone–methanol for 10 min at −20°, and stained with rhodamine B-conjugated immunoglobulin from hamster anti-SV40 tumor serum, as described by Graessmann *et al.*[26] When pTKx1-loaded RSVE were used, cells were first grown for 36 hr in DMEM medium supplemented with 10% fetal calf serum and then transferred to HAT selective medium. TK$^+$ colonies appeared after 3–4 weeks of growth.

may represent an unspecific background seen always when cells are stained with fluorescently labeled anti-T antigen antiserum.

As sialoglycoproteins and sialoglycolipids, serving as receptors of Sendai virus particles, are present in a wide variety of animal cell lines, RSVE may be used almost as a "universal" fusogenic syringe. Indeed, RSVE can be used to transfer SV40-DNA to cells that are resistant to infection by the intact SV40. Table V shows that F 1′ 1-4 cells, an established rat cell line, that do not synthesize SV40-T antigen after incubation with intact SV40 (Table V),[26] are able to synthesize this protein after incubation with SV40-DNA-associated RSVE. About 1% of the "infected" (fused with RSVE) F 1′ 1-4 cells show the appearance of nuclei-associated T antigen, as was assayed with fluorescently labeled anti-T antigen antiserum. It should be mentioned that attempts to transform these cells by coprecipitation of DNA with calcium phosphate have failed.[26] F 1′ 1-4 cells are resistant to infection by intact SV40, since they probably lack receptors of these viruses. This was established by experiments showing that RSVE can be used to transfer membrane components from TC7 cells to the F 1′ 1-4.[22] After such fusion-mediated implanation of membrane fragments from SV40-susceptible cells into F 1′ 1-4, the latter became susceptible to infection by intact SV40. These experiments demonstrate, probably for the first time, transplantation of functional receptors for SV40.

Table V also shows that RSVE can be used to transfer the *HSV-TK* gene (*Herpes simplex* thymidine kinase gene) to LTK^- cells with relatively high efficiency. This was demonstrated by the ability of pTKx1-loaded RSVE to transform LTK^- to LTK^+ cells. The LTK^+ cells, after cultivation in HAT medium, showed an eightfold increase in the activity of the enzyme thymidine kinase, as compared with the activity of the enzyme in control untransformed cells (LTK^-) (not shown).

Conclusions and Comments

Figure 4 illustrates schematically the main steps of the method of using RSVE as a carrier for the introduction of functional DNA molecules and specific genes into cells in culture. From the results described above and from previous work using RSVE as a carrier of macromolecules,[20,37] it appears that the method allows (*a*) simultaneous injection of macromolecules, including DNA, into large numbers of cells; (*b*) introduction of DNA molecules into a broad range of cell lines that have previously been shown to be resistant to transformation by DNA-calcium phosphate complexes or by intact viruses. The only requirement for fusion-mediated injection by RSVE is the presence of sialic acid residues on cell surface glycoproteins or glycolipids.[34]

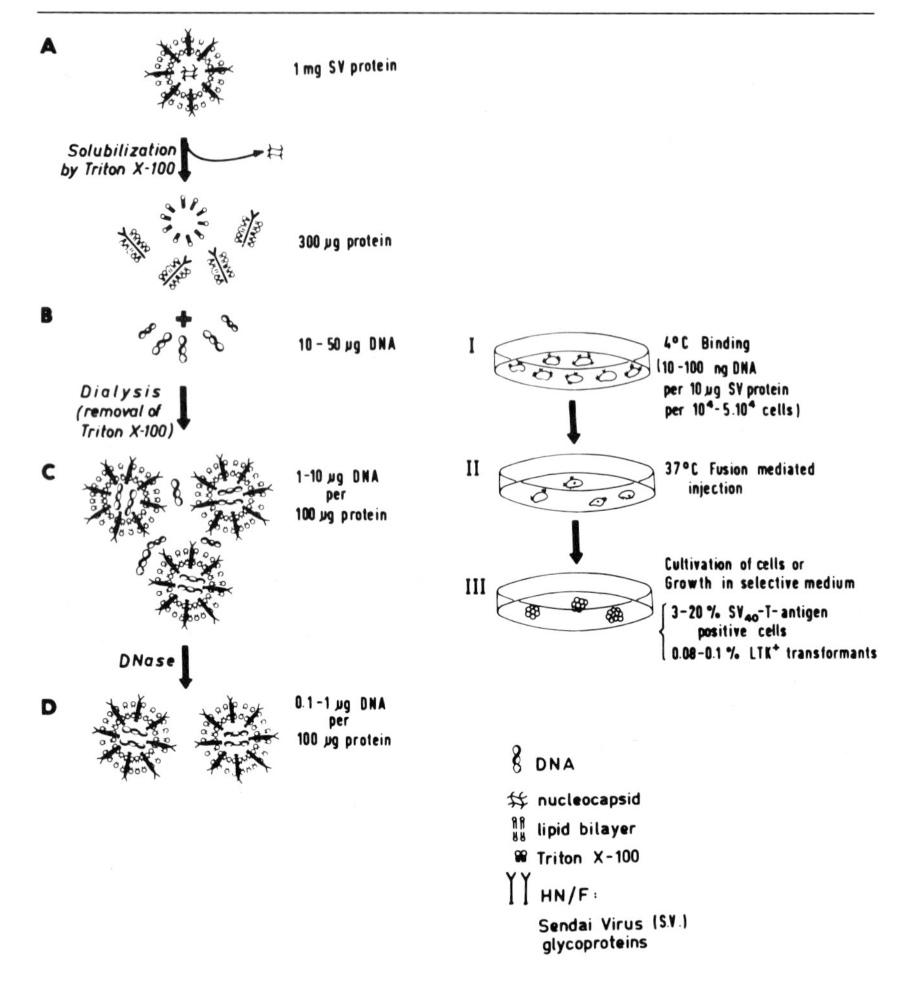

FIG. 4. A schematic illustration of trapping of DNA molecules within reconstituted Sendai virus envelopes (RSVE) and the use of DNA-loaded RSVE for transfer of DNA into cells.

It is anticipated that in the near future various aspects of the method will be studied and modified, such as the phospholipids' composition of the RSVE or a better characterization of their internal volume, thus allowing a better trapping efficiency. However, in all cases the method will be based on two steps: (*a*) trapping of DNA molecules within viral envelopes during their reconstitution; and (*b*) fusion of the loaded reconstituted vesicles with plasma membranes of recipient cells. The presence of the virus F protein (fusion factor)[15,20,31] in the reconstituted vesicles promotes their fusion with cell plasma membranes without the need for an external fusogenic agent such as polyethylene glycol. This may, on the one hand, render this method more useful than other vesicle-mediated microinjection techniques;[11–13] on the other hand, it may allow, eventually, the use of such vesicles for the delivery of macromolecules, especially DNA molecules, into specific cells in the living organism.

A schematic summary of the method is also given in Fig. 4. The calculations made are based on the use of four different species of radiolabeled DNA molecules and of two specific genes (*SV40-DNA* and *HSV-TK*), whose transformation of cells was followed quantitatively. It can be seen that about 10 μg of viral vesicles, containing 0.1–1 μg of DNA, are required to promote synthesis of T antigen in 3% and 20% of TC7 and CV1 cells, respectively. About the same amount of vesicles (10 μg of protein) and of DNA (0.1–1 μg) were required to transform about 0.1% of LTK^- into LTK^+ by fusion-mediated injection of pTKx1. It appears that the same amount of DNA used by the calcium phosphate coprecipitation method produced 0.5–5%[13] T antigen-positive cells and 0.001–0.01% LTK^+ cells.[5,7] Based on these numbers, it appears that frequency of transformation obtained by the method described here, namely, using RSVE as a vehicle for the introduction of DNA, is 5 to 15-fold higher than that obtained with the calcium phosphate method. However, we believe that the main advantage of this system is the possibility that RSVE will allow transfer of functional DNA molecules into cells that were not transformed by previous methods. So far, this was demonstrated with two different cell lines and with two genes. DNA-loaded RSVE have been able to induce SV40-T antigen synthesis in F 1′ 1-4 with SV40-DNA (Table V) and to transform hamster TK^- cells (E36TK^- derivative of V79) to TK^+ cells with *HSV-TK* (unpublished results). Both cell lines are resistant to transformation by the calcium phosphate coprecipitation technique (Table V).[26,38]

Theoretically, it can be assumed that transformation of cells or transfer of functional DNA molecules by DNA-loaded RSVE should be as efficient as transformation by intact DNA animal viruses. Reaching this goal, however, will necessitate significant improvement of the current method, especially modification of the steps involved in trapping the functional DNA. This might be achieved by using reconstitution systems that will permit better trapping of macromolecules such as the one used for obtaining large unilamellar liposomes[39] or by addition of specific phospholipids[40]

[38] L. Siminovitch, *in* "Genes, Chromosomes and Neoplasia" (F. E. Arrighi, P. N. Rao, and E. Stubblefield, eds.), p. 157. Raven, New York, 1981.

[39] F. J. Szoka and D. Papahadjopoulos, *Proc. Natl. Acad. Sci. U.S.A.* **75,** 4194 (1978).

[40] R. M. Hoffman, L. B. Margolis, and L. D. Bergelson, *FEBS Lett.* **39,** 365 (1978).

or proteins that will interact specifically with the DNA and force it to concentrate within the RSVE formed. The possibility of diluting the functional DNA with carrier-nonfunctional DNA should not be excluded. Carrier-nonfunctional DNA has been shown to be useful in transforming cells by the calcium phosphate coprecipitation methods.[5] Using carrier DNA may significantly increase the efficiency of cell transformation by DNA-loaded RSVE. This may be due to the introduction of DNA sequences present in the carrier-DNA that are required for the efficient reconstitution-insertion of the added DNA into the cell chromosomal DNA[10] or by decreasing the amount of the functional DNA required.

Acknowledgments

This work was supported by grants (to A. L.) from the March of Dimes Birth Defects Foundation; from the National Council for Research and Development, Israel; and from the G.S.F., Munich, Germany; and by a grant (to A.G.) from the Deutsche Forschungsgemeinschaft, No. GR 384-8.

[32] Liposomes as Carriers for Intracellular Delivery of Nucleic Acids

By ROBERT M. STRAUBINGER and DEMETRIOS PAPAHADJOPOULOS

A wide variety of molecules can be encapsulated within the aqueous interior of liposomes[1] and can be delivered to cells in a biologically active form. Advantages of liposomes as a carrier for introducing RNA and DNA molecules into cells are simplicity of preparation and long-term stability, low toxicity, and ability to protect encapuslated nucleic acids from degradation. As the pathway for DNA uptake may be mechanistically distinct from methods of gene transduction such as needle microinjection,[2,3] calcium phosphate precipitation,[4–6] and others,[7–12] liposomes may maximize expression in existing transformation and transfection systems and

[1] In this chapter, "liposome" and "phospholipid vesicle" are used synonymously. The nomenclature for specific types of liposomes follows the convention suggested at the New York Academy of Sciences meeting on Liposomes in Biology and Medicine [see "Liposomes in Biology and Medicine" (D. Papahadjopoulos, ed.), *Ann. N.Y. Acad. Sci.* **308** (1978)].

[2] M. Capecchi, *Cell* **22,** 479 (1980).

[3] W. Anderson, L. Killos, L. Sanders-Haigh, P. Kretschmer, and E. Diakumakos, *Proc. Natl. Acad. Sci. U.S.A.* **77,** 5399 (1980).

[4] F. Graham and A. Van der Eb, *Virology* **52,** 456 (1973).

[5] M. Wigler, S. Silverstein, L. Lee, A. Pelecer, Y. Cheng, and R. Axel, *Cell* **11,** 223 (1977).

[6] A. Loyter, G. A. Scangos, and F. H. Ruddle, *Proc. Natl. Acad. Sci. U.S.A.* **79,** 422 (1982).

ISBN 0-12-182001-7

extend the ability for genetic manipulation to cell types that do not respond to other methods for introducing nucleic acids into cells. Since some methods of liposome preparation allow efficient capture of solutes of very high molecular weight, purified genes and conceivably chromosomes can be encapsulated and introduced into cells. The ability to alter such parameters as liposome size, surface charge, bilayer fluidity, and stability affords the opportunity to adapt the carrier to a wide range of experimental conditions. Specific target cells may be recognized by liposomes bearing covalently bound antibodies,[13-18] and glycolipids[19,20] or lectins[21] may be used to improve the overall magnitude of liposome binding to cells. Finally, liposomes may be useful for *in vivo* gene transfer, should there be an advantage to such a course of therapy for inherited diseases.

Although liposomes can be made by many procedures and composed of a variety of amphipathic molecules,[22,23] the goal of encapsulation and delivery to cells[24] of DNA, RNA, and large macromolecular assemblies imposes certain constraints on the choice of method and materials. Thus, the present chapter will be restricted in its review of current methods for preparation of DNA-containing liposomes and their separation from unencapsulated material and will focus on the conditions that favor cellular uptake and expression of entrapped nucleic acids.[25-27]

---

[7] J. McCutchen and J. Pagano, *J. Natl. Cancer Inst.* **41,** 351 (1968).

[8] G. Milman and M. Herzberg, *Somatic Cell Gene.* **7,** 161 (1981).

[9] L. M. Sompayrac and K. J. Danna, *Proc. Natl. Acad. Sci. U.S.A.* **78,** 7575 (1981).

[10] A. Hinnen, J. Hicks, and G. Fink, *Proc. Natl. Acad. Sci. U.S.A.* **75,** 1929 (1978).

[11] N. Stow and N. Wilkie, *J. Gen. Virol.* **33,** 447 (1976).

[12] O. McBride and R. Athwall *In Vitro* **12,** 777 (1977).

[13] T. D. Heath, R. Fraley, and D. Papahadjopoulos, *Science* **210,** 539 (1980).

[14] T. D. Heath, J. A. Montgomery, J. R. Piper, and D. Papahadjopoulos, *Proc. Natl. Acad. Sci. U.S.A.* **80,** in press (1983).

[15] T. D. Heath, B. A. Macher, and D. Papahadjopoulos, *Biochim. Biophys. Acta* **640,** 66 (1981).

[16] A. C. Huang, L. Huang, and S. J. Kennel, *J. Biol. Chem.* **255,** 8015 (1980).

[17] F. Martin, W. Hubbel, and D. Papahadjopoulos, *Biochemistry* **20,** 4429 (1981).

[18] L. Leserman, J. Barbet, R. Kourilsky, and J. Weinstein, *Nature (London)* **288,** 602 (1980).

[19] M. Jonah, E. A. Cerny, and Y. E. Rahman, *Biochim. Biophys. Acta* **541,** 321 (1978).

[20] M. Mauk, R. Gamble, and J. Baldeschwieler, *Science* **207,** 309 (1980).

[21] F. Szoka, K. E. Magnusson, J. Wojcieszyn, Y. Hou, Z. Derzko, and K. Jacobson, *Proc. Natl. Acad. Sci. U.S.A.* **78,** 1685 (1981).

[22] D. Deamer and P. Uster, *in:* "Introduction of Macromolecules into viable Mammalian Cells" (R. Baserga, C. Croce, and G. Roueza, eds.), p. 205. Liss, New York, 1980.

[23] F. Szoka and D. Papahadjopoulos, *Ann. Rev. Biophys. Bioeng.* **9,** 467 (1980).

[24] R. Fraley and D. Papahadjopoulos, *Curr. Top. Microbiol. Immunol.* **96,** 171 (1982).

[25] R. Fraley, S. Subramani, P. Berg, and D. Papahadjopoulos, *J. Biol. Chem.* **255,** 10431 (1980).

Methods Used for Encapsulation of Nucleic Acids

Multilamellar Vesicles (MLV)

These vesicles are perhaps the simplest to prepare: a thin film of phospholipid is deposited on the walls of a glass tube and then hydrated with a solution of the material to be encapsulated. Vigorous agitation usually is necessary to resuspend most of the lipid. The heterogeneous population of particles produced (1.0–5.0 μm diameter) are composed of multiple, concentric lamellae separated by aqueous layers.[28] As a result, the ratio of entrapped volume to lipid is comparatively low[29] (3–7 μl/μmol), as is the efficiency of entrapping gene-sized macromolecules. Furthermore, the fact that MLV formation may damage nucleic acids, and that nucleases may digest some MLV-associated material,[30,31] suggests that the DNA is not completely encapsulated by this method and that formation of MLV involves significant interaction of lipids and nucleic acids. Thus, although multilamellar vesicles have been used to entrap metaphase chromosomes[32] and deliver them to cells, they are not well suited to such an application.

Small Unilamellar Vesicles (SUV)

Extended sonication of multilamellar vesicles produces a relatively homogeneous population of small (0.025 μm in diameter), unilamellar[33,34] liposomes. The material to be entrapped is added to a suspension of preformed MLV. As sonication disrupts the large vesicles, external solute is entrapped and smaller vesicles form. While this phenomenon has been exploited to load SUV with small DNA fragments,[35] the rigors of sonication make it unlikely that nucleic acids are encapsulated intact. Moreover, the

[26] R. Fraley, R. M. Straubinger, G. Rule, E. L. Springer, and D. Papahadjopoulos, *Biochemistry* **20,** 6978 (1981).

[27] R. Fraley, S. Dellaporta, and D. Papahadjopoulos, *Proc. Natl. Acad. Sci. U.S.A.* **79,** 1859 (1982).

[28] A. D. Bangham, M. Standish, and J. Watkins, *J. Mol. Biol.* **13,** 238 (1964).

[29] D. Papahadjopoulos and W. Vail, *Ann. N.Y. Acad. Sci.* **308,** 259 (1978).

[30] R. Hoffman, P. Margolis, and L. Bergelson, *FEBS Lett.* **93,** 365 (1978).

[31] P. Lurquin, *Nucleic Acids Res.* **6,** 3773 (1979).

[32] A. Mukherjee, S. Orloff, J. Butler, T. Triche, P. Lalley, and J. Schulman, *Proc. Natl. Acad. Sci. U.S.A.* **75,** 1361 (1978).

[33] D. Papahadjopoulos and W. Miller, *Biochim. Biophys. Acta* **135,** 624 (1967).

[34] D. Papahadjopoulos and J. C. Watkins, *Biochim. Biophys. Acta* **135,** 639 (1967).

[35] T.-K. Wong, C. Nicolau, and P. Hofschneider, *Gene* **10,** 87 (1980).

small internal aqueous volume[36] of SUV (0.2–0.5 μl/μmol) make them unsuitable for efficient encapsulation of high-molecular weight nucleic acids.

Large Unilamellar Vesicles (LUV)

Unilamellar and oligolamellar vesicles that capture a comparatively large aqueous volume (up to 14 μl/μmol[23]) are produced by a number of methods. Most commonly used for nucleic acid encapsulation are Ca^{2+}-EDTA chelation,[37,38] ether injection,[39–41] detergent dialysis,[42] and reverse-phase evaporation (REV).[25,43,44] Of these methods, the REV procedure is best suited to encapsulating nucleic acids and will be described below in detail. The liposomes produced are relatively homogeneous in size (0.4 μm in diameter[45]), capture 30–50% of a highly concentrated DNA solution without appreciable degradation,[25] and can be made from a wide variety of lipid mixtures. Further, the preparation can be scaled to virtually any volume, so that high ratios of DNA to lipid may be achieved with microgram quantities of DNA. Since the efficiency of encapsulation is independent of solute molecular weight, there are few restrictions on the size of macromolecules that may be entrapped; DNA of approximately 10^8 daltons has been encapsulated in a biologically active form.[46]

Liposome Constituents

Lipids

Lipids of high purity can be obtained from a number of commercial suppliers, but purity should be checked by thin-layer chromatography in

[36] C. Huang, *Biochemistry* **8,** 344 (1969).

[37] D. Papahadjopoulos, W. Vail, K. Jacobson, and G. Poste, *Biochim. Biophys. Acta* **394,** 483 (1975).

[38] T. Wilson, D. Papahadjopoulos, and R. Taber, *Cell* **17,** 77 (1979).

[39] D. Deamer and A. Bangham, *Biochim. Biophys. Acta* **443,** 629 (1976).

[40] M. J. Ostro, D. Giacomoni, and S. Dray, *Biochem. Biophys. Res. Commun.* **76,** 836 (1977).

[41] R. Fraley, C. S. Fornari, and S. Kaplan, *Proc. Natl. Acad. Sci. U.S.A.* **76,** 3348 (1979).

[42] H. Enoch and P. Strittmatter, *Proc. Natl. Acad. Sci. U.S.A.* **76,** 145 (1979).

[43] F. Szoka and D. Papahadjopoulos, *Proc. Natl. Acad. Sci. U.S.A.* **75,** 145 (1978).

[44] M. Schaefer-Ridder, Y. Wang, and P. H. Hofschneider, *Science* **215,** 166 (1982).

[45] F. Szoka, F. Olson, T. Heath, W. Vail, E. Mayhew, and D. Papahadjopoulos, *Biochim. Biophys. Acta* **601,** 559 (1980).

[46] Plant protoplast transformation by liposome-encapsulated Ti plasmid of *Agrobacterium tumefaciens;* S. Dellaporta, R. Fraley, K. Giles, D. Papahadjopoulos, A. Powell, M. Thomashow, E. Nester, and M. Gordon, unpublished observations, 1981.

several solvent systems.[47] Common contaminants such as lyso compounds, fatty acids, and metal ions may allow formation of liposomes, although the vesicles produced may be unstable, cytotoxic, or of a different electrostatic charge than expected for the pure lipid. For example, products of cholesterol oxidation are toxic to some cells,[48] and peroxidation of polyunsaturated fatty acids yields products that may damage DNA.[49] It is recommended that all lipid products be stored under argon or nitrogen and at low temperature (−70°), conditions especially important for the many naturally occuring lipids that contain readily peroxidized polyunsaturated fatty acids.

Acidic lipids, such as phosphatidylglycerol and phosphatidylserine, have been used for encapsulation and intracellular delivery of nucleic acids,[25–27,38,44,46,50] and the most common source of difficulty encountered with these lipids arises from the fact that they sometimes are supplied as the acid, rather than as the sodium salt. In addition, these lipids also may contain significant quantities of divalent cations. Since low pH and divalent cations alter significantly the physical properties of the acidic lipids,[51] it is important to ensure that they have been washed with EDTA[52] and NaCl.[53,54]

The concentration of lipids is especially important in the REV procedure, and the method of Bartlett[55] is convenient for lipid phosphorous determination. Colorimetric, enzymic, and a variety of other methods can be used for quantitation of cholesterol[56] and other lipids.[47]

Encapsulation of DNA (REV Method)

The reverse-phase evaporation method[43] derives its name from the fact that a stable water-in-oil emulsion is produced first by brief sonication

[47] M. Kates, *Techniques in Lipidology, in* "Laboratory Techniques in Biochemistry and Molecular Biology" (T. S. Work and E. Work, eds.). North-Holland Publ., Amsterdam, 1972.

[48] H. W. Chen, A. A. Kandutsch, and C. Waymouth, *Nature (London)* **251,** 419 (1974).

[49] D. D. Pietronigro, W. B. G. Jones, K. Katy, and H. B. Demopoulos, *Nature (London)* **267,** 78 (1977).

[50] G. Dahl, R. Azarnia, and R. Werner, *Nature (London)* **289,** 683 (1981).

[51] D. Papahadjopoulos, W. Vail, W. Pangborn, and G. Poste, *Biochim. Biophys. Acta* **448,** 265 (1976).

[52] Ethylenediaminetetraacetate.

[53] D. Papahadjopoulos, K. Jacobson, S. Nir, and T. Isac, *Biochim. Biophys. Acta* **311,** 340 (1973).

[54] D. Papahadjopoulos and H. Kimelberg, *in* "Progress in Surface Science" (S. G. Davidson ed.), Vol. 4, p. 141. Pergamon, Oxford, 1973.

[55] G. Bartlett, *J. Biol. Chem.* **234,** 466 (1959).

[56] W. Gamble, M. Vaughan, H. S. Kruth, and J. Avigan, *J. Lipid Res.* **19,** 1068 (1978).

of phospholipid plus an aqueous buffer in a larger volume of organic solvent, and the phases are reversed subsequently by evaporation of the organic solvent, leaving a stable oil-in-water emulsion (liposomes). The success of the method depends on a number of parameters, including the type of organic solvent, the type of phospholipid and its solubility in the organic solvent, and on a critical ratio of aqueous phase to lipid phase (in excess organic solvent). The original procedure described[43] used 1.5 ml of aqueous phase and 50 or 100 μmol of phospholipid in 5 ml of ether. However, since restricting the volume of the aqueous phase and increasing the concentration of DNA both reduces the amount of material required and enhances the probability that each vesicle produced will contain DNA, we routinely scale down the preparation to use 0.175 ml of aqueous phase with 5 μmol of lipid and 0.6 ml of ether. The quantities can be reduced to as little as 0.035 ml of aqueous and 1.0 μmol of lipid without adversely affecting vesicle formation. In adapting the procedure to volumes other than those given here, the aqueous volume determines the amount of lipid and ether added to maintain the aqueous : lipid : solvent ratio given below.

Reagents

Lipid (stored in organic solvent), 5–50 μmol
Diethyl or isopropyl ether, 1–10 ml (freshly distilled or H_2O-washed)
DNA in Ca^{2+}- and Mg^{2+}-free, isotonic buffer (pH 7–7.5)

Equipment

Rotary evaporator[57]
Bath-type sonicator[58]

Procedure

Five micromoles of lipid (usually phosphatidylserine and cholesterol in a 1 : 1 mole ratio) is transferred to a 10 × 100-mm screw-cap glass tube with a Teflon-lined cap. While the larger, standard rotary evaporator glassware can be used, the small tube reduces the amount of material lost during such small-scale preparations. The dimensions of the screw-cap tube are chosen so that it can be inserted into the much larger evaporation tube commonly used on rotary evaporators. A small amount of glycerol or H_2O is added to the larger tube to ensure thermal contact between the

[57] Preferably with the bleed valve attached to a supply of inert gas.

[58] E.g., Laboratory Supply Co., Hicksville, New York, Model TS-80-80-IRS; manufacturer's specified output is 80 W at 80 KHz.

sample tube and the evaporator bath, since solvent evaporation can cause a significant drop in sample temperature.

A thin film of lipid is deposited on the walls of the tube as the original solvent is removed by rotary evaporation. Once the lipid has dried, 0.6 ml of diethyl or isopropyl ether is added to resuspend the lipid. It is essential to ensure that the ether is peroxide-free, especially in the case of isopropyl ether. Redistillation or extraction with aqueous bisulfite are two ways to eliminate peroxide from the ether. A benefit of the aqueous extraction method is that the ether remains hydrated, thus aiding resuspension of well-dried lipids. Often, the choice of ether is arbitrary. However, since the phospholipid must be in the liquid-crystalline state for vesicle formation, lipids with a high transition temperature may require warming above the boiling point of diethyl ether. Thus, the higher boiling point of isopropyl ether makes it the solvent of choice for such cases.

DNA is added to the ether–lipid mixture in 0.175 ml of aqueous buffer[59]; only the solubility of high molecular weight DNA (approximately 4.0–10.0 mg/ml) limits the concentration at which it can be encapuslated. The buffer should have a pH in the range of 7 to 8; Tris,[60] Tes,[61] HEPES,[62] or phosphate[63] are acceptable. If the acidic phospholipids are used in the preparation, the buffer must not contain divalent cations at millimolar concentrations, since these phospholipids will precipitate. EDTA (pH 7 to 8) may be included, and may add to the long-term stability of the encapsulated DNA. Since unilamellar vesicles are osmotically sensitive, the buffer must also be isotonic with respect to the final buffer in which the incubation with cells will take place. If high osmolality is required, such as in plant protoplast culture, it is recommended that sucrose, mannitol,[27,64] or other sugars be used instead of NaCl or KCl for raising the tonicity. Negatively charged liposomes aggregate in high concentrations of Na^+ or K^+. Finally, since the captured volume is a useful indicator of the efficiency of encapsulation, a small amount of a radioactive tracer, such as sucrose, inulin, or polyadenylic acid, is included in the buffer. After separation of the liposomes from unentrapped material, the percentage of original radioactivity remaining with the liposomes corresponds to the percentage of the DNA captured.

Brief sonication of the two-phase system generates a stable emulsion that will not coalesce if allowed to stand for 20–30 min. A bath-type soni-

[59] E.g., 150 m*M* NaCl, 10 m*M* Tris, 0.1 m*M* EDTA; pH 7.4.

[60] Tris(hydroxymethyl)aminomethane.

[61] *N*-Tris(hydroxymethyl)methyl-2-aminoethane sulfonic acid.

[62] *N*-2-(hydroxyethyl)piperazine-*N*′-2-ethane sulfonic acid.

[63] E.g., phosphate-buffered saline (PBS), Ca^{2+}- and Mg^{2+}-free: 137 m*M* NaCl, 3 m*M* KCl, 17 m*M* Na_2HPO_4, 1 m*M* KH_2PO_4; pH 7.4.

[64] E.g., 400 m*M* mannitol, 50 m*M* NaCl, 5 m*M* Tris, 0.1 m*M* EDTA; pH 7.4.

cator allows the sonication to be done in a closed tube purged with inert gas. Thus the preparation remains sterile, the aerosols generated are contained, and lipid peroxidation is avoided. Although the original method calls for 2–5 min of sonication, the small volumes used in DNA encapsulation can be emulsified in as little as 5–7 sec of sonication. Low-power or poorly tuned bath sonicators may necessitate much longer irradiation and may cause nicking or shearing of large DNA molecules.

The bulk ether is removed by rotary evaporation under low vacuum (300–400 mm Hg); if a vacuum gauge is not available, the pressure is held at the point at which the sample bubbles in a controlled way. When a viscous gel has formed, the sample tube is removed from the evaporator, vortexed briefly, and then returned to a higher vacuum (500–600 mm Hg). The purpose of vortexing is to promote breakdown of the gel without resorting to high vacuum (>650 mm Hg), since stable gels may break down suddenly at low pressure, ejecting the sample from the tube. The mixture may froth and bubble until the gel breaks and the slightly turbid vesicle suspension is formed. During the frothing, the vacuum is regulated so that material is not lost from the sample tube. Most residual ether can be removed in several minutes at high vacuum (700–750 mm Hg), although care must be taken to ensure that small-volume preparations do not evaporate. An additional volume of buffer may be added to the sample after the gel breaks (e.g., 0.3 ml) and is recommended to prevent dehydration. At this point, encapsulation of the material is complete, and the preparation may be stable for months under an inert atmosphere at 4°.

Separation of Liposomes from Free Material

Although some applications may not require that free and entrapped material be separated, there are a number of reasons for doing so. Separation allows one to determine the efficiency of encapsulation and is necessary when one is interested in maximizing delivery of the encapsulated material under specific conditions. In addition, some methods of separation allow reclamation of the 50–65% of the material that is not encapsulated. Gel chromatography is a common method for resolution of free and entrapped material since large liposomes are excluded by most gels. Sepharose (2B or 4B) or BioGel (A1.5m) exclude LUV, although prior nuclease digestion is required to separate liposomes from nucleic acids of 200,000 daltons or more. Undesirable aspects of column chromatography are degradation of unencapsulated material, sample dilution, difficulty in maintaining sterility, and lipid adsorption to polyacrylamide[65] or agarose gels.[36]

[65] F. J. Martin, unpublished observation, 1979.

Since a large proportion of LUV will pellet if centrifuged at 120,000 *g* for 30 min, differential centrifugation is a method of sterile separation that allows reclamation[44,66] of unentrapped nucleic acid. However, good separation requires several washes, and 20–40% of the phospholipid may be lost in the supernatant.

The most common method used in this laboratory is flotation of the vesicles on discontinuous gradients.[15,25] A variety of media are acceptable, including metrizamide, Ficoll, and dextran. The latter two are useful for negatively charged liposomes but will cause neutral vesicles to flocculate. In addition, Ficoll and dextran precipitate in ethanol and are retained by most dialysis tubing, thus requiring dilution and centrifugation for reclamation of unentrapped DNA. Metrizamide is useful with liposomes of most compositions and is dialyzable, although it will penetrate vesicles in the presence of an osmotic gradient and prevent flotation. Ficoll and dextran are less likely to penetrate than is metrizamide.

With the method described below, resolution of entrapped from free DNA is complete and rapid. The technique is adaptable to multiple samples, amenable to the maintenance of sterility, and does not dilute the sample. For encapsulations using large quantitites of DNA in which some lipid loss is acceptable, both pelleting and flotation can be used. The sample first is diluted in the original encapsulation buffer and centrifuged once (20 min at 120,000 *g*, SW 50.1 Beckman rotor). The supernatant can be ethanol-precipitated directly, and the pellet is resuspended in a small volume and floated on a gradient.

Reagents

Metrizamide, 10% and 5%[67]
or Dextran,[68] 20% and 10%
or Ficoll, 25% and 10%
Original encapsulation buffer

Procedure

The gradient solutions usually are made in the original encapsulation buffer; with metrizamide it is especially important to verify that the solutions are isotonic with respect to the internal space of the liposomes. Since metrizamide undergoes reversible, concentration-dependent di-

[66] T. Wilson, D. Papahadjopoulos, and R. Taber, *Proc. Natl. Acad. Sci. U.S.A.* **74,** 3471 (1977).

[67] All percentages are weight/volume and the volumes given assume the centrifuge tube to be that of the SW 50.1 rotor, with capacity of 5.5 ml.

[68] Average molecular weight 40,000.

merization, it is important to verify that the 10% solution becomes isotonic upon dilution with the volume of buffer in which the vesicles will be added.

The liposome solution is diluted to a volume of 0.5 ml and mixed with 1.0 ml of the more concentrated gradient solution. Three milliliters of the less concentrated solution are overlaid, and 0.5 ml of buffer forms the top gradient step. The gradient is centrifuged at 35,000 rpm for 30 min at 20°. If metrizamide is used, the speed is increased to 45,000 rpm, with no brake and minimal acceleration. The liposomes can be pipetted from the buffer–polymer interface in a volume of 0.5–1.0 ml. While a small quantity of gradient material may be included when the vesicle band is pipetted, those mentioned here generally are regarded as nontoxic.

Liposome–Cell Interaction

General Principles

The interest in liposomes as carriers of macromolecules is based on their perceived potential to enclose and protect diverse materials of biological interest and to deliver them, functionally intact and in significant quantities, to the interior of large numbers of various cell types. While the continuous progress in liposome technology has achieved many of the prerequisite goals, the prospect of realizing the full potential of liposomes depends on understanding and exploiting the mechanisms in nature by which they and a variety of molecules and macromolecular structures gain entry to cells. Two common mechanisms, endocytosis and membrane-membrane fusion, should be efficient pathways for the uptake of liposome contents, especially considering the freedom one has in choosing liposome constituents and thereby choosing the physicochemical properties of the carrier created.

To summarize a great deal of work on liposome–cell interaction, it appears that the best vesicle lipid composition includes high mole percentages of negatively charged phospholipid (customarily phosphatidylserine) and cholesterol. Such vesicles bind avidly to cells, although some may leak their contents or exchange lipids with the plasma membrane. While the binding site is unknown, 10^3 to 10^4 liposomes bound will saturate the capacity of cells to internalize liposome contents. Thus, nonloaded vesicles compete with DNA delivery by loaded vesicles. Although delivery can be relatively efficient, it is statistically infrequent. As a result, it is difficult to determine the mechanism by which very large, liposome-entrapped molecules gain access to the cell interior. A growing body of evidence[69,70] suggests that liposomes do not fuse with the plasma membrane without some sort of perturbation such as polyethyleneglycol treatment[21] or perhaps the inclusion of certain viral proteins thought to promote membrane fusion.[71,72] Recent studies[73] show that negatively charged liposomes are endocytosed in coated pits and proceed intracellularly through the well-defined coated vesicle endocytic pathway.[74] It may be that liposomes subvert existing pathways by which cells process endocytosed macromolecules, and thereby deliver their contents.

Lipid Composition and DNA Delivery

It has been found empirically[25,26] that negatively charged vesicles generally are superior to positively charged or neutral vesicles in functional delivery of DNA. Either by plaque assay on permissive cells (CV-1) or by SV40 T-antigen expression in semipermissive cells,[74a] SV40 DNA in phosphatidylserine liposomes is at least 3-fold more infectious than positively charged sterylamine vesicles and over 10-fold more infectious than DNA in vesicles of the neutral phospholipid phosphatidylcholine. Manipulations in incubation conditions, which are discussed below, make DNA in negatively charged vesicles 10^3 times more infectious than DNA in neutral vesicles. Where the comparison has been made, this generalization on the lipid dependence of delivery has been found to hold for a variety of animal cells and also for plant protoplasts.[27]

The variation in DNA delivery among liposomes of different electrostatic charge corresponds to the variation seen in the amount of cell-associated vesicle lipid. Thus, more negatively charged liposomes adhere to cells than do vesicles of other charge, and infectivity of encapsulated DNA can be explained on the basis of the capacity to bind cells. While the difference in cell-associated lipid between phosphatidylserine and phosphatidylcholine is only 2-fold at lipid-to-cell ratios above the point at which the cellular capacity for functional delivery saturates,[26,38,44] the superiority of phosphatidylserine is comparatively greater (10-fold) in the range of lipid-to-cell ratios that are relevant to DNA delivery (100 nmol/10^6 cells).

There is considerable variability in DNA delivery among specific neg-

[69] F. Szoka, K. Jacobson, and D. Papahadjopoulos, *Biochim. Biophys. Acta* **551,** 295 (1979).

[70] D. K. Struck, D. Hoekstra, and R. E. Pagano. *Biochemistry* **20,** 4093 (1981).

[71] T. Uchida, J. Kim, M. Yamaizumi, Y. Miyake, and Y. Okada, *J. Cell Biol.* **80,** 10 (1979).

[72] A. Helenius, J. Kartenbeck, K. Simons, and E. Fries, *J. Cell Biol.* **84,** 404 (1980).

[73] R. M. Straubinger, K. Hong, D. S. Friend, and D. Papahadjopoulos, *Cell* **32,** in press (1983).

[74] J. L. Goldstein, R. W. G. Anderson, and M. S. Brown, *Nature (London)* **279,** 679 (1979).

[74a] E. g., L. HeLa, 3T6; R. Fraley and R. Straubinger, unpublished observations, 1980.

atively charged lipids, and the differences can be related to differences in retention of aqueous contents upon interaction with cells.[26,38] Phosphatidylserine is the best single lipid for DNA infectivity; phosphatidylglycerol vesicles bind to cells as well as those of phosphatidylserine but can lose 95% of their contents on exposure to cells.

In addition to animal[69,75] and plant[27,76] cells, serum components[77–79] and a variety of soluble proteins promote the leakage of vesicle contents. Cholesterol has been shown to reduce the rate of efflux.[77–81] The effect of cholesterol on vesicle leakage is relevant to DNA delivery,[26] for the addition of cholesterol to phosphatidylglycerol vesicles causes a 9-fold decrease in loss of aqueous contents and a 10-fold increase in the infectivity of DNA. The effect of cholesterol addition is not nearly as dramatic in the case of phosphatidylserine liposomes, owing to the fact that phosphatidylserine forms relatively nonleaky vesicles.

There have been a number of efforts to correlate functional delivery of nucleic acids with cell-association of vesicle lipid and contents. From them, it was possible to estimate the comparative significance[82] of several of the processes mentioned earlier. Using ^{3}H-labeled phospholipid as a marker for vesicle lipid, ^{32}P poliovirus RNA as a marker for vesicle contents, and the appearance of progeny poliovirus as a marker for intracellular delivery,[38] it was found that approximately 10^4 vesicles per cell[83] must be added in order to ensure expression of the encapsulated RNA by every cell. Approximately 10% of the vesicle lipid became cell-associated, while only 1% of the RNA was cell-bound. Thus, each cell bound or internalized about 10^3 vesicles when a saturating concentration of vesicles was applied, and only 100 of those vesicles retained their contents. From the infectivity of the RNA, it was apparent that the vesicles that retained their contents were sufficient to ensure infection of virtually all the cells in a vesicle-exposed population, whether of primate (HeLa) or of rodent (L, CHO) origin, and that infectivity of the entrapped RNA was compara-

[75] J. Van Renswoude and D. Hoekstra, *Biochemistry* **20,** 540 (1981).

[76] P. Lurquin, *Nucleic Acids Res.* **12,** 3773 (1981).

[77] G. Gregoriadis and C. Davis, *Biochem. Biophys. Res. Commun.* **89,** 1287 (1979).

[78] G. Scherphof, F. Roerdink, M. Waite, and J. Parks, *Biochem. Biophys. Acta* **542,** 296 (1978).

[79] T. M. Allen and L. G. Cleland, *Biochim. Biophys. Acta* **597,** 418 (1980).

[80] D. Papahadjopoulos, S. Nir, and S. Ohki, *Biochim. Biophys. Acta* **266,** 561 (1972).

[81] E. Mayhew, Y. Rustum, F. Szoka, and D. Papahadjopoulos, *Cancer Treat. Rep.* **63,** 1923 (1979).

[82] D. Papahadjopoulos, T. Wilson, and R. Taber, *In Vitro* **16,** 49 (1980).

[83] Roughly 100 nmol of phospholipid per 10^6 cells; it can be calculated that there are about 4.4×10^8 unilamellar liposomes per nanomole of lipid, assuming a mean vesicle diameter of 400 nm, a bilayer thickness of 3.5 nm, and an area of 7 nm^2 per phospholipid headgroup.

ble to that of a high-titer stock of whole poliovirus (one plaque per 100 particles).

Enhancing Delivery

Dimethyl sulfoxide (DMSO), glycerol, and polyethylene glycol (PEG) enhance effective delivery of liposome-entrapped DNA.[26] PEG (44% w/v for 90 sec) and DMSO (25% v/v for 4 min) increase the infectivity of entrapped SV40 DNA by a factor of 10. Glycerol has been the most effective agent for increasing DNA delivery, especially for negatively charged liposomes (100 to 200-fold enhancement). The effect is dose- and time-dependent and useful with a variety of cell types (L, 3T6, HeLa). In addition, glycerol enhances both the transient[84] and stable[44,85] expression of the Herpes thymidine kinase gene, so the effect is not limited to SV40 DNA. However, viability after glycerol treatment varies with cell type and should be checked.

The effect of similar treatments to increase the delivery and expression of encapsulated nucleic acids is also seen with plant protoplasts. Although glycerol is toxic to *Nicotiana* protoplasts, treatment with PEG or polyvinyl alcohol increases the functional delivery of liposome-encapsulated tobacco mosaic virus RNA by 200-fold or more.[27] Qualitatively similar results were obtained for delivery of the Ti plasmid of *Agrobacterium tumefaciens,* using transformation to phytohormone independence and octopine-positive phenotype as indications of plasmid expression. Preliminary experiments indicated that the plasmid was integrated in some of the transformed colonies.[46]

Mechanism of Delivery

While many studies show that liposome delivery of large molecules occurs without exposure of the contents to the external medium, such observations offer no definitive evidence for the mechanism of the delivery event. Although it is not inconsistent with the hypothesis of liposome fusion with the plasma membrane, the same result could be explained by endocytosis of liposomes and intracellular processing of the contents through a pathway that allows 1% of the internalized material to escape lysosomal degradation.

Agents which inhibit a variety of cellular functions offer an indication of one mechanism by which liposomes may deliver DNA.[26,86] Short incu-

[84] R. Straubinger, unpublished observations, 1981.
[85] R. Fraley and L. Rall, unpublished observations, 1980.
[86] R. Straubinger, unpublished observations, 1981.

bations of cytochalasin B, colchicine, chloroquine, or azide and 2-deoxyglucose do not alter the basal level or polyethylene glycol enhancement of SV40 DNA delivery, although electron microscope examination shows liposomes bound to coated pits and in intracellular vesicles.[73] The effect of glycerol is quite sensitive to azide and deoxyglucose and partially to cytochalasin B. This and several other lines of evidence suggest that glycerol increases liposome uptake by promoting a process that resembles macropinocytosis. Chloroquine, ammonium chloride, and several other agents[87] with lysosomotropic activity[88] enhance the glycerol effect, so lysosomal degradation may be the ultimate fate of some liposome-delivered DNA. The lysosomotropic amines may exert their effect during the initial period of liposome internalization by inhibiting phagolysosome function, thus increasing the probability that intact DNA may escape to the cytoplasm. Alternatively, they may inhibit autophagic[89,90] processing of intracellular DNA and increase the probability of nuclear access.

Liposome–Cell Incubation

The following is a general summary of the protocol for the incubation of liposomes with cells. It is used for both monolayer and suspension cultures; the only difference in dealing with suspension cells is that washes usually are performed by low speed centrifugation. It is assumed that the composition of the liposomes includes acidic phospholipids.

Procedure

Cells in growth medium are washed twice before the addition of liposomes in order to remove serum and, if necessary, to assure that divalent cations are reduced to submillimolar concentrations. Calcium- and magnesium-free Tris-[91] or phosphate-buffered saline commonly are used, with just enough divalent cations added to keep adherent cells attached to the substrate. Serum-free medium may be substituted for the buffers mentioned and may improve vesicle delivery by improving the metabolic state of the cells. If necessary, EDTA should be added to reduce the divalent cation concentration. If chloroquine is to be used, it is added to the cells 30–60 min before adding vesicles and is maintained in all buffers until the

[87] D. K. Miller and J. Lenard, *Proc. Natl. Acad. Sci. U.S.A.* **78,** 3605 (1981).

[88] C. de Duve, T. de Barsy, B. Poole, A. Trouet, P. Tulkens, and F. van Hoof, *Biochem. Pharmacol.* **23,** 2495 (1974).

[89] D. W. Stacey and V. G. Allfrey, *J. Cell Biol.* **75,** 807 (1977).

[90] L. Marzella, J. Ahlberg, and H. Glaumann, *J. Cell Biol.* **93,** 144 (1982).

[91] TBS: 137 m*M* NaCl, 5 m*M* KCl, 1 m*M* Na_2HPO_4, 25 m*M* Tris; pH 7.4

final postliposome washes. Usually it is most convenient to add the drug to the original growth medium as a concentrated solution at physiological pH. Although chloroquine is more toxic to some cells than to others, most can tolerate a final concentration of 100 μM.

Liposomes are added to cells in a small volume in order to maximize cell–vesicle interaction; 200 μl may be sufficient for 5×10^6 cells. For most cells studied, the capacity for liposome uptake saturates at about 100–200 nmol (negatively charged) phospholipid per 10^6 cells, and there is likely to be little advantage to adding more material. In addition, high concentrations of liposomes (>500 μmol/ml), particularly those made with charged lipids, can be toxic to both animal and plant cells.

For negatively charged liposomes, the interaction relevant to nucleic acid delivery takes place in the first 30 min of incubation.[25,38] Longer incubations may be undesirable because of effects on cellular metabolism. At the end of the incubation, cells are washed twice with drug-free medium or buffer and replated in the usual growth medium.

If polyethylene glycol or glycerol is to be used, it is added just before the postincubation washes. The usual range of PEG concentrations that cells will tolerate is 25–50% (w/v) for 60–90 sec, although toxicity is a dose- and time-dependent effect. There is no known advantage to PEG of any particular molecular weight or supplier, though PEG of 6000 daltons or greater is difficult to wash away and may be toxic for that reason. For glycerol, the useful range of concentrations is 15–35% (v/v, buffer or medium) for 4 min or more. The effectiveness and the toxicity of glycerol both increase with the concentration and the length of exposure. At the end of the period of exposure, the glycerol (or polyethylene glycol) is diluted slowly, and cells are washed three times with medium or buffer before return to the normal growth medium.

Present and Future Prospects

Although specific successes with liposome-encapsulated nucleic acids demonstrate the value of this approach, the method is still evolving. Specific efforts are directed toward increasing the amount of material delivered intracellularly and extending the ability for genetic manipulation to a variety of cell lines for which no method of gene transduction has been very successful. One exciting development in liposome technology is the ability to couple to liposomes ligands that recognize specific cell-surface components. With the proper monoclonal antibody or peptide ligand, it may be possible to direct liposomes and their contents to specific cellular targets with great efficiency. In addition, insight into the ways in which cells process liposomes and their contents suggests that several aspects of cellular metabolism may be exploited to increase the efficiency of intracellular delivery. For example, as coated-pit endocytosis is probably the predominant mechanism by which many cells take up liposomes,[73] pH-dependent fusion of acidic liposomes with endosomal membranes is a mechanism by which liposome contents could gain access to the cytoplasm. Thus the liposome would provide a route by which endocytosed material could escape the endocytic vacuole, similar to that suggested for Semliki Forest virus[72] and vesicular stomatitis virus.[92] It is expected that these and other such developments will provide a realistic basis for important applications of liposome-mediated transfer of nucleic acids.

[92] D. K. Miller and J. Lenard, *J. Cell. Biol.* **84,** 430 (1980).

[33] *Agrobacterium* Ti Plasmids as Vectors for Plant Genetic Engineering

By KENNETH A. BARTON and MARY-DELL CHILTON

Agrobacterium tumefaciens, a pathogenic soil bacterium, incites crown gall disease in a wide variety of dicotyledonous plants.[1] Virulence is conferred on the bacterium by genes carried on large plasmids, the Ti (tumor-inducing) plasmids, which range in size from 160 to 240 kilobases.[2–5] Although details of the mechanism of tumor induction are unknown, the end result is that a segment of the Ti plasmid, called T-DNA, is inserted into the nuclear DNA of the host plant.[6–11] Genetic information

[1] M. DeCleene and J. De Ley, *Botan. Rev.* **42,** 389 (1976).

[2] I. Zaenen, N. Van Larebeke, H. Teuchy, M. Van Montagu, and J. Schell, *J. Mol. Biol.* **86,** 109 (1974).

[3] N. Van Larebeke, G Engler, M. Holsters, S. Van den Elsacker, I. Zaenen, R. A. Schilperoort, and J. Schell, *Nature* (*London*) **252,** 169 (1974).

[4] B. Watson, T. C. Currier, M. P. Gordon, M.-D. Chilton, and E. W. Nester, *J. Bacteriol.* **123,** 255 (1975).

[5] T. C. Currier and E. W. Nester, *J. Bacteriol.* **126,** 157 (1976).

[6] M.-D. Chilton, M. H. Drummond, D. J. Merlo, D. Sciaky, A. L. Montoya, M. P. Gordon, and E. W. Nester, *Cell* **11,** 263 (1977).

[7] M.-D. Chilton, R. K. Saiki, N. Yadav, M. P. Gordon, and F. Quetier, *Proc. Natl. Acad. Sci. U.S.A.* **77,** 2693 (1980).

[8] L. Willmitzer, M. DeBeuckeleer, M. Lemmers, M. Van Montagu, and J. Schell, *Nature* (*London*) **287,** 359 (1980).

[9] N. S. Yadav, K. Postle, R. K. Saiki, M. F. Thomashow, and M.-D. Chilton, *Nature* (*London*) **287,** 458, (1980).

[10] M. F. Thomashow, R. Nutter, K. Postle, M.-D. Chilton, F. R. Blattner, A. Powell, M. P. Gordon, and E. W. Nester, *Proc. Natl. Acad. Sci. U.S.A.* **77,** 6448 (1980).

[11] P. Zambryski, M. Holsters, K. Kruger, A. Depicker, J. Schell, M. Van Montagu, and H. Goodman, *Science* **209,** 1385 (1980).

ISBN 0-12-182001-7

in T-DNA is functional in the transformed plant cell, producing polyadenylated transcripts[12–16] that are translatable *in vitro*.[17–19] T-DNA gene products are apparently responsible for both hormone autotrophy of crown gall cells[20,21] and the synthesis of novel metabolites called opines.[18,19,22–25] Opines are simple derivatives of amino acids, keto acids, and sugars and are specificially catabolized by *Agrobacterium*, an ability conferred by inducible genes on the Ti plasmid.[26–28] The biological rationale for T-DNA insertion into the host plant genome appears to be the production of opines as a nutrient source for the pathogen. *Agrobacterium tumefaciens* thus emerges as an example of genetic engineering in nature, for it assures a supply of specific metabolites by altering metabolic pathways of the host plant to its own ends.

The *A. tumefaciens* Ti plasmids and the Ri (root-inducing) plasmids of *A. rhizogenes*,[29] a closely related pathogen, are the only agents known to insert foreign DNA into higher plant cell genomes. Ti and Ri plasmids in this sense resemble the DNA transforming viruses of animal cells that are

[12] M. H. Drummond, M. P. Gordon, E. W. Nester, and M.-D. Chilton, *Nature* (*London*) **269,** 535 (1977).
[13] A. Ledeboer, Ph.D. Thesis, University of Leiden, The Netherlands (1978).
[14] W. B. Gurley, J. D. Kemp, M. J. Albert, D. W. Sutton, and J. Callis, *Proc. Natl. Acad. Sci. U.S.A.* **76,** 2828 (1979).
[15] F.-M. Yang, A. L. Montoya, D. J. Merlo, M. H. Drummond, M.-D. Chilton, E. W. Nester, and M. P. Gordon, *Mol. Gen. Genet.* **177,** 704 (1980).
[16] F.-M. Yang, J. C. McPherson, M. P. Gordon, and E. W. Nester, *Biochem. Biophys. Res. Commun.* **92,** 1273 (1980).
[17] J. C. McPherson, E. W. Nester, and M. P. Gordon, *Proc. Natl. Acad. Sci. U.S.A.* **77,** 2666 (1980).
[18] J. Schroder, G. Schroder, H. Huisman, R. A. Schilperoort, and J. Schell, *FEBS Lett.* **129,** 166 (1981).
[19] N. Murai and J. D. Kemp, *Proc. Natl. Acad. Sci. U.S.A.* **79,** 86 (1982).
[20] G. Ooms, P. J. Hooykaas, G. Moleman, and R. A. Schilperoort, *Gene* **14,** 33 (1981).
[21] D. J. Garfinkel, R. B. Simpson, L. W. Ream, F. W. White, M. P. Gordon, and E. W. Nester, *Cell* **27,** 143 (1981).
[22] A. Petit, S. Delhaye, J. Tempe, and G. Morel, *Physiol. Veg.* **8,** 205 (1970).
[23] G. Bomhoff, P. M. Klapwijk, H. C. M. Kester, R. A. Schilperoort, J. P. Hernalsteens, and J. Schell, *Mol. Gen. Genet.* **145,** 177 (1976).
[24] J. Tempe and J. Schell, *in* "Translation of Natural and Synthetic Polynucleotides" (A. B. Legocki, ed.), p. 416. University of Agriculture, Poznan, 1977.
[25] J. Leemans, C. Shaw, R. Deblaere, H. DeGreve, J. P. Hernalsteens, M. Maes, M. Van Montagu, and J. Schell, *J. Mol. Appl. Genet.* **1,** 149 (1981).
[26] A. L. Montoya, M.-D. Chilton, M. P. Gordon, D. Sciaky, and E. W. Nester, *J. Bacteriol.* **129,** 101 (1977).
[27] A. Petit and J. Tempe, *Mol. Gen. Genet.* **167,** 147 (1978).
[28] P. M. Klapwijk, M. Oudshoorn, and R. A. Schilperoort, *J. Gen. Microbiol.* **102,** 1 (1977).
[29] M.-D. Chilton, D. A. Tepfer, A. Petit, C. David, F. Casse-Delbart, and J. Tempe, *Nature* (*London*), **295,** 432 (1982).

proving to be useful as gene vectors in mammalian systems.[30–32] Ti and Ri plasmids appear promising as vectors for engineering new genes into higher plants. There appears to be no stringent size limitation on DNA that T-DNA can transport into plants: a Tn7 transposon (14 kb) inserted in T-DNA was found intact in infected plant cells.[33] However, transpositional insertion is not a method of general utility for introduction of new DNA into the T-DNA region of Ti plasmid vectors. Methods for site-specific insertion of desired DNA at specific restriction endonuclease cleavage sites in T-DNA have therefore been developed. In this chapter we describe one such method[34] and some variations on that approach.[35] An alternative procedure that requires the isolation of Ti plasmids constitutive for transfer (Tra^c) has been described elsewhere.[25] The chapter mentions additional technical problems facing the genetic engineer who wishes to use T-DNA as a vector for higher plants.

Methods

Direct introduction of a desired DNA fragment into a specific restriction site in T-DNA of the Ti plasmid is not feasible by the normal route of restriction endonuclease cleavage and ligation, for several reasons.

1. The enormous size of the vector, the Ti plasmid, and its consequent large number of restriction endonuclease cleavage sites (Fig. 1A), make it impossible to achieve cleavage at a specific site to full-length linear plasmids.
2. Reclosure of circular DNA of such enormous size would be prohibitively slow.
3. Because the Ti plasmid does not replicate in *Escherichia coli,* the resulting recombinant plasmid, if it could be constructed, would have to be selected by transformation of *Agrobacterium,* a procedure that occurs at prohibitively low frequency[36] for recombinant DNA work. Alternatively, one could use a Ti plasmid that could replicate in *E. coli* (a Ti : RP_4 cointegrate, for example[25]), but such a

[30] C. Thummel, R. Tjian, and T. Grodzicker, *Cell* **23,** 825 (1981).

[31] J. T. Elder, R. A. Spritz, and S. M. Weissman, *Annu. Rev. Genet.* **15,** 295 (1981).

[32] H. M. Temin, *Cell* **28,** 3 (1982).

[33] J. P. Hernalsteens, F. Van Vliet, M. DeBeuckeleer, A. Depicker, G. Engler, M. Lemmers, M. Holsters, M. Van Montagu, and J. Schell, *Nature (London)* **287,** 654 (1980).

[34] A. J. M. Matzke and M.-D. Chilton, *J. Mol. Appl. Genet.* **1,** 39 (1981).

[35] K. Barton, A. de Framond, and M.-D. Chilton, in preparation.

[36] M. Holsters, D. DeWaele, A. Depicker, E. Messens, M. Van Montagu, and J. Schell, *Mol. Gen. Genet.* **163,** 181 (1978).

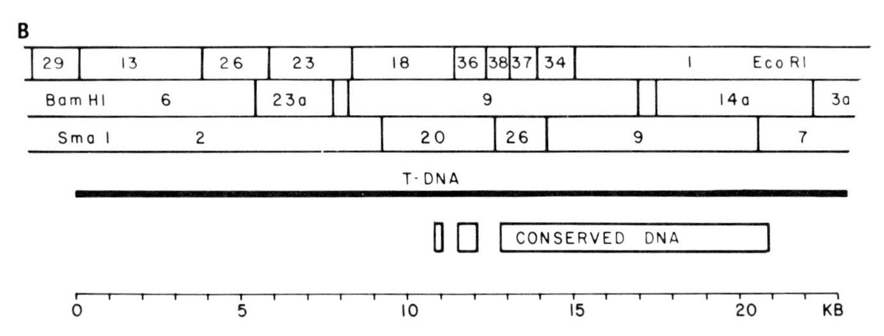

FIG. 1. (A) Cleavage pattern of pTiT37 with restriction endonuclease *Bam*HI. The *Bam*HI fragments of pTiT37, a 190-kilobase resident plasmid of *Agrobacterium tumefaciens* strain A208, have been assigned numbers with respect to relative size following electrophoretic separation on a 0.7% agarose gel. The T-DNA is found in fragments 3a, 6, 9, and 23a. (B) Restriction map of the T-DNA of pTiT37. Data supporting the relative positioning of restriction sites for the T region of pTiT37 are published elsewhere.[34] The T-DNA extends from *Eco* fragment 29 on the left into fragment *Bam*3a on the right. Once such a restriction map has been generated for Ti or Ri plasmids, and a transcriptional[55] and functional[21] map of the T region is available, a suitable target site for gene insertion may be selected. The *Sma*I restriction site in *Bam* fragment 14a is an example of a potential target site.

plasmid would be even larger. Transformation of *E. coli* by such giant plasmids is also extremely inefficient.

To circumvent all these difficulties, one can use an indirect method for spite-specific insertion into the Ti plasmid (see Figs. 2 and 3).

1. A preselected target fragment (part of T-DNA containing the target restriction site) is subcloned from the Ti plasmid into a standard *E. coli* cloning vehicle.

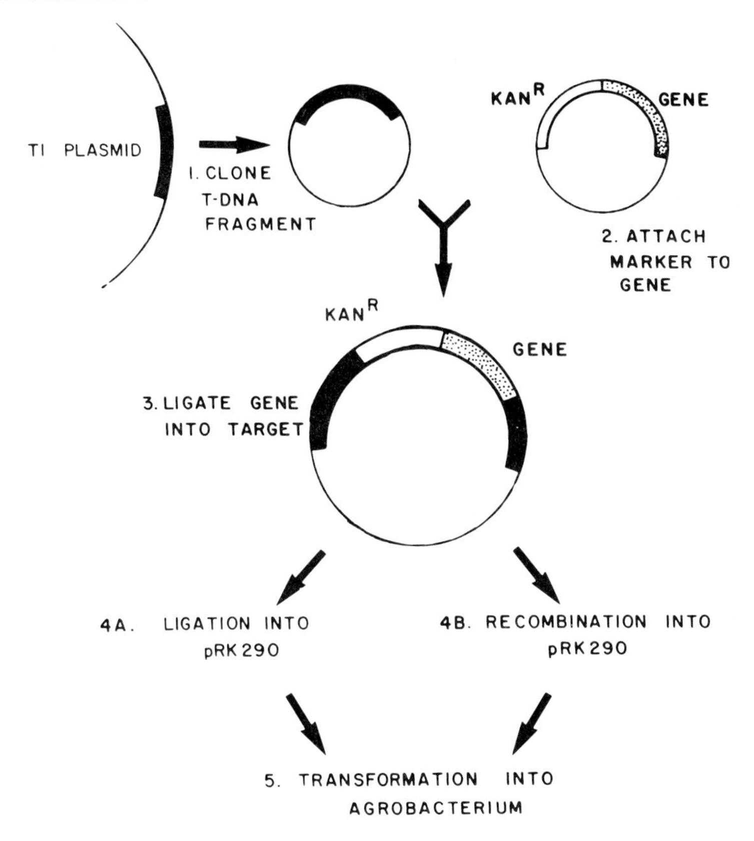

FIG. 2. Manipulations required for use of Ti plasmids as plant engineering vectors.[34] 1, T-DNA surrounding a preselected target restriction site is subcloned into a plasmid vehicle capable of replication in *E. coli;* 2, the eukaryotic gene to be transferred into plants is covalently associated with a bacterial selective marker such as an antibiotic resistance element; 3, the eukaryotic gene with attached marker is ligated into the subcloned target site; 4, the engineered target fragment, until now used only on vectors limited to replication in *E. coli* (for the purpose of biological containment), is cointegrated with a wide host range plasmid; 5, the engineered subclone is then transformed into *Agrobacterium*.

2. The gene to be transferred into plants is covalently attached to a selectable antibiotic resistance marker.
3. The new gene plus the marker can now be introduced into the target site on the T-DNA subclone by routine recombinant DNA techniques, creating an "engineered" T-DNA fragment.
4. The engineered T-DNA fragment is either subcloned or cointegrated into a wide host range plasmid to allow its introduction into *Agrobacterium*.

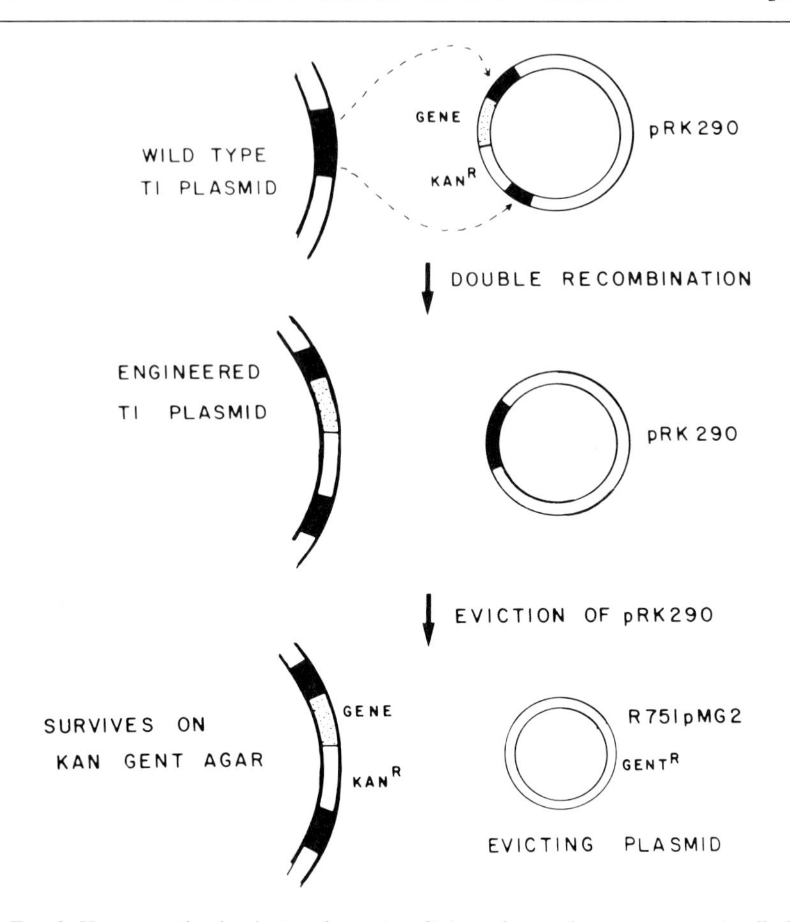

FIG. 3. Homogenotization in *Agrobacterium*.[34] A marker exchange occurs naturally between a wide host plasmid (pRK290) containing an engineered T-DNA fragment and the homologous sequences of DNA located on the resident Ti plasmid of *Agrobacterium*. Selection for the exchange, a result of a double homologous recombination between the two plasmids, is provided using plasmid incompatibility characteristics. A third plasmid (R751.pMG2), which is incompatible with pRK290, is conjugated into *Agrobacterium*, and selection is maintained for the new plasmid as well as for the marker associated with the eukaryotic gene. The desired marker exchange onto the Ti plasmid provides for cell survival, despite eviction of pRK290.

5. Incorporation of the engineered fragment into the Ti plasmid by homologous recombination will occur spontaneously at low frequency. Such recombinants can be recovered by several techniques.

We will describe a plasmid incompatibility selection first used by Ruvkin and Ausubel[37] for similar experiments in *Rhizobium meliloti*.

The Target Fragment

The T-DNA region of octopine-type[38,39] and nopaline-type[40] (Fig. 1B) Ti plasmids has been defined by thorough analysis of many transformed plant cell lines. For the purpose of genetic engineering, a target fragment should be chosen that is reliably and reproducibly found to be a part of T-DNA. The variable boundaries of octopine T-DNA[38,39] make the choice more challenging than is the case for nopaline T-DNA, which is substantially longer and more predictable.[40] Additional factors influence the choice of target fragment and site. First, if insertion interrupts a T-DNA gene, the engineered Ti plasmid could have altered biological properties.[20,21] It could in principle become completely avirulent, although only two examples of this sort have been reported, both involving insertion of RP4, a large block of DNA.[41] Altered virulence properties may be an advantage, presuming that the ultimate goal of the genetic engineer is regeneration of healthy plants from the engineered transformed cells. Second, it may be desirable to insert the novel DNA element in a region of T-DNA that is extensively transcribed. In this way one might hope to exploit promoters on T-DNA to bring about extensive transcription of the inserted DNA. Likely target sites for this purpose would be within the nopaline or octopine synthase genes, which have now been mapped.[18,19,21,25,33,41] Elimination of octopine or nopaline synthase by insertion mutations does not exert a significant effect on virulence.[21,33,41] Third, a final important consideration in choice of a target fragment is the restriction endonuclease cleavage sites found within it. For the strategy presented here, a fragment free from either *Eco*RI, *Bam*HI, or *Bgl*II sites is most convenient. In addition, the target fragment should possess a centrally located unique restriction endonuclease cleavage site, one not present on its cloning vector. If the most convenient central site on the target fragment is one shared by common cloning vectors (*Eco*RI, *Hin*dIII, *Bam*HI), the

[37] G. B. Ruvkin and F. Ausubel, *Nature* (*London*) **289,** 85 (1981).

[38] M. F. Thomashow, R. Nutter, A. L. Montoya, M. P. Gordon, and E. W. Nester, *Cell* **19,** 729 (1980).

[39] M. De Beuckeleer, M. Lemmers, G. De Vos, L. Willmitzer, M. Van Montagu, and J. Schell, *Mol. Gen. Genet.* **183,** 283 (1981).

[40] M. Lemmers, M. De Beuckeleer, M. Holsters, P. Zambryski, J. P. Hernalsteens, M. Van Montagu, and J. Schell, *J. Mol. Biol.* **144,** 353 (1980).

[41] M. Holsters, B. Silva, F. Van Vliet, C. Genetello, M. De Block, P. Dhaese, A. Depicker, D. Inze, G. Engler, R. Villarroel, M. Van Montagu, and J. Schell, *Plasmid* **3,** 212 (1980).

corresponding site in the vector should be annihilated in advance to allow construction of a T-DNA clone with a unique site. Techniques for such constructions have been described in this series.[42] The T-DNA fragment should be as large as possible and must have at least one kilobase of T-DNA flanking each side of the foreign gene insert for a useful frequency of double crossing over,[21] the genetic exchange that will place the engineered trait on the Ti plasmid.

The New Plant Gene and Bacterial Selective Marker

We will assume that the reader has chosen a trait to be engineered that is likely to be expressed in a higher plant cell nucleus. Because of the significant differences in the structural elements of eukaryotic and prokaryotic genes, we will not assume expression of the engineered trait in *A. tumefaciens* or *E. coli*. In order to follow the fate of the eukaryotic gene through the construction of a series of bacterial plasmids, we therefore attach to it a selectable trait that expresses in the bacterial host. Few antibiotic resistances have proved to be useful for stringent selection in *Agrobacterium*, which is resistant to a wide range of antibiotics. Kanamycin resistance, conferred by the neomycin phosphotransferase gene from transposon Tn*5*, is one trait that has been useful[34] in the selection of rare recombinants in *Agrobacterium*. A variety of DNA fragments carrying kanamycin resistance can be excised from Tn*5* using different restriction endonucleases.[43] It is convenient to use the plasmid ColEl::Tn*5* as a source of such fragments[34] because it contains no other interfering drug resistances (such as ampicillin, tetracycline, or chloramphenicol resistance) that could give rise to undesired transformants in the subsequent cloning steps.

For attachment of the selectable marker to the eukaryotic gene, a plasmid containing the latter is opened with a restriction enzyme at a unique site located outside any functional regions of the gene (5′ to the putative promoter region or 3′ to the polyadenylation site). A DNA fragment containing kanamycin resistance is then ligated into this site to yield a recombinant plasmid for which one can exert direct positive selection: it will encode both kanamycin resistance and the drug resistances of the original cloning vehicle. It is important that this step be designed in a way that will be compatible with subsequent recombinant DNA steps. The eukaryotic gene and kanamycin resistance marker must next be ligated into the target site subclone at its unique central cleavage site. While it is possible to carry out such ligations with nonhomologous restriction endonuclease

[42] This series, Vol. 68.

[43] S. J. Rothstein, R. A. Jorgensen, K. Postle, and W. S. Reznikoff, *Cell* **19**, 795 (1980).

sites by using blunt-ended DNA or synthetic linkers, it is most convenient to use homologous sticky-end ligations wherever possible.

Insertion of the Foreign DNA into the Target Site

Once the donor foreign gene has been genetically linked to an antibiotic resistance element, the two can be moved together with positive selection into the subcloned T-DNA target site. The plasmid containing the foreign gene/kanamycin resistance construct is digested with appropriate restriction enzymes to produce a DNA fragment containing the two traits. The T-DNA subclone is opened enzymically at the unique central site, and the foreign gene–kanamycin resistance construct is ligated into position. The reaction is most convenient if compatible restriction sites have been chosen, although blunt ends or synthetic linkers can also be used. Direct selection for the desired recombinant plasmid is achieved by using kanamycin resistance and a second antibiotic resistance found on the T-DNA subclone. The resulting plasmid contains an "engineered" T-DNA fragment with the foreign gene located at the chosen site of the subcloned T-DNA.

Host-Range Extension

Cloning vehicles derived from the plasmid ColEI, such as pBR325, do not replicate in *Agrobacterium*. Although a variety of wide host range plasmids are available, their generally large size, limited antibiotic resistances, or low copy number make them less desirable than the ColEI derivatives for the above restrictions and ligations. Where biological containment is a concern, ColEI vectors provide an additional advantage over conjugative wide host vectors. However, to obtain replication of the engineered plasmid in *Agrobacterium*, it is necessary to ligate or cointegrate the plasmid into a wide host vehicle.

Recloning into pRK290, a Wide Host Range Plasmid. It is possible to excise the engineered T-DNA fragment and ligate it into a unique site in a wide host range plasmid, such as pRK290,[44] which replicates in *Agrobacterium*[21,34] and *Rhizobium*.[37] Unique *Eco*RI and *Bgl*II sites in pRK290 will accept engineered T-DNA fragments that have compatible sticky ends. It is useful to remember that *Bgl*II, *Bam*HI, and *Bcl*I are compatible in ligations. We note also that it is possible to insert the entire engineered T-DNA fragment with its cloning vector into pRK290 by simply linearizing the plasmid at a unique site in the vector.[34] The strategy is simple for en-

[44] G. Ditta, S. Stanfield, D. Corbin, and D. Helinski, *Proc. Natl. Acad. Sci. U.S.A.* **77**, 7347 (1980).

gineered fragments having no *Eco*RI sites, since the *Kan*^{r} gene from Tn*5* possesses no *Eco*RI sites.[43] The single *Bam*HI site in Tn*5* must be eliminated at an earlier stage if *Bam*HI recloning is to be employed in this step. We have found that blunt-end ligations into pRK290 are not practical; however synthetic linkers may be of use for this reaction.[45] Wide host vehicles other than pRK290 offer versatility in restriction sites and antibiotic resistances. However, variations have been observed in the incompatibility characteristics of these plasmids, so the subsequent eviction process could provide additional complications.

A Wide Host Range Shuttle Vector. While it is possible to ligate the engineered T-DNA fragment into a wide host vehicle, careful planning in earlier reactions is required to provide the necessary unique restriction sites to open the engineered plasmid for the reaction. Further, the large sizes of the engineered subclone and the wide host plasmid make such a reaction quite inefficient. Undoubtedly, as genes to be engineered become more complex, difficulties in finding the appropriate unique sites for this final reaction will become greater. To circumvent this problem, we have designed a "shuttle" mechanism that allows host range extension of ColEI-derived plasmids without restrictions or ligations.[35]

A tetracycline-resistant "shuttle" plasmid, pLG17, was constructed by ligating a small fragment of pBR325 into the wide host plasmid pRK290. This plasmid resides in a polA1 mutant of *E. coli*, DS989.[46] The DNA polymerase I activity of this polA1 mutant is present at low level at 30°, but is absent when cells are incubated at 43°. We exploit the fact that replication from the ColEI origin requires DNA polymerase I functions, while that from the pRK290 origin relies on the chromosomal polymerase. Extension of the host range of ColEI derivative plasmids, such as the engineered T-DNA subclone, is accomplished in two stages. First, we transform the T-DNA clone into DS989 containing pLG17 at the permissive temperature, selecting for kanamycin resistance. Next, the transformed cells are incubated at 43° in the presence of kanamycin, to select for bacteria containing cointegrates of the T-DNA subclone with the shuttle plasmid. The cointegration occurs by recombination between the homologous regions of the two plasmids. Cointegrates formed by this method appear to be quite stable and can be isolated by either "minipreps"[34] or through CsCl gradients for transformation into *Agrobacterium*.

Transformation into Agrobacterium

The engineered T-DNA fragment, now attached to a wide host range vector by one of the above procedures, is next introduced by transforma-

[45] R. J. Rothstein, L. F. Lau, C. P. Bahl, S. A. Narang, and R. Wu, this series, Vol. 68, p. 98.

[46] W. Tacon and D. Sherratt, *Mol. Gen. Genet.* **147,** 331 (1976).

tion into *Agrobacterium tumefaciens* containing an intact Ti plasmid. Because the efficiency of transformation is very low, recombinant plasmids should always be amplified by a step of growth in *E. coli* before attempting to transform *Agrobacterium*. Transformation of *Agrobacterium* with plasmid DNA is carried out as described by Holsters *et al.*,[36] using either highly purified plasmid or miniprep DNA. Begin the transformation procedure with a 2-ml inoculum grown for 24 hr at 30° in YEP (1% yeast extract, 1% peptone, and 0.5% NaCl). Use this culture to inoculate 100 ml of YEP, which is grown with shaking for 4–6 hr at 30°. Harvest the bacteria by centrifugation, and wash once with 10 m*M* NaCl. Resuspend the pellet in 1 ml of YEP. Combine 0.2 ml of bacterial suspension with 1 μg or more of plasmid DNA and freeze immediately in a Dry Ice–ethanol bath. Thaw after 5 min, and incubate at 37° for 25 min. Add 0.5 ml of prewarmed YEP, and continue incubation for an additional 1–2 hr. Plate 0.1 ml of bacteria on nutrient agar (Difco) containing appropriate antibiotics and incubate at 30°. Various *Agrobacterium* strains should be individually assayed for drug sensitivity prior to initiation of engineering experiments to ensure the feasibility of the host–marker combination. We now routinely transform strain A208 (host of pTiT37) with selection for resistance to 100 μg of kanamycin per milliliter with colonies appearing after 48 hr of incubation. After the transformation process, confirm the presence of both the original Ti plasmid and the new transforming plasmid in the bacterium by analyses on gels following a rapid alkaline lysis.[34]

Homogenotization

Transformation of *Agrobacterium* strain A208 with an engineered subclone will result in bacterial cells containing a single resident Ti plasmid (pTiT37) and a multicopy wide host plasmid, with regions of DNA homology between the two types of plasmids providing sites for recombination (see Fig. 3). In a population of such bacteria, cells will exist in which a single recombination has occurred between the plasmids, creating a large cointegrate. A subsequent recombination can result in exchanges between the two plasmids. Selection for these events can be obtained using plasmid incompatibilities.[37]

Plasmids of the same incompatibility group coexist in a bacterial cell only if selection is maintained for both plasmids. Eviction of the engineered wide host plasmid is accomplished by conjugation of A208 (containing the engineered subclone) with *E. coli* strain HB101 containing the conjugative gentamycin-resistant plasmid R751.pMG2.[34] This plasmid is of the same incompatibility group (Inc PI) as pRK290 and pLG17 and can therefore cause the eviction of the original engineered plasmid. Conjugation is carried out by preparing fresh 2-ml overnight cultures of the strains in L broth (1% tryptone, 0.5% yeast extract, and 1% NaCl) at 30° for A208 and 37° for HB101. Dilute the two cultures 1:10 with fresh L broth, and incubate for 2 hr at the original tempertures. Combine 0.2 ml of each culture, and add 0.2 ml of fresh L broth. Incubate the mixture at 30° without agitation for 3–5 hr. Plate 10 μl per plate on nutrient agar, selecting for resistance to kanamycin (50–100 μg/ml) and gentamycin (50 μg/ml). After 72 hr, up to 100 colonies per plate will appear. Select single colonies and purify by three successive clonings on selective plates. the homogenotization can yield undesired recombinants; it is therefore essential to confirm the Ti plasmid structure in the putative homogeneote. Because of the multiplicity of bands in a Ti plasmid restriction pattern (Fig. 1), it is best to confirm the plasmid structure by analysis of Southern blot hybridizations, using either rapid lysis DNA or purified plasmids. Southern analysis should show the absence of the wild-type fragment and presence of the engineered fragment[34]; pRK290 or pKB71 should not be detected.

Inoculation of Plant Tissue

Although it is also possible to inoculate intact plants using live bacteria, inoculation of excised stem segments of tobacco may be conveniently carried out *in vitro*,[47] where a high level of biological containment is provided. For stem segment inoculation of tobacco, plants should have 6–10 leaf nodes at time of harvest. After cutting the stem near soil level, trim off leaves and disinfect the stem for 15 min in 1% sodium hypochlorite containing a dilute surfactant (0.1% SDS). Rinse the tissue thoroughly by transferring through several changes of sterile distilled water. Cut the stem into sections, each containing one leaf node, and place basal end up in petri plates on Murashige and Skoog medium[48] (without sucrose or phytohormones) solidified with 0.9% agar. Using a sterile syringe needle, puncture the upper surface of the segment repeatedly until sap flows freely and inoculate with *Agrobacterium* using a sterile toothpick loaded with bacteria. Seal the plate with Parafilm, and incubate at 25° until a crown of callus appears at the top of the stem segment (about 2 weeks). Excise the growth, and transfer to a fresh petri plate containing the same medium plus 200 μg of carbenicillin per milliliter. The crown gall cells will continue to proliferate and will be cured of inciting bacteria by the presence of the antibiotic. After two or three passages, the antibiotic may safely be omitted from the medium.

[47] A. C. Braun, *Cancer Res.* **16,** (1956).

[48] T. Murashige and F. Skoog, *Physiol. Plant.* **15,** 473 (1962).

Regeneration of Transformed Plants

Although there may be practical applications for growing transformed plant cells in culture to produce various metabolites, the ultimate goal of most engineering experiments as described above will be regeneration of modified but healthy plants. Applications of this approach are severely limited by the science of plant tissue culture; regeneration of most agronomically important crops, even from unaltered tissue cultures, has not been accomplished. The regeneration of plants from neoplastic crown gall cultures promises to be more complex.[49,50] However, it is now feasible to alter regions of the T-DNA responsible for the tumorous phenotype without affecting the ability of the remaining T-DNA to integrate into plant nuclear DNA.[21,51,52]

Site-specific insertions into the Ti plasmid can be exploited to determine the functional map of the T-DNA.[21,25] This approach is especially powerful after T-DNA transcripts in tumor cells have been physically mapped.[53-55] Thus far there appear to be no genes on the T-DNA required for integration, although T-DNA genes do affect tumor morphology, host range, and regenerative capability.[20,21,36,52] An adaptation of the site-specific insertion technique described here can be used to delete any undesired genes within T-DNA. In this way one could construct a vector that integrates, but does not initiate oncogenic growth.[52]

The Ri (root-inducing) plasmid of *Agrobacterium rhizogenes* may provide a promising alternative to the Ti plasmid as a plant genetic engineering vector. There is evidence that *A. rhizogenes* infection of plant cells results from transfer of Ri plasmid T-DNA to the plant cell in a manner similar to that of T-DNA transfer during crown gall infection.[29] However, cells transformed by *A. rhizogenes* proliferate in culture as roots rather than a neoplasm. Regeneration of plants from such an organ culture may be possible with plant species where regeneration from callus or crown gall cells has not yet been successful. Thus, as further experience is gained in use of the Ti and related Ri plasmids as genetic engineering vectors, the problem of regeneration of plants may be surmounted.

[49] A. Binns, H. N. Wood, and A. C. Braun, *Differentiation* **11,** 175 (1978).

[50] G. J. Wullems, L. Molendijk, G. Ooms, and R. Schilperoort, *Cell* **24,** 719 (1981).

[51] L. Otten, H. De Greve, J. P. Hernalsteens, M. Van Montagu, O. Schieder, J. Straub, and J. Schell, *Mol. Gen. Genet.* **183,** 209 (1981).

[52] K. A. Barton, A. Binns, A. J. M. Matzke, and M.-D. Chilton, *Cell,* in press.

[53] S. B. Gelvin, M. F. Thomashow, J. C. McPherson, M. P. Gordon, and E. W. Nester, *Proc. Natl. Acad. Sci. U.S.A.* **79,** 76 (1982).

[54] L. Willmitzer, G. Simons, and J. Schell, *EMBO J.* **1,** 139 (1982).

[55] M. W. Bevan and M. D. Chilton, *J. Mol. Appl. Genet.* **1,** 539 (1982).

[34] Isolation of Bacterial and Bacteriophage RNA Polymerases and Their Use in Synthesis of RNA *in Vitro*

By M. CHAMBERLIN, R. KINGSTON, M. GILMAN, J. WIGGS, and A. DEVERA

Bacterial RNA polymerases have been the subject of biochemical interest since their initial isolation, both because of their central role in gene expression and because of the complex series of reactions they carry out. Purified RNA polymerase holoenzymes can display a high degree of specificity *in vitro,* transcribing selectively from promoter sites on DNA templates and responding efficiently to transcriptional termination sites also encoded in the DNA. Hence they offer the possibility of assembling well-characterized systems with which factors that affect gene expression can be studied *in vitro*.

Certain very simple phage RNA polymerases also display this ability to use distinct promoter and terminator sites efficiently on DNA templates *in vitro*. The best-studied examples are the RNA polymerases specified by phages T7, T3, and SP6, which are single subunit proteins of molecular weight about 100,000. Because of their simple structure and high degree of specificity, these enzymes offer a system in which the exact molecular interactions that control protein–nucleic acid interactions at promoter sites can be probed.

Both bacterial and bacteriophage RNA polymerases are also useful in the synthesis and manipulation of genetic sequences *in vitro*. Under appropriate conditions, purified bacterial RNA polymerases can give rise to discrete RNA species that may be of value in studies of *in vitro* RNA splicing, translation, or for use as nucleotide sequence probes or in cloning. The small bacteriophage RNA polymerases may prove particularly useful as tools for preparation of such RNAs since they are highly specific for recognition of specific promoter and terminator sites in DNA, and since they synthesize RNA at an extremely rapid rate *in vitro* for extended periods.

A number of reviews have considered the structure and physical properties of bacterial and bacteriophage RNA polymerases in some detail[1–4] as well as their interaction with DNA sequences involved in the initiation

[1] R. R. Burgess, *in* "RNA Polymerase" (R. Losick and M. Chamberlin, eds.), p. 69. Cold Spring Harbor Laboratory, Cold Spring Harbor, New York, 1976.

[2] M. Chamberlin, *in* "The Enzymes" (P. D. Boyer, ed.), 3rd ed., Vol. 15, p. 61. Academic Press, New York, 1982.

METHODS IN ENZYMOLOGY, VOL. 101

ISBN 0-12-182001-7

and termination of transcription,[5-11] and their role in regulation of gene expression.[11-13] In addition, monographs on RNA polymerases[14,15] and on promoter structure and function[16] have appeared. We focus here on the preparation of these enzymes, the measurement of their activities and specificity, and their application in synthesis of discrete RNA transcripts. In addition to reviewing many of the current methods in general use in this area, we have provided detailed procedures for certain methods in use in our laboratory that we have found to be of value in the preparation or assay of different RNA polymerases.

Isolation of Bacterial RNA Polymerases

Normal and Small-Scale Preparations of RNA Polymerase

The method of preparation is important for bacterial RNA polymerases since the subunit content, specificity, purity, degree of contamination by inhibitors and the fraction of RNA polymerase protein that is active, can all vary with different methods of isolation. Many different fractionation procedures have been employed and are reviewed elsewhere.[1,2] Probably the best current procedure is based on the precipitation of the enzyme with polyethyleneimine (polymin P). The method was introduced by Zillig[17] and has been modified by Burgess and Jendrisak[18]

[3] M. Chamberlin and T. Ryan, *in* "The Enzymes" (P. D. Boyer, ed.), 3rd ed., Vol. 15, p. 87, Academic Press, New York, 1982.

[4] A. Ishihama, *Adv. Biophys.* **14,** 1 (1980).

[5] W. Gilbert, *in* "RNA Polymerase" (R. Losick and M. Chamberlin, eds.), p. 193. Cold Spring Harbor Laboratory, Cold Spring Harbor, New York, 1976.

[6] S. Adhya and M. Gottesman, *Annu. Rev. Biochem.* **47,** 967 (1978).

[7] M. Rosenberg and D. Court, *Annu. Rev. Genet.* **13,** 319 (1980).

[8] H. Bujard, *TIBS* **5,** 274 (1980).

[9] U. Siebenlist, R. B. Simpson, and W. Gilbert, *Cell* **20,** 269 (1980).

[10] T. Platt, *Cell* **24,** 10 (1981).

[11] R. Losick and J. Pero, *Cell* **25,** 582 (1981).

[12] R. Doi, *Arch. Biochem. Biophys.* **214,** 772 (1982).

[13] C. Yanofsky, *Nature (London)* **289,** 751 (1981).

[14] R. Losick and M. Chamberlin (eds.), "RNA Polymerase." Cold Spring Harbor Laboratory, Cold Spring Harbor, New York, 1976.

[15] S. Osawa, H. Ozeki, H. Uchida, and T. Yura (eds.), "Genetics and Evolution of RNA Polymerase tRNA and Ribosomes." Univ. of Tokyo Press, Tokyo, 1980.

[16] R. Rodriguez and M. Chamberlin (eds.), "Promoters, Structure and Function." Praeger, New York, 1982.

[17] W. Zillig, K. Zechel, and H. Halbwachs, *Hoppe-Seyler's Z. Physiol. Chem.* **351,** 221 (1970).

[18] R. R. Burgess and J. J. Jendrisak, *Biochemistry* **14,** 4634 (1957).

for use with the *Escherichia coli* RNA polymerase on a moderately large scale (0.5 kg of cells). The procedure is rapid and reproducible, and the total recovery of *E. coli* RNA polymerase based on analysis of the $\beta' + \beta$ content of cell extracts and final fractions is good. The final fractions contain the polymerase subunits β', β, α, and σ together with the ω polypeptide, as the only major polypeptide components. These fractions are suitable for many applications and are generally free from nuclease contamination. However the sigma content of such preparations is often low (0.3–0.5 equivalent), and this may be a consideration for some applications as discussed below.

The use of polymin P also has proved to be very effective in obtaining highly purified RNA polymerases from a large number of other bacterial species including *Mycobacterium, Rhodospirillum, Salmonella, Bacillus,* and *Clostridium,* as well as T4-infected *E. coli* and the blue-green alga *Anacystis nidulans.*[19–23] The fact that these different RNA polymerases behave similarly in certain fractionation steps probably reflects their common subunit structure, large size, and DNA binding properties. It should be cautioned, however, that the actual solution conditions needed to stabilize the different kinds of bacterial RNA polymerases can vary quite substantially.[1,19,22]

Certain specific fractionation steps have also been valuable in the fractionation of different bacterial RNA polymerases, for removal of particular contaminants, or for separating different forms of the enzyme. These include chromatography on DNA cellulose or DNA agarose[1,24] and on Blue Dextran columns.[25] Noteworthy here is heparin–agarose column chromatography[26,27] which can give a rapid efficient purification of the normal *Bacillus subtilis* and *E. coli* RNA polymerases,[27] as well as the *B. subtilis* σ^{28} polymerase.[28] The exact method of preparation of the heparin–agarose appears to be quite critical.[27]

For rapid preparation of small amounts of *E. coli* RNA polymerase from many bacterial strains, a modified form of the polymin procedure

[19] J. Wiggs, J. Bush, and M. Chamberlin, *Cell* **16,** 97 (1979).

[20] M. Schachner and W. Seifert, *Hoppe-Seyler's Z. Physiol. Chem.* **352,** 734 (1971).

[21] F. Herzfeld and W. Zillig, *Eur. J. Biochem.* **24,** 242 (1971).

[22] K. Stetter and W. Zillig, *Eur. J. Biochem.* **48,** 527 (1974).

[23] C. Murray and J. C. Rabinowitz, *J. Biol. Chem.* **256,** 5153 (1981).

[24] B. Alberts and G. Herrick, this series, Vol. 21, p. 198.

[25] S. Halling, F. Sanchez-Anzaldo, F. Fukuda, R. Doi, and C. F. Meares, *Biochemistry* **16,** 2880 (1977).

[26] H. Sternbach, R. Englehardt, and A. G. Lezius, *Eur. J. Biochem.* **60,** 51 (1975).

[27] B. Davison, T. Leighton, and J. C. Rabinowitz, *J. Biol. Chem.* **254,** 9220 (1979).

[28] J. L. Wiggs, M. Z. Gilman, and M. J. Chamberlin, *Proc. Natl. Acad. Sci. U.S.A.* **78,** 2762 (1981).

has been suggested.[29] We have devised an alternative procedure that employs stepwise elution of the enzyme from a heparin–agarose column after adsorption directly from a cell extract. The method is applicable to small amounts of cells (1–2 g wet weight), and has been used successfully in our laboratory with *E. coli*, T4-infected *E. coli*, *B. subtilis*, and other gram-positive species. The procedure has also been used successfully elsewhere with *Caulobacter* (K. Amemiya and L. Shapiro, personal communication). Because it is very rapid and involves a minimal amount of manipulation, it is attractive as a method for isolation of mutant RNA polymerases that may be quite labile.

Heparin–Agarose Procedure for Small-Scale Isolation of *E. coli* RNA Polymerase

Cells may be freshly grown or frozen as cell paste. Volumes given are for 1 g wet weight of cells; these should be adjusted proportionately for larger or smaller amounts. Lysis buffer contains 10 m*M* Tris-HCl, pH 8, 2 m*M* EDTA, 0.3 m*M* dithiothreitol, 7.5% glycerol, 23 μg of phenylmethylsulfonyl fluoride (PMSF) per milliliter, and 150 μg of lysozyme (Sigma, Grade I) per milliliter.

Cells are suspended in 3 ml of freshly prepared lysis buffer. The suspension is blended for 3 min at low speed in a small Waring blender cup or VirTis mixer and allowed to stand for 30 min at 4–6°. Sodium deoxycholate (50 μl of a 4% w/v solution) is added to give a final concentration of 0.05%. The suspension is blended for 30 sec at low speed and then is allowed to stand for 30 min at 8–12°. The lysate at this point becomes extremely viscous owing to release of cellular DNA. The DNA is then sheared by two 30-sec treatments at high speed in the blender. Two milliliters of lysis buffer without lysozyme are added, and the diluted lysate is blended at high speed for two more 30-sec periods. It is important to keep the lysate cold (below 12°) by chilling the blender or container in an ice bath between and after periods of high speed mixing.

The sheared lysate is centrifuged for 40 min at 10,000 rpm in a Sorvall SS34 rotor in a refrigerated centrifuge. A slightly turbid, amber supernatant fluid (4–5 ml) is recovered. Concentrated NaCl (4 *M*) is added to the extract to give a final concentration of 0.6 *M* NaCl, and the extract is kept on ice for 15–30 min while the heparin–agarose is prepared.

Heparin–agarose, prepared and stored in a 50% suspension as described by Davison *et al.*[27] is washed and resuspended four times in buffer I (10 m*M* Tris pH 8, 10 m*M* $MgCl_2$, 1 m*M* EDTA, 0.3 m*M* dithiothreitol, 7.5% glycerol). This is conveniently accomplished in 12-ml conical centri-

[29] C. Gross, F. Engbaek, T. Flammang, and R. Burgess, *J. Bacteriol.* **128**, 382 (1976).

fuge tubes in a clinical centrifuge; 5 sec at top speed is sufficient to sediment the heparin–agarose. After the final wash, the heparin–agarose is resuspended to twice its packed volume in buffer I, and a volume exactly twice that of the high-salt cell extract is transferred to a clean centrifuge tube. The heparin–agarose is packed again by centrifugation, and the excess buffer is removed. The high-salt extract is added to the packed heparin–agarose, and the gel is rapidly resuspended. Note that, by mixing equal volumes of heparin–agarose equilibrated in buffer I and crude extract of 0.6 *M* NaCl, the final NaCl concentration obtained is 0.3 *M*.

The heparin-agarose suspension is agitated gently in the cold for 60 min to allow RNA polymerase binding, and the suspension is then poured into a column chosen so that the height-to-diameter ratio of the settled bed is near 4:1. The suspension is allowed to stand for several minutes with the column outlet closed so that air bubbles may rise out, then the heparin–agarose is packed with the column flowing. The column is washed with 10–20 volumes of buffer I + 0.3 *M* NaCl. Flow rates for this step may be as high as one-third column volume per minute. RNA polymerase is eluted with 3–4 column volumes of buffer I + 0.6 *M* NaCl. The activity peak is generally recovered in one column volume or less.

Heparin-agarose-purified fractions of normal *E. coli* RNA polymerase are stable for 3–4 days at 0°. For long-term storage, active fractions are dialyzed into buffer containing 10 m*M* Tris, pH 8, 10 m*M* $MgCl_2$, 0.1 m*M* EDTA, 0.1 m*M* dithiothreitol, 100 m*M* NaCl, and 50% glycerol for storage at −20° or −80°.

We have also prepared cell extracts using sonication or by alumina grinding of frozen cells. Sonication gave satisfactory results, with yields and specific activities of RNA polymerase that were only slightly lower than by the blender procedure. However the exact length of time employed for sonication appeared to be important, and lower activities were obtained with longer or shorter periods of sonication. Recoveries with alumina grinding were lower, perhaps because of losses of RNA polymerase on rapidly sedimenting DNA components. We would also expect that cell extracts prepared with the French pressure cell would be suitable for this procedure, although we have not tested such extracts with *E. coli*.

The entire purification including cell lysis requires 5–6 hr. Between 0.3 and 0.5 mg of RNA polymerase is recovered from 1 g of cells. Quantitative RNA polymerase assays using T7 DNA (see below) show that heparin–agarose enzyme fractions have specific activities in the range of 50–100 mU/mg total protein (1 mU of enzyme incorporates 1 nmol of $[^{32}P]$CMP per minute under standard conditions) corresponding to 10–20% active RNA polymerase molecules. Densitometric scanning sodium dodecyl sulfate (SDS)–polyacrylamide gels indicates that these fractions are roughly 30% RNA polymerase subunits by protein, and that RNA polymerase core enzyme is approximately 30% saturated with sigma subunit. The overall recovery of RNA polymerase activity is about 150% of that estimated in cell extracts.

Heparin–agarose-purified RNA polymerase gives transcriptional parameters that are the same as those of highly purified holoenzyme in quantitative assays.[30] These fractions are also free of detectable ribonuclease activity and hence are suitable for use in transcriptional analysis using RNA gels.

Heparin-Agarose Procedure for Small-Scale Isolation of RNA Polymerase from Other Bacteria

The procedure described above has also been used successfully, with minor modifications, to isolate RNA polymerases from *B. subtilis* grown under a variety of conditions, and from *B. stearothermophilus, Lactobacillus casei, L. plantarum,* and *Clostridium pasteurianum*.[31] The initial adsorption and elution with 0.3 *M* NaCl were eliminated for these preparations since, in the case of *B. subtilis,* this leads to loss of minor forms of RNA polymerase with different sigma factors. One example is *B. subtilis* σ^{28} RNA polymerase, which elutes from heparin–agarose at a lower salt concentration than the normal RNA polymerase.[28]

In general, frozen cells (2 g) are suspended in 6 ml of homogenization buffer (10 m*M* Tris-HCl pH 8.0, 10 m*M* $MgCl_2$, 1 m*M* EDTA, 0.3 m*M* dithiothreitol, 7.5% (v/v) glycerol, 50 m*M* NaCl, and 0.5 m*M* PMSF) by pestling the mixture at 0°. The cell suspension is lysed by three passes through a French pressure cell at 10,000–12,000 psi. The cell lysate is then centrifuged at 10,000 rpm in the Sorvall SS34 rotor for 40 min.

A portion (0.25–0.5 ml) of the supernatant fluid containing 10–20 mg of total protein is applied to a 1.5-ml heparin–agarose column previously equilibrated with buffer containing 10 m*M* Tris-HCl pH 8.0, 10 m*M* $MgCl_2$, 1 m*M* EDTA, and 0.3 m*M* dithiothreitol. This step can be scaled up to use the entire cell extract fraction by use of a larger heparin–agarose column. After adsorption of the sample for 60 min at 4°, the column is eluted with 5 ml of 10 m*M* Tris-HCl, pH 8.0, 10 m*M* $MgCl_2$, 1 m*M* EDTA, 0.3 m*M* dithiothreitol, 7.5% (v/v) glycerol, and 0.05 *M* NaCl, followed by 5 ml of the same buffer containing 0.6 *M* NaCl instead of 0.05 *M* NaCl. A flow rate of approximately 15 ml/hr is maintained. One-milliliter fractions are collected and assayed for protein and RNA polymerase activity using appropriate templates. The peak RNA polymerase fractions

[30] M. Chamberlin, W. Nierman, J. Wiggs, and N. Neff, *J. Biol. Chem.* **254,** 10061 (1979).

[31] J. Wiggs, Ph.D. Thesis, University of California, Berkeley, 1981.

are used immediately or dialyzed overnight against a buffer containing 10 mM Tris-HCl pH 8.0, 10 mM $MgCl_2$, 0.1 mM EDTA, 0.1 mM dithiothreitol, 0.1 M NaCl, and 50% (v/v) glycerol, for storage at −80°.

In general all the bacterial cell extracts we have tested using this procedure show similar chromatographic behavior on heparin–agarose. A large protein peak with little or no RNA polymerase protein or activity can be washed off with 0.05 M NaCl, whereas a second peak containing RNA polymerase activity elutes with 0.6 M NaCl. For each of the species we have tested, the recovery of protein from the column was complete, whereas the recovery of RNA polymerase activity varied from 200% to 2000%, as compared with the cell extract. Since the specific activities of the final peak RNA polymerase fractions were relatively constant when poly[d(A-T)] was used as a template (from 150 to 450 units/mg, where 1 unit is 1 nmol of AMP incorporated per hour[19]), this result suggests that essentially the same amount of total RNA polymerase is recovered from each extract and the variation in the percentage of activity recovered relative to the input is the result of variable amounts of RNA polymerase inhibitors in the different cell extracts. The specific activities of the peak RNA polymerase fractions from different bacterial species vary considerably with the DNA template, as expected from their different promoter recognition properties.[19]

The heparin–agarose procedure generally gives RNA polymerase fractions that are pure enough to use directly for the synthesis and analysis of RNA transcripts. However, since many DNases and RNases can potentially interfere with such assays, it is important to do appropriate control studies where new bacterial strains are used. In one case with *Caulobacter crescentus*, a potent DNA restriction endonuclease elutes together with the RNA polymerase (K. Amemiya and L. Shapiro, personal communication).

Isolation of Holoenzyme and Core RNA Polymerase

Most purification procedures give RNA polymerase fractions that contain both core polymerase and RNA polymerase holoenzyme; that is, sigma is not present in stoichiometric amounts. Furthermore, there are multiple sigma factors[11,28,32–34] in *B. subtilis*, although one species (σ^{55}) predominates. These combine with *B. subtilis* core polymerase to give

[32] M. Z. Gilman, J. L. Wiggs, and M. J. Chamberlin, *in* "Promoters, Structure and Function" (R. Rodriguez and M. Chamberlin, eds.). Praeger, New York, 1982.

[33] W. G. Haldenwang and R. Losick, *Proc. Natl. Acad. Sci. U.S.A.* **77,** 7000 (1980).

[34] W. G. Haldenwang, N. Lang, and R. Losick, *Cell* **23,** 615 (1981).

several distinct kinds of RNA polymerase holoenzymes with different promoter specificities.[11,12,35,36]

Since the transcriptional properties, particularly the promoter specificity, of core RNA polymerase and an RNA polymerase holoenzyme are quite different, it is frequently desirable that holoenzyme be prepared free of core polymerase. Chromatography on single-stranded DNA agarose columns[37] or on phosphocellulose in 50% glycerol[38] separates the *E. coli* enzyme into fractions containing predominantly holoenzyme and core polymerase, respectively, as judged by SDS–gel analysis of the fractions. However, the former method can give holoenzyme fractions containing as little as 0.6 equivalent of σ,[39] and there are detectable amounts (<5%) of an activity having the properties of core polymerase in some holoenzyme preparations prepared by the latter method as well.[40] These small amounts of core polymerase may well be serious contaminants for experiments intended to study the specific properties of an RNA polymerase holoenzyme.

Escherichia coli RNA polymerase holoenzyme can also be reconstituted by adding purified σ subunit[39] to partially σ-saturated holoenzyme or core polymerase preparations.[18] Although reconstitution is rapid, it is not a simple reaction. In the absence of Mg^+ ion an enzyme form is obtained which differs in its properties from normal holoenzyme.[41] Since it cannot be assumed *a priori* that holoenzyme reconstituted *in vitro* is identical to that isolated directly, the possibility of differences should be kept in mind if reconstituted holoenzyme is used.

The release of *E. coli* σ subunit from RNA polymerase fractions upon chromatography on phosphocellulose or BioRex-70 columns provides a useful way of purifying that factor.[1,39] *E. coli* σ is a very stable protein and can also be isolated in active form directly by eluting σ containing bands from SDS–polyacrylamide gels.[42,43]

Escherichia coli core RNA polymerase is usually obtained by phosphocellulose or BioRex-70 chromatography, which releases the σ subunit from holoenzyme.[39] However, there are generally small traces (<1%) of

[35] M. Gilman, J. Wiggs, and M. Chamberlin, *Nucleic Acids Res.* **9,** 5991 (1981).
[36] C. Moran, N. Lang, and R. Losick, *Nucleic Acids Res.* **9,** 5979 (1981).
[37] C. Nusslein and B. Heyden, *Biochem. Biophys. Res. Commun.* **47,** 282 (1972).
[38] N. Gonzalez, J. Wiggs, and M. Chamberlin, *Arch. Biochem. Biophys.* **182,** 404 (1977).
[39] P. Lowe, D. Hager, and R. R. Burgess, *Biochemistry* **18,** 1344 (1979).
[40] J. Kaguni and A. Kornberg, *J. Biol. Chem.* **257,** 5437 (1982).
[41] R. Fisher and T. Blumenthal, *J. Biol. Chem.* **255,** 11056 (1980).
[42] K. Weber and D. Kutter, *J. Biol. Chem.* **246,** 4505 (1971).
[43] D. Hager and R. Burgess, *Anal. Biochem.* **109,** 76 (1980).

σ in these preparations and because σ can recycle during transcription, a major part of the activity of such fractions can be due to holoenzyme when duplex DNA templates are used.[44] This has given rise to several reports that core RNA polymerase can show promoter-specific transcription. We have found that *E. coli* core polymerase can be freed of the last traces of σ by chromatography on poly(rC) cellulose columns, as outlined below. These enzyme fractions show no detectable transcription from the strong T7 A promoters but do transcribe actively from single-strand breaks and ends.

Isolation of Sigma-Free *E. coli* Core Polymerase by Poly(rC)-Cellulose Chromatography

Almost any purified sample of *E. coli* RNA polymerase can be used as starting material. We have used enzyme prepared by the method of Burgess and Jendrisak[18] as well as core polymerase fractions from the phosphocellulose–glycerol chromatography procedure of Gonzalez *et al.*[38] with equally good results. The RNA polymerase sample (2–5 mg of protein) is diluted in buffer A if necessary to give a solution containing less than 10% glycerol and less than 20 m*M* NaCl (buffer A is 10 m*M* Tris, pH 8, 10 m*M* $MgCl_2$, 10 m*M* 2-mercaptoethanol, 1 m*M* EDTA, and 5% glycerol). If input protein concentration will be below 0.2 mg/ml after dilution, it is better to dialyze the starting material into buffer A containing 20 m*M* NaCl.

The diluted enzyme is passed through a 5-ml poly(rC) cellulose column [poly(rC) cellulose contains 8 mg of poly(rC) per gram; it is prepared according to Carmichael[45] and is equilibrated at 4° overnight with buffer A containing 20 m*M* NaCl]. The column is washed with two column volumes of buffer A containing 20 m*M* NaCl and then with 50 column volumes of buffer A containing 200 m*M* NaCl. Core polymerase is eluted with buffer A containing 400 m*M* NaCl.

Peak core polymerase fractions are located by UV absorption and are assayed with d(A-T) copolymer or DNase-treated DNA as template.[46] Fractions are dialyzed into storage buffer (10 m*M* Tris, pH 8, 10 m*M* $MgCl_2$, 100 m*M* NaCl, 0.1 m*M* DTT, 0.1 m*M* EDTA, and 50% glycerol) for storage at −80°. The protein peaks eluting at 20 and 200 m*M* NaCl both contain sigma as well as traces of core subunits. The 400 m*M* eluate should be sigma-free as shown both by its low activity on intact DNA templates[46] and by SDS gel analysis of subunits.

[44] R. Tjian, and D. Stinchcomb, and R. Losick, *J. Biol. Chem.* **250,** 8824 (1975).

[45] G. Carmichael, *J. Biol. Chem.* **250,** 6160 (1975).

[46] D. Hinkle, J. Ring, and M. Chamberlin, *J. Mol. Biol.* **70,** 197 (1972).

For bacterial RNA polymerases other than the *E. coli* enzyme, isolation of holoenzyme, core polymerase, or sigma is often more difficult. For several other bacterial RNA polymerases the predominant σ subunit is not released by chromatography on phosphocellulose,[47,48] and this is true also of several of the minor *B. subtilis* σ subunits.[49,50] This makes isolation of these factors or of core polymerases difficult and there are as yet no general methods available. However, like the *E. coli* sigma subunit,[42,43] the minor *B. subtilis* σ factors can be eluted in active form from SDS–polyacrylamide gels.[28–34] The ease of renaturation of σ factors has also allowed isolation of active σ^{37} from *B. subtilis* RNA polymerase by chromatography of the urea-dissociated enzyme on phosphocellulose.[33]

Some bacterial RNA polymerases are quite unstable and dissociate even during purification procedures that are perfectly suitable for the *E. coli* RNA polymerase.[19,22] This can severely limit the choice of purification steps for a particular RNA polymerase.

Detection and Measurement of RNA Polymerase Activity

Qualitative and Quantitative Assay Procedures

While the homogeneity of a purified RNA polymerase preparation is easily monitored by SDS–gel electrophoresis of the protein, the definition and quantitative determination of enzymic activity is extremely difficult. The presence of active enzyme in a preparation is revealed by its ability to direct DNA-dependent incorporation of labeled nucleotide into acid-precipitable material. In general, such procedures employ nucleoside triphosphates labeled in the nucleoside residue with ^{3}H or ^{14}C or in the α-phosphate with ^{32}P and follow their conversion to an acid-insoluble form (RNA) in a 10- or 20-min incubation. These are useful procedures for following total RNA polymerase activity during purification or where the enzyme is used as a preparative reagent.[1,2]

Because of the complex series of steps that are involved in synthesis of a single RNA chain, the amount of incorporation in such an assay generally bears no simple relationship to the fraction of RNA polymerase molecules which are active in a preparation, nor will it normally distinguish among enzyme fractions which may differ greatly in their specificity. In addition, because of the intricacy of the partial reactions carried

[47] J. Johnson, M. Debacker, and J. Boezi, *J. Biol. Chem.* **246,** 1222 (1971).
[48] K. Amemiya, C. W. Wu, and L. Shapiro, *J. Biol. Chem.* **252,** 4157 (1977).
[49] W. Haldenwag and R. Losick, *Nature (London)* **282,** 256 (1979).
[50] J. Jaehning, J. Wiggs, and M. Chamberlin, *Proc. Natl. Acad. Sci. U.S.A.* **76,** 5470 (1979).

out by RNA polymerase in the transcription cycle, no single parameter can be used to describe the "activity" of an enzyme preparation.[30]

A Quantitative Assay Procedure for *E. coli* RNA Polymerase with Bacteriophage T7 DNA

Background. One approach to the quantitative characterization of bacterial RNA polymerase preparations involves the use of a DNA template bearing a single well-defined transcriptional unit, where the values of different transcriptional parameters can be assayed by following labeled nucleotide incorporation during a single cycle of transcription. The basic features of the procedures have been set forth using bacteriophage T7 DNA and *E. coli* RNA polymerase.[30,51] Here there is a single strong transcriptional unit used during free RNA chain initiation *in vitro*, and the rates of all of the different steps in the transcriptional cycle are known from separate studies. The procedure measures the amount of active *E. coli* RNA polymerase in the preparation together with values of several of the parameters for the different steps of the transcriptional cycle including (*a*) the rate of promoter location plus chain initiation; (*b*) the mean rate of RNA chain elongation; and (*c*) the efficiency of chain termination at the strong early T7 termination signal. We have revised the procedure using ionic conditions that correspond more closely to those reported *in vivo*. These include 4 m*M* Mg^{2+} and 120 m*M* KCl,[52] 4 m*M* spermidine,[53] and increased concentrations of the nucleoside triphosphates,[54,55] including ATP (2.7 m*M*), UTP (1.4 m*M*), GTP (1.1 m*M*), and CTP (0.7 m*M*). These latter concentrations are not unreasonable in view of the high K_s values for chain elongation by *E. coli* RNA polymerase *in vitro*.[56] These conditions give a somewhat improved efficiency of RNA chain elongation (25 nucleotides per second), and termination at the early T7 terminator is increased to about 95%. These conditions also increase the amount of active polymerase measured in some enzyme preparations.

The changed conditions necessitate the following modifications to the previous procedure.[30]

1. Rifampin must be used in place of heparin to block reinitiation of RNA chains. Heparin fails to prevent recycling of *E. coli* RNA po-

[51] S. Rosenberg, T. Kadesch, and M. Chamberlin, *J. Mol. Biol.* **155,** 31 (1982).
[52] J. Lusk, R. Williams, and E. Kennedy, *J. Biol. Chem.* **243,** 2618 (1968).
[53] C. Tabor and H. Tabor, *Annu. Rev. Biochem.* **45,** 285 (1976).
[54] C. Matthews, *J. Biol. Chem.* **247,** 7430 (1972).
[55] I. Villadsen and O. Michelsen, *J. Bacteriol.* **130,** 136 (1977).
[56] R. Kingston, W. Nierman, and M. Chamberlin, *J. Biol. Chem.* **256,** 2787 (1981).

lymerase under the new conditions, possibly owing to complexing with spermidine.

2. The spermidine concentration must never approach 10 m*M* when DNA is present, since it efficiently precipitates DNA at this concentration. Hence DNA should be added last before enzyme.
3. Average elongation rates of 25 nucleotides per second are observed under the new conditions, hence some enzyme molecules reach the T7 terminator after 4.5 min. Time points in the assay have been readjusted to fit this reaction rate.

Reagents. RNA polymerase preparations are stored in 50% glycerol at -20 to $-80°$[38] and, if necessary, are diluted prior to use with a new diluting solution (81 RP diluent) containing 10 m*M* Tris, pH 8, 10 m*M* 2-mercaptoethanol, 10 m*M* KCl, 0.1 m*M* EDTA, 0.4 mg of bovine serum albumin per milliliter, 0.1% Triton, and 5% glycerol. Dilution is carried out by adding enzyme (at least 5 μl) to a clean tube with a chilled pipette and slowly adding an appropriate volume of chilled 81 RP diluent with constant mixing. We see much less variability in repeated experiments with this diluent as compared to that used previously,[38] and the diluted enzyme is up to 20-fold more stable to heat inactivation as well.

Reagent mixtures needed for the standard RNA polymerase assay:

81 solution B: 0.38 ml of distilled water, 0.2 ml of 1 *M* Tris-Cl, pH 8, 0.3 ml of 2 *M* KCl, 0.02 ml of 1 *M* $MgCl_2$, 0.05 ml of 1 *M* 2-mercaptoethanol, and 0.05 ml of 400 m*M* spermidine

Nucleoside triphosphate mix: ATP, UTP, and GTP containing 27 m*M* ATP, 14 m*M* UTP, and 11 m*M* GTP

Carrier solution: 50 m*M* sodium pyrophosphate and 50 m*M* EDTA with 0.5 mg/ml of Sigma yeast RNA

Other special reagents: ATP, which we find must normally be purified to remove an inhibitor that reduces the amount of active *E. coli* RNA polymerase when ATP concentrations over 1.5 m*M* are used. Purification is accomplished by chromatography on DEAE or Dowex-1 (Cl form), and the ATP recovered is subsequently desalted by barium precipitation and restored to the potassium form.[57]

Bovine serum albumin, (BSA) which must be purified free from nuclease in some cases. We purify such preparations of BSA by passage of a 5 mg/ml solution through a BioRex 70 column in 0.01 *M* sodium acetate, pH 5.5, followed by concentration and dialysis against 2 m*M* Tris, pH 8.

Procedure. To carry out a quantitative *E. coli* RNA polymerase assay,

[57] I. R. Lehman, M. Bessman, E. Simms, and A. Kornberg, *J. Biol. Chem.* **233,** 163 (1958).

a reaction mixture is prepared in a tube at 0° containing 0.2 ml of 81 solution B, 0.1 ml of nucleoside triphosphate mix (ATP, UTP, and GTP), 0.1 ml of 7 m*M* [α-^{32}P]CTP (specific activity 5–20 cpm/pmol), 0.1 ml of 2 m*M* T7 DNA, and water to give a final volume of 1.0 ml after addition of enzyme. A sample (100 μl) is removed as a zero time control and is mixed with 200 μl of carrier solution (50 m*M* sodium pyrophosphate and 50 m*M* EDTA with 0.5 mg of Sigma yeast RNA per milliliter). Nucleic acids are then precipitated by adding 2.5 ml of an ice-cold solution of 10% trichloroacetic acid, and the mixture is kept on ice prior to filtration.

A sample of RNA polymerase is added to the reaction mixture (5–20 μl, containing from 0.2 to 10 mU), and the solution is mixed well. This mixture can be kept on ice up to 10 min without affecting the subsequent reaction. RNA synthesis is initiated by transferring the tube to a water bath at 30° (zero time) with about 10 sec of gentle shaking to facilitate thermal equilibration. At 1.5 min, 20 μl of a solution of rifampicin (1 mg/ml) are added to the reaction. For *E. coli* RNA polymerase, samples (100 μl) are taken at 1, 2, 3, 4, 10, 15, 20, and 25 min and are diluted and precipitated as for the control. After standing for at least 10 min in cold trichloroacetic acid, the samples of labeled RNA are collected by filtration onto Whatman GF/C filters (previously soaked with 0.1 *M* sodium pyrophosphate). Each filter is washed with about 35 ml of an ice-cold solution containing 1 *M* HCl and 0.1 *M* sodium pyrophosphate and then with 10 ml of ethanol. After drying, the samples are counted in a toluene-based mix in a scintillation counter or in a thin-window counter.

Calculations. The concentration of active *E. coli* RNA polymerase in the sample is determined from the rate of CMP incorporation in the initial linear phase of the reaction (phase II[30]) given that all active RNA polymerase in the reaction is engaged in RNA chain elongation at an average of 25 nucleotides per second. The activity of a preparation can be expressed either as concentration of active RNA polymerase, or in terms of milliunits per milliliter, where 1 mU is 1 nmol of CMP incorporated per minute when all active RNA polymerase molecules are elongating T7 RNA chains. By the latter convention, homogeneous and fully active *E. coli* RNA polymerase (M_r ~450,000; see Chamberlin[2]) has a maximum specific activity of 825 mU per milligram of protein under the new conditions used here.

Several other important parameters can be determined from such a quantitative assay, including the rate of promoter binding plus RNA chain initiation, the rate of RNA chain elongation, and the efficiency of termination at the T7 phage early terminator; the interested reader is referred elsewhere for more detailed discussions of these calculations.[30,51]

Single-Point Assay. Where many enzyme samples must be assayed, as in following an enzyme purification step, a fixed time point assay gives a more convenient although somewhat less precise measure of the concentration of active RNA polymerase. Reaction solutions are prepared containing 10–40 μl of water, 20 μl of 81 solution B, 10 μl of T7 DNA (2 m*M* in DNA nucleotide), 10 μl of [α-^{32}P]CTP, and 10 μl of nucleoside triphosphate mix (ATP, UTP, GTP). Reaction solutions are kept at 0°. Samples of RNA polymerase (0.02 to 1 mU) are added in 10–30 μl; the volume of enzyme sample plus water is adjusted to give a final reaction volume of 100 μl. RNA synthesis is initiated by transfer of the tubes to a 30° bath and is terminated after 3.5 min by adding 100 μl of a solution containing 50 m*M* sodium pyrophosphate and 50 m*M* EDTA with 0.5 mg of Sigma yeast RNA per milliliter, followed by 2.5 ml of ice-cold 10% trichloroacetic acid. Samples are kept at least 10 min on ice and are then filtered to determine acid-insoluble labeled RNA as for the previous procedure.

For *E. coli* RNA polymerase there is a 30-sec lag in RNA synthesis; hence, the initial rate of RNA synthesis in phase II[30] is calculated for each sample assuming a 3-min total reaction period. This procedure gives only a single determination of the concentration of active RNA polymerase. For other bacterial RNA polymerases or with altered reaction conditions, the lag must be determined empirically using the standard assay procedure, and the total reaction time required may have to be adjusted taking account of the lag and the duration of linear phase II synthesis.

Protein concentrations of *E. coli* RNA polymerase holoenzyme in terms of weight of amino acids are determined by ultraviolet absorption using the extinction coefficient $\epsilon_{1\%}^{280} = 6.5$.[1,2] As noted by Burgess,[1] these values must be corrected for light scattering for some enzyme samples. This correction can be substantial (over 50%) and cannot be ignored. As a general rule this correction is less than 5% if the absorbance at 320 nm is less than 2% of that at 280 nm.[38] It is often more convenient to determine the protein concentration by the Lowry procedure[58] after precipitation of the enzyme with trichloroacetic acid to remove interfering substances such as salt, mercaptans, and glycerol. With BSA as standard, 1 mg/ml of RNA polymerase as determined by Lowry method at 280 nm has a true protein concentration of 0.8 mg/ml. The extinction coefficient of BSA is $\epsilon_{1\%}^{280} = 6.6$.[59]

Quantitative Assay Procedures for Other RNA Polymerases and Using Other Templates

The assumptions on which the assay procedure is based appear to be valid in most cases also for bacterial RNA polymerases from a wide range of taxonomically different bacterial strains.[19] This follows from the obser-

[58] O. Lowry, N. Rosenbrough, A. Farr, and R. Randall, *J. Biol. Chem.* **193,** 265 (1951).
[59] S. Leach and H. Scheraga, *J. Am. Chem. Soc.* **82,** 4790 (1960).

vation that these diverse bacterial RNA polymerases all transcribe T7 DNA selectively and generally employ the same promoter and terminator sites used by the *E. coli* enzyme. RNA polymerases from *Rhodosporillum rubrum, Mycobacterium smegmatis, Caulobacter crescentus, Azotobacter vinelandii,* and *Bacillus subtilis* give single-cycle transcriptional curves with T7 DNA similar to those of the *E. coli* polymerase, although the exact values of the assay parameters differ significantly among the different enzymes.[19]

In principle, a quantitative assay for an RNA polymerase can be devised with any template and RNA polymerase for which the transcriptional parameters fill the requirements set forth above. Practically speaking, the following major difficulties must be considered.

1. The promoter site or sites used must be strong enough to give rapid initiation, and there must be no competing promoter sites or tight binding sites that cause loss of enzyme.
2. The transcriptional unit must be reasonably long, to allow a phase of transcription in which all of the active polymerase is growing chains. The rate of elongation can be reduced by lowering the temperature or the substrate concentrations,[60] and this will help to lengthen the elongation phase. However, these conditions can also slow promoter complex formation and/or chain initiation and may extend the initiation phase unacceptably. In addition, low nucleotide concentrations can lead to extensive "pausing" of transcriptional complexes and give nonlinear kinetics of chain elongation.[61]
3. There must be no premature termination. This latter requirement has proved to be a major problem in devising quantitative assay procedures with cloned promoters. For example, the vector pBR322[62] bears a partially effective transcriptional terminator located near 3100 bp on the standard map of pBR322, which interrupts what would otherwise be a very satisfactory transcription unit read from the *tet* promoter or from promoters on fragments cloned into the *Eco*RI site or neighboring sites.[2] The existence of partial terminators on templates is especially a problem with *B. subtilis* RNA polymerase, which appears to respond much more strongly to such sites than the *E. coli* enzyme.[60]

Detection and Assay of Promoter-Specific RNA Polymerases or RNA Polymerases of Novel Specificity

The quantitative assay procedures discussed above generally depend on having an enzyme of defined promoter specificity. However, the speci-

[60] N. Neff and M. Chamberlin, *Biochemistry* **19,** 3005 (1980).
[61] G. Kassavetis and M. Chamberlin, *J. Biol. Chem.* **256,** 2777 (1981).
[62] J. G. Sutcliffe, *Cold Spring Harbor Symp. Quant. Biol.* **43,** 77 (1979).

ficity and the efficiency of utilization of promoter and terminator sites can be changed for different forms of bacterial RNA polymerase[28,32–34] or even for RNA polymerase mutants.[60,63] Hence the investigator is often faced with the problem of determining whether an RNA polymerase fraction is specific for promoter sites and, if so, whether it has the specificity of the normal *E. coli* holoenzyme or is significantly different. This information can be essential for purification of RNA polymerases of novel specificity or in characterization of purified fractions. Ideally one would like an assay procedure that gives a quantitative measurement of the amount and specificity of each distinct RNA polymerase species present in a fraction. In practice, the specificity of such enzyme species is usually not known *a priori,* and such assays are not available. There is also often a large background of transcription by other forms of RNA polymerase in the extract.

One approach to this problem is to monitor the transcriptional specificity of RNA polymerase fractions by transcribing well defined templates and analyzing the RNA transcripts by polyacrylamide gel electrophoresis. This is most effective where there is a promoter site on the template that is coupled with a strong rho-independent terminator to give a transcription unit of well defined size that can be sharply resolved by gel analysis.[19] Alternatively, the DNA template can be cleaved at a site downstream from the promoter using an appropriate restriction endonuclease. This gives a discrete "runoff" transcript terminated at the cleavage site.[49,64–67]

Although this procedure generally does not give a quantitative analysis of the amount of RNA polymerase specific for a given promoter, it can be highly effective as a qualitative test for such an activity. Advantages of the procedure are that the presence of promoter-specific enzyme can often be picked up against a large background of nonspecific transcription or transcription from different promoters. This depends on the size and strength of the transcriptional unit. Short transcription units (up to 500 nucleotides) are advantageous since the resolution of the gel system is better than for larger fragments, and even the presence of small amounts of contaminating nucleases may not preclude detection of the transcript. Furthermore, the procedure need not depend on *a priori* knowledge of the promoter location.[34]

An extension of this approach involves using a template that bears a collection of different promoter sites, having different transcriptional properties. Here, by following transcription from each promoter, one can

[63] C. Yanofsky and V. Horn, *J. Bacteriol.* **145,** 1334 (1981).
[64] B. Meyer, D. Kleid, and M. Ptashne, *Proc. Natl. Acad. Sci. U.S.A.* **72,** 4785 (1975).
[65] S. Gilbert, H. de Boer, and M. Nomura, *Cell* **17,** 211 (1979).
[66] R. Young and J. Steitz, *Cell* **17,** 225 (1979).
[67] B. Davison, C. Murray, and J. Rabinowitz, *J. Biol. Chem.* **255,** 8819 (1980).

monitor not only large changes in promoter specificity (promoter recognition identity) but also detect changes that affect the relative utilization of promoters that share a common specificity (promoter strength). One such promoter test system has been described using bacteriophage T7 DNA as template.[19]

Two other useful procedures to detect and analyze transcription from specific promoters or transcription units should be mentioned. The presence of transcripts derived from a particular region of a DNA template in a mixture of *in vitro* transcripts can be detected by cleavage of the template using appropriate restriction endonucleases, followed by gel electrophoresis to separate DNA fragments and hybridization of labeled transcripts to the fragments by the Southern method.[68] Alternatively, RNA chains can be terminally labeled with [γ-^{32}P] nucleotides and subsequently cleaved with an appropriate ribonuclease to generate a mixture of 5′-γ-^{32}P-labeled oligonucleotides. These can be separated and the amount of transcription from particular promoter sites can be analyzed.[69] Both of the above procedures are applicable even where there is enough ribonuclease or deoxyribonuclease in the enzyme fractions to preclude gel analysis of the intact RNAs. However, each has its own weaknesses; random transcription by core polymerase will be detected in the first procedure and must be distinguished from selective transcription. The use of γ-^{32}P-labeling requires rather active and highly purified polymerase fractions as well as high specific activity nucleotides to get sufficient incorporation to study.

Purification of RNA Polymerases from T7-like Bacteriophages

Bacteriophage T7 induces synthesis of its own DNA-dependent RNA polymerase soon after infection,[70] and similar enzymes are induced by a variety of morphologically similar bacteriophages (for reviews, see Chamberlin and Ryan[3], Hausmann,[71] and Kruger and Schroeder[72]). These enzymes are of interest to the biochemist for several reasons. First, they are single-subunit enzymes, M_r around 100,000, which are highly selective for utilization of specific promoter sites *in vitro*. Thus they offer the potential to study the physical and chemical aspects of RNA polymerase–promoter interactions with a much simpler protein than the multisubunit bacterial RNA polymerases. A second, more practical property of the phage RNA

[68] E. M. Southern, *J. Mol. Biol.* **98,** 503 (1975).

[69] J. Miller and R. Burgess, *Biochemistry* **17,** 2054 (1978).

[70] M. Chamberlin, J. McGrath, and L. Waskell, *Nature (London)* **228,** 227 (1970).

[71] R. Hausmann, *Curr. Top. Microbiol. Immunol.* **75,** 77 (1976).

[72] D. Kruger and C. Schroeder, *Microbiol. Rev.* **45,** 9, (1981).

polymerases is their ability to synthesize RNA rapidly and for extended periods of time *in vitro,* which makes them reagents of potential value to those who wish to prepare specific RNA species *in vitro* in large amounts.

The major drawbacks to use of T7 RNA polymerase in biochemical studies have been the low yields of the protein available from T7-infected cells. The protein is not abundant in infected cells, and it also appears to be quite unstable at low protein concentrations under the conditions used in most laboratories. These factors combined to make it difficult to prepare highly active T7 RNA polymerase except on a rather large scale. This problem is likely to be alleviated when the cloned T7 polymerase gene[73] is placed in suitable vectors to allow hyperexpression of the polymerase. In addition, we have reported on the isolation and properties of a similar RNA polymerase from *Salmonella* phage SP6, which is more stable than the T7 enzyme[74] and can be purified in a single step for purposes of *in vitro* transcription, as we describe below.

Purification Procedures

A number of bacteriophages specify DNA-dependent RNA polymerase similar to the T7 enzyme; these are all morphologically similar to T7 and often show genetic homology as well.[71] The T7 RNA polymerase is the most thoroughly studied of these and will be taken as representative of the general class, although individual phage polymerases may show significant differences.

A variety of procedures has been used to purify T7 RNA polymerase; however, there is still no really satisfactory method that gives high yields of homogeneous and active polymerase. The original method of Chamberlin *et al.*[70] for purification of T7 polymerase employed streptomycin sulfate precipitation to remove nucleic acids, precipitation and extraction with ammonium sulfate, followed by column chromatography on DEAE-cellulose and phosphocellulose. The peak fractions from phosphocellulose were over 90% T7 polymerase protein as judged by SDS-gel analysis.[70] However, the yields and specific activities obtained by this method are relatively low, and there is often significant variation in the early steps. Niles *et al.*[75] introduced a modified procedure in which nucleic acids and T7 polymerase are precipitated from the extract with polyethyleneimine and the T7 polymerase is then extracted from the precipitate with salt. These fractions were fractionated with ammonium sulfate, and then by column chromatography on phosphocellulose, DEAE cellulose,

[73] S. Stahl and K. Zinn, *J. Mol. Biol.* **148,** 481 (1981).
[74] E. Butler and M. Chamberlin, *J. Biol. Chem.* **257,** 2772 (1982).
[75] E. Niles, S. Conlon, and W. Summers, *Biochemistry* **13,** 3904 (1974).

and hydroxyapatite, respectively. This procedure gives good yields of enzyme activity through the phosphocellulose step and is quite reproducible.[76] The phosphocellulose fractions are only of moderate specific activity and contain contaminating peptides, but they give normal amounts of the large T7 transcripts as measured by RNA gel analysis of the products and are quite adequate for transcriptional analysis, or preparation of specific labeled RNAs. Such fractions have been kept at $-20°$ in 50% glycerol solutions without substantial loss in activity for several years.[76] Further purification of these fractions by heparin–agarose chromatography gives enzyme of very high specific activity,[76] but these fractions have been somewhat less stable, possibly owing to the low protein concentrations involved.

An alternative modification of the Niles *et al.*[75] procedure was reported to give homogeneous T7 polymerase.[77] However these fractions are apparently contaminated with a single-strand specific endonuclease, and gel analysis suggests that as much as 30–50% of the protein can be in peptides other than T7 polymerase.[78] Chromatography of these fractions on T7 DNA cellulose may give homogeneous enzyme,[78] but the capacity of the column is said to be quite low.

Similar procedures are generally applicable for the purification of T3 RNA polymerase. McAllister and his collaborators[79] have isolated the T3 RNA polymerase using the polyethyleneimine procedure through the phosphocellulose step, followed by chromatography on heparin–agarose and phosphocellulose. An alternative procedure for purification of T3 RNA polymerase[80] takes advantage of the fact that the enzyme in cell extracts is easily sedimented with cell debris, probably owing to binding to ribosomes.[81] It can be eluted from the pellet with salt solutions and subsequently fractionated by column chromatography. The procedure is reported to give reasonable recoveries (ca 20%) of enzyme of good specific activity (ca 600,000 units/mg), which gives a single band on SDS–polyacrylamide gels.

Isolation of the RNA polymerase from *Salmonella typhimurium* infected with phage SP6 has proved to be somewhat easier than for the T7 enzyme, owing in part to the greater stability of SP6 enzyme. After removal of nucleic acids with streptomycin and ammonium sulfate precipitation, the enzyme is chromatographed on phosphocellulose, Blue Dextran-Sepharose, and BioGel P200, respectively, to give a homogeneous

[76] G. Kassavetis and M. Chamberlin, *J. Virol.* **29,** 196 (1979).
[77] J. Oakley, J. Pascale, and J. Coleman, *Biochemistry* **14,** 4684 (1975).
[78] R. Strothkamp, J. Oakley, and J. Coleman, *Biochemistry* **19,** 1074 (1980).
[79] J. Bailey and W. McAllister, *Nucleic Acids Res.* **8,** 5071 (1980).
[80] P. Chakraborty, P. Sarkar, H. Huang, and U. Maitra, *J. Biol. Chem.* **248,** 6637 (1973).
[81] J. Dunn, F. Bautz, and E. Bautz, *Nature (London) New Biol.* **230,** 94 (1971).

protein fraction. This gives yields of up to 30% overall of SP6 RNA polymerase activity and specific activities of 700,000 units/mg.[74]

For studies where SP6 polymerase is to be used for *in vitro* RNA synthesis and the enzyme need not be homogeneous, a procedure has been used that gives acceptable yields of active enzyme in a single fractionation step.

Procedure for Single-Step Isolation of SP6 RNA Polymerase. The method depends on the high affinity of the SP6 RNA polymerase for the dye Cibacron Blue,[82] which can be used as an affinity chromatography agent. However, in our procedure elution is with a high-salt buffer, as we have not been able to obtain effective elution of the polymerase with substrates. In our trials, Cibacron Blue F3GA dye was coupled to Sepharose 4B using a modification of the procedure of Heyns and Demoor[83] to produce an affinity matrix with a very high capacity for SP6 RNA polymerase. It is likely that commercially available column matrices such as Bio-Rad Affigel Blue would be equally suitable for this purpose; however, the capacity of these materials for SP6 polymerase may vary slightly. SP6-infected cells were grown and lysed exactly as described for the normal purification procedure.[74] Fraction I (20–30 ml; 15–20 mg/ml protein) is adjusted to a final concentration of 0.25 *M* NaCl and passed slowly through a Cibachron Blue–Sepharose column (2 ml total bed volume). The column is washed with 25 ml of buffer II containing 0.25 *M* NaCl (buffer II contains 10 m*M* KPO_4, pH 7.9, 10 m*M* 2-mercaptoethanol, 0.1 m*M* EDTA, and 10% glycerol), and then with 15 ml of buffer II containing 1 m*M* each of ATP and GTP. The column is washed with buffer II containing 0.25 *M* NaCl until the absorbancy of the effluent at 280 nm is negligible, and the enzyme is eluted with 1 *M* KPO_4, pH 7.9, containing 1 m*M* DTT, 0.05 m*M* EDTA, and 10% glycerol; peak fractions are dialyzed against buffer II containing 0.2 *M* KCl and 50% glycerol for storage at −20°. The enzyme elutes in a broad peak over about 10–15 ml. SDS–gel electrophoresis of SP6 polymerase fractions show one major band at about 96,000 M_r together with a number of lower M_r contaminants, suggesting that the enzyme is about 10–20% pure. Specific activities are around 45,000 units/mg, about 10% of those obtained for the full purification.[74]

Assay of Bacteriophage RNA Polymerases

T7 RNA polymerase is usually assayed by following incorporation of a radioactively labeled nucleoside triphosphate into acid-insoluble material in the presence of T7 DNA as template. The reaction shows an absolute

[82] S. Thompson and E. Stellwagen, *Proc. Natl. Acad. Sci. U.S.A.* **73,** 361 (1976).
[83] W. Heyns and P. Demoor, *Biochim. Biophys. Acta* **358,** 1 (1974).

requirement for the four ribonucleoside triphosphates, Mg^{2+}, and T7 DNA.[70,84] T7 DNA can be replaced by other duplex DNA templates bearing specific T7 polymerase promoter sites (see below) or by synthetic polynucleotides such as $(dG)_n \cdot (dC)_n$, $(dI)_n \cdot (dC)_n$, or poly(dC). The rate of synthesis is optimal between pH 7.7 and 8.3. The rate of synthesis is highly sensitive to reaction temperature[84] and falls off rapidly below 37°; there is about a twofold reduction at 30°. The may be due, in part, to a requirement for DNA strand separation in a rate controlling step,[85] but the rate falls off nearly as rapidly with single-stranded poly$(dC)_n$ as template,[84] suggesting that other steps or temperature-dependent changes in enzyme conformation are also involved.

The rate and also the extent of T7 RNA synthesis are affected by sulfhydryl-reactive agents, such as *p*-chloromercuribenzoate; hence a thiol such as dithiothreitol or 2-mercaptoethanol is included in the reaction solution. Similarly, the reaction, especially with early enzyme fractions, shows an enhancement or even complete dependence on the addition of BSA.[70] This may be due to the high sensitivity of the T7 polymerase to inhibition by polyanionic compounds.[86]

Under optimal conditions the reaction continues at constant rate for at least an hour after a short (10–15 sec) lag. However, this extended period of synthesis involves many cycles of transcription initiation, elongation, and termination for each active polymerase (see below). The longest of the transcription units controlled by a strong (class III) T7 promoter is 12,000 bases,[87] corresponding to a transit time for T7 RNA polymerase of 60 sec at 200 nucleotides per second,[88] and the average transit time for class III transcriptional units is 20–25 sec. Hence each active T7 RNA polymerase must repeat the transcription cycle about three times each minute.

An apparent consequence of this extensive recycling during transcription with T7 DNA is that the reproducibility of assays falls off rapidly at times over 5–10 min (M. Chamberlin and J. Ring, unpublished studies); hence, for assays, reaction times of no more than 5–10 min should be used. This loss of reproducibility probably results because T7 RNA polymerase is not highly stable at 37°, so that small variations in reaction conditions, glassware, etc., affect the lifetime of enzyme released during the recycling process. Since loss of a small fraction of enzyme during recycling leads to a decreasing geometrical progression (see the section on the use of isolated RNA polymerases, below), there can be a dispropor-

[84] M. Chamberlin and J. Ring, *J. Biol. Chem.* **248,** 2235 (1973).
[85] J. Oakley, R. Strothkamp, R. Sarris, and J. Coleman, *Biochemistry* **18,** 528 (1979).
[86] M. Chamberlin and J. Ring, *J. Biol. Chem.* **248,** 2245 (1973).
[87] A. Carter, C. Morris, and W. McAllister, *J. Virol.* **37,** 636 (1981).
[88] M. Golomb and M. Chamberlin, *J. Biol. Chem.* **249,** 2858 (1974).

tionate effect on the extent of recycling at longer reaction times depending on small variations in the reaction.

The definition of a unit of T7 RNA polymerase activity has generally been based on measurement of the rate of reaction in a 10-min incubation under specified reaction conditions with T7 DNA as template.[70,85] One unit is the amount of enzyme needed to give a rate of incorporation of 1 nmol of labeled substrate per hour under these conditions.

It would be highly desirable to have a quantitative RNA polymerase assay involving a single transcriptional cycle, similar to that developed for bacterial RNA polymerases.[30] Such an assay for T7 RNA polymerase cannot be devised with T7 phage DNA, as it contains far too many transcription units and these vary in transit time from about 3 to 60 sec. Since chain initiation, or establishment of a normal rate of chain elongation, requires about 10–15 sec, it is clearly impossible to separate chain elongation from initiation and termination with T7 DNA. Such a separation should be possible with cloned T7 polymerase promoters inserted in large cloning vectors. In principle, if there were no chain termination and transcription were initiated on an intact circular DNA, RNA chain elongation would be continued more or less indefinitely and a quantitative measure of the concentration of active RNA polymerase could be obtained simply from the rate of incorporation and the elongation rate.

Unfortunately, although cloned phage polymerase promoters are available,[89] currently available vectors contain several partially effective terminator sites.[89] Hence termination and recycling begins with these templates shortly after the reaction starts, and there is still the same problem of separating the different reaction steps. In view of the importance of obtaining a quantitative T7 RNA polymerase assay, it would be useful to attempt to develop better DNA templates for these assays by systematically attempting to remove *in vitro* termination sequences from the cloning vectors.

The biochemical properties and assay procedures for other T7-like phage RNA polymerases are generally similar to those of the T7 polymerase, and these have been studied for the T3 RNA polymerase,[80,81,90–93] *Pseudomonas* phage gh-1,[94] and *Salmonella* phage SP6.[74,95] These differ-

[89] W. McAllister, C. Morris, A. Rosenberg, and F. Studier, *J. Mol. Biol.* **153**, 527 (1981).

[90] J. Dunn, W. McAllister, and E. Bautz, *Eur. J. Biochem.* **29**, 500 (1972).

[91] W. McAllister, H. Kupper, and E. Bautz, *Eur. J. Biochem.* **34**, 489 (1973).

[92] R. Salvo, P. Chakraborty, and W. Maitra, *J. Biol. Chem.* **248**, 6647 (1973).

[93] P. Chakraborty, P. Bandyopadhyay, H. Huang, and U. Maitra, *J. Biol. Chem.* **249**, 6901 (1974).

[94] H. Towle, J. Jolly, and J. Boezi, *J. Biol. Chem.* **250**, 1723 (1975).

[95] G. Kassavetis, E. Butler, D. Roulland, and M. Chamberlin, *J. Biol. Chem.* **257**, 5779 (1982).

ent enzymes all show a requirement for their own specific class of promoter sites and generally will not use heterologous templates.[3] However, all of the T7-like RNA polymerases studied thus far will use the synthetic polynucleotides $(dG)_n \cdot (dC)_n$ or poly(dC) as templates. These templates lack specific promoter sequences and transcription on such templates probably reflects simply the general catalytic activity of the polymerase in a reaction where specific promoter binding has been bypassed.

Use of Isolated RNA Polymerases to Obtain Extensive Transcription *in vitro*

The Problem of Bacterial RNA Polymerases

In many instances the investigator wishes to prepare large amounts of RNA *in vitro* using a particular DNA template and a preparation of RNA polymerase. The goal may be use of the RNA for sequence analysis, mapping of a transcriptional start site or terminator, or simply preparation of a highly labeled hybridization probe. Generally one wants the best possible yield of intact RNA with the expenditure of the least possible amount of DNA template and labeled nucleotide, since both of the latter may be difficult to obtain or expensive. However most *in vitro* transcriptional systems have been optimized to give favorable initial rates of transcription. These need not be those that give good yields of RNA.

We have carried out some studies on the factors that affect extensive transcription of T7 ΔD111 DNA to begin to define some of the important variables that affect yields of RNA *in vitro*. T7 ΔD111 DNA and *E. coli* RNA polymerase holoenzyme were chosen, since this template has a single strong promoter (T7 A1) that gives rise to a discrete RNA of 6100 nucleotides[19] and since the parameters that govern the rate and extent of the transcription steps are well studied with this template.[30] Furthermore, use of the quantitative assay[30] allows determination of the exact concentration of active *E. coli* RNA polymerase in each reaction, under a variety of altered reaction conditions.

Under the ionic conditions described above for the quantitative RNA polymerase assay [these include 4 m*M* Mg^{2+}, 120 m*M* KCl, 4 m*M* spermidine, together with "*in vivo*" levels of nucleotides including CTP (0.7 m*M*), UTP (1.7 m*M*), GTP (1.1 m*M*) and ATP (2.7 m*M*), and 0.3 m*M* T7 DNA nucleotide], over 50% of the limiting nucleotide (CTP) is converted to acid-insoluble material in a 6-hr synthesis with a total RNA polymerase concentration of 36 μg/ml (Table I). Analysis of the products in a long-term reaction at 36 μg of RNA polymerase per milliliter by thin-layer chromatography suggests that at the limit of incorporation none of the α-

TABLE I
REQUIREMENTS FOR EXTENSIVE T7 RNA SYNTHESIS[a]

Conditions	CMP incorporated (nmol/ml)	Total CTP used (%)
Complete system	400	57
Omit KCl	300	<0.1
Increase KCl (320 m*M*)	5	0.7
Omit spermidine	260	37
Add 1/10 × ATP, UTP, GTP, CTP*	400	57
Add 1/10 × T7 D111 DNA	280	40
Add 1/10 × T7 D111 DNA, 18 hr	400	57

[a] Reaction conditions in the complete system are those given in the text for quantitative *E. coli* RNA polymerase assay with [α-^{32}P]CTP and 36 μg/ml of *E. coli* RNA polymerase holoenzyme[38] which was 25% active by the quantitative assay procedure with T7 DNA.[30] Reactions were run for 6 hr at 30° and analyzed for incorporation of [^{32}P]CMP into RNA.[30]

^{32}P-labeled substrate remains; virtually all the label is found in polynucleotide except for small amounts present as CMP,[32] which may be due to a slight contamination of the enzyme with a triphosphatase. Addition of more nucleoside triphosphate substrates after 16 hr restores synthesis; hence under these optimal conditions with T7 DNA, the enzyme remains active through many transcriptional cycles and for many hours.

At 18 μg of RNA polymerase per milliliter a similar limit is reached; however, 15–18 hr are required for maximal synthesis. Concentrations of RNA polymerase over 36 μg/ml do not increase the yield of RNA, and there is a slow progressive decline in the yield at polymerase concentrations over 70μg/ml. Concentrations of polymerase below 18 μg/ml give proportionately lower yields of RNA.

The optimal level of incorporation corresponds to formation of at least 1.6 m*M* RNA nucleotides or over 0.5 mg of RNA per milliliter; RNA gel analysis shows that most of this is intact T7 A1 RNA. Thus there is no intrinsic kinetic or thermodynamic barrier to *in vitro* synthesis of very substantial amounts of RNA, given appropriate conditions. However, these high yields depend on a narrow range of reaction conditions and on several specific features of the T7 ΔD111 DNA template. Extensive synthesis decreases sharply at KCl concentrations over 200 m*M*, and at 300 m*M* synthesis is almost completely blocked (Table I). The decrease in yield at higher KCl concentration may be related to the fact that the strength of binding between RNA polymerase and the T7 A1 promoter is decreased up to 10^{12}-fold by a 10-fold increase in salt concentration.[96] This

[96] T. Kadesch, S. Rosenberg, and M. Chamberlin, *J. Mol. Biol.* **155,** 1 (1982).

suggests that efficiency of interaction of the enzyme with the available promoter sites is a critical factor in obtaining high yields of RNA *in vitro*. This view is also supported by studies of the effect of DNA concentration on extensive RNA synthesis (Table I). Reduction of the DNA concentration by 10-fold in the standard reaction reduces the yield of RNA in a 6 hr reaction, but synthesis continues to the normal limit after 15–18 hr. However, reduction of the DNA concentration by 100-fold reduces the yield of T7 RNA to below 1 nmol/ml under all the conditions we have tested, even after 24 hr of reaction.

An explanation for the dramatic reduction in yield of T7 A1 RNA at low DNA concentrations comes from studies on the effect of DNA concentration on the concentration of active RNA polymerase using the quantitative RNA polymerase assay (Table II). As the T7 DNA concentration is reduced below the normal, optimal concentration of 320 n*M* (8 μM in T7 A1 promoter concentration), there is a precipitate reduction in the amount of active enzyme that can initiate a round of T7 RNA synthesis. Thus, even though the rate of binding of RNA polymerase to the strong T7 A1 promoter is very rapid, at promoter concentrations below 0.8μM almost all the active enzyme added is inactivated prior to forming a stable promoter complex.

In similar kinds of experiments, we have followed transcription from

TABLE II

EFFECT OF DNA CONCENTRATION ON THE AMOUNT OF ACTIVE *Escherichia coli* RNA POLYMERASE IN AN *in Vitro* TRANSCRIPTION REACTION WITH T7 DNA[a]

DNA concentration (nM)	Active RNA polymerase (%)
320	100
32	31
3.2	4
0.32	<1

[a] Assays were carried out under conditions described originally for the quantitative RNA polymerase assay with T7 DNA,[30] except that enzyme was preincubated with the DNA for 10 min to eliminate the initial (phase I) lag in synthesis. Each reaction contained 0.2 μg of *E. coli* RNA polymerase holoenzyme.[38] Under standard conditions (320 n*M* T7 DNA) 12% of the protein added was active; this corresponds to a rate of incorporation of 64 mU/mg, under the conditions of Chamberlin *et al.*[30]

the *E. coli rrnB* promoters on plasmid pKK3535[97] under the standard conditions used in Table I; yields after 6 hr have never exceeded 1 nmol/ml CMP incorporated (J. Levin, unpublished observations). Again, it is likely that this reflects the fact that the *E. coli* ribosomal promoters are much weaker than those of T7 *in vitro,* and this leads to greatly reduced yields of RNA, not simply to slower accumulation of transcripts.

Reduction of the nucleoside triphosphate concentration lowers the extent of incorporation substantially (Table I). However the yield of T7 RNA relative to the amount of CTP added is unchanged. Thus for purposes of obtaining labeled RNA of high specific activity, suboptimal concentrations of substrates may be preferable.

What general conclusions can be drawn from these experiments that may be applicable for other templates? First, it is likely that extensive synthesis such as we see with T7 ΔD111 DNA depends on extensive recycling of RNA polymerase molecules. Under our optimal conditions and using 18 μg of RNA polymerase per milliliter, we obtain a yield of up to 500 nmol/ml of CMP incorporated. Hence each active RNA polymerase molecule added to the reaction has synthesized over 25 T7 A1 RNA chains under normal conditions. In such a reaction the ultimate yield is highly sensitive to the fraction of RNA polymerase that is recycled at each step. For example, if 5% of the enzyme is lost at each cycle, then after three rounds of synthesis the number of RNA chains, N, per active RNA polymerase added, will be

$$N = 1 + 0.95 + (0.95)^2$$

This is a geometrical progression[98] that reaches a limit after synthesis is completed of

$$\text{limit } N = 1/(1 - r)$$

where r is the fraction of enzyme surviving each cycle. For the example given above, r must be about 0.96 to give 25 chains per polymerase, and even for the complete system shown in Table I (400 nmol/ml CMP; 12 chains per polymerase) r must be about 0.92. If r drops to 0.5, only two chains will be obtained per polymerase; hence r is a critical factor.

At what steps in transcription is loss of active RNA polymerase likely to occur? Elongating polymerase complexes are very stable, and there is no evidence for random inactivation–termination during normal elongation.[30] Hence, during extensive synthesis, inactivation is most likely dur-

[97] R. Kingston and M. Chamberlin, *Cell* **27,** 523 (1981).

[98] CRC Handbook of Chemistry and Physics, 47th ed., p. A235 (1966).

ing the initial phase of promoter binding and then again after termination–release and prior to rebinding at the promoter. A number of different factors can lead to inactivation at the following steps.

1. While promoter binding is very efficient at high polymerase concentrations (over 1 μg/ml), at lower concentrations there can be substantial losses of activity.[99] Since release from a terminator is not synchronous, concentrations of recycling free enzyme are likely to be very low. This may account for the need for rather high enzyme concentrations for extensive synthesis; much of the polymerase may simply play a protective role.

2. Binding of polymerase to ends or breaks in DNA or to the RNA product competes directly with promoter binding[46] and can reduce the enzyme concentration. This is suppressed by concentrations of KCl over 100 m*M*, but these levels of KCl can weaken promoter–polymerase interactions substantially for weak promoters.

3. RNA polymerase is subject to thermal inactivation; at elevated polymerase concentrations a half-time of about 20 min is seen at 30°.[100] This is not significant for strong promoter sites and elevated enzyme concentrations where binding is complete in 10–15 sec.[96,99] However, for weaker promoter sites where minutes are required, this can be a major factor.

From this analysis we conclude that there are probably very narrow limits under which extensive RNA synthesis can be obtained with bacterial RNA polymerases *in vitro*. In a worst-case situation with a weak (salt-sensitive) promoter on a short restriction fragment, it may not be possible to obtain even 0.1 chain per active polymerase added. This situation could be improved only if bacterial RNA polymerases are found with much greater stability to inactivation, or if reagents are devised that give a comparable stabilization of the *E. coli* enzyme.

In such cases two alternative procedures for obtaining extensive *in vitro* transcription can be considered. Where the DNA to be transcribed can be manipulated genetically, the sequence to be transcribed can be inserted next to a strong bacterial promoter, using DNA cloning technology. Vectors are now available that contain strong promoters near strong, rho-independent terminators.[101] Insertion of a sequence in such a transcription unit should allow transcription of supercoiled DNA with production of discrete transcripts and a large increase in *in vitro* promoter

[99] D. Hinkle and M. Chamberlin, *J. Mol. Biol.* **70,** 187 (1972).

[100] R. C. Williams and M. Chamberlin, *Proc. Natl. Acad. Sci. U.S.A.* **74,** 3740 (1977).

[101] R. Gentz, A. Langner, A. Chang, S. Cohen, and H. Bujard, *Proc. Natl. Acad. Sci. U.S.A.* **78,** 4936 (1981).

strength and hence RNA yield. An alternative approach that employs bacteriophage RNA polymerase promoters is described in the next section.

Use of Phage RNA Polymerases for Extensive Synthesis of RNA in Vitro

While extensive RNA synthesis with bacterial RNA polymerases has rather stringent requirements, synthesis with the T7-like phage RNA polymerases under normal reaction conditions often continues for long periods.[84] Since the rate of chain elongation for these enzymes is about 10-fold more rapid than for the bacterial polymerases, this leads directly to extensive RNA synthesis. As discussed above in the section on assay of bacteriophage RNA polymerases, this synthesis involves extensive recycling because of the rapid elongation rate and the presence of short transcription units on normal phage DNA templates, and in a 60-min reaction with T7 DNA template the active T7 RNA polymerase will have recycled a minimum of 180 times. Because of the considerations discussed in the preceding section (The Problem of Bacterial RNA Polymerases), this could occur only if there is virtually no loss of enzyme in each cycle; clearly the phage polymerases are recycled much more efficiently than the bacterial enzymes, which accounts for their ability to make large amounts of RNA. This may be due to the fact that as compared to the bacterial polymerases, the phage enzymes bind poorly to sites that might compete for promoter binding, such as the RNA product, single-strand breaks in templates, or template ends.[86]

This ability of phage RNA polymerases to synthesize large amounts of RNA *in vitro* is not useful for transcription of DNAs that do not contain cognate promoter sites for the particular phage polymerase. However, several groups have now prepared cloning vectors that bear a strong phage promoter site upstream of a cloning site. One such vector pSP6-R7, prepared by Eugene Butler and Peter Little in Tom Maniatis' laboratory, bears an SP6 RNA polymerase promoter upstream of a poly-linker cloning site in pBR322 DNA. Under normal SP6 polymerase assay conditions, transcription of pSP6-R7 continues for over 60 min at almost constant rate. In one experiment 13 units of fraction IV SP6 polymerase (0.5 μg) converted about 10% of the total CTP in a 1-ml reaction to RNA in 60 min, a yield of about 0.16 μmol of RNA transcripts.

Unfortunately, many DNA sequences contain partially effective terminators for the SP6 RNA polymerase, which makes it difficult to obtain complete transcripts of cloned sequences with such a vector. One solution to this problem may be a vector prepared by Michael Green and Tom Maniatis (Harvard) in which a poly(A) sequence has been inserted just downstream of the cloning site in pSP6-R7 followed by a unique restriction site that can be used to linearize the template. Complete runoff transcripts of an insert can be selected from the products by chromatography on oligo(dT) cellulose columns. Development of such vectors should greatly facilitate the problem of obtaining large quantitities of RNA *in vitro* from defined genetic sequences.

Acknowledgments

Many of the methods described here have been worked out in collaboration with colleagues at Berkeley, both in our own and other laboratories. We are grateful for their contributions. We particularly thank Jon Narita and Daisy Roulland, for their studies on SP6 RNA polymerase, and Tom Kadesch and Steve Rosenberg for their work on the stability and DNA binding properties of *E. coli* RNA polymerase.

This work was supported by a research grant (GM12010) from the National Institute of General Medical Sciences.

[35] *In Vitro* Transcription: Whole-Cell Extract

By JAMES L. MANLEY, ANDREW FIRE, MARK SAMUELS, and PHILLIP A. SHARP

Three classes of nuclear DNA-dependent RNA polymerases have been identified in eukaryotic cells.[1] RNA polymerase I catalyzes the synthesis of rRNA precursors, RNA polymerase II transcribes primarily the genes that give rise to mRNA, and RNA polymerase III transcription results in the production of tRNAs, 5 S RNA, and several other RNAs of unknown function. It has been clear for many years that in order to study the mechanisms of transcription as well as to identify the factors and nucleotide sequences that control and regulate gene expression, cell free systems that accurately and specifically transcribe exogenously added DNA are required. Early attempts at achieving this aim, which utilized purified RNA polymerases, were unsuccessful. In the last several years, however, *in vitro* systems have been developed in which accurate transcription by all three types of RNA polymerases can be obtained. Two basic approaches have been successful. In one, purified RNA polymerase is supplemented with cell extracts that contain factors required for accurate transcription.[2] This method has been used primarily for RNA polymerase II-mediated transcription and is described in this volume [36]. The

[1] R. G. Roeder, *in* "RNA Polymerase" (R. Losick and M. Chamberlin, eds.), p. 285. Cold Spring Harbor Laboratory, Cold Spring Harbor, New York, 1976.

[2] P. A. Weil, D. S. Luse, J. Segall, and R. G. Roeder, *Cell* **18,** 469 (1979).

ISBN 0-12-182001-7

other approach is to prepare concentrated cell lysates that contain not only the factors required for transcription, but also sufficient amounts of RNA polymerase so that addition of purified enzyme is not required. For RNA polymerase I[3] and III[4] such systems can be simply prepared from cytoplasmic extracts, because sufficient amounts of these polymerases and their required factors leak out of the nucleus at isotonic salt concentrations. RNA polymerase II, on the other hand, remains almost entirely within the nucleus. Thus, to obtain extracts containing this activity, a whole cell extract must be prepared.[5] We describe here the preparation and properties of such an extract, which contains all the factors and enzymic activities necessary for accurate and specific transcription, not only by RNA polymerase II, but also by RNA polymerases I and III.

Most experiments to date have utilized extracts prepared from human cells that grow in suspension culture (HeLa or KB). Such extracts show quite broad species specificities for RNA polymerases II and III. Polymerase III genes from virtually all higher eukaryotes that have been tested are accurately transcribed in HeLa lysates. Whole-cell extracts do not seem able to transcribe yeast polymerase II genes accurately, but have been shown to be capable of transcribing *Drosophila*[6] and chicken[7] polymerase II genes as well as many such genes from higher eukaryotes and their viruses.

Synthesis of mature RNA molecules requires additional enzymes and factors other than those needed to bring about accurate transcriptional initiation (e.g., processing enzymes). Soluble HeLa lysates appear to contain virtually all the enzymes required for tRNA processing, including the splicing enzymes.[8,9] Although work with RNA polymerase I systems is just beginning, soluble extracts appear to contain at least one processing enzyme.[10,11] RNA polymerase II transcripts synthesized *in vitro* are efficiently capped and methylated at their 5′ ends.[2,5] Although one report in the literature claims that an extract efficiently spliced RNA,[12] we have not

[3] I. Grummt, *Proc. Natl. Acad. Sci. U.S.A.* **78,** 727 (1981).

[4] G. J. Wu, *Proc. Natl. Acad. Sci. U.S.A.* **75,** 2175 (1978).

[5] J. L. Manley, A. Fire, A. Cano, P. A. Sharp, and M. L. Gefter, *Proc. Natl. Acad. Sci. U.S.A.* **77,** 3855 (1980).

[6] R. Morimoto, unpublished observations.

[7] B. Wasylyk, C. Kedinger, J. Corden, O. Brison, and P. Chambon, *Nature (London)* **285,** 366 (1980).

[8] D. N. Standring. A. Venegas, and W. J. Rutter, *Proc. Natl. Acad. Sci. U.S.A.* **78,** 5963 (1981).

[9] F. Laski, A. Fire, U. L. RájBhandary, and P. A. Sharp, unpublished observations.

[10] I. Grummt, E. Roth, and M. R. Paule, *Nature (London)* **296,** 173 (1982).

[11] K. G. Miller and B. Sollner-Webb, *Cell* **27,** 165 (1981).

[12] B. Weingartner and W. Keller, *Proc. Natl. Acad. Sci. U.S.A.* **78,** 4092 (1981).

observed in a variety of experiments either splicing or creation of polyadenylated 3′ termini in the cell-free system described here.

Methods

Preparation of Extract

Extracts are prepared by modification of a procedure originally described by Sugden and Keller,[13] who used the method as a first step in RNA polymerase purification. We have used HeLa cells almost exclusively. These cells are easy to obtain in large quantities, and the resultant extracts are relatively free of nuclease activity at 30°. Lysates with transcriptional activity have been prepared from a few other cells lines; most other cell lines and tissues, however, have yielded extracts without detectable levels of transcription or with high levels of nuclease. Cells are grown in suspension culture in Eagle's minimal essential medium supplemented with 5% horse serum to a density of 4 to 8 × 10^5 cells/ml. The cell density appears not to be crucial, although we have obtained slightly more active extracts with cells harvested at the lower end of the range indicated.

The following operations are carried out at 0–4°.

1. Cells are harvested by centrifugation and washed twice with phosphate-buffered saline.
2. The volume of the resultant cell pellet is determined, and the cells are resuspended in four packed-cell volumes (PCV) of 0.01 *M* Tris-HCl (pH 7.9), 0.001 *M* EDTA, and 0.005 *M* dithiothreitol (DTT) (6 × 10^8 cells yield approximately 2 ml of packed cells). At this point, the cells should visibly swell.
3. After 20 min, the cells are lysed by homogenization in a Dounce homogenizer with eight strokes using a "B" pestle.
4. Four PCV of 0.05 *M* Tris-HCl (ph 7.9), 0.01 *M* $MgCl_2$, 0.002 *M* DTT, 25% sucrose, and 50% glycerol are added, and the suspension is gently mixed. With continued gentle stirring, one PCV of saturated $(NH_4)_2SO_4$ is added dropwise. After this addition, the highly viscous lysate is gently stirred for an additional 30 min. Stirring must be very gentle to prevent shearing of the DNA, which would interfere with its removal in the next step. Nuclear lysis can be detected by increased viscosity after approximately half the $(NH_4)_2SO_4$ has been added. Occasionally, lysates appear clumpy and only slightly viscous, rather than extremely viscous

[13] B. Sugden and W. Keller, *J. Biol. Chem.* **248,** 3777 (1973).

and uniform as usually observed. We have obtained active extracts from both types of lysates, although more reproducibly from the latter.

5. The extract is carefully poured into polycarbonate tubes and centrifuged at 45,000 rpm in a SW 50.2 rotor for 3 hr.

6. The supernatant is decanted so as not to disrupt the pellet (the last 1 or 2 ml are left behind), and protein and nucleic acid are precipitated by addition of solid $(NH_4)_2SO_4$ (0.33 g/ml of solution). After the $(NH_4)_2SO_4$ is dissolved, 1 *N* NaOH [0.1 ml/10 g solid $(NH_4)_2SO_4$] is added and the suspension stirred for an additional 30 min.

7. The precipitate is collected by centrifugation at 15,000 *g* for 20 min (the supernatant should be completely drained off), and resuspended with 5% of the volume of the high-speed supernatant with 0.025 *M* HEPES (adjusted to pH 7.9 with NaOH), 0.1 *M* KCl, 0.012 *M* $MgCl_2$, 0.5 m*M* EDTA, 2 m*M* DTT, and 17% glycerol.

8. The suspension is dialyzed against two changes of 50–100 volumes each of the resuspension buffer for a total of 8–12 hr. The volume of the solution increases 30–50% during dialysis. The conductivity of a 1:1000 dilution of dialyzed extract into distilled H_2O (23°) should be 12–14 μmho.

9. The dialyzate is centrifuged at 10,000 *g* for 10 min to remove insoluble material. The supernatant is divided into small aliquots (0.2–0.5 ml), quick frozen in liquid nitrogen or powdered dry ice, and stored at −80°. Extract can be thawed and quick frozen several times without loss of activity and retains full activity at −80° for at least a year.

10. Lysates contain between 15 and 30 mg of protein per milliliter and up to 2 mg of nucleic acid per milliliter. We routinely start with 1–50 liters of cells. One liter of cells should yield about 2 ml of whole-cell extract (WCE), or enough for 100–400 assays. More concentrated extracts are desirable because with these the same optimal protein concentration can be obtained in reaction mixtures with a smaller volume of lysate. In this manner, the salt concentration in the *in vitro* reaction mixture can be lowered (high salt severely inhibits transcription; see below). Attempts to obtain more concentrated extracts by resuspending the pellet in a smaller volume after precipitation, or by tying the dialysis bag tightly (to reduce expansion during dialysis), have not been reproducibly successful owing to increased protein precipitation during dialysis. Likewise, dialysis against buffer containing lower salt concentrations results in less active lysates, again as a result of increased protein precipitation.

The Transcription Reaction

Reactions can be done in volumes of a few microliters or more. Analytical reactions are conveniently performed in 20 μl. A typical reaction mix might contain the following: 30–60% whole-cell extract in its dialysis buffer, 0.2–1.5 μg of template DNA (see below for effects of DNA and extract concentrations), 50 μM ATP, 50 μM GTP, 50 μM CTP, 5 μM UTP, 5 mM creatine phosphate, and 10 μCi of [α-^{32}P]UTP [commercial preparations of aqueous nucleotides can be obtained at a high enough concentration (~10 mCi/ml) and a sufficient specific activity (>200 Ci/mmol) to be added directly to the transcription]. After incubation for 30–120 min, the reactions can be extracted directly or placed at −80° for up to a week before extraction.

Extraction of RNA and Resolution of Products by Gel Electrophoresis

To terminate transcription, 200 μl of stop buffer [7 M urea, 100 mM LiCl, 0.5% sodium dodecyl sulfate (SDS), 10 mM EDTA, 250 μg/ml tRNA, 10 mM Tris-HCl (pH 7.9)] and 300 μl of PCIA (phenol–chloroform–isoamyl alchohol, 1 : 1 : 0.05, water saturated and buffered with 20 mM Tris, pH 7.9) are added, the tubes are blended in a vortex mixer and centrifuged at 12,000 g for 15 min. The aqueous phase (discarding interface) is extracted once more with PCIA and once with chloroform and then pooled with 200 μl of 1.0 M ammonium acetate and precipitated with 900 μl of ethanol. The pellet is washed with ethanol and resuspended in 20 μl of 10 mM Na_2HPO_4 (pH 6.8)–1 mM EDTA; to this is added 50 μl of 1.4 M deionized glyoxal–70% dimethyl sulfoxide–10 mM Na_2HPO_4 (pH 6.8)–1 mM EDTA–0.04% bromophenol blue. After 1 hr at 50°, 25 μl of the samples are loaded on 1.4% agarose gels (run in 10 mM PO_4–1 mM EDTA).[14] This extraction procedure removes most of the free nucleotides from the RNA preparation.

For some techniques, larger reaction volumes are necessary. The above extraction protocol can be scaled up with modifications as follows: After removal of the first aqueous phase, three reaction volumes of stop buffer are added to the first organic phase, and the mixture is again homogenized and spun at 12,000 g (for 1 min). The organic phase is removed, and an equal volume of chloroform is added. After brief homogenization and centrifugation, the aqueous phase can be easily removed. The two aqueous phases are then pooled and reextracted once with PCIA, and twice with chloroform. Any precipitate at the interface of these extractions should be discarded. After the first ethanol precipitation, the pellet is resuspended in 200 μl of 0.2% SDS and 1 mM EDTA. An equal volume of 2 M ammonium acetate is added, and nucleic acid is reprecipitated with ethanol. The pellet is washed with ethanol and can be resuspended in the buffer of choice.

For analysis by hybridization and Sl nuclease digestion,[15] it is impor-

[14] G. K. McMaster and G. C. Carmichael, *Proc. Natl. Acad. Sci. U.S.A.* **74,** 4835 (1977).
[15] P. A. Sharp, A. J. Berk, and S. M. Berget, this series, Vol. 65, p. 750.

tant first to remove the template DNA. The pellet is resuspended in 0.3 *M* sodium acetate (pH 5.2) and reprecipitated and washed with ethanol. The dried pellet is resuspended in 100 μl of 10 m*M* Tris-HCl (pH 7.5), and 100 m*M* NaCl. RNase-free DNase (treated with iodoacetate[16]) to 50 μg/ml, and $MgCl_2$ to 10 m*M* are added. After 5 min at 37°, 100 μl of 10 m*M* EDTA, 0.2% SDS, 150 m*M* NaCl are added, and the mixture is extracted with PCIA and chloroform. The final aqueous is reprecipitated with 0.25 ml of ethanol as above, and the pellet is resuspended in 60 μl of 0.2% Sarkosyl–1 m*M* EDTA (pH 8.0) and stored at −20°.

Sizing and Mapping *in Vitro* RNA

In general, RNA polymerase II does not terminate transcription *in vitro*. However, distinct length RNA products can be generated by the "runoff" assay. This method uses, as template, DNA molecules that have been cleaved by a restriction enzyme that cuts downstream from a putative transcription start site. RNA polymerases that transcribe this DNA will stop or fall off when they reach the end of the DNA. If a substantial number of enzymes initiate transcription at the same site, then a population of molecules of a discrete size will be produced; such a population will migrate as a band on gel electrophoresis. DNA segments that have been cleaved by different restriction enzymes are used as templates in separate reaction mixtures; the transcription start site can be deduced by comparison of the sizes of the RNAs produced. This technique has been widely used for promoter mapping with *in vitro* transcription systems.

An example of the technique is shown in Fig. 1. The RNAs were transcribed from recombinant plasmids containing the adenovirus late promoter and various segments of the long (30 kb) late transcription unit. Several points are exemplified by this experiment. The *in vitro* system is capable of synthesizing very long RNAs, up to 7–8 kb, and hence contains little nuclease activity. Also, the glyoxal method of analyzing RNA is sensitive over a wide range of sizes. Plots of log molecular weight vs migration are linear for transcripts from 0.2 kb to over 5 kb.

Analysis of runoff transcripts is a simple, sensitive, and accurate method for determining the structure of *in vitro* synthesized RNA. However, it does have some limitations. The WCE contains relatively high levels of nucleic acid. Since most of this is 18 S and 28 S rRNA, it is impossible to load more than 25–50% of the RNA obtained from a 20-μl reaction mix onto a standard size gel slot (6 mm × 3 mm) without producing severe overloading of the gel in the regions occupied by these

[16] S. B. Zimmerman and G. Sandeen, *Anal. Biochem.* **14,** 269 (1966).

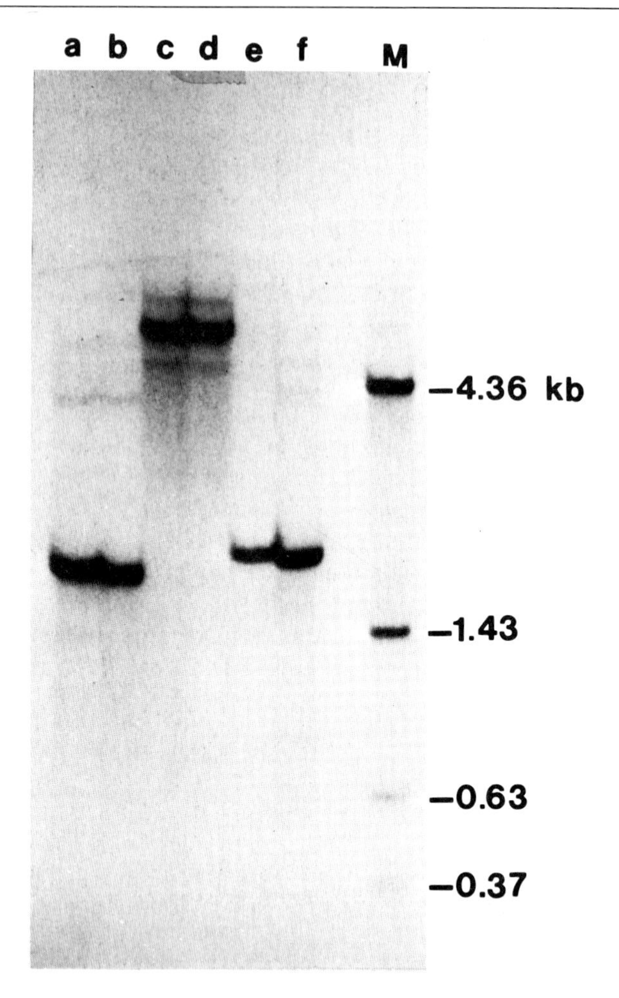

FIG. 1. Analysis of RNA runoff products by glyoxalation and agarose gel electrophoresis. Recombinant plasmids containing the adenovirus late promoter and various segments of the late transcription unit were constructed (R. Jove and J. L. Manley, unpublished), cleaved with restriction enzymes, and used as templates for *in vitro* transcription. DNAs were cleaved so that runoff transcripts of 1.8 kb (lanes a and b), 7.0 kb (lanes c and d), and 1.95 kb (lanes e and f) would be produced. Size markers (lane M) were also produced by *in vitro* transcription (see text).

RNAs. An additional problem is that specific transcripts produced by very weak promoters can sometimes be obscured by nonspecific initiation or termination, or by exogenous nucleic acids labeled by end labeling activity in the extract (particularly rRNA and its breakdown products). This latter activity is insensitive to amanitin and actinomycin D and can thus be distinguished from *de novo* RNA synthesis. Use of radioactive GTP as tracer produces the least end labeling. Use of CTP or ATP produces high levels of tRNA labeling by enzymes exchanging the 3′-terminal CCA. A third limitation of analyzing runoff RNAs by denaturation and agarose gel electrophoresis is that end points of transcripts can be mapped only to within ~20 nucleotides at best. To map the 5′ ends of RNAs more precisely, short runoff transcripts can be analyzed on polyacrylamide sequencing gels.

The above analysis can be extended, and some of the problems circumvented, by using several variants on the technique of hybridization and Sl nuclease digestion. By using labeled RNA and a nonradioactive DNA probe, the problem of spurious RNA labeling can be eliminated, since RNA not complementary to the DNA probe is destroyed by the nuclease. Use of nonradioactive RNA and a DNA probe 5′ end-labeled 50–200 nucleotides downstream from the promoter allows resolution of ±1 nucleotide. The structure of the 5′ end of *in vitro* synthesized RNA can also be studied by classical RNA fingerprinting techniques.[2,5]

Transcriptional Activity

Extracts made from different cell preparations can vary in activity over a 5- to 10-fold range, with about two in three extracts exhibiting activity within 2-fold of the observed maximum. Extracts should be compared for their activity using a runoff assay from a standard polymerase II promoter, such as the major late promoter of adenovirus 2. With optimal DNA and extract concentrations, a good extract (20 μl) will yield 10^6 dpm, or 20 ng of a 2200 nucleotide runoff transcript from the Ad2 late promoter, in 1 hr ([α-^{32}P]UTP at 100 Ci/mmol). This represents the synthesis of one RNA molecule per 10 DNA template molecules present. However, the extract may actually be utilizing a smaller fraction of templates with multiple rounds of initiation per active template.

DNA and Extract Concentrations

Titrations both of DNA and of extract yield nonlinear responses. At a constant extract concentration, measuring runoff transcription as a function of DNA concentration yields (*a*) a threshold DNA concentration below which no transcription occurs; and (*b*) an inhibitory effect of high DNA concentration.[5] The requirement for a minimal DNA concentration is nonspecific; i.e., by using a concentration of a promoter specific DNA that is below the threshold, carrier DNA such as pBR322 or *Escherichia coli* DNA can be added to stimulate specific transcription. The duplex alternating copolymers poly[d(I-C)]:poly[d(I-C)] and poly[d(A-T)]:poly [d(A-T)] will also act as carrier DNA, thereby demonstrating a total lack of sequence specificity in the bulk DNA requirement.[17] A further advantage of these copolymers as carrier DNA is that the transcribed RNA products of the carrier poly[d(I-C)]:poly[d(I-C)] and poly[d(A-T)]:poly[d(A-T)] contain only two nucleotides. Thus, poly[d(I-C)]:poly[d(I-C)] carrier in a reaction containing [α-^{32}P]UTP yields no radioactive background. The key aspect of bulk DNA dependence is that at a fixed total DNA concentration, the molar yield of transcripts per promoter is constant and independent of the source of carrier. In general, specific competition between promoter-containing fragments is not observed.

A critical dependence of transcription upon extract concentration is also observed.[5] In fact, DNA concentration dependence and extract protein concentration dependence are not independent.[18] Specific transcription can be obtained in a range of 4–18 mg of extract protein per milliliter. At low extract concentration the DNA optima tend to be much lower (in the range of 10 μg/ml). There is still a bulk DNA dependence, but it is less steep and the threshold concentrations are lower. At a high extract concentration the DNA titration becomes sharper, and the threshold becomes higher. Under such conditions it is often necessary to use 60 μg/ml of DNA in order to see any transcription. Thus, for each new extract it is necessary to do careful DNA and extract titrations, to determine optimal conditions.

For a given promoter, very short runoff transcripts (<300 n) have a higher optimum DNA concentration than longer runoff transcripts.[18] This effect can be taken into account by measuring the synthesis of different length runoff products from the same promoter. No length dependence has been observed with runoff products between 400 and 4000 nucleotides.

To further complicate matters, the ratio of activity from two promoters can vary as much as 20-fold over a range of DNA and extract concentration.[17–20] An example of this is shown in Fig. 2 where the relative activities of an early and a late Ad2 promoter are compared in an uninfected extract. The ratio of these activities varies 10-fold at different DNA concentrations. Comparison of promoter strengths in different extracts must thus be cautiously controlled and interpreted, a crucial point in assaying for regulatory phenomena.

[17] U. Hansen, D. J. Tenen, D. M. Livingston, and P. A. Sharp, *Cell* **27,** 603 (1981).

[18] A. Fire, C. C. Baker, J. L. Manley, E. B. Ziff, and P. A. Sharp, *J. Virol.* **40,** 703 (1981).

[19] D. Rio, A. Robbins, R. Myers, and R. Tjian, *Proc. Natl. Acad. Sci. U.S.A.* **77,** 5706 (1980).

[20] D. C. Lee and R. G. Roeder, *Mol. Cell. Biol.* **1,** 635 (1981).

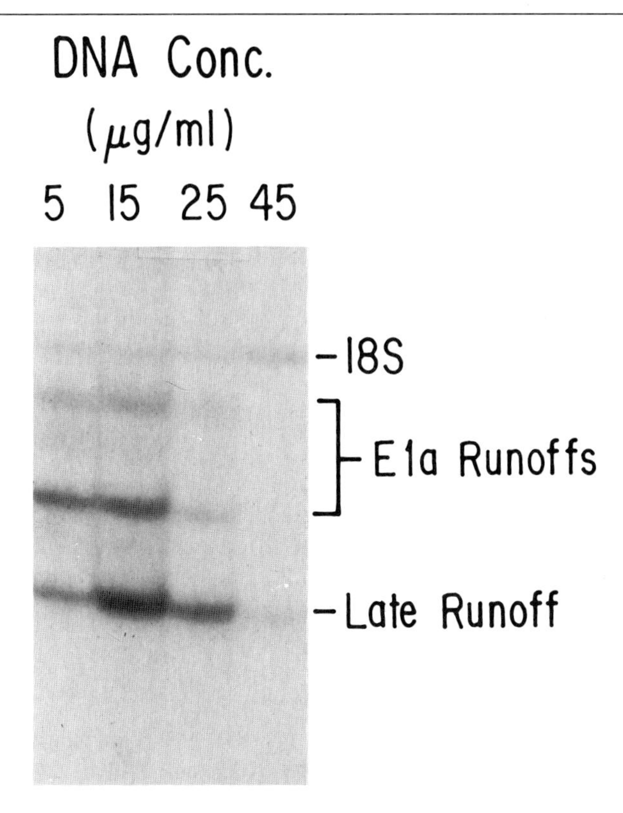

FIG. 2. Relative transcriptional activities of two adenovirus promoters as a function of bulk DNA concentration. Each reaction contained, per milliliter, 4 μg of a plasmid containing the adenovirus Ela promoter cleaved to give a 1220 n runoff and 1 μg of a plasmid containing the late promoter cleaved to give a 974 n runoff. Bulk DNA concentration was increased by addition of poly[d(I-C)]:poly[d(I-C)]. The transcription products were resolved on glyoxal gels as described above.

Reaction Conditions

One unusual feature of the WCE is the temperature dependence of the reaction. Transcription is routinely done at 30°, where the *in vitro* synthesized RNA product is stable for 8 hr. Increasing the temperature to 37° greatly enhances the rate of RNA degradation; RNA made at 30° is degraded within 10 min after shifting to 37°. Transcription assayed at 23° yields the expected Arrhenius effect.

Specific transcription in the WCE is highly sensitive to ionic strength. Concentrations of KCl and NaCl above 60 m*M* significantly inhibit the reaction; concentrations in the 30–40 m*M* range are optimal. Reactions can also be performed in 15–30 m*M* $(NH_4)_2SO_4$. The divalent cations

Ca^{2+}, Zn^{2+} and Mn^{2+} inhibit transcription and 0.5 m*M* EDTA is added to control their effect and the effect of other heavy metal contaminants. Reactions are done at pH 7.9, which is optimal for purified RNA polymerase II.[1]

Even after extensive dialysis, most WCEs seem to contain a free pool of 1 μM nucleotides.[18] The extract also contains creatine kinase (CPK) and other kinases and phosphatases so that the β and γ phosphates in nucleotides are labile.[21] For example, [γ-^{32}P]ATP will rapidly exchange label with other triphosphates. Label at the α position of the nucleotide triphosphates does not exchange in the WCE, thus allowing RNA to be uniquely labeled with ^{32}P in the α position of each triphosphate. Addition of 5 m*M* creatine phosphate to the reaction mix ensures charging of the triphosphates and allows reduction in triphosphate concentrations, thus permitting the use of higher specific activities.[21] Extracts from some cell lines tested lack CPK activity and the enzyme must be added exogenously to maintain nucleotide concentrations.[22] One must also recall that $(NH_4)_2SO_4$ inhibits creatine kinase. In the presence of creatine phosphate, concentrations of UTP, CTP, and GTP as low as 5 μM saturate specific transcription; higher concentrations (up to 500 μM) do not inhibit specific transcription.[21] Because of endogenous pools, the transcription reaction is not fully dependent on addition of these three nucleotides.[18] A higher concentration of ATP is required for optimal activity (50 μM); ATP concentrations above 500 μM inhibit the reaction.[21]

The dialyzed extract also contains sufficient *S*-adenosylmethionine (SAM) to methylate the 5′ ends of the *in vitro* transcripts.[5] Internal methylation has not been studied, however. Addition of exogenous SAM does not affect the reaction.[18,23]

Time Course

The rate of elongation in the WCE is approximately 300 nucleotides/min.[23] The rate *in vivo* is 10-fold higher, but one must recall the difference of 7° in temperature between the two. After DNA template, nucleotides, and extract are mixed, there is about a 5-min lag before specific transcription commences. The lag cannot be eliminated by preincubation of extract alone or with nucleotides, but is eliminated by preincubation of extract together with DNA (in the absence of nucleotides).[24] This suggests that the lag represents the time required for assembly on the DNA of fac-

[21] H. Handa, R. J. Kaufman, J. L. Manley, M. L. Gefter, and P. A. Sharp, *J. Biol. Chem.* **256,** 478 (1981).

[22] N. Crawford, unpublished results.

[23] R. Jove, unpublished results.

[24] M. Samuels and A. Fire, unpublished results.

tors required for initiation. The rate of accumulation of runoff transcripts is approximately linear for over an hour after the initial lag period.[5]

Some Other Properties of the WCE

The preparation procedure for the WCE was originally designed for solubilization of RNA polymerase II from mammalian cells. A standard 20 μl reaction mix typically contains 2–3 units of RNA polymerase II.[5] Under optimal conditions, at most one in ten polymerase II molecules gives rise to a specific transcript in a 1-hr reaction. Supplementation with excess purified polymerase has no significant effect on the WCE.[22] Endogenous RNA polymerase II in the WCE is inhibited by α-amanitin at 0.5 μg/ml; addition of a purified mutant enzyme resistant to α-amanitin reconstitutes specific transcription.[6,24,25]

RNA polymerase II preferentially initiates transcription at the termini of DNA fragments and at internal nicks.[26] The enzyme is also capable of end labeling DNA fragments with [α-^{32}P]NTPs to yield full-length labeled molecules, which are resistant to RNase digestion.[24] These reactions are each sensitive to α-amanitin (0.5 μg/ml), and are suppressed *in vitro* by the addition of a 110,000-dalton ADP-ribosyltransferase that may blockade nicks and ends.[27] This 110,000 dalton protein is present in large quantities in the WCE (up to 0.1% of total WCE protein).

The WCE should contain most of the soluble proteins in the cell. Most of these are of no concern; however, some can interfere with interesting experiments. Most extracts have high levels of topoisomerase type I and II activities as well as DNA ligase. Thus DNA topologies can change rapidly in the reaction mix, preventing, for instance, studies of supercoiled DNA. The extract also contains RNA polymerases I and III.[13] Their contribution to the background pattern can be assessed with α-amanitin. Most template DNAs do not contain promoters for these enzymes, and their contribution to background incorporation is small. Some genomic clones contain dispersed repetitive elements, which often contain polymerase III genes.

Partial Fractionation of the WCE

A number of inhibitory activities can be removed by fractionation on phosphocellulose, yielding a more efficient transcription extract.[24,28]

[25] C. J. Ingles, *Proc. Natl. Acad. Sci. U.S.A.* **75,** 405 (1978).
[26] M. K. Lewis and R. R. Burgess, *J. Biol. Chem.* **255,** 4928 (1980).
[27] E. Slattery, J. D. Dignam, and R. G. Roeder, unpublished results.
[28] T. Matsui, J. Segall, P. A. Weil, and R. G. Roeder, *J. Biol. Chem.* **255,** 11992 (1980).

In Vitro Transcription Studies Using a Whole-Cell Extract (WCE)

Template	Comments	References*
Adenovirus	Almost all *in vivo* promoters are recognized *in vitro*	2, 5, 7, 12, 18, 20
	Detailed 5′ terminal analysis, dependence on upstream sequences	18, 20, *a–d*
	Inactivation of transcription in WCEs of poliovirus-infected cells	*e*
	Changes in transcriptional pattern in WCEs of adenovirus-infected cells	18
Globins	α-Globin and β-globin genes are recognized *in vitro*, dependence on upstream sequences	*f–i*
	Mutant α- and β-thalassemia globin genes are transcribed *in vitro*	*g, j–l*
SV40	Early and late promoters are recognized *in vitro*	19, 21
	Detailed 5′-terminal analysis of RNA from early promoter	17, *m, n*
	In vitro inhibition of transcription by T antigen	17, 19, *o*
	Cell-free translation of *in vitro* synthesized RNA	*p*
Conalbumin and ovalbumin	Promoters are recognized *in vitro*, dependence on upstream sequences, effects of altering TATA box	7, *a, q, r*
	Transcription in a homologous system	29
Type C retroviruses	Promoters in the long terminal repeat (LTR) of several RNA tumor viruses are recognized *in vitro*, dependence on upstream sequences	*s–u*
Fibroin	Promoter is recognized *in vitro* by WCEs of HeLa cells and of silk worm glands, dependence on upstream sequence in both extracts	*v, w*
Herpes simplex virus	Early promoters are recognized *in vitro* in uninfected cell WCEs	*x*

Template	Comments	References*
Histone H2A	Promoter is recognized *in vitro,* dependence on upstream sequences using linear or circular DNA template	*y*
Adeno-associated virus	Identification of a new promoter, detailed 5′ terminal analysis	*z*

* *Key to references:* Numbers refer to text footnotes. Letters refer to the following.

a. J. Corden, B. Wasylyk, A. Buchwalder, P. Sassone-Corsi, C. Kedinger, and P. Chambon, *Science* **209,** 1406 (1980).
b. S.-L. Hu and J. L. Manley, *Proc. Natl. Acad. Sci. U.S.A.* **78,** 820 (1981).
c. O. Hagenbuchle and U. Schibler, *Proc. Natl. Acad. Sci. U.S.A.* **78,** 2283 (1981).
d. D. J. Mathis, R. Elkaim, C. Kedinger, P. Sassone-Corsi, and P. Chambon, *Proc. Natl. Acad. Sci. U.S.A.* **78,** 7383 (1981).
e. N. Crawford, A. Fire, M. Samuels, P. A. Sharp, and D. Baltimore, *Cell* **27,** 555 (1981).
f. D. S. Luse and R. G. Roeder, *Cell* **20,** 691 (1980).
g. N. J. Proudfoot, M. H. M. Shander, J. L. Manley, M. L. Gefter, and T. Maniatis, *Science* **209,** 1329 (1980).
h. C. A. Talkington, Y. Nishioka, and P. Leder, *Proc. Natl. Acad. Sci. U.S.A.* **77,** 7132 (1980).
i. G. C. Grosveld, C. K. Shewmaker, P. Jat, and R. A. Flavell, *Cell* **25,** 215 (1981).
j. R. A. Spritz, P. Jagadeeswaran, P. V. Choudary, P. A. Biro, J. T. Elder, J. K. DeRiel, J. L. Manley, M. L. Gefter, B. G. Forget, and S. M. Weissman, *Proc. Natl. Acad. Sci. U.S.A.* **78,** 2455 (1981).
k. S. H. Orkin, S. C. Goff, and R. L. Hechtman, *Proc. Natl. Acad. Sci. U.S.A.* **78,** 5041 (1981).
l. S. H. Orkin and S. C. Goff, *J. Biol. Chem.* **756,** 9782 (1981).
m. D. J. Mathis and P. Chambon, *Nature (London)* **290,** 310 (1981).
n. P. Lebowitz and P. K. Ghosh, *J. Virol.* **41,** 449 (1982).
o. R. M. Myers, D. C. Rio, A. K. Robbins, and R. Tjian, *Cell* **25,** 373 (1981).
p. C. L. Cepko, U. Hansen, H. Handa, and P. A. Sharp, *Mol. Cell. Biol.* **1,** 919 (1981).
q. S. Tsai, M.-J. Tsai, and B. W. O'Malley, *Proc. Natl. Acad. Sci. U.S.A.* **78,** 879 (1981).
r. B. Wasylyk and P. Chambon, *Nucleic Acids Res.* **9,** 1813 (1981).
s. T. Yamamoto, B. deCrombrugghe, and I. Pastan, *Cell* **22,** 787 (1980).
t. M. C. Ostrowski, D. Benard, and G. L. Hager, *Proc. Natl. Acad. Sci. U.S.A.* **78,** 4485 (1981).
u. L. A. Fuhrman, C. Van Beveren, and I. M. Verma, *Proc. Natl. Acad. Sci. U.S.A.* **78,** 5411 (1981).
v. Y. Tsujimoto, J. Hirose, M. Tsuda, and Y. Suzuki, *Proc. Natl. Acad. Sci. U.S.A.* **78,** 4838 (1981).
w. M. Tsuda and Y. Suzuki, *Cell* **27,** 175 (1981).
x. R. J. Frink, K. G. Draper, and E. K. Wagner, *Proc. Natl. Acad. Sci. U.S.A.* **78,** 6139 (1981).
y. R. Grosschedl and M. L. Birnstiel, *Proc. Natl. Acad. Sci. U.S.A.* **79,** 297 (1982).
z. M. R. Green and R. G. Roeder, *Cell* **22,** 231 (1980).

After dialysis to remove Mg^{2+} and dilution to 40 mM KCl, a WCE is chromatographed on phosphocellulose (Whatman P-11) yielding a breakthrough fraction, and two higher salt washes (0.35 and 1.0 M KCl). Reconstitution of the breakthrough and the dialyzed 1.0 M wash with purified RNA polymerase II in optimal ratios yields a mixture capable of specifically transcribing DNA at 10 times the efficiency of the original WCE. Tsai *et al.* have used a similar protocol to remove inhibitors from an extract of hen oviduct.[29]

Summary of Results

Since the first demonstration of transcription by RNA polymerase II in a soluble system and development of the WCE procedure a number of investigators have studied transcription of promoters using these systems. The preceding table represents a moderately comprehensive listing of such studies.

Acknowledgments

The work of R. Jove, H. Handa, U. Hansen, C. Cepko, N. Crawford, and A. Cano in helping to characterize the WCE and the advice and guidance of M. Gefter are gratefully acknowledged. This work was supported by Grant PCM78-23230 from the National Science Foundation, by Public Health Service Grants CA26717 (Program Project Grant) to P. A. S. and partially by CA14051 Center for Cancer Biology at MIT (Core). J. L. M. gratefully acknowledges support from Public Health Service Grant GM28983.

[29] S. Y. Tsai, M.-J. Tsai, L. E. Kops, P. P. Minghetti, and B. W. O'Malley, *J. Biol. Chem.* **256,** 13055.

[36] Eukaryotic Gene Transcription with Purified Components

By JOHN D. DIGNAM, PAUL L. MARTIN, BARKUR S. SHASTRY, and ROBERT G. ROEDER

In eukaryotic organisms the large ribosomal RNAs, messenger RNAs, and some low molecular weight RNAs (5 S RNA, tRNA, and some small viral RNAs) are transcribed by three structurally and functionally distinct enzymes, RNA polymerases I, II, and III, respectively (see reviews [1-3]). Although these enzymes are structurally complex, highly purified preparations of RNA polymerases I, II, and III are unable to execute accurate

[1] R. G. Roeder, *in* "RNA Polymerase" (R. Losick and M. Chamberlin, eds.), p. 285. Cold Spring Harbor Laboratory, New York, 1976.

[2] P. Chambon, *Annu. Rev. Biochem.* **44,** 613 (1975).

[3] M. R. Paule, *Trends Biochem. Sci.* **6,** 128 (1981).

ISBN 0-12-182001-7

transcription initiation on appropriate DNA templates.[1,2,4–9] In contrast, it has been shown that a variety of purified class III RNA polymerases mediate accurate transcription of class III genes (5 S, tRNA, adenovirus VA) in chromatin or nuclear templates, indicating the presence of additional components that are required for accurate transcription and remain stably associated with complex nucleoprotein templates.[10,10a] More recently, a number of less purified soluble systems have been developed and shown to be appropriate for analysis of transcription by all three classes of RNA polymerases.[6–9,11–13] These systems can accurately and selectively initiate transcription on specific genes in intact viral DNAs or in cloned fragments of viral or cellular genes. The first of these systems demonstrated accurate initiation and termination for genes transcribed by RNA polymerase III (5 S, tRNA, and VA-RNA)[8,11–13] and consisted of a purified DNA template (either viral DNA or a cloned viral or cellular gene) and a crude extract derived from cultured cells,[11,12] oocyte nuclei,[13] or whole oocyte.[8,14] Although these systems did not require exogenous RNA polymerase III (which was present in the extract with other putative factors), another reconstituted system was shown to require an exogenous (purified) RNA polymerase III, in addition to a crude soluble extract (from immature oocytes).[8] Subsequently, soluble systems were developed for genes transcribed by RNA polymerase II and consisted of a cloned cellular or viral DNA template and either a soluble extract from cultured cells supplemented with purified RNA polymerase II[9] or a whole cell extract containing endogenous RNA polymerase II.[15] Similar soluble systems have been reported for the transcription of ribosomal RNA gene promoters by RNA polymerase I.[16,17] What is apparent, at least from

[4] R. G. Roeder, R. H. Reeder, and D. D. Brown, *Cold Spring Harbor Symp. Quant. Biol.* **35,** 727 (1970).
[5] C. S. Parker and R. G. Roeder, *Proc. Natl. Acad. Sci. U.S.A.* **74,** 44 (1977).
[6] J. A. Jaehning, P. S. Woods, and R. G. Roeder, *J. Biol. Chem.* **252,** 8762 (1977).
[7] C. S. Parker, J. A. Jaehning, and R. G. Roeder, *Cold Spring Harbor Symp. Quant. Biol.* **42,** 577 (1978).
[8] S.-Y. Ng, C. S. Parker, and R. G. Roeder, *Proc. Natl. Acad. Sci. U.S.A.* **76,** 136 (1979).
[9] P. A. Weil, D. S. Luse, J. Segall, and R. G. Roeder, *Cell* **18,** 469 (1979).
[10] J. A. Jaehning and R. G. Roeder, *J. Biol. Chem.* **252,** 8753 (1977).
[10a] V. A. Sklar and R. G. Roeder, *Cell* **10,** 405 (1977).
[11] G. J. Wu, *Proc. Natl. Acad. Sci. U.S.A.* **75,** 2175 (1978).
[12] P. A. Weil, J. A. Segall, B. Harris, S.-Y. Ng, and R. G. Roeder, *J. Biol. Chem.* **254,** 6163 (1979).
[13] E. H. Birkenmeier, D. D. Brown, and E. J. Jordon, *Cell* **15,** 1077 (1978).
[14] D. R. Engelke, S.-Y. Ng, B. S. Shastry, and R. G. Roeder, *Cell* **19,** 717 (1980).
[15] J. L. Manley, A. Fire, A. Cano, P. A. Sharp, and M. L. Gefter, *Proc. Natl. Acad. Sci. U.S.A.* **77,** 3855 (1980).
[16] I. Grummt, *Proc. Natl. Acad. Sci. U.S.A.* **78,** 727 (1981).
[17] K. G. Miller and B. Sollner-Webb, *Cell* **27,** 165 (1981).

some studies of RNA polymerase II and III,[7,12] is that some components are provided in these extracts that enables the RNA polymerase accurately and selectively to transcribe the corresponding genes.

Establishing these soluble systems was an essential first step in the analysis of cellular components that give RNA polymerases the specificity and selectivity that they exhibit *in vivo,* but not in the highly purified state. On the one hand, they have allowed the deduction of gene sequences important for recognition by RNA polymerases and/or accessory factors (see review[18]). On the other hand, they have provided a starting point for the separation, isolation, and analysis of those factors required (in addition to the previously characterized RNA polymerases) for accurate transcription initiation *in vitro.* In this chapter we describe some of the procedures for the isolation of these auxiliary transcription components. Protocols are presented for the purification of a 5 S gene-specific transcription factor for RNA polymerase III (designated TFIIIA) from frog ovaries.[14] The last section presents a general protocol for the isolation of transcription factors for both RNA polymerase III and RNA polymerase II from cultured human cells.[19,20]

Materials and Methods

Materials. Ammonium sulfate is obtained from Schwarz-Mann, HEPES from Calbiochem, and nucleotide triphosphates are from P-L Biochemicals. Other reagents are obtained from either Sigma or Mallinckrodt. DEAE-cellulose (DE-52) and phosphocellulose (P-11) are from Whatman, and BioRex 70 is from Bio-Rad. All ion exchange resins are precycled according to the manufacturer's instructions prior to equilibration with starting buffer. Guanosine triphosphate (α-^{32}P, greater than 600 Ci/mol) is obtained either from New England Nuclear or Amersham. Bovine serum albumin (Pentex) is obtained from Miles. Single-stranded DNA cellulose (1.0–1.5 mg of DNA per milliliter of gel) is prepared as described by Alberts.[21]

Preparation of RNA Polymerase II. Human[19] and calf thymus[22] RNA polymerase II are prepared according to published procedures. Calf thymus RNA polymerase II is interchangeable with the human enzyme and is

[18] N. Heintz and R. G. Roeder, *in* "Genetic Engineering" Vol. 4, p. 57. Academic Press, New York, 1982.

[19] T. Matsui, J. Segall, P. A. Weil and R. G. Roeder, *J. Biol. Chem.* **255,** 11992 (1980).

[20] J. Segal, T. Matsui, and R. G. Roeder, *J. Biol. Chem.* **255,** 11986 (1980).

[21] B. Alberts and G. Herrick, this series, Vol. 21, p. 198.

[22] H. G. Hodo III and S. P. Blatti, *Biochemistry* **16,** 2334 (1977).

used routinely for assaying chromatographic fractions in complementation assays.

Conditions for Transcription with Class III Templates. Transcription assays are performed as described previously,[8,12] using plasmids containing *Xenopus laevis* tRNA or 5 S RNA genes or whole adenovirus DNA as templates (containing the adenovirus VA genes). Reaction mixtures (usually 50 μl) contain: 12.5 m*M* HEPES–NaOH (pH 7.7) or 40 m*M* Tris-HCl (pH 7.9), 10% glycerol, 70 m*M* KCl or 25 m*M* $(NH_4)_2SO_4$, 0.5 m*M* dithiothreitol (DTT), 5 m*M* $MgCl_2$, 600 μM ATP, 600 μM CTP, 600 μM UTP, 25 μM [α-^{32}P]GTP (4 mCi/mol), 10 μg of DNA per milliliter, and an aliquot of the appropriate crude extract or column fraction. (The concentrations given for buffer, glycerol, and salts include the contribution from the extract or column fraction.) Reactions are initiated by the addition of the extract or fractions. Mixtures are incubated for 60 min at 30°, the reactions are terminated and the RNA isolated as described below.

Conditions for Transcription with Class II Templates. Reaction conditions are as described previously[9] using as templates plasmids containing fragments of viral or cellular DNA. The plasmid DNAs are cleaved at a restriction site downstream from the promoter so that correct initiation and elongation generates an RNA that corresponds in size to the distance from the transcription start site to the restriction site. We have routinely used the clone pSmaF that contains the Adenovirus major late promoter[12] for assays of column fractions; when cleaved with *Sma*I and employed for specific transcription, this template generates an RNA transcript of 536 nucleotides. Transcription reactions contain in 50 μl: 12.5 m*M* HEPES–NaOH (pH 7.9), 12.5% glycerol, 60 m*M* KCl, 0.33 m*M* DTT, 0.13 m*M* EDTA, 8–12 m*M* $MgCl_2$, 600 μM ATP, 600 μM CTP, 600 μM UTP, 25 μM [α-^{32}P]GTP (8 mCi/mol), 10–20 μg of DNA per milliliter, and an aliquot of the appropriate crude extract or column fraction (the concentrations given for buffer, glycerol, and salts include the contributions from the extract or column fraction). The reaction mixtures are incubated 60 min at 30°, the reaction is terminated, and the RNA is isolated. Since the Mg^{2+} optimum varies for different templates and the DNA optimum varies for different extracts, these conditions must be optimized for each DNA template. For Adenovirus 2 major late (pSmaF), the Mg^{2+} optimum is 12 m*M* and the DNA optimum is 10–20 μg/ml. Although nuclear extracts contain sufficient endogenous RNA polymerase II, 100 units (1 unit equals 1 pmol of UTP incorporated in 10 min) of calf thymus or human RNA polymerase II are added to assays of column fractions. Reactions are initiated by the addition of extract (20–30 μl) or fractions (up to 30 μl).

Isolation of RNA from Reaction Mixtures. After incubation of reaction mixtures, 50 μl of a solution containing 0.5% SDS, 0.1 *M* sodium acetate (pH 5.5), and 1 mg of carrier RNA per milliliter are added. The samples are vortexed and extracted with 200 μl of 50% phenol, 50% chloroform. After separation of the phases by centrifugation, the aqueous phase is extracted with 200 μl of chloroform, then mixed with 250 μl of 90% ethanol, 0.1 *M* sodium acetate (pH 5.5). After freezing the samples on Dry Ice, the RNA is collected by centrifugation and the pellet dried.

Gel Electrophoresis and Autoradiography. RNA samples are dissolved in 15 μl of deionized 50% formamide containing 0.1% xylene cyanole and 0.1% bromophenol blue. Samples are applied to 12% acrylamide gels (for 5 S RNA, tRNA, and VA RNA) or 4% acrylamide gels containing 7 *M* urea (for longer transcripts obtained with class II templates). The electrophoresis buffer contains 0.09 *M* Tris, 0.09 *M* borate, and 0.0028 *M* EDTA. Electrophoresis is for 1 hr at 300 V for polymerase III transcripts or 2 hr at 300 V for polymerase II transcripts utilizing gels 14 cm wide, 12 cm long, and 1 mm thick. Gels are dried (when employed, urea is removed by soaking the gel in water for 20 min), and exposures are obtained using either Kodak XAR5 (1–2 hr) or BB5 (8–12 hr) X-ray film and using Dupont Cronex intensifying screens.

Isolation of a 5 S Gene-Specific Transcription Factor from *Xenopus laevis* Ovary

Overview

The dual 5 S RNA gene system in *Xenopus* provides an attractive model system not only for the analysis of simple eukaryotic genes, but also for the analysis of transcriptional regulation of developmentally regulated genes in eukaryotes. These genes have been isolated, cloned, and characterized by Brown and co-workers.[23,24] Studies of *in vitro* deletion mutants[25,26] have further shown that a region in the center of the gene is necessary and sufficient for the correct initiation of transcription of these genes in cell-free systems. To further understand the nature and mechanism of action of the factors responsible for 5 S gene transcription, our laboratory has emphasized the fractionation of transcriptionally competent extracts from various *Xenopus* tissues; this has led to the identification of at least three separate factors that are required for 5 S gene transcription, in addition to RNA polymerase III.[27] One of these factors was

[23] L. J. Korn and D. D. Brown, *Cell* **15,** 1145 (1978).
[24] R. C. Peterson, J. L. Doering, and D. D. Brown, *Cell* **20,** 131 (1980).
[25] S. Sakonju, D. F. Bogenhagen, and D. D. Brown, *Cell* **19,** 13 (1980).
[26] D. F. Bogenhagen, S. Sakonju, and D. D. Brown, *Cell* **19,** 27 (1980).
[27] B. S. Shastry, S.-Y. Ng, and R. G. Roeder, *J. Biol. Chem.* **257,** 12979 (1982).

previously purified to homogeneity and shown to be a 37,000-dalton polypeptide that is specifically required for 5 S genes and interacts with an intragenic region of 5 S genes.[14] Since this region is identical to the promoter site identified by Brown and collaborators,[25,26] this factor must be an initiation factor. This protein has also been shown to be associated with 5 S RNA in 7 S ribonucleoprotein storage particles in the oocyte, and this second function accounts for the exceptionally high levels of this protein in oocytes. Because the other transcriptional factors (IIIB and IIIC) have not been purified, this section will focus on the 5 S gene specific factor TFIIIA, and describes two methods of purification from *Xenopus* ovaries. The first method involves gentle chromatographic fractionation of extracts from large, mature ovaries, containing a mixture of oocytes at various developmental stages. The second method involves the isolation of ribonucleoprotein particles from immature ovaries followed by DE-52 chromatography of the particles and separation from RNA and elution of TFIIIA under denaturing conditions.

The first procedure, although lengthy, rigorously avoids denaturing conditions. The second procedure has the advantages of high yield and rapidity, but employs denaturing conditions that may alter the tertiary structure of the protein and possibly some of its functional properties. While the proteins isolated by these procedures appear to be structurally and functionally similar, as yet we do not know whether they are identical in all respects. Both proteins are active in reconstituted systems, but it is not known whether they have the same specific activity, nor is it known whether the primary structure is identical in both proteins. In view of the potential for differences between the proteins isolated by these two procedures, we have presented both protocols.

Purification of Transcription Factor TFIIIA from Mature Ovary

Materials. All the stock solutions and water used are sterile components.

Buffer A: 20 m*M* HEPES (pH 7.5), 20% Glycerol (v/v), 5 m*M* $MgCl_2$, 0.1 m*M* EDTA, 1.0 m*M* DTT

Buffer B: Equivalent to buffer A with 0.5 m*M* Spermine in place of 5 m*M* $MgCl_2$

Buffer C: 50 m*M* Tris-HCl (pH 7.9), 25% glycerol, 50 m*M* KCl, 0.1 m*M* EDTA, 2 m*M* DTT

Buffer D: Barth's modified saline HEPES (BMSH)[28], pH adjusted to 7.6 with NaOH, 88 m*M* NaCl, 2 m*M* Kcl, 1 m*M* $CaCl_2$, 1 m*M*

[28] R. A. Laskey, A. D. Mills, and N. R. Norris, *Cell* **10,** 237 (1977).

$MgSO_4 \cdot 7\ H_2O$, 0.5 m*M* Na_2HPO_4, 2.0 m*M* $NaHCO_3$, 15 m*M* HEPES

Animals: Adult *Xenopus laevis* are obtained from the South African Snake Farm (P.O. Box 6, Fish Hoek, Cape Province, South Africa) and maintained at 22° as previously described.[29]

Preparation of Unfertilized Egg Extract for Complementation Assay. The assay for the transcription factor (TFIIIA) is based on the observation[14] that S100 extracts of unfertilized *Xenopus* eggs do not actively support *in vitro* 5 S RNA synthesis but do support tRNA gene transcription. However, a combination of egg extract and an aliquot (1 μg of protein) of an extract derived from ovaries can direct the accurate transcription of 5 S RNA genes. The egg extract apparently provides all the other components, including RNA polymerase III, necessary for selective transcription of 5 S RNA genes.[14] This egg complementation assay greatly facilitates the analysis and provides a valuable system to detect functional TFIIIA during or after its purification. The following procedure describes the preparation of this extract.

Unfertilized eggs are obtained from adult (4–5 inch) female *Xenopus laevis* by the following procedure.[23,25] The dorsal lymph sacs of several female frogs are injected with 800–1000 units of human chorionic gonadotropin, and eggs are collected overnight in tanks containing buffer D, to which additional sodium chloride is added to give 110 m*M*. The eggs are completely dejellied by gentle swirling in 2% cysteine hydrochloride in water made pH 7.8 with sodium hydroxide. Dejellied eggs are rinsed four times with buffer D (400–500 ml) and three times with buffer C (50–100 ml), and all the floating bad eggs are removed by decantation. The packed volume (about 10 ml per frog) is measured, and the eggs are transferred immediately to a 40-ml Kontes glass homogenizer (precooled in ice). The eggs are homogenized in 2 volumes of buffer C (precooled) with 10 strokes of a loose-fitting Teflon pestle. The homogenate is centrifuged at 3000 *g* for 20 min at 4°. The yellow yolk that floats is removed with a spatula, and the supernatant is decanted to a cold beaker and centrifuged at 100,000 *g* for 90 min at 4°. The supernatant is aspirated slowly without mixing the loosely sedimented pellet and is stored as 50-μl aliquots at −80°. This extract is stable for several months without freezing and thawing. Generally 20 μl of egg extract and 1–5 μl of the fraction to be tested are used in the assay.

[29] J. B. Gurdon and H. R. Woodland, *in* "Handbook of Genetics" Vol. 4, (R. C. King, ed.), Vol. 4, pp. 35–50. Plenum, New York, 1971.

Isolation of TFIIIA[14]

Step 1. Preparation of Crude Ovarian Extract. About 100 g of mature ovaries (predominantly large oocytes) are dissected from 3–10 adult female *Xenopus laevis* and rinsed once in cold 0.15 *M* sodium chloride, 0.015 *M* sodium citrate, pH 7.0 (1 × SSC) and once in cold buffer C and blotted to remove the excess buffer. (All the subsequent steps are carried out at 4°.) Ovaries are transferred to a cold beaker and minsed with scissors in 50 ml of buffer C. The ovaries are then dispersed in another 150 ml of buffer C with 10 strokes of a Teflon pestle in a Kontes glass homogenizer, followed by 1 stroke with a glass pestle, A. The homogenate is filtered through one layer of cheese cloth and rehomogenized with five strokes of a B pestle. The homogenate is centrigued at 3000 *g* for 20 min. The yellow yolk is removed, and the supernatant is recentrifuged at 100,000 *g* at 4° for 90 min. The supernatant is aspirated slowly. This extract, which is designated S100, is assayed and an aliquot is frozen at −80°. About 125 ml of S100 (10 mg of protein per milliliter) is obtained by this procedure. [Note: It is advisable to avoid the accumulation of blood during the dissection of ovaries, since yields are lower when blood-contaminated ovaries are used. Avoid mixing the loosely sedimented pellet with the supernatant, since this contaminating material may cause problems in subsequent steps.]

Step 2. Ammonium Sulfate Precipitation. To 120 ml of S100 is added 0.3 g of ammonium sulfate per milliliter over 30 min with constant slow stirring. After complete addition, the solution is stirred an additional 30 min and centrifuged at 100,000 *g* for 1 hr at 4°. The supernatant is discarded, the side of the tube is slowly rinsed with buffer A, and the rinse is discarded. The pellet is resuspended in buffer A to a final ammonium sulfate concentration of 50 m*M* (volume was 90–100 ml). This extract can be stored at −80°. About 900–1000 mg of protein are obtained from this step.

Step 3. DEAE-Cellulose Chromatography. All chromatography is performed in columns with a height-to-width ratio of 2–3:1. Columns are equilibrated immediately prior to use with the buffer in which the sample is applied.

The ammonium sulfate fraction (95 ml, step 2) is applied to an 80-ml DEAE-cellulose column that has been previously equilibrated with buffer A containing 50 m*M* ammonium sulfate at a flow rate of 2 column volumes per hour. The column is washed with equilibrating buffer (3 column volumes) at a flow rate of 2000 ml/hr. The bound protein is step-eluted with buffer A containing 125 m*M* ammonium sulfate. After determining the absorption at 280 nm, 3–4 peak fractions are pooled (about one-fifth of input protein is recovered in the step fraction).

Step 4. DEAE-Sephadex A-25 Chromatography. The ammonium sulfate concentration of combined fractions (step 3) is readjusted to 125 m*M*, and the fraction is applied to an 80-ml DEAE-Sephadex A-25 column equilibrated with buffer A containing 125 m*M* ammonium sulfate. After the measurement of optical density of 280 nm of the flowthrough fraction, 3–4 fractions across the peak are combined (95% of the input protein is in the breakthrough fraction) and dialyzed for 5 hr against buffer B containing 100 m*M* ammonium sulfate. This chromatographic step removes residual nucleic acids and RNA polymerase III, which would otherwise interfere in the subsequent purification. [Note: Long dialysis may cause aggregation and subsequent precipitation of the factor and hence is to be avoided.]

Step 5. Adsorption and Elution on Phosphocellulose (P-11). The dialyzate from step 3 is applied at 50 ml/hr to a 25-ml phosphocellulose column equilibrated with buffer B containing 100 m*M* ammonium sulfate. The column is washed with equilibrating buffer (2 column volumes), and the bound material was eluted with a 5 column volume linear 95–700 m*M* ammonium sulfate gradient in buffer B. The size of each gradient fraction is about one-fifth of the column volume. Gradient fractions are dialyzed against buffer A containing 50 m*M* ammonium sulfate and frozen at −80° after removing aliquots for assays; TFIIIA is located in the fraction using the egg complementation assay. Since the protein content of gradient fractions is small, optical density measurements are not possible at this step. [Note: After this step of purification the gradient fractions are collected in plastic tubes, since the pure protein sticks to glass.]

Step 6. Adsorption and Elution on BioRex-70 (BR-70). The peak active fractions that elute between 300 and 600 m*M* ammonium sulfate are thawed, combined, and applied at 12 ml/hr to a 4-ml BioRex-70 column previously equilibrated with buffer A containing 50 m*M* ammonium sulfate. The column is washed with 2 column volumes of equilibrating buffer, and the bound protein is eluted with a 5 column volume linear 50–500 m*M* ammonium sulfate gradient. The fractions are about 0.8 ml (one-fifth of the column volume). The gradient fractions are assayed for transcription factor activity by complementation with the egg extract, and the active fractions are frozen individually at −80°, where the activity is stable for several months.

Purity and Recovery. Sodium dodecyl sulfate (SDS) gel electrophoresis is used to analyze the transcription factor obtained by this procedure. The preparations are routinely about 95% pure. About 500 μg of purified protein are routinely obtained from 100 g of ovaries.

Purification of Transcription Factor TFIIIA from Immature Ovaries

Transcription factor TFIIIA detected by direct assay (above) is shown to be equivalent to a protein bound to 5 S ribosomal RNA in immature oocytes in the form of 7 S RNP particles.[30,31] This second method of purification is based on this observation. The procedure is essentially as described by Picard and Wegnez[32] as modified by Pelham and Brown[30].

Materials

Buffer A: 15% sucrose (w/v), 50 m*M* Tris-HCl (pH 7.5), 5 m*M* $MgCl_2$, 25 m*M* KCl, 0.25 m*M* DTT
Buffer B: 30% sucrose (w/v), 50 m*M* Tris-HCl (pH 7.5), 5 m*M* $MgCl_2$, 25 m*M* KCl, 0.25 m*M* DTT
Buffer C: 50 m*M* Tris-HCl (pH 7.5), 5 m*M* $MgCl_2$, 25 m*M* KCl, 0.25 m*M* DTT
Buffer D: 0.15 *M* sodium chloride, 0.015 *M* sodium citrate, pH adjusted to 7.0
Buffer E: 20 m*M* Tris-HCl (pH 7.5), 0.2 *M* KCl, 0.25 m*M* DTT
Buffer F: 50 m*M* Tris-HCl (pH 7.5), 0.25 m*M* DTT
Buffer G: 50 m*M* Tris-HCl (pH 7.5), 7 *M* urea, 2.5 m*M* DTT
Buffer H: 20 m*M* Tris-HCl (pH 7.5), 20% glycerol (v/v), 50 m*M* KCl, 2 m*M* $MgCl_2$, 2 m*M* DTT
Buffer I: 10 m*M* Tris-HCl (pH 7.5), 0.25 m*M* DTT
Animals. Young *Xenopus laevis* (3–5 cm) are obtained from Nasco, Fort Atkinson, Wisconson, and are maintained at 22° as previously described.[29]

Isolation of TFIIIA

Step 1. Preparation of Immature Ovary Extracts. Twenty ovaries from young *Xenopus laevis* are rinsed with 10 ml of buffer D, then 10 ml of buffer C, and weighed (1.2 g). Ovaries are transferred to a precooled 7-ml Dounce homogenizer and homogenized in 3 ml of buffer C with 10 strokes of a glass pestle B. The homogenate is centrifuged at 20,000 *g* for 15 min at 4°, and the supernatant (3–5 ml) is collected.

Step 2. Preparation of 7 S Ribonucleoprotein Particles. Sucrose gradients of 15–30% are prepared in a series of 12-ml polyallomer tubes using solutions A and B. The supernatant from step 1 is layered (0.5 ml per tube) on the top of the sucrose gradient. The gradients are centrifuged in a Beckman SW 40 rotor at 36,000 rpm for 26 hr at 4°; alternatively, a

[30] H. R. B. Pelham and D. D. Brown, *Proc. Natl. Acad. Sci. U.S.A.* **77,** 4170 (1980).
[31] B. M. Honda and R. G. Roeder, *Cell* **22,** 119 (1980).
[32] B. Picard and W. Wegnez, *Proc. Natl. Acad. Sci. U.S.A.* **76,** 241 (1979).

Beckman VTi50 rotor may be used, in which case the samples are centrifuged 6 hr at 50,000 rpm. The gradients are fractionated (0.6-ml fractions), and the 7 S RNP particles are located by their absorbance at 260 nm. Under these conditions of centrifugation, the 7 S RNP particles form a prominent peak approximately half-way from the top of the gradient.

Step 3. Purification of 7 S RNP Particle Protein. The three peak fractions from each gradient are combined, adjusted to 5 m*M* EDTA and loaded onto a 1.0-ml DEAE-cellulose column previously equilibrated with buffer I. The column is washed with 10 ml of buffer E, and then 5 ml of buffer F. The protein is eluted with buffer G. The protein-containing fractions are pooled and dialyzed for 6–8 hr against buffer H and frozen in small aliquots at −80°.

Comments. This second method is more rapid and efficient in terms of yield for a given amount of starting material. One gram of tissue yields 700–1000 μg of purified protein. The protein isolated by this method is at least 90% pure on SDS-polyacrylamide gel electrophoresis and is functionally similar to the protein isolated from mature oocytes.[30,31]

Isolation of Transcription Factors for RNA Polymerase II and III from HeLa Cells

Soluble Extracts

To date, cell-free systems for *in vitro* transcription derived from cultured cells have been of two types: the first reported was an S100 extract,[11] which is a postribosomal supernatant prepared at 150 m*M* KCl; the second was a "whole cell extract," which is a high-salt extract of a cellular homogenate,[15] an approach originally employed for the quantitative solubilization and purification[33,34] of eukaryotic RNA polymerase II. The former method, which was originally developed in this laboratory, continues to be successfully employed[35] and is advantageous for some studies because it lacks endogenous RNA polymerase II. Because we encountered difficulties in reproducibly preparing active extracts by the former procedure, we developed a method for the extraction of the transcription components from nuclei. It is noteworthy that these nuclei are isolated at low ionic strength (0.01 *M* KCl), which appears to favor the retention of most of the components for both RNA polymerase II and III within them. These transcription components are then extracted from the nuclei with 0.42 *M* NaCl. We have found that this procedure, in addition to being simple and reproducible, achieves significant separation of the transcription factors from contaminating cytoplasmic and nuclear compo-

[33] B. Sugden and W. Keller, *J. Biol. Chem.* **248,** 3777 (1973).

[34] R. G. Roeder and W. J. Rutter, *Proc. Natl. Acad. Sci. U.S.A.* **65,** 675 (1970).

[35] D. Mathis and P. Chambon, *Nature (London)* **290,** 310 (1981).

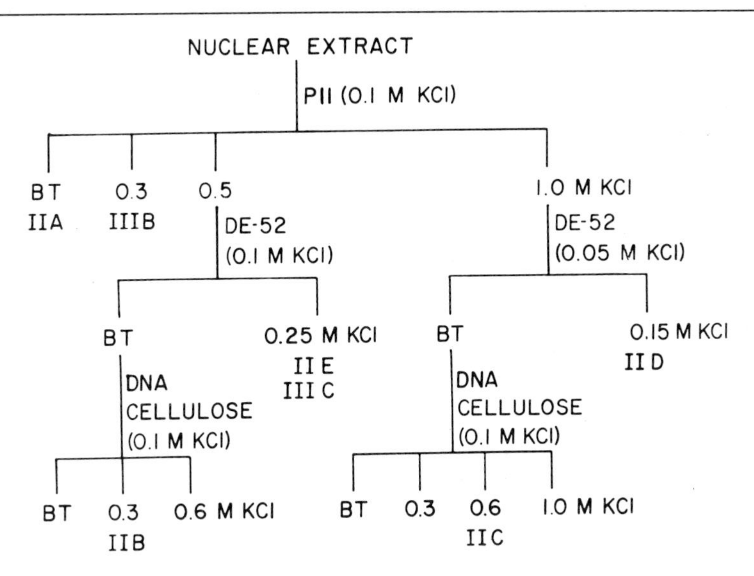

FIG. 1. Schematic representation of chromatographic resolution of class III and class II gene transcription factors from HeLa cell extract. The putative transcription factors present in the various fractions are indicated in the figure (see text for details).

nents. Using these extracts, we have separated four RNA polymerase II transcription components (designated TFIIA, TFIIB, TFIIC, and TFIID) previously identified in an S100[19] and an additional component (designated TFIIE), which was not detected previously. As reported previously for the S100 fractionation,[20] two RNA polymerase III transcription components (designated TFIIB and TFIIIC) can also be resolved from this extract, but the 5 S RNA gene specific transcription component (designated TFIIIA) remains in the soluble S100 fraction. The human 5 S gene specific factor, TFIIIA, appears to be functionally equivalent to the frog TFIIIA[20] and is also not required for tRNA transcription. Only components TFIIIB and TFIIIC and RNA polymerase III are required for the transcription of tRNA and Adenovirus VA genes. Figure 1 summarizes the separation of these components, the details of which are described in the accompanying text.

Extract Preparation

Phosphate-buffered saline (PBS)
Buffer A: 10 m*M* HEPES (pH 7.9), 1.5 m*M* $MgCl_2$, 10 m*M* KCl, 0.5 m*M* DTT, 0.5 m*M* PMSF
Buffer B: 0.3 *M* HEPES (pH 7.9), 1.4 *M* KCl, 0.03 *M* $MgCl_2$,
Buffer C: 20 m*M* HEPES (pH 7.9), 25% glycerol, 0.42 *M* NaCl, 1.5 m*M* $MgCl_2$, 0.2 m*M* EDTA, 0.5 m*M* PMSF, 0.5 m*M* DTT

Buffer D: 20 m*M* HEPES (pH 7.9), 20% glycerol, 0.1 *M* KCl, 0.2 m*M* EDTA, 0.5 m*M* DTT, 0.5 m*M* PMSF

DTT and PMSF are added to the buffers just before use. PMSF should be dissolved in ethanol or isopropanol as a 0.25 *M* stock solution. It is necessary to add the PMSF drop by drop with stirring to the buffers so that it goes into solution.

Procedure. Cells are harvested from tissue culture media by centrifugation (1000 *g* for 10 min) and suspended in 5 volumes of cold PBS. All subsequent operations are done at 0–4°. The cells are collected again by centrifugation (1000 *g* for 10 min) and suspended in five volumes of buffer A, allowed to stand on ice for 10 min, and collected by centrifugation (1000 *g* for 10 min). The cells are suspended in two volumes of buffer A, homogenized by 10 strokes of a glass Dounce homogenizer (B pestle), and checked for cell lysis microscopically (lysis should be greater than 90%). The homogenate is spun for 10 min at 1000 *g*, the supernatant is carefully decanted, and the nuclear pellet is spun for 20 min at 25,000 *g*. The supernatants from the two spins are combined, 0.11 volume of buffer B is added, and the preparation is centrifuged for 60 min at 100,000 *g*. The 100,000 *g* supernatant from this step is dialyzed for 4–5 hr against 20 volumes of buffer D (this fraction is designated S100). While the S100 fraction is being prepared, the nuclear pellet obtained after the 25,000 *g* spin is homogenized in 2.5 ml per 10^9 cells of buffer C by 10 strokes of a Dounce homogenizer (B pestle) and stirred for 30 min. This material is spun for 30 min at 25,000 *g*, and the supernatant is dialyzed for 4–5 hr against 50 volumes of buffer D (this fraction is designated the nuclear extract). Both the S100 and nuclear extract are clarified after dialysis by centrifugation at 25,000 *g* for 30 min to remove precipitated material and lipid. The S100 and nuclear extract are quick frozen in liquid nitrogen and stored at −80°. These preparations are stable for at least 4 months under these conditions. Starting from 20 g of HeLa cells, one obtains approximately 20 mg of protein per 10^9 cells in the nuclear extract and 40 mg of protein per 10^9 cells in the S100.

It is noteworthy that the salt concentration used for extraction of the nuclei is critical since extracts prepared from nuclei at 0.3 *M* and 0.6 *M* NaCl are significantly less active, and extracts prepared at 0.2 or 0.8 *M* NaCl are inactive.

Partial Resolution of Transcription Components for RNA Polymerases II and III

The following procedure is based on the isolation of components from 35 g of cells, but can be adjusted for different amounts of starting material. The first chromatographic step (phosphocellulose) gives partial resolution of components of polymerases II and III.

Approximately 250 mg of nuclear extract protein (40 ml, 25 mg per milliliter of bed volume) is loaded onto a 10-ml bed volume column of phosphocellulose (1.8 cm $\times$ 4 cm) equilibrated in buffer D at 30 ml/hr. Fractions corresponding to 20% of the bed volume (2 ml) are collected. The column is washed with buffer D at the same flow rate until the A_{280} is below 0.5 (approximately 2 column volumes). The breakthrough fraction contains component TFIIA. The column is then step eluted with 3 column volumes of buffer D containing 0.3 *M* KCl. The 0.3 *M* KCl fraction contains component TFIIIB. Components TFIIB, TFIIE and TFIIIC are eluted with 2.5 column volumes of buffer D containing 0.5 *M* KCl and can be resolved from one another in subsequent chromatographic steps. Components TFIIC and TFIID are eluted with 2.5 column volumes of buffer D containing 1.0 *M* KCl and 0.25 mg of BSA per milliliter. These two components are separable by chromatography on DEAE-cellulose. While fractions corresponding to 1–2 column volumes can be pooled based on ultraviolet absorbance (greater than 1.0 at 280 nm), some side fractions should be retained and assayed for activity. The step fractions are dialyzed against 100 volumes of buffer D in small-diameter dialysis tubing for 2 hr; it is important to avoid long dialyses, particularly for component TFIID, which is less stable than the other components. The recovery of protein from the material applied to the column is 50–60% in the breakthrough (at 4.0–6.0 mg/ml), 15–20% in the 0.3 *M* KCl step fraction (at 3.0–4.0 mg/ml), 7–10% in the 0.5 *M* KCl step fraction (at 2.0–3.0 mg/ml), and 4–5% in the 1.0 *M* KCl step fraction (at 1.0–2.0 mg/ml). The fractions may be quick frozen at this point or may be used immediately for the subsequent chromatographic steps. The column fractions are assayed by complementation. Generally, 5 μl of the breakthrough fraction (TFIIA), 10 μl of the 0.5 *M* KCl step fraction (TFIIB and TFIIE), 10 μl of the 1.0 *M* KCl step fraction (TFIIC, and TFIID), and 100 units of calf thymus RNA polymerase II are active for the transcription of the Adenovirus 2 major late promoter (pSmaF). Addition of the 0.3 *M* KCl step fraction inhibits transcription of class II templates. Although activity can be obtained without additional RNA polymerase II, addition of more enzyme greatly increases the signal. Since the addition of too much RNA polymerase II generates a high background of random transcription, several amounts of enzyme should be tested.

For the transcription of class III genes, the combination of the 0.3 *M* KCl step fraction (TFIIIB) and the 0.5 *M* KCl step fraction (TFIIIC) is active. Adenovirus VA and *Xenopus* tRNA gene transcription can be reconstituted with 5 μl each of these two fractions. The transcription of 5 S genes (*Xenopus* or human) requires an additional component (TFIIIA)[20] that is absent in the nuclear extract but can be supplied by addition of 5 μl of the S100 fraction. Since the 0.3 and 0.5 *M* KCl step fractions contain endogenous RNA polymerase III, it is not necessary to add additional enzyme.

Resolution of TFIIB from TFIIE and TFIIIC. The 0.5 *M* KCl step fraction from phosphocellulose (20 mg of protein) is applied to a column of DEAE-cellulose equilibrated in buffer D at 7 mg of protein per milliliter of bed volume (the bed volume is 2.5–3 ml) at 2 column volumes per hour and is washed with 2 column volumes of buffer D. The breakthrough fraction contains component TFIIB; fractions with absorbance at 280 nm over 0.1 are pooled. Components TFIIE and TFIIIC can be eluted with two column volumes of buffer D containing 0.25 *M* KCl. The breakthrough fraction contains 40–50% of the applied protein (at 0.6–0.8 mg/ml), and the 0.25 *M* KCl step contains 20–30% of the applied protein (at 2.0–3.0 mg/ml). The fractions are assayed for RNA polymerase II components by using 7 μl of the breakthrough fraction (TFIIB) with 7 μl of the 0.25 *M* KCl step fraction to complement the phosphocellulose breakthrough (TFIIA) fraction and 1.0 *M* KCl step fraction (TFIIC and TFIID). The RNA polymerase III component in the 0.25 *M* KCl step fraction, TFIIIC, can be detected by complementing 5 μl of the 0.3 *M* KCl step fraction of phosphocellulose with 5 μl of the DEAE 0.25 *M* KCl step fraction. Although not detailed here, components TFIIE and TFIIIC can be separated from one another by gel filtration on agarose 1.5 m.

Component TFIIB can be purified further on single-stranded DNA cellulose. The breakthrough fraction from DEAE cellulose (6.0 mg) is applied to a 1.0 ml bed volume column of single-stranded DNA cellulose at 10 ml/hr, and 0.3 ml fractions were collected. The column is washed with 5.0 ml of starting buffer and eluted with 5.0 ml of buffer D containing 0.3 *M* and 0.6 *M* KCl. The breakthrough fraction contains 60–70% of the applied protein (at 0.6–0.8 mg/ml), the 0.3 *M* KCl step fraction contains 15–20% of the applied protein (at 2.0–3.0 mg/ml), and the 0.6 *M* KCl step fraction contains 5–10% of the applied protein (at 1.0–2.0 mg/ml). Fractions with absorbance at 280 nm over 0.1 are pooled. The 0.3 *M* KCl step fraction contains TFIIB, which can be assayed by complementing 5.0 μl of this fraction with 5.0 μl of the phosphocellulose breakthrough fraction (TFIIA), 10 μl of the phosphocellulose 1.0 *M* step fraction (TFIIC and TFIID) and 5.0 μl of the DEAE-cellulose 0.25 *M* step fraction (TFIIE).

Resolution of Components TFIIC and TFIID on DEAE-Cellulose. The separation of these components is achieved on DEAE-cellulose. The 1 *M* KCl step fraction from phosphocellulose is brought to 0.05 *M* KCl either by dilution with an equal volume of buffer D lacking KCl or by a brief dialysis. The sample (approximately 10 mg of protein) is applied to a

2.5 ml of DEAE-cellulose column (1 cm × 2 cm) equilibrated in buffer D containing 0.05 *M* KCl, and 0.5-ml fractions are collected. Component TFIIC is in the breakthrough fraction. The column is washed with three column volumes of starting buffer, and component TFIID is eluted with three column volumes of buffer D containing 0.15 *M* KCl. Fractions with absorbance at 280 nm over 0.1 are combined. The breakthrough fraction contains 5–10% of the applied protein (0.2–0.5 mg/ml), and the 0.15 *M* KCl step contains 30–40% of the applied protein (2–3 mg/ml). The fractions are assayed by complementing the phosphocellulose breakthrough (TFIIA) and 0.5 *M* KCl step fractions (TFIIB, TFIIE) with 5.0 μl of the DEAE cellulose breakthrough fraction (TFIIC), 5.0 μl of the DEAE-cellulose 0.15 *M* KCl step fraction, and 100 units of calf thymus RNA polymerase.

While component D is required to obtain specific transcription, component TFIIC serves only to suppress a background of random transcription that is initiated at nicks in the template (E. Slattery, unpublished). We have identified component TFIIC as poly(ADPribose)polymerase. Although the enzymic activity of poly(ADPribose)polymerase (TFIIC) does not appear to be involved in transcription initiation or chain elongation, its reported property of binding to nicks in DNA[36] explains its ability to suppress transcription initiated at nicks. Component TFIIC can be purified further by chromatography on single-stranded DNA cellulose. The breakthrough fraction from DEAE-cellulose is brought to 0.1 *M* KCl and applied to a 0.5-ml column of single-stranded DNA cellulose equilibrated in buffer D containing 0.1 mg of BSA per milliliter. The column is eluted successively with 3 ml of buffer D containing 0.1 *M*, 0.3 *M*, 0.6 *M*, and 1 *M* KCl, all containing 0.1 mg of BSA per milliliter. TFIIC (polyADPribose polymerase) elutes at 0.6 *M* KCl and contains predominantly a single polypeptide (in addition to serum albumin) of 110,000 molecular weight.

Isolation of TFIIIA from the HeLa Cell S100 Fraction. As described above, the nuclear extract contains factors that are sufficient to allow specific and accurate transcription of class II genes. It has also been determined that the nuclear extract is sufficient to allow specific and accurate transcription of tRNA and VA-RNA genes, which are transcribed by RNA polymerase III. However, the nuclear extract is unable to transcribe 5 S RNA genes. The human transcription factor IIIA, which is analogous in function to *Xenopus* TFIIIA, is either absent or inactive in the nuclear extracts. However, when the low-salt S100, derived from the nuclear ex-

[36] Y. Yoshihara, T. Hashida, Y. Tanaka, N. Matsunami, A. Yamaguchi, and T. Kamiya, *J. Biol. Chem.* **256,** 3471 (1981).

tract preparation, is mixed with the nuclear extract, specific and accurate transcription of the 5 S gene is reconstituted. Therefore, we have concluded that most of the active human TFIIIA is present in the low-salt S100 and only low amounts in the nuclear extract. There is variability in activity among nuclear extracts and their associated S100s. This apparently relates to the partitioning of TFIIIB and TFIIIC. Usually most of the TFIIIB and TFIIIC is found in the nuclear extract. Sometimes as much as half of the TFIIIB and TFIIIC is found in the S100, the remaining half being associated with nuclear extract. The basis for this partitioning has not been extensively studied.

To separate the TFIIIA from any remaining TFIIIB and TFIIIC, the S100 is chromatographed over phosphocellulose. The conditions of this column are the same as described above for the nuclear extract. TFIIIA is found in the breakthrough fraction, along with about 80% of the protein. Any contaminating TFIIIB is eluted with a 0.3 *M* KCl step, and any TFIIIC present is eluted with a 0.5 *M* KCl step. There appears to be no functional difference between the TFIIIB or TFIIIC derived from nuclear extracts or S100s.

Concluding Remarks

Our goal in these studies has been to reconstruct *in vitro* transcription regulatory events observed *in vivo*. To this end we have concentrated on the purification of the protein components involved so that we can examine protein–protein and protein–nucleic acid interactions that mediate events in transcription. The utility of this approach has already been demonstrated for the 5 S RNA gene-specific factor from *Xenopus*.[14] Although most of the transcription components for RNA polymerases II and III have not as yet been purified to homogeneity, *in vitro* reconstitution with at least partially purified components should provide insight into the basic processes of transcription and transcription regulation.

[37] Bacterial *in Vitro* Protein-Synthesizing Systems

By GLENN H. CHAMBLISS, TINA M. HENKIN, and JUDITH M. LEVENTHAL

In vitro protein-synthesizing systems can be useful in the analysis of gene expression at the levels of transcription and translation. In this chapter we describe the details of systems using cell extracts of *Escherichia*

ISBN 0-12-182001-7

coli or *Bacillus subtilis* for either mRNA-dependent[1] or DNA-dependent[2] protein synthesis *in vitro*.

Preparation of Cell Fractions

Bacillus subtilis 168T$^+$ (obtained from P. Schaeffer) and *E. coli* MRE 600 (ribonuclease I$^-$, obtained from W. McClain) are grown at 37° with vigorous aeration in CHT50 medium[3] (0.125% casein hydrolyzate, 0.2% $(NH_4)_2SO_4$, 1.4% K_2HPO_4, 0.6% KH_2PO_4, 0.1% sodium citrate, 0.5% glucose, 0.5 m*M* $CaCl_2$, 0.8 m*M* $MgSO_4$, 50 μg of tryptophan per milliliter). Growth is monitored with a Bausch & Lomb Spectronic 20 spectrophotometer, and cells are harvested in mid-exponential phase (A_{570} = 0.5). Cultures are poured over ice, and the cells are collected by centrifugation in a refrigerated Sharples centrifuge. Cells are washed twice with buffer I (0.01 *M* Tris-HCl, pH 7.5, 0.015 *M* $MgCl_2$, 1.0 *M* KCl, 5 m*M* EDTA, 10% glycerol, 0.01 *M* 2-mercaptoethanol) and once with buffer II (buffer I with 0.05 *M* KCl), then frozen rapidly and stored at −80°. Cells are kept on ice throughout the harvesting and washing procedure. All buffers are autoclaved then stored at 4°, and 2-mercaptoethanol is added just before use.

Frozen pellets of *B. subtilis* or *E. coli* cells are thawed on ice and resuspended in buffer III [0.01 *M* Tris-HCl, 0.015 *M* $MgCl_2$, 0.06 *M* NH_4Cl, 10% glycerol, 5 m*M* 2-mercaptoethanol, 3.45 m*M* phenylmethylsulfonylfluoride (PMSF, Sigma), 0.2 m*M* diisopropyl fluorophosphate (DFP, Sigma), pH 7.0]. Addition of protease inhibitors to buffer III is carried out as follows.

1. Heat the buffer to 37°.
2. While stirring the heated buffer III, add PMSF dissolved in ethanol (0.6 g of PMSF in 25 ml of ethanol per liter of buffer III), submerging the tip of the pipette beneath the surface of the liquid to prevent precipitation and to avoid release of toxic fumes.
3. After PMSF addition, adjust pH to 7.0.
4. Cool buffer to 25°, then add 2-mercaptoethanol (0.35 ml/liter).
5. Readjust the pH to 7.0
6. Add DFP (1 ml of a 0.2 *M* stock in isopropanol per liter).
7. Make final adjustment of pH to 7.5 at 4°.

Addition of inhibitors takes place immediately before buffers are to be used. Resuspension volume is 1.5 ml of buffer III per gram wet weight of

[1] L. Legault-Demare, and G. H. Chambliss, *J. Bacteriol.* **120,** 1300 (1974).
[2] J. M. Leventhal, and G. H. Chambliss, *Biochim. Biophys. Acta* **564,** 162 (1979).
[3] C. Anagnostopoulos and J. Spizizen, *J. Bacteriol.* **81,** 741 (1961).

cells; this volume is used in all later resuspension steps unless otherwise indicated. The cells are broken by two passages through a chilled French pressure cell (Aminco) at 16,000 psi. Cell extracts are kept chilled at all times. The lysate is stirred for 30 min to shear DNA, and is then centrifuged in a refrigerated Sorvall centrifuge (SS 34 rotor) at 13,000 rpm for 20 min. The supernatant is carefully decanted and centrifuged again. The final supernatant (S-30 extract) is fractionated into a ribosomal pellet and a high speed supernatant (S-150) by centrifugation at 45,000 rpm for 4.1 hr at 4° (Beckman L5-75B ultracentrifuge, 50.2 Ti rotor). The supernatant (S-150) is dialyzed for 4 hr against four 1-liter changes of buffer III, divided into small aliquots, frozen rapidly, and stored at −80°. Immediately after centrifugation the ribosomal pellet is resuspended in buffer III by gentle agitation. When the ribosomal pellet is fully resuspended, deoxyribonuclease I (Sigma) is added to 1 μg/ml to decrease endogenous activity in the DNA-dependent translation system, and ribosomes are stirred for 30 min at 4°. The ribosome suspension is centrifuged at 10,000 rpm for 10 min to remove membrane material, and the suspension is ultracentrifuged at 45,000 rpm for 4.1 hr (50.2 Ti rotor). The ribosome pellet is resuspended in buffer IV (buffer III in which the NH_4Cl concentration is increased to 1.0 *M*) and agitated gently for 16 hr. The ribosome suspension is clarified by centrifugation at 10,000 rpm for 10 min, and the ribosomes are collected from this supernatant by ultracentrifugation.

The supernatant fluid from this step contains the initiation factors (IFs), which are precipitated by addition of ammonium sulfate (47 g/100 ml) and stirring for 1 hr at 4°. The precipitate is collected by centrifugation at 13,000 rpm for 20 min and resuspended in buffer L (0.02 *M* Tris-HCl, 0.02 M NH_4Cl, 5 m*M* EDTA, 5 m*M* $MgCl_2$, 5% glycerol, 0.2 m*M* DFP, 5 m*M* 2-mercaptoethanol, pH 7.0), using 5 ml per 100 g wet weight of cells. The crude initiation factor preparation is then dialyzed for 12 hr against four 1-liter changes of buffer L and centrifuged at 13,000 rpm for 45 min to remove insoluble material; the supernatant is divided into small aliquots and rapidly frozen at −80°. The ribosomal pellet is resuspended in buffer III, layered over 10-ml cushions of 30% sucrose (gradient grade, Sigma) in buffer III, and centrifuged at 45,000 rpm for 16 hr (50.2 Ti rotor). The brown layer that forms on top of the clear ribosomal pellet is scraped off and discarded, and the ribosomal pellet is resuspended in buffer III to a concentration of 1000 A_{260} units/ml. The ribosomal suspension is divided into small aliquots and rapidly frozen at −80°.

Ribosomal subunits are prepared basically according to Cannon and Bott[4] from crude 70 S ribosomes (i.e., from the first ultracentrifugation

[4] J. G. Cannon and K. F. Bott, *Mol. Gen. Genet.* **174,** 149 (1979).

step described here in the preparation of washed 70 S ribosomes). This pellet is gently resuspended in buffer SUB I (0.02 *M* Tris-HCl, 0.02 *M* MgAc, 0.1 *M* NH_4Cl (pH 7.8) with 6 m*M* 2-mercaptoethanol and 0.2 m*M* DFP added just before use), using 0.5 ml of buffer per gram wet weight of cells broken. The ribosome suspension is clarified by centrifugation at 12,000 rpm for 10 min, and the supernatant is dialyzed against four 1-liter changes of buffer SUB II (buffer SUB I with 0.2 m*M* MgAc) for a total of 4 hr. This suspension is then carefully layered over freshly prepared, chilled 35-ml 10 to 30% sucrose gradients made up in buffer SUB II and centrifuged in a SW 28 rotor at 23,000 rpm for 13 hr. One milliliter fractions are collected into tubes kept on ice, and the absorbance at 260 nm is monitored (Gilford 250 spectrophotometer). Peak fractions are pooled, and the MgAc concentration is raised to 20 m*M*. Subunits from the pooled fractions are pelleted in the ultracentrifuge at 45,000 rpm for 8 hr, resuspended in buffer SUB I, and dialyzed against buffer SUB II. The sucrose gradient centrifugation is repeated, and the final 30 S and 50 S pellets are resuspended to a concentration of 200 A_{260} units/ml. DFP must be added to all buffers immediately before use.

RNA Extraction

All glassware and buffers are autoclaved before use. Cells are grown at 37° in aerated CHT50 medium and harvested in mid-exponential growth phase by pouring over ice sprinkled with solid $MgCl_2$ and NaN_3 to bring the final concentrations to 1 m*M* $MgCl_2$ and 10 m*M* NaN_3. Cells are collected in a refrigerated centrifuge, and the pellets are frozen at −80°, then lyophilized. The cells are broken by dry rupture in a dental amalgamator (Wig-L-Bug, Crescent Dental Mfg. Co.) for eight 1-min bursts with 30-sec cooling periods between bursts. Each breakage capsule contains 100 mg of dried cells, 100 mg of glass beads (75–150 μm) and a pestle to aid in grinding the cells. The broken cells and glass beads are then added directly to freshly prepared water-saturated phenol on ice (15 ml of phenol per 1-gram dry weight of cells). Buffer VI (0.01 *M* Tris-HCl, pH 7.5, 0.01 *M* KCl, 5 m*M* $MgCl_2$, and 0.01 *M* NaN_3 added after autoclaving) and Macaloid (National Lead Co.) are added (15 ml of buffer VI and 50 mg of Macaloid per gram dry weight of cells), and the mixture is gently shaken at 37° for 10 min, then centrifuged at 10,000 rpm for 15 min. The aqueous upper layer is transferred to a sterile flask on ice, and the phenol layer is reextracted with an equal volume of buffer VI by shaking at 37° for 10 min. After centrifugation, the aqueous layers are pooled, and solid NaCl is added to 0.1 *M*. The RNA is precipitated by addition of two volumes of cold absolute ethanol and storage at −20° for at least 3 hr. The

RNA precipitate is collected by centrifugation and dissolved in NaAc buffer (0.1 *M* acetic acid, adjusted to pH 5.0 with NaOH), using 15 ml per gram dry weight of cells. This solution is extracted with an equal volume of water-saturated phenol and centrifuged; the aqueous layer is removed and reextracted with phenol until no white material appears at the interface between the two layers. After the final extraction, the aqueous layer is transferred into a clean flask and solid NaCl is added to 2 *M*. The RNA is ethanol-precipitated, and the precipitate is collected as before. The pellet is washed three times with TNE buffer (0.1 *M* Tris-HCl, pH 7.15, 0.1 *M* NaCl, 70% ethanol), then dissolved in TNH buffer (0.01 *M* Tris-HCl, pH 7.0, 0.1 *M* NH_4Cl) and centrifuged at 12,000 rpm for 40 min to remove insoluble material. The RNA is then precipitated with ethanol once again, and the final pellet is resuspended in sterile water to a concentration of 750 A_{260} units per milliliter and stored in sterile vials at −20°.

DNA Extraction

We have used DNA from a variety of sources to prime the coupled transcription–translation *in vitro* system and have found no significant differences in the activities of DNA preparations made using different extraction procedures. A general method for plasmid DNA isolation can be found in Guerry *et al.*[5] A procedure for isolation of phage DNA can be found in Groves *et al.*[6]

In Vitro Protein Synthesis

RNA-Dependent Protein Synthesis

The *in vitro* translation assay is described as a set of solutions (which can be made ahead and stored as indicated) and mixes (which are made up at 4° on the day the assay is to be done). The labeled amino acid is [^{14}C]phenylalanine, which can be used with natural mRNA or polyuridylic acid [poly(U)] as template; this can be altered to suit the particular requirements of an experiment. A higher specific activity is required for autoradiography of *in vitro* synthesized products, so [^{35}S]methionine or a mixture of ^{14}C-labeled amino acids is used. The assay as written is designed for 20 individual reactions (i.e., 20 tubes); it can be scaled up for larger experiments with no difficulty.

Solutions

Tris AA: 19 amino acids (each 0.3 m*M*) omitting phenyalanine (the la-

[5] D. Guerry, D. J. LeBlanc, and S. Falkow, *J. Bacteriol.* **116,** 1064 (1973).
[6] D. J. Groves, G. A. Wilson, and F. E. Young, *J. Bacteriol.* **120,** 219 (1974).

beled amino acid), made up in sterile 0.3 *M* Tris-HCl, pH 7.4. Store at −20°.

MgAc, 0.5 *M* Autoclave, then store at 4°.

NH_4Ac, 3 *M* Autoclave, then store at 4°.

Dithiothreitol 0.1 *M* (DTT) Store at −20°.

Cold Phe: 3.2 m*M* phenylalanine. Store at −20°.

Tris-HCl, 0.01 *M* pH 7.4. Autoclave, then store at 4°.

EDTA, 0.5 *M* pH 7.5. Autoclave, then store at 4°.

Diisopropyl fluorophosphate, 0.2 *M* in isopropanol (DFP). Store at −20°.

Energy: 0.1 *M* ATP, 0.25 *M* PEP (phosphoenolpyruvate, Sigma), 0.005 *M* GTP, neutralized with NH_4OH. Store at −20°.

Pyruvate kinase, 5000 units/ml (Sigma). Store at 4°.

Folinic acid, 5 m*M*. Store at −20°.

[^{14}C]Phe: [^{14}C]phenylalanine, diluted in water to 5 μCi/ml, to give 14 cpm/pmol in each assay tube. Store at −20°.

Mixes

Inhibitors:

0.070 ml	Tris-HCl
0.020 ml	EDTA
0.010 ml	DFP
0.100 ml	

Preincubation mix (PM):

Tris AA	0.250 ml
MgAc	0.0125 ml
NH_4Ac	0.025 ml
DTT	0.025 ml
Cold Phe	0.025 ml
Inhibitors	0.010 ml
Tris-HCl	0.277 ml
	0.625 ml

Reaction mix (RM) for 20 tubes:

Tris AA	0.500 ml
MgAc	0.050 ml
NH_4Ac	0.050 ml
DTT	0.050 ml
Energy	0.050 ml
Pyruvate kinase	0.005 ml
Folinic acid	0.050 ml
DFP	0.001 ml
[^{14}C]Phe	0.050 ml
Crude IFs	0.200 ml
Tris-HCl	0.500 ml
	1.506 ml

Preincubation (PRE): for 20 tubes

PM	0.250 ml	
70 S ribosomes	0.020 ml	(ribosomes at 1000 A_{260} units/ml)
Buffer III	0.080 ml	(buffer in which ribosomes are diluted, without inhibitors)
S-150	0.150 ml	
	0.500 ml	

Pre is incubated at 37° for 15 min just before addition to assay mixture. For assays using ribosomal subunits, 0.067 ml of 30 S subunits and 0.033 ml of 50 S subunits (both at 200 A_{260} units/ml) are mixed and incubated at 37° for 20 min, after which the PM and S-150 are added and incubated at 37° for 15 min as with 70 S ribosomes.

Assay Procedure. All solutions and mixes are kept on ice at all times except during specific incubations. Ribosomes, S-150, and IFs are thawed just before use, and are not refrozen or reused.

Each assay tube contains: 0.075 ml of RM; 0.010 ml of RNA (750 A_{260} units/ml); 0.015 ml of Tris-HCl; 0.025 ml of PRE (total 0.125 ml). The Tris in the assay mix can be replaced by antibiotic solutions made up in Tris, or by other substances the effects of which are under investigation, and is increased to replace the RNA when endogenous mRNA activity is measured.

Assays are incubated at 37° for 30 min. The tubes are placed on ice, and 2 ml of 5% trichloroacetic acid (TCA) is added per tube. The mixture is heated at 90° for 15 min, and the precipitate is collected on glass fiber filters and counted in a liquid scintillation counter.

When poly (U) is used as template, 100 μg is used per assay, and the IFs in the RM are replaced by spermidine (0.0625 *M* in 0.01 *M* Tris-HCl, pH 7.4).

DNA-Dependent Protein Synthesis

The coupled transcription-translation system uses the same basic solutions and mixes as the translation system, with some modifications that stimulate transcription.

Solutions. All solutions used in the translation system are needed.

PEP: 0.375 *M* phosphoenolpyruvate (Sigma) neutralized with NH_4OH. Store at −20°.

XTP: 17 m*M* GTP, 17 m*M* UTP, 17 m*M* CTP. Store at −20°.

KAc: 1.25 *M* potassium acetate. Store at 4°.

[^{3}H]UTP. Add to RM in place of [^{14}C]Phe to give 4.2 μCi of [^{3}H]UTP per tube when mRNA synthesis is to be monitored.

Mixes. PM and inhibitor mixes are unchanged.

PRE for 20 tubes:

PM	0.250 ml
70 S ribosomes	0.050 ml
S150	0.200 ml
	0.500 ml

RM for 20 tubes:

Tris AA	0.500 ml
MgAc	0.050 ml
NH_4Ac	0.050 ml
KAc	0.100 ml
DTT	0.050 ml
Energy	0.050 ml
Folinic acid	0.050 ml
XTP	0.020 ml
PEP	0.020 ml
Pyruvate kinase	0.005 ml
DFP	0.001 ml
[^{14}C]Phe	0.050 ml
Crude IFs	0.200 ml
Tris-HCl	0.355 ml
	1.501 ml

Assay Procedure. The assay is carried out as described for the translation system. Each assay tube contains 0.075 ml of RM, 0.010 ml of DNA (170 μg/ml), 0.015 ml of Tris, 0.025 ml of PRE (total 0.125 ml).

Amino acid incorporation is determined by measuring ^{14}C label remaining in material insoluble in hot TCA when [^{14}C]phenylalanine is used. Nucleotide incorporation is measured as the amount of ^{3}H label remaining in material insoluble in unheated TCA when [^{3}H]UTP is used.

The optimal concentrations of various components in the system are routinely checked. Initiation factor and S-150 preparations may vary, and the optimum amount to be added should be determined for each preparation; the system is also extremely sensitive to magnesium concentration. The optimum concentrations may be different in the RNA-dependent and DNA-dependent protein synthesis systems, and should be determined for each system. Cell extracts can be stored as long as 1 year with no decline in activity.

[38] Translation of Exogenous mRNAs in Reticulocyte Lysates

By WILLIAM C. MERRICK

In 1976, Pelham and Jackson first reported on the preparation of an active protein-synthesizing system for the translation of exogenous mRNAs.[1] By simply treating reticulocyte lysates with a Ca^{2+}-dependent nuclease and subsequently inactivating the nuclease with a chelator of Ca^{2+}, EGTA, they were able to convert the most efficient eukaryotic cell-free protein synthesis system[2] into one that would readily translate a variety of different mRNAs. In general this system is deemed superior to a Krebs II ascites or wheat germ cell-free system or frog oocytes because of the greater mRNA sensitivity (due to low backgrounds and high synthetic rates), low nuclease activity and/or low protease activity.[1] The availability of such an active translation system has been significant for molecular biologists. More than ever before it has become possible to assay for small amounts of mRNAs in order to monitor the purification of an mRNA, to quantitate levels of a specific mRNA during a developmental period, or to characterize recombinant DNA clones.

Solutions and Reagents

Phenylhydrazine, 2.5%: Phenylhydrazine (25 g) is dissolved in 1000 ml of H_2O that has been deoxygenated by bubbling with N_2. This solution is neutralized to pH 7 with 1 *N* NaOH. Single-use aliquots are frozen and protected from decomposition by storage in amber, airtight containers at −20°. Once thawed for use, residual amounts of the 2.5% phenylhydrazine solution are discarded, not refrozen for subsequent use.

Heparin solution, 500 units/ml

Physiological saline: 140 m*M* NaCl, 5 m*M* KCl, 7.5 m*M* magnesium acetate. This solution should be autoclaved to reduce the possibility of nuclease contamination.

Amino acid mixture: This solution contains all the nonradioactive amino acids at a concentration of 1 m*M* except for glutamate, glutamine, aspartate, asparagine, alanine, and glycine, which would be 2 m*M*.

Hemin hydrochloride solution: A 1 m*M* solution of hemin is prepared

[1] H. R. B. Pelham and R. J. Jackson, *Eur. J. Biochem.* **67,** 247 (1976).

[2] T. Hunt and R. J. Jackson, "Modern Trends in Human Leukemia" (R. Neth, R. C. Gallo, S. Spiegelman, and F. Stohlman, eds.), p. 300. Lehmanns Verlag, Munich, 1974.

ISBN 0-12-182001-7

as described by Ranu and London.[3] Initially 6.5 mg of hemin hydrochloride (Calbiochem-Behring) is dissolved in 0.25 ml of 1 *M* KOH. Subsequently, the solution is brought up to 10 ml final volume by the addition of 0.55 ml of distilled H_2O, 0.1 ml of 1 m*M* Tris-HCl (pH 7.9), 8.9 ml of ethylene glycol, and 0.2 ml of 1 *M* HC1. The resulting 1 m*M* solution of hemin can be stored at $-20°$ and is stable for more than a year.

Creatine phosphokinase solution: Ten milligrams of creatine phosphokinase (Sigma Chemical Co.) are dissolved in 1 ml of a buffer that is 20 m*M* HEPES–KOH (pH 7.5) and 50% glycerol. This enzyme solution is stable for more than 6 months when stored at $-20°$.

Master cocktail: To simplify the addition of small volumes of items that will normally be common to each assay, a number of stable reagents are mixed together to be subsequently added to the protein synthesis reaction mixture. The generation of a master cocktail not only saves time, but it also safeguards against the omission of a single component. The master cocktail is made up as follows:

100 μl	amino acid mixture
20 μl	100 m*M* GTP
50 μl	100 m*M* ATP
200 μl	2 *M* KCH_3CO_2, pH 7.5
10 μl	1 *M* Mg $(CH_3CO_2)_2$
75 μl	1 *M* HEPES–KOH, pH 7.5
X μl	radiolabeled amino acid (to be from 10 to 80 μ*M* in the master cocktail, depending on expense)
(545-*X*) μl	H_2O
1000 μl	

It should be noted that depending on the amount of radiolabeled amino acid added to the master cocktail, it is possible that the rate and/or extent of protein synthesis will be limited by this component (especially if one uses high specific activity [^{35}S]methionine or [^{3}H]leucine: sp. act. 100 to 1500 Ci/mmol).

Micrococcal nuclease: Dissolve in H_2O at a concentration of 1 mg/ml and store frozen in aliquots.

Procedures

Preparation of Lysate

Immature New Zealand white rabbits (2.5–5 lb) are injected with 2.5% phenylhydrazine (0.10 ml/lb) for 5 consecutive days to cause reticu-

[3] R. S. Ranu and I. M. London, this series, Vol. 60, p. 459.

locytosis. During the last 1 or 2 days, the rabbits will become quite anemic and usually develop dark, watery stools. Subsequently the rabbits are allowed to recover for 2 days, then are bled on day 8. A few minutes prior to bleeding, the animals are anesthetized by an intraperitoneal injection of 30 mg of sodium pentobarbital. Care should be taken not to overdose the animals with the pentobarbital, as this may cause death of the animal prior to bleeding. Blood is taken by cardiac puncture with a No. 14 Huber-point needle (Becton-Dickinson) on a 60-ml plastic disposable syringe that contains 0.1 ml of heparin solution. Approximately 60–80 ml of blood can be obtained from each animal; however, some care in bleeding must be taken to avoid puncturing the heart too often, which leads to either loss of blood or death of the animal prior to removal of a satisfactory volume of blood.

After the blood has been withdrawn, it is transferred to chilled, plastic centrifuge tubes (250 ml) that contain 0.5 ml of the heparin solution. To avoid clotting and to increase cooling, the bottle is swirled gently with each fresh addition of blood and then kept on ice. After all the animals have been bled, the reticulocytes are collected by centrifugation in a Sorvall GSA rotor (8000 *g*, 10 min at 4°). The plasma is discarded and the reticulocytes are washed three times with chilled physiological saline. Washing of the cells involves (*a*) gentle but complete resuspension of the reticulocytes in 3–4 volumes of chilled physiological saline; (*b*) collection of the cells by centrifugation (8000 *g*, 10 min at 4°); (*c*) aspiration of the wash solution and the buffy coat (leukocytes), which sediment on the top of the reticulocyte pellet. Finally, the reticulocytes are osmotically lysed by the addition of chilled, distilled water (1.5–2.0 volumes of H_2O per volume of packed cells). Lysis is maximized by stirring the cells in the distilled water for 3–5 min on ice. Subsequently, the lysate is cleared of cell debris and mitochondria by centrifugation for 20 min at 16,000 *g* at 4°. The supernatant is carefully decanted off, taking care not to include any cell debris (the pellet formed is not very firm and does not remain attached to the bottom of the centrifuge tube). The supernatant is then immediately frozen in 0.5–2 ml aliquots and stored at the vapor temperature of liquid nitrogen.

Characterization of Reticulocyte Lysate

Prior to treatment with nuclease, it is worthwhile to check several parameters of the reticulocyte lysate to ensure that subsequent time and expense are not wasted on inactive lysate. Five relatively easily checked characteristics are as follows: (*a*) the A_{415} should be about 800–1000, which corresponds to a hemoglobin concentration of 200–250 mg/ml; (*b*) high speed centrifugation (100,000 *g* for 3 hr) should yield about 20–

25 A_{260} units of polysomes per milliliter of lysate; (*c*) using intact lysate, a 50-μl reaction mixture that is one-half lysate should direct the incorporation of about 300–400 pmol of radiolabeled leucine into hot trichloroacetic acid-precipitable product in 30 min; (*d*) in the absence of added hemin, protein synthesis should proceed at about 5–10% of the rate indicated above; (*e*) protein synthesis should be linear for the first 30 min and continue for 45–60 min. Although the list of characteristics may seem excessive, it should be noted that preparing active lysate is what is most important. The above characteristics provide a useful yardstick to judge the preparation. Another possibility is to compare the preparation with one of the commercially available "reticulocyte lysate translation kits" (see other considerations).

Assay for Hemoglobin Synthesis

The standard protein synthesis assay would contain the following components in a total of 100 μl: 20 μl of master cocktail, 50 μl of reticulocyte lysate, 1 μl of creatine phosphokinase solution, 5 μl of 100 mM creatine phosphate, 2 μl of 1 mM hemin hydrochloride solution, X μl of poly(A)$^+$ mRNA (0.3–5 μg), (22-X) μl of H_2O

For most assays the variable of interest will be the added mRNA. In such instances all the ingredients except the mRNA and lysate should be mixed together. Next, this mixture is then added to the unthawed lysate (ratio, 28 μl of mixture : 50 μl lysate) and the lysate is thawed immediately. This thawing of the lysate in the presence of hemin minimizes any loss of activity due to apparent hemin deficiency in the lysate.[3] As should be apparent, in order to maximize the use of available lysate, aliquots should be made of appropriate size depending on need (100–1000 μl). The unused material should be discarded. Once the ingredients have been mixed on ice, they are transferred to a 30° water bath and incubated for the desired period of time. Reactions are terminated by dilution of a small aliquot, 2–10 μl, into 1 ml of H_2O and protein, then quantitated as (*a*) hot, TCA-precipitable radioactivity; or (*b*) alkali resistant, cold TCA-precipitable radioactivity (see below). For the initial untreated lysate, the investigator should optimize the assay conditions for added hemin (5–30 μM) and for added $Mg(CH_3CO_2)_2$ (0–4 mM). In most instances, those conditions optimal for globin mRNA translation will be similar to the conditions optimal for the translation of other mRNAs.

Hot TCA-Precipitable Radioactivity. To the 1 ml of distilled H_2O that contains 2–20 μl of the 100 μl reaction mixture is added in equal volume of 20% TCA. The solution is heated at 90° for 20 min to degrade aminoacyl-tRNA and then cooled on ice for 10 min. Precipitated radioactive protein is collected by vacuum filtration using glass fiber filters (Whatman GF/C) or nitrocellulose filters (type HA, Millipore). Glass fiber filters tend to have lower backgrounds when sulfur-containing or aromatic radiolabeled amino acids are used, but nitrocellulose filters give slightly better retention (background problems can be reduced by inclusion of 1 mM of the appropriate unlabeled amino acid in the TCA solution and rinsing of the filter prior to sample collection). The tube containing the 2-ml of sample is poured onto the filter and then rinsed three times with 5% TCA. The filters are then dried under an infrared lamp and transferred to scintillation vials; counting scintillant is added, and then radioactivity is determined using a liquid scintillation spectrometer.

Alkali-Resistant, TCA-Precipitable Radioactivity. This method of determination is primarily used for those few proteins that are not stable to hot TCA (i.e., collagen) or when ^3H-labeled amino acids are used (the precipitated hemoglobin will markedly quench ^3H but has little effect on ^{35}S or ^{14}C). In this instance, 0.5 ml of 1 M NaOH containing 0.5 M H_2O_2 (to decolorize) is added to the 1-ml sample (aliquot plus 1 ml of H_2O), and the mixture is incubated at 37° for 15 min. Then 1 ml of cold 25% TCA is added, and the sample is mixed and kept on ice for 10 min. Subsequently the radiolabeled, precipitated protein is collected by vacuum filtration, and radioactivity is quantitated as described above for hot TCA-precipitable radioactivity.

Treatment of Lysate with Micrococcal Nuclease

Having established that one has an active reticulocyte lysate, the next step is to remove the endogenous globin mRNA. This is done by treatment with a calcium-dependent RNase, micrococcal nuclease. Hemin (2–4 μl per 100 μl of lysate, as optimized in hemoglobin synthesis) is added to thawing lysate. This solution is then made 1 mM in $CaCl_2$ by the addition of 1 μl of 100 mM $CaCl_2$ per 100 μl of lysate and 10 μg/ml micrococcal nuclease by the addition of 1 μl of 1 μg/ml micrococcal nuclease per 100 μl of lysate. The lysate is then incubated for varying lengths of time and the action of the nuclease is stopped by the addition of the calcium chelator EGTA (1 μl of 100 mM EGTA per 100 μl of lysate). Subsequently the remaining ingredients may be added and protein synthesis tested. Usually, incubation of the lysate with nuclease for about 10 min gives optimal results (i.e., a low background and full activity). Underdigestion leaves residual globin mRNA, and consequently high backgrounds, in the absence of added mRNA. Overdigestion yields low backgrounds but the level of activity in the presence of added mRNA is reduced. Using the conditions indicated above, satisfactory lysate should be obtained with nuclease digestion times of 5–20 min.

Translation of Exogenous mRNA

The first mRNA to be tested should be one that is known to be active in directing protein synthesis. Having access to such an mRNA depends on whether or not one has a colleague from whom one might borrow such an mRNA. Lacking this, the preparation of globin mRNA using oligo(dT) cellulose is quite routine.[4,5] The advantage of using globin mRNA as an original test mRNA is that it should yield exactly the same incorporation as the untreated lysate, and the proteins (α and β globin) are readily characterized. Globin mRNA should saturate the assay system at about 20 μg/ml.[1] The major disadvantage in using globin mRNA is that one cannot determine how well larger mRNAs will be translated.

Having established an active, mRNA-dependent lysate, the mRNA is now to be analyzed. Listed below are a few considerations that should be reviewed prior to testing the mRNA.

1. It takes about 2 min to synthesize an intact globin chain at 30°. Using this ratio it will take about 10–15 min to synthesize a 75,000 dalton peptide, 20–30 min for a 150,000 dalton peptide, and 40–60 min for a 300,000 dalton peptide.

2. The reticulocyte tRNAs are ideally suited for the translation of globin mRNA, as the amino acid acceptance distribution very nearly matches the amino acid composition of the α and β globin chains.[6] Quite often the addition of a more general tRNA (i.e., liver tRNA) or tRNA from the same tissue will greatly enhance the translation of a specific mRNA. This is readily understandable for proteins of unique amino acid composition, such as collagen (high in glycine, serine, and proline), but some apparently normal proteins may also be dramatically responsive to added tRNA. A good example is the synthesis of fatty acid synthetase, where the presence of calf liver tRNA increased fatty acid synthetase production nine-fold while total protein synthesis was stimulated only slightly.[7] The preparation of any tRNA should include phenol extraction, ethanol precipitation, and Sephadex G-100 chromatography[8] to ensure that additional mRNA or mRNA fragments are not added.

3. The purity of the mRNA is not critical; however, poly(A)$^+$ mRNA should be used. A total RNA preparation that is approximately 90–95% rRNA translates poorly, usually no better than 30–50% of maximum. A poly (A)$^+$ mRNA preparation that is 20–60% rRNA translates well. The

[4] A. Krystosek, M. L. Cawthon, and D. Kabat, *J. Biol. Chem.* **250,** 6077 (1975).

[5] H. Aviv and P. Leder, *Proc. Natl. Acad. Sci. U.S.A.* **69,** 1408 (1972).

[6] D. W. E. Smith and A. L. McNamara, *Science* **171,** 577 (1971).

[7] A. G. Goodridge, S. M. Morris, Jr., and T. Foldflam, this series, Vol. 71, p. 139.

[8] W. C. Merrick, this series, Vol. 60, p. 108.

presence of some rRNA may help as an alternative substrate for nucleases and will aid in any precipitation steps as carrier RNA.

4. For most translations, one will use either [^{35}S]methionine (sp. act. 1000 Ci/mmol) or [^{35}S]cysteine (sp. act. 600 Ci/mmol) because of the very high specific activities and the higher energy of ^{35}S relative to ^{3}H (0.167 MeV vs 0.0186 MeV). However, although [3,4,5-^{3}H]leucine, [4,5,-^{3}H]isoleucine, or [4,5,-^{3}H]lysine are of lower specific activity (sp. act. about 120 Ci/mmol, 105 Ci/mmol, and 80 Ci/mmol, respectively), these amino acids constitute 6–10% of the amino acid composition of most proteins compared to the average 1% contribution of methionine or cysteine. The difference in specific activity between ^{35}S- and ^{3}H-labeled amino acids can thus be negated by the increased abundance of the three ^{3}H-labeled amino acids indicated above. This will not, however, compensate for the difference in the energy of decay between the two isotopes.

5. As indicated above, in general those conditions that are optimal for the translation of total poly(A)$^{+}$ mRNA (or even globin mRNA) should be useful in identifying the optimal conditions for translation of specific mRNAs. However, if mRNA is available, it would be worthwhile to titrate spermidine (0, 0.5, 1.0, and 1.5 m*M*), $Mg(CH_3CO_2)_2$ (1.0, 1.5, 2.0, 2.5, and 3.0) and KCH_3CO_2 (50, 100, 150, and 200 m*M*) for the specific enhancement of translation of the mRNA under study.

Analysis of Translation Products

A major part of using nuclease-treated lysates is the subsequent analysis of translation products, because only for a pure mRNA will TCA-precipitable radioactivity reflect the optimal translation conditions for a given mRNA. There are two basic methodologies used to characterize translation products, and they are presented below.

1. *Polyacrylamide Gel Electrophoresis and Autoradiography*. Aliquots (2–10 μl) of sample reaction mixtures are mixed with 15 μl sample buffer [60 m*M* Tris-HCl, pH 6.8, 10% glycerol, 1% sodium dodecyl sulfate (SDS), 1% dithiothreitol, and 0.002% bromophenol blue] and then heated to 100° for 5 min to achieve uniform coating of the proteins with SDS. Subsequently, the 20–25-μl sample is subjected to electrophoresis in 7.5–12% polyacrylamide slab gels (100 × 140 × 0.75 mm; 20 wells per slab) as described by Laemmli.[9] The advantage of this gel system is that it is easy to prepare, readily reproducible and has high resolving power due to the discontinuous buffer system. The main variable is the percentage of acrylamide that yields best resolution of the molecular weight product expected.

[9] U. K. Laemmli, *Nature* (*London*) **227**, 680 (1970).

After electrophoresis, the proteins are visualized by staining with Coomassie Blue in 50% methanol, 7.5% acetic acid and destained with 5% methanol, 7.5% acetic acid.[10] Once the gel is destained, it is soaked in 10% acetic acid, 1% glycerol for 30–60 min and then dried onto filter paper (Whatman 3 MM or Bio-Rad filter paper backing). The dried gel is then exposed to X-ray film (Kodak X-Omat AR). Usually 1000 cpm of a ^{14}C- or ^{35}S-labeled protein can be visualized in 24 hr. However, for a mixture where it is not possible to anticipate the number of discrete polypeptides, it may take several days to visualize all the radiolabeled bands.

There are two common alternatives to the above procedure. The first is not to stain the gel, as staining is useful only in visualizing proteins used as molecular weight standards. In this case, after electrophoresis the gel is soaked in cold 20% TCA for 60 min to fix proteins, rinsed in the 10% acetic acid, 1% glycerol solution for 30–60 min, and then dried and exposed to X-ray film as above. The second alternative is the use of fluorography instead of autoradiography. In essence, this technique involves impregnating the gel with scintillation chemicals and then exposure to X-ray film.[11] In this process, the film is exposed by energy in the radioactive decay via the light emitted by the scintillation chemicals rather than by a photoconversion induced directly by the radioactive decay. This type of analysis is mandatory if ^{3}H-labeled amino acids are used, owing to the low energy in the ^{3}H decay, and gives about a 10-fold enhancement if ^{14}C- or ^{35}S-labeled amino acids are used. While greater sensitivity always sounds good, the use of the Bonner and Laskey procedure[11] or a commercial substitute (i.e., EN^{3}HANCE, New England Nuclear) is expensive, a bit messy, and involves the use of hazardous solvents (i.e., dimethyl sulfoxide).

2. *Immunoprecipitation*. For many specific mRNAs to be studied, the protein for that mRNA and an antibody against the protein are available. It is possible to monitor the synthesis of a specific protein in an unfractionated mRNA population by immunoprecipitation. In this case, aliquots of the reaction mixture are diluted into an Eppendorf centrifuge tube that contains 1 ml of 150 m*M* NaCl, 10 m*M* EDTA, 20 m*M* Tris-HCl, pH 7.5, and 1% Nonidet P-40. To this solution is added 1 μl of a 200 m*M* solution of phenylmethylsulfonyl fluoride in absolute ethanol (phenylmethylsulfonyl fluoride is a general protease inhibitor that breaks down in H_2O and consequently must be added fresh; it is stable as a 200 m*M* solution in ethanol when stored at 4°). To quantitate total protein synthesis, a small aliquot can be taken, and hot TCA-precipitable radioactivity determined

[10] K. Weber and M. Osborne, *J. Biol. Chem.* **244,** 4406 (1969).

[11] W. M. Bonner and R. A. Laskey, *Eur. J. Biochem.* **46,** 83 (1974).

as above. To the remaining solution is added 10 μg of carrier protein and an appropriate amount of antibody to effect precipitation (this must be determined independently). The immunoprecipitate is allowed to form by incubating the tubes for 1 hr at room temperature and then overnight at 4°. The immunoprecipitate is collected by centrifugation for 5 min (Beckman or Brinkmann microfuge) and washed four times with 1 ml of 0.15 *M* NaCl, 1% Nonidet P-40 with vigorous mixing to resuspend the pellet uniformly. The pellet is then solubilized in SDS sample buffer as described above. At this point an aliquot is taken for determination of radioactivity. This radioactivity represents both completed chains as well as shorter peptides released by the EDTA present in the original buffer (see above). To quantitate the radioactivity in complete chains vs total radioactivity, it is necessary to analyze the immunoprecipitate by SDS polyacrylamide slab gel electrophoresis as described above. Such an analysis will indicate not only the percentage of radioactivity that represents complete chain synthesis, but also check the specificity of the immunoprecipitation process. As incomplete products will probably be randomly distributed throughout the low molecular weight region of the gel, it may be necessary to cut the dried gel into strips to quantitate the relative amount of radioactivity as full-length product.

As may be appreciated, the use of immunoprecipitation and SDS polyacrylamide gel electrophoresis should greatly increase the ability to identify and quantitate a specific radiolabeled peptide from the mixture. This is especially true for proteins that are minor representatives of the total due either to low mRNA abundance or relatively high molecular weight (all lysates are biased to yield a disproportionate amount of low molecular weight peptides because of low levels of nuclease or protease in the lysate and the greater length of time required to synthesize high molecular weight polypeptides). This procedure also allows resolution from more abundant proteins of similar molecular weight. However, as this process is more time consuming and requires additional materials, it is worthwhile to attempt identification just by SDS polyacrylamide gel electrophoresis[9] or as an alternative two-dimensional polyacrylamide gel electrophoresis.[12]

Other Considerations

Use of "The Kit." There are several commercial sources of nuclease-treated reticulocyte lysate that are guaranteed active and are usually accompanied by a proven test mRNA. If one is well-funded or plans only a limited series of translation experiments, the commercial materials (espe-

[12] P. H. O'Farrell, *J. Biol. Chem.* **250**, 4007 (1975).

cially from New England Nuclear) will readily prove cost-effective as the translation materials (lysate, mRNA, salts, etc.) are about the same price as the amount of radiolabeled amino acid required for the translations. In addition, most companies will give volume discounts (up to 40%) for an individual or group. Although optimal conditions for the test mRNA are provided, it is still necessary to ensure that these conditions are optimal for the mRNA being examined (most especially for high molecular weight polypeptides).

Messenger RNA Studies. The nuclease-treated lysate provides an excellent system for the examination of competition between different mRNAs for translation, especially as translation occurs at *in vivo* rates. Such studies can be extended by the use of inhibitors of initiation (i.e., ediene), elongation (i.e., cycloheximide), or initiation of capped mRNAs (i.e., m^7GDP).[13-15] The major asset of this system is the rapid rate of translation, which is usually 8- to 15-fold greater than most *in vitro* translation systems. The major limitation is that this is a system that would normally synthesize only α and β globin and as such is not prepared for many cellular responses (i.e., new mRNA synthesis, stimulation by hormones, cell division, viral infection, etc.).

Posttranslation Studies. Nuclease-treated lysates may be used to examine posttranslational modification. A number of studies, only a very few of which are referenced,[16-21] have examined *in vitro* the posttranslational modification of secreted proteins or proteins synthesized in the cytosol, but destined to be in mitochondria or chloroplasts. While it has not been demonstrated that the reticulocyte lysate system will be compatible with chloroplasts, it has been possible to demonstrate processing with microsomal membranes[16,17,19] or uptake into mitochondria of proteins synthesized *in vitro*.[20,21] At present a commercial kit for processing proteins with microsomal membranes is available (New England Nuclear), and by the time this chapter is published, one for the mitochondrial uptake of *in vitro* synthesized proteins will probably also be available.

[13] T. Brendler, T. Godefroy-Colburn, R. D. Carlill, and R. E. Thach, *J. Biol. Chem.* **256,** 11747 (1981).

[14] T. Godefroy-Colburn and R. E. Thach, *J. Biol. Chem.* **256,** 11762 (1981).

[15] D. Skup and S. Millward, *Proc. Natl. Acad. Sci. U.S.A.* **77,** 152 (1980).

[16] R. C. Jackson and G. Blobel, *Proc. Natl. Acad. Sci. U.S.A.* **74,** 5598 (1977).

[17] J. A. Majzoub, P. C. Dee, and J. F. Habener, *J. Biol. Chem.* **257,** 3581 (1982).

[18] A. R. Grossman, S. G. Bartlett, G. W. Schmidt, J. E. Mullet, and N. Chua, *J. Biol. Chem.* **257,** 1558 (1982).

[19] G. Rogers, J. Gruenebaum, and I. Boime, *J. Biol. Chem.* **257,** 4179 (1982).

[20] S. M. Gasser, A. Ohashi, G. Daum, P. Bohni, J. Gibson, G. A. Reid, T. Yonetani, and G. Schatz, *Proc. Natl. Acad. Sci. U.S.A.* **79,** 267 (1982).

[21] T. Morita, S. Miura, M. Mori, and M. Tatibana, *Eur. J. Biochem.* **122,** 501 (1982).

[39] Translational Systems Prepared from the Ehrlich Ascites Tumor Cell

By EDGAR C. HENSHAW and RICHARD PANNIERS

Cell-free protein-synthesizing systems that can translate exogenous mRNA have been prepared from the Ehrlich cell[1-3] and many other cultured cell lines.[4-7] In general none of these systems support as many rounds of initiation as the reticulocyte system. However, they have certain advantages. Perhaps most important is the ready availability of the cells. Laboratories that are routinely growing cells in suspension culture may find this source of cells more rapid, convenient, and reliable than cardiac puncture of phenylhydrazine-treated rabbits. As described below, the cultured cell systems can be made to translate exogenous mRNA species preferentially compared to endogenous messages by degrading endogenous mRNA with micrococcal nuclease or by addition of excess exogenous mRNA to cell extracts that have been preincubated under conditions that allow runoff of ribosomes from endogenous mRNA. Once the optimal parameters for the system have been determined, a system with predictable characteristics can be generated reliably under standardized conditions.

Solutions

MOPS-buffered saline: NaCl, 154 m*M*, 3-[*N*-morpholino]propanesulfonic acid (MOPS), 20 m*M* (pH 7.2)

Hypotonic lysis buffer: MOPS, 10 m*M* (pH 7.2), KCl, 10 m*M*, magnesium acetate ($MgAc_2$), 1.5 m*M*, dithiothreitol (DTT), 0.5 m*M*, ethylenediaminetetraacetic acid (EDTA), 0.1 m*M*

Medium A: KCl, 2.0 *M*, $MgAc_2$, 50 m*M*

Ethylene glycol bis(2-aminoethyl ether)-*N*,*N'*-tetraacetic acid (EGTA), neutralized with KOH, 50 m*M*

G-25 elution medium: KCl, 80 m*M*, MOPS, 10 m*M* (pH 7.2), $MgAc_2$,

[1] K. L. Eggen and A. J. Shatkin, *J. Virol.* **9,** 636 (1972).
[2] W. D. Ensminger and E. C. Henshaw, *Biochem. Biophys. Res. Commun.* **52,** 550 (1973).
[3] S. Pestka, J. McInnes, E. A. Havell, and J. Vilcek, *Proc. Natl. Acad. Sci. U.S.A.* **72,** 3898 (1975).
[4] M. B. Matthews and A. Korner, *Eur. J. Biochem.* **7,** 328 (1970).
[5] I. M. Kerr and E. M. Martin, *J. Virol.* **7,** 438 (1971).
[6] H. Aviv, I. Boime, and P. Leder, *Proc. Natl. Acad. Sci. U.S.A.* **68,** 2303 (1971).
[7] L. A. Weber, E. M. Feman, and D. Baglioni, *Biochemistry* **14,** 5315 (1975).

ISBN 0-12-182001-7

2.5 m*M*, DTT, 0.5 m*M*, EGTA, 0.5 m*M*. EGTA is omitted in the ribosome runoff method.

Energy mix: MOPS, 50 m*M* (pH 7.2), ATP, 12 m*M*, GTP, 3 m*M*, phospho(enol)pyruvate (PEP), 30 m*M*, pyruvate kinase, 400 μg/ml, DTT, 5 m*M*

Energy and amino acid mix: MOPS, 50 m*M* (pH 7.2), ATP, 12 m*M*, GTP, 3 m*M*, PEP, 30 m*M*, pyruvate kinase, 400 μg/ml, 19 amino acids, 400 μM, DTT, 5 m*M*

Salts-energy-amino acid mix: KCl, 400 m*M*, $MgAc_2$, 12.5 m*M*, MOPS, 50 m*M* (pH 7.2), ATP, 12 m*M*, GTP, 3 m*M*, PEP, 30 m*M*, pyruvate kinase, 400 μg/ml, 19 amino acids, 400 μM, DTT, 5 m*M*, spermidine, 4 m*M*

Protocol

The basic protocol for translating exogenous mRNA has three steps: (*a*) preparation of a cell extract by hypotonic swelling and Dounce homogenization of cells; (*b*) preincubation of the extract with micrococcal nuclease to degrade endogenous mRNA, or preincubation under protein-synthesizing conditions to allow ribosome runoff; and (*c*) incubation with the added mRNA and a labeled amino acid to synthesize labeled proteins. In some systems the nuclease-treated or runoff extracts are also passed over Sephadex G-25 to remove small molecules and create defined salt conditions, before the final incubation. We describe here the preparation of Ehrlich cell systems; the same methods and considerations are expected to apply to other cultured cell lines.

The Nuclease-Treated System

The development by Pelham and Jackson[8] of a technique for degrading endogenous mRNA has allowed preparation of greatly improved translation systems that have low endogenous background incorporation and therefore have much higher sensitivity for detecting translation of exogenous mRNA species. In this technique the calcium-dependent micrococcal nuclease is used to degrade endogenous mRNA and is then inactivated by EGTA, which has a higher specificity for Ca^{2+} than for Mg^{2+}. Two nuclease-treated systems will be described. In the simplest, calcium and nuclease are incubated with the extract; after this preincubation, EGTA is added, and mRNA and a labeled amino acid are added immediately for the final incubation.[9] In a slightly more complex version, the nuclease-

[8] H. R. B. Pelham and R. J. Jackson, *Eur. J. Biochem.* **67**, 247 (1976).

[9] M. Centrella and J. Lucas-Lenard, *J. Virol.* **41**, 781 (1982).

treated extract is first desalted by gel filtration before the final incubation; it may be stored frozen until needed.

Unfiltered, Nuclease-Treated System

Preparation of Cell Extract (S-10). Small volumes of cells are harvested conveniently by batchwise centrifugation. Larger volumes of suspension may be harvested by continuous flow centrifugation or may be concentrated by filtration before batch centrifugation. Whereas rapid chilling is necessary for "freezing" of polyribosomes and preservation of *in vivo* polyribosome patterns for analytical purposes,[10] it seems unlikely to be required for the present purpose, since the extracts themselves will be preincubated at 20° or 37°. Therefore, we use the Sorvall continuous-flow apparatus without prior chilling of the cells. If rapid chilling is preferred, the cell suspension is poured over ice in centrifuge bottles. Because ice can interfere with subsequent decanting, only sufficient ice is used (about one-third volume) so that the ice is melted by the time of the decanting step. Cells are sedimented by centrifugation at 1000 *g* for 5 min.

The cell pellets are resuspended in and resedimented from MOPS-buffered saline at 0–4°. (For hemin-treated preparations, discussed below, hemin should be added to this and subsequent media.) The pellets are then resuspended in 1 volume of hypotonic lysis buffer and are allowed to swell for at least 10 min on ice. It is quite important to allow a sufficient swelling period. Cells are lysed by 20 strokes of a tight-fitting Dounce homogenizer. Cells are monitored by light microscopy to confirm lysis. A wet mount of the cell preparation under a cover slip and stained with a drop of toluidine blue or methylene blue allows resolution of nucleus and cytoplasm. (Detergent lysis gives higher ribosome yields and cleaner nuclei, but our standard detergent treatment with Triton X-100 plus deoxycholate,[10] or with Triton X-100 alone, does not produce active lysates.) The concentrations of KCl and Mg^{2+} are raised by the addition of 0.05 ml of medium A per 1.0 ml of hypotonic lysis buffer used. Nuclei and mitochondria are removed by sedimentation at 15,000 *g* for 10 min, and the supernatant solution is retained as the S-10 fraction. For the Ehrlich cell this preparation usually contains 100–150 A_{260} units/ml, and the yield is 6 ml per 10^9 cells. The S-10 fraction is generally carried immediately through the nuclease step (below). However, it can be frozen in aliquots and stored in liquid nitrogen and be nuclease-treated on the day of use.[9]

Preincubation of the S-10 Fraction with Micrococcal Nuclease. To each 1.0 ml of freshly prepared S-10 fraction, 10 μl of 100 m*M* $CaCl_2$ are

[10] W. J. W. VanVenrooij, E. C. Henshaw, and C. A. Hirsch, *J. Biol. Chem.* **245,** 5947 (1970).

added to bring the concentration of added Ca^{2+} to 1 mM. Micrococcal nuclease is added to a predetermined concentration (see below), and the extract is incubated at 20° for 15 min. Nuclease action is stopped by the addition of 40 μl of 50 mM EGTA–KOH, pH 7.2, to bring EGTA to 2 mM, and by chilling on ice. The preparation can be stored frozen in liquid nitrogen for many months.

Translation of Added mRNA. Incubations are set up in convenient volumes, frequently 25 μl, 50 μl, or 100 μl, containing, per 1.0 ml of final volume, 0.5 ml of nuclease-treated S-10, 0.1 ml of the energy and amino acid mix (containing 19 amino acids minus the labeled amino acid), and the remaining amino acid as [^{35}S]methionine or a ^{14}C-labeled amino acid. Because of the presence of endogenous amino acids, large amounts of the labeled amino acid may be required, typically 100 μCi of [^{35}S]methionine per milliliter of final volume. Incubation is at 37° for 30–60 min.

Incorporation is measured by spotting 10- μl samples of the incubation onto squares or circles of Whatman No. 3 filter paper, numbered by pencil and mounted by a pin on a styrofoam board. The filters are dropped into a beaker of 5% TCA, 1% casamino acids. Filters are rinsed twice in this solution and then boiled in it for 10 min. The TCA is decanted; the filters are washed three times more with 95% ethanol, dried, and counted.

Optimization of the Concentration of Micrococcal Nuclease. The amount of nuclease to be added to the preincubation should be determined for each batch of nuclease. Fifty-microliter fractions of freshly prepared S-10 are brought to 1 mM $CaCl_2$ and are incubated at 20° in a range of nuclease concentrations generally 1–50 μg/ml, two tubes per concentration. After 15 min, tubes are chilled and EGTA is added to 2 mM. Messenger RNA is added to one tube of each pair. Ten microliters of energy and amino acid mix, 0.5 μCi of [^{14}C]leucine, and water to bring the volume to 100 μl are added to each incubation; the tubes are incubated at 37° for 30 and 60 min. Duplicate 20-μl samples are processed on filters, and radioactivity is determined as described above (Fig. 1).

Use of too little nuclease results in high endogenous backgrounds, as measured in the samples without added mRNA. Too much nuclease presumably damages ribosomal RNA and leads to poor stimulation by added mRNA. Thus, the background is reduced at the expense of activity. A level of nuclease is chosen consonant with the stringency of the background requirements.

Gel-Filtered Nuclease-Treated Systems

Preparation of Extracts, Nuclease Treatment, and Gel Filtration. The S-10 fraction is prepared as described above and is immediately nuclease

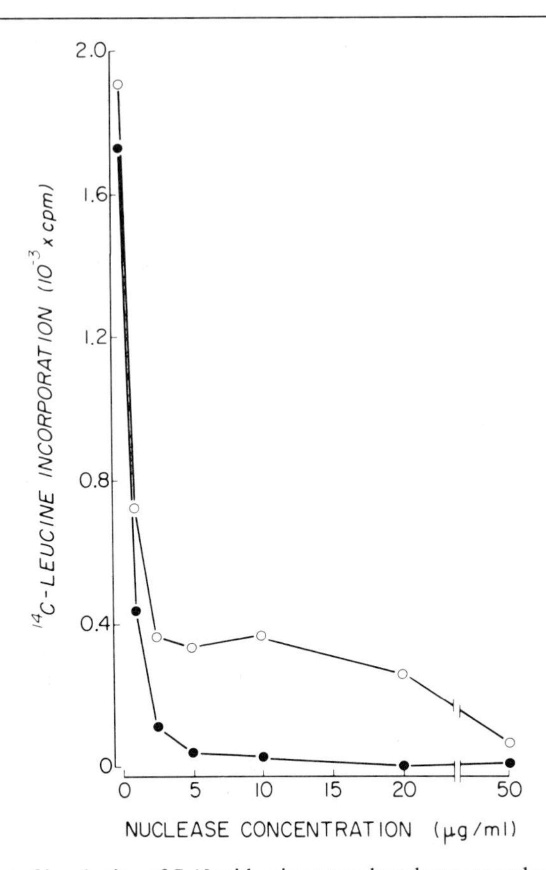

FIG. 1. Effect of incubation of S-10 with micrococcal nuclease on endogenous synthesis and synthesis using exogenous mRNA. The S-10 was incubated as described. The micrococcal nuclease was stated to have 7000 units/mg. ●——●, Endogenous synthesis; ○——○, 0.5 μg of globin mRNA was added per 100 μl incubation. Incubation was for 60 min; sample size was 20 μl; and 5 μCi of [^{14}C]leucine were used per milliliter.

treated (i.e., without freezing), also as described. After 15 min at 20°, EGTA is added to 2 m*M* and the preparation is layered on a column of Sephadex G-25 equilibrated with G-25 elution medium. Elution is with the same buffer; 2 ml of preincubated S-10 can be desalted per 10-ml column bed volume. The peak of excluded material is collected by reference to the A_{260} curve or more simply by visual inspection of the effluent. The active fractions are detectably translucent. The A_{260} of the preparation is measured; the G-25 desalting step usually causes a two- to threefold dilution. The preparation is frozen in aliquots and is stored in liquid nitrogen.

Translation of Added mRNA. In 1.0 ml final volume, the translation system contains 0.5 ml of the thawed, gel-filtered S-10, 80 m*M* KCl,

10 mM MOPS, 2.5 mM $MgAc_2$, 1.2 mM ATP, 0.3 mM GTP, 5 mM PEP, and 40 μg of pyruvate kinase per milliliter (alternatively 5 mM phosphocreatine and 100 μg/ml creatine phosphokinase), 0.4 mM sperimidine, 0.5 mM DTT, and 19 amino acids (minus the labeled amino acid) at 40 μM. These concentrations are attained by adding, per 1.0 ml final volume, 0.1 ml of salts–energy–amino acid mix. With 5 μCi of [^{14}C]leucine per milliliter and 5 μg of globin mRNA per milliliter, incorporation is usually in the range of 7000–9000 Cpm for 20-μl sample in 30 min.

The Ribosome-Runoff System

In this system polyribosomes are reduced to monomeric ribosomes and free mRNA by a preincubation allowing ribosome runoff. The preparation is gel-filtered and incubated with mRNA. In some cell lines or tissues RNA may be destroyed by endogenous nucleases during the preincubation, but in the Ehrlich cell, at least, a large amount of endogenous mRNA is preserved.

Preparation of Extracts, Preincubation, and Gel Filtration. The S-10 fraction is prepared as described. To each 0.9 ml of S-10, 0.1 ml of energy mix is added immediately (i.e., without freezing). This preparation is incubated for 30 min at 37° to allow ribosome runoff. Amino acid incorporation generally ceases by about 15 min. (We have incubated for up to 70 min with little alteration in results.) It is prudent to confirm the completeness of runoff by sucrose gradient analysis at this point. The preparation is centrifuged at 15,000 g to remove a small amount of particulate matter that forms during incubation at 37°. The chilled preparation is desalted by Sephadex G-25 chromatography as described, except that EGTA is omitted from the elution medium. This preparation is stored in liquid nitrogen and is stable for at least 2 years.

Translation of Added mRNA

Translation is as described for the gel-filtered, nuclease-treated system.

Discussion

Initiation in the Cultured Cell System

The major weakness of the Ehrlich cell and other similar cultured-cell systems compared to reticulocyte or wheat germ systems is the early failure of the initiation process. Inclusion of hemin prolongs initiation in the reticulocyte system and has been reported to do so in HeLa cell extracts,[7] presumably owing to the presence of a hemin-controlled protein kinase, which phosphorylates eukaryotic initiation factor 2. We have not been able to produce any consistent improvement in initiation by adding hemin at any stage in the preparation of cell-free systems from the Ehrlich cell. However it would seem reasonable for anyone setting up a new system to evaluate the effect of adding hemin to the lysing medium, as described by Weber *et al.*[7]

Adenine derivatives, at millimolar concentrations (adenine, 2,6-diaminopurine, cAMP, 2-aminopurine), phosphorylated glycolytic intermediates (glucose 6-phosphate, fructose 6-phosphate, fructose 1,6-bisphosphate), and elevated levels of GTP prolong initiation in the reticulocyte system.[7,11,12] The mechanisms of these effects remain unproved. We have not found adenine, glucose 6-phosphate, or GTP concentration above 0.3 m*M* to be useful in the Ehrlich cell system, but they might be in extracts prepared from some other cell lines.

The effects of the addition of a crude preparation of Ehrlich cell or reticulocyte initiation factors (i.e., the ribosomal KCl wash) to the gel-filtered, ribosome runoff S-10 system are variable, depending upon the particular KCl wash and S-10 extract used, the mRNA used, and the ionic conditions. Incorporation may be stimulated or even inhibited. The increased incorporation that is sometimes possible with added KCl wash does not seem to justify the increased effort required, except perhaps in special circumstances. In general we would recommend switching to the reticulocyte system if increased activity is required.

Optimization of the Mg^{2+}, Spermidine, and KCl Concentrations

As noted, the nuclease concentration must be optimized (Fig. 1). However, the useful range is fairly broad, and, once an optimum has been found for a particular nuclease preparation, results are sufficiently consistent so that it does not seem necessary to reoptimize for each S-10 preparation.

For the unfiltered S-10 preparations, the Mg^{2+} optimum is also broad enough so that as long as a concentration on the higher side of the peak is used, the optimum need be measured only once. However, with the gel-filtered S-10 fraction the optimum varies slightly among preparations and is so sharp that it probably should be determined for each filtered S-10 preparation (Fig. 2). A difference of 0.5 m*M* can cause a large difference in incorporation.

[11] H. Giloh (Freudenberg) and J. Mager, *Biochim. Biophys. Acta* **414,** 293.

[12] S. Legon, A. Brayley, T. Hunt, and R. J. Jackson, *Biochem. Biophys. Res. Commun.* **56,** 745 (1974).

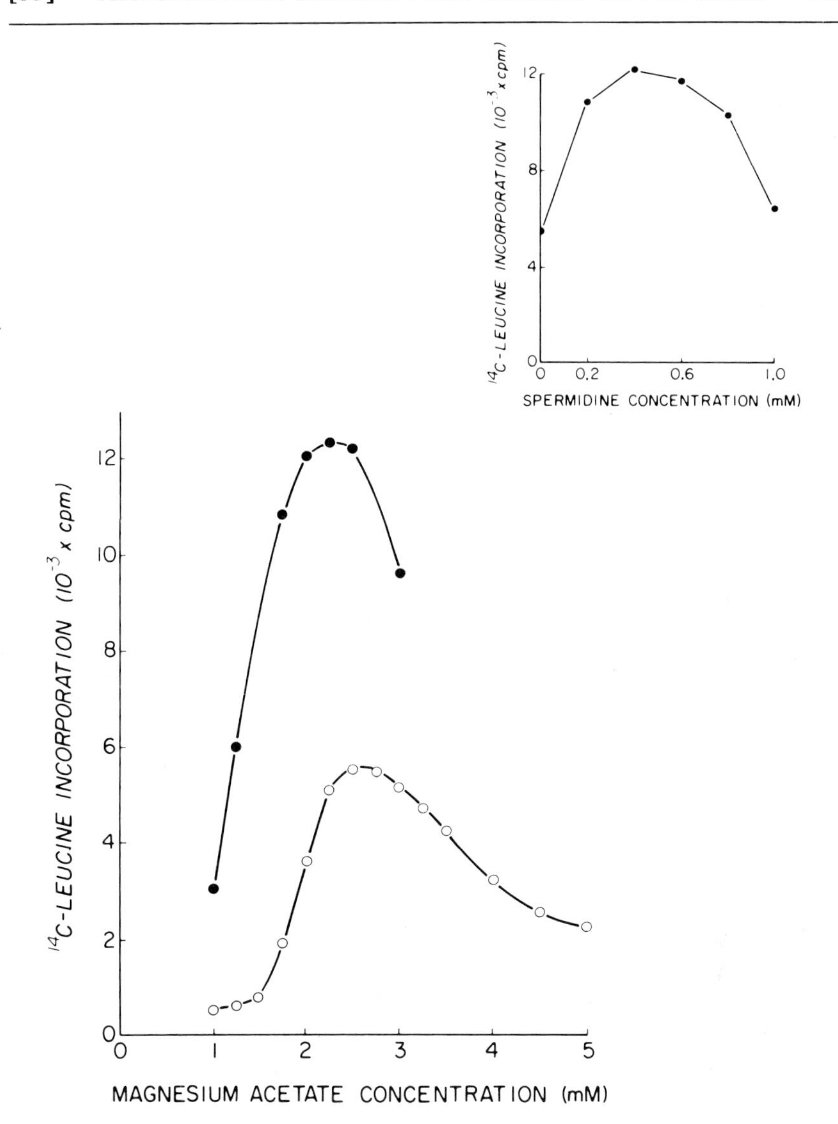

FIG. 2. [^{14}C]Leucine incorporation as a function of spermidine and magnesium concentration, in the ribosome runoff system. ○——○, No spermidine; ●——●, plus 0.4 m*M* spermidine. Inset: Incorporation as a function of spermidine concentration. Magnesium acetate = 2.5 m*M*.

After gel filtration, spermidine stimulates incorporation above that possible with optimum Mg^{2+} alone (Fig. 2). Spermidine also decreases the requirement for Mg^{2+}, so that the curve of incorporation versus the Mg^{2+} concentration should be measured in the presence of spermidine. The spermidine curves have been sufficiently reproducible so that we have used 0.4 m*M* without determining the optimum for each preparation (Fig. 2, inset). Spermine stimulates about as well as spermidine but has a lower optimum concentration. Since the polyamines are very inhibitory at supraoptimal concentrations, the optimum must certainly be determined for any new system.

The KCl optimum is sufficiently broad so that, once determined, it need not be determined for each new preparation (Fig. 3). The Cl^- anion inhibits initiation at high concentrations so that substitution of potassium acetate for KCl has been recommended.[13] When potassium acetate is substituted for KCl, the concentration curve shifts and a higher potassium concentration can be used. This may increase initiation somewhat and should perhaps be recommended for those setting up a new system.

Gel Filtration

The gel filtration step is required in the ribosome runoff system in order to reactivate the system. It is not a necessity in the nuclease treated system, and is not used in the original procedure developed for reticulocytes by Pelham and Jackson. The step does have the advantages that it removes unlabelled amino acid, allowing the use of considerably less of the added isotope, and it allows precise control of the ionic conditions. Filtration is recommended because of the large increase in incorporation.

The "Energy Generating System" and the Buffer

Two systems, creatine phosphate plus creatine phosphokinase, and PEP plus pyruvate kinase, can be used more or less interchangeably. We prefer the latter because it can also convert GDP to GTP directly, whereas the former depends upon the presence of nucleoside diphosphate kinase and ATP to regenerate GTP. In addition, some commercial preparations of creatine phosphate are contaminated with pyrophosphate,[14] which is a potent inhibitor of the aminoacyl tRNA synthetases.[15] Baglioni and Weber[16] have reported contamination of some lots of creatine phosphokinase with nuclease and suggested using fructose 1,6-bisphosphate

[13] L. A. Weber, E. D. Hickey, P. A. Maroney, and C. Baglioni, *J. Biol. Chem.* **4007** (1977).

[14] T. J. Wheeler and J. M. Lowenstein, *J. Biol. Chem.* **254,** 1484 (1979).

[15] J. D. Dignam and M. P. Deutscher, *Biochemistry* **18,** 3165 (1979).

[16] C. Baglioni and L. A. Weber, *FEBS Lett.* **88,** 37 (1978).

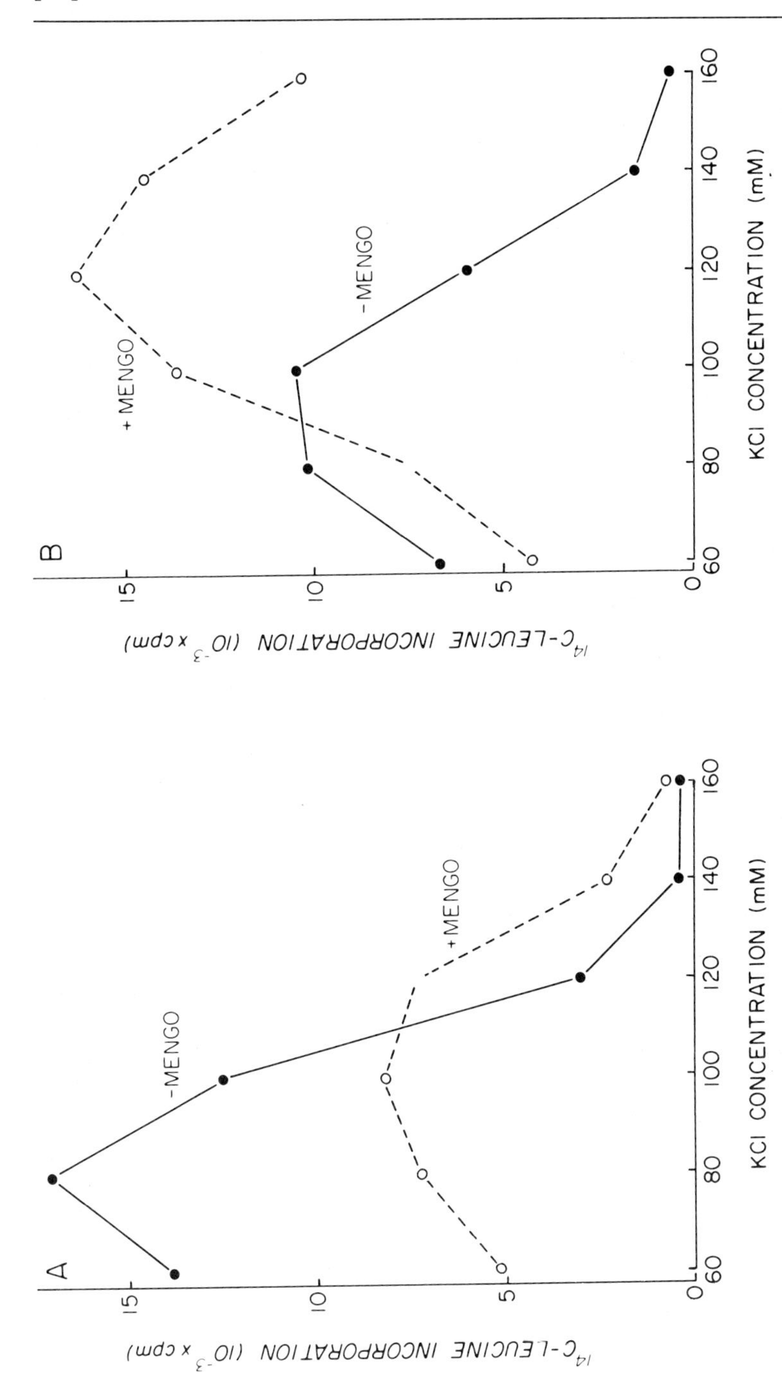

FIG. 3. [^{14}C]Leucine incorporation in the ribosome runoff system as a function of KCl and Mg^{2+} concentration, for endogenous mRNA and with added mengovirus RNA. Panel A: Mg^{2+} = 2.5 m*M*. Panel B: Mg^{2+} = 3.5 m*M*. ●——●, endogenous synthesis; ○——○, plus added mengovirus RNA.

and the endogenous glycolytic pathway enzymes to regenerate ATP. We have no experience with this in the Ehrlich cell; however, there may be some cell lines in which the endogenous glycolytic enzymes are present in insufficient quantity to prevent increasing levels of ADP and GDP. It is not clear why incorporation ceases in cell-free systems after a time, but conversion of ATP and GTP to the diphosphates is suggested as one cause.[11] This is supported by the reactivation of the rundown system by gel filtration; we have serially reactivated the same Ehrlich cell preparation by gel filtration after two sequential incubations, indicating low molecular weight components as limiting in this case.[17]

MOPS, HEPES, and tris(hydroxymethyl)aminomethane (Tris) are all mildly inhibitory at high concentrations. In the Ehrlich cell system, all three are equivalent at the concentrations used. However, Tris is undesirable because its pK is high and the pH changes markedly with temperature in the range of 0 to 37°. We use MOPS because it has been less expensive than HEPES. However, most laboratories use HEPES, and HEPES might therefore be preferred on the basis of the wider experience with it.

Sulfhydryl Reagent

We have not found any effect of adding DTT to the unfiltered preparations. However, gel-filtered systems become more sensitive to the need for sulfhydryl reagents. Increasing the concentration above 0.5 mM had no effect. Scheele and Blackburn have recommended inclusion of 2 mM DTT in all steps to protect the endogenous ribonuclease inhibitor.[18] They reported that addition of a ribonuclease inhibitor isolated from human placenta increases translation of large mRNA species in the wheat germ systems. They found an excess of the inhibitor existing in the free form in reticulocyte and Krebs cell extracts and attributed the absence of measurable ribonuclease activities and the lack of effect of added inhibitor in these systems to the endogenous inibitor. High-molecular-weight proteins (200,000) are synthesized in the Ehrlich and reticulocyte systems in our hands, but we have not examined the addition of this inhibitor. We have incubated the Ehrlich cell S-10 for up to 70 min at 30° and find only a small reduction in incorporation on endogenous mRNA between 30-min and 70-min incubations, suggesting good endogenous inhibition of ribonuclease.

Temperature of Incubations

Many cell-free systems are incubated at temperatures in the 25–30° range. This has been recommended for the reticulocyte system because the hemin-regulated inhibitor is activated much more rapidly at 37°. This

[17] W. D. Ensminger and E. C. Henshaw, unpublished observations.
[18] G. Scheele and P. Blackburn, *Proc. Natl. Acad. Sci. U.S.A.* **76,** 4898 (1979).

effect has not been noted in the Ehrlich cell system. Incorporation is more rapid at 37° than at 30°; it ceases sooner, but at about the same level. We chose 37° because of the advantage of speed. In a new system we would recommend comparing total incorporation at 25° (where the hemin-regulated inhibitor is inactive) and 37°.

Freezing and Thawing

In any preparation, sequential cycles of freezing and thawing seem to have progressively more damaging effects. Our gel-filtered preparations lose only 25% or less of their activity when they are frozen the first time, but may lose 90% if they are thawed and refrozen. We also have the impression that freezing and thawing sensitizes the preparations to gel filtration. Therefore we have designed the procedures described to have only one freezing and thawing step. However, crude preparations seem to be more resistant to freezing and thawing than gel-filtered preparations. The reticulocyte lysate can be frozen when obtained, thawed for nuclease treatment, and refrozen for storage and will maintain good activity. It is possible that if convenience dictated two freezing steps for the cultured cell system, they could be performed safely.

Comparison of the Nuclease Treated and Ribosome Runoff Systems

The nuclease-treated system is certainly the most generally useful of the two because of its wide applicability. It is slightly more complicated because of the nuclease step, which requires careful calibration. The nuclease also reduces activity to some extent. However, all mRNA species can be assayed directly in the system, with low backgrounds.

In the ribosome runoff system initiation ceases during the preincubation, but endogenous mRNA is still present. Gel filtration of the rundown system reactivates initiation, presumably through the removal of small molecules such as GDP. Endogenous mRNA is then utilized when the system is incubated under incorporating conditions. The optimum Mg^{2+} and KCl concentrations for utilization of endogenous mRNA are 2.0–2.5 m*M* and 80–90 m*M*, respectively (Fig. 3). Thus, at these ionic concentrations the effect of added mRNA may be obscured by endogenous synthesis. However, the system may be useful under several conditions: (*a*) if the optimal ionic conditions for translation of added mRNA are different from those for endogenous mRNA; (*b*) if an antibody is available for precipitation of the protein product of interest; (*c*) if the products are to be examined by gel electrophoresis. The first situation is illustrated by mengovirus RNA, which can be translated at substantially higher KCl and $MgAc_2$ concentration than endogenous mRNA (Fig. 3).

It is evident from Fig. 3 that mengovirus RNA inhibits endogenous

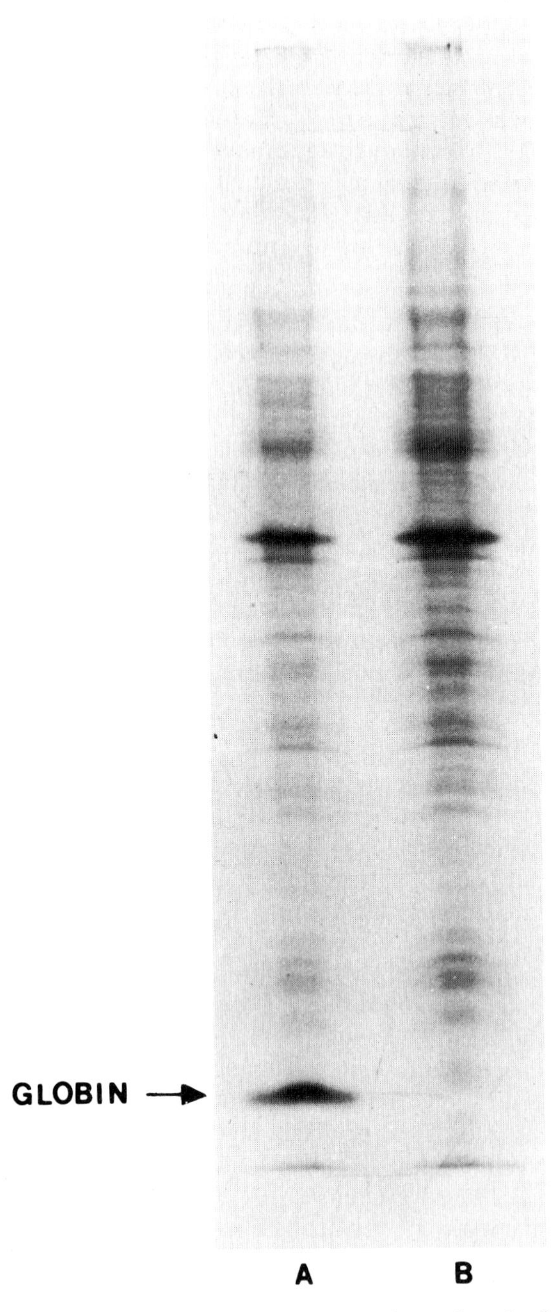

FIG. 4. Autoradiogram of SDS–polyacrylamide gel electrophoresis of translation products of the ribosome runoff system. The ribosome runoff system was incubated for 60 min with 40 μCi of [^{35}S]methionine, per milliliter, specific activity 25 Ci/mmol; KCl concentration, 80 mM; $MgAc_2$, 2.5 mM. Ten-microliter samples were analyzed on SDS gels. Autoradiography was for 28 hr. Lane A: With 5 μg/ml added globin mRNA; lane B: no added RNA.

synthesis at KCl concentrations of 80 m*M* and is itself most actively translated at about 110 m*M* and 3.5 m*M* $MgAc_2$. Thus, increasing KCl concentration to 130 m*M* and $MgAc_2$ to 3.5 m*M* produces a system in which translation is virtually confined to the exogenous mRNA. The differential effect of salt appears to depend primarily upon the structure of the mRNA translated rather than its presence as endogenous versus exogenous mRNA, since the KCl concentration curves of exogenous Ehrlich cell mRNA and globin mRNA are similar to endogenous mRNA rather than exogenous mengovirus RNA.

Translation of messages that can be analyzed by SDS–polyacrylamide gel electrophoresis is illustrated in Fig. 4. In this case, the synthesis of globin can be detected readily after the addition of 5 μg of globin mRNA per milliliter to the runoff system, because the globin band is readily resolved from endogenous proteins. The availability of an antibody for the product also obviously allows detection of the message despite the background. In our hands total incorporation in the ribosome runoff system is usually at least twice as great as in the nuclease-treated system.

Acknowledgments

The authors gratefully acknowledge the expert technical assistance of Eileen Canfield. This work was supported by NIH Grants CA-21663 and CA-11198.

[40] Preparation of a Cell-Free System from Chinese Hamster Ovary Cells That Translates Natural and Synthetic Messenger Ribonucleic Acid Templates[1]

By KIVIE MOLDAVE and ITZHAK FISCHER

A number of eukaryotic cell-free systems that translate exogenous mRNAs, from such diverse cells as reticulocytes, ascites, wheat germ, embryonic chick muscle, yeast, several cultured cell lines, etc., have been reported.[2–11] The preparation of a cell-free system from cultured Chinese hamster ovary cells that actively and accurately translates mes-

[1] This work was supported in part by Research Grants AM 15156 and AG 00538 from the National Institutes of Health, and NP 88 from the American Cancer Society. The authors thank Mr. Robert L. Jones, Mrs. Eva Mack, Mr. Wayne Sabo, and Drs. Isaac Sadnik, James S. Hutchison, Stewart Laidlaw, and E. Thalia David for their many contributions.

[2] M. B. Mathews and A. Korner, *Eur. J. Biochem.* **17,** 328 (1970).

[3] M. H. Schrier and T. Staehelin, *J. Mol. Biol.* **73,** 329 (1973).

[4] R. G. Crystal, N. A. Elson, and W. F. Anderson, this series, Vol. 30, p. 101.

[5] S. M. Heywood and A. W. Rourke, this series, Vol. 30, p. 669.

ISBN 0-12-182001-7

senger RNAs, reported previously[12,13] is detailed below. Exponentially growing cells are homogenized in hypotonic buffer; the unbroken cells, debris, nuclei, and mitochondria are removed by centrifugation. The supernatant fraction contains polysomes, ribosomes, ribosomal subunits, aminoacyl-tRNA synthetases, tRNA, and translational factors; it incorporates radioactive amino acids into protein when incubated with ATP, GTP, an energy-generating system, and amino acids. When endogenous mRNA is degraded by preincubating the postmitochondrial extract with micrococcal nuclease,[10] polysomes are no longer evident and the ribonucleoprotein particles exist as 40 S and 60 S ribosomal subunits and 80 S ribosomes; synthesis of protein with nuclease-treated extract requires exogenous mRNA in addition to the components listed above for protein synthesis with endogenous mRNAs.

Reagents

Cell culture media: Eagle's minimal essential medium/alpha modified (KC Biological, Inc., Lenexa, Kansas), 100.3 g; adenosine, 0.1 g; deoxythymidine, 0.1 g; $NaHCO_3$, 22 g; and glass-distilled H_2O to 10 liters. Powdered MEM/alpha modified, with L-glutamine and without $NaHCO_3$, is available packaged in quantities sufficient for 10 liters. The solution is sterilized by filtration under pressure with 5% CO_2–95% air, through a Millipore Twin-90 unit (0.22 μm pore size), and stored at 4°. Immediately prior to use, add 10% (v/v) fetal calf serum or 5% fetal calf serum and 5% calf serum. Sera are usually heat-inactivated at 55° for 30 min.

Solution A: Tris-HCl buffer, 0.035 *M*; NaCl, 0.15 *M*; glucose, 0.011 *M*; pH 7.3

Solution B: Tris-HCl buffer, 0.01 *M*; KCl, 0.015 *M*; $Mg(OAc)_2$, 0.0015 *M*; 2-mercaptoethanol, 0.006 *M*; pH 7.3

Solution C: HEPES–KOH buffer, 0.2 *M*; KCl, 1.2 *M*; $Mg(OAc)_2$, 0.05 *M*; 2-mercaptoethanol, 0.06 *M*; pH 7.3

Solution D: HEPES–KOH buffer, 0.02 *M*; KCl, 0.12 *M*; $Mg(OAc)_2$, 0.0015 *M*; 2-mercaptoethanol, 0.006 *M*; pH 7.3

[6] R. T. Schimke, R. E. Rhoads, and G. S. McKnight, this series, Vol. 30, p. 694.

[7] G. Schutz, M. Beato, and P. Feigelson, this series, Vol. 30, p. 701.

[8] L. Villa-Komaroff, M. McDowell, D. Baltimore, and H. Lodish, this series, Vol. 30, p. 709.

[9] A. Marcus, D. Efron, and D. Weeks, this series, Vol. 30, p. 749.

[10] H. R. B. Pelham and R. J. Jackson, *Eur. J. Biochem.* **67,** 247 (1976).

[11] E. Gasior, F. Herrera, I. Sadnik, C. S. McLaughlin, and K. Moldave, *J. Biol. Chem.* **254,** 3965 (1979).

[12] I. Fischer, S. M. Arfin, and K. Moldave, *Biochemistry* **19,** 1417 (1980).

[13] I. Fischer and K. Moldave, *Anal. Biochem.* **113,** 13 (1981).

Sephadex G-25 medium: About 10 g of Sephadex (4–6 ml bed volume per gram of dry gel) are boiled for 30 min in distilled H_2O, and then allowed to cool to 4°. The gel is poured into a glass column carefully, avoiding channeling or air bubbles, and 50 ml of solution D are then passed through at 4°. The column can be used several times by washing with 50 ml of 0.6–1 *M* KCl followed by 50 ml of solution D.

Incubation media (-Leu), (-Met), or (-Phe): HEPES–KOH, pH 7.2, 0.18 *M*; dithiothreitol, 0.01 *M*; ATP, 0.009 *M*; GTP, 0.0018 *M*; and 19 nonisotopic amino acids, 0.001 *M* (excluding the isotopic amino acid, for example, Leu, Met, or Phe, to be used in the second step of the incubation). Tyrosine and cysteine are added separately to the incubation media; 2.0 and 1.2 mg, respectively, per 10 ml of solution.

KOAc, 1 *M*

$Mg(OAc)_2$, 0.06 *M*

Creatine phosphate, 0.3 *M*

Creatine phosphokinase, 12 mg/ml

$CaCl_2$, 0.005 *M*

Micrococcal nuclease, 4000 units/ml (Boehringer Mannheim, from *Staphylococcus aureus*). It is necessary to test each batch of nuclease, since the batches vary in activity.

EGTA, 0.005 *M*; ethylene glycol bis(β-aminoethyl ether)-*N,N,N',N'*-tetraacetic acid, neutralized with 2 *M* NH_4OH or KOH

Trichloroacetic acid, 10%; casamino acids (Difco), 2%

Trichloroacetic acid, 5%; casamino acids, 2%

$Mg(OAc)_2$, 0.1 *M*

Poly(U), 10 mg of polyuridylic acid per milliliter

Preparation of the Cell-Free System

Unless otherwise indicated, all steps were carried out at 2°–4°, and sterile glassware is used throughout. One to three liters of Chinese hamster ovary cells (for example CHO, GAT^-) are grown in cell culture media[14] in spinner flasks at 34°, for at least three generations (doubling time approximately 20 hr) to a density of about 5×10^5 cells/ml. The cell suspension is rapidly added to an equal volume of frozen, crushed isotonic buffered salts solution (solution A) and centrifuged at 2000 *g* for 5 min at 2°. Essentially similar results are obtained when the cell culture is centrifuged directly at room temperature to obtain the cells. The sedimented cells are washed four times by repeated resuspension and centrifugation with 50–100 ml of cold solution A, and the packed cell volume is

[14] L. H. Thompson, D. J. Lofgren, and G. M. Adair, *Cell* **11,** 157 (1977).

noted. The cells are washed once with a hypotonic buffered salts solution (solution B) resuspended (in three times the volume of packed cells noted above) in solution B, and homogenized in a tight-fitting Dounce glass–glass homogenizer; approximately 40 strokes are adequate to obtain a good homogenate, but, owing to the variation in homogenizers, the process is monitored by examining in the microscope for breakage every 10–20 strokes. It is possible to achieve 70 to 80% lysis. One-tenth volume of solution C is then added, and the homogenate is centrifuged at 30,000 g for 20 min to obtain the postmitochondrial fraction, which is removed by careful aspiration of the supernatant, avoiding the pellet and the lipid layer at the meniscus. The supernatant solution is passed through a column of Sephadex G-25 medium (20 × 1.6 cm) equilibrated with solution D, and eluted with the same solution. Two-milliliter effluent fractions are collected, and those with absorbance at 260 nm of over 15 A_{260} units/ml are pooled. Aliquots of this S-30 preparation, about 0.04 ml, are frozen into beads by dropping from a Pasteur pipette into liquid nitrogen; in some cases, 0.1-ml aliquots are placed in plastic microfuge tubes or vials and quick frozen in liquid nitrogen. The extracts thus prepared are stable for over 3 months in a liquid nitrogen storage container.

The S-30 preparation containing endogenous polysomes incorporates radioactive amino acids extensively into protein in the absence of added mRNA when the reaction is carried out with buffered salts, nucleotides, an energy-generating system, and a mixture of amino acids. The addition of mRNA (globin mRNA) to the incubations containing S-30 extract stimulates both the initial rate and the total incorporation over twofold. The 5′-cap analog, 7-methylguanylate, which inhibits initiation of protein synthesis,[11,15–20] decreases amino acid incorporation in the S-30 preparation by about 50%; the initial rate of incorporation (for the first 40-min of incubation) is similar with and without 7-methylguanylate but protein synthesis stops much earlier when the analog is present. These results[13] indicate that the cell-free extract contains a pool of ribosomal subunits that are functional in initiation, that initiation-competent ribosomal subunits are generated in the course of the incubation as a consequence of the ribosome cycle, and that endogenous mRNA is recycled as reinitiation of pro-

[15] G. W. Both, A. K. Banarjee, and A. J. Shatkin, *Proc. Natl. Acad. Sci. U.S.A.* **72,** 1189 (1975).

[16] S. Muthukirshnan, G. W. Both, Y. Furiuchi, and A. J. Shatkin, *Nature* (*London*) **225,** 33 (1975).

[17] D. Canaani, M. Revel, and Y. Groner, *FEBS Lett.* **64,** 326 (1976).

[18] E. D. Hickey, L. A. Weber, and C. Baglioni, *Proc. Natl. Acad. Sci. U.S.A.* **73,** 19 (1976).

[19] L. A. Weber, E. R. Fernan, E. D. Hickey, M. C. Williams, and C. Baglioni, *J. Biol. Chem.* **251,** 5657 (1976).

[20] R. Kaempfer, H. Rosen, and R. Israeli, *Proc. Natl. Acad. Sci. U.S.A.* **75,** 650 (1978).

tein synthesis takes place on open initiation sites in polysomal mRNA as they become available.

Translation of Natural mRNA

Translation of exogenous natural and synthetic mRNAs with the postmitochondrial extract is carried out with a two-step incubation. The first step involves the degradation of endogenous mRNA in the S-30 fraction with micrococcal nuclease plus Ca^{2+}; the second step measures incorporation of radioactive amino acid protein when EGTA is added to chelate the calcium and mRNA is added as a template.

Between 5 and 18 μl of postmitochondrial extract (50–300 μg of protein) are incubated with 5 μl of incubation media (–Leu), 3 μl of 1 *M* KOAc, 2 μl of 0.06 *M* $Mg(OAc)_2$, 3 μl of creatine phosphate, 2 μl of creatine phosphokinase, 2.5 μl of $CaCl_2$, 2.5 μl of micrococcal nuclease, and solution D to a volume of 0.038 ml. After 8–10 min at 20°, the reaction mixtures are placed in ice (2°) and the following components are added: 5 μl of EGTA, 4 μl of radioactive amino acid such as [^{3}H]leucine (final concentration, 1 to 2×10^{-5} *M*), and 3 μl of mRNA (0.2–0.4 μg; for example, reticulocyte poly(A)$^+$ RNA[21,22]) or H_2O. The final concentrations of various reaction components, in a total volume of 0.05 ml, are as follows: HEPES–KOH buffer, 25 m*M*; potassium acetate (and potassium chloride), 100 m*M*; magnesium acetate, 2.9 m*M*; dithiothreitol, 1 m*M*; ATP, 0.9 m*M*; GTP, 0.18 m*M*; creatine phosphate, 18 m*M*; 19 nonisotopic amino acids, 0.1 m*M* each; calcium chloride, 0.25 m*M*; EGTA, 0.5 m*M*. These concentrations include amounts of various components such as K^+ and Mg^{2+} introduced with other incubation additions, such as postmitochondrial extract.

Incubations are for 1–3 hr at 30°. At the end of the incubation period, 1 ml of cold 10% trichloroacetic acid is added, the reaction tubes are placed on ice for 20–30 min, then heated in a water bath at 90° for 15 min and cooled. The acid-insoluble protein is collected on glass-fiber filters, washed three times with 5 ml of 5% trichloroacetic acid, dried under an infrared lamp, and counted in a scintillation counter.

Translation of Synthetic Polynucleotides (Polyuridylic Acid)

This assay is similar to the one described above for natural mRNA, except that the incubation media (–Phe) is used, and the concentration of

[21] A. Krystosek, M. L. Cawthon, and D. Kabat, *J. Biol. Chem.* **250,** 6077 (1975).
[22] H. Aviv and P. Leder, *Proc. Natl. Acad. Sci. U.S.A.* **69,** 1408 (1972).

Mg^{2+} is higher, 8.5 m*M* instead of 2.9 m*M*. Postmitochondrial extract (5–10 μl) is incubated with 5 μl of incubation media (−Phe), 3 μl of 1 *M* KOAc, 4 μl of 0.1 *M* $Mg(OAc)_2$, 3 μl of creatine phosphate, 2 μl of creatine phosphokinase, 2.5 μl of $CaCl_2$, 2.5 μl of nuclease, and solution D to a volume of 0.036 ml. After a period of 8–10 min at 20°, 5 μl of EGTA, 4 μl of [^{3}H]phenylalanine, and 5 μl of poly(U) or H_2O are added. The solution is incubated at 30° for 60 min, then the hot acid-insoluble fractions are prepared and counted.

Characteristics of the Nuclease-Treated Cell-Free Translational System[12,13]

Sucrose gradient centrifugation of the S-30 extract preincubated with nuclease reveals that all the polysomes are degraded to ribosomes and ribosomal subunits. The ability to incorporate amino acids into protein in the fully supplemented system (see the table) indicates that the nuclease-treated extract contains functional aminoacyl-tRNA synthetases, tRNAs,

EFFECT OF VARIOUS INCUBATION COMPONENTS ON PROTEIN SYNTHESIS IN THE POSTMITOCHONDRIAL EXTRACT DEPLETED OF mRNA

Incubation components	cpm of [^{3}H]leucine incorporated
Complete system[a]	122,300
Complete system,[a] − mRNA	1,500
Complete system,[a] − ATP and GTP	7,200
Complete system,[a] − creatine phosphate	2,800
Complete system,[a] − amino acids	4,800
Complete system[b]	192,200
Complete system,[b] − poly(U)	2,200

[a] The complete system contained nuclease-treated postmitochondrial extract, Incubation media (-Leu): KOAc, $Mg(OAc)_2$, creatine phosphate, creatine phosphokinase, EGTA, [^{3}H]leucine, and reticulocyte poly(A)$^+$ RNA, as described in the text; $CaCl_2$ and nuclease from the first incubation step were also present. In these experiments 1.4 μ*M* [^{3}H]leucine (22,000 cpm/pmol) was used. Incubations were carried out for 60 min.

[b] The complete system contained nuclease-treated postmitochondrial extract, Incubation media (-Phe), $Mg(OAc)_2$ to 8.5 m*M*, [^{3}H]phenylalanine, poly(U), and the other components described above and in the text. In these experiments 1.4 μ*M* [^{3}H]phenylalanine (5500 cpm/pmol) was used. Incubations were carried out for 120 min.

ribosomes and ribosomal subunits, and translational factors; however, as shown below, protein synthesis is markedly dependent with this preparation on the addition of template as well as nucleotides, an energy source, and amino acids. Incorporation is dependent on the time of incubation up to about 3 hr, and on the concentration of mRNA added, although relatively high levels of some mRNAs tend to inhibit the system.

A number of natural mRNAs are translated in the CHO system described here. For example, poly(A)-containing RNAs from CHO cells, reticulocytes, yeast, mouse melanoma and mouse neuroblastoma cells, as well as RNA from brome mosaic virus, markedly stimulate protein synthesis. Gel electrophoretic analysis (in SDS) of the products synthesized by the mRNA-depleted S-30 extract in the presence of globin mRNA indicates that the major product migrates as a single sharp band in the region of the gel corresponding to a myoglobin marker. When poly(A)$^+$ RNA from CHO cells is used as a template and the products are analyzed by one-dimensional SDS–gel electrophoresis, at least 60 bands are detected which correspond, with few exceptions, to those seen in labeled preparations from intact CHO cells. These results indicate that the cell-free system translates faithfully essentially all the mRNAs that are translated in the intact cells, and that it probably carries out most of the appropriate posttranslational modifications properly. Also, polypeptides of relatively high molecular weight are synthesized *in vitro*.

In incubations with exogenous mRNAs, all the reactions involved in the initiation of protein synthesis, polypeptide chain elongation, polysome formation, termination, and release can be detected. Procedures for the analysis of these intermediary reactions in translation have been described.[12,13]

[41] Preparation of a Cell-Free Protein-Synthesizing System from Wheat Germ

By CARL W. ANDERSON, J. WILLIAM STRAUS, and BERNARD S. DUDOCK

Crude extracts of commercial wheat germ are known to be capable of translating a wide variety of messenge RNAs. The basic wheat germ cell-free system used in most laboratories was initially described by Roberts and Paterson.[1] However, several significant improvements have been in-

[1] B. E. Roberts and B. M. Paterson, *Proc. Natl. Acad. Sci. U.S.A.* **70,** 2330 (1973).

ISBN 0-12-182001-7

troduced since that publication, such as the use of polyamines and the substitution of acetate for chloride ions in buffering systems.[2,3] The purpose of this communication is to describe in detail the preparation of a wheat germ cell-free system capable of synthesizing high molecular weight polypeptides and to enumerate some of the difficulties that may be encountered in preparing and using such a system. It should be noted that cell-free protein-synthesizing systems have been prepared also from wheat embryos.[4] Such systems are reported to have lower endogenous ribonuclease activity, but have not been as popular owing to the added procedure of isolating the embryos.

Preparation of Wheat Germ Extract

Materials. Wheat germ extract is commercially available from Bethesda Research Laboratories (BRL) (Gaithersburg, Maryland) and Miles Laboratories (Elkhart, Indiana), but can be easily and inexpensively prepared in any modern laboratory. Untoasted wheat germ can be obtained in small quantities (ca 1 pound) from General Mills (contact Director, Quality Control, P.O. Box 113, Minneapolis, Minnesota 05426; telephone 612-540-2354). General Mills has been setting aside parts of certain lots for research. We have not tested a sufficient number to comment on the relative activity of different lots. Alternatively, untoasted wheat germ can be purchased from local health food stores; however, only two or three of over a dozen lots tested gave satisfactory activity. Wheat germ can be stored for several years in a vacuum desiccator over silica gel at 4°. Exposure to moisture, heat, or freezing will result in a severe loss of activity.

Prior to extraction, assemble the following materials:

Acid-washed sand or powdered glass (oven sterilized)

Medium-sized mortar (ca. 10 cm in diameter) and pestle, rubber policeman, and spatula

U.S. Standard Sieve Series No. 25 screen (opening = 0.71 mm)

Corex centrifuge tubes, 30 ml (new, unscratched); refrigerated centrifuge

Sephadex G-25 medium or fine column (2.5 × 30 cm) equilibrated with "column buffer," set up in a cold room at 4°; fraction collector and tubes

Spectrophotometer

[2] K. Marcu and B. Dudock, *Nucleic Acids Res.* **1,** 1385 (1974).

[3] J. F. Atkins, J. B. Lewis, C. W. Anderson, and R. F. Gesteland, *J. Biol. Chem.* **250,** 5688 (1975).

[4] A. Marcus, D. Efron, and D. P. Weeks, this series, Vol. 30, p. 749.

Plastic microcentrifuge tubes, 500 μl (ca 150 per extract)
Liquid nitrogen in a small Dewar flask
Stock solutions:

a. 1.0 *M* HEPES buffer, adjusted to pH 7.60 (at 20°) with KOH. Hepes is not stable to autoclaving and will support growth of mold and bacteria. Store as 5-ml aliquots at −20°.
b. 1.0 *M* KOAc (acetate), adjusted to pH 7 with acetic acid. Autoclave in 100-ml bottles.
c. 0.1 *M* $Mg(OAc)_2$, adjusted to pH 7.0 with KOH if necessary. Autoclave in 100-ml bottles.
d. 0.1 *M* $CaCl_2$. Autoclave in 100-ml bottles.
e. 1.0 *M* dithiothreitol (DTT). Store at −70° in 1.0 ml aliquots.

Extraction buffer (make fresh, keep ice cold):

	Volume	*Final conc.*
1.0 *M* HEPES, pH 7.6	1.0 ml	20 mM
1.0 *M* KOAc	5.0 ml	100 mM
0.1 *M* $Mg(OAc)_2$	0.5 ml	1 mM
0.1 *M* $CaCl_2$	1.0 ml	2 mM
1.0 *M* DTT	0.05 ml	1 mM
Distilled H_2O to final vol. 50.0 ml		—

Column buffer (make fresh, keep at 4°):

1.0 HEPES, pH 7.6	10.0 ml	20 mM
1.0 *M* KOAc	60.0 ml	120 mM
0.1 *M* $Mg(OAc)_2$	25.0 ml	5 mM
1.0 *M* DTT	0.5 ml	1 mM
Distilled H_2O to final vol. 500.0 ml		—

The extraction should be performed as rapidly as possible and all materials be kept at 0–4°. It is preferable to work in a cold room. All glassware should be scrupulously clean and oven-sterilized to minimize contamination from exogenous nucleases. Plastic gloves should be worn throughout. If nuclease contamination is a problem, then placental ribonuclease inhibitor[5] can be used to reduce degradation of the RNA.[6]

Extraction Procedure. Sieve about 13 g of wheat germ with moderate vigor for 3–5 min. The germ is retained on the screen while about 20% of the weight (mostly endosperm and chaff) passes through.

Vigorously grind 10 g of sieved wheat germ with an equal weight of acid-washed sand or powdered glass for 1 min in a prechilled mortar.

[5] P. Blackburn, G. Wilson, and S. Moore, *J. Biol. Chem.* **252,** 5904 (1977).
[6] G. Scheel and P. Blackburn, *Proc. Natl. Acad. Sci. U.S.A.* **76,** 4898 (1979).

Note that overgrinding will result in a loss of activity. Then mix in 20 ml of cold "extraction buffer" (see above) in 4–5 ml aliquots, with a rubber policeman or spatula, to obtain a thick paste. Transfer the paste with a spatula to a 30-ml Corex centrifuge tube and centrifuge at 30,000 *g* for 10 min at 2° (e.g., 16,000 rpm in a Sorvall SS-34 rotor). After centrifugation, remove the brownish supernatant (~12 ml) with a Pasteur pipette, taking care to avoid contamination from the yellow surface layer of fatty material and from the pellicle that covers the sand. A $\frac{1}{800}$ dilution of this supernatant into water should give approximate spectrophotometric readings at A_{260} of 0.5, and at A_{280} of 0.4.

Low molecular weight material is then removed from the extract by gel filtration. Load the extract on a Sephadex G-25 column (see above) and collect 4-ml fractions at a flow rate of 2–3 ml/min. The extract can be observed as a light brown band that elutes in the void volume. A yellow band is retained on the column and if desired, can be eluted in 1–2 column volumes. The void volume peak should be evaluated spectrophotometrically by diluting 10-μl aliquots of each fraction into 2 ml of water and determining the A_{260} values. The void volume should contain about one-third of the applied optical density at 260 nm. Pool appropriate portions of the void volume peak (i.e., discard the last third of the void volume peak) and centrifuge (15,000 *g* for 10 min at 2°) to remove any insoluble material. Note that the concentration of pooled extract can be changed by adding column buffer. We find it convenient to adjust the A_{260} to 100 for uniformity between different extracts.

The pooled extract is immediately aliquoted into 500-μl plastic tubes (e.g., 100–200 μl/tube) and dropped into a small Dewar flask of liquid nitrogen. The frozen tubes are then transferred to an appropriately labeled box and stored in liquid nitrogen or in a freezer at −70°. Extracts are stable at −70° for at least 6 months and often up to several years. From 10 g of wheat germ, we obtain about 25 ml of extract. The preparation can be scaled up or down as required.

Wheat Germ Protein Synthesis

Materials. Assemble reaction tubes, water bath (30°), wheat germ extract, mRNA, any test compounds, solutions, and assay materials (see below). The reactions can be conveniently performed in 1.5-ml plastic microcentrifuge tubes. The wheat germ extract, mRNA, and solutions should be kept ice cold prior to incubation.

The required stock solutions are listed below.

Hepes, 1.0 *M*, pH 7.60 (at 20°), adjusted with 5 *M* KOH, (do not autoclave)

$Mg(OAc)_2$, 0.25 *M*, adjusted to pH 7 with KOH, autoclaved
KOAc, 1.0 *M*, autoclaved (KOAc is very hydroscopic), adjusted to pH 7.0 with acetic acid
Stock solutions, 0.1 *M*, each of 20 amino acids (see Table I)
Working amino acid mixture (minus radioactive methionine, leucine, or other amino acids as required), 5 m*M* each amino acid ($\frac{1}{20}$ dilution of 0.1 *M* stock)
Creatine phosphokinase (CPK) (Sigma), 10 mg/ml in H_2O, 40–170 IU/mg
Creatine phosphate (CP) disodium salt (Sigma), 0.4 *M*, adjusted to pH 7
ATP, 0.1 *M*, adjusted to pH 7 with KOH
GTP + $Mg(OAc)_2$ (each 40 m*M*), adjusted to pH 7
DTT, 0.1 *M*
Spermine tetrahydrochloride, 1.5 mg/ml, adjusted to pH 7

All solutions are stored at −20° in aliquots of 1–5 ml in glass vials or polypropylene plastic tubes. It may prove to be convenient to premix several components.

Procedure. Decide on the amount and type of radioactive label and lyophilize (if necessary). It is not usually necessary to lyophilize radioac-

TABLE I
AMINO ACIDS FOR STOCK SOLUTIONS

Acid	Molecular weight	0.1 *M* (mg/ml)
L-Alanine	89.09	8.9
L-Arginine-HCl	210.7	21.1
L-Asparagine-H_2O	150.10	15.0 suspension
L-Aspartic acid	133.10	13.3 suspension
L-Cysteine	121.16	12.1
L-Glutamic acid	147.13	14.7 suspension
L-Glutamine	146.16	14.6
Glycine	75.07	7.5
L-Histidine-HCl	209.6	21.0
L-Isoleucine	131.17	13.1
L-Leucine	131.17	13.1
L-Lysine-HCl (hydrate)	182.7	18.3
L-Methionine	149.12	14.9
L-Phenylalanine	165.19	16.5
L-Proline	115.13	11.5
L-Serine	105.09	10.5
L-Tryptophan	204.22	20.4 suspension
L-Threonine	119.12	11.9
L-Tyrosine	181.19	18.1 suspension
L-Valine	117.15	11.7

tive amino acids unless they are stored in acid (HCl) or unless large volumes are to be used. In general, [^{14}C]leucine is used for kinetic measurements and [^{35}S]methionine is used when products are to be analyzed by polyacrylamide gel electrophoresis. Typical assays contain: [^{14}C]leucine (ca 300 mCi/mmol, 100 μCi/ml) 1.5 μl/assay = 0.15 μCi/assay (0.5 nmol/assay = 20μM); [^{35}S]methionine (ca 350 Ci/mmol, 5 mCi/ml) 3 μl assay = 15 μCi/assay (42.9 pmol/assay = 1.7 μM).

Note that if substantially less than 15 μCi per assay of [^{35}S]methionine is used, cold methionine should be added to provide approximately 2 μM methionine. For maximum incorporation of isotope it is worthwhile to titrate the [^{35}S]methionine to see that it is not inhibitory at high concentrations.

Write up the protocol, thaw components (on ice), and set up reaction tubes in an ice water bath. Assemble the "master mix" (see Table II) in the tube containing the radioactive label, remembering to dilute the stock solution of 40 mM GTP + Mg^{2+} 1:20. Mix stock solutions before making additions, since some components (i.e., GTP, amino acids) are suspensions. Sufficient water may be added to the "master mix" to minimize pipetting of the reaction components into the individual reaction tubes.

Add to the individual reaction tubes appropriate volumes of water and master mix, any test solutions, and mRNA (approximately 0.1–1.0 μg) in order. Approximately 3 μg of rRNA or tRNA should be added to the background control tube to protect endogenous protein synthesis from endogenous nucleases so as to provide an adequate indication of the endogenous background (this RNA should be tested for inhibition of translation). Be careful to minimize bubble formations when pipetting and to mix tubes gently before incubation. Then incubate each tube for 90 min at 30°.

Assays for Protein Synthesis. If the entire sample is to be assayed for incorporated radioactivity, add 50 μl of bovine serum albumin (5 mg/ml) and 100 μl of 2 M KOH to each tube at the end of the incubation period (to provide carrier during precipitation and to hydrolyze aminoacylated tRNA). Incubate for 15 min at 30°, and then add 1.5 ml of cold 5% trichloroacetic acid (TCA) containing amino acids (20 mM each) or 2 g of "casamino acids" (Difco Laboratories) per liter and put on ice for 10 min. Filter the reaction mixtures through glass fiber filters (Whatman GF/C), and thoroughly rinse both the tubes and filters with additional TCA. The filters are then labeled (e.g., with a No. 1 pencil) and dried for 10–15 min under a heat lamp. Dried filters may then be placed in shell-vials and counted for radioactivity in about 3 ml of toluene-based scintillation fluid, such as Econofluor (NEN). Discard TCA wash in radioactive waste.

If samples are to be analyzed by polyacrylamide gel electrophoresis,[7]

[7] U. K. Laemmli, *Nature (London)* **227,** 680 (1970).

TABLE II
COMPONENTS FOR PROTEIN SYNTHESIS[c] BY WHEAT GERM EXTRACTS

Master mix[a]	Number of 25-μl reactions					Concentration		
	10 (μl)	20 (μl)	30 (μl)	40 (μl)	50 (μl)	Extract	Component	Total
1. 1.0 *M* HEPES, pH 7.6[b]	3.5	7	10.5	14	17.5	6	14	20 m*M*
2. 0.1 *M* ATP	2.5	5	7.5	10	12.5	—	1	1 m*M*
3. 0.4 *M* CP (creatine phosphate)	5	10	15	20	25	—	8	8 m*M*
4. 10 mg/ml CPK (creatine phosphokinase)	1	2	3	4	5	—	40	40 μg/ml
5. 1.5 mg/ml spermine, pH 7	5	10	15	20	25	—	30	30 μg/ml
6. 0.1 *M* DTT	4.25	8.5	12.75	17	21.25	0.3	1.7	2 m*M*
7. 2 m*M* GTP + Mg^{2+} (1/20 of stock)	2.5	5	7.5	10	12.5	—	20	20 μ*M*
8. 5 m*M* each, 19 amino acid mix	1.25	2.5	3.75	5	6.25	—	25	25 μ*M*
9. 0.25 *M* $Mg(OAc)_2$	1	2	3	4	5	1.5	1	2.5 m*M*
10. 1.0 *M* KOAc	21	42	63	84	105	36	84	120 m*M*
11. Wheat germ extract ($A_{260} \simeq 100$)	75	150	225	300	375	—		
Volume of "Master Mix"	122	244	366	488	610			
Total assay volume	250	500	750	1000	1250			
Remaining volume	128	256	384	512	640			
Additional components (for remaining volume)								
12. Radioactive amino acid								
13. Messenger RNA								
14. Other test substances								
15. Water								

[a] Master mix optimized for translation of total cytoplasmic RNA from adenovirus-infected cells. Optimal translation of other RNAs may require slight adjustments of some concentrations.

[b] Tris-acetate, 1.0 *M* pH 7.2, will substitute for HEPES buffer.

[c] Incubate individual reactions at 30° for 90 min.

remove 2–5-μl aliquots for analysis as above (scintillation counting). The remainder of the sample is incubated with 5 μl per assay of 200 μg/ml RNase A, 200 m*M* EDTA for 15 min at 30° prior to being precipitated with 0.5 ml of 90% aqueous acetone for about 30 min at 0–4°, and centrifuged (2000 rpm, 10 min at 4° in a clinical centrifuge). The supernatant, which is discarded, contains most of the unincorporated amino acid, and the precipitate is dissolved in 50 μl of standard SDS sample buffer. It is essential that the sample be well suspended in SDS sample buffer before it is heated. This is conveniently accomplished by placing the samples in an ultrasonic cleaning bath for 30 min prior to heating. Samples may be stored frozen at −20° until they are assayed. For best resolution do not apply more than 10 μl to a well of a standard 25-slot slab gel. It is desirable to stain gels or to soak gels in destaining solution to remove residual unincorporated radioactive amino acids prior to drying for autoradiography.

If the protein under study is soluble in acetone, alternative procedures are to dissolve the sample in an equal volume of sample buffer directly, or to precipitate the sample with 10% TCA and wash three times with ethanol–ether (1:3) before dissolving in sample buffer. If ethanol–ether washes are not thorough, residual TCA will be revealed by the sample turning yellow during heating. This is a pH change of bromophenol blue and is reversible.

When the reaction product is to be subjected to chemical or sequence analysis, it should be noted that the wheat germ extract can contribute significant quantities of endogenous amino acids, presumably from aminoacylated tRNA.[8] Furthermore, terminal amino acids in the polypeptide chains may be partially acetylated. Acetylation can be prevented by pretreating the reaction mixture with citrate synthetase and oxaloacetate as described by Palmiter.[9]

Troubleshooting. There are many ways in which a protein synthesis assay can fail, either totally or partially. Components should be reasonably fresh and handled and stored carefully. This is particularly true of the wheat germ extract and the mRNA. If it is not possible to obtain satisfactory results, then it may be necessary to test each individual component. It is preferable to test components one at a time in a system that is known to function, since it is otherwise difficult to identify the problem if two or more components are bad.

Commercially available untoasted wheat germ may range in activity from very good to no activity at all. Extracts should be tested occasion-

[8] C. W. Anderson, *in* "Genetic Engineering: Principals and Methods," Vol. 4 (J. K. Setlow and A. Hollaender, eds.), p. 147–167. Plenum, New York, 1982.

[9] R. D. Palmiter, *J. Biol. Chem.* **252,** 8781 (1977).

ally with an mRNA that is known to be efficiently translated. The 9 S RNA (globin mRNA) from rabbit reticulocytes (BRL) makes a useful standard with which to compare different extracts or different translation conditions. For maximal activity, the optimal amount of extract per 25-μl reaction should be determined. With extracts at A_{260} = 100, we have usually obtained satisfactory activity at 7.5 μl of extract per 25-μl reaction mixture. One cannot exceed 12 μl of extract per assay without exceeding the optimal Mg^{2+} concentration of 2.5 m*M* for most eukaryotic messages.

Most wheat germ extracts have very low endogenous activity. If necessary, endogenous activity may be further reduced by a preincubation step with creatine phosphokinase[1] or by treatment with micrococcal nuclease.[10] In practice we have generally found either treatment to be unnecessary (the overall activity is reduced as much or more than the endogenous activity). Even with extracts that have a relatively high endogenous activity as determined by the incorporation of radioactive amino acids, few if any endogenous protein "bands" are observed on SDS–polyacrylamide gels.

Pure mRNAs are extremely sensitive to degradation and should be stored frozen at −20° in small aliquots or as an alcohol precipitate (do not use a frost-free refrigerator). Plastic gloves should be worn when handling RNA, and all solutions and glassware or plasticware must be nuclease free.

With regard to radioactive amino acids, [^{14}C]leucine is reasonably stable and can be stored frozen for several years without appreciable loss of activity. [^{35}S]Methionine is very sensitive to oxidation and should be stored frozen at −70° in small aliquots in 25 m*M* mercaptoethanol. It should be used as soon as possible, certainly within 3 months of manufacture. The quality of [^{35}S]methionine can be readily checked in a few hours by thin-layer chromatography on cellulose sheets using *n*-butanol–pyridine–acetic acid–water (300:60:200:240) as a solvent. [^{35}S]Methionine should be mixed with a solution of cold methionine (ca 10 m*M*) to reduce oxidation during spotting and to provide a marker that can be detected by ninhydrin staining.

The assay system has a very sharp pH optimum. Furthermore, the pH of HEPES buffer changes significantly upon dilution. The value given above is for the 1.0 *M* stock solution, which is made from the acid by titration to pH 7.6 with 5 *M* KOH. The actual pH of the reaction mix is considerably below pH 7.6 (i.e., pH 6.9–7.0). One should make up a series of HEPES solutions between pH 7.2 and pH 8 to determine the opti-

[10] H. R. B. Pelham and R. J. Jackson, *Eur. J. Biochem.* **67**, 247 (1967).

mum for each new extract. HEPES should be stored frozen in small aliquots; it supports microbial growth, and it is not stable to autoclaving. One should also remember that the buffer supplies some potassium that has not been accounted for in the concentrations given above. Note that 1.0 *M* Tris-acetate buffer, pH 7.20 (at 20°) can be substituted for HEPES, pH 7.6, without apparent loss of activity.

The system has a very broad potassium optimum (using KOAc) with a maximum around 120 m*M* with Ad2 RNA. The potassium optimum reported by different groups (using different mRNAs) is often quite different. The Mg^{2+} [as $Mg(OAc)_2$] optimum is rather sharp but is close to 2.2 m*M* in the presence of 20 μg of spermine per milliliter for all mRNAs we have tested, including BMV, TMV, Ad2, KB cell, globin, and T7. The Mg^{2+} optimum is dependent on the concentration of polyamine used; spermidine or putrescine at the appropriate concentration will substitute for spermine. ATP, GTP, DTT, CP, and CPK are all somewhat unstable and are suspect when all else fails. Occasionally one may receive a bad lot of CPK.

[42] Preparation of a Cell-Free System from *Saccharomyces cerevisiae* That Translates Exogenous Messenger Ribonucleic Acids[1]

By KIVIE MOLDAVE and EUGENIUSZ GASIOR[2]

A number of laboratories have described cell-free systems from yeast that carry out chain elongation reactions, the translation of polyuridylic acid, or the isolation of factors from yeast that catalyze elongation reactions.[3-9] This chapter describes the preparation of a yeast cell-free sys-

[1] This work was supported in part by research grants from the National Institutes of Health (AM 15156, AG 00538, and GM 27999). The authors acknowledge the many valuable contributions of their colleagues, Drs. Flor Herrera, Isaac Sadnik, Barry Feinberg, Mick F. Tuite, Judith Plesset, and Calvin S. McLaughlin and the technical assistance of Mrs. Eva Mack, Mrs. Pamela Sutherland, Mr. Wayne Sabo, Mr. Igor Jercinovich, and Ms. Cherie J. Nevara.

[2] Permanent address, Department of Molecular Biology, Institute of Microbiology and Biochemistry, University of Maria Curie-Sklodowska, Lublin, Poland.

[3] D. Richter and F. Lipmann, *Biochemistry* **9,** 5065 (1970).

[4] D. Richter and F. Klink, this series, Vol. 20, p. 349.

[5] A. Torano, G. Sandoval, C. Sanjose, and C. F. Heredia, *FEBS Lett.* **22,** 11 (1972).

[6] L. Skogerson and E. Wakatama, *Proc. Natl. Acad. Sci. U.S.A.* **73,** 73 (1976).

[7] L. Spermulli and J. Ravel, *Arch. Biochem. Biophys.* **172,** 261 (1976).

ISBN 0-12-182001-7

tem that translates exogenous natural and synthetic mRNA templates and carries out all the reactions involved in polypeptide chain initiation, elongation, and termination, as reported previously.[10-13] A number of strains of *Saccharomyces cerevisiae*, such as diploid SKQ2N, haploid A364A and several temperature-sensitive mutants derived from it, have been used to prepare active translational systems. Spheroplasts are prepared from growing yeast cells and allowed to incubate for a short period of time in fully supplemented media; they are then lysed, and the remaining intact cells, nuclei, mitochondria, and polysomes are removed by centrifugation. The postpolysomal extract is treated with nuclease to remove residual endogenous mRNA. This preparation incorporates amino acids into proteins when it is incubated at 20° with mRNA, ATP, GTP, an energy-generating system, and amino acids. The use of actively growing cells, the preparation of spheroplasts and incubation in growth-promoting media, removal of the heavy polysome fraction, degradation of residual mRNA with nuclease, and incubation temperatures no higher than 20° have been found to be critical for optimal protein synthesis.

Reagents

YM-1 media: succinic acid, 10 g; NaOH, 6 g; yeast extract, 5 g; adenine, 0.01 g; uracil, 0.01 g; Bactopeptone, 10 g; yeast nitrogen base without amino acids, 6.7 g; water to 900 ml. Autoclave. Add 10 g of glucose in 100 ml of H_2O, autoclaved separately. Final pH, 5.8. Yeast extract, Bactopeptone, and yeast nitrogen base are available from Difco Laboratories, Detroit, Michigan.

Sorbitol, 1 *M*

Glusulase (β-glucuronidase/arylsulfatase, Boehringer Mannheim)

YEPD-agar: Bactopeptone, 10 g; yeast extract, 10 g; adenine, 0.01 g; uracil, 0.01 g; agar, 20 g; H_2O to 900 ml. Autoclave and add 40 g of glucose in 100 ml H_2O (autoclaved separately) prior to pouring plates.

Sorbitol, 1.2 *M*

YM-5 media plus $MgSO_4$: succinic acid, 10 g; NaOH, 6 g; yeast ex-

[8] L. Skogerson and D. Engelhardt, *J. Biol. Chem.* **252,** 1471 (1977).

[9] L. Skogerson, this series, Vol. 60, p. 676.

[10] E. Gasior, F. Herrera, I. Sadnik, C. S. McLaughlin, and K. Moldave, *J. Biol. Chem.* **254,** 3965 (1979).

[11] E. Gasior, F. Herrera, C. S. McLaughlin, and K. Moldave, *J. Biol. Chem.* **254,** 3970 (1979).

[12] F. Herrera, E. Gasior, C. S. McLaughlin, and K. Moldave, *Biochem. Biophys. Res. Commun.* **88,** 1263 (1979).

[13] M. F. Tuite, J. Plesset, K. Moldave, and C. S. McLaughlin, *J. Biol. Chem.* **255,** 8761 (1980).

tract, 1 g; adenine, 0.01 g; uracil, 0.01 g; Bactopeptone, 2 g; yeast nitrogen base with amino acid, 6.7 g; $MgSO_4$, 48 g; H_2O to 700 ml. Autoclave. Add 200 ml of 2 *M* $MgSO_4$ and 10 g of glucose in 100 ml, autoclaved individually. Final pH, 5.8.

Solution A: HEPES [4-(2 hydroxyethyl)-1-piperazineethanesulfonic acid]–KOH buffer, 0.02 *M*; KOAc, 0.1 *M*; $Mg(OAc)_2$, 0.002 *M*; dithiothreitol, 0.002 *M*; pH 7.4. When solution A is used in incubations, glycerol to 20% is added.

2-Mercaptoethanol, 10 m*M*; EDTA (ethylenediaminetetraacetic acid, disodium salt), 2.5 m*M*; 1:1

Incubation media A: HEPES–KOH buffer, 0.4 *M*; $Mg(OAc)_2$, 0.04 *M*; dithiothreitol, 0.04 *M*; pH 7.4

KOAc, 4 *M*

ATP–GTP mixture: ATP, 0.020 *M*; GTP 0.004 *M*

Creatine phosphate, 0.5 *M*

Creatine phosphokinase, 8 mg/ml

$CaCl_2$, 0.0125 *M*

Amino acid mixture: 19 nonisotopic amino acids, 0.0016 *M*, excluding the isotopic amino acid (for example, Leu, Met) to be used in the second step of the incubation. Stocks of individual amino acids, except for cysteine and tyrosine, are used to make the amino acid mixture. Stocks are 50 m*M* for each amino acid except glutamic acid and phenylalanine, which are 25 m*M*. Cysteine and tyrosine are added separately, in solid form, to the amino acid mixture after it is made up.

Micrococcal nuclease, 0.5 mg/ml (Boehringer Mannheim, from *Staphylococcus aureus*). It is necessary to test each batch of nuclease, since the batches vary in activity.

EGTA, 0.0125 *M*; ethylene glycol bis(β-aminoethyl ether)-*N,N,N',N'*-tetraacetic acid, neutralized with 2 *M* KOH

Trichloroacetic acid, 10%; casamino acids (Difco), 0.2%

Trichloroacetic acid, 5%; casamino acids, 0.2%

$Mg(OAc)_2$, 0.4 *M*

Polyuridylic acid, 10 mg/ml

Preparation of the Cell-Free System

Unless otherwise indicated, all steps are carried out at 2–4°, and sterile glassware is used throughout the procedure. Yeast cultures of 1–40 liters have been used successfully. In most cases, cells are grown in 1.5-liter batches at 20%, in YM-1 media, to a density of 1.0 unit at 660 nm. The cells are obtained by centrifugation at 3000 *g* for 5 min at 2°. For very large culture volumes, the use of a continuous-flow centrifuge is recommended. Approximately 2–2.5 g wet weight of cells are obtained per liter of culture. The cells are washed with cold (2°) sterile water and resuspended in 1 *M* sorbitol (40–50 ml of sorbitol per liter of original cell culture). The cell suspension in sorbitol is incubated with glusulase (0.4–0.8 ml per liter of original cell culture) at 20° for 1 hr, to form spheroplasts.[14] The extent to which cells are converted to spheroplasts can be checked by serial dilution of a small aliquot (0.05 ml) of the suspension 20,000-fold with sterile distilled water and plating 0.5 ml on YEPD agar plates. The spheroplasts are obtained from the glusulase suspension by centrifugation, washed once with 1.2 *M* sorbitol, and resuspended in YM-5 media[15] containing 0.4 *M* $MgSO_4$ (80–120 ml of YM-5 per liter of original cell culture). This suspension is incubated for 60–90 min at 20°, with occasional swirling. The incubated spheroplasts are obtained by centrifugation, resuspended in solution A (1.5–2 ml per liter of original cell culture), and homogenized with 15–30 strokes in a tight Dounce glass–glass homogenizer. The degree of breakage is monitored microscopically; occasionally, more than 30 strokes may be necessary.

The lysate is centrifuged at 27,000–30,000 *g* for 15 min, and the supernatant is carefully aspirated, avoiding the flocculent material at the bottom of the tube and the lipid layer at the top of the solution. The supernatant (S-30) is centrifuged at 100,000 *g* for 30 min after reaching speed, to sediment the polysomes. The polysome pellet is saved and used for the preparation of yeast mRNA. The postpolysomal fraction, a modified S-100 preparation, is passed through a Sephadex G-25 medium column (25 × 1.5 cm, for 2–3 ml of supernant, obtained from 5 liters of original cell culture), previously equilibrated with solution A containing 20% glycerol. Elution is carried out with the same solution, and 0.5–1.0-ml fractions are collected. Fractions with absorbance values at 260 nm above 30 A_{260} units/ml are pooled. About 4–12 A_{260} units are obtained per liter of original cell suspension, containing between 10 and 15 mg of protein per milliliter, and the $A_{260}:A_{280}$ ratio is 1.5–1.7. Aliquots of 0.25–0.5 ml are quick-frozen in liquid nitrogen in small plastic centrifuge tubes or vials; also, aliquots of about 0.04 ml can be frozen into beads by dropping into liquid nitrogen. Some frozen preparations of the modified S-100 extract have been stored for over 2 years at −70° to −80° with little or no loss of activity.

With some yeast strains, particularly with some haploid temperature-sensitive mutants, where the activity of the resulting cell-free system is

[14] H. T. Hutchison and L. H. Hartwell, *J. Bacteriol.* **94,** 1697 (1967).

[15] L. H. Hartwell, *J. Bacteriol.* **93,** 1662 (1967).

low, the following modification is used: The cells sedimented from the growing cultures are washed once with cold distilled H_2O, resuspended in 10 m*M* 2-mercaptoethanol–2.5 m*M* EDTA (20 ml per liter of original cell culture) and incubated at 20° for 30 min; cells are then recovered by centrifugation, resuspended in 1 *M* sorbitol (40–50 ml sorbitol per liter of original cell culture) and treated with glusulase to produce spheroplasts, as described above.

Translation of Natural mRNA

Translation of exogenous natural and synthetic mRNAs with the postpolysomal extract is carried out with a 2-step incubation. The first serves to degrade residual mRNA in the modified S-100 fraction with micrococcal nuclease plus calcium,[16] and the second measures the incorporation of radioactive amino acid into protein when nuclease action is stopped by adding EGTA to chelate the Ca^{2+} and exogenous mRNA is added as a template for translation.

Between 5 and 24 μl of modified S-100 extract (0.1–0.4 mg of protein) are incubated with 2.5 μl of incubation media A, 1.25 μl of amino acid mixture (−Leu), 2.0 μl of 4 *M* KOAc, 1.25 μl of ATP–GTP mixture, 2.5 μl of 0.5 *M* creatine phosphate, 1.25 μl of creatine phosphokinase, 1 μl of 0.0125 *M* $CaCl_2$, 1.25 μl of micrococcal nuclease, and solution A containing 20% glycerol to a final volume of 0.037 ml. After 10 min at 20°, the reaction mixtures are placed in ice (2°), and the following components are added: 3 μl of EGTA, 5 μl of [^{3}H]leucine (final concentration, 2×10^{-5} *M*), and 5 μl of mRNA, such as polysomal RNA or polysomal poly(A)$^+$ mRNA[17,18] (200–250 A_{260} units/ml), or H_2O. If yeast poly(A)$^+$ RNA purified on oligo(dT) cellulose columns is used, the addition of 10–20 μg of yeast or *E. coli* tRNA is also required for maximal activity. The reaction mixtures, in a total volume of 0.05 ml, are incubated at 20° for 1–2 hr. At the end of the incubation period, 1 ml of cold 10% trichloroacetic acid is added, the reaction tubes are placed in ice for 20–30 min, then heated in a water bath at 90° for 15 min and cooled. The acid-insoluble protein is collected on glass-fiber filters, washed three times with 5 ml of 5% trichloroacetic acid, dried under an infrared lamp, and counted in a scintillation counter.

[16] H. R. B. Pelham and R. J. Jackson, *Eur. J. Biochem.* **67,** 247 (1976).

[17] B. M. Gallis, J. P. McDonnell, J. E. Hopper, and E. T. Young, *Biochemistry* **14,** 1038 (1975).

[18] H. Aviv and P. Leder, *Proc. Natl. Acad. Sci. U.S.A.* **69,** 1408 (1972).

Translation of Synthetic Polynucleotides (Polyuridylic Acid)

This assay is similar to the one described above for natural mRNA, except that the amino acid mixture does not contain phenylalanine and the $Mg(OAc)_2$ concentration is increased to 13 m*M*. Five to 20 μl of post-polysomal extract are incubated with 2.5 μl of incubation media A, 2.0 μl of KOAc, 1.25 μl of 0.4 *M* $Mg(OAc)_2$, 1.25 μl of ATP–GTP mixture, 2.5 μl of creatine phosphate, 1 μl of $CaCl_2$, 1.25 μl of nuclease, and solution A–20% glycerol to a final volume of 0.037 ml. After 10 min at 20°, 3 μl of EGTA, 5 μl of [^{3}H]phenylalanine (final concentration, 1 to 2×10^{-5} *M*), and 5 μl of poly(U) or H_2O are added. The reaction mixtures are incubated at 20° for 60 min, and the hot trichloroacetic acid-insoluble fractions are then prepared and counted.

Characteristics of the Cell-Free Translational System[10,13]

The postmitochondrial fraction of yeast contains all of the required translational factors, aminoacyl-tRNA synthetases, tRNAs, ribosomal subunits, ribosomes (and/or monosomes) and polysomes containing at least a dozen ribosomes per polynucleotide chain. After centrifugation for a short period of time at 100,000 *g*, the supernatant no longer contains polysomes, but all the other translational components are present. Chromatography of this preparation through a molecular sieve column and degradation of residual mRNA with nuclease yields a preparation that is stringently dependent for maximal activity on mRNA, nucleotides and a nucleoside triphosphate-generating system, and amino acids (see the table). Incorporation in the presence of mRNA is linear for 1–2 hr, and in some cases incorporation with yeast poly(A)$^+$ mRNA is over 500 times greater than in the absence of mRNA. In addition to the RNA prepared from the polysome pellet obtained from centrifugation at 100,000 *g*, several nucleic acids tested markedly stimulate protein synthesis in this system. Whole cell RNA from yeast, rabbit reticulocyte poly(A)$^+$ RNA (globin mRNA), and RNAs from brome mosaic virus, tobacco mosaic virus, and turnip yellow mosaic virus are actively and faithfully translated *in vitro*. Also, the use of acetate salts of K and Mg^{2+}, rather than the chloride equivalents result in a significant increase in protein synthesis.

Incubations with exogenous mRNAs have provided evidence that all the intermediary reactions that are involved in the initiation, elongation, and termination of protein synthesis are carried out in this cell-free translating system.[11] One- and two-dimensional gel electrophoresis of the products synthesized *in vitro* with yeast polysomal mRNA, as compared to

EFFECT OF VARIOUS INCUBATION COMPONENTS ON PROTEIN SYNTHESIS IN THE POSTPOLYSOMAL EXTRACT DEPLETED OF mRNA

Incubation components	cpm of [^{3}H]leucine incorporated
Complete system[a]	898,800
Complete system,[a] − mRNA	3,300
Complete system,[a] − ATP and GTP	330,000
Complete system,[a] − creatine phosphate	10,300
Complete system,[a] − creatine phosphokinase	56,800
Complete system,[a] − amino acids	170,600
Complete system,[b]	614,400
Complete system,[b] − poly(U)	5,800

[a] The complete system contained nuclease-treated postpolysomal extract, buffered salts-dithiothreitol (incubation media A), amino acid mixture (-Leu), KOAc, ATP, GTP, creatine phosphate, creatine phosphokinase, glycerol, EGTA, [^{3}H]leucine, and yeast polysomal poly(A)$^+$ RNA, as described in the text; $CaCl_2$ and nuclease from the first incubation step were also present. In these experiments 1.4 μM [^{3}H]leucine (22,000 cpm/pmol) was used, and incubations were carried out for 30 min.

[b] The complete system contained nuclease-treated postpolysomal extract, incubation media (-Phe), $Mg(OAc)_2$ to 13 mM, [^{3}H]phenylalanine, poly(U), and the other components described above and in the text. In these experiments 1.4 μM [^{3}H]phenylalanine (20,000 cpm/pmol) was used. Incubations were carried out for 60 min.

those synthesized *in vivo,* indicate that essentially all the mRNAs are translated faithfully, both qualitatively and quantitatively; further, the results suggest that posttranslational modifications appear to operate normally in this cell-free system.

[43] Methods Utilizing Cell-Free Protein-Synthesizing Systems for the Identification of Recombinant DNA Molecules

By JACQUELINE S. MILLER, BRUCE M. PATERSON, ROBERT P. RICCIARDI, LAWRENCE COHEN, and BRYAN E. ROBERTS

Recombinant DNA procedures can be used to generate libraries of fragments of genomic DNA or DNA copies of mRNAs. These libraries represent most of the cell's genetic information in the form of recombinant molecules, many of which contain part or all of the sequences encoding information for a particular polypeptide. In order to study a specific gene or family of genes, the investigator is often confronted with the te-

METHODS IN ENZYMOLOGY, VOL. 101

ISBN 0-12-182001-7

dious task of identifying a subset of sequences within a large number of recombinant molecules. Identification of these recombinants can be made either by utilizing a specific probe for the sequence of interest or by employing a procedure which specifically identifies the encoded product(s).

In practice the availability of purified probes has been restricted to those mRNAs that represent the bulk of the messenger RNA population, such as RNAs encoding globin, ovalbumin, and parathyroid hormone. In order to obtain probes for less abundant mRNAs, a method for the fractionation of cellular RNAs is required. RNAs can be separated by electrophoresis in methylmercury hydroxide agarose gels, and the enriched fractions containing RNAs encoding specific polypeptides are used to prepare probes. These radiolabeled probes are used to screen libraries to identify clones containing homologous sequences. The recombinant DNAs that contain these sequences can be further characterized to define which specific polypeptide they encode. The basis of this identification relies on the fact that each recombinant DNA sequence will hybridize stably with its corresponding mRNA. This can be utilized to select a specific mRNA from the total mRNA population and determine the polypeptide it encodes in a cell-free protein synthesizing system. Alternatively, the specific hybridization of the DNA and mRNA sequences can be utilized directly to define the encoded polypeptide based upon the arrest of the translation of a specific polypeptide when the hybridized mRNA is translated in a cell-free system.

These methods used in concert constitute a general approach for screening DNA libraries and defining those recombinants containing sequences coding for specific polypeptides. Moreover, these methods can also be utilized to determine the size and arrangement of mRNA within a DNA fragment and to define precisely the genomic location of coding and noncoding sequences of individual mRNAs. This refined topological information is essential for deducing the pathways of synthesis of mature mRNAs and also for studying the regulation of expression of specific genes and gene families in eukaryotic organisms.

The methodological details of this approach, in its current and most widely used format, are described. Each procedure is outlined, from the preparation of stock solutions to a detailed description of the manipulations, followed by appropriate comments and useful troubleshooting hints.

Isolation of Messenger RNA and the Preparation of Cell-Free Protein-Synthesizing Systems

The success of the procedures described in this chapter requires the preparation of translatable mRNAs and the availability of efficient cell-

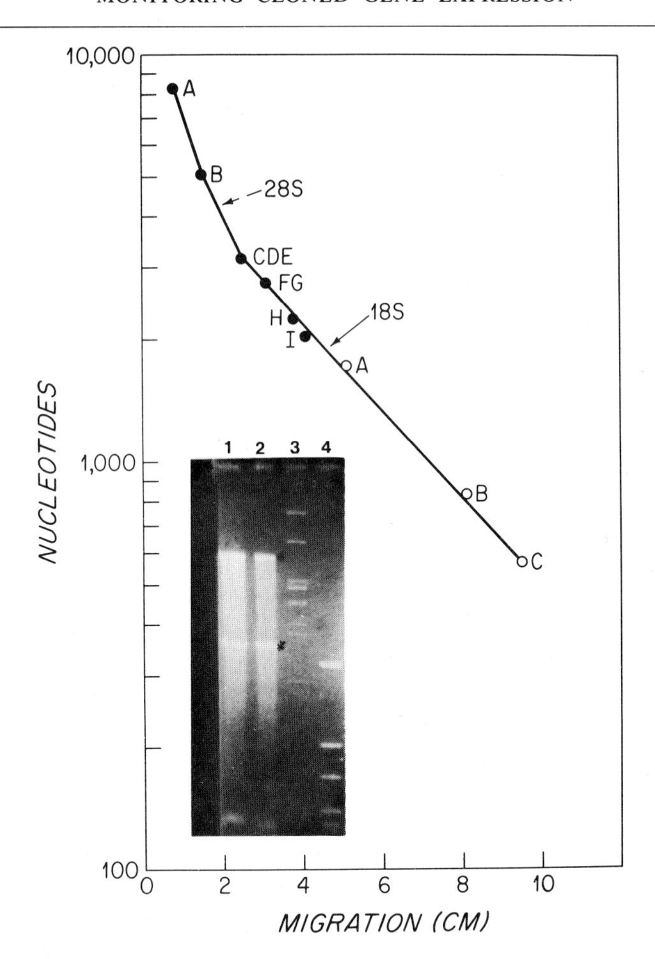

FIG. 1. The inset is a photograph of a 1.1% low-melt agarose gel containing methylmercury hydroxide. Lanes 1 and 2 contain 50 and 25 μg, respectively, of total HeLa cytoplasmic RNA; the asterisk denotes the 28 S and 18 S ribosomal RNAs. Lanes 3 and 4 contain DNA markers of known molecular weight, which are respectively the *Hin*dIII digest of Adenovirus 2 DNA and the *Hae*III digest of SV40 DNA. The size of the RNA in each fraction was determined by direct comparison to the DNA fragments, and the arrows on the graph indicate the position of the 28 S and 18 S ribosomal RNAs.

free protein synthesizing systems. The appropriate choice of RNA isolation method will depend on the specific characteristics of the organism under investigation, such as the ease with which it can be lysed, the abundance of ribonucleases, and the ease of separating ribonucleic acids from deoxyribonucleic acids. Recently, RNA isolation procedures were comprehensively reviewed and described[1–3] and therefore will not be discussed further. In addition the procedures for the preparation of two commonly used eukaryotic cell-free protein synthesizing systems, wheat germ[4] and rabbit reticulocyte lysate,[5,6] have been fully described. Moreover these systems are available from commercial vendors.

Fractionation of mRNAs in Methylmercury Hydroxide Agarose Gels

Methylmercury hydroxide is a potent denaturant of nucleic acids, and its incorporation into agarose gels permits the fractionation of RNA or DNAs according to size[7] (Fig. 1). Either total mRNA or poly(A)$^+$ enriched RNA can be fractionated in sufficient quantities by analytical gel electrophoresis to obtain a gel fraction highly enriched for a particular mRNA species. The fraction of interest is defined by slicing the gel lane, efficiently eluting the RNA from the individual slices, translating a portion of each, and fractionating the polypeptides synthesized on an SDS–polyacrylamide gel[8] (Figs. 2 and 3). The fractions containing the polypeptide of interest can be assayed by direct visualization of the autoradiograph (Fig. 3), by immunoprecipitation with specific antibodies, or by an appropriate bioassay. Often the fractionated RNA will direct substantial synthesis of polypeptides not easily detected in the translation of total unfractionated RNA. These data define the size of the mRNA encoding the specific polypeptide and identify those fractions wherein the specific mRNA is enriched between 20- to 100-fold (Fig. 4). This enriched RNA fraction is then used to synthesize a single-stranded cDNA probe for the subsequent screening of a cloned cDNA library that is representative of the mRNA sequences transcribed in the cell under investigation (Fig. 5).

This approach was successfully utilized to isolate variant cDNA clones from a cDNA library representing genes expressed in cultures of embryonic chick muscle cells. These include cDNA clones for the actin gene family (α, β, and γ), vimentin, glyceraldehyde-3′-phosphate dehydrogenase, pyruvate kinase, and the myosin light chains LC1 and LC3.

[1] B. M. Paterson and B. E. Roberts *in* "Gene Amplification and Analysis" (J. G. Chirikjian and T. S. Papas, eds.), p 417. Elsevier/North-Holland, Amsterdam, 1981.

[2] J. M. Taylor, *Annu. Rev. Biochem.* **48,** 681 (1979).

[3] J. S. Miller, B. E. Roberts, and B. M. Paterson, *in* "Genetic Engineering" (in press).

[4] B. E. Roberts and B. M. Paterson, *Proc. Natl. Acad. Sci. U.S.A.* **70,** 2230 (1973).

[5] H. R. B. Pelham and R. J. Jackson, *Eur. J. Biochem.* **67,** 247 (1976).

[6] See appropriate section in this volume.

[7] J. M. Bailey and N. Davidson, *Anal. Biochem.* **70,** 75 (1976).

[8] J. S. Miller, R. P. Ricciardi, B. E. Roberts, B. M. Paterson, and M. B. Mathews, *J. Mol. Biol.* **142,** 455 (1980).

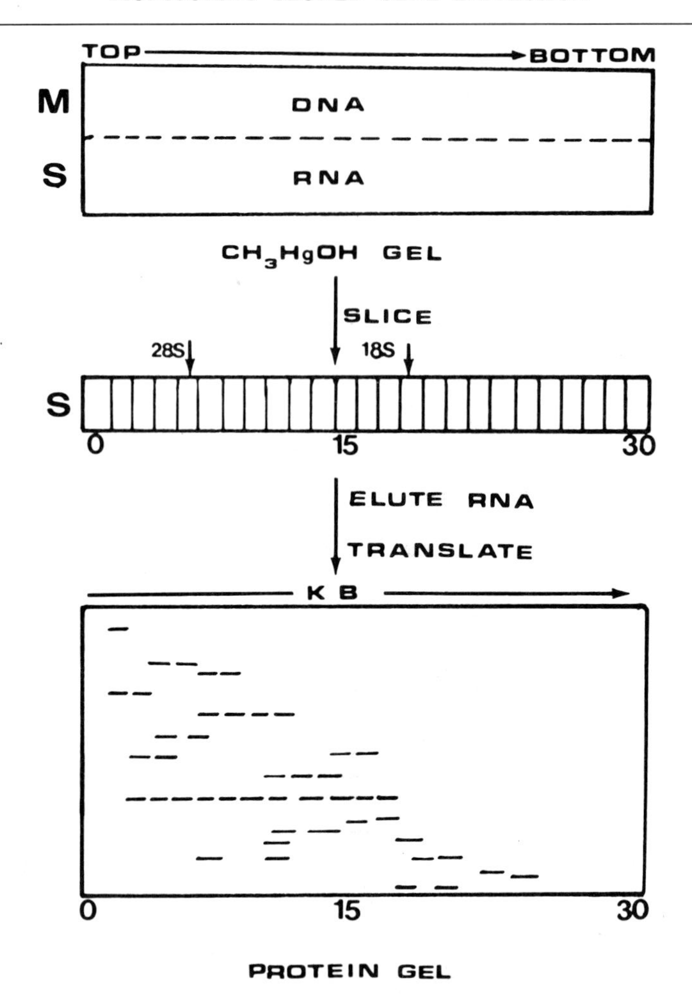

FIG. 2. Isolation and identification of mRNAs fractionated on methylmercury hydroxide gels.

Alternatively, the enriched RNA fractions can be used directly to prepare double-stranded cDNA that can be cloned into PBR322 using standard procedures. This approach was successfully applied in the cloning of rat actin and tubulin cDNAs.[9]

Stock Solutions

NH_4OAc, 5.0 *M* (Millipore filtered)

NH_4OAc, 0.5 *M*; 2.5 m*M* EDTA (Millipore filtered)

[9] I. Lemischka, J. Schwarzbauer, R. Hynes, and P. Sharp, unpublished data.

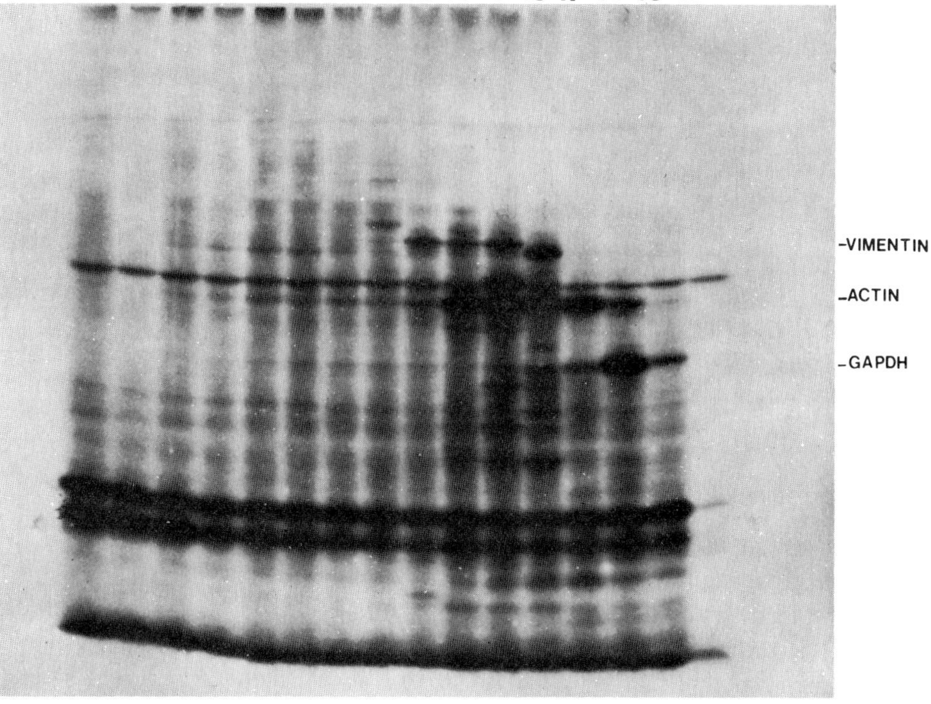

FIG. 3. Fluorograph of an SDS–polyacrylamide gel visualizing the [^{35}S]methionine-labeled polypeptides encoded by the enriched mRNA fractions isolated from a methylmercury agarose gel. The fractions derive from the portion of the agarose gel that includes the 18 S ribosomal RNA (see Fig. 4).

Dithiothreitol (DTT), 0.1 *M*, freshly made

CH_3HgOH, 1.0 *M*, in water (Alpha Products, Ventron Division)

Buffers

10× Running buffer (Millipore filtered): 0.5 *M* boric acid, 0.05 *M* $Na_2B_4O_7$, 0.1 *M* Na_2SO_4, 0.01 *M* Na_2EDTA. When diluted to 1×, pH should be 8.35.

2× Sample buffer: 2× running buffer, 0.025 *M* CH_3HgOH, 20% glycerol, 0.01% bromophenol blue

Boehringer-Mannheim calf liver tRNA, 10 mg/ml in sterile distilled, water

Redistilled phenol was shaken with an equal volume of a solution containing 0.1 *M* NaCl and 0.5% SDS, then equilibrated with 0.5 *M* NH_4OAc, 0.001 *M* EDTA, 0.002 *M* mercaptoethanol.

Gel Preparation. The percentage of agarose included in the gel determines the nucleic acid size range that is optimally resolved. For example, a 1.1% gel is used to resolve mRNAs larger than 2 kb whereas a 1.8% gel is ideal for mRNAs smaller than 1 kb. The gels can be run either in a vertical or horizontal orientation. In this outline, the preparation of a horizontal gel 20 cm long, 13 cm wide, and 3–4 mm thick is described. First, 1× running buffer with 1% agarose is melted in a microwave oven, and wicks are poured and allowed to solidify. Then 200 ml of 1× running buffer is supplemented with 1.1–1.8% low-melt agarose, melted in a microwave, cooled to 70° and, *in a fume hood,* made up to a final concentration of 12.5 m*M* with CH_3HgOH. The gel is poured in the fume hood; a comb with 10 equally spaced teeth 0.7 cm wide and 0.1 cm thick is positioned appropriately, and the gel is allowed to solidify. Then the comb is removed, the 1 cm of gel surrounding the slots is excised, the comb is replaced, and new slots are formed using 1× running buffer containing 1% normal agarose and 12.5 m*M* CH_3HgOH. The reason for this is that slots made with low melt agarose tend to collapse. The replacement with normal agarose ensures good slot formation for convenient sample loading.

Gel Running. An equal volume of 2× sample buffer is added to the RNA samples and the DNA markers, and the samples are loaded. The sample is electrophoresed into the gel at 100 V using an ISCO Model 493 power pack set on constant voltage, then the slots are filled with running buffer and the entire gel covered with Saran Wrap to prevent desiccation. The gel is run at 40–50 V, constant voltage, overnight until the bromphenol blue front migrates 14–16 cm.

At the conclusion of the electrophoresis, the lanes containing DNA and RNA markers are cut out of the gel and soaked in two changes of 250 ml of 0.5 *M* NH_4OAc for 20 min each, to remove the methylmercury hydroxide. Then the markers are stained by soaking the gel for 1 hr in ethidium bromide (1 μg/ml) in 0.5 *M* NH_4OAc. The bands were visualized by UV photography, and their migration distance is recorded by aligning the gel next to a ruler (Fig. 2). The gel is fragile, so care should be exercised in its handling.

Elution of RNA from the Gel Matrix. The portions of the gel containing the fractionated RNAs are soaked in two changes each of 100 ml of freshly prepared 0.1 *M* DTT until the gel appears totally clear, that is, approximately 30–60 min. The individual lanes are sliced out with a sterile razor blade and placed on Whatman 3 MM paper mounted on the stage of a Mickle automatic gel slicer, and carefully aligned. These manipulations of the fragile gel slice can be facilitated by wetting the gel and the paper with 0.1% SDS in 0.5 *M* NH_4OAc. The gel is then sliced into 1-mm-thick slices, and every three slices are pooled into a 1.5 ml Eppendorf tube. These slices may be kept frozen at −80° or processed immediately.

To elute the RNA from the agarose matrix, the following procedure is followed. To each gel sample, 500 μl of 0.5 M NH_4OAc, 2.5 mM EDTA was added, and the tubes placed four at a time into the microwave oven for 20 sec, vortexed, then microwaved again for 20 sec. At this point the gel is totally liquefied and 500 μl of the equilibrated phenol and 5 μl of 10% SDS are added. Placement of the tubes within the microwave oven is important for efficient melting and varies owing to the uneven distribution of emission from the source. So the placement of the samples and the time of melting should be determined experimentally for the individual microwave oven. The liquefied gel sample and the phenol are vortexed and spun 5 min in a microfuge to separate the phases; and the aqueous layer is transferred to new tubes. The inclusion of the SDS causes the interface to compact tightly and permits the complete removal of the aqueous layer. The aqueous layer is finally extracted twice with chloroform : isoamyl alcohol (24 : 1 v/v). Then 10 μg of calf liver tRNA are added as carrier and 2.5 volumes of cold 100% ethanol. The samples are precipitated overnight and spun for 10 min in a microfuge in the cold room; the pellet is washed twice with 1.5 ml of cold 70% ethanol and dried under vacuum; a portion of each sample is translated in the rabbit reticulocyte lysate (Figs. 3 and 4).

Comments. Methylmercury hydroxide is a very toxic compound and therefore must be handled with caution. All steps in which the compound is present, including pouring, running, and slicing the gel, should be done *in a hood* calibrated for adequate air flow. True touch (thick) gloves should be worn on top of a second pair of gloves. All pipettes, tips, gel fragments, tubes, etc., that have come in contact with the methylmercury hydroxide should be placed in a plastic bag in a hood and disposed of by procedures for volatile toxic chemicals.

The Use of Enriched RNA Preparations To Screen cDNA Libraries for Specific Genes

Poly(A)$^+$ RNA (50 μg) from embryonic chick muscle cells was fractionated on a 1.5% agarose gel containing CH_3HgOH. The RNAs were eluted, 5% of each sample was translated, and the [^{35}S]methionine-labeled polypeptides were fractionated on an SDS-polyacrylamide gel and visualized by autoradiography. A portion of a typical profile of the polypeptides encoded by the RNAs isolated through the 18 S region of the gel is shown in Fig. 3. The sequences encoding α- and β-actin were resolved in fractions 14 and 17, respectively. This was confirmed by two-dimensional gel analysis of the cell-free products encoded by these fractions (Fig. 4). The two vimentin mRNAs are separated in fractions 13 and 15, and glyceral-

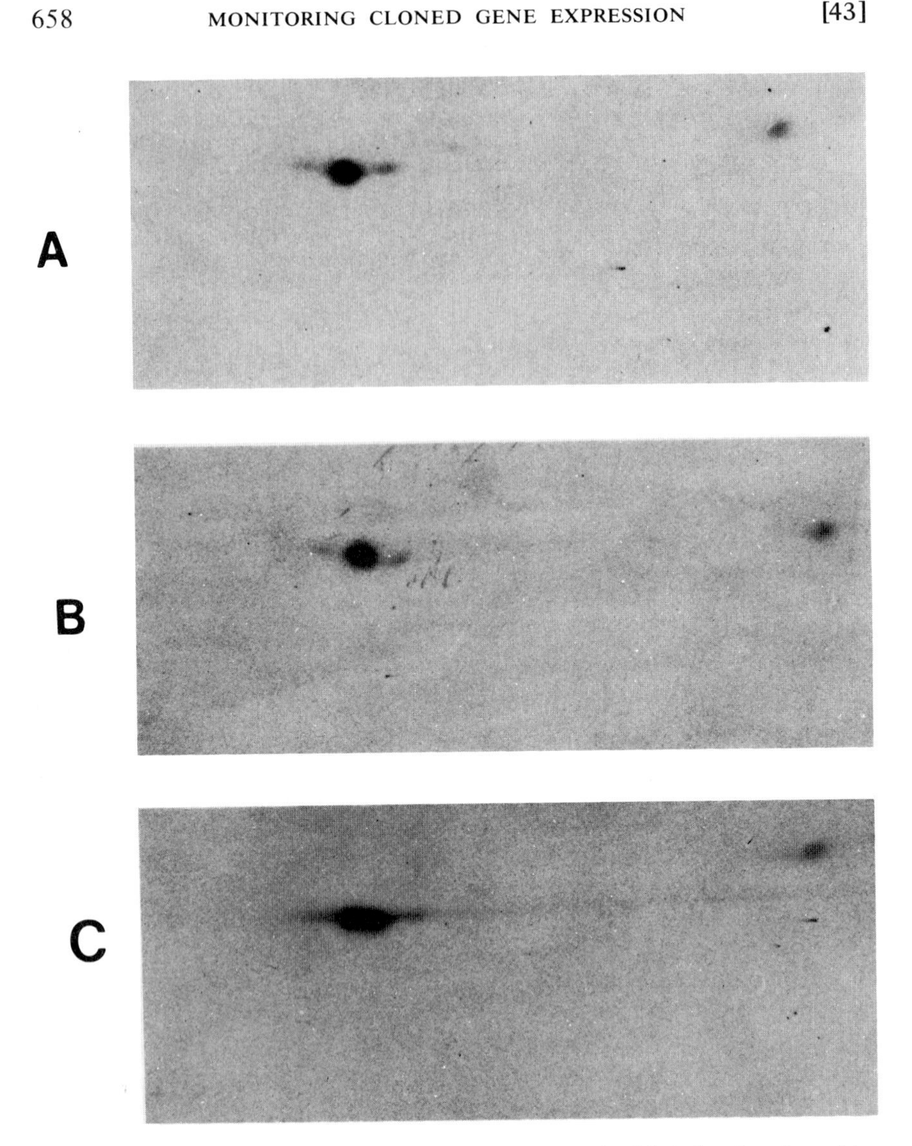

FIG. 4. Two-dimensional gel analysis of the [^{35}S]methionine-labeled actin polypeptides encoded by the RNAs in methylmercury gel fractions 14 and 17 (see Fig. 3). (A) Fraction 14 contains predominantly β-actin. (B) Fraction 17 contains predominantly α-actin. (C) A mixture of fractions 14 and 17 demonstrates the presence of both α- and β-actin.

dehyde-3′-phosphate dehydrogenase is greatly enriched in fraction 18 (Fig. 3). These RNA preparations were used to screen cDNA libraries to isolate recombinant DNAs complementary to these specific sequences.

In each case the 20% of the enriched RNA fraction was used to synthesize a single-stranded cDNA probe using reverse transcriptase.[10] The inclusion of a single ^{32}P-labeled deoxynucleotide triphosphate resulted in the total incorporation of between $10 \times$ and 20×10^6 cpm at a specific activity of 10^8 cpm per microgram of cDNA. This incorporation can be enhanced by including more than one radiolabeled deoxynucleotide triphosphate. These cDNA probes were used in a protocol for screening large numbers of colonies using very small amounts of probe and hybridization solution. This procedure was developed in Dave Hogness' laboratory at Stanford and is an adaptation of the original Grunstein and Hogness screening method. Bacterial colonies containing recombinant plasmids are grown on a Millipore filter, lysed, affixed to the filter, and probed in a standard hybridization reaction.

The colonies to be screened are transformants grown on the appropriate selective media plate. These colonies are toothpicked onto a storage plate and then spread within a grid square of a 4.5 cm Millipore filter (with grid lines) resting on the surface of a nutrient agar petri plate. The filter is marked with dots, using a black government-issue ballpoint to indicate the start of columns and rows. The dots serve also as visual aids in the alignment of the subsequent autoradiogram. It is also convenient to use the same type of matrix on both the Millipore filter and under the storage plate. A convenient matrix for this size Millipore filter is 10×9 colonies.

Both the storage and Millipore filter-bearing plates are incubated at 37° to permit colony growth. The filter-bearing plates are incubated long enough to ensure good growth. The growth of the colonies can be facilitated by using thick nutrient plates and transferring the filter to a new plate after 12 hr of growth. In order to prevent contaminants it was not necessary to sterilize the Millipore filters; however, sterile forceps are used to transfer the filters, and colony picking and spreading was carried out under a hood. An oblique light source also aids in visualizing those grids already wetted by the spreading.

To prepare the Millipore filters bearing adequate-sized colonies for hybridization the following procedure is adopted. Place three squares of filter paper in three large petri dishes; soak the first in 0.5 *M* NaOH; the

[10] Reviewed by A. Efstratiadis and L. Villa-Komaroff, *in* "Genetic Engineering: Principles and Methods" (A. Hollaender and J. Setlow, eds.), Vol. 1. Plenum, New York, 1979.

second in 1 *M* Tris-HCl, pH 7.4; and the third in 0.5 *M* Tris-HCl, pH 7.4–1.5 *M* NaCl. The Millipore filters are placed colony side up on the first (NaOH) filter for 10–20 min, during which time the cells lyse and the DNA denatures. During this period the colonies take on a wet appearance, but the remainder of the filter does not appear wet. The filters are transferred to the second filter pad for neutralization for 2–5 min, and then to the third pad to complete neutralization and to adjust the salt concentration for 10–20 min. Thereupon the lysate is pulled down onto the Millipore filter by placing it on a filtration device colony side up and applying a vacuum for 1–2 min. The colonies can wash off if this step is omitted. Then the filters are placed on a clean nonadsorbent surface, and a solution of proteinase K (2 mg/ml in 1× SSC) is pipetted to just cover the colony lysates. The solution is retained on the filter by surface tension, and digestion is allowed to continue for 30 min at room temperature. The filter is then returned to the filtration apparatus, washed once with ethanol and three times with chloroform, and allowed to dry on the filter tower. Finally the filters are placed immediately in a lightly weighted paper towel and dried at 80° under vacuum for 2–4 hr. The filters need the light weight, $\frac{1}{4}$–$\frac{1}{2}$ lb, to avoid wrinklings.

Ten filters can be conveniently hybridized in 10 ml of 50% formamide, 5× SSC; 250 g of denatured herring sperm DNA per milliliter contain 500,000 cpm of the appropriate cDNA probe per milliliter at 37° overnight. The filters are washed four times in 2× SSC for 30 min each at 25°, blotted dry, and subjected to autoradiography. An example of a successful screen is shown in Fig. 5 wherein clear positive colonies are in evidence, using ^{32}P-labeled cDNA from methylmercury gel fraction 14 to define potential actin recombinants and from fraction 17 to define actin recombinants.

Hybrid Selection

The recombinant molecules that hybridized to the cDNA probes were recovered from the storage plate and further characterized to determine the polypeptide that they encode utilizing hybrid selection. This identification depends on the formation of a stable hybrid between the recombinant DNA sequence and its corresponding mRNA. This affects the purification of a specific mRNA from the plethora of cellular mRNAs. The polypeptide encoded by this selected mRNA is determined by translation in a cell-free protein-synthesizing system (as diagrammed in Fig. 6).

Hybridization of the RNA to DNA immobilized on a solid support provides a convenient method for separation of the DNA-mRNA hybrid from the remainder of the mRNAs. Several procedures for hybrid selection

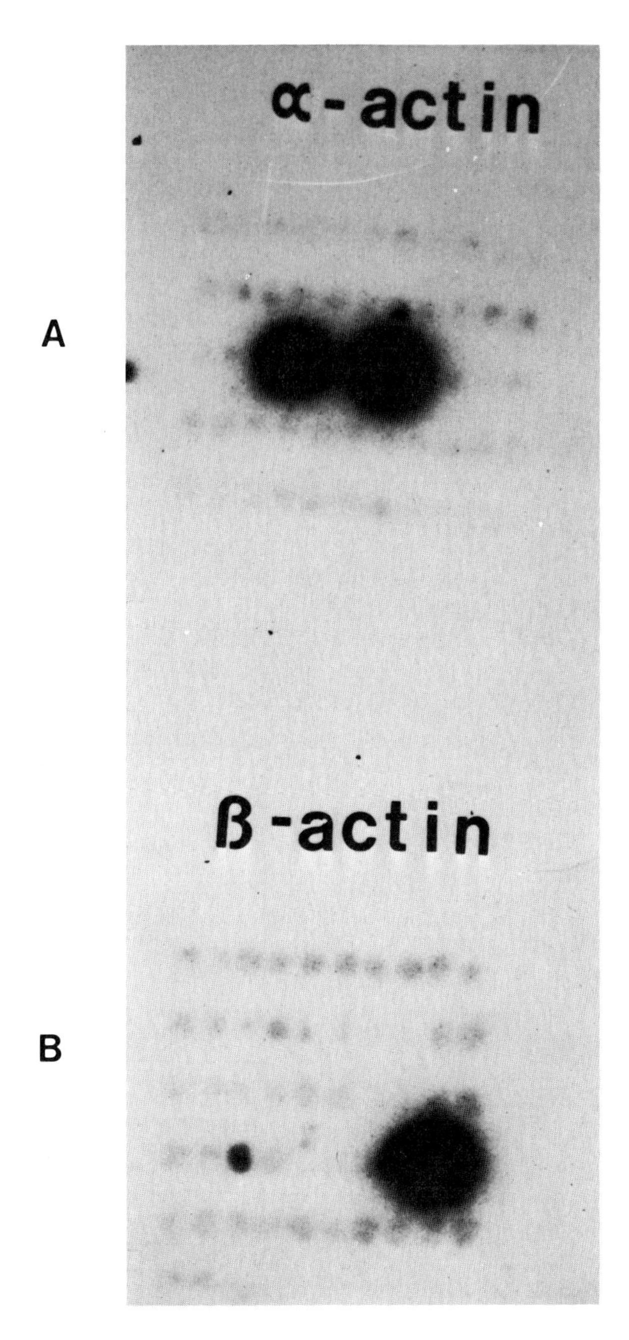

FIG. 5. Autoradiograph of Millipore filters bearing recombinant DNAs hybridized with ^{32}P-labeled cDNA prepared from enriched RNA fractions. (A) The enriched RNA from methylmercury gel fraction 17 contained sequences encoding α-actin. (B) The enriched RNA from fraction 14 contained sequences encoding β-actin.

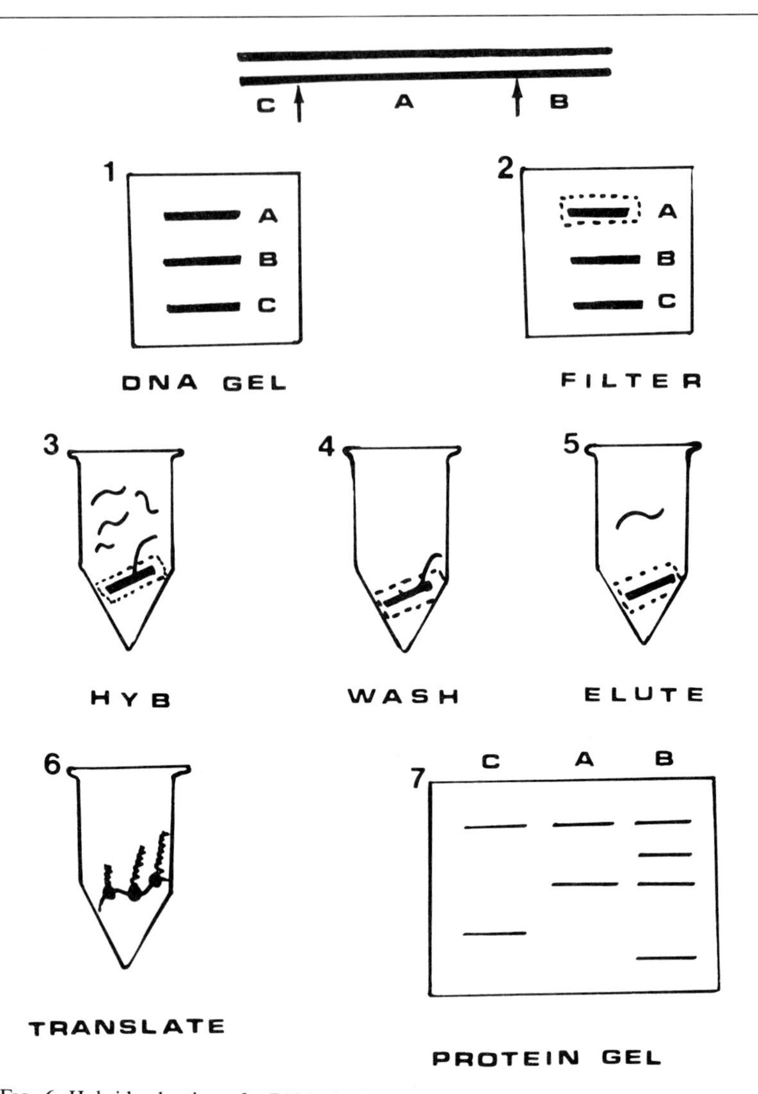

FIG. 6. Hybrid selection of mRNAs by DNA immobilized on a solid support.

using DNA immobilized on solid supports have been described.[11-20] In this section, methods that we have found to be the most efficient and reliable are outlined.

Hybrid Selection with DNA Immobilized on Nitrocellulose Filters

Solutions for Hybridization in Buffer Containing Formamide

Stock solutions

Formamide 100% (Matheson, Coleman and Bell)

0.1 *M* Piperazine-*N*,*N*′-bis(2-ethanesulfonic acid) (PIPES) pH 6.4 (with NaOH), 4 *M* NaCl (Millipore filtered)

20× SSC: 3 *M* NaCl, 0.3 *M* sodium citrate (Millipore filtered)

SDS, 10% (Millipore filtered)

EDTA 2 m*M*, pH 7.9 (Millipore filtered)

tRNA (10 mg of calf liver per milliliter; from Boehringer-Mannheim).

Hybridization buffer: 65% formamide, 0.01 *M* PIPES (pH 6.4), 0.4 *M* NaCl

Wash solution: 1× SSC, 0.5% SDS

Elution buffer: 1 m*M* EDTA, pH 7.9

Deionize formamide by stirring for 40–60 min with mixed-bed resin [Bio-Rad Ag501 × 8 −(D)]. Remove mixed-bed resin by filtration through Whatman 3 MM paper. Alternatively, formamide may be recrystallized.[21] Store in aliquots at −80°.

Solutions for Hybridization in Aqueous Buffer

Stock solutions

NaAc, 4 *M*, pH 6.0 (Millipore filtered)

EDTA, 0.01 *M*, pH 7.9 (Millipore filtered)

[11] C. L. Prives, H. Aviv, B. M. Paterson, B. E. Roberts, S. Rozenblatt, M. Revel, and E. Winocour, *Proc. Natl. Acad. Sci. U.S.A.* **71,** 302 (1971).

[12] R. P. Ricciardi, J. S. Miller, and B. E. Roberts, *Proc. Natl. Acad. Sci. U.S.A.* **76,** 4927 (1979).

[13] W. Buttner, Z. Veres-Molnar, and M. Green, *Proc. Natl. Acad. Sci. U.S.A.* **71,** 2971 (1974).

[14] M. M. Harpold, P. R. Dobner, R. M. Evans, and F. C. Bancroft, *Nucleic Acids Res.* **5,** 2039 (1978).

[15] M. McGrogan, D. J. Spector, C. J. Goldenberg, D. Halbert, and H. J. Raskas, *Nucleic Acids Res.* **6,** 593 (1979).

[16] B. E. Noyes and G. R. Stark, *Cell* **5,** 301 (1975).

[17] T. Y. Shih and M. A. Martin, *Biochemistry* **13,** 3411 (1974).

[18] E. Gilboa, C. L. Prives, and H. Aviv, *Biochemistry* **14,** 4215 (1975).

[19] P. Venetianer and P. Leder, *Proc. Natl. Acad. Sci. U.S.A.* **71,** 3892 (1974).

[20] M. L. Goldberg, R. P. Lifton, G. R. Stark, and J. G. Williams, this series, Vol. 68, p. 206.

[21] J. Casey and N. Davidson, *Nucleic Acids Res.* **4,** 1539 (1977).

Hybridization buffer: 0.4 *M* NaAc, pH 6.0; 0.001 *M* EDTA, pH 7.9
Wash solution: 0.5× SSC, 0.5% SDS

Application of DNA to Nitrocellulose Membranes. Prior to application it is essential to linearize circular DNAs. This is accomplished by digestion with a restriction enzyme, followed by extraction with phenol, then chloroform–isoamyl alcohol extraction followed by ethanol precipitation to remove contaminating proteins and salts. Alternatively, the circular DNA can be fragmented by boiling in 0.01 *M* Tris-HCl, pH 7.9, 0.001 *M* EDTA for 10 min. This procedure is more economical; however, the fragmentation is size dependent so the results should be monitored by gel electrophoresis prior to application onto nitrocellulose.

DNA can be applied to filters by any one of the following procedures.

1. Boil the DNA for 60 sec and freeze quickly in a Dry Ice–ethanol bath. Thaw and immediately apply to a square of dry nitrocellulose. Allow to dry, then wash the filter in 5× SSC for 30 min.[12]

2. Apply DNA solution to dry nitrocellulose, air-dry 30–60 min, and denature the DNA with alkali by laying the filter over Whatman No. 1 paper (three layers) soaked in 0.5 *N* NaOH, 1.5 *M* NaCl. Repeat three times for 1 min each time, blotting filter lightly on 3 MM paper between each time. Neutralize in the same manner with 2 *M* Tris-HCl (pH 7.4), 2× SSC. Wash in a similar fashion in 2× SSC, then immerse in 2× SSC for 30 min.[1]

3. Soak nitrocellulose filter in deionized water for 60 min, then in 1 *M* NH_4OAc, pH 5.5. Denature the DNA in solution in 0.4 *N* NaOH at room temperature for 10 min, then place on ice. When ready to apply, add an equal volume of cold 2 *M* NH_4Ac. Solution may then be applied to the nitrocellulose either by spotting 10 μl at a time, drying between applications, or under vacuum at a rate of 100 μl/15 sec.[22]

4. DNA digested with restriction enzymes are fractionated by agarose gel electrophoresis. The gel is stained, and the position of each fragment is indicated by cutting small notches on either side of the band. The fragments are then transferred to a nitrocellulose membrane by the procedure of Southern.[23] Transfer of large fragments of DNA is enhanced by soaking the gel in 0.25 *M* HCl for 15 min prior to denaturation. The position of each fragment on the nitrocellulose is indicated by marking the filter in ink through the notches in the gel prior to separating the agarose from the membrane. The filter is washed in 10× SSC for 30 min to remove any adhering agarose. After air drying and baking (see below), each fragment is excised for use in the hybrid selection procedure.[12]

[22] F. C. Kafatos, C. W. Jones, and A. Efstratiadis, *Nucleic Acids Res.* **7,** 1541 (1979).
[23] E. M. Southern, *J. Mol. Biol.* **94,** 51 (1975).

In all instances the filter is then blotted dry with Whatman 3 MM paper, allowed to air dry overnight, and then baked for 2 hr under vacuum at 70°.

The amount of DNA that can be applied to the nitrocellulose depends upon the method of application. In procedures 1 and 2 the DNA is concentrated, generally at 0.5 mg/ml. A maximum of 5–10 μg of DNA is recommended for immobilization per square centimeter of nitrocellulose using these procedures. However, the capacity of nitrocellulose for DNA has been demonstrated to be considerably higher.[24] If the concentration of DNA is maintained at 25 μg/ml, as described in procedure 3, at least 50 μg of DNA can be applied to 1 cm^2 of nitrocellulose with no apparent loss of DNA during the hybridization reaction.

The initial concentration of DNA bound to the filter and that retained subsequent to each selection can be monitored by including trace amounts of radiolabeled DNA along with the recombinant DNA to be characterized. For this purpose DNA digested with *Hin*dIII was radiolabeled using [γ-^{32}P]ATP and polynucleotide kinase[25] or by using reverse transcriptase to introduce radiolabeled deoxynucleotides at the staggered ends formed by the restriction enzyme.[10] A trace quantity of the ^{32}P-labeled DNA was mixed with a known concentration of the recombinant DNA so that the DNA affixed to the filter could be monitored by Cerenkov counts.

Hybridization Conditions. The optimal conditions for hybridization vary as a function of the sequence of the nucleic acids involved and also the abundance of the mRNA of interest. The variable in the hybridization reaction include temperature (37–50°), time (2–12 hr), formamide concentration (50–70%) and the amount of RNA included (30–1200 μg per 100-μl reaction). The following conditions, optimized for the selection of mRNA, made late during Adenovirus infection, have been used successfully in the selection of numerous other mRNAs.

Filters (1 cm^2) containing the DNA are wetted with hybridization solution without RNA, cut in 12 small squares with a sterile razor blade, and placed into a microfuge tube using a sterile syringe needle. Hybridization buffer (100 μl) (65% formamide, 0.01 *M* PIPES, pH 6.4, 0.4 *M* NaCl), containing 30 μg of total cytoplasmic RNA is added, and the reaction is incubated at 48° for 2 hr. The supernatant is removed and may be precipitated for translation by diluting the formamide to 25%. The filters are washed by repeating the following procedure 10 times: 1 ml of wash buffer at 60° is added, the reaction is vortexed, and the wash buffer is re-

[24] P. S. Thomas, *Proc. Natl. Acad. Sci. U.S.A.* **77**, 5201 (1980).
[25] A. Maxam and W. Gilbert, *Proc. Natl. Acad. Sci. U.S.A.* **74,** 560 (1977).

moved by aspiration. Three rinses in 2 m*M* EDTA follow. The selected RNA is then eluted by boiling for 60 sec in 300 μl 1 m*M* EDTA containing 10 μg of tRNA, followed by quick freezing in an ethanol–Dry Ice bath. The supernatant is removed to a new tube, made 0.2 *M* NaAc (pH 5.5), and precipitated with 3 volumes of 100% ethanol at −20° overnight. The RNA is then pelleted in a microfuge for 10 min, washed twice with 70% EtOH (−20°), and dried under vacuum. Translation components may be added directly to the RNA, and translation products fractionated in an SDS–polyacrylamide gel.[26] An example of hybrid selection is shown in Fig. 7.

Hybridization reactions can be carried out in the absence of formamide. Conditions of 0.4 *M* NaAc (pH 6.0), 0.001 *M* EDTA at 60° for 4 hr have been used very successfully to select mRNAs from late in Adenovirus 2 infection. Higher backgrounds have been observed using these conditions; however, this can be eliminated by increasing the stringency of the wash conditions.

Hybrid Selection with DNA Immobilized onto Diazobenzyloxymethyl Paper

Stock Solutions

NaOH, 1 *M*
Sodium acetate, 2 *M*, pH 5.5 (Millipore filtered)
HCl, 1 *M*
Formamide, 100%
SDS, 10% (Millipore filtered)
NaCl, 5 *M* (Millipore filtered)
Tris-HCl, 1 *M*, pH 8.0 (Millipore filtered)
EDTA, 0.5 *M* (Millipore filtered)
20× SET: 2.3 *M* NaCl, 0.4 *M* Tris-HCl (pH 7.8)
20 *M* EDTA (Millipore filtered)

Buffers

Hybridization buffer: 50% formamide, 0.1% SDS, 0.6 *M* NaCl, 0.08 *N* Tris-HCl (pH 7.8), 0.004 *M* EDTA

[26] U. K. Laemmli, *Nature (London)* **227,** 680 (1970).

FIG. 7. Hybrid selection of Adenovirus 2 mRNA by recombinant DNAs containing Adenovirus 2 DNA. The [^{35}S]methionine labeled cell-free polypeptides are directed by: (1) total mRNA isolated from HeLa cells after 24 hr of infection with Adenovirus 2; (2) specific mRNAs selected from the total RNA by a recombinant molecule containing Adenovirus 2 DNA, map units 31.5–37.3; (3) specific mRNA selected from the total RNA by a recombinant molecule containing Adenovirus 2 DNA, map units 41.0–51.0; (4) no added RNA.

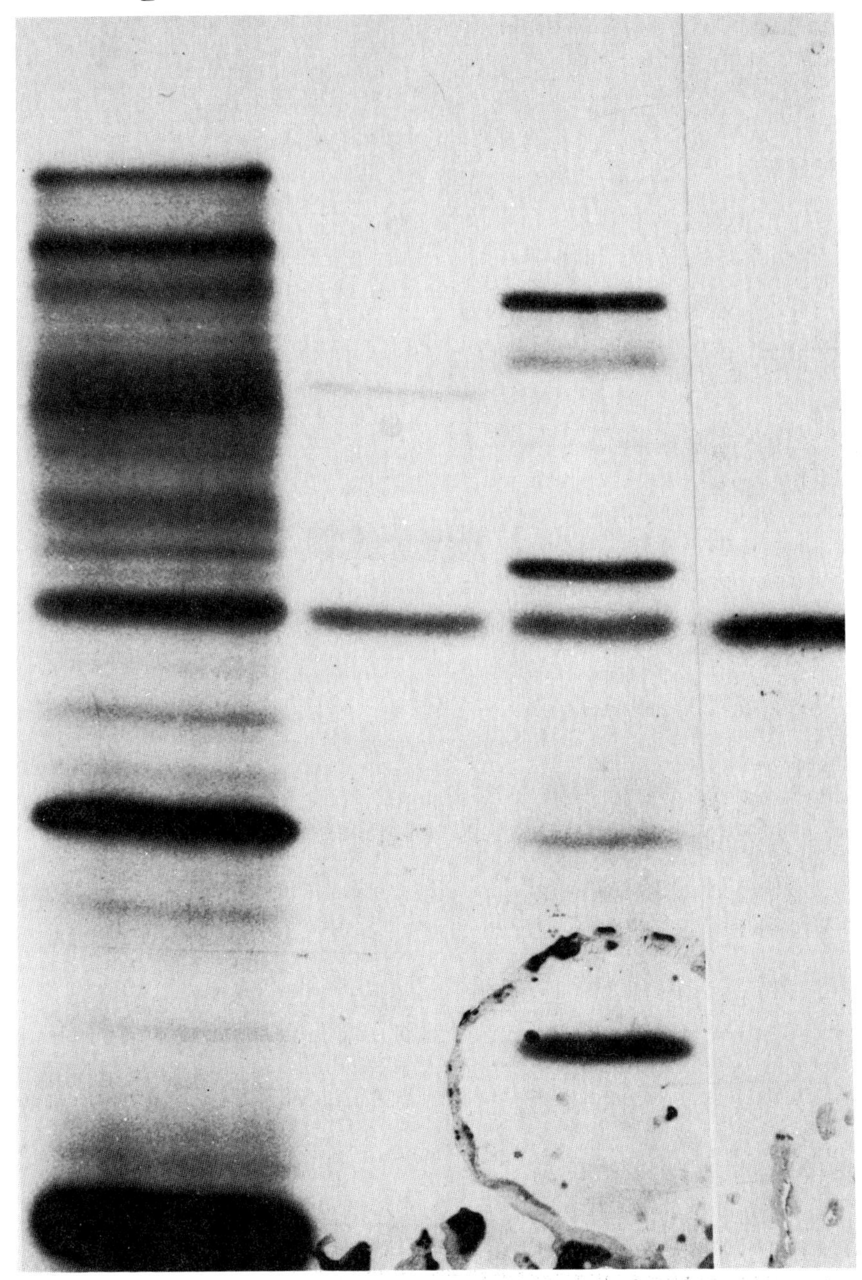
1
2
3
4

Wash solution: 50% formamide, 0.2× SET, 0.1% SDS
Elution solution: 99% formamide, 0.01 *M* Tris-HCl (pH 7.8)

Formamide should be treated as described in the preceding section.

Preparation of DBM Paper. Procedures for the preparation of DBM paper have been described in detail previously.[27,28] Immediately prior to the application of the DNA (within 5–10 min), the paper is activated by washing twice with ice-cold distilled water, followed by two washes with 20 m*M* NaOAc, pH 4.0. The following procedure is a modification of that described by Barnett *et al.*[29]

Application of DNA. The DNA to be applied (5–10 μg/cm^2) is dissolved in 100 μl of sterile, double-distilled water; 50 μl of 1 *M* NaOH are added, and the solution is boiled for 3–5 min to fragment the DNA. The solution is then neutralized by the addition of 30 μl of 2 *M* NaAc, pH 5.5, and 40 μl of 1 *M* HCl. The DNA is precipitated by the addition of 2.5 volumes of ice-cold ethanol (−80° for 60 min), resuspended in 25 μl of sterile distilled water, denatured by boiling for 1 min, and quick frozen in a Dry Ice–ethanol bath. The DNA is then applied directly to the DBM paper. The filter is allowed to air dry and placed in a closed microfuge tube overnight. The filter is then washed three times with sterile, double-distilled water, four times with 0.4 *N* NaOH, four times with water, and rinsed in hybridization buffer. A sterile razor is used to cut the filter into small squares that are replaced into the microfuge tube.

Hybridization conditions are as follows: Hybridization buffer (100 μl) containing 50–70 μg of total cytoplasmic RNA is added to the tube, and hybridization is carried out for 18 hr at 37°. The supernatant is removed and may be precipitated for translation by diluting the 200 μl of sterile, double-distilled water and adding 2.5 volumes of 100% ethanol. The filters are then washed by adding 1 ml of wash buffer at 37°, vortexing, and aspirating the supernatant. This is repeated 10 times. The RNA is eluted by incubating the filters in 300 μl of elution buffer (containing 15 μg of tRNA) at 65° for 5 min. The eluent is transferred to a new tube, brought to 0.2 volume of 100% ethanol. The RNA is then translated, and the products are analyzed as described above.

Comments on the Application of Hybrid Selection

The choice of the solid support matrix for use in hybridization selections is, generally, dependent upon the requirements of the investigator. The covalent binding of the DNA to the DBM paper provides a more per-

[27] J. C. Alwine, D. J. Kemp, and G. R. Stark, *Proc. Natl. Acad. Sci. U.S.A.* **74,** 5350 (1977).

[28] G. M. Wahl, M. Stern, and G. R. Stark, *Proc. Natl. Acad. Sci. U.S.A.* **76,** 3683 (1979).

[29] T. Barnett, C. Pachel, J. P. Gergen, and P. C. Wensink, *Cell* **21,** 729 (1980).

manent record. The filters are sturdy and can be reused an indefinite number of times without apparent loss of signal. In contrast, the basis of DNA immobilization on nitrocellulose filters is undefined; however, filters have been used repeatedly in hybrid selection assays with no apparent diminution of signal. Nitrocellulose membrane requires no preparation prior to DNA application and is generally of uniform quality. Furthermore it demonstrates a higher binding capacity for DNA than does DBM and is more sensitive in DNA and RNA blotting analyses.[24] Once the DNA is bound, both types of solid supports can be stored indefinitely at 4°.

Optimal hybridization conditions will depend on both the sequence and abundance of the mRNA of interest. The parameters to be varied to ascertain these optima include the concentrations of formamide and input RNA and the time and temperature of the hybridization. A convenient method of assessing the appropriate temperature for selection involves the use of a temperature programmer that defines a linear decrease in the temperature of the hybridization reaction. In addition the stringency of the washing conditions should be optimized both to retain the mRNA of interest and to minimize nonspecific hybridization. Examples of hybrid selection conditions that have been optimized for the selection of different abundance classes of mRNAs are as follows:

High-abundance class (e.g., late mRNAs of Adenovirus 2, actin and tubulin): 65% formamide, 0.01 *M* PIPES–NaOH (pH 6.4), 0.4 *M* NaCl, 0.3 mg/ml total cytoplasmic RNA at 48° for 2 hr

Low-abundance class (e.g., Adenovirus 5 early mRNAs and Epstein-Barr virus mRNAs): 50% formamide, 0.01 *M* PIPES NaOH (pH 6.4), 0.4 *M* NaCl, 12 mg/ml total cytoplasmic RNA at 37° for 12 hr

In both cases the wash conditions were 1 × SSC, 0.5% SDS at 60°.

Failure of hybrid selection assays is most often due to degradation of the RNA by components contaminating the formamide. This can be easily ascertained by incubating the RNA under the appropriate hybridization conditions, then diluting the formamide concentration to less than 25% and precipitating the RNA with ethanol. A fraction of the precipitated RNA is then translated, and the cell-free products are compared directly with those synthesized by the untreated RNA. Evidence of RNA degradation requires the deionization of new formamide and the preparation and filtration of newly made buffers. Furthermore, throughout all manipulations the investigator must wear gloves to prevent the introduction of RNases.

Central to successful hybrid selection is an efficient cell-free translation system.[30] The addition of small amounts of tRNA has been shown to

[30] See appropriate sections in this volume.

stimulate translation; therefore, it is recommended that calf-liver tRNA be added to the elution buffer in the selection procedure. However, it should be noted that amounts of tRNA in excess of 25 μg per 25-μl translation reaction will cause inhibition.

Hybrid selection has many applications: (*a*) the determination of the arrangement of mRNA sequences within a specific DNA; (*b*) the purification of specific mRNAs as templates for the preparation of cDNA recombinant molecules; (*c*) the determination of the kinetics of RNA synthesis by the electrophoretic fractionation of selected RNA labeled with ^{32}P[31]; (*d*) the isolation of specific polypeptides for detailed analysis, e.g. tryptic digestion[32]; (*e*) the identification of cDNA and genomic recombinant molecules containing sequence information for specific polypeptides.

In this chapter we outline the application of hybrid selection to the identification of recombinant molecules encoding specific polypeptides. This can be done by using hybrid selection in concert with methylmercury hydroxide agarose gel electrophoresis as described above. Alternatively, hybrid selection can be used to screen large numbers of recombinant DNA molecules directly. Plasmid DNAs isolated by a rapid preparation procedure are linearized and applied in a gridwork pattern on a nitrocellulose filter. By using DNA application procedure 3, several different plasmids may be applied to the same 1-cm^2 region. The nitrocellulose is then placed in a plastic, sealable bag, and the hybridization solution (300 μl to 1 ml, depending on the size of the filter) containing the RNA (the amount depending on the volume of the hybridization and the abundance of the mRNA of interest) is added, the bag is sealed, and hybridization is carried out. Subsequent to hybridization, the filter is washed in a tray under the conditions described above. Individual filters containing the mRNA:DNA hybrids are excised from the gridwork with a sterile razor blade and placed in separate microfuge tubes; the RNAs are eluted as described above. Translation of the RNAs permits identification of those recombinant DNA molecules containing sequences for the desired polypeptide.

Hybrid Arrest of Translation

An alternative method for the identification of recombinant DNA molecules containing part or all of a sequence coding for a specific polypeptide is hybrid arrest of translation.[33,34] The basis for this technique is the

[31] D. Biswas, S. Hanes, and B. Brennessel, *Proc. Natl. Acad. Sci. U.S.A.* **79,** 66 (1982).

[32] J. Morgan, unpublished results.

[33] B. M. Paterson, B. E. Roberts, and E. L. Kuff, *Proc. Natl. Acad. Sci. U. S.A.* **74,** 4370 (1977).

[34] N. D. Hastie and W. A. Held, *Proc. Natl. Acad. Sci. U.S.A.* **75,** 1217 (1978).

observation that when sequences within mRNA essential for translation are in hybrid with their complementary DNA, translation of its encoded polypeptide in a cell-free system is blocked. Upon heat dissociation of this complex, complete translational activity is reinstated. In addition, whereas hybrid selection can be used to define the genomic location of all the constituent sequence-specific mRNA, hybrid arrest of translation can be used to identify those sequences within the mRNA that are required for the translation. Duplex formation involving the entire translated sequence or that portion of the mRNA sequence corresponding to the N terminus of the polypeptide abolishes its translation. Duplex formation involving the C-terminal portion of the coding sequence may result in the synthesis of the truncated polypeptide.[33] These variations are diagrammed in Fig. 8. Thus, these two techniques when used in concert, permit the detailed localization of coding and noncoding sequence of an RNA transcript along the DNA.

Stock Solutions

Deionized Formamide (Matheson, Coleman and Bell)

NaCl, 4 *M* (Millipore filtered)

0.4 *M* Piperazine-*N*,*N'*-bis(2-ethanesulfonic acid)–NaOH, pH 6.4 (Millipore filtered)

EDTA, 0.5 *M*, pH 7.9 (Millipore filtered)

Procedure

Plasmid DNA is linearized either by digestion with restriction enzymes or by boiling the DNA in 0.01 *M* Tris-HCl, pH 7.9, 0.001 *M* EDTA for 10 min. Purified DNA, free from contaminating proteins and salts, is critical to the success of this procedure, therefore the DNA is extracted sequentially with phenol and chloroform–isoamyl alcohol (24 : 1). The DNA and total cytoplasmic RNA are ethanol-precipitated with 25 μg of carrier tRNA. Two different procedures for hybridization have been used successfully.

Procedure 1. The precipitate is dissolved in 2–3 μl of sterile double-distilled water, boiled for 30 sec, and quick frozen in a Dry Ice–ethanol bath. The reaction is spun for 10 sec, 22 μl of 80% formamide, 0.01 *M* PIPES (pH 6.4), 0.4 *M* NaCl is added and the reaction is incubated for 2 hr at 48°.[33]

Procedure 2. The RNA–DNA precipitate is dissolved in 4.5 μl 0.1 *M* Pipes. 40 μl of 100% formamide are added, and the solution is boiled by 2 min to denature the DNA. The reaction is quenched in a Dry Ice–ethanol bath and 5 μl of 4 *M* NaCl, 0.4 *M* PIPES, pH 6.4, 0.01 *M* EDTA is added. Hybridization is carried out for 3 hr at 37°.[35]

Each DNA preparation to be used in the hybrid-arrested translation

[35] J. A. Cooper, R. Wittek, and B. Moss, *J. Virol.* **37,** 284 (1981).

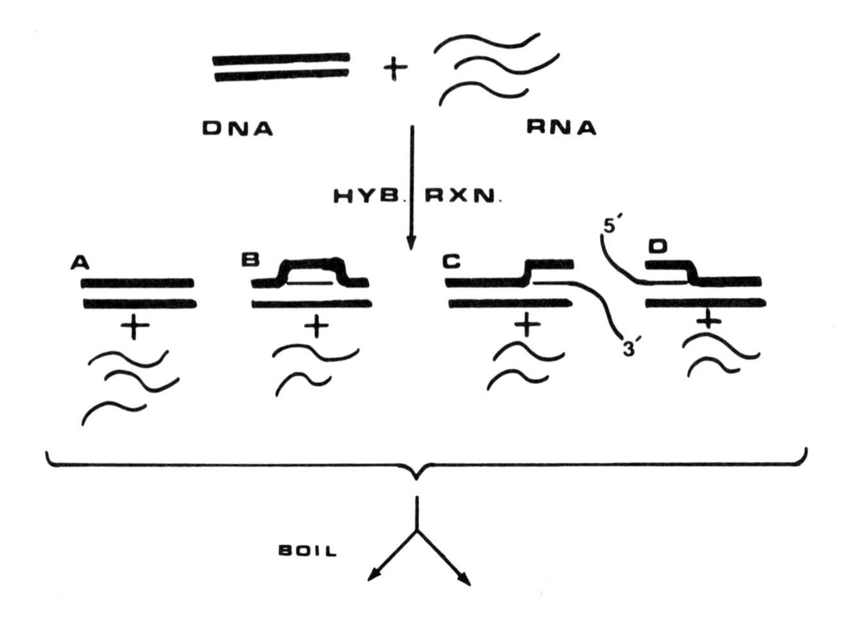

TRANSLATE

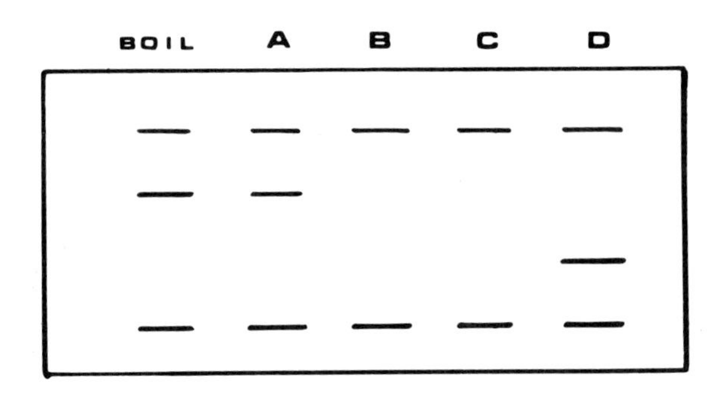

PROTEIN GEL

FIG. 8. Hybrid-arrested translation.

reaction should be titrated to determine the DNA concentrations that result in arrest of translation without causing inhibition of translation. The RNA concentration should be twice that required to give adequate incorporation of $[^{35}S]$methionine into polypeptides for clear visualization overnight by autoradiography.

The hybridization reaction is terminated by the addition of 200 μl of ice cold, sterile, double-distilled water. One half of the reaction is transferred to a new tube, boiled for 60 sec, then quick frozen in a Dry Ice–ethanol bath. Both reactions (arrested and heat dissociated) are adjusted to 0.2 *M* sodium acetate, pH 5.5, and ethanol-precipitated with 2.5 volumes of 100% ethanol and dried under vacuum; translation components are added directly to the pellet. The $[^{35}S]$methionine-labeled cell-free products are fractionated on an SDS–polyacrylamide gel, and the products of the arrest reaction are compared to those of the heat-dissociation reaction.

Comments on the Application of Hybrid Arrest of Translation

In contrast to hybrid selection, which identifies recombinant DNA molecules by hybridization to coding or noncoding sequences, hybrid arrest is limited to the identification of those clones containing sequences essential for translation of the encoded polypeptide.

Since hybrid arrest is a subtractive assay, its application is limited to those mRNAs whose encoded polypeptide is both obvious and clearly resolved from adjacent polypeptides in an SDS–polyacrylamide gel. This limitation can be circumvented if antibodies for the specific polypeptide of interest are available.

After translation of the RNAs in the hybrid arrest and heat-dissociated reaction, the translation reaction is diluted 10-fold in 0.05 *M* Tris-HCl (pH 7.4), 0.005 m*M* EDTA, 0.15 *M* NaCl, and specific polypeptides are immunoprecipitated by standard procedures. These immunoprecipitates may then be fractionated on an SDS–polyacrylamide gel.

For the purposes of mapping RNA transcripts along the DNA, hybrid-selected RNA can be utilized in a hybrid-arrest assay instead of total cytoplasmic RNA. This permits unambiguous resolution of unaltered, truncated, or arrested polypeptides on an SDS–polyacrylamide gel and defines the coding and noncoding sequences of an mRNA on the DNA.[8]

Summary

Herein we outline three methods that, when coupled with cell-free protein synthesis, permit the identification of recombinant DNA molecules encoding specific polypeptides. RNAs enriched by fractionation on methylmercury hydroxide agarose gels are used to prepare sequence-specific probes. Recombinant DNA clones thus identified are further characterized as to their encoded polypeptides by either hybrid selection or hybrid arrest of translation.

[44] Prokaryotic Coupled Transcription–Translation

By HUI-ZHU CHEN and GEOFFREY ZUBAY

In prokaryotes transcription and translation are usually coupled; i.e., translation starts shortly after transcription has begun and long before transcription is completed. Investigation of genes with attenuators (e.g., the *trp* operon) indicate that the lead ribosome is not far behind the RNA polymerase and may even modulate its movement.[1] In view of this situation, its seems likely that the most realistic cell-free conditions for studying transcription and translation should involve a system in which the two events are coupled. In 1967 J. DeVries and I demonstrated synthesis of an N-terminal fragment of the enzyme β-galactosidase in such a coupled system.[2] The conditions used in this system have undergone some changes to improve the quantity and quality of the synthesized products, and the variety of applications has expanded. In this chapter, preferred conditions are described and a number of experiments are reviewed to illustrate the versatility of the system.

The basic components of the system include a cell-free extract of *Escherichia coli* called an S-30, the cofactors and substrates necessary for RNA and protein synthesis, and a source of DNA to direct the synthesis. It is possible to synthesize most bacterial and viral RNAs and proteins in readily detectable amounts. All aspects of transcription and translation can be studied under conditions believed to be as close to conditions inside the cell as one can hope to attain. The cell-free system possesses a number of advantages over whole cells for analytical purposes. Small molecules required for synthesis may be controlled quantitatively by simple addition. Macromolecules required for synthesis can be determined by *in vitro* complementation if suitable mutants are available. The crude cell-free system made from the S-30 extract has a number of advantages

[1] C. Yanofsky, *Nature (London)* **289,** 751 (1981).

[2] J. K. DeVries and G. Zubay, *Proc. Natl. Acad. Sci. U.S.A.* **57,** 1010 (1967).

[3] G. Zubay, *Annu. Rev. Genet.* **7,** 267 (1974).

METHODS IN ENZYMOLOGY, VOL. 101

ISBN 0-12-182001-7

over more fractionated systems; it is more efficient in terms of rates, yield, and longevity, and it is more likely to contain most if not all of the factors required for transcription and translation.

Basic Methodology

A summary of all the procedures needed to make the standard coupled system is given below.

Growth of Cells. The bacterial cells used in preparation of S-30s are grown at 28° to mid-log phase in a New Brunswick microferm fermentor in the following medium (per liter of distilled water): KH_2PO_4 (anhydrous), 5.6 g; K_2HPO_4 (anhydrous), 28.9 g; yeast extract, 10 g; thiamin, 10–15 mg; and 40 ml of 25% glucose added after autoclaving. Fermentors containing 10 liters of medium are inoculated with 1 liter of culture grown overnight on the same medium. About 4 hr after inoculation, cells are collected without otherwise interrupting the growth process in the fermentors. Cells from the fermentor are chilled to 1° by passage through a copper coil immersed in an ice-water bath and collected in a Lourdes continuous-flow centrifuge at a rate of 100 ml/min. The yield of cells is about 10 g of wet paste per liter of medium. The cells are removed from the rotor, flattened into pancakes about $\frac{1}{8}$ inch thick, and frozen at −90° in a Revco ultradeep-freeze. The cells are stored in this manner overnight.

Preparation of S-30 Extract. The S-30 extract is prepared with minor modifications by the method of Nirenberg.[4] Fifty grams of frozen cells are allowed to soften at 4° for 30 min and homogenized in a Waring blender with 500 ml of buffer I (0.01 *M* Tris-acetate, pH 8.2, 0.014 *M* magnesium acetate, 0.06 *M* potassium chloride, and 0.006 *M* 2-mercaptoethanol). The suspension is centrifuged for 30 min at 10,000 rpm in a large Servol rotor. The sediment of cells is resuspended in 65 ml of buffer II (buffer I containing 0.001 *M* dithiothreitol in place of 2-mercaptoethanol). The cell suspension is lysed in an Aminco pressure cell at pressures of 4000–8000 psi. Immediately after lysis, 1 μmol of dithiothreitol per milliliter is added to the lysate. No deoxyribonuclease is added to the lysate. After two 30-min centrifugations at 30,000 *g*, the resulting supernatant is mixed with 8.0 ml of a solution containing: 6 mmol of Tris-acetate, pH 8.2, 0.06 mmol of dithiothreitol, 0.17 mmol of $Mg(OAc)_2$, 0.6 μmol of the 20 amino acids, 0.048 mmol of ATP, 0.54 mmol of Na_3PEP, and 0.16 mg of pyruvate kinase. The mixture is incubated in a light-protected vessel at 37° for 80 min and dialyzed for 18 hr at 4° against buffer III with one change (buffer II containing 0.06 *M* potassium acetate in place of the po-

[4] M. W. Nirenberg and J. H. Matthaei, *Proc. Natl. Acad. Sci. U.S.A.* **47**, 1588 (1961).

tassium chloride). The S-30 extract is rapidly frozen in small portions and stored at −90°. The S-30 is thawed at 4°, then used immediately.

Incubation Procedure. The incubation mixture contains (per milliliter): 44 μmol of Tris-acetate (pH 8.2); 1.37 μmol of dithiothreitol; 55 μmol of KOAc; 27 μmol of NH_4OAc; 14.7 μmol of $Mg(OAc)_2$; 7.4 μmol of $Ca(OAc)_2$; 0.22 μmol each of 20 amino acids; 2.2 μmol of ATP; 0.55 μmol of GTP, CTP, and UTP; 21 μmol trisodium phosphoenolpyruvic acid; 100 μg of *E. coli* tRNA; 27 μg of pyridoxine HCl; 27 μg of TPN; 27 μg of FAD; 11 μg of *p*-aminobenzoic acid; 27 μg of folic acid; 16 mg

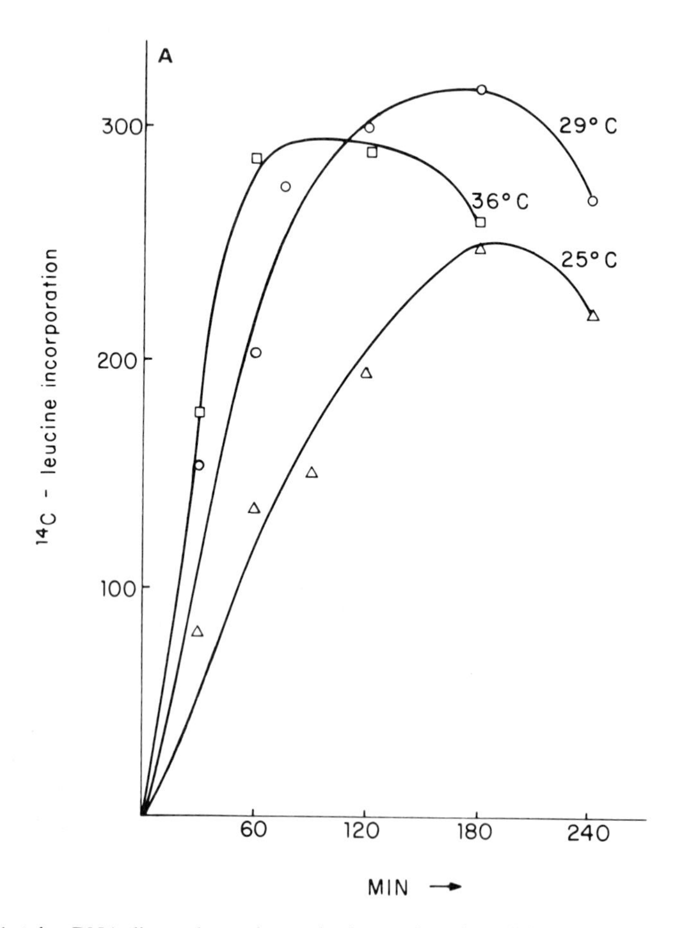

FIG. 1. λ *lac* DNA-directed protein synthesis as a function of time at different temperatures. In (A) gross peptide synthesis is measured by trichloroacetic acid precipitation of [^{14}C]leucine labeled protein. In (B) β-galactosidase synthesis is measured by a colorimetric assay using *O*-nitrophenyl-β-galactoside as substrate. See Zubay *et al.*[6] for details.

polythylene glycol 6000. The amounts given for the calcium and magnesium salts are only approximate. In practice the ratio is kept constant and the total divalent cation concentration is optimized for each S-30. The above ingredients are incubated for 3 min at 37° with the DNA (usually 50 μg), with shaking, before 6.5 mg of S-30 extract protein is added. Incubations with shaking are allowed to continue for 1–2 hr at 37°. The shaking should be as vigorous as possible without producing bubbles or splashing. During the incubation, a viscous pellet is formed in the bottom of the tube

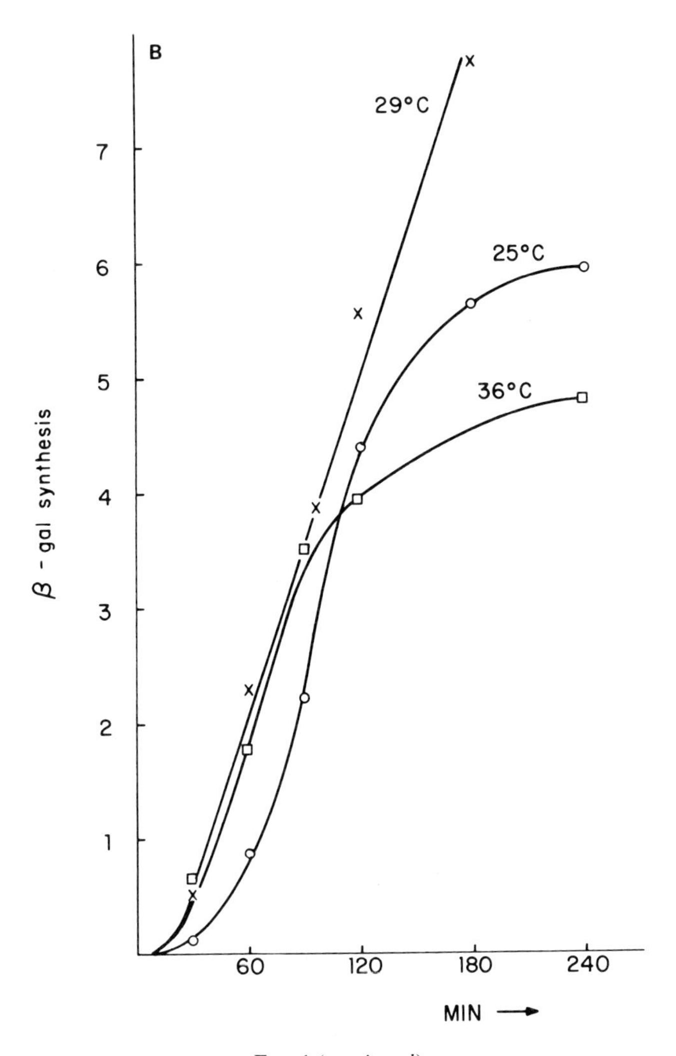

FIG. 1 (*continued*)

or vessel. At the termination of the incubation, the precipitate is gently resuspended and aliquots for assay are removed.

Troubleshooting. Under optimal conditions protein synthesis occurs at a linear rate for 2–3 hr (Fig. 1) before gradually subsiding. Any deviation from the procedures described above or poor quality control of the ingredients can result in a drastic reduction in yields. One of the most sensitive factors is the divalent cation concentration. In practice it has been found desirable to determine the optimum concentration on each S-30 by testing the amount of synthesis around the designated concentration of Ca^{2+} and Mg^{2+} indicated above.

Analysis of RNA and Protein Products. RNA and protein product may be analyzed by gel electrophoresis of *in vitro* radioactively labeled products. Standard procedures for so doing have been described elsewhere in this series. Sometimes it is desirable to assay an RNA or protein products by a specific biochemical or biological assay. For example, β-galactosidase made in the coupled system can be assayed by a simple colorimetric method (total incubation time in the assay is 3 min or longer depending upon amount of enzyme made); colicin E1 (ColE1) factor synthesized in the coupled system has been assayed by killer activity (unpublished data of our laboratory); tyrosine suppressor transfer RNA has been assayed by its ability to suppress *amber* triplet termination in a comparable coupled system.[5]

Some Applications of the Coupled System

With the coupled system it is possible to synthesize most bacterial and viral proteins in readily detectable amounts. The main question for the future is not whether a particular RNA or protein can be made in the cell-free situation, but, rather, what can be learned by studying its synthesis. One of the most useful applications of the coupled system has been the detection of factors involved in the regulation of transcription and translation. Once factors have been detected and studied in the coupled system they can be purified, if necessary, for further characterization, using the coupled system as an assay to guide the purification. Factors are usually detected by their ability to stimulate or inhibit protein synthesis. Small molecule factors are amenable to direct study, as they can be added to or deleted from the cell-free system at will. Macromolecular factors, present in cell-free extracts prepared from wild-type bacterial strains can be studied through a technique in which the cell-free extract (S-30) is prepared from mutant strains defective in the factor(s), and the system is reconstituted with fractionated portions of cell-free extract from a normal strain.

[5] G. Zubay, L. Cheong, and M. Gefter, *Proc. Nalt. Acad. Sci. U.S.A.* **68,** 2195 (1971).

Studies of Factors Affecting Gene Expression

The versatility of the coupled system can be illustrated by studies made on the *lac* operon. The *lac* operon has been one of the most intensively studied and best understood inducible gene clusters. The operon contains three structural genes, which code for β-galactosidase (β-gal), lactose permease, and galactoside transacetylase, as well as a promoter and an operator. The first protein hydrolyzes lactose to its constituent monosaccharides, the second concentrates lactose from outside the cell, and the third catalyzes the acetylation of β-galactosides. The structural genes for these three proteins are adjacent to one another and the controlling elements are located at one end of the gene cluster. In most of the coupled system studies of the *lac* operon the level of β-gal synthesis has been used as an index of gene activity. This assumption seems justifiable in retrospect, but it does not eliminate the possibility of purely translational control mechanisms.

Small molecules of two classes exert a profound effect on the enzyme yield as a result of what appears to be a gene-regulating function.[6] The first class includes inducers of the *lac* operon, of which isopropylthiogalactopyranoside (IPTG) is one of the most potent known. In systems isogenic except for the *lac* operon repressor, IPTG stimulates only that system containing the repressor. If repressor is absent, maximum activity is obtained without IPTG. The interaction of IPTG with the repressor is believed to result in the release of the latter from the operator. The amount of IPTG required to induce β-gal synthesis depends upon the amount of repressor present; more IPTG is required at higher concentrations of repressor. Quantitative studies have led to the proposal that the binding of two inducer molecules per repressor molecule is required for maximum derepression. The second class of small molecules regulating enzyme yield are those associated with catabolite repression of which cyclic 3′,5′-AMP is a prime example. Cyclic AMP (cAMP) is usually added to the coupled system, as its omission results in much lower enzyme activities. The synthesis of galactoside transacetylase has also been shown to depend upon the presence of cAMP *in vitro*.

The effects of both IPTG and cAMP appear to be localized at the level of initiation of transcription. Several experiments support this proposition. First, in a modified coupled system transcription is initiated in the absence of translation by leaving out all amino acids. After a set time, transcription is stopped, either by adding DNase or rifampicin, and simultaneously translation is begun by adding amino acids. The stimulatory ef-

[6] G. Zubay, D. A. Chambers, and L. C. Cheong, *in* "The Lactose Operon" (J. R. Beckwith and D. Zipser, eds.), p. 375. Cold Spring Harbor Laboratory, Cold Spring Harbor, New York, 1970.

fect of IPTG and cAMP is confined to the transcription phase of the synthesis. Second, mRNA reisolated from the coupled system may be used to stimulate enzyme synthesis in the absence of added DNA. In this system, neither IPTG or cAMP have any effect. Finally when DNA is used in a purified transcription system containing only those proteins necessary for transcription (RNA polymerase holoenzyme) or for regulating transcription (*lac* repressor and CAP), both IPTG and cAMP have a specific stimulatory effect on *lac* mRNA synthesis.

In addition to studying the conditions of synthesis, the coupled system may be used as an assay tool for regulatory protein purification. Historically, the *lac* repressor was first isolated by W. Gilbert and B. Müller-Hill using the binding between the repressor and [^{14}C]IPTG as the assay during purification. However, the DNA-directed system provides an assay for repressor with much greater specificity and greater sensitivity. The inducer binding assay is limited by the affinity constant of the inducer for repressor, which is about 10^6 liters mol^{-1}, whereas the assay provided by the coupled system is limited by the affinity constant of the repressor for the DNA, which is about 10^{10} liters mol^{-1}. This makes the latter assay more sensitive by a factor of $10^{10}/10^6$ or 10^4. The coupled system assay also allows for a more comprehensive characterization of the repressor. Thus in this assay the repressor is characterized by its ability to inhibit β-gal synthesis and the reversal of this effect by inducer. In the inducer binding assay only the ability of repressor to bind inducer is measured. For these reasons it seems likely that the coupled system provides a superior probe for isolation of most gene-regulating proteins. This is exemplified by the work on CAP, described below.

The coupled system was used to isolate the cAMP activator protein called CAP.[7] For this purpose an S-30 extract was prepared from a CAP mutant that fails to synthesize appreciable β-gal unless a fraction from wild-type cells containing CAP is added. The stimulation of β-gal synthesis by CAP is completely dependent upon the presence of cAMP. Using this assay, CAP can be monitored in crude extracts and extracts purified to varying degrees. It has been claimed that CAP could be isolated using a binding assay of CAP to labeled cAMP. Because of the low affinity constant between CAP and cAMP and the binding of cAMP to other components of the cell, this latter approach is not sufficiently sensitive or selective for detecting CAP in crude extracts.

In the coupled system a number of studies on the gene activity as a function of the concentration of various regulatory components have been made. Such studies have helped in determining the mechanism of turning the operon on and off.[8]

[7] W. Gilbert and B. Müller-Hill, *Proc. Natl. Acad. Sci. U.S.A.* **56,** 1891 (1966).
[8] G. Zubay, this series, Vol. 65, p. 856.

The compound guanosine tetraphosphate (ppGpp), known as magic spot, influences *lac* operon expression. This compound rises in concentration in amino acid-starved cells. The most notable correlated effect of this rise is the cessation of tRNA and rRNA synthesis. As in the case of factors that control cAMP levels, the changes in metabolism surrounding these circumstances are too complex to distinguish direct effects from indirect ones on gene expression in whole cells. *In vitro* the direct effects of ppGpp on the expression of a number of genes has been studied[9–11]; rRNA synthesis is severely inhibited paralleling the *in vivo* results, whereas tRNA is not. The lack of inhibition to tRNA synthesis *in vitro* is puzzling; one possibility is that the ppGpp effect on tRNA synthesis *in vivo* (at least for tRNA genes that do not occur as part of an rRNA cistron) is indirect. The effects of ppGpp on a number of structural genes was also examined using an assay for the protein products. Several operons were found to be stimulated by ppGpp (for example, the arabinose operon, the lactose operon, and the trytophan operon). The pleiotropic effects of ppGpp in the cell-free system have led to the hypothesis that ppGpp direct interacts with RNA polymerase and changes its structure; the structural change in the polymerase either enhances or diminishes the interaction between polymerase and promoter according to the structure of the promoter. In a purified system containing *lac* operon DNA, RNA polymerase, CAP, cAMP and the substrates necessary for RNA synthesis ppGpp stimulates gene expression severalfold (unpublished results from our laboratory). Unlike the studies on CAP, no mutants resistant to ppGpp effects have been isolated, and until this is done the *in vitro* results cannot be related to the physiologically significant site of ppGpp action. The *in vitro* studies with the small molecule effectors ppGpp and cAMP have been greatly aided by the fact that both of these compounds are stable *in vitro*. This is in contrast to the *in vivo* situation, where ppGpp and cAMP have rapid turnover times, steady-state concentrations being maintained by balancing the synthetic rates against the degradative rates.

Negative supercoiling of the DNA template has a large stimulatory effect on the level of *lac* operon expression. Negative supercoiling of DNA in *E. coli* is catalyzed by DNA gyrase, an essential enzyme for cell survival. Antibodies such as coumermycin and novobiocin that specifically inhibit DNA gyrase facilitate investigations of the effects of negative su-

[9] H.-L. Yang, G. Zubay, E. Urm, G. Reiness, and M. Cashel, *Proc. Natl. Acad. Sci. U.S.A.* **71,** 63 (1974).
[10] G. Reiness, H.-L. Yang, G. Zubay, and M. Cashel, *Proc. Natl. Acad. Sci. U.S.A.* **72,** 2881 (1975).
[11] H.-L. Yang, K. Heller, M. Gellert, and G. Zubay, *Proc. Natl. Acad. Sci. U.S.A.* **76,** 3304 (1979).

percoiling *in vivo*. *In vitro* the effects of supercoiling on gene expression can be examined more directly. When λ *lac* DNA is used as a template *in vitro,* it becomes very rapidly ligated and subsequently supercoiled. Expression of the *lac* operon is lowered about five-fold in the presence of novobiocin. This drug prevents supercoiling but not ligation. When a novobiocin-resistant strain is used to make the S-30 extract, the effect of the antibiotic is negligible. This demonstrates the importance of negative supercoiling to optimal expression of the *lac* operon. Other genes have been studied in this way, some genes are much more sensitive to the effects of supercoiling than others, as shown in Table I. No cases have been found where negative supercoiling inhibits expression. Since negative supercoiling exerts a torsional tension favoring the untwisting of the duplex, it is presumed that gyrase facilitates formation of the so-called open promotor complex with RNA polymerase. Mutational change in the *lac* operon promotor to a sequence more favorable for RNA polymerase—promoter complex formation decreases the sensitivity to negative supercoiling as well as the dependence of cAMP and ppGpp (see Table II).

When T4 bacteriophage infects *E. coli,* a complex series of events take place including gross inhibition of host transcription. The coupled system has been helpful in studying inhibition of host transcription using *lac* expression as an index of host activity.[12] It is known that host transcription *in vivo* is inhibited by at least two different mechanisms, one requiring

TABLE I
SENSITIVITY OF DIFFERENT GENES TO SUPERCOILING

		Ratio of synthesis in the absence and the presence of 1 μg/ml of novobiocin
Insensitive		
↓	β-Lactamase in pBR322	1–1.2
	β-Gal in *UV5* λ *lac*	1–1.2
	β-Gal in λ *trp-lac* (i.e., the *lac* operon fused to the *trp* promoter)	1–1.2
	Colicin in ColE1	2–3
	β-Gal in wild-type λ *plac* or λ *dlac*	4–5
	rrn in λ ^{d}*rif18*	4–6
	Tetracycline resistance protein in pBR322	6–10
↓	4.5 S RNA in ColE1	50–100
Ultrasensitive		

[12] R. Mailhammer, H.-L. Yang, G. Reiness, and G. Zubay, *Proc. Natl. Acad. Sci. U.S.A.* **72,** 4928 (1975).

TABLE II
COMPARISON OF THE β-GALACTOSIDASE SYNTHESIZED BY WILD-TYPE λ *plac5* AND MUTANT *UV5* DNA IN THE PRESENCE AND THE ABSENCE OF cAMP, ppGpp, AND NOVOBIOCIN[a]

	Yield of enzyme			Promoter	
	cAMP	ppGpp	Novobiocin	λ*plac*5	*UV5*
1	+	−	−	1.67	3.03
2	+	+	−	3.27	2.46
3	+	−	+	0.42	2.43
4	+	+	+	1.60	2.20
5	−	−	−	0.2	3.30
6	−	+	−	0.1	2.64
7	−	−	+	0.2	2.45
8	−	+	+	0.1	2.35

[a] Synthesis reactions are carried out under standard conditions for 1 hr. The yield of enzyme is expressed in relative amounts based on a standard colorimetric assay. When present the concentrations of cAMP, ppGpp, and novobiocin were 0.5 m*M*, 0.1 m*M*, and 1 μg/ml, respectively. The sequences of the Pribnow boxes in the λ *plac5* and *UV5* promoters are TATGTTG and TATAATG, respectively.

protein synthesis after virus infection and one that does not. The latter mechanism is demonstrable by incubating cells with chloramphenicol at the time of phage infection. Inhibition of host transcription under such conditions is incomplete and increases in the multiplicity of infection increases.

It is known that T4 infection results in modification of the host RNA polymerase. In order to test the effect of the phage-modified polymerase on β-galactosidase synthesis in the coupled system it was necessary to prepare the polymerase from T4-infected cells and add it to the coupled system. At the same time it was necessary to block transcription by the polymerase already present in the S-30 extract. This was done by adding rifampicin to the extract and preparing the T4-modified enzyme from rifampicin-resistant cells. As a control rifampicin-resistant polymerase was also prepared from uninfected mutant cells. The enzyme prepared from the uninfected control cells stimulated β-galactosidase like normal enzyme in the absence of rifampicin. Enzyme isolated from T4 infected cells was completely inactive in stimulating β-galactosidase synthesis. The same T4 modified enzyme was shown to be highly active in T4 DNA-directed protein synthesis. From these results it was concluded that T4 modification alters the ability of RNA polymerase to recognize host promoters.

The more immediate type of inhibition of host transcription observed when T4 infects *E. coli* at a high multiplicity was explored by simply adding T4 DNA to a normal coupled system containing λ *lac* DNA. It was found that T4 DNA at low levels (4 μg/ml) produces 90% inhibition of λ *lac* DNA-directed β-gal synthesis even when the latter DNA is present at a high concentration (53 μg/ml). It is believed that this experiment simulates the *in vivo* situation at a high multiplicity of infection and suggests that T4 DNA has a much higher affinity for available host polymerase than does λ *lac* DNA. The inhibitory effect of T4 DNA can be partially overcome by using higher polymerase concentrations. Thus possible explanations for both types of T4 inhibition of host transcription have been found by appropriate manipulations of the coupled system.

Applications of the Coupled System to Studies Using Linear DNA Templates

In most of the results reported above, λ DNA or suitable derivatives of λ DNA were used as a template to direct RNA and protein synthesis. As indicated earlier λ DNA circularizes rapidly when present in the coupled system. The experiments reported here and elsewhere show that DNA endonuclease activity must be very low under the conditions of incubation. A single nick in the λ DNA would eliminate the effect of supercoiling by permitting relaxation of the DNA. This amazing stability of the DNA came as a great surprise to us.

By contrast, linear DNA is rapidly degraded in the coupled system. For example, normal circular ColE1 DNA is completely stable and highly supercoiled even after a 1-hr incubation in the coupled system. This was shown by gel analysis of the reisolated DNA. If the ColE1 DNA is linearized by *Eco*RI restriction enzyme cleavage prior to incubation in the S-30, it is completely degraded in 1 hr. The exonucleolytic activity present in the S-30 system makes it impossible to assess the expression of linear DNA quantitatively.

In view of the growing possibility of obtaining well-defined linear DNAs with the help of restriction enzymes it seemed likely that it would be worthwhile to have a coupled system in which linear DNA could be studied. For this purpose, a mutant *recB*-strain was used for preparing the S-30.[13] In this S-30 linear DNA is completely stable, showing that the *recB* gene product is involved in the degradation of linear DNA at least *in vitro*. In this connection it is of interest that *in vivo* replication of λ DNA by the rolling-circle mode requires inactivation of the *recB* encoded enzyme. Normally the *recB* enzyme is inactivated by a λ function early in

[13] H.-L. Yang, I. Ivashkiv, H.-Z. Chen, G. Zubay, and M. Cashel, *Proc. Natl. Acad. Sci. U.S.A.* **77**, 7029 (1980).

infection (M. Gottesman, personal communication). Use of the *recB*- S-30 extract has substantially expanded the range of usefulness of the coupled system. Given the restriction map of a particular DNA or chromosome, one can cleave the DNA at precise locations and measure the effects on gene expression. By using a battery of restriction enzymes, it is possible to assign gene locations to particular RNA or protein products as well as to determine when genes are organized into multifunctional clusters or operons.

The effectiveness of the *recB*- S-30 extract was demonstrated by examining two situations which were already well understood from other observations.[13] λ*plac5* DNA has a *Hin*dIII cleavage site close to the upstream side of the *lac* promoter; there are no *Hin*dIII cleavage sites within the first part of the operon including the *Z* gene for β-galactosidase. *Hin*dIII treatment of λ*plac5* DNA reduces β-gal synthesis by 93% in the normal S-30, but by only 35% in the *recB*- S-30. The β-operon was also examined after *Hin*dIII cleavage in the *recB*- S-30. This operon encodes four proteins, two ribosomal proteins, L10 and L7/L12, and two RNA polymerase subunits, β and β′. *Hin*dIII makes a single cut in the operon within the promoter proximal gene that encodes the L10 protein. This cleavage eliminates synthesis of all four proteins, even though the full-length transcript is known to contain at least one other ribosome recognition site before the *rpoBC* genes that encode β and β′. This result indicates that the normal promotor is essential for expression; apparently potential *in vitro* artifacts, such as initiation of transcription from the end of the linear DNA, do not contribute substantially to expression, at least in this instance. The possibility of terminal transcription initiation should be carefully inspected for each template before the generality of this conclusion is accepted.

Use of the *recB*- S-30 raises background synthesis, that is, the synthesis that occurs even in the absence of added template. When radioactively labeled protein synthesized *in vitro* is examined by gel electrophoresis and fluorography, this background synthesis sometimes makes it difficult to detect products synthesized by the added template. Background synthesis has always been a problem with S-30 extracts, but it has been easier to correct for other uses with normal strains. A key step in the preparation of S-30 extracts involves a preincubation. When the extracts are to be used for mRNA-directed protein synthesis, a 30-min preincubation in the presence of DNase removes all DNA and endogenous mRNA, so that the ribosomes can be programmed by adding the mRNA of choice. When the S-30 system was modified to make it suitable for investigating DNA-directed RNA and protein synthesis, a new background problem emerged. This resulted from the transcription of endogeneous fragments of chromosomal DNA that inevitably are found in the S-30. Preincubation times were increased from 30 to 80 min, but this did not completely eliminate the problem. With the S-30 extract made from a *recB*- mutant of *E. coli*, linear as well as circular DNA showed negligible degradation during the course of the usual incubation time for protein synthesis (60–90 min at 37°). New steps can be taken to reduce the relative importance of background synthesis. A fractionation procedure has been used that eliminates endogenous DNA[13+] or the relative importance of background synthesis can be lessened by using more of the test template. Neither of these solutions to the background problem are ideal. The fractionation procedure is laborious, difficult to reproduce, and usually lower in synthetic efficiency. It is not always desirable to add a large amount of test template, especially when it is precious. Another solution to this problem, which produces miraculous results when it works, is reported below.

A simple lowering of the amount of S-30 extract used in synthesis produces a drastic lowering in background synthesis without appreciable effects on the level of synthesis resulting from added DNA. This is illustrated here for the situation where ColE1 DNA is used to direct synthesis. The increased background synthesis when the *recB*- S-30 is used in place of the normal S-30 can be seen by comparing the fluorographs of the gel patterns of the protein products synthesized in the absence of added DNA (compare a and c in Fig. 2 and Table III). In a number of instances, it is difficult to decide whether the proteins made in the *recB*- S-30 result from background synthesis or from the added ColE1 DNA (compare columns c and d in Fig. 2). The seriousness of the problem would obviously be augmented in lower levels of ColE1 DNA had been used. Reduction in background synthesis without significant reduction in the synthesis of proteins encoded by ColE1 DNA can be accomplished merely by lowering the amount of S-30 extract used (see e and f in Fig. 2 and Table III). In these experiments the S-30 concentration was reduced from the usual 6.5 mg of protein to 1 mg of protein per milliliter. Background synthesis is so low that the fluorographs of the gels have to be greatly overexposed before faint traces of background bands can be seen (see Fig. 3). All the bands seen when ColE1 DNA is present during synthesis unambiguously result from ColE1 DNA-directed synthesis. The amount and quality of synthesis is usually improved somewhat by supplementing the S-30 with 60 μl/100 μl of an S-100 extract prepared from the same cells (compare f and h in Table III and Fig. 2). In most of the experiments the incubation

FIG. 2. Fluorographs of L-[^{35}S]methionine-labeled proteins synthesized in the coupled system after polyacrylamide gel electrophoresis. Conditions of synthesis gel electrophoresis and fluorography are described in Table III. Direction of electrophoresis is from top to bottom. A 5- μl aliquot of the appropriate incubation mixture was applied to each gel slot. Labeling is the same as in Table III. Fluorography time was 16 hr.

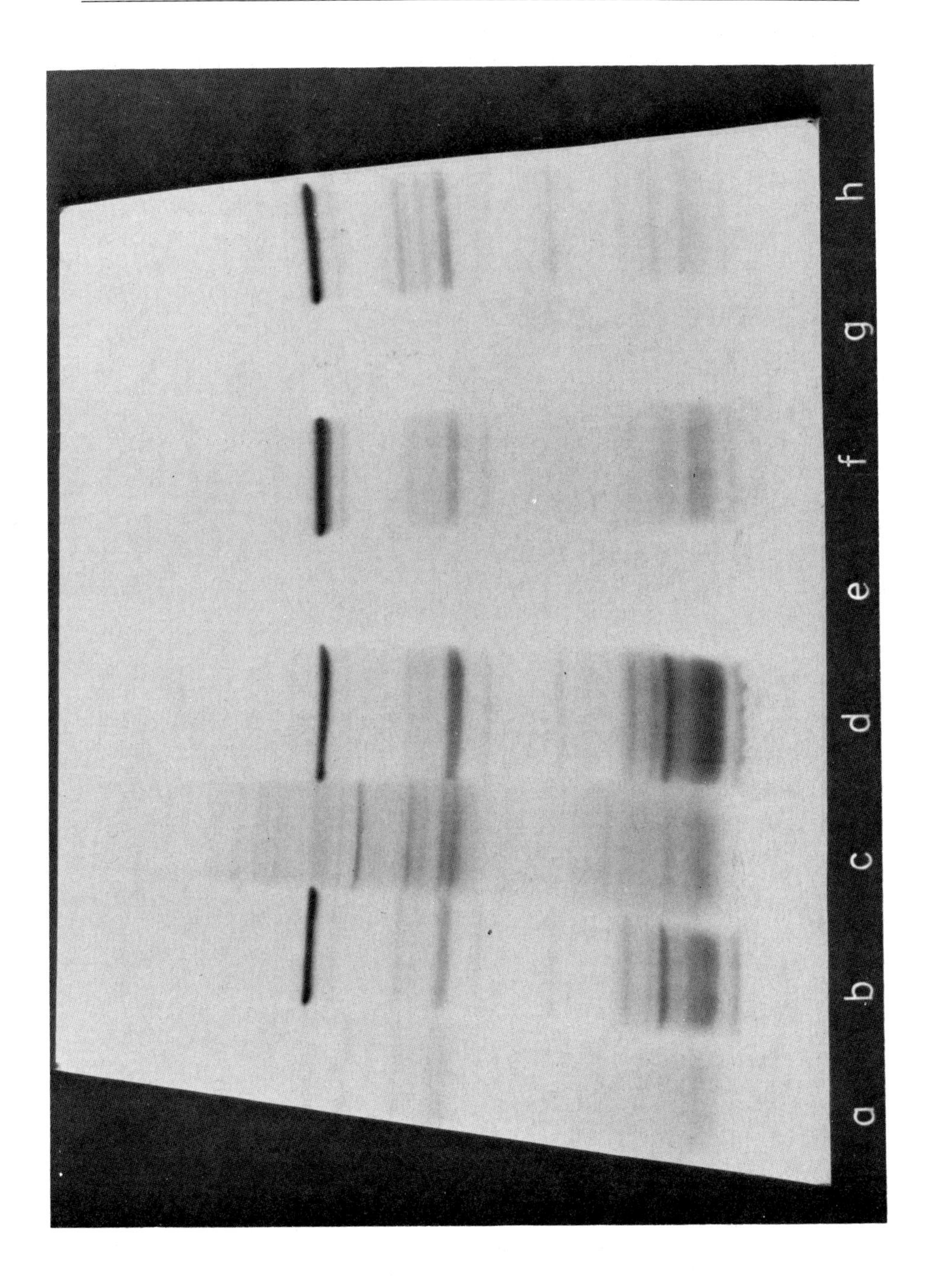
a
b
c
d
e
f
g
h

TABLE III
ColE1 DNA-Directed Protein Synthesis Using Different Kinds and Amounts of S-30 and S-100 Extract[a]

Lane	S-30 system	ColE1	Incorporation (cpm)	Δ
a	S30(2089), 33 μl	−	27,005	—
b	S30(2089), 33 μl	+	65,892	38887
c	S30(CF300), 33 μl	−	69,575	—
d	S30(CF300), 33 μl	+	81,088	11513
e	S-30(CF300), 5 μl	−	12,303	—
f	S-30(CF300), 5μl	+	42,400	30097
g	S-30(CF300), 5 μl + 10 μl of S-100	−	14,540	—
h	S-30(CF300), 5 μl + 10 μl of S-100	+	58,568	44028

[a] Synthesis conditions are described in Methods. A 10-μl aliquot of the incubation mixture was processed and counted in the usual way. Δ is the increase in counts per minute in the presence of added DNA. Strain 2089 is *recB*$^+$; strain CF300 is *recB*$^-$.

mixture contains 5% S-30 (21 mg/ml protein) and 10 per cent S-100 extract (6 mg/ml protein).

It was surprising to find that the proteins of ColE1 could be made with almost equal efficiency at much lower levels of S-30 extract since previous experience had indicated that this was not the case for β-gal synthesis. Two other templates, λ *rifd18* and λ *plac5* DNAs, were examined. The λ *rifd18* DNA carries the genes for several ribosomal proteins and the RNA polymerase subunits β and β′ as well as those of λ. The use of the lower level of S-30 extract does not substantially affect the efficiency of protein synthesis here even of the very large β and β′ proteins (results not shown). In the case where λ*plac5* DNA is employed, use of the lower level of S-30 extract reduces the efficiency of β-gal synthesis by more than 100-fold, verifying previous observations. Apparently β-galactosidase synthesis has a special requirement for one or more factors present in the S-30 in small amounts. This cannot be due to the lack of CAP protein, since the λ *plac5* template (unlike the λ*dlac* DNA) gives a substantial amount of β-galactosidase synthesis in the absence of CAP protein. In this connection it should be pointed out that others have suggested that β-galactosidase synthesis has a special requirement for a protein known as L factor, a product of the *nusA* gene.[14] It is clear that further exploration is necessary to determine those cases in which lower levels of S-30 extract can and cannot be used. When it works, it is preferable to other procedures for minimizing background problems.

[14] J. Greenblatt, S. Adhya, D. I. Friedman, L. S. Baron, B. Redfield, H.-F. Kung, and H. Wiessbach, *Proc. Natl. Acad. Sci. U.S.A.* **77,** 1991 (1980).

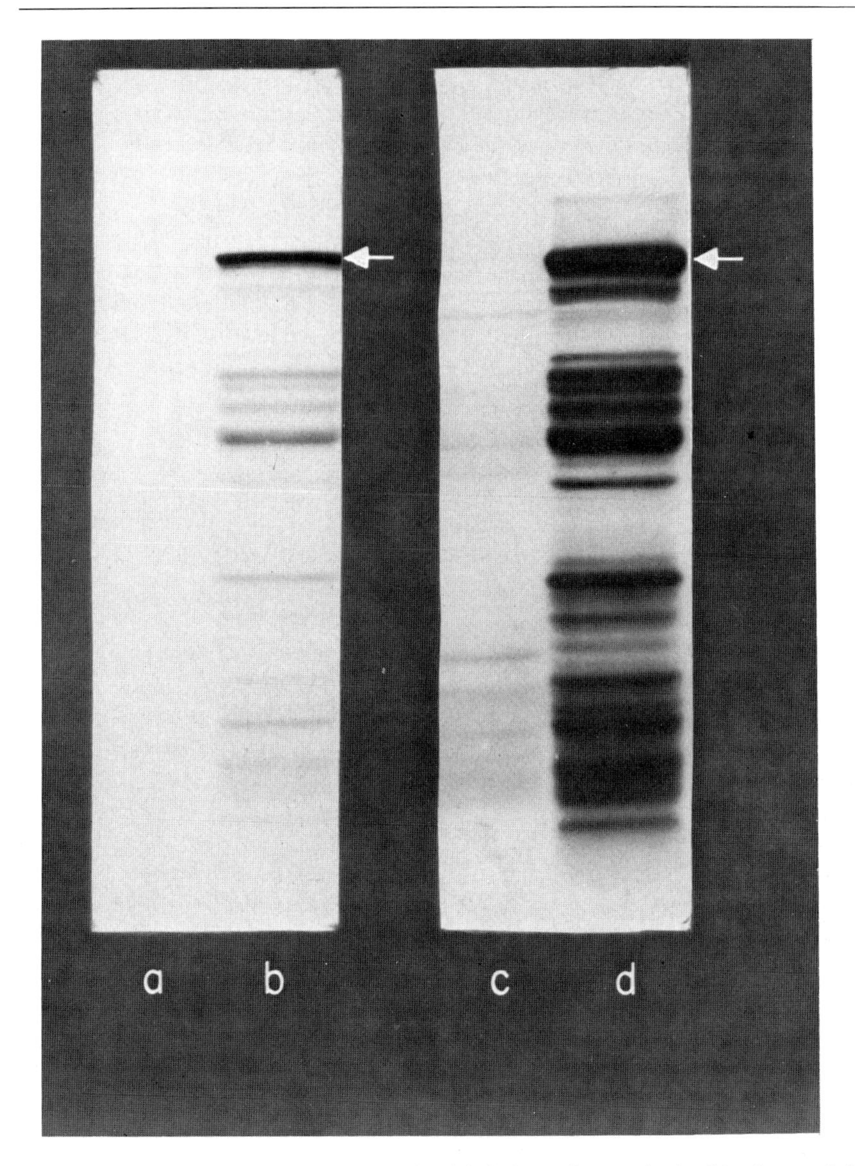

FIG. 3. Fluorographs of L-[^{35}S]methionine labeled proteins synthesized in the coupled system. Conditions used are identical to those described in Fig. 2 unless otherwise indicated. Columns a and c: complete system minus added DNA; columns b and d: complete system with ColE1 DNA; columns a and b: exposed for 16 hr; columns c and d: exposed for 96 hr. Arrows direct attention to the 56,000-dalton colicin protein. In all cases 5 μl of S-30 and 10 μl of S-100 were used per 100 μl of incubation mixture.

Concluding Remarks

In this chapter a description of the coupled transcription translation system has been given. A limited number of examples have been discussed to illustrate the types of questions that can be answered by appropriate manipulations of the system. Little has been said about analysis of RNA made in the system. Suitably labeled RNA made in the coupled system is amenable to standard methods of analysis after extraction. Variants of the coupled system that use reconstructed cytoplasmic extracts have been described by others (for example, see Herrlich and Schweiger[15] and Robakis *et al.*,[16]). For applications where separation of components present in the S-30 is desirable, these procedures should be consulted. The main advantage of the S-30 for most applications is the simplicity of preparation, its completeness, and its high activity.

Acknowledgments

This work was supported by grants from the National Institutes of Health (5 RO1 GM-29217-05) and the National Science Foundation (NSF-PCM-80-21265).

[15] P. Herrlich and M. Schweiger, this series, Vol. 20, p. 537.

[16] N. Robakis, Y. Cenatiempo, L. Meza-Basso, N. Brot, and H. Weissbach, this volume [45].

[45] A Coupled DNA-Directed *in Vitro* System to Study Gene Expression Based on Di- and Tripeptide Formation

By NIKOLAOS ROBAKIS, YVES CENATIEMPO, LUIS MEZA-BASSO, NATHAN BROT, and HERBERT WEISSBACH

Several DNA-directed *in vitro* protein-synthesizing systems have been developed to study the regulation of gene expression.[1,2] Most laboratories use a modification of the original unfractionated system described by Zubay and co-workers,[3] but a highly defined system has also been de-

[1] H.-Z. Chem and G. Zubay, this volume [44].

[2] B. de Crombrugghe, *in*: "Molecular Mechanisms of Protein Biosynthesis" (H. Weissbach and S. Pestka, eds.), p. 603. Academic Press, New York, 1977.

[3] G. Zubay, D. A. Chambers, and L. C. Cheong, *in* "The Lactose Operon" (J. R. Beckwith and D. Zipser, eds.), p. 375. Cold Spring Harbor Laboratory, Cold Spring Harbor, New York, 1970.

ISBN 0-12-182001-7

scribed.[4,5] In most cases, the synthesized complete protein product is assayed by using gel electrophoresis and/or immunoprecipitation techniques—procedures that are long and rather complicated. In some instances, enzymic assay of the product is possible, e.g., β-galactosidase,[3] but this depends on the gene product and the activity of the *in vitro* synthesizing system.

In this chapter we describe a simplified DNA-directed *in vitro* system that measures the formation of the initial dipeptide (or tripeptide) characteristic of the gene product. It is most applicable with templates that contain a limited number of genes, e.g., plasmids, restriction fragments, and where the N-terminal sequence of the protein product is known. Measuring a dipeptide provides specificity and is quantitative, and the product can be easily and quickly assayed. The system can be constructed with five highly purified factors (defined system) or with a ribosomal high-salt wash (RSW[6]) fraction plus RNA polymerase (crude system).

Dipeptide and tripeptide synthesis can be specifically obtained by limiting the acylated tRNA species that are used in the system. Thus, by using fMet-tRNA and the appropriate aminocyl-tRNA species for the second and third amino acid, a specific di- or tripeptide gene product can be selectively synthesized. An important aspect of the measurement of a tripeptide is that it can be used as a rapid, specific, and quantitative assay of the amount of a particular mRNA template present in a mixture of mRNA species. Finally, since di- or tripeptide formation from a DNA template involves accurate transcription and proper initiation of translation, this system is very well suited for studies on the regulation of gene expression.

Reagents

Restriction endonuclease *Hin*cII was purchased from Bethesda Research Laboratories, Inc. (Gaithersburg, Maryland) and *Hin*dIII from New England BioLabs (Beverly, Massachusetts). Unfractionated *Escherichia coli* tRNA, purified $tRNA_f^{Met}$, phosphoenolpyruvate (PEP), and pyruvate kinase (PK) were purchased from Boehringer-Mannheim. Purified tRNA isoacceptor species $tRNA_1^{Ser}$, $tRNA_3^{Ser}$, $tRNA_3^{Ala}$ and $tRNA_4^{Leu}$ were kindly supplied by Dr. B. R. Reid, University of Washington, Seat-

[4] H. F. Kung, B. Redfield, B. V. Treadwell, B. Eskin, C. Spears, and H. Weissbach, *J. Biol. Chem.* **252,** 6889 (1977).

[5] T. Zarucki-Schulz, C. Jerez, G. Goldberg, H. F. Kung, K. H. Huang, N. Brot, and H. Weissbach, *Proc. Natl. Acad. Sci. U.S.A.* **76,** 6115 (1979).

[6] The following abbreviations are used: RSW, ribosomal high-salt wash; PEP, phosphoenolpyruvic acid; DTT, 1,5-dithiothreitol; TLC, thin-layer chromatography; RUBPCase, ribulose bisphosphate carboxylase; PEG, polyethylene glycol; BSA, bovine serum albumin; SA, specific activity; PK, pyruvate kinase.

tle. A 0.25 *M* salt elute from a DEAE-cellulose column was used as the source of the enzymes required to acylate and transformylate the tRNA species.[7] The acylation and transformylation reactions were carried out as described elsewhere.[8-11] The acylated tRNA species were purified as previously described.[12] An *E. coli* RSW was prepared by washing ribosomes with 1.0 *M* NH_4Cl as described previously.[7] The extract was dialyzed against a buffer containing 10 m*M* Tris-acetate, pH 8.2, 14 m*M* Mg^{2+} acetate; 60 m*M* K^+ acetate, and 1 m*M* DTT. High-salt washed 70 S ribosomes and initiation factors IF-1, IF-2, and IF-3 were prepared as described.[13,14] EF-Tu was prepared according to Miller and Weissbach,[15] and EF-G according to Rohrbach *et al.*[16] RNA polymerase was prepared according to the method of Burgess.[17]

N-Formyl-L-methionine, *N*-formyl-L-methionyl-L-alanine, L-methionyl-L-serine, and spermidine were purchased from Sigma Chemical Co. L-Methionyl-L-leucine was obtained from Vega Biochemicals. *N*-Formyl-L-methionyl-L-alanyl-L-leucine was kindly supplied by Dr. W. Danho (Hoffmann-La Roche, Inc.). *N*-Formyl-L-methionyl-L-serine and *N*-formyl-L-methionyl-L-leucine were prepared as described previously.[12] PEG 6000 was obtained from J. T. Baker Chemical Co., precoated thin-layer chromatography plates (silica gel G, 250 μM) from Analtech (Newark, Delaware), and Seakem and Seaplaque agarose from FMC Corporation, Rockland, Maine. L-[^{3}H]Alanine, L-[^{3}H]serine, and L-[^{3}H]leucine were obtained from New England Nuclear, L-[^{35}S]methionine was from Amersham/Searle. All radioactive samples were dissolved in Bray's solution (National Diagnostics) and counted in a liquid scintillation spectrometer.

Plasmids Used

Escherichia coli JF943 containing either plasmid pNF1337 or pNF1341 was kindly supplied by Dr. J. Friesen (University of Toronto, Toronto,

[7] H. F. Kung, C. Spears, and H. Weissbach, *J. Biol. Chem.* **250,** 1556 (1975).

[8] H. W. Dickerman, E. Steers, Jr., B. Redfield, and H. Weissbach, *J. Biol. Chem.* **242,** 1522 (1967).

[9] A. Böck, *Arch. Mikrobiol.* **68,** 165 (1969).

[10] K. L. Roy and D. Soll, *J. Biol. Chem.* **245,** 1394 (1970).

[11] P. Schofield, *Biochemistry* **9,** 1694 (1970).

[12] N. Robakis, L. Meza-Basso, N. Brot, and H. Weissbach, *Proc. Natl. Acad. Sci. U.S.A.* **78,** 4261 (1981).

[13] N. Brot, E. Yamasaki, B. Redfield, and H. Weissbach, *Biochem. Biophys. Res. Commun.* **40,** 698 (1970).

[14] J. W. B. Hershey, J. Yanov, K. Johnston, and J. L. Faukunding, *Arch. Biochem. Biophys.* **182,** 626 (1977).

[15] D. L. Miller and H. Weissbach, *Arch. Biochem. Biophys.* **141,** 26 (1970).

[16] M. S. Rohrbach, M. E. Dempsey, and J. W. Bodley, *J. Biol. Chem.* **249,** 5094 (1974).

[17] R. R. Burgess, *J. Biol. Chem.* **244,** 6160 (1969).

Canada). The plasmid DNA was isolated from this transformant and used to transform *E. coli* strain RRI.[18] Strain RRI was also used as host for plasmids pBR322[19] and pJEA4. Plasmid pJEA4 was prepared (from plasmid pSoe3101)[20] by Dr. J. Erion in this laboratory and has a 2.2 kilobase(kbp) fragment that contains the gene for the large subunit of spinach RUBPCase.[21]

Plasmid pNF1337 was used as a DNA template for most of the studies described here (Fig. 1). This plasmid has a bacterial insert (cloned into pBR322) that starts at codon 106 of ribosomal protein L11, contains all the genetic information for ribosomal proteins L1, L10, and L12, and terminates within the gene coding for the β subunit of RNA polymerase.[22] This gene cluster is inserted into pBR322 at the *Pst*I restriction endonuclease site, and therefore plasmid pNF1337 contains a truncated β-lactamase gene.[22] Since it has been shown that ribosomal protein L1 cannot be synthesized from pNF1337 DNA,[23] the only bacterial genes expressed are *L10, L12,* and the NH_2-terminal fragment of the β subunit. DNA sequence studies of this genetic region[24] have revealed that the N-terminal nascent dipeptides of these proteins are fMet-Ala (L10), fMet-Ser (L12), and fMet-Val (β-subunit). Similar studies of the β-lactamase gene have shown

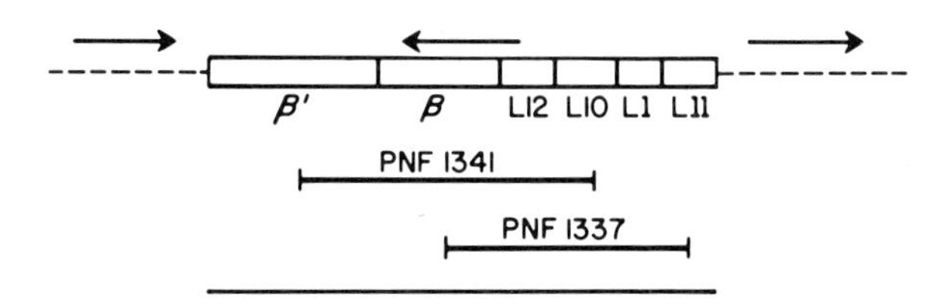

FIG. 1. Map of the bacterial inserts on plasmids pNF1337 and pNF1341. These plasmids contain fragments derived from the transducing phages λ *rif*^{d}*18* by limited digestion with *Pst*I.[22] Dashed lines, plasmid DNA; arrows, direction of transcription. Adapted from Goldberg *et al.*[23]

[18] F. Bolivar and K. Backman, this Series Vol. 68, p. 245.

[19] F. Bolivar, R. Rodriquez, P. J. Greene, M. C. Betlach, H. L. Heyneker, H. W. Boyer, J. Crosa, and S. Falkow, *Gene* **2,** 95 (1977).

[20] J. L. Erion, J. Tarnowski, H. Weissbach, and N. Brot, *Proc. Natl. Acad. Sci. U.S.A.* **78,** 3459 (1981).

[21] Y. Cenatiempo, N. Robakis, L. Meza-Basso, N. Brot, H. Weissbach, and B. R. Reid, *Proc. Natl. Acad. Sci. U.S.A.* **79,** 1466 (1982).

[22] N. P. Fiil, D. Bendiak, J. Collins, and J. D. Friesen, *Mol. Gen. Genet.* **173,** 39 (1979).

[23] G. Goldberg, T. Zarucki-Schulz, P. Caldwell, H. Weissbach, and N. Brot. *Biochem. Biophys. Res. Commun.* **91,** 1453 (1979).

[24] L. E. Post, P. D. Strycharz, M. Nomura, H. Lewis, and P. P. Dennis, *Proc. Natl. Acad. Sci. U.S.A.* **76,** 1697 (1979).

that the N-terminal dipeptide of this enzyme is fMet-Ser.[25] Like pNF1337, pNF1341 also contains a fragment of λ *rif*^{d}*18* DNA, but the insert begins at codon 26 of the *L10* gene and extends through the *L12* and the β subunit genes (Fig. 1).[22] This DNA, therefore, lacks both the promoter and NH_2-terminal fragment of L10.

DNA Fragments and mRNA

Restriction Analysis. A 1.6 kbp DNA fragment containing the entire genes for ribosomal proteins L10 and L12 was prepared by digestion of the plasmid pNF1337 with restriction endonuclease *Hin*cII. This fragment also contains the genetic information for several amino acids at the carboxyl terminus of ribosomal protein L1 and terminates within the leader sequence of the β-subunit of RNA polymerase (Fig. 2). Further restriction digestion of the 1.6 kbp DNA fragment with *Hin*dIII yields two fragments. One of them is 880 bp long and contains the promoter region and coding sequences for approximately 90% of the *L10* gene product. The smaller fragment (740 bp) contains the remainder of the *L10* gene and the entire *L12* gene (Fig. 2). The *L12* gene should not be expressed to any significant extent from this fragment, since the fragment is missing a functional promoter.[23,26,27] mRNA templates were synthesized from plasmid pNF1337, the *Hin*cII fragment, and the *Hin*dIII restriction digest of the 1.6 kbp fragment.

Preparation and Isolation of DNA Fragments. The following procedure was used to obtain the 1.6 kbp *Hin*cII fragment and the 880 bp *Hin-*

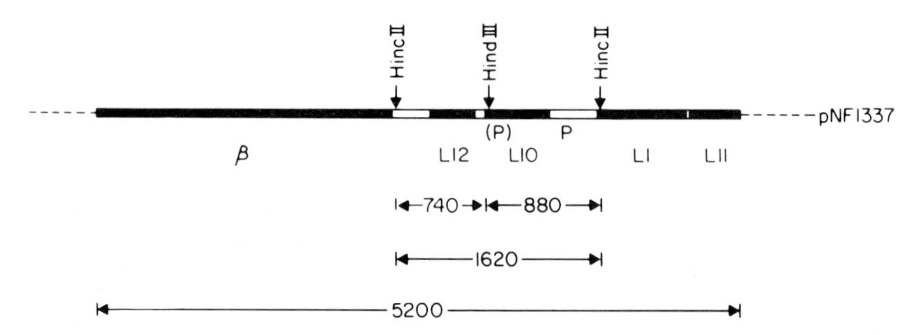

FIG. 2. Partial restriction map of the bacterial DNA inserted into plasmid pBR322. Direction of transcription is from right to left. P, primary promoter; (P), secondary promoter.

[25] J. G. Sutcliffe, *Proc. Natl. Acad. Sci. U.S.A.* **75,** 3737 (1978).
[26] M. Yamamoto and M. Nomura, *Proc. Natl. Acad. Sci. U.S.A.* **75,** 3891 (1978).
[27] T. Linn and J. Scaife, *Nature (London)* **275,** 33 (1978).

dIII fragment. The reaction mixture contained 10 m*M* Tris-HCl, pH 7.9, 1 m*M* dithiothreitol (DTT), 30 m*M* NaCl, 1 m*M* $MgCl_2$, 0.6 mg of DNA per milliliter, one enzyme unit (*Hin*cII) per microgram of DNA, and 100 μg of nuclease-free BSA per milliliter. The extent of digestion was determined by analytical agarose gel electrophoresis (1.2% Seakem). After digestion was complete, the reaction mixture was extracted with an equal volume of a phenol–chloroform mixture (1:1) saturated with electrophoresis buffer (40 m*M* Tris-acetate, pH 7.8, 5 m*M* Na^+ acetate, 1 m*M* EDTA). The aqueous phase was made 0.3 *M* in Na^+ acetate, and the DNA was precipitated with ethanol. The precipitate was dissolved in electrophoresis buffer, glycerol was added (final concentration 25%), and the sample was electrophoresed overnight at 25 mA in a 1.2% Seakem agarose gel. After staining with ethidium bromide, the band containing the 1.6 kbp fragment was excised and the 1.6 kbp fragment was isolated by the procedure of Thuring *et al.*[28] In brief, the agarose band was cut into small pieces (approximatcly 1 × 0.5 × 0.5 cm), frozen at −20°, and then placed individually into a petri dish between two Parafilm sheets. The agarose was pressed by hand between the two Parafilm sheets until fluid was expelled. The droplets were collected and extracted with buffer-saturated phenol; the DNA was precipitated with ethanol. The *Hin*cII 1.6 kbp DNA prepared in this way was incubated with *Hin*dIII in the following buffer: 7 m*M* Tris-HCl, pH 7.4; 60 m*M* NaCl; 7 m*M* $MgCl_2$; 100 μg of BSA and 0.2 mg of DNA per milliliter, and one *Hin*dIII unit per microgram of DNA. The mixture was incubated at 37° for about 1 hr. After complete digestion, the mixture was kept at 4° and could be used for mRNA synthesis without further treatment.

Table I lists the genes present on plasmid pNF1337, the amino-termi-

TABLE I
DIFFERENT PLASMIDS USED AND PEPTIDE PRODUCTS[a]

Plasmid	Gene product	NH_2-terminal sequence	Nucleotide coding sequence	Second codon isoacceptor tRNA
pNF1337	Ribosomal protein L10	fMet-Ala	AUG GCU	$Ala_{3\ or\ 2}$
pNF1337	Ribosomal protein L12	fMet-Ser	AUG UCU	$Ser_{1\ or\ 2}$
pNF1337	β-subunit	fMet-Val	AUG GUU	Val_1
pNF1337, pBR322	β-Lactamase	fMet-Ser	AUG AGU	Ser_3

[a] Adapted, in part, from Cenatiempo *et al.*[21]

[28] R. W. J. Thuring, J. P. M. Sanders, and P. Borst, *Anal. Biochem.* **66,** 213 (1975).

nal peptides for each gene product, the code words for each amino acid (from the DNA sequences), and the tRNA isoacceptor species that are required.

mRNA Preparation. mRNA was prepared by incubating 40 μg of plasmid DNA or a restriction fragment at 37° in the following reaction mixture (final volume 0.5 ml): 50 m*M* Tris-acetate, pH 7.5; [^{3}H]UTP (5000 cpm/nmol), CTP, GTP, ATP, 0.8 mM each; DTT, 1 m*M*; PEP, 25 m*M*; PK, 14 μg/ml; RNA polymerase, 60 μg/ml; magnesium acetate, 10 m*M*; spermidine, 1 m*M*. At the end of the incubation, the reaction mixture was made 0.01% in $NaDodSO_4$ and extracted with phenol saturated with 50 m*M* Tris-HCl, pH 7.5. The aqueous phase was extracted three times with an equal volume of ether, and then concentrated five-fold (Speed-Vac, Savant Instruments). The nucleic acids were precipitated by cold ethanol in the presence of 0.3 *M* sodium acetate, and the precipitate was dissolved in a small volume of sterile water to which three volumes of 4 *M* Na^+ acetate were added. The mixture was left overnight at 4° and then placed at −20° for about 2 hr. The precipitated mRNA was collected by centrifugation and redissolved in sterile water. The DNA fragment in the supernatant could be recovered by overnight dialysis against 5 m*M* Tris-HCl, pH 7.5, followed by concentration and ethanol precipitation as described above. The recovered DNA was active as a template in the dipeptide system (see below).

DNA-Directed Dipeptide Synthesis (fMet-Ala)

The expression of the *L10* gene, as measured by fMet-Ala formation (Table I) is used here as an example, although the system as described could be used for other dipeptides directed by different templates. The steps involved in fMet-Ala formation starting with a DNA template are shown in Fig. 3. The *in vitro* incubation mixture (final volume 35 μl) contains the following components: 30 m*M* Tris-acetate (pH 7.5); 10 m*M* sodium dimethylglutarate (pH 6.0); 36 m*M* ammonium acetate; 2 m*M* DTT; 9.2 m*M* magnesium acetate[29]; 2.9 m*M* ATP, 0.7 m*M* CTP, GTP, and UTP; 28 m*M* PEP; 0.5 μg of PK; 39 m*M* K^+ acetate; 0.8 m*M* spermidine, 3 μl of 50% PEG 6000[30]; 0.3 μg of IF-1, 0.5 μg of IF-2, 0.6 μg of IF-3, 1.0 μg of EF-Tu, 2 μg of RNA polymerase, 0.5–1.0 A_{260} unit of NH_4Cl-washed 70 S ribosomes; 10 pmol of either unlabeled fMet-$tRNA_f^{Met}$ or

[29] The Mg^{2+} concentration for optimal dipeptide synthesis should be titrated when the incubation components are changed (e.g., different ribosomes, tRNA).

[30] Addition of PEG 6000 to the incubation mixture usually increases the yield of dipeptides by twofold. This is probably due to the protection of mRNA against nucleases.

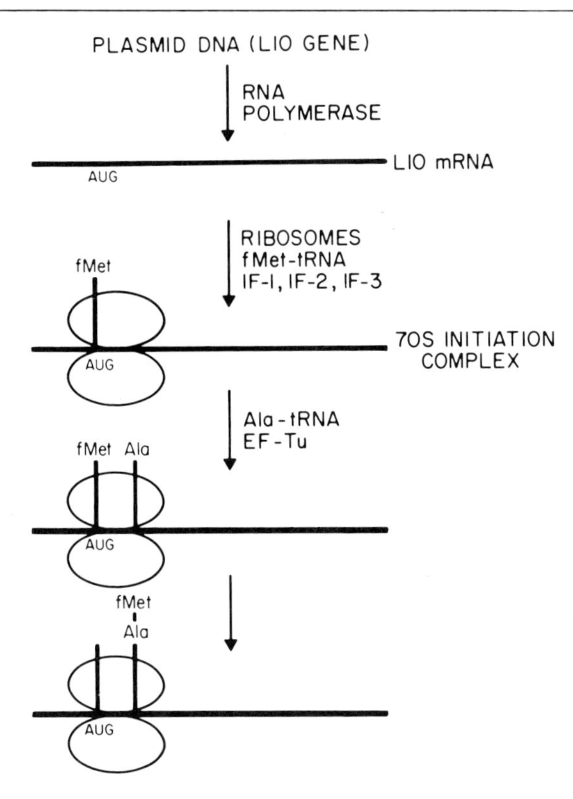

FIG. 3. Outline of steps leading to dipeptide formation in the *in vitro* system. Taken from Robakis *et al.*[2]

f[^{35}S]Met-tRNA$_f^{Met}$ (about 7000 cpm/pmol) and 10 pmol of [^{3}H]Ala-tRNA (about 6000 cpm/pmol) prepared by acylating unfractionated *E. coli* tRNA with [^{3}H]alanine. The reaction is initiated by the addition of 2 μg of plasmid pNF1337 or 0.4 μg of a DNA fragment (see above). When mRNA (0.5 μg) is used as a template, RNA polymerase is omitted from the reaction mixture. The reaction mixture (in 1.5-ml Eppendorf tubes) is incubated at 37° for 1 hr. The reaction is stopped by the addition of 2.5 μl of 1 *N* NaOH, and the mixture is incubated for an additional 10 min at 37° to hydrolyze any peptidyl-tRNA. Two assays have been used to measure dipeptide formation. One is based on TLC, which gives chromatographic identification of the product. The other involves a simple extraction that is rapid and suitable for routine analysis once the dipeptide has been identified by TLC.

Assay by Thin-Layer Chromatography. To the alkalinized reaction mixture prepared as described above, 2 μl of a 10 mg/ml solution of fMet and fMet-Ala are added as carrier. The reaction mixture is acidified by the addition of 4 μl of 2 *N* HCl. The tubes are centrifuged for 2 min in an Eppendorf centrifuge 5414, and the precipitate is discarded. A portion of the supernatant (usually 24 μl) is applied in 3-μl aliquots to a silica gel G thin-layer plate. The solvent system used is a mixture of ethyl acetate–hexanes–acetic acid, 8:3:1 (v/v/v). The plate is developed for 50 min; after drying, the methionine-containing areas are visualized by exposing the plate to iodine vapor in a closed glass tank that contains iodine crystals. The fMet-Ala spot on the silica plate is scraped off into scintillation vials. The silica scraping is extracted with 1 ml of water for about 3 min. Nine milliliters of scintillation fluid are added and the radioactivity is determined.

In this solvent system, fMet, fMet-Ala, and fMet-Ser display R_f values of 0.48, 0.37, and 0.2, respectively, whereas all free amino acids remain at, or close to, the origin. Therefore, the system has also been used to study the synthesis of fMet-Ser (β-lactamase and L12). Usually for thin-layer analysis, both the fMet-tRNA and second aminoacyl-tRNA are labeled but, if the dipeptide does not separate from free fMet (e.g., fMet-Val), only the second aminoacyl-tRNA should be labeled.

Ethyl Acetate Extraction. This method is based on the extraction of the dipeptide from acidic solution into ethyl acetate. It is simpler and faster than the thin-layer method described above. For these incubations, since fMet extracts into ethyl acetate under the conditions used, unlabeled fMet-tRNA is used and only the Ala-tRNA (or other second amino acid) is labeled. The reaction mixture after NaOH treatment (see above) is transferred to a glass test tube and 0.5 ml of 0.5 *N* HCl is added followed by 3 ml of ethyl acetate. The tubes are mixed for a few seconds (Vortex) and then centrifuged at low speed in a bench-top centrifuge for 30 sec. Two milliliters of the ethyl acetate phase are transferred to a scintillation vial, and the radioactivity is measured. To determine the amount of product formed, the extraction coefficient of the different peptides must be known. Under the conditions used, about 75% of fMet-Ala and 30% of fMet-Ser extract into ethyl acetate.

Characteristics of fMet-Ala Synthesis

Figure 4 shows the kinetics of fMet-Ala synthesis in a defined system using plasmid pNF1337 as template. The synthesis of fMet-Ala is linear for at least 45–60 min when plasmid DNA or DNA fragments are used as templates. Similar results were obtained when other dipeptides were synthesized, e.g., fMet-Ser, fMet-Val. Table II shows the dependencies for

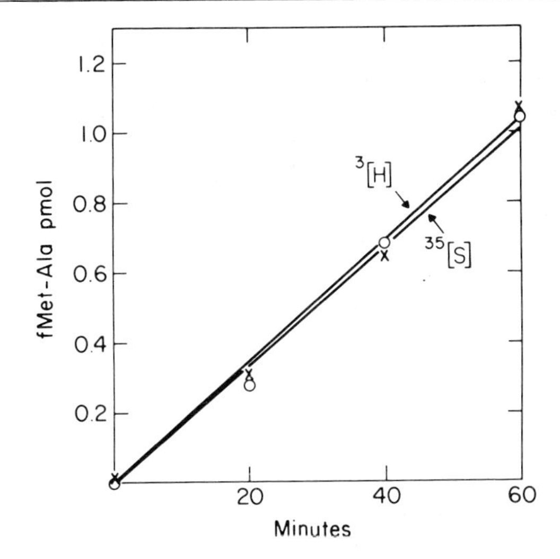

FIG. 4. Kinetics of fMet-Ala synthesis in the defined system. The incubation conditions and thin-layer assay are described in the text. The incubation mixture contained f[^{35}S]Met-tRNA$_f^{Met}$ and [^{3}H]Ala-tRNA. Unfractionated *Escherichia coli* tRNA was used as a source of tRNA^{Ala}. Taken from Robakis *et al.*[12]

TABLE II
COMPONENTS REQUIRED FOR THE SYNTHESIS OF fMet-Ala[a]

Omission	fMet-Ala (pmol)
None	1.7
RNA polymerase	0
Ribosomes	0
IF-1	0.7
IF-2	0
IF-3	0
EF-Tu	0
−1337 DNA + 1341 or pBR322 DNA	0

[a] Details of the incubations are described in the text. Taken from Robakis *et al.*[12]

the formation of fMet-Ala (or other dipeptides) in this system. With the exception of IF-1, there is an absolute requirement for all the factors. Further evidence that fMet-Ala synthesis is due to the expression of the *L10* gene is demonstrated by the absence of any fMet-Ala formation when plasmid pNF1341 (see Fig. 1) or pBR322 is used as a template instead of pNF1337.

The present system can efficiently use small DNA fragments and mRNA instead of plasmid DNA as templates. DNA fragments and mRNA were prepared as described above. The results are shown in Table III.

Separation of Gene Products Containing the Same N-Terminal Dipeptide

The dipeptide assay is not product specific if there are two genes on the template whose protein products begin with the same N-terminal dipeptide. To deal with this problem the dipeptide system can be modified in two ways: (*a*) the use of specific tRNA isoacceptor species; and (*b*) tripeptide formation.

Use of Isoacceptor tRNA Species. Unfractionated tRNA, acylated with a specific amino acid or purified tRNA isoacceptor species, functions well in the system. If the codons for the second amino acid differ so that a different tRNA isoacceptor species is required for each gene product, the dipeptides can be distinguished. An example of how codon specificity has been used to differentiate between the expression of the β-lactamase and *L12* genes is shown here. As seen in Table I, both gene products begin with fMet-Ser, but β-lactamase requires $tRNA_3^{Ser}$ whereas L12 uses $tRNA_1^{Ser}$. The reaction mixtures are essentially as described above for fMet-Ala synthesis, except that instead of [^{3}H]Ala-tRNA, 10 pmol of the purified isoacceptor [^{3}H]Ser-$tRNA_1^{Ser}$ or [^{3}H]Ser-$tRNA_3^{Ser}$ species are

TABLE III
fMet-Ala SYNTHESIS DIRECTED BY DNA FRAGMENTS OR mRNA[a]

Template	fMet-Ala (pmol)
*Hin*cII DNA fragments	3.0
*Hin*dIII DNA fragments	2.0
mRNA prepared from *Hin*cII DNA fragments	1.7
mRNA prepared from *Hin*dIII DNA fragments	1.5

[a] The details of the incubations and the preparation of the DNA fragments and mRNA are described in the text.

added. The results are shown in Table IV. With pBR322 as template, only $tRNA_3^{Ser}$ is active, since β-lactamase is the primary gene product formed. Using pNF1337, both $tRNA_3^{Ser}$ and $tRNA_1^{Ser}$ are active, since both the β-lactamase and L12 genes are expressed. With the *Hinc*II DNA fragment, which directs the synthesis of L12, only $tRNA_1^{Ser}$ is active. Similar results have been obtained with plasmid pJEA4, which contains the gene for the large subunit of ribulose bisphosphate carboxylase and the gene for β-lactamase. Both begin with fMet-Ser, but the fMet-Ser synthesis from the large subunit gene uses $tRNA_1^{Ser}$, whereas β-lactamase synthesis requires $tRNA_3^{Ser}$.[21]

Synthesis of Tripeptides. For tripeptide synthesis, e.g., fMet-Ala-Leu, the N-terminal tripeptide of L10, the following modifications of the dipeptide system are made. Polyethylene glycol is omitted, and 0.1 μg of EF-G, 10 pmol of [^{3}H]LeutRNA$_4^{Leu}$ (codon UUA), and nonradioactive Ala-$tRNA_3^{Ala}$ are added to the standard incubations for dipeptide synthesis. At the end of the incubation 20 μg of unlabeled fMet-Ala-Leu in 2 μl of H_2O are added, and the reaction mixture is treated as described above for dipeptide synthesis. The tripeptide can be separated by TLC using the same solvent system described for fMet-Ala isolation (R_f 0.29). The tripeptide region is scraped from the plate, and radioactivity is determined as described above. The ethyl acetate extraction procedure can also be used. At 37°, tripeptide synthesis is linear for about 40 min, but less tripeptide is formed than dipeptide. This lower tripeptide synthesis is not due to accumulation of the dipeptide, fMet-Ala, since, under the conditions in which the tripeptide is made, there is no synthesis of dipeptide (Table V). Other experiments showed that fMet-Ala-Leu synthesis required all three acylated tRNA species and was markedly stimulated by EF-G.

TABLE IV
UTILIZATION OF TWO $tRNA^{Ser}$ ISOACCEPTING SPECIES IN THE DIPEPTIDE ASSAY[a]

		Synthesis of fMet-Ser	
Template	Protein products containing fMet-Ser	$tRNA_1^{Ser}$ (pmol)	$tRNA_3^{Ser}$ (pmol)
pBR322	β-Lactamase	0	4.0
pNF1337	L12, β-lactamase	2.4	1.6
*Hinc*II fragment	L12	1.2	0

[a] The reactions were carried out as described in the text, using 2 μg of plasmid DNA (pBR322, pNF1337) or 0.2 μg of the 1.6 kbp *Hinc*II DNA fragment. Adapted, in part, from Cenatiempo *et al.*[21]

TABLE V
COMPARISON OF DI- AND TRIPEPTIDE SYNTHESIS[a]

System	fMet-Ala (pmol)	fMet-Ala-Leu (pmol)
Dipeptide system	2.0	0
Tripeptide system	0	0.6

[a] Plasmid 1337 was used as template. Di- and tripeptides were detected and separated by thin-layer chromatography. The tripeptide system contained the components required for dipeptide synthesis plus EF-G and Leu-tRNA$_4^{Leu}$, except that polyethylene glycol 6000 was omitted from the reaction mixtures. Taken from Y. Cenatiempo, N. Robakis, B. R. Reid, H. Weissbach and N. Brot.[31]

Tripeptide Synthesis as a Measure of Functional mRNA

Contrary to dipeptide synthesis, under the conditions used, tripeptide synthesis results in a stable tripeptidyl-tRNA · mRNA · ribosome complex and hence there is no recycling of the mRNA.[31] The higher level of dipeptide formation (Table V) is due to the instability of the dipeptidyl-tRNA · mRNA · ribosome complex and results in a recycling of the mRNA when only dipeptide synthesis occurs. Since tripeptides remain attached to 70 S ribosomes, nitrocellulose filtration can also be used as an assay for the amount of tripeptide formed. In this case, at the end of the incubation, the reaction mixture is diluted to 0.3 ml with cold buffer containing 50 m*M* Tris-HCl pH 7.8, 100 m*M* NH_4Cl, 12 m*M* $MgCl_2$, and 2 m*M* 2-mercaptoethanol and then rapidly passed through a nitrocellulose filter (0.45 μM, Millipore). The filter is washed three times with 3 ml of the same buffer and dissolved in scintillation fluid, and the radioactivity is determined. A comparison of di-and tripeptide formation using two different assays is shown in Table VI. Although there is excellent dipeptide synthesis by the thin layer (or extraction) procedure, no stable complex, that is retained on a filter, is seen. This is due to rapid dissociation of the dipeptide ribosome complex. In contrast, there is reasonable agreement between the thin-layer procedure and the filter assay for tripeptide synthesis. The above results suggested that tripeptide synthesis could also be used as a rapid and qunatitative assay for the amount of a specific functional mRNA present in a mixture of mRNA species, since only those mRNA molecules that can bind to ribosomes and direct the synthesis of a specific tripeptide (active mRNA) would be measured.

[31] Y. Cenatiempo, N. Robakis, B. R. Reid, H. Weissbach, and N. Brot, *Arch. Biochem. Biophys.* **218,** 572 (1982).

TABLE VI
DIPEPTIDE AND TRIPEPTIDE SYNTHESIS BY DIFFERENT ASSAYS[a]

Assay	fMet-Ala (pmol)	fMet-Ala-Leu (pmol)
Thin layer	2.5	0.9
Filter	<0.1	0.8

[a] The assay systems and incubation conditions are described in the text except that polyethylene glycol 6000 was omitted from the reaction mixtures. Adapted, in part, from Y. Cenatiempo, N. Robakis, B. R. Reid, H. Weissbach and N. Brot.[31]

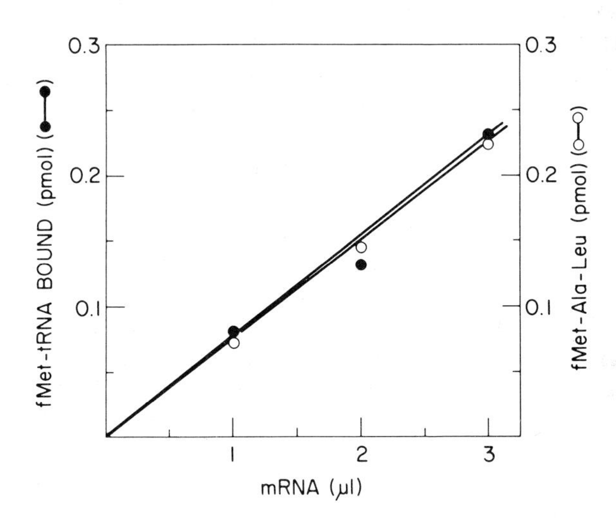

FIG. 5. fMet-tRNA binding to 70 S ribosomes and tripeptide formation in the presence of various amounts of *L10* mRNA. The mRNA (0.5 μg/μl) was prepared as described in the text. For fMet-tRNA binding, the mRNA was incubated with f[^{35}S]Met-tRNA$_f^{Met}$ and 70 S ribosomes in a buffer containing 50 m*M* Tris-HCl, pH 7.4; 100 m*M* NH_4Cl; 6 m*M* magnesium acetate; 2 m*M* 2-mercaptoethanol; 0.6 m*M* GTP; 1.0 μg of IF-2, 0.5 μg IF-3; 10 pmol of f[^{35}S]Met-tRNA$_f^{Met}$, and 1 A_{260} of 70 S ribosomes. After 10 min at 37°, the mixture was diluted with 0.3 ml of cold buffer containing 100 m*M* Tris-HCl, pH 7.6; 200 m*M* NH_4Cl, 10 m*M* magnesium acetate, and 4 m*M* 2-mercaptoethanol, and filtered through a nitrocellulose membrane. The filters were washed with 2 ml of buffer, and radioactivity was determined. The conditions for tripeptide synthesis from *L10* mRNA are described in the text. Taken from Cenatiempo, N. Robakis, B. R. Reid, H. Weissbach, and N. Brot.[31]

To validate this procedure, the *Hin*dIII mRNA transcript containing the L10 mRNA was used. This mRNA preparation directs the synthesis of fMet-Ala but not fMet-Ser. It has only one fMet-tRNA binding site and fMet-tRNA binding should be a measure of the amount of mRNA present. Figure 5 shows the results of a typical experiment comparing fMet-tRNA binding and tripeptide formation using limiting amounts of mRNA. An excellent correlation is obtained between the amount of fMet-RNA bound and the amount of the tripeptide synthesized.

Use of a Ribosomal High-Salt Wash for Di- and Tripeptide Synthesis

The di- and tripeptide system as described above has the advantage that it is highly defined, but it may not be of general use since at least five purified factors are needed. Because of this, the system has been simplified by using an RSW instead of the purified initiation and elongation factors. For these experiments, the reaction mixture is similar to that described above, except that IF-1, IF-2, IF-3, EF-Tu, and EF-G are omitted; they are replaced by an RSW preparation (6 μg of protein). In addition to ribosomes, only RNA polymerase is required, since it is not present in the RSW. The effect of RSW on fMet-Ala synthesis is shown in

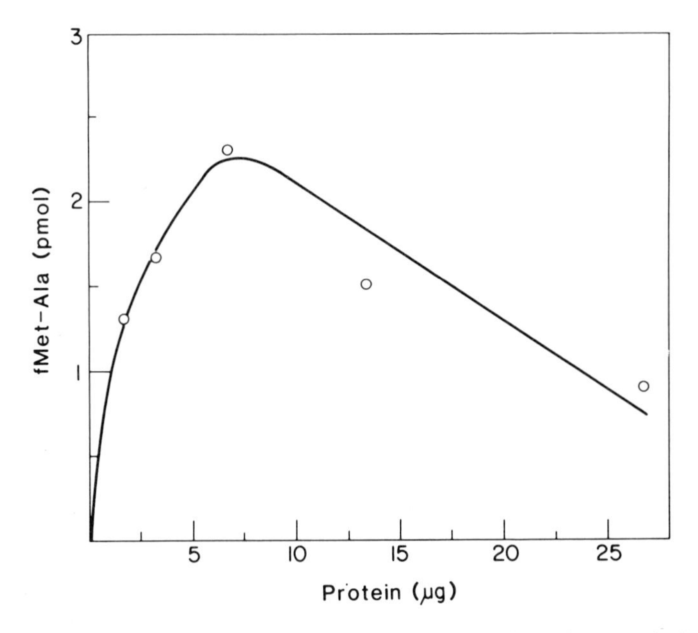

FIG. 6. Effect of ribosomal salt wash protein on fMet-Ala formation. Incubation conditions and assays are described in the text.

Fig. 6. It should be stressed that each RSW preparation must be titrated, since inhibitory components are sometimes present in these extracts. As with the defined system, di- and tripeptides can be synthesized by this system, and it can use plasmid DNA, DNA fragments, or mRNA as templates. Both total *E. coli* tRNA or purified isoacceptor tRNA species can be used as a source of the acylated tRNAs.

Use of the Dipeptide System To Study Regulation of Gene Expression

Di- or tripeptide synthesis can be used to study the regulation of gene expression at either the initiation of transcription or translation, and the system has been used to study the expression of both prokaryotic and chloroplast genes.[12,21] As an example, the effect of L10 on its own synthesis (autogenous regulation) has been investigated using the dipeptide system (fMet-Ala synthesis). Previous results had shown that the inhibition by L10 is at the level of translation.[32] As shown in Fig. 7, L10 inhibits the synthesis of fMet-Ala indicating that the autoregulation occurs before the formation of the first peptide bond.[33] This effect was shown to be specific since ribosomal protein L12 had no effect on the synthesis of fMet-Ala and L10 did not inhibit fMet-Ser formation (L12 or β-lactamase).[12]

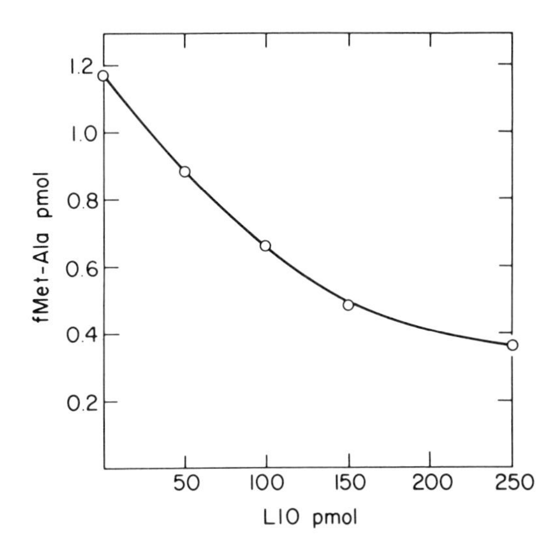

FIG. 7. Effect of protein L10 on fMet-Ala formation. For details see text. Taken from Robakis *et al.*[12]

[32] N. Brot, P. Caldwell, and H. Weissbach, *Proc. Natl. Acad. Sci. U.S.A.* **77,** 2592 (1980).

[33] Recently Johnsen *et al.* [*EMBO J.* **1,** 999 (1982)] reported that L8 (an L10:L12 complex) binds to the leader region of the L10 mRNA and protects a fragment of about 80 nucleotides from RNase digestion.

By examining the partial reactions involved in dipeptide formation (Fig. 3), it has been possible to show that L10 inhibits the formation of the 30 S initiation complex.[34] Transcriptional control of fMet-Ala synthesis by guanosine 5′-diphosphate-3′-diphosphate has also been shown.[12]

Summary

In this report, a simplified coupled DNA-directed *in vitro* system has been described that is based on the formation of the first di- or tripeptide of the gene product. This system is gene specific and quantitative, and the assay (especially the extraction procedure) is very rapid. The fact that both transcription and translation initiation occur in this system makes it ideally suited for studies on the regulation of prokaryotic gene expression. The ideal templates are plasmids, DNA fragments or purified mRNAs that direct the synthesis of a limited number of products with different second amino acids. An essential requirement is that the initial sequence of the protein products be known, although this system could be used to determine the second amino acid in cases where there is some doubt from the DNA sequence as to where a particular protein initiates. A difficulty arises when a plasmid contains more than one gene whose protein products have the same initial dipeptide. One solution to the problem is to measure tripeptide formation if the third amino acid is different. A second procedure, if the code word for the second amino acid differs between the genes, is to use purified isoacceptor tRNA species to distinguish the products. Another important application of tripeptide synthesis is that it can be used as a measure of the amount of active mRNA present in a mixture of mRNAs.

The use of a ribosomal high-salt wash instead of the purified initiation and elongation factors greatly simplifies this system and should make it suitable for routine analysis in most laboratories.

[34] N. Robakis, Y. Cenatiempo, S. Peacock, N. Brot, and H. Weissbach, *in* "Interaction of Translational and Transcriptional Controls in the Regulation of Gene Expression" (M. Grunberg-Manago and B. Safer, eds.), pp. 129–146. Elsevier, Amsterdam, 1982.

Author Index

Numbers in parentheses are reference numbers and indicate that an author's work is referred to although the name is not cited in the text.

A

B

C

E

F

G

H

I

J

K

L

M

N

O

P

Q

R

S

T

U

V

W

X

Y

Z

Subject Index

A

B

C

D

F

G

H

M

R

S

Z